Biogenic Nano-Particles and their Use in Agro-ecosystems

Mansour Ghorbanpour •
Prachi Bhargava • Ajit Varma •
Devendra K. Choudhary
Editors

Biogenic Nano-Particles and their Use in Agro-ecosystems

Springer

Editors
Mansour Ghorbanpour
Department of Medicinal Plants, Faculty of Agriculture and Natural Resources
Arak University
Arak, Iran

Prachi Bhargava
Department of Bioscience & Technology
Shri Ramswaroop Memorial University
Barabanki, Uttar Pradesh, India

Ajit Varma
Amity Institute of Microbial Technology
Amity University
Noida, Uttar Pradesh, India

Devendra K. Choudhary
Amity Institute of Microbial Technology
Amity University
Noida, Uttar Pradesh, India

ISBN 978-981-15-2984-9 ISBN 978-981-15-2985-6 (eBook)
https://doi.org/10.1007/978-981-15-2985-6

This Springer imprint is published by the registered company Springer Nature Singapore Pte Ltd.
The registered company address is: 152 Beach Road, #21-01/04 Gateway East, Singapore 189721, Singapore

Preface

In the present scenario nanoscale devices have emerged with the capability to analyse associated plant diseases, nutrient deficiencies or any other ailments that affect food security in agroecosystems. It has been envisioned that smart delivery systems for agriculture can exhibit benign multifunctional characteristics to avoid habitat-imposed stresses for successful targeting. Nanoparticle-mediated smart delivery systems can examine the effects of delivery of nutrients or bioactive and/or pesticide molecules. It has been investigated that nanoparticles in plants might determine the nutritional status of plants and help in suitable curative measures upon induction in agroecosystems.

To enhance soil and crop productivity, attempts have been made to make nano-fertilizers and their delivery to the crops by deploying nanotechnology. In concomitant with nano-delivery, nano-fertilizer may be defined as nanoparticles that may directly help in supplying essential nutrients for plant growth and soil productivity. Herein, nanoparticles can absorb on to the clay network, thereby preventing fixation through releasing into the soil solution that can be utilized by plants; thus, progression improves soil health and nutrient use efficiency by crops. Furthermore, fertilizer particles can be coated with nanoparticles that make possible the slow and steady release of nutrients, thereby reducing loss of nutrients and enhancing its use efficiency of agri-crops.

Therefore, regardless of the fact that nanotechnology in agroecosystem is mostly speculative at this stage and requires an attention to be continued to have a momentous effect in nanoparticle-mediated delivery systems for agri-inputs to sustenance.

By keeping above views in mind, the present book is designed to explore nanotechnology in agroecosystems with reference to biogenic nanoparticles. Hence, the emphasis has given on:

- Occurrence and diversity of biogenic nanoparticles
- Mechanistic approach involved in the synthesis of biogenic nanoparticles
- Synthesis of nanoparticles by employing photoactivation and their fate in soil ecosystem
- A general idea on potential applications of nanoparticles in the agricultural systems
- Application and biogenic synthesis of gold nanoparticles and characterization
- Impact of biogenic nanoparticles on biotic stress to plants

- Mechanistic approach involves in biogenic nanoparticles on antimicrobial effects and cytotoxicity
- Role of biogenic nanoparticles in plant disease management
- A relevance of biological synthesized nanoparticles on longevity of agricultural crops
- A design and synthesis of nanobiosensor for monitoring of pollutants in water, soil and plant systems
- An applications of nanotechnology in agriculture with special reference to soil, water and plant sciences

In the present book, editors compiled researches in the form of compendium with elaborate description related to 'deployment of nanoparticles in amelioration of agriculture'. It will be one more milestone in addition to the published books that have been edited by our team to resolve the issues around food security with reference to biotechnological implication of microbes in agroecosystems.

Chapter 1 discusses exciting examples of selected applications of nanotechnologies in farm animals to realize the benefits of emerging nano-tools that promise the increase in food production in a sustainable manner.

Chapter 2 emphasizes complete knowledge of the positive and negative impacts of nanoparticles and its associated characteristics and highlights the impact of nanoparticles on the growth and development of plants.

Chapter 3 describes the use of nanobiotechnology in improving crop and food processing and supports the expansion and commercial applications of nanotechnology at large scale through public and private sectors.

Chapter 4 stresses that in agriculture, a positive effect on ecology can be made by using nano-agrochemicals instead of conventional pesticides.

Chapter 5 elaborately describes a great deal of processes for biogenic nanoparticle production schemes suitable for large-scale (industrial-scale) production.

Chapter 6 provides a gist of the traditional insect/pest control strategies and discusses the potentials of nanotechnology as a new tool for insect control.

Chapter 7 emphasizes that there is an augmenting need to evolve sustainable, reliable and eco-friendly procedures to fabricate nanoparticles with broad scope. Biogenic nanoparticles (NPs) synthesized via nanobiotechnology-related processes have the potential to carry hygienic manufacturing technologies. This latest technology can remarkably decrease the contaminants in the environment and danger to humans due to the usage of poisonous solvents and chemicals.

Chapter 8 involves understanding the behaviour of nanoparticles in the soil, influencing soil physicochemical properties and assessing possibility hazards.

Chapter 9 discusses control of plant diseases by using nanoparticles as antimicrobial factors that have become highly prevalent as technological progression, making their production affordable.

Chapter 10 defines the role of nanoparticles in controlling insect pests.

Chapter 11 emphasizes the application of gold NPs in sustainable agriculture.

Chapter 12 specially focuses on the uses of biological and non-biological (both biogenic) agents on the shelf life of agricultural crops.

Chapter 13 describes various methods of NP synthesis followed by a detailed description of the fate of NPs in soil with regard to their impact on the growth of plants and microbes in different prospective.

Chapter 14 emphasizes engineered nanoparticle (ENPs) by using gold, silver and aluminium to test a possible eco-friendly agent and others on the PGPR.

Chapter 15 defines a precise and on-demand application of nano-pesticides or nano-fertilizers that can enhance the productivity and prove protection against several pests without harming the environment.

Chapter 16 elaborately discusses that nanomaterials are designed to be utilized in the field of personal care items, agribusiness, food, pharmaceuticals and biotechnology. It is perceived that nanomaterials may effectively contribute in the analysis, diagnosis and avoidance of diseases such as microbe-based infections or carcinoma and its treatment.

Chapter 17 focuses on the various aspects of nanotechnology and nanoinformatics in the field of agriculture. This includes different approaches and applications of nanotechnology and nanoinformatics especially in the area of agriculture.

Chapter 18 focuses on the aspects of nanotechnology that have revolutionized the agriculture field, leading to better environmental management and sustainable practices.

Chapter 19 focuses on the applications of nanotechnologies in nutrigenomics particularly for the prevention or treatment of certain metabolic disorders.

Chapter 20 describes the development of multifunctional nutraceutical nanocarriers combining several useful properties in one particle that can boost up the efficacy of many therapeutic and diagnostic protocols.

Chapter 21 focuses on the benefits of advances in nanotechnology and probable risks involve along with future approaches.

Chapter 22 highlights particularly ignored or missed case of using nanoparticle producers – microorganisms. Authors briefly discuss, as another concept, that enhancing anti-phytopathogen potential capacity of soil can negatively be affected by NP-synthesizing microorganism that may drastically impair microflora balance and its own biocontrol capacity besides their expected positive advantages in purpose of antimicrobial property.

Chapter 23 highlights the use of nanomaterials in the detection of different pesticides even at lower concentrations using various types of sensors. These nanostructured materials are also able to degrade variety of pesticides to almost harmless or less harmful compounds and will provide a pathway of detection and degradation of pesticides to keep our Earth clean and green for years to come.

Chapter 24 highlights the use of NPs in food science through nanosensors that are developed for smart packaging so as to detect spoilage of food and to release nano-antimicrobials as and when required to extend the shelf life, thereby helping keep the food fresh for a longer period. There is a great demand for nano-enabled packaging of food, beverage and pharmaceutical products in industries due to regularly changing consumption patterns.

Chapter 25 discusses a thorough summary of the prospective uses of distinct biological sources for the synthesis of nanoparticles, types, physiochemical properties and nanoparticle characterizations which exist in multiple forms in agriculture.

Chapter 26 describes the role of nanoparticles in developing biosensors and in detecting, monitoring and diagnosis of pollutant in environment and agriculture.

Chapter 27 especially focuses on synthesis particularly plant-intermediated biosynthesis of metal nanoparticles and their classifications.

Arak, Iran — Mansour Ghorbanpour
Barabanki, Uttar Pradesh, India — Prachi Bhargava
Noida, Uttar Pradesh, India — Ajit Varma
Noida, Uttar Pradesh, India — Devendra K. Choudhary

Contents

Editors and Contributors

About the Editors

Mansour Ghorbanpour is a scientific member at the Department of Medicinal Plants, Faculty of Agriculture and Natural Resources, Arak University, Arak, IRAN. Also, scientific member at Institute of Nanoscience and Nanotechnology, Arak University, Arak, Iran. Presently, Dr. Ghorbanpou is an Associate Professor at the Department of Medicinal Plants, Faculty of Agriculture and Natural Resources, Arak University, Iran. Dr. Ghorbanpour maintains a worldwide network, and has published over 120 research articles, 15 review articles, 40 chapters and 7 books. He is an editor of various international journals such as "*Industrial Crops and Products*", and "*Phyton-International Journal of Experimental Botany*".

Prachi Bhargava has over 16 years of teaching and research experience in microbial molecular biology and is currently working as an Associate Professor at Shri Ramswaroop Memorial University, Lucknow. She has worked as Principle Investigator on various major GOI-sponsored projects, and was recently involved in a project sponsored by SERB, New Delhi. As an active researcher, Dr Bhargava has published numerous research and review articles in respected journals, along with several book chapters.

Ajit Varma is The Group Deputy Vice Chancellor of Amity University and Head of Amity Institute of Microbial Technology. He spent almost three decades as a Senior Professor at JNU, New Delhi (1985–2004). He has published over 300 research articles for respected national and international journals, as well as several major review articles and chapters in books. He has published more than 120 books in the area of microbial technology and has been the series editor for Springer-Verlag's Series on Soil Biology. He has extesive experience of organizing national and international training workshops/symposia and congresses.

Devendra K. Choudhary is a member of the National Academy of Sciences (MNASc), India. He holds a PhD in Microbiology and is currently working as an Associate Professor at Amity University, Noida. Dr. Choudhary has worked as Principle Investigator on major GOI-sponsored projects. He has published over 87 research articles for respected national and international journals, as well as several major review articles and chapters in books. He has also edited more than 10 books sponsored by SpringerNature.

Contributors

Mahsa Abbasi Department of Horticultural Science, Gorgan University of Agricultural Sciences and Natural Resources, Gorgan, Iran

Heba M. M. Abdel-Aziz Botany Department, Faculty of Science, Mansoura University, Mansoura, Egypt

Sakshi Agrahari Department of Biotechnology, Netaji Subhash Institute of Technology, New Delhi, India

Mohammad Akram Department of Plant Protection, Aligarh Muslim University, Aligarh, India

Demet Altindal Fethiye Ali Sıtkı Mefharet Koçman Vocational School, Department of Crop and Animal Production, Organic Farming Programme, Muğla Sıtkı Koçman University, Muğla, Turkey

Nüket Altindal Sivaslı Vocational School, Department of Crop and Animal Production, Programme of Medical and Aromatic Plants, Uşak University, Uşak, Turkey

Rakshit Ameta Department of Chemistry, JRN Rajasthan Vidyapeeth (Deemed to be University), Udaipur, Rajasthan, India

Satish Kumar Ameta Department of Environmental Science, Mewar University, Chittorgarh, Rajasthan, India

Suresh C. Ameta Department of Chemistry, PAHER University, Udaipur, Rajasthan, India

Behnam Asgari Lajayer Department of Soil Science, Faculty of Agriculture, University of Tabriz, Tabriz, Iran

Ashish Department of Bioengineering, Integral University, Lucknow, Uttar Pradesh, India

Yaseen Ayub Department of Chemistry, Governmant Islamia College Civil Lines, Lahore, Pakistan

Anju Bajpai ICAR-Central Institute for Subtropical Horticulture, Lucknow, India

Purushothaman Balakrishnan Department of Biotechnology, Periyar Maniammai Institute of Science & Technology, Thanjavur, Tamil Nadu, India

Abarna Balamurugan Department of Biotechnology, Periyar Maniammai Institute of Science & Technology, Thanjavur, Tamil Nadu, India

Ambreen Bano Department of Biosciences, Integral University, Lucknow, Uttar Pradesh, India

Ömür Baysal Faculty of Science, Department of Molecular Biology and Genetics, Muğla Sıtkı Koçman University, Kötekli/Muğla, Turkey

Jayesh Bhatt Department of Chemistry, PAHER University, Udaipur, Rajasthan, India

Fateme Khadivi Derakhshan Department of Microbiology, College of Science, Urmia Branch, Islamic Azad University, Urmia, Iran

Seema Dua Division of Animal Physiology and Reproduction, ICAR-Central Institute for Research on Buffaloes, Hisar, India

Akhilesh Dubey Department of Biotechnology, Netaji Subhash Institute of Technology, New Delhi, India

Prekshi Garg AMITY Institute of Biotechnology, AMITY University Uttar Pradesh, Lucknow, Uttar Pradesh, India

Mansour Ghorbanpour Department of Medicinal Plants, Faculty of Agriculture and Natural Resources, Arak University, Arak, Iran

Fatemeh Graily-Moradi Department of Plant Protection, Faculty of Agriculture, University of Tabriz, Tabriz, Iran

Anmol Gupta Department of Biosciences, Integral University, Lucknow, Uttar Pradesh, India

Neeraj Gupta Institute of Bioscience and Technology, Shri Ramswaroop Memorial University, Lucknow, Uttar Pradesh, India

Yasmin M. Heikal Botany Department, Faculty of Science, Mansoura University, Mansoura, Egypt

Divya Hiran Department of Home Science, Government Meera Girls College, Udaipur, Rajasthan, India

Kashmiri Jadhav ICAR-Indian institute of Rice Research, Hyderabad, India

Monika Jangid Department of Chemistry, RNT College, Kapasan/Chittorgarh, Rajasthan, India

P. Janhavi Department of Biochemistry, CSIR-CFTRI, Mysuru, India

Neha Kapoor Department of Chemistry, PAHER University, Udaipur, Rajasthan, India

R. C. Kasana ICAR-Central Arid Zone Research Institute, Jodhpur, Rajasthan, India

Kamal Khalkhal Department of Soil Science, Faculty of Agriculture, University of Tabriz, Tabriz, Iran

Mujeebur Rhaman Khan Department of Plant Protection, Aligarh Muslim University, Aligarh, India

Dharmendra Kumar Division of Animal Physiology and Reproduction, ICAR-Central Institute for Research on Buffaloes, Hisar, India

Nitin Kumar Department of Biotechnology, Periyar Maniammai Institute of Science & Technology, Thanjavur, Tamil Nadu, India

Sandeep Kumar ICAR-Central Institute for Subtropical Horticulture, Lucknow, India

Ayda Maadani Mallak Department of Soil Science, Faculty of Agriculture, Ferdowsi University of Mashhad, Mashhad, Iran

G. Ali Mansoori Departments of Bio & Chemical Engineering, University of Illinois at Chicago (M/C 063), Chicago, IL, USA

Nariman Maroufpour Department of Plant Protection, Faculty of Agriculture, University of Tabriz, Tabriz, Iran

Keshav Meghvanshi Department of Biotechnology, School of life sciences, Central University of Rajasthan, Ajmer, India

Reza Mohammadinejad Pharmaceutics Research Center, Institute of Neuropharmacology, Kerman University of Medical Sciences, Kerman, Iran

Mahdieh Mousavi Department of Plant Protection, Faculty of Agriculture, Urmia University, Urmia, Iran

M. Muthukumar ICAR-Central Institute for Subtropical Horticulture, Lucknow, India

J. Natasha Department of Biochemistry, CSIR-CFTRI, Mysuru, India

Rabia Nazir Applied Chemistry Research Centre, Pakistan Council of Scientific and Industrial Research Labs Complex, Lahore, Pakistan

R. Neelam Department of Biochemistry, CSIR-CFTRI, Mysuru, India

Neelam Pathak Department of Biochemistry, Dr. Rammanohar Lohia Avadh University, Ayodhya, Uttar Pradesh, India

R. Pathak ICAR-Central Arid Zone Research Institute, Jodhpur, Rajasthan, India

Afrin Rahim Department of Biotechnology, Periyar Maniammai Institute of Science & Technology, Thanjavur, Tamil Nadu, India

Avinash Kumar Rai Department of Chemistry, PAHER University, Udaipur, Rajasthan, India

Smita Rai Department of Biosciences, Integral University, Lucknow, Uttar Pradesh, India

P. V. Ravindra Department of Biochemistry, CSIR-CFTRI, Mysuru, India

Monika Saini Division of Animal Physiology and Reproduction, ICAR-Central Institute for Research on Buffaloes, Hisar, India

Mojtaba Salouti Biology Research Center, Zanjan Branch, Islamic Azad University, Zanjan, Iran

Naresh L. Selokar Division of Animal Physiology and Reproduction, ICAR-Central Institute for Research on Buffaloes, Hisar, India

M. Mohiraa Shafreen Department of Biotechnology, Periyar Maniammai Institute of Science & Technology, Thanjavur, Tamil Nadu, India

Kumaran Shanmugam Department of Biotechnology, Periyar Maniammai Institute of Science & Technology, Thanjavur, Tamil Nadu, India

Bharti Sharma Department of Bio & Nano Technology, Guru Jambheshwar University of Science & Technology, Hisar, India

Rolee Sharma Department of Biosciences, Integral University, Lucknow, Uttar Pradesh, India

Swati Sharma Department of Biosciences, Integral University, Lucknow, Uttar Pradesh, India

Jayendra Nath Shukla Department of Biotechnology, School of life sciences, Central University of Rajasthan, Ajmer, India

Nidhi Shukla Department of Biotechnology and Bioinformatics, Birla Institute of Scientific Research, Jaipur, Rajasthan, India

R. Soner Silme Center of Biotechnology and Genetic Engineering Research and Practice (BIYOGEM), Istanbul University, Fatih-Istanbul, Turkey

Deepti Singh Institute of Bioscience and Technology, Shri Ramswaroop Memorial University, Lucknow, Uttar Pradesh, India

S. K. Singh ICAR-Central Arid Zone Research Institute, Jodhpur, Rajasthan, India

Mousa Solgi Department of Horticultural Science, Faculty of Agriculture and Natural Resources, Arak University, Arak, Iran

Neha Srivastava AMITY Institute of Biotechnology, AMITY University Uttar Pradesh, Lucknow, Uttar Pradesh, India

Prachi Srivastava AMITY Institute of Biotechnology, AMITY University Uttar Pradesh, Lucknow, Uttar Pradesh, India

Gaurav Srivatava Bhundelkhand University, Jhansi, India

Mina Taghizadeh Department of Horticultural Science, Faculty of Agriculture and Natural Resources, Arak University, Arak, Iran

Lubna Tahir Applied Chemistry Research Centre, Pakistan Council of Scientific and Industrial Research Labs Complex, Lahore, Pakistan

Siddharth Vats Shri Ram Swaroop Memorial University, Lucknow, Uttar Pradesh, India

Kanchan Vishwakarma Amity Institute of Microbial Technology, Amity University, Noida, Uttar Pradesh, India

R. S. Yadav ICAR-Central Arid Zone Research Institute, Jodhpur, Rajasthan, India

Application of Nanotechnology in Agricultural Farm Animals

1

Naresh L. Selokar, Seema Dua, Dharmendra Kumar,
Bharti Sharma, and Monika Saini

Abstract

The agricultural farm animal production system is under a massive burden to provide adequate food to the rapidly growing human population. The emergence of new diseases and climate change has enhanced burdens on the animal production system. Nanotechnology, an application of materials at the nano scale, offers opportunities to improve animal health and production by providing better therapeutics, diagnostics, vaccines and adjuvants, animal feed and additives, and even helping animal reproduction. This chapter discusses the exciting examples of the selected application of nanotechnology in farm animals to realize the benefits of emerging nanotools that promise increase in food production from farm animals in a sustainable manner.

Keywords

Nanotechnology · Agriculture · Farm animals · Food production

1.1 Introduction

Nanotechnology (NT), an application of materials at the ultramolecular level, has expanded its uses to agricultural animal health and their food production. NT aims to produce materials at the nanoscale size with at least one dimension between 1

N. L. Selokar (✉) · S. Dua · D. Kumar · M. Saini
Division of Animal Physiology and Reproduction, ICAR-Central Institute for Research on Buffaloes, Hisar, India

B. Sharma
Department of Bio & Nano Technology, Guru Jambheshwar University of Science & Technology, Hisar, India

M. Ghorbanpour et al. (eds.), *Biogenic Nano-Particles and their Use in Agro-ecosystems*, https://doi.org/10.1007/978-981-15-2985-6_1

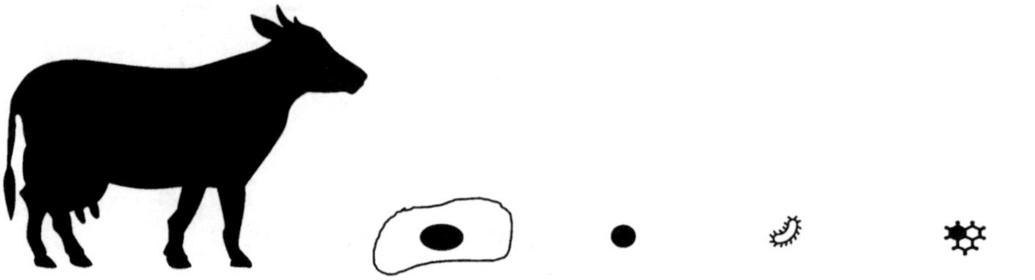

Fig. 1.1 Size of different biological structures in farm animals

to 100 nanometers (Roco 2003), in relation to biological structures of different sizes (Fig. 1.1). These ultrasmall materials have high surface-area ratio, called nanoparticles (NPs). NPs can easily be made from a variety of materials, including metals, flowers, leaves, and chemicals (Charitidis et al. 2014). At the ultrasmall level, NPs acquire unique physical, chemical and biological properties that provide new ways to study the structural, cellular, and molecular mechanisms in mammalian cells (Scott 2005). There are several areas in the agricultural animal production system in which NT can be applied with an aim to improve animal health and food production.

For many decades, the domestic animals, such as cattle, goat, sheep, pig, buffalo, and poultry have been used as a source of food (milk, meat, and their products) for human consumption. It is expected that the human population will be about 9 billion by 2050, and the global agricultural production needs to double or triple to provide sufficient food to the ever-growing human population (Sekhon 2014). Therefore, we need to use modern technologies, including NT, in the agricultural animal production system, to boost the productivity of domestic animals. In the last few years, there is a growing interest to explore NT in animal health and production. Here, we highlight some of the NT applications in agricultural animal production.

1.2 Improvement in Animal Health

Healthy farm animals support high animal productivity, which ensures a regular and safe food supply, less use of antibiotics and vaccines, and regular trade of animal products. Therefore, good health care and prevention of disease outbreaks can save millions of dollars by eliminating the investment of money on disease treatment and its eradication. One of the most noticeable examples is an outbreak of the foot and mouth disease, a devastating viral disease of cattle, buffalo, sheep, goats, and pigs, which costs 11 billion US dollars to the outbreak countries, mainly India and China (Knight-Jones and Rushton 2013). Furthermore, animal products (milk, meat, and their products, including hide) are not accepted by disease-free importing countries like the USA and EU, which also causes heavy loss in animal trade. The NPs are at the same scale as disease-causing viruses (nanometer in size) that would open up the possibility to develop effective diagnosis and treatment strategies to a particular disease (Scott 2005). Hence, NT is considered as an attractive option to improve animal health, drug delivery, and veterinary medicine.

1.2.1 Veterinary Diagnostics

The rapid, accurate, and sensitive detection of a disease-causing pathogen is the basis for effective treatment and eradication of a disease. Recently, the advancements in NT have revolutionized the veterinary diagnostics. For example, iron oxide-based NPs can be used to determine the distribution of a drug in the body using magnetic resonance imaging (MRI) (Soenen et al. 2010). Also, a specific drug can be tagged with fluorescent NPs to detect its localization inside the targeted cells (Ajmal et al. 2015). Furthermore, NP-based diagnostic chips are made available to quickly analyze a large number of samples (Craighead 2006). NP-based diagnostic chips take less time, a small volume of starting material (blood or serum), fewer consumables, and give more accurate results compared to classical laboratory tests, such as polymerase chain reactions and enzyme-linked immunosorbent assays (Bai et al. 2018). A few examples of NP-based veterinary diagnostics are summarized in Table 1.1.

1.2.2 Veterinary Therapeutics and Vaccine Delivery

Investment in the field of research and development activities of veterinary medicine will create ample access to the best drugs and vaccines to treat diseases. NT has opened new avenues in veterinary therapeutics by developing a smart drug delivery system that ensures the efficient delivery of drugs to the target tissues (Scott 2005). The smart drug delivery system ensures maximum absorption and low irritation at

Table 1.1 Some innovative applications of NP-based veterinary diagnostics

Animal species	Nanoparticle-based diagnostic tool	References
Poultry	Quantum dot-based 3D-printed immunoassay to detect avian influenza (H7N9) virus	Xiao et al. (2019)
Pig	Nano-polymerase chain reaction assay to detect the epidemic porcine diarrhea virus	Wanzhe et al. (2015)
Poultry	Gold nanoparticle-coated polyvinylidene difluoride membrane to detect avian flu antibodies in serum	Emami et al. (2012)
Cattle	Nanoparticle-based array of sensors (NA-NOSE) to detect volatile organic compounds linked to bovine tuberculosis in breath	Peled et al. (2012)
Domestic and wild ruminants	Nanoparticle-based biosensor assay to detect the Mycobacterium avium subsp. paratuberculosis (MAP), the causative agent of Johne's disease (JD), in fecal samples	Kumanan et al. (2009)
Chicken and goat	Quantum dot-based fluoroimmunoassay to detect antibodies of chicken Newcastle and goat pox virus in serum	Yuan et al. (2009)

the target site, and also has maximum therapeutic activity for an adequate duration (Sabry et al. 2018). In recent years, several NPs such as polymeric NPs, carbon nanotubes, liposomes, dendrimers, nanoshells, nanopores, and magnetic nanoparticles have been used for targeted delivery of drugs to treat veterinary diseases (reviewed by Muktar et al. 2015).

A wide range of antibiotics such as penicillin, amoxicillin, streptomycin, tetracycline, and gentamycin are frequently used drugs in veterinary medicine. In farm animals, particularly in poultry and pigs, the antibiotics are not only used to kill pathogenic bacteria but also used as growth promoters (Gross 2013). Humans exposed to antibiotics through the consumption of antibiotics-treated milk and meat products may increase the incidence of antibiotics resistance (Bartlett et al. 2013). NT can play an important role in developing effective and nontoxic antimicrobial agents to overcome the excessive use of antibiotics in farm animals.

Vaccines, antigenic components of pathogens, are regularly used to protect animals from the occurrence of a disease state. The vaccine stimulates the body's immune system to produce specific antibodies against a particular pathogen (Pulendran and Ahmed 2011). The efficacy of vaccines is highly dependent upon antigen type, route of delivery, and vaccine composition. With vaccines, the adjuvants (immunological agents) are often injected to augment the body's immune response that provides stronger and long-lasting immunity to a particular disease (Awate et al. 2013). New vaccine candidates like synthetic peptides and recombinant proteins are sensitive to degradation, and commonly use aluminum-based adjuvants could not protect these new classes of vaccines from early degradation (Underwood and van Eps 2012). NPs can be used to engineer the adjuvants in such way that vaccines have longer bioactivity with reduced dose to provoke a specific immune response. Various forms of NPs such as liposomes, polystyrene nanobeads, and immune-stimulating complexes can be used to engineer the adjuvants (Underwood and van Eps 2012). A few examples of NP-based therapeutics in veterinary are summarized in Table 1.2.

1.3 Improvement in Animal Production

Farm animals are reared to produce large quantities of milk, eggs, meat, fiber, and hide at the lowest possible cost. Efficient farm production requires good management practices that include adequate nutrition and good health, and the adaptation of animals to specific production conditions. In the past few decades, the fast-growing demand for animal origin food has massively changed the farm animal production system, which is supported by the mechanization of agriculture and technological developments. Nanotechnologies can also be applied to increase the production potential of farm animals. For example, supplementation of NP-coated feed, called nanofeed, to promote animal growth and yield, and to fortify milk, eggs, and meat products (Damian and Konrad 2018). At present, NT in animal production systems is in its budding stage, there is great potential in the coming years that NT will be extensively used to boost farm animal production.

Table 1.2 Examples of NP-engineered therapeutics and vaccines in veterinary medicine

Animal species	Nanoparticle-based therapeutic and vaccine
Cattle	Liposome-based streptomycin delivery for the treatment of brucellosis disease
	Intra-nasal delivery of ring-shaped nanoparticles for the treatment of respiratory syncytial virus
	Transdermal delivery of liposomes-based diclofenac as an anti-inflammatory and analgesic drug
Sheep	Polystyrene nanobeads to deliver foot and mouth disease vaccine
	DNA chitosan nanospheres to deliver Newcastle disease vaccine
	Liposome-based staphylococcal mastitis vaccine
	Liposome-based bovine leukemia virus vaccine
Horse	Liposome-based delivery of diamidine for the treatment of babesiosis disease
	Micelle-based delivery of ivermectin for the treatment of *Strongylus vulgaris*
	Water-based nanoparticle adjuvant vaccine against *Rhodococcus equi* pneumonia
	Liposome-based delivery of *Toxoplasma gondii* vaccine
Pig	Dendrimer-based delivery of foot and mouth vaccine
	Polymeric *E. coli* fimbriae vaccine

The table is adopted and modified from Underwood and van Eps 2012

1.3.1 Nanofeed

In an animal production system, approximate 40-50% of operating costs are due to animal feed (Wanapat et al. 2015). The feed industry's aim is to increase the efficiency of feed and its additives. Nutritional deficiency significantly decreases the production potential of animals, and nutrient-deficient animals are more prone to disease (Wanapat et al. 2015). The animal feed industry should formulate animal feed with NPs with the following aims: (1) improvement of overall feed efficiency, (2) increased production levels and quality of animal products (milk, eggs, and meat), (3) use of NPs having antioxidant and immune-modulatory properties to improve health, (4) reducing the requirement of antibiotics as growth promoters, as these may have negative effects on human health, and (5) removing unpleasant smells (boar taint) of animals products, particularly in meat of pigs (Hill and Li 2017). Several studies suggested that nanoform of feed can be used to enhance its nutritive value and to promote animal growth (Hill and Li 2017). Some of the examples of nanofeed additives are summarized in Table 1.3.

1.3.2 Nano-Reproduction

Reproduction is a natural way to increase the population of livestock in which sex cells (oocyte and sperm) are united to form an embryo that develops into a new generation of animal. Since many decades, humans have been exploiting the best genotype and phenotype parent animals to produce next-generation animals that are

Table 1.3 Commonly used nanoparticle additives in farm animal feed

Animal species	Nanomaterial	Application
Sheep, goat and poultry	Selenium	Stimulation of rumen microbial and enzyme activity
		Enhancing the semen quality
		Enhancing the immune response
Ruminants, pig and poultry	Zinc	Improving feed conversion ratio that promotes growth
		Enhancing the immune response
Pig	Chromium	Anti-diarrheal
		Improving carcass quality, particularly lean meat production
		Enhancing the immune response
Poultry	Copper	Stimulating the metabolic rate in broilers during embryonic development
		Strengthening immunological biocompatibility
Poultry	Montmorillonite –composite	Reducing aflatoxin toxicity
Farm animals, including poultry	Nano-polystyrene with polyethylene glycol linkers and mannose targeting biomolecules	Binding and removal of food-borne pathogens in animal feed

The table is adopted from Sabry et al. 2018

intended to produce more food (milk and meat) for human consumption. In farm animal reproduction, artificial insemination with frozen semen is the most commonly used method that assists faster and wider multiplication of elite animal production traits.

Recently, NPs have been used to study the physiological interaction between sperm and oocyte to understand the physiology of the fertilization process (Vasquez et al. 2016; Feugang et al. 2015). Furthermore, NP-based magnetic purification method of bull semen has demonstrated separation of damaged and defective sperm, thereby improving fertilization ability both in vitro and in vivo (Odhiambo et al. 2014; Durfey et al. 2019). Recently, Falchi et al. 2018 reviewed the potential applications of NPs as antioxidants and protective agents that are supplemented in semen extender to protect sperm during cryopreservation stress. NT can be coupled with gamete biology and semen cryopreservation methods to enhance the reproduction potential of animals.

1.4 Conclusion

Nanotechnology has great potential to improve agricultural farm animal health and production. Existing research studies have clearly demonstrated the practicality of NT in accurate diagnostic tools, targeted drug delivery, enhanced vaccine response, improved feed efficiency, and augmentation of reproduction in farm animals. Scientists, engineers, and biologists working in the field of nanotechnology need to continue efforts to revolutionize animal health and production, and also require significant budgetary allocation in research and development.

References

Ajmal M, Yunus U, Matin A, Haq NU (2015) Synthesis, characterization and in vitro evaluation of methotrexate conjugated fluorescent carbon nanoparticles as drug delivery system for human lung cancer targeting. J Photochem Photobiol B 153:111–120

Awate S, Babiuk LA, Mutwiri G (2013) Mechanisms of action of adjuvants. Front Immunol 4:114

Bai DP, Lin XY, Huang YF, Zhang XF (2018) Theranostics aspects of various nanoparticles in veterinary medicine. Int J MolSci 19(11):3299

Bartlett JG, Gilbert DN, Spellberg B (2013) Seven ways to preserve the miracle of antibiotics. Clin Infect Dis 56(10):1445–1450

Charitidis CA, Georgiou P, Koklioti MA, Trompeta A-F, Markakis V (2014) Manufacturing nanomaterials: from research to industry. Manuf Rev 1:11

Craighead H (2006) Future lab-on-a-chip technologies for interrogating individual molecules. Nature 442:387–393

Damian K, Konrad W (2018) The use of nanominerals in animal nutrition as a way to improve the composition and quality of animal products. J Chem 2018:5927058

Durfey CL, Swistek SE, Liao SF, Crenshaw MA, Clemente HJ, Thirumalai RVKG, Steadman CS, Ryan PL, Willard ST, Feugang JM (2019) Nanotechnology-based approach for safer enrichment of semen with best spermatozoa. J Anim Sci Biotechnol 10:14

Emami T, Madani R, Rezayat SM, Golchinfar F, Sarkar S (2012) Applying of gold nanoparticle to avoid diffusion of the conserved peptide of avian influenza nonstructural protein from membrane in Western blot. J Appl Poult Res 21:563–566

Falchi L, Khalil WA, Hassan M, Marei WFA (2018) Perspectives of nanotechnology in male fertility and sperm function. Int J Vet Sci Med 6(2):265–269

Feugang JM, Youngblood RC, Greene JM, Willard ST, Ryan PL (2015) Self-illuminating quantum dots for non-invasive bioluminescence imaging of mammalian gametes. J Nanobiotechnol 13:1–16

Gross M (2013) Antibiotics in crisis. Curr Biol 23(24):R1063–R1075

Hill EK, Li J (2017) Current and future prospects for nanotechnology in animal production. J Anim Sci Biotechnol 8:26

Knight-Jones TJD, Rushton J (2013) The economic impacts of foot and mouth disease - what are they, how big are they and where do they occur? Prev Vet Med 112(3-4):161–173

Kumanan V, Nugen SR, Baeumner AJ, Chang YF (2009) A biosensor assay for the detection of Mycobacterium avium subsp paratuberculosis in fecal samples. J Vet Sci 10(1):35–42

Muktar Y, Bikila T, Keffale M (2015) Application of nanotechnology for animal health and production improvement: a review. World Appl Sci J 33:1588–1596

Odhiambo JF, DeJarnette JM, Geary TW, Kennedy CE, Suarez SS, Sutovsky M, Sutovsky P (2014) Increased conception rates in beef cattle inseminated with nanopurified bull semen. Biol Reprod 91:1–10

Peled N, Ionescu R, Nol P, Barash O, Mccollum M, Vercauteren K, Koslow M, Stahl R, Rhyan J, Haick H (2012) Detection of volatile organic compounds in cattle naturally infected with Mycobacterium bovis. Sens Actuators B Chem 171–172:588–594

Pulendran B, Ahmed R (2011) Immunological mechanisms of vaccination. Nat Immunol 12(6):509–517

Roco M (2003) Nanotechnology: convergence with modern biology and medicine. Curr Opin Biotechnol 14(3):337–346

Sabry MIE, McMillin KW, Sabliov CM (2018) Nanotechnology considerations for poultry and livestock production systems – a review. Ann Anim Sci 18(2):319–334
Scott NR (2005) Nanotechnology and animal health. Rev Sci Tech 24(1):425–432
Sekhon BS (2014) Nanotechnology in agri-food production: an overview. Nanotechnol Sci Appl 7:31–53
Soenen SJH, Himmelreich U, Nuytten N, Pisanic TR, Ferrari A, De Cuyper M (2010) Intracellular nanoparticle coating stability determines nanoparticle diagnostics efficacy and cell functionality. Small 6:2136–2145
Underwood C, van Eps AW (2012) Nanomedicine and veterinary science: the reality and the practicality. Vet J 193(1):12–23
Vasquez ES, Feugang JM, Willard ST, Ryan PL, Walters KB (2016) Bioluminescent magnetic nanoparticles as potential imaging agents for mammalian spermatozoa. J Nanobiotechnol 14:1–9
Wanapat M, Cherdthong A, Phesatcha K, Kang S (2015) Dietary sources and their effects on animal production and environmental sustainability. Anim Nutr 1(3):96–103
Wanzhe Y, Yanan L, Peng L, Qinye S, Limin L, Jiguo S (2015) Development of a nanoparticle-assisted PCR assay for detection of porcine epidemic diarrhoea virus. J Virol Meth 220:18–20
Xiao M, Huang L, Dong X, Xie K, Shen H, Huang C, Xiao W, Jin M, Tang Y (2019) Integration of a 3D-printed read-out platform with a quantum dot-based immunoassay for detection of the avian influenza A (H7N9) virus. Analyst 144(8):2594–2603
Yuan P, Ma Q, Meng R, Wang C, Dou W, Wang G, Su X (2009) Multicolor quantum dot-encoded microspheres for the fluoroimmunoassays of chicken Newcastle disease and goat pox virus. J Nanosci Nanotechnol 9(5):3092–3098

2 Nanoparticles in Plant Growth and Development

Sakshi Agrahari and Akhilesh Dubey

Abstract

Over time, nanotechnology has enabled a wide range of applications in the agricultural field due to the distinctive properties of nanoparticles, including high surface area, reactivity, agglomeration, penetration capability, size and structure. Nanoparticles have been by far advantageous for plant growth, development and protection. Nanoparticles bestow specificity in pesticide delivery, enhanced nutrient supply, managing pathogenicity, increasing photosynthetic capacity and germination rate. Apart from beneficial impacts on plants, there have been instances of toxicity and bioaccumulation of nanoparticles, which led to a few setbacks. Thus, it is necessary to have a complete knowledge of the positive and negative impacts of nanoparticles and to study all their characteristics in detail. This chapter highlights the impact of nanoparticles on the growth and development of plants.

Keywords

Nano-particles · Distinctive properties · Impact on plants · Bioaccumulation · Toxicity · Agriculture · Nanotechnology · Metals · Plant growth · Uptake and translocation · Carbon nanotubes · Graphene · Fertilizers · Phytotoxicity

2.1 Introduction

"Nanotechnology is a novel, innovative, interdisciplinary scientific approach of designing, developing, manipulating and application of materials at nano-scale" (Ali, Muhammad, et al. 2014). "Nano" signifies one-billionth unit, therefore

S. Agrahari · A. Dubey (✉)
Department of Biotechnology, Netaji Subhash Institute of Technology, New Delhi, India

M. Ghorbanpour et al. (eds.), *Biogenic Nano-Particles and their Use in Agro-ecosystems*, https://doi.org/10.1007/978-981-15-2985-6_2

nanotechnology involves substances quantified as billionth of a meter. Ten hydrogen atoms placed side by side cover a distance equivalent to a nanometre. Although nanotechnology involves the science of minute components, it also encompasses a much wider range of disciplines, including expertise from physics, biology, chemistry and various other disciplines.

Nanoparticles (NPs) of sizes less than 100 nm belong to an intermediate zone between an atom and its bulk components, having the capability to alter a material's physicochemical properties, i.e. exceptional reactivity, sensitivity and conductivity. (Mishra and Kumar 2009). Nanotechnology has been widely applied in the agricultural field, as it is quite a challenge to feed the increasing population, beyond 7 billion, as well as to simultaneously provide adequate nutrients.

Among the discrete molecules and their respective bulk components, there exists a transitional zone where nanoparticles lie, with properties that are novel from those of its bulk as well as molecular equivalents (Singh et al. 2015). With the development of nanotechnology, apart from the classical agricultural methods, scientists have tried to make use of the advanced characteristics of nanoparticles to enhance the growth and development of plants.

Nanosized components can be engulfed by bacteria and can penetrate plant cells (Liu et al. 2009a, b), and at high levels of dosage can induce phytotoxicity (Stampoulis et al. 2009). Research on nanotechnology-based agrochemicals has influenced numerous scientists to ponder over the advantages that nanotechnology can bestow upon agricultural crops. Advantages offered by nanotechnology include treatment of plants by nanocides, nutrient maintenance through nano-fertilizers and prevention of diseases (Moraru Carment 2003; Priester et al. 2012). In different areas, the effect and usage of nanomaterial variants such as carbon nanotubes, polymers, metals and nonmetals, quantum dots, magnetic particles, etc. are being studied (Rico et al. 2015).

Nanoparticles have the capability to transform the food and agricultural industries due to their unique characteristics that enhance nutrient adsorption by plants, molecular level disease management and pathogen detection. It involves the operation at a similar level with those of the disease causing particles, which enables the instinctive detection and elimination of those particles (Prasad et al. 2014).

Nanoparticles with characteristics such as enhanced reactivity, small surface to volume ratio, surface structure, agglomeration, etc. have found application in various areas, like cancer therapy, nano-pharmacology, targeted drug delivery, nanomedicine and delivery of agrochemicals. These extraordinary characteristics of nanoparticles have enabled scientists to study their interaction with plants, both in vivo and in vitro.

Nanoparticles are specifically designed and engineered with unique surface and chemical properties. A varying class of nanoparticles have been produced, including metal oxide nanoparticles, magnetic nanoparticles, gold nanoparticle, mesoporous silica nanoparticle, quantum dots, carbon nanomaterials such as carbon nanotubes, fullerenes and graphene (Wang et al. 2016a). Scientists demonstrated that carbon nanotubes have the ability to penetrate seed husks, which enables faster germination of seeds (Zhang et al. 2015a).

Nanoparticles are designed such that they can favour usage of optimal concentrations, regulated release, decline in phytotoxicity and targeted delivery. Pesticides are loaded into the inner core of the mesoporous silica nanoparticles, leading to regulated release as well as protection from photodegradation (Wang et al. 2016a). Quantum dots have a diameter from 2 to 10 nm. They are minute semiconductor particles that are used for cellular imaging and labelling by the production of fluorescence.

Nanoparticles have found important perspective in targeting specific biotic life forms through their unique delivery system and are highly used in the medical field. A similar principle is applied in plants, specifically in tackling phytopathological infections, growth adjuvant and supplementation of nutrients.

Use of nanoparticles is still a fresh and new approach, which needs further study and research for proper understanding and implementation of their properties for the betterment of food and crop as well as in other fields of science. In this chapter, will discuss some of the widely studied effects of nanoparticles on agriculture, which specifically focus on plant growth promotion. Nanotechnology is a niche area which still needs a thorough understanding, but it is sure to expand its boundaries, including agriculture and allied sectors, providing immense benefits.

In food and agricultural fields, nanotechnology has found widespread applications, but still at a budding stage, and thus they require thorough knowledge and guidance so as to indulge in developing "green nanotechnologies" and take into account all the necessary precautions in order to minimize the prospective unfavourable impacts they can pose to the environment and to human health (Mishra and Kumar 2009).

Nanoparticles have gained demand in the agricultural and medical fields due to their unique physicochemical properties, including ability to penetrate, larger surface area and chemically active. With the increase in demand they have also become potential threats to the environment (Borm et al. 2006; Kreyling et al. 2006; Lam et al. 2006; Maynard 2004).

Nano-enabled products have been profoundly used globally due to their immense availability, which can be released into the environment in high quantities as engineered nanoparticles usually displaying properties different from their bulk components (Geisler-Lee et al. 2012).

Nanoparticles have aggregation tendencies and low solubility in water, which limits their access to most living organisms (Maynard et al. 2004; Brant et al. 2005). Certain experiments demonstrated penetration of nanoparticles through skin, food and water or air into animals and humans (Oberdörster et al. 2006; Jain et al. 2007). Nanoparticles can translocate along the food web and can accumulate in higher-level consumers, influencing the extent of toxicity on organisms which belong to distinct trophic levels (Krysanov et al. 2010).

Metal oxide nanoparticles, i.e. titanium oxide and carbon-based fullerene nanoparticles, show microbial toxicity, as well as fullerenes display low mobility in soil and aqueous medium (Lecoanet and Wiesner 2004; Brant et al. 2005). Nanoparticle toxicity is not only associated with size but also with their physicochemical properties.

It is presumed that biologically attainable nanoparticles will materialize in high quantities in the future and will be higher in organisms at higher trophic levels. Accumulation of nanoparticles in organs or tissues leads to increased effects on cells and cellular structures. The issue with translocation of nanoparticles within the food web is that it is inevitable.

Farmers used conventional chemicals for controlling pests and pathogens, which had a drastic impact on the environment as well as on the farmer's economy, because upon application 90% was lost as runoff or into the air (Thul and Sarangi 2015). Release of pesticides and insecticides through a nano-scaled delivery system led to the application of these chemicals only when the need arises (Gruère et al. 2011).

Toxicity and accumulation of nanoparticles have led to the necessity to search for biodegradation pathways for nanoparticles and their impacts on living species, including the natural structures and functions as well as artificial biocenoses (Krysanov et al. 2010).

At present, the knowledge on the providence of nanoparticles into the environment is scarce and their bioaccumulation by living beings and their tissues, is also sparse rather practically absent and specifically whether they have chronic impact or not (Fig. 2.1).

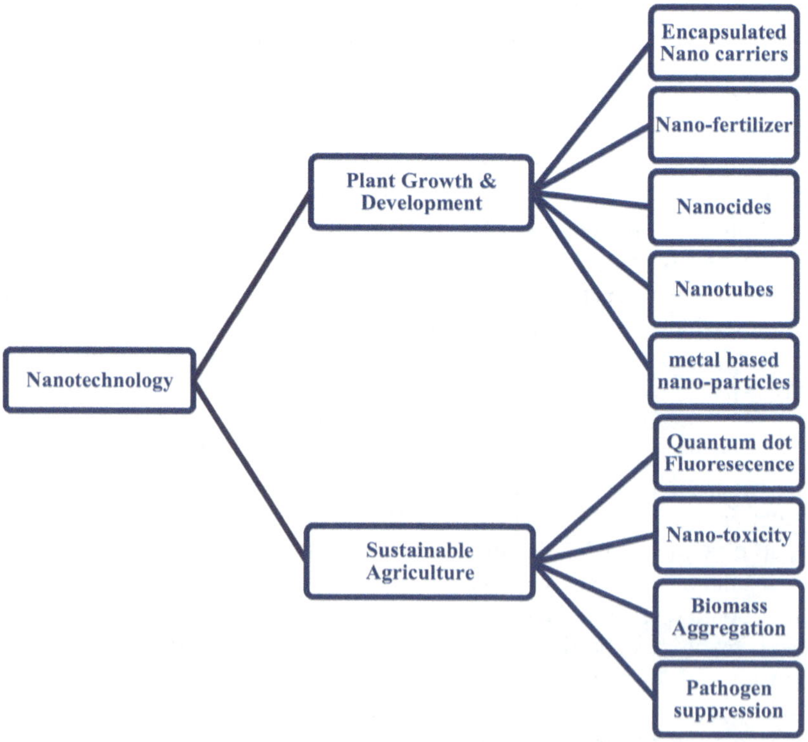

Fig. 2.1 Diversification in the application of nanotechnology (Modified from Ditta et al. 2015)

2.2 Molecular Mechanism of Nanoparticles in Plant Growth and Protection

2.2.1 Mode of Entry and Uptake

Plant transport pathways play a vital role in the entry of nanoparticles into plants as well as into the surrounding environment, which can lead to their accumulation. Engineered nanoparticles were transferred through protoplasts, intact plants and dissected organs (Wang et al. 2016a). Nanoparticle entry into plant cells largely depends on the species of plant and the nanoparticle properties and is obstructed due to its cell wall (Singh et al. 2015). Entry through the plant cell wall occurs either through engulfing, endosome formation or through sieving mechanism. Entry of ENPs occurs either by way of organs and tissues above the ground level, i.e. stomata, hydathodes, cuticles, stigma and trichomes, or by the root tissues, as well as through junctions and injuries (Wang et al. 2016a).

A plant cell wall has pores with diameters in the range of 5–20 nm. Usually, through the cell wall water molecule as well as solute accretion occurs, this is due to the porous polysaccharide fibre matrix of the cell wall (Tripathi et al. 2017). Thus, for efficient entry into the plant cell wall, the size of the nanoparticles should be less than the cell wall pore diameter. The efficient passage of these nanoparticles through the pores enables them to extend towards the plasma membrane. The pore size of the cell wall can even be enlarged upon association with engineered nanoparticles, thereby favouring the uptake of nanoparticles (Nair et al. 2010). Plants grown in soil as well as on sand depicted no uptake or minimal uptake of nanoparticles. Therefore plant cells are grown on growth medium for the uptake of nanoparticles. The low or no uptake of nanoparticles by soil- and sand-grown plants is due to the adhesion of metal oxide nanoparticles (Singh et al. 2015). The growth medium varies with different types of nanoparticle uptake.

The ion channels and carrier proteins which are embedded in the membrane also lead to the transport of nanoparticles across the membranes in the cell. Nanoparticle entry into the plant cells is therefore an active transport process which is regulated by various cellular mechanisms, including signal transfer, plasma membrane regulation and recycling (Tripathi et al. 2017). When applied on the surface of leaves, the entry of nanoparticles is governed either by the stomatal openings, by the trichomes or by cuticular routes which accumulate and are then transported to the varying plant tissues, which is described as a top-down movement.

After penetration into the root's epidermal cell wall and membrane, the nanoparticles enter the vascular bundles through a series of steps. To achieve the crossing of nanoparticles into the cell membrane, it is necessary for the nanoparticles to undergo passive integration from the endodermal apoplast (Tripathi et al. 2017). The penetration into the seed coat occurs by uptake through parenchymatous spaces and is regulated by the aquaporins present in the seed, which thereby enhances liquid diffusion into the cotyledons (Wang et al. 2016a). Entry of nanoparticles through the vascular system or lateral root sites occurs due to incomplete formation or breakdown of Casparian strips, respectively.

In comparison to plants, uptake of nanoparticles by animal cells occurs by the endocytic pathway which includes both clathrin-independent and clathrin-dependent pathways. Fluid phase and caveolae endocytosis and phagocytosis constitute the independent pathway whereas the synthesis of clathrin-coated forms by the formation of sheathed assembly on the plasma membrane constitutes the dependent pathway.

The uptake of zinc oxide nanoparticles was established by the electron microscopic images, which depicted the damage caused to the epidermal as well as cortical cells due to the uptake of nanoparticles. This even led to an injury on the vascular and endodermal cells, resulting in the inhibition of growth of ryegrass.

2.2.2 Nanoparticle-Plant Interactions

Engineered nanoparticles include:

- Carbon nanomaterials – carbon nanotubes (CNTs), single-walled carbon nanotubes (SWCNTs), multi-walled carbon nanotubes (MWCNTs), graphene and fullerenes
- Quantum dots – Cadmium-selenium (CdSe), Cadmium-telerium (CdTe), etc.
- Metal-based nanomaterials – metal oxide, i.e. zinc oxide (ZnO), titanium oxide (TiO_2), copper oxide (CuO) silicon dioxide (SiO_2), etc.; zero valent, i.e. iron (Fe), silver (Ag), gold (Au), etc.; and metal salts, i.e. silicates and ceramics
- Nanopolymers – latex, dendrimers, etc.

These particles have exceptional reactivity and surface areas in comparison to their bulk equivalents (Service 2003) which is the result of eccentric physical and chemical properties. The presence of nanoparticles in the environment and biotic surroundings leads to inevitable interactions with the biotic components, thereby causing physicochemical alterations, like dissolution, incidental coating by biomolecules and redox reactions (Rico et al. 2015).

Plants exhibit intense interactions with the external environment, thus exposure to nanoparticles affects plants strongly. These interactions lead to numerous changes, including anatomical and morphological, the alteration depends largely on the concentration and nature of nanoparticles.

López-Moreno et al. (2010) showed that roots of soybean when germinated upon treating with zinc oxide nanoparticles transformed from +2 to nitrite or acetate, whereas when treated with cerium oxide no such transformation occurred.

Dissolution is widely considered for the transformation of metal-based nanomaterials, which alters their properties and fate in the plant species. Plant roots excrete organic acids, which are necessary for biotransformation of nanoparticles as they promotes dissolution. Transformation of nanoparticles takes place outside the roots as well as after entering the roots of a plant. Metal nanoparticles with variable valences undergo redox reactions in the soil, transforming these particles by their interaction with plant's biogenic redox agents (Rico et al. 2015). Nanoparticles upon interacting with plant cells release reactive oxygen species (ROS) in large amounts due to stress which in turn affects the plant biomolecules. Among nanoparticles, carbon nanotubes specifically induce accumulation of ROS.

2.2.3 Translocation

Research on the uptake of nanoparticles led to the study of their absorption, translocation or transport and accumulation in plants, which is still not very clearly understood. Ongoing research on the transport of nanoparticles suggests that these particles have the ability to move across tissues either intra- or extracellular to the xylem.

Nanoparticles with small sizes are able to pass through the cell wall pores into the cell membrane whereas those larger in size than the pores accumulate outside the cell wall, unable to enter. For example, nano-conjugates of titanium oxide alizarin red (30 nm) can traverse through the cell wall and accumulate in the subcellular portions of roots and leaves of Arabidopsis plant (Kurepa et al. 2010); on the other hand the accumulation of 25 nm titanium oxide on the surface of roots of maize altered the hydraulic conductivity and availability of water, thus reducing transpiration rate and affecting development of plants (Asli and Neumann 2009).

Sabo-Attwood et al. (2012) demonstrated that the absorption of Au nanoparticles takes place in a size selective pattern in tomato seedlings, among which the 18 nm size particles are restricted from entering the pores and thus accumulate on the surface of roots whereas the 3.5 nm Au nanospheres easily traverse into the plant cells.

Zhu et al. (2012) depicted that roots easily take up gold nanoparticles (AuNPs) with positive charge whereas the negatively charged ones translocate from roots into the stems as well as leaves. The sequence of concentration of Au in the roots was AuNPs(+) > AuNPs(0) > AuNPs(−) whereas the reverse sequence was present in the shoots. Au concentration in rice roots followed the above order whereas reversed order for shoots, specifying favourable translocation of Au nanoparticles with negative charge.

Birbaum et al. (2010) suggested that treating maize plants with 37 nm cerium oxide nanoparticles either in the form of aerosols or in suspension form resulted in no internalization or translocation. Wang et al. (2012) provided evidence on the penetration of CuO nanoparticles into the root system of maize plants by the combination of energy dispersive spectroscopy (EDS) and transmission electron microscopy (TEM) of the xylem sap. Engineered nanomaterials upon reaching the xylem tissues are translocated towards the aerial segments. Through the experiments by split roots and observation by high-resolution TEM showed that the translocation of these nanoparticles can also occur from shoots into the roots by phloem, thus reducing nanoparticles from the copper (Cu) (II) to the copper (Cu) (I) state (Zhang et al. 2015b).

2.3 Effect of Nanoparticles

Nanoparticles with the emerging advancement have shown varying beneficial effects upon plants. They have been implemented for conversion of waste to energy, production of by-products via nano-bioprocessing, usage as nano-fertilizers and nano-pesticides for regulated delivery.

These nanoparticles play an important role in plant growth and development, which renders them widely used. Different nanoparticles bestow a different effect on plants as every particle interacts with plants in a different pattern, including alteration in morphology as well as physiology. The chemical composition, reactivity, size and other properties of these nanoparticles determine its function upon interaction with plants, which can be positive or negative (Rico et al. 2015).

Copper, gold, zinc oxide, cerium oxide, titanium oxide and silver nanoparticles are some of the widely used and synthesized metal nanoparticles. Apart from these, manganese, cobalt-ferric oxide and ferric oxide are also used. Among the engineered nanoparticles, both metal and carbon-based nanoparticles have the ability to accumulate ROS, which affects the macromolecules in plants and thus leads to stress.

Carbon-based nanoparticles have distinctive chemical, electrical, mechanical and thermal characteristics which render them their significant functions (Singh et al. 2015). Single-walled carbon nanotubes are responsible for the transport of DNA and other molecules, such as dye, across plant cells as well as from the outer surrounding into the plant cells. They are regarded as nanotransporters. On the other hand, MWCNTs play a vital role in the enhancement of water uptake with the uptake of nutrients, thereby augmenting germination of seeds and growth of plants. Studies reveal that carbon nanotubes promote accumulation of ROS and lead to peroxidation of lipid molecules in the root tips of seedling and cultures of cells (Siddiqui et al. 2015). The rate of germination of seeds is elevated due to the perforation of nanoparticles into the cell wall, which strengthens uptake of water.

2.3.1 Effect on Photosynthesis

For plant growth, hardly 2–4% of the radiation energy is used. Photosynthesis provides an easy detection parameter for the stress produced due to the living and non-living factors. Nanoparticles disturb the photochemical fluorescence, photosynthetic activity and efficiency and the quantum yield, causing oxidative stress to plants.

Govorov and Carmeli (2007) conducted an experiment in which nAg and nAu were bound to the chlorophyll of the reaction centre of PSI forming a unique hybrid system, leading to two contradictory effects on photosynthetic efficiency, i.e. the nanoparticles had plasmon resonance effect, which enhanced the chlorophyll's light absorption efficiency and hence led to a decrease in the photosystem's quantum yield which was due to a tenfold increase in the transfer of energy upon chlorophyll enhancement (Siddiqui et al. 2015).

Scientists observed that there was three times increase in the activity of photosynthesis upon supplementing single-walled carbon nanotubes into chloroplasts compared to those lacking nanoparticles (Giraldo et al. 2014). Treating with SWCNTs increased the transport of electrons and enhanced the ability of plants to recognize signal molecules, such as nitric oxide (Siddiqui et al. 2015). The enhancement of photosynthetic activity of plants by SWCNTs is due to its transport and irrevocable concentration within the chloroplast's lipid layer. The chloroplast-SWCNTs complex increases transport rate in leaves in vivo by a process supplemented with photo absorption.

Noji et al. (2011) deduced through his experiments that, the activity of reaction responsible for the generation of photosynthetic oxygen can be stabilized by the formation of a complex between the PSII and the nano-mesoporous silica compound (SBA), thus depicting the transport of electrons from water, due to light-mediated reaction, to quinine. This complex is expected to render properties needed for the development of artificial photosynthetic system and photo biosensors (Siddiqui et al. 2015).

The effect of Au nanoparticles upon fluorescence quenching of chlorophyll *a* of PSII in soybean leaves was analysed. The extracted chlorophyll was blended with varying concentrations and sizes of Au nanoparticles and their absorbance and fluorescence spectra at 538 nm and 625–800 nm was noted, respectively. This led to the conclusion that with increasing concentration of Au nanoparticles, the fluorescence quenching increased with increase in absorbance, whereas at the largest size of Au nanoparticles, the absorbance was lowest. Depicting that size of nanoparticles suppresses the fluorescence, thus lower the size of Au nanoparticles, higher will be the fluorescence quenching. The suppression of fluorescence was due to the high surface area which favours increase in adsorption of chlorophyll and enhancing electron transfer from chlorophyll to nanoparticles.

Metal nanoparticles, such as nano-anatase titanium oxide refine the rate of photosynthesis, transpiration rate and water conductivity in plants, upon their exogenous application. The photocatalysing feature of titanium oxide refines the absorption and transformation of light as well as encourages assimilation of carbon dioxide (Siddiqui et al. 2015). Carbon assimilation is enhanced by the activation of rubisco, the most abundant enzyme responsible for the carboxylation reaction occurring during the light-mediated photosynthesis pathway, and hence promotes plant growth.

The function of nano-anatase largely depends on its high thermal conductivity, high surface area and photocatalytic activity. The chloroplast of spinach upon treatment led to enhanced absorption of light by chlorophyll *a*, oxygen evolution rate, quantum yield and transfer of electrons in PSII. However, in *Ulmus elongata* the foliar application of 0.1–0.4% *n*TiO2-A when exposed to 800 and 1600 μmol $m^{-2}s^{-1}$light intensity led to decrease in quantum yield of photosystem II (PSII), electron transfer and fluorescence quenching but increase in water loss and non-photochemical quenching.

Similarly, the effect on photosynthesis by zinc oxide nanoparticles varied with the species of plant, though the concentration and stage of growth were among the other parameters upon which its activity depended. ZnO NPs (24 ± 3 nm) showed that after 20 days, there was reduction in photosynthetic activity, chlorophyll concentration and stomatal conductivity at 800 mg/kg in corn but no change in the other stages at 400 mg/kg (Du et al. 2017).

Among all the photosynthetic pigments, chlorophyll *a* has more sensitivity towards photodegradation and is used as an indicator of toxicity by nanoparticles, in comparison to other parameters. The photosynthetic efficiency can be determined by the chlorophyll *a*/*b* ratio. The ratio determines the availability of nutrients, specifically N and light to plants. Comparison of nanoparticle-treated plants with that of control elaborates that the ratio of chlorophyll in ferric oxide and cobalt ferric oxide nanoparticles decreased. These observations were in contrast to chloroplasts treated with cerium oxide nanoparticles, which have the ability to eliminate ROS, thereby protecting chloroplast and improving photosynthesis. However, these nanoparticles alter the stomatal openings and modify the microstructure of chloroplast, which adversely effects photosynthesis (Du et al. 2017).

Therefore, it is necessary to have detailed knowledge about the interaction of nanoparticles with plants' photosynthetic system, which will hence determine how nanoparticles promote anti-oxidant defence and oxidative stress in plants. It has been shown that ions released by the nanoparticles lead to stress generation and induce accumulation of ROS.

2.3.2 Effect on Seed Germination

The initial stages of growth and development of plants begins with seed germination, leading to elongation of roots and emergence of shoot. Depending on the concentration and on the plant species, nanoparticles have varying effects on seed germination. Metal oxide nanoparticles and carbon-based nanoparticles exhibit diversifying effects on seed germination, root elongation and shoot growth. Scientists have shown through their experiments that different nanoparticles augment both positive and negative effects upon different plant species. Upon exposure to metal oxide nanoparticles, the effect on seed germination as well as elongation can be inhibitory, neutral or promoting.

Among the metal-oxide nanoparticles, TiO_2 nanoparticles have the ability to enact as photo catalyst, thereby inducing redox reaction which enhances germination of seeds with the initiation of growth of plumule and radicle in the seedlings of canola (Crabtree 1998). TiO_2 NPs specifically favour vigour of aged seeds and formation of chlorophyll as well as enhance the activity of rubisco enzyme, which promotes photosynthesis and growth of plants (Siddiqui et al. 2015).

The effect of CuO nanoparticles in rice exhibited inhibitory patterns on elongation and germination of seeds specifically with size greater than 50 nm. This pattern was also seen in soybean, chickpea, maize, cucumber and Indian mustard seeds. These inhibitory patterns were studied by various scientists, including Da Costa and Sharma (2016), Wang et al. (2015), Adhikari et al. (2012).

Similarly, ZnO nanoparticles at lower concentrations exhibited neutral or promoting effects on seed germination of soybean seeds and inhibitory effects at higher concentrations, depicting a dosage-dependent pattern (Stampoulis et al. 2009; Ghodake et al. 2011; Lee et al. 2010; Yang et al. 2015). In a case studied by Helaly et al. (2014), supplementing ZnO nanoparticles with MS media stimulated plantlet regeneration and somatic embryogenesis, and enhanced the activity of superoxide, peroxidase, dismutase and synthesis of proline and catalase, which led to an increase in biotic stress tolerance. Wang et al. (2012) showed that CeO_2 nanoparticles led to no change in the germination of rice and tomato seeds, whereas there was significant reduction in the germination of seeds of cucumber and corn (Du et al. 2017).

Ag and Fe nanoparticles with zero valence at varying concentrations inhibit germination of seeds, which is observed at the incubation process of seeds instead of the soaking process. AgNPs had no effect on germination but caused alteration in the cell wall's chemical composition, which confirmed that the impact of nanoparticles was up to cellular and molecular levels. Among metal-based nanoparticles, AuNPs have shown no impact on germination of barley seeds, rather they led to lower biomass production and stimulating impact on growth. However, a contradictory observation by Savithramma et al. (2012) and Gopinath et al. (2014) in *Boswellia ovalifoliolata* and in *Gloriosa superb*, respectively, claimed that AuNPs did improve the germination of seeds.

When silicon oxide nanoparticles (up to $8gL^{-1}$) were applied exogenously to seedlings, it was observed that they enhanced growth of the seedlings as well as the quality, including diameter of root collar, quantity of seedling's lateral roots, mean height and root length. This exogenous application led to improved germination of tomato seeds, germination index, fresh and dry weight of seedlings, seed vigour index and utilization of nutrients which increases the parameters for germination making them available for the seeds thereby maintaining the growth medium's pH and conductivity.

Apart from enhancing germination and quality of seeds, SiO_2 nanoparticles also stimulated chlorophyll synthesis, which is effective for crop growth and yield (Haghighi et al. 2012; Li et al. 2012). Under salinity constraints, silicon oxide nanoparticles improve accumulation of proline, antioxidant enzymes, free amino acids which improve abiotic stress tolerance by plants (Kalteh et al. 2014; Shalaby et al. 2016).

Apart from metal oxide nanoparticles, multi-walled carbon nanotubes have been found to enhance seed germination. They specifically prompt the uptake efficiency of macronutrients especially Ca and Fe and water, which will in turn enhance the growth and development of plants. Like other nanoparticles, even MWCNTs have the ability to penetrate the seed coat, which triggers the germination of barley, corn and soybean seeds when added to a sterilized agar medium. This penetration was due to the regulatory effects of MWCNTs on the expression of genes encoding the proteins of the water channel.

Khodakovskaya et al. (2012) showed that upon upregulating the marker genes responsible for formation of cell wall (NtLRX1), cell division (CycB) and transport of water (aquaporin, NNtPIP1), they accelerated the tobacco cell growth in culture.

This proved that MWCNTs play a vital role as regulators of the growth and germination of seeds (Ditta and Arshad 2016). MWCNTs upon primary uptake and aggregation in roots, can improve the activity of peroxidase and dehydrogenase enzymes, which stimulate the growth of roots and shoots. Followed by accumulation, these MWCNTs translocate from the roots towards the leaves, which induces expression of genes (Smirnova et al. 2012).

Similarly, Graphene accelerates germination of seeds and specifically decreases the time duration of germination. The rate of germination of seeds treated with graphene showed exceptional increase for the first few days as compared to untreated seeds (Zhang et al. 2015a).

2.3.3 Root and Shoot Growth

Nanoparticles not only affect photosynthesis or seed germination, they also affect root and shoot growth. They have the ability to enhance or inhibit the root and shoot length. Different nanoparticles have differing impacts on root and shoot growth, including those discussed so far.

Roots of *Vigna radiata* and *Cicer arietinum* seedlings, upon absorbing zinc oxide nanoparticles, boosted the length of roots and shoots and their biomass as well (Mahajan et al. 2011). Gruyer et al. (2013) found that depending on the species of plant, Ag nanoparticles can induce and inhibit the elongation of roots. In case of barley, the length of roots increased whereas in lettuce it was inhibited. The enhancement of root growth in *Crocus sativus* occurs due to blockage of ethylene signalling. As compared to $AgNO_3$, AgNPs increased the length of roots in maize, barley and cabbage (Siddiqui et al. 2015).

Apart from the plant species, the morphology of nanoparticles also plays a vital role in root growth. Syu et al. (2014) demonstrated the effect on root growth and the physiological changes in Arabidopsis seedlings when subjected to Ag nanoparticles with three different morphologies, among which the decahedral morphology exhibited the highest level of promotion of root growth, whereas there was no effect on root growth in the case of spherical one, rather stimulated accumulation of anthocyanin at high levels. Ag nanoparticles also inhibited elongation of root by activating aminocyclopropane-1-carboxylic acid (ACC) in Arabidopsis seedlings and lowered the expression of ACC oxidase 2 as well as ACC oxidase 7, implicating that perception and synthesis of ethylene was inhibited by Ag nanoparticles (Siddiqui et al. 2015).

Metal-based nano-particles including silicon, palladium, high levels of copper, low levels of gold and mixture of gold and copper led to a positive impact on seedlings growth and ratio of shoot to root, while cerium oxide nanoparticles effected only root elongation of lettuce at 2000 mg/L concentration. Seeds of parsley when treated with nano-anatase had stimulated root and shoot length, germination and chlorophyll content of seeds. In pumpkin, elongation of roots occurs when exposed to iron oxide nanoparticles.

ZnO nanoparticles promote elongation of roots in soybean. In *Cyamopsis tetragonoloba*, the biomass of plant, root and shoot length, synthesis of protein and chlorophyll and other parameters of growth improved on exposure with ZnO nanoparticles (Singh et al. 2015). In radish and rape plant, the growth of roots decreased upon incubation in a suspension of Zn nanoparticles. However, this kind of inhibition was not seen in suspensions of ZnO nanoparticles because of seed coat's selective permeability.

In a range of plants, carrot, cucumber, cabbage and corn growth declined as pure alumina nanoparticles (13 nm) reduced the elongation of roots without causing any modifications. In the presence of Cu nanoparticles, seed germination in lettuce led to an increase in the ratio of shoot to root in comparison to plants in the absence of nanoparticles (Nair et al. 2010).

Supplementation of nutrient medium having protein in which infusorium *Tetrahymena pyriformis* was cultured with nanotubes, it had unexpected growth simulation and increased nanotube concentration (Zhu et al. 2006). This unexpected simulation was presumed to be due the binding between protein and nanotube supplements, thus increasing protein penetration into cells and enhancing growth (Krysanov et al. 2010).

Carbon-based nanoparticles, like carbon nanotubes and graphene, have also shown varying impact on roots and shoots of plants. In onion and cucumber, elongation of roots was induced by carbon nanotubes, as well as formation of nanotube sheets on the surface of roots of cucumber upon interaction with fCNTs and CNTs. But these nanotubes were unable to enter the roots. These nanotubes had no effect on cabbage and carrot plants. Elongation of roots was inhibited by fCNTs in lettuce and by CNTs in tomato, while tomato being highly sensitive to CNTs. Scientists demonstrated that at concentrations of 0.5, 0.9 and 0.16 gL^{-1}, SWCNTs enhanced the growth of roots in onion and cucumber seeds.

Among, carbon-based nanoparticles, graphene showed exceptional effects on seeds. In an experiment on tomato seedlings exposed to graphene, it was observed that on the 19th day, the seedlings exposed to graphene had stems up to 17% longer compared to the control seedlings and longer length roots of up to 12.5% compared to control (Zhang et al. 2015a).

2.3.4 Effect on Nutrient Delivery

In agriculture, the requirement for nutrients is fulfilled by the use of fertilizers as the soil lacks most of the macro- and micronutrients. Almost 35–40% of the overall productivity of crops is dependent on fertilizers and most fertilizers directly affect the growth of plants. At present, the nutrient utilization efficiency is quiet low, as approx. 50–70% of nitrogen provided through conventional fertilizers is lost. In the continually increasing population, the demand for food is increasing, which increases the need for macronutrients by crops.

To decrease the loss of nutrients, new systems were exploited for the delivery of nutrients which involved porous nanoscale plant parts that led to the reduction in nitrogen loss. Upon encapsulating fertilizers into nanoparticles the uptake of nutrients can be increased. Use of nano-fertilizers and nano-composites instead of the conventional fertilizers is an exquisite breakthrough in science as these have a slow nutrient release rate, which continues throughout the growth of crops, enabling the crops to utilize nutrients without wasting them and prevents water pollution (Singh et al. 2015).

Nano-fertilizers are composed of nanosized macronutrients such as N, P, K, Ca, Mg and S, which are needed in high demand by crops or to supplement the activities of chemical fertilizers. N is the chief nutrient for the growth of all plants, which is released slowly through urea-coated zeolite chips. Similarly, in the soft wood cavities of *Gliricidia sepium* nanoparticles of hydroxyapatite, a derivative of urea was encapsulated and observed for the slow and feasible release of nitrogen. Zeolites have a crystal structure with honeycomb-like layers and occur as natural mineral groups, supplying nutrients slowly on demand (Manjunatha et al. 2016).

Fertilizers are encapsulated within nanoparticles through three ways (Naderi and Danesh-Shahraki 2013).

1. Encapsulation of nutrients within nanoporous components.
2. Thin coating of polymer film.
3. Delivering nanoscale dimensions in the form of particles or emulsifiers.

Phosphorus is one of the essential components in most metabolites and is involved in almost all the metabolic processes, which is supplied through conventional fertilizers. Crops take up only 20% of the available phosphorus while the rest 80% accumulates in soil and water bodies due to runoff leading to eutrophication. Use of nanotechnology increases the efficiency of phosphorus utilization and eliminates environmental menace.

In greenhouse conditions, soybean (*Glycine max*) showed 33% increase in the rate of growth and 20% yield of seeds in comparison to chemical phosphatic fertilizers due to continuous Ca and P supply (Singh et al. 2015). Upon foliar application of nano-fertilizers, there was prominent increase in the yield of crops (Tarafdar et al. 2012a, b). Yield of 80 kg ha^{-1} of cluster bean and pearl millet was obtained through foliar application of 640 mg ha^{-1} (40 ppm concentration) nanophosphorus under an arid environment (Manjunatha et al. 2016).

Compared to chemical P fertilizers, administration of nanoparticles elevated the rate of growth and germination of seeds by 33% and 20%, respectively, indicating that soybean roots absorbed hydroxyapatite nanoparticles as an implicit P source. Nano- and sub-nano-composites control the release of nutrients from the fertilizer capsule (Singh et al. 2015).

The nutrient utilization efficiency (NUE) increases up to three times as well as provides ability to tolerate stress by the use of nano-fertilizers. Combining nano-fertilizers with nano-devices releases N and P fertilizer and the uptake by plants in a synchronized manner, thereby eliminating the undesirable loss of nutrients and preventing interaction with soil, air, microorganisms and water (Manjunatha et al. 2016). With respect to nutrient utilization, scientists demonstrated that nano-composites containing macronutrients, micronutrients, amino acids and mannose upon application influenced uptake and utilization (Ali et al. 2014). Iron chelated nano-fertilizers showed increased photosynthesis, adsorption and surface area expansion in leaves (Singh et al. 2015).

2.3.5 Effect on Rhizospheric Environment

Soil is an omnipresent habitat for a wide range of microbes interacting with the biotic components, specifically rhizosphere, and with each other. At the rhizospheric site, a complex association between root and associated microbes occurs with a high diversity of microbes. Microbes present in soil are involved in the productivity of crops, functions of the ecosystem and maintenance of soil health (Mishra and Kumar 2009).

Effective molecular techniques and certain biochemical processes were developed by microbes far before detected by plants to detoxify, efflux and accumulate metal ions. Microbes have the ability of volatilizing metal ions in order to eliminate acute toxicity (De Souza et al. 2000).

Rhizobacteria exhibiting propitious effects on growth of plants are termed as plant-growth-promoting rhizobacteria (PGPR). PGPR are soil-borne, free-living bacteria isolated from the rhizosphere, and upon application to seeds or crops enhance plant growth. PGPRs are involved in controlling plant pathogens, nutrient cycle, growth of seedlings and many other ecosystem functions. PGPR are associated with asymbiotic fixation of nitrogen, production of phytohormones, i.e. IAA (Indole-3-Acetic Acid), gibberellins, cytokinins, phosphate solubilization and production of siderophores, which help in the growth of plants (Mishra and Kumar 2009).

Bacterial taxa are altered in a dose-dependent manner in which some taxa increase in proportion while others decrease, resulting in reduction in diversity. The application of nanoparticles directly on land or through treated biosolids with mobile nanoparticles interacts with soil microbes. Microbes can absorb and accumulate nanomaterials effectively and initiate mobilization through food chain and altering taxa with diverse populations, i.e. bacteria, plants, fishes, within the food web (Holden et al. 2013). Soil bacteria and fungi help plants to take up nutrients easily from the soil (Thul and Sarangi 2015).

Uptake of cerium oxide nanoparticles into the roots and nodules led to the elimination of nitrogen fixation and impairing growth of soybean. Zinc oxide, titanium dioxide and silicon dioxide nanoparticles relay toxic impact upon bacteria, which intensifies in the presence of light (Thul and Sarangi 2015).

Fortner et al. (2005) demonstrated the inhibitory effect of C_{60} fullerene aggregates on *Escherichia coli* (gram negative) and *Bacillus subtilis* (gram positive) which were grown on rich and minimal media and under both aerobic and anaerobic conditions, respectively. It was observed that at concentrations above 0.4 mg/L, complete inhibition of bacterial growth occurred in both the cultures which were subjected to both the absence and presence of oxygen and light conditions, while at concentrations up to 2.5 mg/L it was observed that no inhibition occurred in rich media, which can be because of precipitation of C_{60} or by protein coating in the media (Mishra and Kumar 2009).

Nyberg et al. (2008) depicted that there was no effect of C_{60} fullerene nanoparticles on anaerobic microbes. Fullerenes inhibit growth of bacteria mostly found in soil and water, which can be due to the antioxidant property of fullerenes leading to generation of ROS which disrupts membrane lipids and other biomolecules specifically DNA. They adsorb vitamins, minerals and trace elements found in soil which limit the growth of bacteria indirectly, ultimately leading to adverse impacts of nanoparticles on the environment.

R. metallidurans cells grown in medium containing $AuCl^{4-}$ (50 mM) showed toxicity to gold as 90% of the cells died after 4 h however, increased after 72 h of inoculation, depicting that *R. metallidurans* possesses toxicity resistance to $AuCl^{4-}$ and can adapt to high concentrations of gold. Initially precipitation of gold was significant by *R. metallidurans* but after incubation of 8 h, 3 mM gold was precipitated.

Silver nanoparticles have size in the range of 1–50 nm, the surface area of such nanoparticles is larger as compared to their volume. The large surface area provides an increase in reactivity and toxicity towards various microorganisms and bacteria. In some cases, the usage of silver nanoparticles leads to antibiotic resistance amid toxic or lethal bacteria. Various ecosystem processes controlled by bacterial species are endangered due to the use of silver nanoparticles, as they act as a strong bactericide (Elechiguerra et al. 2005).

Antibacterial activity towards *Pseudomonas aeruginosa, Salmonella paratyphi, Shigella* strains and *Klebsiella pneumoniae* PGPs (Plant Growth Promoters) was experimented using copper oxide nanoparticles (80–160 nm) (Mahapatra et al. 2008). Apart from this, it was interesting to find that copper/copper oxide nanoparticles were synthesized by *Serratia*, a gram-negative bacterium which dies in the process (Mishra and Kumar 2009).

It is presumed that iron- and copper-based nanoparticles produce free radicals by reacting with peroxides found in the environment. These radicals are known to be highly toxic to microorganisms. ZnO and magnesium oxide nanoparticles are potent microbe killers and act as food preservative as well (Mishra and Kumar 2009).

Upon evaluation of the impact of various nanoparticles, including fullerenes, aluminium, silver, gold, etc. on PGPR, it was suggested that nanoparticles effectively degraded phytostimulatory bacteria found in soil and also caused ecotoxicity (Table 2.1).

2.3.6 Toxicity

Application of gentle nanoparticles has been commercially approved by the Food and Drug Administration (FDA) but accumulation of nanoparticles like heavy metal nanoparticles, nanopesticides, etc. can lead to mild as well as severe nanotoxicity, which is not rationalized with respect to its bulk counterparts (Dubey et al. 2018).

Table 2.1 Impact of various nanoparticles on different plant species (Siddiqui et al. 2015; Singh et al. 2015; Manjunatha et al. 2016)

S.No	Nanoparticles	Plant species	Optimal concentration	Impact
1	Silver	*Boswelia ovalifoliolata*	10–30 μg/ml	Enhancement in seed germination and growth
2	Gold	*Cucmis sativus, Lactuca sativa*	62 μg/ml	Germination index elevates considerably
3	Selenium	*Nicotinia tabacum*	0.1 mg/gm	Initiation of callus and micro-shoot formation
4	Aluminium	*Raphanus raphanistrum, Brassica napus*	2 mg/ml	Ameliorated root growth
5	Alumina	*Lemna minor*	0.3 mg/ml	Elongation of root length
6	Titanium oxide	*Lycopersicum esculantum*	0.05–0.2 mg/ml	Overall rate of photosynthesis and conductivity of water increased
		Triticum aestivum	1 mg/ml	Rise in amount of chlorophyll
7	Ferrous oxide	*Glycine max*	0.5–0.75 mg/ml	Quality and yield refinement
8	Cobalt (II,III) oxide	*Raphanus sativus*	5 mg/ml	Enhanced growth of root
9	Zinc oxide	*Gycine max*	0.5 mg/ml	Enhanced root growth
		Arachis hypogeal	1 mg/ml	Elevated shoot and root growth and increased yield
		Cicer arietinum L.	0.15 mg/ml	Considerable increase in shoot and dry weight
10	Multi-walled carbon nanotubes	*Lycopersicum esculantum*	50 μg/ml 200 μg/ml	Increase in flower number and boosted plant height
11	Carbon nanotubes	*Lycopersicum esculantum*	40 μg/ml	Improvement in seedling germination and growth
12	Silicon dioxide	*Arabidopsis thaliana*	0.4 mg/ml	Elevation in root length
			2 mg/ml and 4 mg/ml	Reduction in root length

Nanoparticles, upon direct application can have disastrous impact, whereas upon diffusing into the apoplectic intercellular space, results in membrane adsorption and incorporation (Nowack and Bucheli 2007). Transport of compounds with negative charge occurs due to a negatively charged plant cell surface. Transfer into apoplast is interfered because of the Casparian strips, which act as a barrier to the flow, thus favouring transport into xylem symplastically (Thul and Sarangi 2015).

The production and usage of nanoparticles is increasing, leading to increase in congregation into the environment. Toxicity and environmental harm due to nanoparticles, including both direct and indirect exposures, is an ongoing debate (Brayner 2008; Panda et al. 2011; Dubey et al. 2018).

2.3.6.1 Pathogen Suppression

Crop improvement and conservation was mainly achieved through pesticide application, plant breeding and maintaining regular sanitation. The yield and quality of crops decline due to pathogens, thus there was high usage of the traditional crop improvement techniques. The traditional techniques were expensive as well as led to increase in resistance with time. Scientists found that the use of nanoparticles not only led to preservation of crop quality but also enhanced the productivity of crops (Emamifar et al. 2010; Bouwmeester et al. 2009).

Toxicity cognate with nanoparticles was utilized to tackle microbial pathogenicity in plants. Suppression of pathogens due to the antimicrobial properties associated with nanoparticles has shown to increase product quality and yield.

Antimicrobial function of nanoparticles occurs through five general mechanisms. specifically of metal nanoparticles (Lemire et al. 2013; Zeng et al. 2007):

1. Alteration of membrane protein function and permeability due to liberation of toxic ions
2. Impairment in uptake of nutrients and transport system in membranes
3. Rise in genotoxicity and cell death due to the interaction of toxic ions with DNA
4. Oxidative stress and damage of cellular components as well as DNA due to the production of reactive oxygen species (ROS)
5. Interference in metabolic pathways, which alters energy generation, membrane properties and protein oxidation

The antimicrobial properties of nanoparticles are presumed to aid unique and improved antimicrobial actions which are dosage and temperature dependent. Macromolecule oxidation or microbial membrane interaction of nanoparticles exert antimicrobial functions, which either damage or alter the permeability of membranes. Upon entry into bacterial cell, nanoparticles produce reactive oxygen species (ROS), which oxidize macromolecules, hampering important processes of cell leading to cell apoptosis (Musee et al. 2011). Nanoparticles such as silver, titanium oxide, fullerene C_{60} and single-walled carbon nanotubes particularly depict antimicrobial functions on bacterial monocultures (Morones et al. 2005; Lyon et al. 2006).

Nanoparticle toxicity of Au, Ag, Fe and fullerene C_{60} towards bacterial pathogens including *E. coli*, *B. subtilis* and *A. tumefaciens* was evaluated, among which Ag nanoparticles exhibited vigorous bactericidal activity upon every tested strain while fullerenes also showed inhibitory action on growth of all the strains. Among all the pathogenic microbes, fungi has been shown to cause highly detrimental infections and diseases in plants. Zn, Si, Ti and Cu nanoparticles showed strong antifungal properties and it was seen that the growth of fungal pathogen *A. niger* was inhibited by ZnO and zinc titanate ($ZnTiO_3$) nanoparticles (Ruffolo et al. 2010). A comparative study of the antifungal impact of sulphur nanoparticles with their bulk counterpart against *A. niger* showed increase in growth inhibition due to sulphur nanoparticles (Choudhury et al. 2010).

Si nanoparticles organize themselves as biomineralized silicon dioxide, which imparts fungal resistance (Wang et al. 2001). Two rice varieties, the resistant Nongda

18 and the susceptible Mongolian, were tested for inhibitory effects when treated with and without Si nanoparticles showed inhibition of growth of fungal pathogen *M. grisea* while there was no effect on the control ones (Dubey et al. 2018).

Composition of Ag-Si nanoparticle as an effective alternative for expensive fungicides was sought to be environment friendly, feasible as well as advantageous to humans and offered resistance from pathogens, including *Pythium* spp., *Blumeria* spp., *Colletotrichum* spp., *Sphaerotheca* spp., *Botrytis* spp., *Magnaporthe* spp., *Rhizoctonia* spp. *and Phytophthora* spp. (O'Neill et al. 2003; Shankar et al. 2003; Yau et al. 2004).

2.3.6.2 Regulated Delivery of Pesticides

Application of agrochemicals on crops is usually done through suspensions or sprays. These agrochemicals are mostly leached or degraded by either microbes or by photolytic or hydrolytic mechanism, causing loss of chemicals as well as pollution of soil and ground water. In order to eliminate this problem, designing of nanoencapsulated agrochemicals was highly important as they have the necessary characteristics, including high effectiveness, stability and solubility, on demand release, target-specific activity and minimal toxicity (Boehm et al. 2003; Green and Beestman 2007; Thul and Sarangi 2015).

Silica nanoparticles with hydrophobic surface modification were used for controlling various agricultural pests (Rahman et al. 2009) while mesoporous silica nanoparticles with surface functions specifically manipulated the expression of genes at the cellular level through delivery of DNA and regulatory proteins in a controlled manner (Torney et al. 2007).

It was observed by Kaunisto et al. (2013) that using a polymer matrix, which has swelling and dissolution properties, influenced transport pathways, thereby altering the release conditions. The polymer matrix were prepared from polyethylene glycol or polyvinylpyrrolidone nanospheres. Liposome nanopolymers as regulated delivery units for monitored insecticide release were first prepared and used by Bang et al. (2009). Moreover, the haphazard use of pesticide increased the bioaccumulation and resistance in pathogens and pests with reduction in biodiversity of soil, leading to declining nitrogen fixation and decline in pollinators with diminishing bird habitats (Ghormade et al. 2011; Thul and Sarangi 2015).

Varying compositions of insecticides based on polyethylene glycol led to systemic release at rates lower than the commercially produced ones with imidacloprid, thiram and carbofuran (Adak et al. 2012; Pankaj et al. 2012; Kaushik et al. 2013). In insects, blockage of neurotransmission through avermectin, a pesticide which inhibits chloride ion channel, having 6 h half-life upon exposure to ultraviolet rays gets inactivated in the field, which can be avoided by the controlled release up to 30 days due to encapsulation of avermectin by NP carriers (Thul and Sarangi 2015).

Nanocrystals of polylactic acid and cellulose were organized as a network of nanofibers to which thiamethoxam at a concentration of 50% was added, resulting in the decline of whitefly within 9 days monitoring in a glass house experiment

Table 2.2 Toxicity caused by nanoparticles on diverse microbes (Thul and Sarangi 2015; Ditta et al. 2015)

S.No	Nanoparticles	Microbial species	Toxicity
1	Silver	*Escherichia coli*	Bactericidal activity inhibits growth of bacteria
		Staphylococcus aureus	
		Salmonella typhimurium	
2	Ferrous oxide	*Trifolium repens*	Decrease in biomass of mycorrhizal clover
3	Zinc oxide	*Rhizobiales, Bradyrhizobium.*	Reduction in bacterial communities
		Pseudomonas putida	Hinders growth of bacteria
		Bacillus subtilis	Synthesis of ROS causes mild toxicity
4	Aluminium oxide	*Bacillus cereus, Pseudomonas stutzeri*	Response to transcription by microbes decline
5	Titanium dioxide	*Bradyrhizobiaceae*	Bacterial community diminishes
		Escherichia coli	ROS production leads to amiable toxicity
6	Cerium oxide	*Rhizobium, Azorhizobium*	Decline in nitrogen fixation ability
7	Silicon dioxide	*Escherichia coli*	ROS production leads to toxicity
8	Single-walled carbon nanotubes	*Rhizobium leguminosarum*	Alters the cellular morphology of bacteria

(Xiang et al. 2013). Nano-formulations associated with active compounds were observed effective compared to the commercial compounds for use in agriculture (Thul and Sarangi 2015).

2.3.6.3 Physiological and Biochemical Changes in Plants

Generation of reactive oxygen species due to the application of nanoparticles leads to the peroxidation of lipid molecules (Cabiscol et al. 2000), which remarkably impacts the biochemical as well as molecular properties of the membrane, including permeability, fluidity, osmotic stress susceptibility and loss in uptake of nutrients. Osmotic stress was identified due to soil and water, which activates an array of metabolic functions, in turn alleviating metal stressors (Chinnusamy et al. 2004; Thul and Sarangi 2015) (Table 2.2).

2.3.7 Accumulation of Nanoparticles

Nanoparticles have been found to accumulate in plants and the surrounding environment, as plants bear large size, higher leaf area and are immobile in nature, rendering them highly susceptible to exposure from varied nanoparticles present in the environment (Dietz and Herth 2011).

The initiation of nanoparticle mobilization within the food chain due to the adsorption and accumulation of nanoparticles by soil microbes upon direct application onto land or biosolids having mobile nanoparticles can lead to alteration in the community's food web and multiple populations, including plant, microbes and fishes (Thul and Sarangi 2015).

The size of nanoparticles contributes to the contamination of the surrounding environment, depicting distinctive physicochemical characteristics, such as high surface area, energy and surface confinement, resulting in alteration in environmental behaviour and increasing toxicity drastically, compared to their bulk components. The surface properties are the main reason for toxicity of ENPs (Engineered Nano-Particles), which can be inhibited using surface functioning (Geisler-Lee et al. 2012).

Environmental conditions alter the surface coating of nanomaterials, which can amend or simulate toxicity in microbes (Suresh et al. 2013). Due to high demand for nanoparticle-derived consumer goods, soil and water are highly potent for contamination (Thul and Sarangi 2015).

2.3.7.1 In Plants

Accumulation in plants occurs mainly through the roots harbouring most of the nanoparticles they are exposed to and cause toxicity. In terrestrial plants, most of the soil is exposed to nanoparticles, leading to leaching of nanoparticles, release into subsurface, contamination of land due to biosolid applications and discharge into wastewaters (Pokhrel and Dubey 2013; Hai et al. 2013). Therefore plants can effectively decide the fate of nanoparticles and their environmental transport by aggregating them into their biomass (Navarro et al. 2008; Ma et al. 2010; Anjum et al. 2013).

The final consumers of products produced by plants in the ecosystem are humans and animals. Metal nanoparticles like AgNPs upon accumulation would be transported into humans and animals through plant products (Cheng et al. 2011; Kim et al. 2011; Yin et al. 2011). AgNPs-treated *Lemna paucicostata*, an aquatic plant, and *Loliummultiflorum*, common grass seedlings, led to toxicity whether they were placed in a petri plate or in an aquatic environment (Geisler-Lee et al. 2012).

The problem of concern is the transport of nanoparticles into the food chain, which happens through consumption of edible plants. Using synchrotron X-ray fluorescence (μ-XRF) and X-ray absorption near edge structure (μ-XANES) analysis of *Cucumis sativus,* a garden vegetable, was conducted, which demonstrated titanium dioxide nanoparticle translocation from roots to the fruit without biotransformation (Servin et al. 2013; Thul and Sarangi 2015).

It is difficult to completely explain the phytotoxicity caused due to AgNPs just by the release of Ag ions. Plants subjected to Ag NP and silver nitrate ($AgNO_3$) suspension indicated that higher accumulation of silver occurred in AgNP suspensions compared to $AgNO_3$ due to uptake of AgNP by plants. In the case of *Brassica juncea*, there was no accumulation of Ag when exposed to AgNP.

Ultra-small anatase TiO_2 nanoparticles upon entering plant cells get accumulated in the sub-cellular sections, including vacuoles and root nuclei, leading to reorientation and exclusion of microtubules, which thereby inhibit elongation of roots in

Arabidopsis (Kurepa et al. 2010; Wang et al. 2011). Similarly, effect of ZnO nanoparticles was also observed in *Arabidopsis*, which showed reduced accumulation of biomass in shoots and roots and decrease in chlorophyll content instead of carotenoid content (Wang et al. 2016b).

Through μ-XRF imaging, it was established that Ce and Zn traverse between tissues with water flow during transpiration, which led to their bioaccumulation (Zhao et al. 2013; Thul and Sarangi 2015).

Hu et al. (2014) suggested that upon exposure to ZnO nanoparticles upto 7 days, their accumulation and dissipation led to aggregation of zinc in leaves and roots of *Salvinia natans*. Whereas Zhai et al. (2014) demonstrated that in the cytoplasm, cellular organelles, root cells and leaf cells, there was accumulation of Au nanoparticles, which was confirmed by transmission electron microscopy and measured by inductively coupled plasma mass spectrometry (ICP-MS) (Thul and Sarangi 2015).

2.3.7.2 In Soil and Water Bodies

Aquatic invertebrates, including copepods (*Amphiascus tenuiremis*), amoebae (*Entamoeba histolytica*), cladocerans (*Daphnia magna*), infusoria (*Tetrahymena pyriformis* and *Stylonychia mytilus*), absorb carbon nanoparticles from food, but it is still unspecified if these nanoparticles penetrate internal organs (Oberdörster et al. 2006; Templeton et al. 2006; Zhu et al. 2006, 2009; Elías et al. 2007; Roberts et al. 2007).

It was experimentally observed that the existence of nanoparticles within aquatic organisms leads to reduced fertility, abnormalities in behaviour, physiological alterations and increased rate of mortality (Lovern and Klaper 2006; Templeton et al. 2006; Krysanov et al. 2010).

The penetration ability of carbon nanoparticles into aquatic organisms depends upon their structure and modification. In the embryos of zebrafish (*Danio rerio*), the unmodified C_{60} fullerenes penetrated through chorion, whereas hydroxylated derivatives ($C_{60}(OH)_{24}$) could not (Isaacson et al. 2007). Similarly, when subjected to copper nanoparticles, the NPs accumulated at a concentration of 1.9 fold in the gills after 48 h as compared to the controls (Griffitt et al. 2007).

Daphnias were subjected to titanium dioxide, aluminium oxide and zinc oxide nanoparticles for 48 h, which accumulated in the gut (Zhu et al. 2009). This was due to the fact that the nanoparticles aggregated within 12 h but their excretion was delayed due to which most of the nanoparticles were retained in the body for another 72 h.

Zhao et al. (2012) demonstrated the concurrence of zinc oxide nanoparticles with zinc-dissolved species which was liberated constantly into the soil, leading to replenishment of zinc ions, which were rummaged by roots in comparison to alginate-treated soil, thus promoting aggregation of zinc in the tissues of corn (Thul and Sarangi 2015).

Quantum dots have the ability to fluoresce, which makes them efficient in monitoring the penetration and translocation pattern within living organisms. Quantum dots have superseded in reaching not only to the gut but to various other parts of daphnia (*D. magna*) (Ingle et al. 2008; Thul and Sarangi 2015).

References

Adak T, Kumar J, Shakil NA, Walia S (2012 Mar 1) Development of controlled release formulations of imidacloprid employing novel nano-ranged amphiphilic polymers. J Environ Sci Health B 47(3):217–225

Adhikari T, Kundu S, Biswas AK, Tarafdar JC, Rao AS (2012 Jun 1) Effect of copper oxide nano particle on seed germination of selected crops. J Agri Sci Technol 2(6A):815

Ali MA, Rehman I, Iqbal A, Din S, Rao AQ, Latif A, Samiullah TR, Azam S, Husnain T (2014 May 25) Nanotechnology, a new frontier in agriculture. Adv Life Sci 1(3):129–138

Anjum NA, Gill SS, Duarte AC, Pereira E, Ahmad I (2013 Sep 1) Silver nano-particles in soil–plant systems. J Nanopart Res 15(9):1896

Asli S, Neumann PM (2009 May) Colloidal suspensions of clay or titanium dioxide nano-particles can inhibit leaf growth and transpiration via physical effects on root water transport. Plant Cell Environ 32(5):577–584

Bang SH, Yu YM, Hwang IC, Park HJ (2009 Dec 1) Formation of size-controlled nano carrier systems by self-assembly. J Microencapsul 26(8):722–733

Birbaum K, Brogioli R, Schellenberg M, Martinoia E, Stark WJ, Günther D, Limbach LK (2010 Oct 21) No evidence for cerium dioxide nanoparticle translocation in maize plants. Environ Sci Technol 44(22):8718–8723

Boehm AL, Martinon I, Zerrouk R, Rump E, Fessi H (2003 Jan 1) Nanoprecipitation technique for the encapsulation of agrochemical active ingredients. J Microencapsul 20(4):433–441

Borm PJ, Robbins D, Haubold S, Kuhlbusch T, Fissan H, Donaldson K, Schins R, Stone V, Kreyling W, Lademann J, Krutmann J (2006 Dec) The potential risks of nanomaterials: a review carried out for ECETOC. Part Fibre Toxicol 3(1):11

Bouwmeester H, Dekkers S, Noordam MY, Hagens WI, Bulder AS, De Heer C, Ten Voorde SE, Wijnhoven SW, Marvin HJ, Sips AJ (2009 Feb 1) Review of health safety aspects of nanotechnologies in food production. Regul Toxicol Pharmacol 53(1):52–62

Brant J, Lecoanet H, Wiesner MR (2005 Oct 1) Aggregation and deposition characteristics of fullerene nano-particles in aqueous systems. J Nanopart Res 7(4–5):545–553

Brayner R (2008 Feb 1) The toxicological impact of nano-particles. Nano Today 3(1–2):48–55

Cabiscol CE, Tamarit SJ, Ros SJ (2000) Oxidative stress in bacteria and protein damage by reactive oxygen species. Int Microbiol 3(1):3–8

Cheng Y, Yin L, Lin S, Wiesner M, Bernhardt E, Liu J (2011 Mar 1) Toxicity reduction of polymer-stabilized silver nano-particles by sunlight. J Phys Chem C 115(11):4425–4432

Chinnusamy V, Schumaker K, Zhu JK (2004 Jan 1) Molecular genetic perspectives on cross-talk and specificity in abiotic stress signalling in plants. J Exp Bot 55(395):225–236

Choudhury SR, Nair KK, Kumar R, Gogoi R, Srivastava C, Gopal M, Subhramanyam BS, Devakumar C, Goswami A (2010 Oct 4) Nanosulfur: a potent fungicide against food pathogen, Aspergillus niger. AIP Conf Proceed 1276(1):154–157

Crabtree RH (1998 Dec 11) A new type of hydrogen bond. Science 282(5396):2000–2001

Da Costa MV, Sharma PK (2016 Mar 1) Effect of copper oxide nano-particles on growth, morphology, photosynthesis, and antioxidant response in Oryza sativa. Photosynthetica 54(1):110–119

De Souza MP, Pilon-Smits EA, Terry N (2000) The physiology and biochemistry of selenium volatilization by plants. In: Phytoremediation of toxic metals: using plants to clean-up the environment. Wiley, New York, pp 171–190

Dietz KJ, Herth S (2011 Nov 1) Plant nanotoxicology. Trends Plant Sci 16(11):582–589

Ditta A, Arshad M, Ibrahim M (2015) Nanoparticles in sustainable agricultural crop production: applications and perspectives. In: Nanotechnology and plant sciences. Springer, Cham, pp. 55–75

Ditta A, Arshad M (2016 Apr 1) Applications and perspectives of using nanomaterials for sustainable plant nutrition. Nanotechnol Rev 5(2):209–229

Du W, Tan W, Peralta-Videa JR, Gardea-Torresdey JL, Ji R, Yin Y, Guo H (2017 Jan 1) Interaction of metal oxide nano-particles with higher terrestrial plants: physiological and biochemical aspects. Plant Physiol Biochem 110:210–225

Dubey A, Mishra V, Kumar S, Ahmed SU, Goswami M (2018 Apr 17) Nanotoxicity on human and plant pathogenic microbes and aquatic organisms. In: Environmental toxicity of nanomaterials. CRC Press, Boca Raton, pp 241–280

Elechiguerra JL, Burt JL, Morones JR, Camacho-Bragado A, Gao X, Lara HH, Yacaman MJ (2005 Jun) Interaction of silver nano-particles with HIV-1. J Nanobiotechnol 3(1):6

Elías AL, Carrero-Sánchez JC, Terrones H, Endo M, Laclette JP, Terrones M (2007 Oct 1) Viability studies of pure carbon-and nitrogen-doped nanotubes with Entamoeba histolytica: from amoebicidal to biocompatible structures. Small 3(10):1723–1729

Emamifar A, Kadivar M, Shahedi M, Soleimanian-Zad S (2010 Oct 1) Evaluation of nanocomposite packaging containing Ag and ZnO on shelf life of fresh orange juice. Innovative Food Sci Emerg Technol 11(4):742–748

Fortner JD, Lyon DY, Sayes CM, Boyd AM, Falkner JC, Hotze EM, Alemany LB, Tao YJ, Guo W, Ausman KD, Colvin VL (2005 Jun 1) C60 in water: nanocrystal formation and microbial response. Environ Sci Technol 39(11):4307–4316

Geisler-Lee J, Wang Q, Yao Y, Zhang W, Geisler M, Li K, Huang Y, Chen Y, Kolmakov A, Ma X (2012 May 1) Phytotoxicity, accumulation and transport of silver nano-particles by Arabidopsis thaliana. Nanotoxicology 7(3):323–337

Ghodake G, Seo YD, Lee DS (2011 Feb 15) Hazardous phytotoxic nature of cobalt and zinc oxide nano-particles assessed using Allium cepa. J Hazard Mater 186(1):952–955

Ghormade V, Deshpande MV, Paknikar KM (2011 Nov 1) Perspectives for nano-biotechnology enabled protection and nutrition of plants. Biotechnol Adv 29(6):792–803

Giraldo JP, Landry MP, Faltermeier SM, McNicholas TP, Iverson NM, Boghossian AA, Reuel NF, Hilmer AJ, Sen F, Brew JA, Strano MS (2014 Apr) Plant nanobionics approach to augment photosynthesis and biochemical sensing. Nat Mater 13(4):400

Gopinath K, Gowri S, Karthika V, Arumugam A (2014 Sep 1) Green synthesis of gold nano-particles from fruit extract of Terminalia arjuna, for the enhanced seed germination activity of Gloriosa superba. J Nanostruct Chem 4(3):115

Govorov AO, Carmeli I (2007 Mar 14) Hybrid structures composed of photosynthetic system and metal nano-particles: plasmon enhancement effect. Nano Lett 7(3):620–625

Green JM, Beestman GB (2007 Mar 1) Recently patented and commercialized formulation and adjuvant technology. Crop Prot 26(3):320–327

Griffitt RJ, Weil R, Hyndman KA, Denslow ND, Powers K, Taylor D, Barber DS (2007 Oct 24) Exposure to copper nano-particles causes gill injury and acute lethality in zebrafish (Danio rerio). Environ Sci Technol 41(23):8178–8186

Gruère G, Narrod C, Abbott L (2011 Feb) Agricultural, food, and water nanotechnologies for the poor. International Food Policy Research Institute, Washington, DC

Gruyer N, Dorais M, Bastien C, Dassylva N, Triffault-Bouchet G (2013) Interaction between silver nano-particles and plant growth. In: International symposium on new technologies for environment control, energy-saving and crop production in greenhouse and plant 1037, 2013 Oct 6, pp 795–800

Haghighi M, Afifipour Z, Mozafarian M (2012) The effect of N-Si on tomato seed germination under salinity levels. J Biol Environ Sci 6(16):87–90

Hai ZH, Cheng PE, Jian-jun YA, Ji-yan SH (2013 Mar 1) Eco-toxicological effect of metal-based nano-particles on plants: research progress. Yingyong Shengtai Xuebao 24(3)

Helaly MN, El-Metwally MA, El-Hoseiny H, Omar SA, El-Sheery NI (2014 Apr) Effect of nano-particles on biological contamination of'in vitro'cultures and organogenic regeneration of banana. Aust J Crop Sci 8(4):612

Holden PA, Nisbet RM, Lenihan HS, Miller RJ, Cherr GN, Schimel JP, Gardea-Torresdey JL (2013 Oct 5) Ecological nanotoxicology: integrating nanomaterial hazard considerations across the subcellular, population, community, and ecosystems levels. Acc Chem Res 46(3):813–822

Hu C, Liu X, Li X, Zhao Y (2014 Jan 1) Evaluation of growth and biochemical indicators of Salvinia natans exposed to zinc oxide nano-particles and zinc accumulation in plants. Environ Sci Pollut Res 21(1):732–739

Ingle TM, Alexander R, Bouldin J, Buchanan RA (2008 Sep 1) Absorption of semiconductor nanocrystals by the aquatic invertebrate Ceriodaphnia dubia. Bull Environ Contam Toxicol 81(3):249–252

Isaacson CW, Usenko CY, Tanguay RL, Field JA (2007 Dec 1) Quantification of fullerenes by LC/ESI-MS and its application to in vivo toxicity assays. Anal Chem 79(23):9091–9097

Jain AK, Kumar Mehra N, Lodhi N, Dubey V, Mishra DK, Jain PK, Jain NK (2007 Jan 1) Carbon nanotubes and their toxicity. Nanotoxicology 1(3):167–197

Kalteh M, Alipour ZT, Ashraf S, Aliabadi MM, Nosratabadi AF (2014) Effect of silica nanoparticles on basil (Ocimum basilicum) under salinity stress. J Chem Health Risks 4:49–55

Kaunisto E, Tajarobi F, Abrahmsen-Alami S, Larsson A, Nilsson B, Axelsson A (2013 Mar 12) Mechanistic modelling of drug release from a polymer matrix using magnetic resonance microimaging. Eur J Pharm Sci 48(4–5):698–708

Kaushik P, Shakil NA, Kumar J, Singh MK, Singh MK, Yadav SK (2013 Aug 3) Development of controlled release formulations of thiram employing amphiphilic polymers and their bioefficacy evaluation in seed quality enhancement studies. J Environ Sci Health B 48(8):677–685

Khodakovskaya MV, De Silva K, Biris AS, Dervishi E, Villagarcia H (2012 Feb 29) Carbon nanotubes induce growth enhancement of tobacco cells. ACS Nano 6(3):2128–2135

Kim E, Kim SH, Kim HC, Lee SG, Lee SJ, Jeong SW (2011 Mar 1) Growth inhibition of aquatic plant caused by silver and titanium oxide nano-particles. Toxicol Environ Heal Sci 3(1):1–6

Kreyling WG, Semmler-Behnke M, Möller W (2006 Oct 1) Health implications of nano-particles. J Nanopart Res 8(5):543–562

Krysanov EY, Pavlov DS, Demidova TB, Dgebuadze YY (2010 Aug 1) Effect of nano-particles on aquatic organisms. Biol Bull 37(4):406–412

Kurepa J, Paunesku T, Vogt S, Arora H, Rabatic BM, Lu J, Wanzer MB, Woloschak GE, Smalle JA (2010 Mar 10) Uptake and distribution of ultrasmall anatase TiO_2 alizarin red S nanoconjugates in Arabidopsis thaliana. Nano Lett 10(7):2296–2302

Lam CW, James JT, McCluskey R, Arepalli S, Hunter RL (2006 Jan 1) A review of carbon nanotube toxicity and assessment of potential occupational and environmental health risks. Crit Rev Toxicol 36(3):189–217

Lecoanet HF, Wiesner MR (2004 Aug 15) Velocity effects on fullerene and oxide nanoparticle deposition in porous media. Environ Sci Technol 38(16):4377–4382

Lee CW, Mahendra S, Zodrow K, Li D, Tsai YC, Braam J, Alvarez PJ (2010 Mar) Developmental phytotoxicity of metal oxide nano-particles to Arabidopsis thaliana. Environ Toxicol Chem Int J 29(3):669–675

Lemire JA, Harrison JJ, Turner RJ (2013 Jun) Antimicrobial activity of metals: mechanisms, molecular targets and applications. Nat Rev Microbiol 11(6):371

Li B, Tao G, Xie Y, Cai X (2012) Physiological effects under the condition of spraying nano-SiO_2 onto the Indocalamus barbatus McClure leaves. J Nanjing Fores Univ (Natural Sciences Edition) 36(4):161–164

Liu Q, Chen B, Wang Q, Shi X, Xiao Z, Lin J, Fang X (2009a Feb 3) Carbon nanotubes as molecular transporters for walled plant cells. Nano Lett 9(3):1007–1010

Liu S, Wei L, Hao L, Fang N, Chang MW, Xu R, Yang Y, Chen Y (2009b Nov 6) Sharper and faster "nano darts" kill more bacteria: a study of antibacterial activity of individually dispersed pristine single-walled carbon nanotube. ACS Nano 3(12):3891–3902

López-Moreno ML, de la Rosa G, Hernández-Viezcas JÁ, Castillo-Michel H, Botez CE, Peralta-Videa JR, Gardea-Torresdey JL (2010 Apr 12) Evidence of the differential biotransformation and genotoxicity of ZnO and CeO2 nano-particles on soybean (Glycine max) plants. Environ Sci Technol 44(19):7315–7320

Lovern SB, Klaper R (2006 Apr) Daphnia magna mortality when exposed to titanium dioxide and fullerene (C60) nano-particles. Environ Toxicol Chem Int J 25(4):1132–1137

Lyon DY, Adams LK, Falkner JC, Alvarez PJ (2006 Jul 15) Antibacterial activity of fullerene water suspensions: effects of preparation method and particle size. Environ Sci Technol 40(14):4360–4366

Ma X, Geiser-Lee J, Deng Y, Kolmakov A (2010 Jul 15) Interactions between engineered nano-particles (ENPs) and plants: phytotoxicity, uptake and accumulation. Sci Total Environ 408(16):3053–3061

Mahajan P, Dhoke SK, Khanna AS (2011) Effect of nano-ZnO particle suspension on growth of mung (Vigna radiata) and gram (Cicer arietinum) seedlings using plant agar method. J Nanotechnol 2011

Mahapatra O, Bhagat M, Gopalakrishnan C, Arunachalam KD (2008 Sep 1) Ultrafine dispersed CuO nano-particles and their antibacterial activity. J Exp Nanosci 3(3):185–193

Manjunatha SB, Biradar DP, Aladakatti YR (2016) Nanotechnology and its applications in agriculture: a review. J Farm Sci 29(1):1–3

Maynard AD, Baron PA, Foley M, Shvedova AA, Kisin ER, Castranova V (2004 Jan 1) Exposure to carbon nanotube material: aerosol release during the handling of unrefined single-walled carbon nanotube material. J Toxic Environ Health A 67(1):87–107

Mishra VK, Kumar A (2009 Sep 1) Impact of metal nano-particles on the plant growth promoting rhizobacteria. Dig J Nanomater Biostruct 4:587–592

Moraru Carment IN (2003) A new frontier in food science. Carment I. Moraru, Panchapakesan Chithra P., Huand Oingrong et al. Food Technol 57(12):24–29

Morones JR, Elechiguerra JL, Camacho A, Holt K, Kouri JB, Ramírez JT, Yacaman MJ (2005 Aug 26) The bactericidal effect of silver nano-particles. Nanotechnology 16(10):2346

Musee N, Thwala M, Nota N (2011) The antibacterial effects of engineered nanomaterials: implications for wastewater treatment plants. J Environ Monit 13(5):1164–1183

Naderi MR, Danesh-Shahraki A (2013 May 16) Nanofertilizers and their roles in sustainable agriculture. Int J Agric Crop Sci 5(19):2229

Nair R, Varghese SH, Nair BG, Maekawa T, Yoshida Y, Kumar DS (2010 Sep 1) Nanoparticulate material delivery to plants. Plant Sci 179(3):154–163

Navarro E, Baun A, Behra R, Hartmann NB, Filser J, Miao AJ, Quigg A, Santschi PH, Sigg L (2008 Jul 1) Environmental behavior and ecotoxicity of engineered nano-particles to algae, plants, and fungi. Ecotoxicology 17(5):372–386

Noji T, Kamidaki C, Kawakami K, Shen JR, Kajino T, Fukushima Y, Sekitoh T, Itoh S (2011 Dec 20) Photosynthetic oxygen evolution in mesoporous silica material: adsorption of photosystem II reaction center complex into 23 nm nanopores in SBA. Langmuir 27(2):705–713

Nowack B, Bucheli TD (2007 Nov 1) Occurrence, behavior and effects of nano-particles in the environment. Environ Pollut 150(1):5–22

Nyberg L, Turco RF, Nies L (2008 Feb 14) Assessing the impact of nanomaterials on anaerobic microbial communities. Environ Sci Technol 42(6):1938–1943

O'Neill MA, Vine GJ, Beezer AE, Bishop AH, Hadgraft J, Labetoulle C, Walker M, Bowler PG (2003 Sep 16) Antimicrobial properties of silver-containing wound dressings: a microcalorimetric study. Int J Pharm 263(1–2):61–68

Oberdörster E, Zhu S, Blickley TM, McClellan-Green P, Haasch ML (2006 May 1) Ecotoxicology of carbon-based engineered nano-particles: effects of fullerene (C60) on aquatic organisms. Carbon 44(6):1112–1120

Panda KK, Achary VM, Krishnaveni R, Padhi BK, Sarangi SN, Sahu SN, Panda BB (2011 Aug 1) In vitro biosynthesis and genotoxicity bioassay of silver nano-particles using plants. Toxicol In Vitro 25(5):1097–1105

Pankaj, Shakil NA, Kumar J, Singh MK, Singh K (2012 Jul 1) Bioefficacy evaluation of controlled release formulations based on amphiphilic nano-polymer of carbofuran against Meloidogyne incognita infecting tomato. J Environ Sci Health B 47(6):520–528

Pokhrel LR, Dubey B (2013 May 1) Evaluation of developmental responses of two crop plants exposed to silver and zinc oxide nano-particles. Sci Total Environ 452:321–332

Prasad R, Kumar V, Prasad KS (2014) Nanotechnology in sustainable agriculture: present concerns and future aspects. Afr J Biotechnol 13(6):705–713

Priester JH, Ge Y, Mielke RE, Horst AM, Moritz SC, Espinosa K, Gelb J, Walker SL, Nisbet RM, An YJ, Schimel JP (2012 Sep 11) Soybean susceptibility to manufactured nanomaterials with evidence for food quality and soil fertility interruption. Proc Natl Acad Sci 109(37):E2451–E2456

Rahman A, Seth D, Mukhopadhyaya SK, Brahmachary RL, Ulrichs C, Goswami A (2009 Jan 1) Surface functionalized amorphous nanosilica and microsilica with nanopores as promising tools in biomedicine. Naturwissenschaften 96(1):31–38

Rico CM, Peralta-Videa JR, Gardea-Torresdey JL (2015) Chemistry, biochemistry of nanoparticles, and their role in antioxidant defense system in plants. In: Nanotechnology and plant sciences. Springer, Cham, pp 1–17

Roberts AP, Mount AS, Seda B, Souther J, Qiao R, Lin S, Ke PC, Rao AM, Klaine SJ (2007 Apr 15) In vivo biomodification of lipid-coated carbon nanotubes by Daphnia magna. Environ Sci Technol 41(8):3025–3029

Ruffolo SA, La Russa MF, Malagodi M, Rossi CO, Palermo AM, Crisci GM (2010 Sep 1) ZnO and ZnTiO 3 nanopowders for antimicrobial stone coating. Appl Phys A 100(3):829–834

Sabo-Attwood T, Unrine JM, Stone JW, Murphy CJ, Ghoshroy S, Blom D, Bertsch PM, Newman LA (2012 Jun 1) Uptake, distribution and toxicity of gold nano-particles in tobacco (Nicotiana xanthi) seedlings. Nanotoxicology 6(4):353–360

Savithramma N, Ankanna S, Bhumi G (2012) Effect of nano-particles on seed germination and seedling growth of Boswellia ovalifoliolata an endemic and endangered medicinal tree taxon. Nano Vision 2(1):2

Service RF (2003 Apr 11) American Chemical Society meeting. Nanomaterials show signs of toxicity. Science (New York, NY) 300(5617):243

Servin AD, Morales MI, Castillo-Michel H, Hernandez-Viezcas JA, Munoz B, Zhao L, Nunez JE, Peralta-Videa JR, Gardea-Torresdey JL (2013 Sep 27) Synchrotron verification of TiO_2 accumulation in cucumber fruit: a possible pathway of TiO_2 nanoparticle transfer from soil into the food chain. Environ Sci Technol 47(20):11592–11598

Shalaby TA, Bayoumi Y, Abdalla N, Taha H, Alshaal T, Shehata S, Amer M, Domokos-Szabolcsy É, El-Ramady H (2016) Nano-particles, soils, plants and sustainable agriculture. In: Nanoscience in food and agriculture. Springer, Cham, pp 283–312

Shankar SS, Ahmad A, Sastry M (2003 Jan 1) Geranium leaf assisted biosynthesis of silver nanoparticles. Biotechnol Prog 19(6):1627–1631

Siddiqui, Manzer H., et al. "Role of nanoparticles in plants." Nanotechnology and Plant Sciences. Springer, Cham, 2015. 19-35.

Singh A, Singh NB, Hussain I, Singh H, Singh SC (2015) Plant-nanoparticle interaction: an approach to improve agricultural practices and plant productivity. Int J Pharm Sci Invent 4(8):25–40

Smirnova E, Gusev A, Zaytseva O, Sheina O, Tkachev A, Kuznetsova E, Lazareva E, Onishchenko G, Feofanov A, Kirpichnikov M (2012 Jun 1) Uptake and accumulation of multiwalled carbon nanotubes change the morphometric and biochemical characteristics of Onobrychis arenaria seedlings. Front Chem Sci Eng 6(2):132–138

Stampoulis D, Sinha SK, White JC (2009 Nov 19) Assay-dependent phytotoxicity of nano-particles to plants. Environ Sci Technol 43(24):9473–9479

Suresh AK, Pelletier DA, Doktycz MJ (2013) Relating nanomaterial properties and microbial toxicity. Nanoscale 5(2):463–474

Syu YY, Hung JH, Chen JC, Chuang HW (2014 Oct 1) Impacts of size and shape of silver nanoparticles on Arabidopsis plant growth and gene expression. Plant Physiol Biochem 83:57–64

Tarafdar JC, Agrawal A, Raliya R, Kumar P, Burman U, Kaul RK (2012a Aug 1) ZnO nanoparticles induced synthesis of polysaccharides and phosphatases by Aspergillus fungi. Adv Sci Eng Med 4(4):324–328

Tarafdar JC, Raliya R, Rathore I (2012b Dec 1) Microbial synthesis of phosphorous nanoparticle from tri-calcium phosphate using Aspergillus tubingensis TFR-5. J Bionanosci 6(2):84–89
Templeton RC, Ferguson PL, Washburn KM, Scrivens WA, Chandler GT (2006 Dec 1) Life-cycle effects of single-walled carbon nanotubes (SWNTs) on an estuarine meiobenthic copepod. Environ Sci Technol 40(23):7387–7393
Thul ST, Sarangi BK (2015) Implications of nanotechnology on plant productivity and its rhizospheric environment. In: Nanotechnology and plant sciences. Springer, Cham, pp 37–53
Torney F, Trewyn BG, Lin VS, Wang K (2007 May) Mesoporous silica nano-particles deliver DNA and chemicals into plants. Nat Nanotechnol 2(5):295
Tripathi DK, Singh S, Singh S, Pandey R, Singh VP, Sharma NC, Prasad SM, Dubey NK, Chauhan DK (2017 Jan 1) An overview on manufactured nano-particles in plants: uptake, translocation, accumulation and phytotoxicity. Plant Physiol Biochem 110:2–12
Wang L, Wang Y, Zhou Y, Duan X, Li M, Zhang F (2001) Relationship between nanostructure SiO_2 and occurrence of plant fungi. J Huazhong (Central China) Agri Univ 20(6):593–597
Wang S, Kurepa J, Smalle JA (2011 May) Ultra-small TiO_2 nano-particles disrupt microtubular networks in Arabidopsis thaliana. Plant Cell Environand 34(5):811–820
Wang Z, Xie X, Zhao J, Liu X, Feng W, White JC, Xing B (2012 Apr 4) Xylem-and phloem-based transport of CuO nano-particles in maize (Zea mays L.). Environ Sci Technol 46(8):4434–4441
Wang S, Liu H, Zhang Y, Xin H (2015 Mar) The effect of CuO NPs on reactive oxygen species and cell cycle gene expression in roots of rice. Environ Toxicol Chem 34(3):554–561
Wang P, Lombi E, Zhao FJ, Kopittke PM (2016a Aug 1) Nanotechnology: a new opportunity in plant sciences. Trends Plant Sci 21(8):699–712
Wang X, Yang X, Chen S, Li Q, Wang W, Hou C, Gao X, Wang L, Wang S (2016b Jan 12) Zinc oxide nano-particles affect biomass accumulation and photosynthesis in Arabidopsis. Front Plant Sci 6:1243
Xiang C, Taylor AG, Hinestroza JP, Frey MW (2013 Jan 5) Controlled release of nonionic compounds from poly (lactic acid)/cellulose nanocrystal nanocomposite fibers. J Appl Polym Sci 127(1):79–86
Yang Z, Chen J, Dou R, Gao X, Mao C, Wang L (2015 Dec) Assessment of the phytotoxicity of metal oxide nano-particles on two crop plants, maize (Zea mays L.) and rice (Oryza sativa L.). Int J Environ Res Public Health 12(12):15100–15109
Yau CP, Wang L, Yu M, Zee SY, Yip WK (2004 Mar 1) Differential expression of three genes encoding an ethylene receptor in rice during development, and in response to indole-3-acetic acid and silver ions. J Exp Bot 55(397):547–556
Yin L, Cheng Y, Espinasse B, Colman BP, Auffan M, Wiesner M, Rose J, Liu J, Bernhardt ES (2011 Feb 22) More than the ions: the effects of silver nano-particles on Lolium multiflorum. Environ Sci Technol 45(6):2360–2367
Zeng F, Hou C, Wu S, Liu X, Tong Z, Yu S (2007 Jan 9) Silver nano-particles directly formed on natural macroporous matrix and their anti-microbial activities. Nanotechnology 18(5):055605
Zhai G, Walters KS, Peate DW, Alvarez PJ, Schnoor JL (2014 Jan 15) Transport of gold nanoparticles through plasmodesmata and precipitation of gold ions in woody poplar. Environ Sci Technol Lettand 1(2):146–151
Zhang M, Gao B, Chen J, Li Y (2015a Feb 1) Effects of graphene on seed germination and seedling growth. J Nanopart Res 17(2):78
Zhang P, Ma Y, Zhang Z (2015b) Interactions between engineered nanomaterials and plants: phytotoxicity, uptake, translocation, and biotransformation. In: Nanotechnology and plant sciences. Springer, Cham, pp 77–99
Zhao L, Hernandez-Viezcas JA, Peralta-Videa JR, Bandyopadhyay S, Peng B, Munoz B, Keller AA, Gardea-Torresdey JL (2012 Dec 20) ZnO nanoparticle fate in soil and zinc bioaccumulation in corn plants (Zea mays) influenced by alginate. Environ Sci: Processes Impactsand 15(1):260–266

Zhao L, Sun Y, Hernandez-Viezcas JA, Servin AD, Hong J, Niu G, Peralta-Videa JR, Duarte-Gardea M, Gardea-Torresdey JL (2013 Nov 23) Influence of CeO_2 and ZnO nano-particles on cucumber physiological markers and bioaccumulation of Ce and Zn: a life cycle study. J Agric Food Chem 61(49):11945–11951

Zhu Y, Ran T, Li Y, Guo J, Li W (2006 Aug 30) Dependence of the cytotoxicity of multi-walled carbon nanotubes on the culture medium. Nanotechnology 17(18):4668

Zhu X, Zhu L, Chen Y, Tian S (2009 Jan 1) Acute toxicities of six manufactured nanomaterial suspensions to Daphnia magna. J Nanopart Res 11(1):67–75

Zhu ZJ, Wang H, Yan B, Zheng H, Jiang Y, Miranda OR, Rotello VM, Xing B, Vachet RW (2012 Nov 2) Effect of surface charge on the uptake and distribution of gold nano-particles in four plant species. Environ Sci Technol 46(22):12391–12398

3 Use of Nanotechnology in Quality Improvement of Economically Important Agricultural Crops

Anju Bajpai, Kashmiri Jadhav, M. Muthukumar, Sandeep Kumar, and Gaurav Srivatava

Abstract

Increased crop production while maintaining sustainability in agroecosystems is the thrust of modern technologies. Biotic and abiotic stresses are the major challenges faced by agriculture in times of climate change, emerging pests and diseases and crop behaviour. Hence, development of agricultural technology becomes necessary to cope with these challenges. Though nanobiotechnology has demonstrated envious benefits in agriculture and food processing, its advancements remain at the laboratory and pilot scale. The Indian government has supported Nano Mission in last decade and supports expansion and commercial applications of nanotechnology by encouraging participation of private sector investments. The concerns related to the availability and synthesis of nanomaterials, toxicity and health hazards associated with nanoparticles, besides regulatory framework, are major constrains in their application from the Indian perspective. Hence, the current chapter focuses on the use of nanotechnology in quality improvement of economically important agricultural crops.

Keywords

Nanofertilizer · Nanosensor · Nanoemulsion · Nanopesticide · Nanopackaging · Nanoequipment

---

A. Bajpai (✉) · M. Muthukumar · S. Kumar
ICAR-Central Institute for Subtropical Horticulture, Lucknow, India

K. Jadhav
ICAR-Indian institute of Rice Research, Hyderabad, India

G. Srivatava
Bhundelkhand University, Jhansi, India

M. Ghorbanpour et al. (eds.), *Biogenic Nano-Particles and their Use in Agro-ecosystems*, https://doi.org/10.1007/978-981-15-2985-6_3

3.1 Introduction

Nanotechnology is the study of nanoscale (1–100 nm) materials known as nanoparticles (NPs), exhibiting unique and novel physical, chemical and biological properties (Li et al. 2001). Over the past decade, the use of nanotechnology in the agricultural field has gained momentum by delivering robust applications (Tarafdar 2012; Kah and Hofmann 2014). These include nanotechnological applications in plant disease resistance and plant growth that have overwhelmingly remoulded the agricultural sector (Ghormade et al. 2011; Chowdappa and Gowda 2013). The major applications of nanotechnology in agriculture are schematically represented in Fig. 3.1 and they include the following:

- Improved seed germination using nanoformulations
- Formulation of nanofertilizers for balanced crop nutrition
- Development of nanoherbicides for weed control
- Development of efficient nanopesticides for pest and disease control
- Management of post-harvest diseases and improving post-harvest quality
- Diagnostic devices based on nanosensors for monitoring agroecosystems
- Improved agricultural engineering using nanotechnology in the field of Agricultural machinery

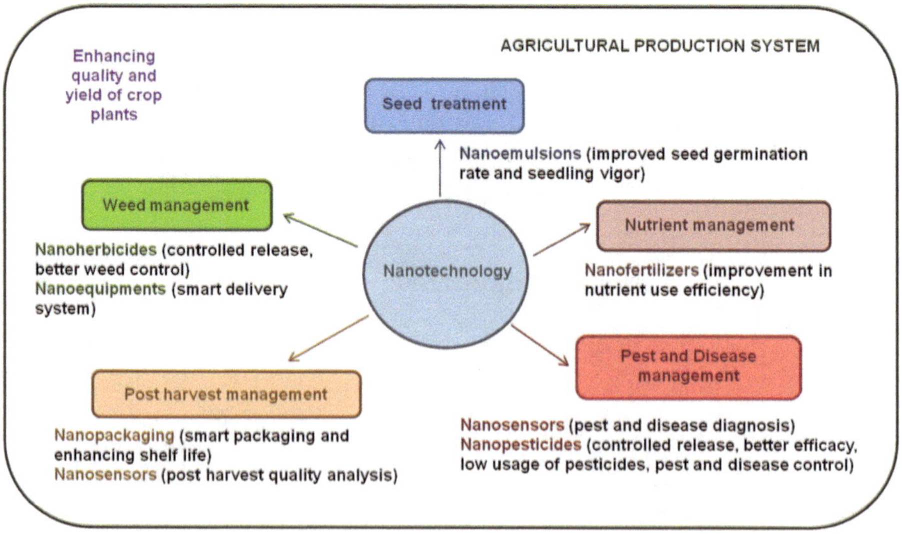

Fig. 3.1 Overview of nanotechnology applications for enhancing quality and yield of crop plants

3.2 Nanoparticles in Agriculture and Their Fabrication

The emergence of nanotechnology has revolutionized the scientific world because of its novelty, high growth and pertinent broad impacts. These NPs have flexible physical properties with a large surface area to volume ratio and a strong affinity for proteins. Polymeric nanoparticles have been most commonly extensively used in making nanoproducts or formulations for potential use in agriculture. These polymeric nanoparticles are defined as particles of diameter <1 μm and made up of either biodegradable or non-biodegradable polymers from synthetic or natural source. Mostly pesticides or fungicides are either integrated in the matrix or attached to the surface.

3.3 Polymers Used as Nanocarriers

Polymers that are used as nanocarriers for formulating nano-agrochemicals are classified as follows:

A. On the basis of occurrence in nature, polymers are classified as natural and synthetic polymers:
   (i) Natural polymers: Natural polymers are obtained naturally and can be directly used. Examples include alginate, gelatin, albumin and chitosan.
   (ii) Synthetic polymers: Synthetic polymers are chemically synthesized and do not occur in nature. Some examples are polylactides (PLA), polylactide-co-glycolide (PLGA), poly epsilon-caprolactone (PECL), poly ethylene glycol (PEG) and poly butyl cyanoacrylate (PBCA).
B. On the basis of biodegradability, polymers are classified into biodegradable and non-biodegradable polymers:
   (i) Biodegradable polymers: These polymers get degraded easily in the environment by the influence of microbial action. The most commonly used biodegradable polymers are polylactides (PLAs), poly cyanoacrylate and poly (D, L-lactides).
   (ii) Non-biodegradable polymers: These polymers do not get degraded by microbial interventions and are hard to dispose and hazardous to the environment. Examples are polyethene and polystyrene.

3.4 Nanoparticles for Seed Germination and Plant Growth

Nanoemulsions of chemicals have been reported to significantly influence seed germination and growth in many crop plants. Use of nanoformulations of plant growth regulators (plant growth promoting substances) in the form of nanoemulsions as

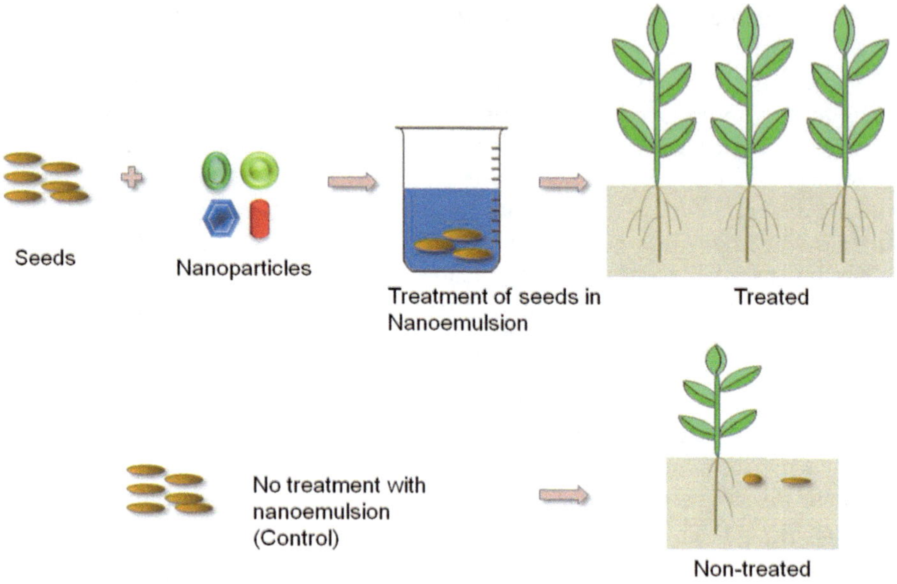

Fig. 3.2 Nanoemulsion seed treatment improves seed germination rate, breaks seed dormancy and improves seedling vigour

well as nanofertilizers improves the rate of seed germination, influences breaking seed dormancy and improves seedling vigour (Fig. 3.2). Use of various nanoparticles, such as of Ag (Almutairi and Alharbi 2015), FeS_2, ZnO and Co nanoparticles (Hoe et al. 2018), SiO_2 (Karunakaran et al. 2015), has shown promising results in seed germination and plant growth. Nanoparticles are known to penetrate the seed coat and exert a beneficial effect on the process of seed germination through increased water absorption, nitrate reductase enzyme levels, seed antioxidant systems, reduced antioxidant stress through H_2O_2 and superoxide radicals by increased enzyme activity (superoxide dismutase, ascorbate peroxidase, guaiacol peroxidase and catalase). As defined earlier, the principal factors of nanoparticles are their increased surface area and quantum effects, which enable easy absorption and smart delivery of nutrients. Also, aspect ratio of the size of nanoparticles helps in competitive binding with receptors and in subsiding reactive oxygen species. Some types of nanoparticles have been found to be well known for their antimicrobial properties, which results in controlling seed-borne pathogens (Table 3.1).

3.5 Nanofertilizers

Recent applications of nanotechnology in agriculture have successfully demonstrated the utility of nanomaterials as a potential plant growth regulators, but practical application of nanomaterial-based fertilizers on agricultural lands requires a

Table 3.1 Effect of nanoparticles on seed germination in crop plants

Nanoparticles	Uses	References
Mixture of nanoscale SiO_2-TiO_2	Increase nitrate reductase activity in soybean (*Glycine max*), thus improving seed growth	Lu et al. (2002)
Single-walled (SWCNTs) and multi-walled (MWCNTs) carbon nanotubes	Penetrate tomato seeds and increase the germination rate by increasing the seed water uptake	Khodakovskaya et al. (2009)
NanoTiO2	Promote photosynthesis and nitrogen metabolism, and thus greatly improve growth of spinach	Yang et al. (2007)
Nano-ZnCuFe-oxide followed by nano-FeO and nano-ZnO	Suspension of nanoparticles affects the growth of mung (*Vigna radiata*) seedlings	Dhokev et al. (2013)
FeS_2 NPs	Increase the production of spinach and different crops	Srivastava et al. (2014)
SiO_2 NPs	Have a significant impact on the seed germination potential in tomato (*Lycopersicum esculentum)*	Siddiqui and Whaibi (2014)
Silver (AgNPs)	Have significant effects on seed germination and seedling growth of corn, watermelon and zucchini plants	Almutairi and Alharbi (2015)
ZnO NPs	For seedling emergence and germination of wheat (*Triticum aestivum*)	Hussain et al. (2018)

suitable substrate to effectively disperse the nanomaterials (Kumar et al. 2018). Nanofertilizers relates to nanoformulations of fertilizers like N, P and K along with other essential minerals and micronutrients. They are modified fertilizers synthesized by chemical, physical or biological methods involving nanotechnology with the aim to improve their attributes and composition as well as enhance the productivity of crop plants (Singh et al. 2017). Their formulations are made with the prime motto of achieving the following objectives: (i) to reduce element volatilization, (ii) increase nutrient uptake, (iii) improve nutrient use efficiency, (iv) to reduce fertilizer dosage, (v) improve soil microflora and fauna and (vi) improve water-holding capacity. In simpler terms, nanofertilizer application is expected to improve the yield and nutritive value of crop plants, thereby resulting in improved plant health. Encapsulation of fertilizers with nanomaterials is the most common method employed till date for developing nanofertilizers. Fertilizer delivery through nanoparticle can be done in three ways: (i) encapsulation in nonporous material, (ii) coating of polymer around fertilizers and (iii) nutrient itself in nano form (Naderi and Shahraki 2013). The most commonly used nanocarriers of nanofertilizers are chitosan and oleylamine along with surfactants such as gluconic acid and cellulose. In order to stabilize and prevent agglomeration, polymers like polyethylene glycol (PEG), poly N-vinyl-2pyrrolidone (PVP), poly (methyl-methacrylate) (PMMA) and poly (methacrylic acid) (PMAA) are also used in nanofertilizer commercial formulations (Grillo et al. 2015).

3.6 Classification of Nanofertilizers

Four classes of nanofertilizers are currently available on the market:

1. Nitrogen nanofertilizers
2. Potash nanofertilizers
3. Zinc nanofertilizers
4. Nanoporous zeolite

Among these, zeolites are most unique as they individually serve as a fertilizer as well as can be used as a material for nanoformulation of other nutrients. Zeolites at nanoscale are used to slow down release of fertilizers by virtue of their property to form pores which can hold cations, small molecules and water within a crystalline state. They can hold cationic nutrients (e.g. NH^{4+} and K^{+}) as well as anionic forms such as SO_4^{2-}, NO_3^- and PO_4^{3-} with surface modification by the cationic surfactant hexadecyltrimethylammonium bromide (CTAB). Zeolite NPs have been employed in delivering nitrogen, phosphorous, potassium and sulphur (Table 3.2) (Subramanian et al. 2015; Manikandan and Subramanian 2014).

Nanofertilizers are applied to the soil or sprayed as foliar sprays. Soil applied nanofertilizers facilitate easy movement in the soil, increasing the release of nutrients and penetration into the roots, thereby enabling better uptake and translocation in the plant system, while foliar spray nanofertilizers increase chlorophyll formation and dry matter production, which consequently improve plant growth (Dhokev et al. 2013). They are less toxic to humans and animals (Leon-Silva et al. 2018). Movement of nanoparticles in the plants is determined by the mode of application of nanoparticles. A nanoparticle having better translocation ability via phloem is more likely to get a good distribution in the plant. This indicates that if the application of nanoparticle is done via foliar spraying, then maximum efficient usage of the chemicals is possible. For nanofertilizers, wherein the nutrients move through roots, soil application is advised, which allows better uptake via xylem (Fig. 3.3).

Table 3.2 Nanofertilizers and nanoformulations for plant nutrition

S. No.	Nanomaterial	Nutrient	References
1	Surface-modified zeolite	Phosphate	Rayalu et al. (2006)
2	Nanocomposites containing organic polymer intercalated in the layers of kaolinite clays	Wide range of nutrient	Liu et al. (2006)
3	Nanocoating of sulphur with chitosan and other nanocomposites	NPK	Wilson et al. (2008)
4	Surface-modified zeolite	Sulphate	Li and Zhang (2010)
5	Montmorillonite	Nitrogen	Bortolin et al. (2013)
6	Zeolite	Nitrogen	Komarneni (2010) and Manikandan and Subramanian (2014)

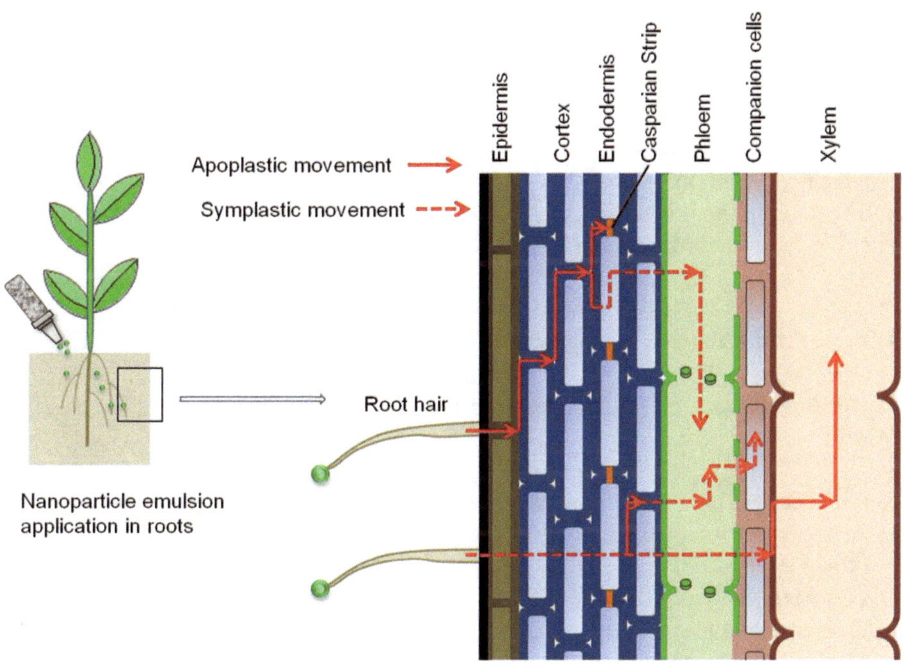

Fig. 3.3 Mechanism of movement of nanofertilizer into the root system of plant

In root-feeding, the movement of nanoparticles in plants tissues occurs either by apoplastic or symplastic action. In the apoplastic movement, the nanoparticles move through the extracellular spaces, cell walls of the adjacent cell and xylem vessels (Sattelmacher 2001). Apoplastic movement allows nanoparticles to move towards the central cylinder of root and into the vascular tissues for further movement towards the aerial part through xylem by following the transpiration system (Larue et al. 2012; Zhao et al. 2012; Sun et al. 2014). In the symplastic movement, the particles move between the cytoplasm of adjacent cells through plasmodesmata and sieve plates (Roberts and Oparka 2003). In the symplastic pathway, nanoparticles have to cross the Casparian strip to reach inside the phloem. Sun et al. (2014) and Lv et al. (2015) showed that some nanomaterials get stuck and accumulate at the Casparian strip. Distribution of nanoparticles in non-photosynthetic tissues and organs takes place with the help of the sieve tube elements in the phloem (Wang et al. 2012; Raliya et al. 2016).

On foliar application, nanoparticles follow a lipophilic or hydrophilic pathway (Schönherr 2002). Lipophilic diffusion of nanoparticles takes place through cuticular waxes, while in hydrophilic pathway nanoparticles move via polar aqueous pores present in the cuticle and stomata. Foliar application of nanoparticles is size limiting because the diameter of cuticular pores are approximately 2 nm (Eichert and Goldbach 2008). So nanoparticles of size more than 2 nm cannot get past and are restricted from foliar application.

3.7 Benefits of Nanofertilizers

1. Field trials with nanofertilizers in sunflower have demonstrated grain yields increase by 50% and in cucumber by 25%.
2. Increase of about 10% in protein and sugar content in many fruits and vegetables have been demonstrated.
3. The overall health of the plant is enhanced, as the host defence system is enhanced indirectly by nutrients.

3.8 Advantages of Nanofertilizers

The following are the advantages of nanofertilizers as described by Leon-Silva (2018):

1. Reduces fertilizer requirements and lowers the cost of production
2. Reduces nutrient loss through leakage and seepage in soil, especially in the case of soil application
3. Eco-friendly synthesis
4. Economical and easily available custom-made products
5. Controlled release of plant nutrients and improved nutrient uptake
6. Improved soil nutrient status and bioavailability of essential nutrients
7. Less negative impacts and toxicity
8. Improved product quality

3.9 Nanoherbicides

Weeds are plants out of place in a cultivated ecosystem of agricultural crop plants, which compete with them for nutrients, water and sunlight. Heavy weeds in the cultivated fields of crop plants affect the growth, alter the developmental stages and ultimately account for 30–40% yield loss. Weeding is an intercultural operation, wherein removing unwanted parasitic weeds is done manually. Herbicides are an alternative to the laborious and time-consuming weeding operations, and the mode of action of any herbicide depends on the moisture content, time of application, persistence and availability. In order to increase the efficiency of herbicides, nano-encapsulation has been attempted, which could facilitate controlled release and delivery when sufficient moisture is available. Nanoherbicides are formulated mainly using nanoencapsulation techniques and a few successful examples of nano-encapsulated herbicides are presented in Table 3.3.

Table 3.3 Nanoherbicides for effective weed management

S. No	Nanomaterial	Herbicide	Purpose	References
1	MnO_2 core shell protected with bilayer polymers	Herbicide	Controlled release of active ingredient after receiving rainfall	Chinnamuthu and Kokiladevi (2007)
2	Manganese carbonate core nanoshell coated with water soluble polymer sodium poly styrene sulfonate and poly allylamine hydrochloride	Pendimethalin	Release of active ingredient on receipt of rainfall for rainfed cropping system	Kanimozhi and Chinnamuthu (2012)
3	Poly(epsilon-caprolactone) nanoparticles	Atrazine	Control weeds and reduce damage to the environment	Pereira et al. (2014)
4	Herbicide-loaded pectin nanoparticles	Metsulfuron methyl	Low herbicide loading (6.30%) and high encapsulation efficiency (63 to 65%)	Kumar et al. (2017)

3.10 Advantages of Nanoherbicides

1. Herbicide-loaded nanoparticles reduced the use and rate of herbicide application.
2. Improved efficacy and environmental safety.
3. Low cytotoxic effect on cell lines and therefore pose lesser risk to animals and humans.

3.11 Nanopesticides

Nanopesticides are pesticides formulated in nano to micro scales with the aim of slow and steady release, such that the active ingredient is completely used for the control of pests and diseases. In general, polymer-based nanoencapsulated formulations are widely used for developing nanopesticides using different polysaccharides such as chitosan, alginates, starch and polyesters (e.g. poly-ε-caprolactone, polyethylene glycol). The first polymer nanopesticide formulation was reported in the 1970s for controlled release of biocides. Nanobiocides, also grouped in the category of nanopesticides, are nanoformulations of microbial or living agents used for killing other living entities. Some successful examples of nanopesticides that have been reported are ZnO and Ag NPs. Foliar application of nanopesticides works in the mode as explained earlier in the mechanism of movement and translocation of nanoparticles in the plant system (Fig. 3.4).

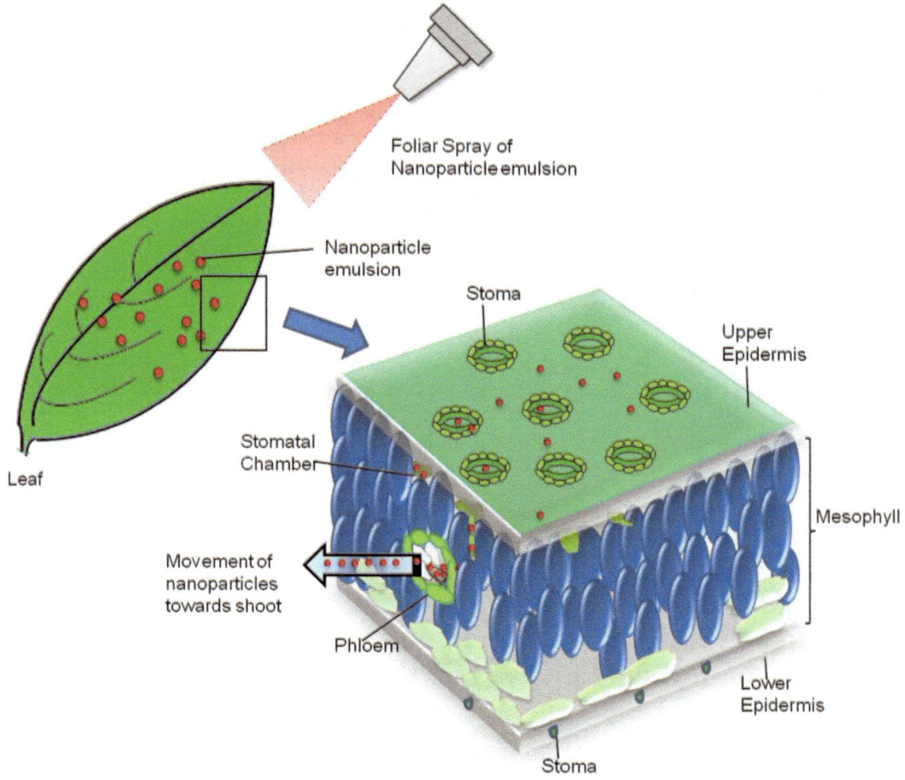

Fig. 3.4 Mechanism of movement of nanoparticles from foliar spray of nanopesticides into the aerial shoot system through leaves

3.12 Advantages of Nanopesticides

The following are the advantages summarized from the results obtained in the application of nanopesticides at laboratory and field trials (Table 3.4);

1. Reduces pesticide requirements and application rates, thereby reducing costs of cultivation
2. Controlled and slow release by smart delivery system, resulting in enhanced efficiency
3. Prolonged availability of active ingredient for better control of insect pests
4. Reduced environmental contamination and easily biodegradable
5. Reduced toxicity risks to animals, humans and natural flora and fauna

Table 3.4 Nanopesticides and their examples in pest and disease management of different crop plants

S. No.	Nanoformulation	Purpose	Crop	Pest management[a]	References
I Nanopesticides, mode of operation and control of insect pests of different crops					
1	Solid lipid nanoparticles of essential oil from *Artemisia arborescens*	Reduce the rapid evaporation of essential oil	Citrus *Quercus suber*	*Aphis gossipy Bemisia tabaci* and *Lymantria dispar*	Lai et al. (2006)
2	Polyethylene glycol polymer coated NPs: Essential oil from garlic	Slow and persistent release of the active componentsß	Cereals, beans, nuts	*Tribolium castaneum* insect in stored products	Yang et al. (2009)
3	Alumina NPs	Significant mortality of pest	Cereals, beans, nuts	*S. oryzae* L. and *Rhyzopertha dominica* (F.),	Stadler et al. (2010)
4	Silver nanoparticles	High efficacy and mortality	Rice	*Sitophilusoryzae*	Debnath et al. (2011)
5	Nanosilica	Control of neonates by affecting their feeding preference	Tomato	*Spodoptera littoralis*	EL-bendary and El-Helaly (2013)
6	Imidachloprid – sodium alginate	Low pesticide load and prolonged availability	Bhendi	Leafhoppers, Jassids (sucking pests)	Kumar et al. (2014)
II Nanofungicides, mode of operation and control of fungal pathogens of different crops					
1	Porous hollow silica NPs – validamycin	Antibacterial, controlled delivery system of water-soluble bactericide	Rice, potato	Rice sheath blight, black scurf on seed potatoes, damping-off, *Rhizoctonia solani*	Liu et al. (2006)
2	Silver nanoparticles	Inhibition of spore and germ tube formation	Cereals	*B. sorokiniana* and *M. grisea* in vitro studies	Jo et al. (2009)
3	Silver nanoparticles	Antifungal	Pear, apple, grapes etc.	*Phoma glomerata*	Gajbhiye et al. (2009)
4	Fungicide – ZnO	Antifungal	All crops	*Botrytis cinerea* and *Penicillium expansum*	He et al. (2011)

(continued)

Table 3.4 (continued)

S. No.	Nanoformulation	Purpose	Crop	Pest management[a]	References
5	Zinc oxide	Deformation in mycelia mats and prevention of development of conidia and conidiophore	Fruit crops	*Botrytis cinerea* and *Penicillium expansum*	He et al. (2011)
6	AgNPs – *Bacillus* sp.	Antifungal	Many crops	*Fusarium oxysporum*	Gopinath and Velusamy (2013)
7	Thiram – ZnO nanoparticles	Multipurpose	Different crops	Antifungal	Xue et al. (2014)
8	Green Bs AgNPs	Antifungal	Wheat	Spot blotch disease, *Bipolaris sorokiniana*	Mishra et al. (2014b)
9	AgNPs – *Serratia* sp.	Antifungal	Wheat	*Bipolaris sorokiniana*	Mishra et al. (2014a)
10	AgNPs – *Brassica rapa*	Antifungal	Tree species	Wood-rotting pathogens	Narayanan and Park (2014)
11	AgNPs – *Calotropis procera*	Antimicrobial activity	Different crops	Many pathogens	Mohamed et al. (2014)

[a]Pests here include insect pests as well as disease-causing pathogens

3.13 Nanosensors and Their Applications

Nanosensors are modified sensors used to detect diseases, moisture and nutrient condition for determining optimum usage of pesticides, water requirements, and fertilizer dosages (Rai and Ingle 2012). These nanosensors are developed using top-down lithography, molecular self-assembly and bottom-up assembly approaches. The potential use of nanosensors is in the detection of pathogens causing diseases in crop plants (Table 3.5), identification of pests and for post-harvest quality assessment. Nanosensors primarily consist of three components: (i) bioreceptor, (ii) transducer and (iii) detector. Nanosensors are devices of nanoscale in which the biodetector measures an analyte by selective binding, signal is generated due its interaction with the bio-element present in the bioreceptor, the signal is processed into useful metrics by transducer and then detected by the detector and analysed (Fig. 3.5). Different types of nanosensors work on the same principle with a common basic workflow but for modifications in their bioreceptor component.

Nanomaterials with mechanical, electronic, photonic, and magnetic properties prove advantageous in the development of sensors and diagnostic tools. E-nose is a nanotechnology-based sensor which is used to assess pest or mechanical damage in cucumber,

Table 3.5 Nanosensors in pathogen detection, plant disease monitoring and management

S. No.	Nanosensors	Pathogen	Disease	Crop	References
1	Fluorescent silica nanoparticles (FSNP)	Bacterial spot disease	Bacterial spot disease	Tomato	Yao et al. (2009)
2	Gold nanoparticles based	*Tilletia indica (fungus)*	Karnal bunt disease	Wheat	Singh et al. (2010)
3	Luminescent semiconductor nanocrystals (QDs)	Beet necrotic yellow veinvirus (BNYVV)	Rhizomania	Beet	Safarpour et al. (2012)
4	Gold nanoparticles	Triteza virus	Tristeza	Citrus	Shojaei et al. (2016)

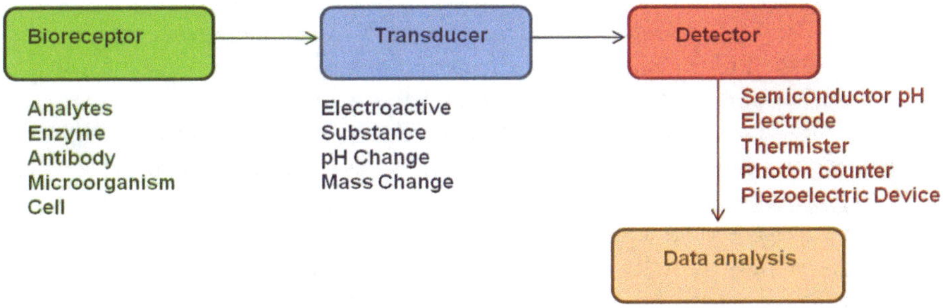

Fig. 3.5 Schematic representation of the working principle of a nanosensor

pepper and tomato leaves by sensing differences in the profiles of volatile organic compounds (VOCs) emitted from both damaged and undamaged plants. This nanosensor was successfully demonstrated for precisely discriminating spider-mite-infested cucumbers from healthy plants. Similarly, e-nose-based sensors have been used to identify plants infected by powdery mildew in tomato. Nanobiosensors can detect soil condition and provide advisory for optimum chemical fertilizer dosages. Nanosensors can be linked to a global positioning system (GPS) for real-time monitoring of disease, and distributed throughout the field to monitor soil conditions and crop health.

3.14 Classification of Nanosensors

Liu (2003) has classified nanosensors based on their mechanism and material used in their development into the following five classes:

1. Nanostructured materials, e.g. porous silicon
2. Nanoparticles-based sensors
3. Nanoprobes
4. Nanowire nanosensors
5. Nanosystems: cantilevers, nano-electromechanical systems (NEMS)

3.15 Advantages of Nanosensors

The following are the advantages of nanosensors listed out by Lu and Bowles (2013):

1. High sensitivity and selectivity
2. Near-real-time detection
3. Low cost and portability

3.16 Nanotechnology for Post-harvest Improvement

Nanotechnology has found a significant foothold in production technology and post-harvest shelf life enhancement through the control of microorganisms, packaging materials (films), etc. Nanotechnology-led packaging allows control of the influence of gases and UV, increased strength, quality and aesthetic value enhancement. Use of multiple chips with nanobiosensors has led to automated storage control. Application of chitosan has gained importance for controlling pre- and post-harvest diseases for improvement of produce quality and environment safety due to low toxicity, initially in tomato grey mould resistance (*Botrytis cineraria*). Since, newer nanoparticles have been used for fungal control towards increased post-harvest life of banana, carrot, tomato, onion, etc. Newer range of packaging coverages have evolved, such as TiO2 and silver ions that are suited owing to physical and chemical stability, low cost, ease of availability and non-toxicity. These have also attracted the interest of several researchers, because of their photoactivity, semiconductor photocatalysis, nanocrystallites and antibacterial activity. Development of nano-TiO2 with light catalysing capability, nano-Ag with quanta and large external area effect have shown ability to absorb and decompose ethylene. Fuji apples with nano-SiOx/chitosan and Green tea have demonstrated better-quality maintenance through nanopacking. Technologies for using biological molecules, such as sugars or proteins food biosensors, are at their infancy. Such biosensors can serve as detectors of food pathogens/contaminants and as devices to track food products. Delivery and controlled-release systems of nutraceuticals in food grains through nanotechnology are another feasibility. Some application of nanobiosensors include: virus recognition using antibody sensor arrays on self-assembled nanoscale block copolymer patterns, detection of food-borne toxins with multifunctional nanoparticles, development and characterization of nanocomposite materials for the detection of pore-forming toxins and molecular imprinted polymers for plant and insect virus recognition.

Table 3.6 Microbial-based nanoformulations and their use in agriculture

Sources	Nanoformulation	Applications	References
AgNPs	*Trichoderma viride*	Vegetable and fruit preservation	Fayaz et al. (2009)
AgNPs	*Aspergillus niger*	Antifungal and antibacterial	Jaidev and Narasimha (2010)
AgNPs	*Spirulina platensis*	Bactericidal	Mala et al. (2012)
AgNPs	Cow's milk	Antifungal	Lee et al. (2013)
ZnNPs	*Aspergillus fumigatus*	Enhanced P-mobilizing enzymes and gum production in cluster bean	Raliya and Tarafdar (2013)

3.17 Microbial Nanoformulations in Quality Enhancement

Nanoparticles are not only used to control insect pests and diseases but also for improving quality-related traits in cereals, pulses and post-harvest quality of fruits and vegetables. Spoilage by phytopathogens can be reduced by use of biocontrol agents which improve quality. Some microbial biocontrol agents are formulated using nanotechnology and have been successfully demonstrated in quality improvement in different crops (Table 3.6).

3.18 Nanopackaging Technology

Smart packaging is a system of packaging which involves use of nanotechnology for packing perishable fruits and vegetables with reduced space, minimal processing and at the same time enhance their shelf life for better marketability and distant marketing. In India, lack of cold storage accounts for major post-harvest losses. Post-harvest losses can be minimized by use of ethylene blockers, delaying ripening processes by manipulation of ethylene biosynthesis, wax coating, etc. An alternate to these strategies is smart packaging systems which involve the use of nanofilm, called as nanopackaging technology. When applied to fruits and vegetables, nanofilms was found to emit a chemical vapor which can extend the shelf life of vegetables by upto 21 days without any loss of quality. This nanomaterial contains a key ingredient known as hexanal, a synthetic version of that found usually in beans and cucumbers. Attempts have been made to use hexanal in natural plant fibres and wax coatings so as to enhance shelf life of fruits and vegetables. Hexanal inhibits phospholipase-D in the peel of fruits, which is involved in the deterioration of fruit quality (Jincy et al. 2017). In Tamil Nadu Agricultural University, Coimbatore, India, nano-based hexanal technology was developed to extend the shelf life of mangoes.

This was prepared by subjecting a polymer to high-intensity voltage to convert the solution into a nano-fibre. Using the electrospinning technique, a nano-fibre was developed with polyvinyl alcohol and cyclodextrin as a sheath and the hexanal in the core, thereby providing a unique advantage of high surface to mass ratio and regulating the release of hexanal vapor. This technology was demonstrated to increase the shelf life of mango fruit by 2–3 weeks. In India, trials are under way using hexanal nanofilm technology for extending the shelf life of mangoes, especially in Tamil Nadu (Anusuya et al. 2016). The TERI report (2017) had emphasized reduced nutrient losses and agrochemicals usage, by ensuring smart delivery of using nanotechnology. Furthermore DBT (2019) has formulated draft guidelines for evaluation of nano-agri input (NAIPs) and nano-agriproducts (NAPs) in India, that is aimed at providing uniform guidelines to support quality commercialization of nanotechnology-based innovations that are safe for human health and environment. Very few nanotechnology applications have yielded commercialization of products due to inconsistent national legislative frameworks, limited regulatory guidelines and a lack of public licensing initiatives. Food and Agriculture Organization, the Organisation for Economic Co-operation and Development (OECD) and the Australian Pesticides and Veterinary Medicines Authority (APVMA) have argued for prioritizing regulatory frameworks to reap the potential benefits.

References

Almutairi, Alharbi A (2015) Effect of silver nanoparticles on seed germination of crop plants. J Adv Agric 4:280–285

Anusuya P, Nagaraj R, Janavi GJ, Subramanian KS, Paliyath G, Subramanian J (2016) Pre-harvest sprays of hexanal formulation for extending retention and shelf life of mango fruits. Sci Hortic 211:231–240

Bortolin A, Aouada FA, Mattoso LH, Ribeiro C (2013) Nanocomposite PAAm/methyl cellulose/montmorillonite hydrogel: evidence of synergistic effects for the slow release of fertilizers. J Agric Food Chem 61(31):7431–7439

Chinnamuthu C, Kokiladevi E (2007) Weed management through nanoherbicides. In: Chinnamuthu CR, Chandrasekaran B, Ramasamy C (eds) Application of nanotechnology in agriculture, vol 10. Tamil Nadu Agricultural University, Coimbatore, pp 978–971

Chowdappa P, Gowda S (2013) Nanotechnology in crop protection: status and scope. Pest Manag Hortic Ecosyst 19(2):131–151

Debnath N, Das S, Seth D, Chandra R, Bhattacharya SC, Goswami A (2011) Entomotoxic effect of silica nanoparticles against *Sitophilus oryzae* (L.). J Pest Sci 84(1):99–105

Dhokev S, Mahajan P, Kamble R, Khanna A (2013) Effect of nanoparticles suspension on the growth of mung (*Vigna radiata*) seedlings by foliar spray method. Nanotechnol Dev 3(1):e1. https://doi.org/10.4081/nd.2013.e1

Eichert T, Goldbach HE (2008) Equivalent pore radii of hydrophilic foliar uptake routes in stomatous and astomatous leaf surfaces further evidence for a stomatal pathway. Physiol Plant 132:491–502. https://doi.org/10.1111/j.1399-3054.2007.01023.x

El-bendary H, El-Helaly A (2013) First record nanotechnology in agricultural: Silica nano-particles a potential new insecticide for pest control. Appl Sci Rep 4(3):241–246. E-ISN: 2310–2940

Fayaz AM, Balaji K, Girilal M, Kalaichelvan PT, Venkatesan R (2009) Mycobased synthesis of silver nanoparticles and their incorporation into sodium alginate films for vegetable and fruit preservation. J Agric Food Chem 57:6246–6252

Gajbhiye M, Kesharwani J, Ingle A, Gade A, Rai M (2009) Fungus-mediated synthesis of silver nanoparticles and their activity against pathogenic fungi in combination with fluconazole. Nanomedicine 5(4):382–386. https://doi.org/10.1016/j.nano.2009.06.005

Ghormade V, Deshpande MV, Paknikar KM (2011) Perspectives for nano-biotechnology enabled protection and nutrition of plants. Biotechnol Adv 29(6):792–803

Gopinath V, Velusamy P (2013) Extracellular biosynthesis of silver nanoparticles using *Bacillus* sp. GP-23 and evaluation of their antifungal activity towards *Fusarium oxysporum. Spectrochim Acta A Mol Biomol Spectrosc* 106:170–174. https://doi.org/10.1016/j.saa.2012.12.087

Grillo R, Rosa AH, Fraceto LF (2015) Engineered nanoparticles and organic matter: a review of the state-of-the-art. Chemosphere 119:608–619

He L, Liu Y, Mustapha A, Lin M (2011) Antifungal activity of zinc oxide nanoparticles against *Botrytis cinerea* and *Penicillium expansum*. Microbiol Res 166:207–215. https://doi.org/10.1016/j.micres.2010.03.003

Hoe PT, Mai NC, Lien LQ, Ban NK, Minh CV, Chau NH et al (2018) Germination responses of soybean seeds under Fe, ZnO, Cu and Co nanoparticle treatments. Int J Agric Biol 20:562–1568

Hussain A, Ali S, Rizwan M, Zia Ur Rehman M, Javed MR, Imran M, Chatha SAS, Nazir R (2018) Zinc oxide nanoparticles alter the wheat physiological response and reduce the cadmium uptake by plants. Environ Pollut 242(PtB):1518–1526

Jaidev LR, Narasimha G (2010) Fungal mediated biosynthesis of silver nanoparticles, characterization and antimicrobial activity. Colloids Surf B: Biointerfaces 81(2):430–433

Jincy M, Djanaguiraman M, Jeyakumar P, Subramanian KS, Jayasankar S, Paliyath G (2017) Inhibition of phospholipase D enzyme activity through hexanal leads to delayed mango (*Mangifera indica* L.) fruit ripening through changes in oxidants and antioxidant enzymes activity. Sci Hortic 218:316–325

Jo YK, Kim BH, Jung G (2009) Antifungal activity of silver ions and nanoparticles on phytopathogenic fungi. Plant Dis 93(10):1037–1043. https://doi.org/10.1094/PDIS-93-10-1037

Kah M, Hofmann T (2014) Nanopesticide research: current trends and future priorities. Environ Int 63:224–235

Kanimozhi V, Chinnamuthu C (2012) Engineering core/hallow shell nanomaterials to load herbicide active ingredient for controlled release. Res J Nanosci Nanotechnol 2(2):58–69

Karunakaran G, Suriyaprabha R, Rajendran V, Kannan N (2015) Effect of contact angle, zeta potential and particles size on the in vitro studies of Al2O3 and SiO2 nanoparticles. IET Nanobiotechnol 9(1):27–34. https://doi.org/10.1049/iet-nbt.2013.0067

Khodakovskaya M, Dervishi E, Mahmood M, Xu Y, Li Z, Watanabe F (2009) Carbon nanotubes are able to penetrate plant seed coat and dramatically affect seed germination and plant growth. ACS Nano 3:3221–3227

Komarneni S (2010) Potential of nanotechnology in environmental soil science. In: 9th international conference of the East and Southeast Asia federation of soil science societies

Kumar S, Bhanjana G, Sharma A, Sidhu MC, Dilbaghi N (2014) Synthesis, characterization and on field evaluation of pesticide loaded sodium alginate nanoparticles. Carbohydr Polym 101:1061–1067

Kumar S, Bhanjana G, Sharma A, Dilbaghi N, Sidhu MC, Kim KH (2017) Development of nanoformulation approaches for the control of weeds. Sci Total Environ 15(586):1272–1278. https://doi.org/10.1016/j.scitotenv.2017.02.138.

Kumar R, Ashfaq M, Verma N (2018) Synthesis of novel PVA–starch formulation-supported Cu–Zn nanoparticle carrying carbon nanofibers as a nanofertilizer: controlled release of micronutrients. J Mater Sci 53:7150–7164. https://doi.org/10.1007/s10853-018-2107-9

Lai F, Wissing SA, Müller RH, Fadda AM (2006) Artemisia arborescens L essential oil-loaded solid lipid nanoparticles for potential agricultural application: preparation and characterization. AAPS PharmSciTech 7(1):E10–E18. https://doi.org/10.1208/pt070102

Larue C, Veronesi G, Flank AM, Surble S, Herlin-Boime N, Carrière M (2012) Comparative uptake and impact of TiO2 nanoparticles in wheat and rapeseed. J Toxicol Environ Health A 75:722–734

Lee KJ, Park SH, Govarthanan MA, Hwang PH, Seo YS, Cho M, Lee WH, Lee JY, Kannan SK, Taek B (2013) Synthesis of silver nanoparticles using cow milk and their antifungal activity against phytopathogens. Mater Lett 105:128–131. https://doi.org/10.1016/j.matlet.2013.04.076

León-Silva S, Fernández-Luqueño F, López-Valdez F (2018) Engineered nanoparticles: are they an inestimable achievement or a health and environmental concern? In: López-Valdez F, Fernández-Luqueño F (eds) Agricultural nanobiotechnology: modern agriculture for a sustainable future. Springer Nature, Cham, pp 183–212. https://doi.org/10.1007/978-3-319-96719-6_10

Li Z, Zhang Y (2010) Use of surfactant-modified zeolite to carry and slowly release sulfate. Desalin Water Treat 21(1–3):73–78

Li LS, Hu J, Yang W, Alivisatos AP (2001) Band gap variation of size- and shape-controlled colloidal CdSe quantum rods. Nano Lett 1:349

Liu Y (2003) Nanosensors, http://www.slideserve.com/kim-johnston/nanosensors dt. 3 Oct 2016.

Liu F, Wen LX, Li ZZ, Yu W, Sun HY, Chen JF (2006) Porous hollow silica nanoparticles as controlled delivery system for water-soluble pesticide. Mater Res Bull 41:2268–2275

Lu J, Bowles M (2013) How will nanotechnology affect agricultural supply chains? Int Food Agribus Manag Rev 16(2):21–42

Lu CM, Zhang CY, Wen JQ, Wu GR, Tao MX (2002) Research of the effect of nanometer materials on germination and growth enhancement of Glycine max and its mechanism. Soybean Sci 21:168–172

Lv J, Zhang S, Luo L, Zhang J, Yangc K, Christie P (2015) Accumulation, speciation and uptake pathway of ZnO nanoparticles in maize. Environ Sci Nano 2:68–77

Mala R, Arunachalam P, Sivsankari M (2012) Synergistic bactericidal activity of silver nanoparticles and ciprofloxacin against phytopathogens. J Cell Tissue Res 12:3249–3254

Manikandan A, Subramanian K (2014) Fabrication and characterisation of nanoporous zeolite based N fertilizer. Afr J Agric Res 9(2):276–284

Mishra S, Singh BR, Singh A, Keshwani C, Naqvi AH, Singh HB (2014a) Biofabricated silver nanoparticles act as a strong fungicide against *Bipolaris sorokiniana* causing spot blotch disease in wheat. PLoS One 9(5):e97881

Mishra S, Singh BR, Singh A, Keswani C, Naqvi AH et al (2014b) Biofabricated silver nanoparticles act as a strong fungicide against *Bipolaris sorokiniana* causing spot blotch disease in wheat. PLoS One 9(5):e97881. https://doi.org/10.1371/journal.pone.0097881

Mohamed N, Mady I, Wael AM, Ahmed S, Shoreit AAFMSM (2014) Antimicrobial activity of latex silver nanoparticles using Calotropis procera. Asian Pac J Trop Biomed 4(11):876–883. https://doi.org/10.12980/APJTB.4.201414B216

Naderi MR, Danesh-Shahraki A (2013) Nanofertilizers and their roles in sustainable agriculture. Int J Agric Crop Sci 5:2229–2232

Narayanan K, Park H (2014) Antifungal activity of silver nanoparticles synthesized using turnip leaf extract (Brassica rapa L.) against wood rotting pathogens. Eur J Plant Pathol 140(2):185–192. https://doi.org/10.1007/s10658-014-0399-4

Pereira AE, Grillo R, Mello NF, Rosa AH, Fraceto LF (2014) Application of poly (epsilon-caprolactone) nanoparticles containing atrazine herbicide as an alternative technique to control weeds and reduce damage to the environment. J Hazard Mater 268:207–215

Rai M, Ingle A (2012) Role of nanotechnology in agriculture with special reference to management of insect pests. Appl Microbiol Biotechnol 94(2):287–293. https://doi.org/10.1007/s00253-012-3969-4

Raliya R, Tarafdar J (2013) ZnO nanoparticle biosynthesis and its effect on phosphorous-mobilizing enzyme secretion and gum contents in Clusterbean (*Cyamopsis tetragonoloba L.*). Agric Res 2(1):48–57. https://doi.org/10.1007/s40003-012-0049-z

Raliya R, Franke C, Chavalmane S, Nair R, Reed N, Biswas P (2016) Quantitative understanding of nanoparticle uptake in watermelon plants. Front Plant Sci 26(7):1288. https://doi.org/10.3389/fpls.2016.01288

Rayalu SS, Bansiwal AK, Meshram SU, Labhsetwar N, Devotta S (2006) Fly ash based zeolite analogues: versatile materials for energy and environment conservation. Catal Surv Jpn 10:74

Roberts AG, Oparka KJ (2003) Plasmodesmata and the control of symplastic transport. Plant Cell Environ 26:103–124. https://doi.org/10.1046/j.1365-3040.2003.00950.x

Safarpour H, Safarnejad R, Tabatabaei M, Mohsenifar A, Rad F, Shahryari F (2012) Detection of Polymyxa betae; the transmitting agent of rhizomania disease of sugar beet, with quantum dots FRET-based biosensor. Can J Plant Pathol 34(4):507–515

Sattelmacher B (2001) The apoplast and its significance for plant mineral nutrition. New Phytol 149:167–192

Schönherr J (2002) A mechanistic analysis of penetration of glyphosate salts across astomatous cuticular membranes. Pest Manag Sci 58:343–351. https://doi.org/10.1002/ps.462

Shojaei TR, Salleh MA, Sijam K, Rahim RA, Mohsenifar A, Safarnejad R, Tabatabaei M (2016) Detection of Citrus tristeza virus by using fluorescence resonance energy transfer-based biosensor. Spectrochim Acta A Mol Biomol Spectrosc 169:216–222. https://doi.org/10.1016/j.saa.2016.06.052

Siddiqui MH, Al-Whaibi MH (2014) Role of nano-SiO2 in germination of tomato (*Lycopersicum esculentum* seeds mill.). Saudi J Biological Sci 21(1):13–17. https://doi.org/10.1016/j.sjbs.2013.04.005

Singh S, Singh M, Agrawal V, Kumar A (2010) An attempt to develop surface plasmon resonance based immunosensor for Karnal bunt (Tilletia indica) diagnosis based on the experience of nano-gold based lateral flow immuno-dipstick test. Thin Solid Films 519:1156–1159. https://doi.org/10.1016/j.tsf.2010.08.061

Singh MD, Chirag G, Patidar OP, Meena HM, Prakasha G, Vishwajith (2017) Nano-fertilizers is a new way to increase nutrients use efficiency in crop production. IJAS 97:3831–3833

Srivastava G, Das CK, Das A, Singh SK, Roy M, Kim H, Sethy N, Kumar A, Sharma R, Singh S, Philip D, Das M (2014) Seed treatment with iron pyrite (FeS2) nanoparticles increase the production of spinach. RSC Adv 4:58495–58504. https://doi.org/10.1039/C4RA06861K

Stadler T, Buteler M, Weaver DK (2010) Novel use of nanostructured alumina as an insecticide. Pest Manag Sci 66(6):577–579

Subramanian KS, Manikandan A, Thirunavukkarasu M, Rahale CS (2015) Nano-fertilizers for balanced crop nutrition. In: Rai M et al (eds) Nanotechnologies in food and agriculture, vol 3, 1st edn. Springer, Cham, pp 69–80

Sun D, Hussain HI, Yi Z, Siegele R, Cresswell T, Kong L (2014) Uptake and cellular distribution, in four plant species, of fluorescently labeled mesoporous silica nanoparticles. Plant Cell Rep 33:1389–1402

Tarafdar J (2012) Perspectives of nanotechnological applications for crop production. NAAS News 12:8–11

Wang Z, Xie X, Zhao J, Liu X, Feng W, White JC (2012) Xylem and phloem-based transport of CuO nanoparticles in maize (*Zea mays* L.). Environ Sci Technol 46:4434–4441

Wilson MA, Tran NH, Milev AS, Kannangara GSK, Volk H, Lu GQM (2008) Nanomaterials in soils. Geoderma 146(1–2):291–302

Xue J, Luo Z, Li P, Ding Y, Cui Y, Wu Q (2014) A residue-free green synergistic antifungal nanotechnology for pesticide thiram by ZnO nanoparticles. Sci Rep 4:5408. https://doi.org/10.1038/srep05408

Yang F, Liu C, Gao F et al (2007) Role of biosynthesized nanoparticles (NPs) in agriculture. Biol Trace Elem Res 119:77. https://doi.org/10.1007/s12011-007-0046-4

Yang FL, Li XG, Zhu F, Lei CL (2009) Structural characterization of nanoparticles loaded with garlic essential oil and their insecticidal activity against Tribolium castaneum (Herbst) (Coleoptera: Tenebrionidae). J Agric Food Chem 57(21):10156–10162

Yao KS, Li SJ, Tzeng KC, Cheng TC, Chang CY, Chiu C, Liao C, Hsu JJ, Lin ZP (2009) Fluorescence silica nanoprobe as a biomarker for rapid detection of plant pathogens. Adv Mater Res 79:513–516

Zhao L, Peralta-Videa JR, Ren M, Varela-Ramirez A, Li C, Hernandez-Viezcas JA (2012) Transport of Zn in a sandy loam soil treated with ZnO NPs and uptake by corn plants: electron microprobe and confocal microscopy studies. Chem Eng J 184:1–8

Agriculture and Nanoparticles

4

Nüket Altindal and Demet Altindal

Abstract

Herbal, animal and agricultural activities that have been applied to meet human needs in harmony with nature throughout human history have not harmed the ecosystem and have not caused environmental problems. However, the current ecosystem balance continues to deteriorate as a result of classical agricultural practices to get more products from the unit area to meet the food needs of the rapidly growing population. Therefore, new approaches to agricultural production and techniques such as nanotechnology are needed. In this context, nanoparticles that form the basis of nanotechnology have emerged as a versatile platform for solving the problems encountered. Nanoparticles have the potential in agricultural applications to be used in plant nutrition, plant and animal breeding and in the fight against herbicides and harmful insects.

Keywords

Agriculture · Nanoparticle · Field · Plant · Nanotechnology

Headings

1. The effects of nanoparticles on the growth and development of plants are important.
2. Nanoparticular fertilizers can increase the effective use of plant nutrients.

N. Altindal (✉)
Sivaslı Vocational School, Department of Crop and Animal Production, Programme of Medical and Aromatic Plants, Uşak University, Uşak, Turkey

D. Altindal
Fethiye Ali Sıtkı Mefharet Koçman Vocational School, Department of Crop and Animal Production, Organic Farming Programme, Muğla Sıtkı Koçman University, Muğla, Turkey

M. Ghorbanpour et al. (eds.), *Biogenic Nano-Particles and their Use in Agro-ecosystems*, https://doi.org/10.1007/978-981-15-2985-6_4

3. Nanotechnology applications in agriculture aim to reduce excessive use of chemical-containing plant protection products.
4. Nanoparticles are important in the increase of agricultural productivity.

4.1 Introduction

The need for food to feed the population of billions of people around the world is increasing. Global climate changes, rapid population growth, and inadequate food production have begun to worry all countries. The adverse effects of excessive chemical use in agriculture for the purpose of controlling microorganisms that are responsible for virus and fungi infections, which cause diseases before and after harvesting, are still unsolved. Therefore, research is focused on the search for alternatives that may be environmentally friendly and may be effective for microbial control. Therefore, the importance of agriculture has gradually increased and the use of new techniques has become extremely important. Nowadays, traditional agriculture techniques have reached saturation, and modern agricultural technologies aimed at increasing efficiency have also ignored environmental factors especially in developing countries and have caused a damaged ecosystem. In this context, people have needed to mention crop productivity increase and sustainable environmental management; thus, novel, big, friendly and cost-effective approaches to agriculture and environment are needed for drought conditions. Today, many types of research are being carried out on the use of nanotechnology, which can help with overcoming the issues mentioned. Therefore, agricultural nanotechnology can play a fundamental role as an environmentally friendly, cost-effective green technology for sustainable agriculture. Historically, nanotechnology was developed for industrial applications half a century ago, and then the use of nanotechnology in agriculture attracted attention (Mukhopadhyay and Sharma 2013). The applications of nanotechnology in agriculture aim to reduce the excessive use of chemical-containing plant protection products and to increase nutrient intake and productivity with effective fertilization.

At least one dimension of nanoparticles which originate from nanotechnology have a minimum size of ≤100 nm and are made of metals such as silver, copper, carbon and silicon (0.1–100 nm diameter). Because of their extremely small dimensions, nanoparticles are used in many areas such as medicine, environment and energy and the food industry. They are derived from biopolymers such as proteins and carbohydrates, which have a low negative impact on human health and the environment. Biosynthesized nanoparticles are obtained from plant extracts with various analytical techniques.

Because plants contain alkaloids, flavonoids, phenols, terpenoids, alcohols and sugars, their extracts are used in the synthesis of gold nanoparticles (Siddiqi and Husen 2016).

Increasing the use areas and applications of nanomaterials is thought to cause a significant increase in their interactions with the environment and lead to very different environmental behaviours and effects of their distinctly different physicochemical properties, such as the higher surface-to-volume ratio of nanoparticles, electronic structure, intermediate surface reactivity, etc. (Ma and Wang 2010).

Nanoparticles with different concentrations can have both positive and negative biological effects. Physical nanoparticles have a less toxic effect than chemical nanoparticles (Taran et al. 2017).

Nanoparticles are effective in the fight against herbicides and harmful insects and have the ability to provide effective use of water and fertilizers in agricultural production, and are important in the increase of agricultural productivity. It is also used in the production of insecticides, insect-repellents (Bhattacharyya et al. 2010), herbicides, and fungicides (Worrall et al. 2018). However, in molecular studies, nanoparticles provide significant contributions in many areas, such as the development of a gene transferred transgenic plant such as Medicago truncatula, Zea mays, Nicotiana tabacum, Gossypium hirsutum and Oryza sativa L. that is resistant against stress factors, developing defence mechanisms against diseases caused by pathogens, etc. (Aras et al. 2015). The use of nanoparticles as biosensors in the diagnosis of plant disease has also been among the research subjects (Elmer and White 2018).

Nanoparticles can pass through the cell wall, membrane and ultimately penetrate the double membranes of chloroplasts. Thus, by injecting genes directly into plant chloroplasts, DNA transmission to the cells is possible. In a study, with gene transfer, new plants with the desired characteristics were obtained as a result of genetic modification. In some plants such as cotton, because in vitro regeneration is difficult, troublesome and complex, DNA was injected into pollen through nanoparticles, and direct transgenic seeds were obtained as a result of pollination (Zhao et al. 2018). In another study, to produce transgenic seeds of cotton directly without tissue culture, transgenic seeds were obtained by transferring the plasmid DNA loaded magnetic nanoparticles carrying functional genes to the pollens (Zhang et al. 2018). However, it was reported that when soybean was exposed to cerium dioxide nanoparticles (CeO_2-NPs), the nanoparticles caused mutations by showing a toxic effect on DNA and genes in soybean (López- Moreno et al. 2010). Another significant breakthrough of nanotechnology to ensure improvement in agricultural production is the development of insect-resistant varieties as a result of DNA transfer through nanoparticles in plants.

Nanoparticles, which have gained extensive usage, participate in the ecosystem but their environmental impact is not fully known. The number of studies that determine the effects of nanoparticles on plants that form the basis of the ecosystem is few. Nanoparticles sometimes adversely affect the ecosystem by causing toxicities on living organisms, environment and plants, i.e. by producing free radicals or reactive oxygen derivatives that can cause oxidative stress in organisms. Nano-pesticides that are very effective at even low doses and cause environmental pollution, are placed in the leaf and flower parts of plants during air transport, blocking stoma, causing a layer on the stigma that prevents pollen germination and affecting the transport of water, food and assimilation products adversely by entering the conductive tissues.

Although research has reported harmful effects on human health and environment, nanoparticles have innovative features in the economic field. Today, with the development of nanotechnology, nanoparticle availability for agriculture has increased. Because of their small sizes, wide and reactive surface areas, they can be used as bactericide, fungicide, nanofertilizers and used in the diagnosis of plant diseases and agricultural chemicals.

Pesticides used in agriculture are not very efficient, and also pollute both terrestrial and aquatic environments as a result of their widespread use. Therefore, in agriculture, a positive effect on ecology can be made by using nano-agrochemicals instead of conventional pesticides. Nano-agrochemicals containing polymeric nanoparticles, silver ions, gold nanoparticles and iron oxide nanoparticles are used as pesticides (Al-Samarrai 2012).

4.2 Effects of Nanoparticles on Plant Development

The effect of many nanoparticles on the growth and metabolic functions of plants varies according to plant species; research examining their effects on plant growth and development have been conducted, with both positive and negative results.

In agriculture, rapid and homogeneous seed germination and seedling emergence are important in terms of yield. In recent years, numerous nanoparticles have been applied as pretreatment agents to wheat (Taran et al. 2017; Li et al. 2019; Jhanzab et al. 2019), corn (Mahakham et al. 2016) and spinach (Srivastava et al. 2014) seeds to stimulate seed germination, seedling growth and stress tolerance. Mahakham et al. (2017) reported that silver nanoparticles (AgNPs) induced seed germination and starch mobilization in rice, and their work will shed light on the future of nano-priming for sustainable agricultural practices and seed industry. However, extensive studies on the physiological and molecular mechanisms of nano-priming effects on seed germination are insufficient. Therefore, further research is needed specifically to determine the effect of nanoparticles on seed germination.

It has been reported that in the (Bt)-transgenic and non-transgenic cotton plant, $Si0_2$NPs applications inhibit the growth of plants, and also that the nanoparticles are located in the xylene tissue of transgenic cotton roots, and this condition is risky for human health (Le et al. 2014) Copper oxide (CuO) nanoparticles have been reported to prevent the growth of wheat crops grown in the sand, changing the structure of the roots (Tang et al. 2016). However, starch-based nanoparticles are biodegradable and can also be used in food packaging technology (Aldao et al. 2018), which are not toxic to plants, animals or the environment, may be an alternative to the chemicals in agriculture and are suitable for sustainable agriculture (Marchiol 2018). A study concluded that carbon-based nanoparticles (CNP_S) created a physiological response in the mung bean and positively influenced its growth (Li et al. 2016).

Research has reported that zinc oxide nanoparticles (ZnONPs) are harmless at low concentrations, stimulate certain enzymes in plants, inhibit diseases (Singh et al. 2017), increase yield under cadmium (Cd) stress by increasing wheat development and photosynthesis (Hussain et al. 2018), and can be used in the fight against rice bacterial leaf blight (*Xanthomonas oryzae* pv. oryzae) (Ogunyemi et al. 2019).

Since antique times, silver and its salts have been used, and the effect of silver nitrate particles in plants is important because of their easy distribution in the environment. Silver nanoparticles have been tested for antimicrobial effects against many diseases caused by pathogens in animals and plants and are also plant growth stimulators. A study indicated that copper oxide (CuO) and titanium dioxide (TiO_2) nanoparticles in the leaves of rose plant caused an increase in the zeatin riboside (ZR) phytohormone, thereby the nanoparticles had antifungal effects against the *Podosphaera pannosa* pathogen causing powdery mildew, and that it could be used as a new plant protection strategy (Hao et al. 2019).

In a study conducted in greenhouse conditions, silver nanoparticles (AgNPs) applied to *Triticum aestivum, Brasica juncea* and *Vespertilio sinensis* plants increased the length of shoot and root in plants (Mehta et al. 2016).

Small-size chitosan nanoparticles (ChNPs) can be used in agriculture, genetic engineering, food industry, environmental pollution control, water treatment, paper production and so on. In the conducted studies, it was reported that ChNPs inhibited the growth of *Fusarium oxysporum* in vitro (Oh et al. 2019) and Cu-chitosan nanoparticles increased germination rate, shoot and root length, number of roots, seedling length and wet and dry weight (Saharan et al. 2016).

Silicon dioxide nanoparticles (SiO_2NPs) are also used in agriculture. It has been reported that these nanoparticles have an important effect on seed germination potential in tomatoes and can be used as a fertilizer source in sustainable agriculture (Siddiqui and Al-Whaibi 2014). Lack of water in drought conditions in agricultural production is an important problem. Hydrogels containing silicon dioxide nanoparticles (SiO_2NPs) can help conserve water in agricultural soils (Pathak and Kumar 2017).

In a study of barley (*Hordeum sativum distichum*), it was revealed that copper oxide nanoparticles (CuO-NPs) reduce the number of chloroplasts but increase the size of chloroplast (Rajput et al. 2018).

Titanium dioxide nanoparticles (TiO_2-NPs) can be used in nano-agriculture, but limited studies on photosynthesis are available (Dias et al. 2018). Abiotic stress factors in plants such as drought, salinity and heavy metals affect the development, germination and some physiological developments in plants. The application of TiO_2-NPs can be a promising approach in preventing the adverse effects of wheat seed germination and cadmium (Cd) stress in plant development (Faraji and Sepehri 2018).

Due to excessive and irregular use of chemical fertilizers in agriculture, many problems have occurred, such as atmospheric and groundwater pollution, decreased soil efficiency and loss of biodiversity. For this reason, instead of traditional fertilizers, environmentally friendly bio- and nano-fertilizers that tend to increase soil efficiency have been obtained with nanotechnology methods. Thus, the increase in the quantities of nitrogen and phosphoric compounds in waters (eutrophication) and groundwater pollution are prevented (Mukhopadhyay and Sharma 2013).

Nano-structured fertilizers can increase the effective use of plant nutrients through mechanisms such as being target-oriented and having slow or controlled release. In recent studies, nano-fertilizers have been reported to have the ability to control plant diseases and to increase the rate of seed germination, seedling growth, photosynthetic activity, nitrogen metabolism, carbohydrate and protein synthesis and also product quality and efficiency (León-Silva et al. 2018; El-Ghamry et al. 2018; Shinde et al. 2018; Hussein et al. 2019).

Due to the nano size of the particles, their permeation to the plant cells is too high and can be effective at very low doses. They increase the benefits of micro and macronutrients, interacting with plants and causing various physiological and morphological changes due to their different physicochemical properties. Thus, they can increase the photosynthetic efficiency of the plants and achieve higher productivity in the unit area. Magnesium hydroxide ($Mg(OH)_2$) nanoparticles help promote seed germination and plant growth in corn (*Zea mays* L.), and therefore such nanoparticles have been expressed to be used as nano-feeders for effective plant development.

4.3 Conclusion

With intensive, irregular application of traditional agriculture, enough yield can be achieved in the desired product, but natural resources are exhausted simultaneously, biodiversity is decreased and ecosystem balance deteriorates due to air pollution, water pollution and soil pollution, leading to irreversible problems. With the application of excessive agriculture such as excessive use of natural resources, faulty farming practices and unconscious and excessive use of chemical drugs in agriculture, it is inevitable to continue the loss of soil. This situation has become critical to global agricultural production. Therefore, the use of nanoparticles in agricultural nanotechnology has gained importance. A number of analytical studies should be performed to determine and characterize the intake, translocation and intracellular biotransformation of nanoparticles in plants, and further studies should be carried out to provide adequate information about the interaction between nanoparticles and plants.

References

Aldao DC, Šárka E, Ulbrich P, Menšíková E (2018) Starch nanoparticles-two ways of their preparation. Czech J Food Sci 36:133–138

Al-Samarrai AM (2012) Nanoparticles as alternative to pesticides in management plant diseases-a review. Int J Sci Res Publ 2(4):1–4

Aras S, Soydam-Aydın S, Fazlıoğlu A, Cansaran-Duman D, Büyük İ, Derici K (2015) RNA interference in plants. Turk Hij Deney Biyol Derg 72(3):255–262

Bhattacharyya A, Bhaumik A, Rani PU, Mandal S, Epidi TT (2010) Nano-particles – a recent approach to insect pest control. Afr J Biotechnol 9(24):3489–3493

Dias MC, Santos C, Pinto G, Silva AMS, Silva S (2018) Titanium dioxide nanoparticles impaired both photochemical and non-photochemical phases of photosynthesis in wheat. Protoplasma 256(1):69–78

El-Ghamry AM, Mosa AA, Alshaal TA, ElRamady HR (2018) Nanofertilizers vs. biofertilizers: new insights. Environ Biodivers Soil Secur 2:51–72

Elmer W, White JC (2018) The future of nanotechnology in plant pathology. Annu Rev Phytopathol 56:111–133

Faraji J, Sepehri A (2018) Titanium dioxide nanoparticles and sodium nitroprusside alleviate the adverse effects of cadmium stress on germination and seedling growth of wheat (*Triticum aestivum* L.). Univ Sci 23(1):61–87

Hao Y, Fang P, Ma C, White JC, Xiang Z, Wang H, Zhang Z, Rui Y, Xing B (2019) Engineered nanomaterials inhibit *Podosphaera pannosa* infection on rose leaves by regulating phytohormones. Environ Res 170:1–6

Hussain A, Ali S, Rizwan R, Rehman MZ, Javed MR, Imran M, Chatha SAS, Nazir R (2018) Zinc oxide nanoparticles alter the wheat physiological response and reduce the cadmium uptake by plants. Environ Pollut 242(Part B):1518–1526

Hussein HS, Shaarawy HH, Hussien NH, Hawash SI (2019) Preparation of nano-fertilizer blend from banana peels. Bull Natl Res Cent 43:26. https://doi.org/10.1186/s42269-019-0058-1

Jhanzab HM, Razzaq A, Bibi Y, Yasmeen F, Yamaguchi H, Hitachi K, Tsuchida K, Komatsu S (2019) Proteomic analysis of the effect of inorganic and organic chemicals on silver nanoparticles in wheat. Int J Mol Sci 20(4):825

Le VN, Rui Y, Gui X, Li X, Liu S, Han Y (2014) Uptake, transport, distribution and bio-effects of SiO_2 nanoparticles in Bt-transgenic cotton. J Nanobiotechnol 12:50

León-Silva S, Arrieta-Cortes R, Fernández-Luqueño F, López-Valdez F (2018) Design and production of nanofertilizers. In: López-Valdez F, Fernández-Luqueño F (eds) Agricultural nanobiotechnology. Springer, Cham, pp 17–31

Li W, Zheng Y, Zhang H, Liu Z, Su W, Chen S, Liu Y, Zhuang J, Lei B (2016) Phytotoxicity, uptake, and translocation of fluorescent carbon dots in mung bean plants. ACS Appl Mater Interfaces 8:19939–19945

Li R, He J, Xie H, Wang W, Bose SK, Sun Y, Hu J, Yin H (2019) Effects of chitosan nanoparticles on seed germination and seedling growth of wheat (*Triticum aestivum* L.). Int J Biol Macromol 126:91–100

López-Moreno ML, de la Rosa G, Hernández-Viezcas JÁ, Castillo-Michel H, Botez CE, Peralta-Videa JR, Gardea-Torresdey JL (2010) Evidence of the differential biotransformation and genotoxicity of ZnO and CeO_2 nanoparticles on soybean (*Glycine max*) plants. Environ Sci Technol 44:7315–7320

Ma X, Wang C (2010) Fullerene nanoparticles affect the fate and uptake of trichloroethylene in phytoremediation systems. Environ Eng Sci 27(11):989–992

Mahakham W, Theerakulpisut P, Maensiri S, Phumying S, Sarmah AK (2016) Environmentally benign synthesis of phytochemicals-capped gold nanoparticles as nanopriming agent for promoting maize seed germination. Sci Total Environ 573:1089–1102

Mahakham W, Sarmah AK, Maensiri S, Theerakulpisut P (2017) Nanopriming technology for enhancing germination and starch metabolism of aged rice seeds using phytosynthesized silver nanoparticles. Sci Rep 7:8263

Marchiol L (2018) Nanotechnology in agriculture: new opportunities and perspectives. In: New visions in plant science. InTechOpen, London, pp 121–141. https://doi.org/10.5772/intechopen.74425

Mehta CM, Srivastava R, Arora S, Sharma AK (2016) Impact assessment of silver nanoparticles on plant growth and soil bacterial diversity. 3 Biotech 6:254

Mukhopadhyay SS, Sharma S (2013) Nanoscience and nanotechnology: cracking prodigal farming. J Bionano Sci 7:1–5

Ogunyemi SO, Abdallah Y, Zhang M, Fuad H, Hong X, İbrahim E, Masum MMI, Hossain A, Mo J, Li B (2019) Green synthesis of zinc oxide nanoparticles using different plant extracts and their antibacterial activity against *Xanthomonas oryzae* pv. Oryzae. Artif Cells Nanomed Biotechnol 47(1):341–352

Oh JW, Chun SC, Chandrasekaran M (2019) Preparation and *in vitro* characterization of chitosan nanoparticles and their broad-spectrum antifungal action compared to antibacterial activities against phytopathogens of tomato. Agronomy 9(1):21

Pathak VM, Kumar N (2017) Dataset on the superabsorbent hydrogel synthesis with SiO_2 nanoparticle and role in water restoration capability of agriculture soil. Data Brief 13:291–294

Rajput V, Minkina T, Fedorenko A, Sushkova S, Mandzhieva S, Lysenko V, Duplii N, Fedorenko G, Dvadnenko K, Ghazaryan K (2018) Toxicity of copper oxide nanoparticles on spring barley (*Hordeum sativum* distichum). Sci Total Environ 645:1103–1113

Saharan V, Kumaraswamy RV, Choudhary RC, Kumari S, Pal A, Raliya R, Biswas P (2016) Cu-chitosan nanoparticle mediated sustainable approach to enhance seedling growth in maize by mobilizing reserved food. J Agric Food Chem 64(31):6148–6155

Shinde S, Paralikar P, Ingle AP, Rai M (2018) Promotion of seed germination and seedling growth of *Zea mays* by magnesium hydroxide nanoparticles synthesized by the filtrate from *Aspergillus niger*. Arab J Chem. https://doi.org/10.1016/j.arabjc.2018.10.001

Siddiqi KS, Husen A (2016) Recent advances in plant-mediated engineered gold nanoparticles and their application in biological system. J Trace Elem Med Biol 40:10–23

Siddiqui MH, Al-Whaibi M (2014) Role of nano-SiO_2 in germination of tomato (*Lycopersicum esculentum* seeds mill.). Saudi J Biol Sci 21(1):13–17

Singh A, Singh NB, Afzal S, Singh T, Hussain I (2017) Zinc oxide nanoparticles: a review of their biological synthesis, antimicrobial activity, uptake, translocation and biotransformation in plants. J Mater Sci 53:185–201. https://doi.org/10.1007/s10853-017-1544-1

Srivastava G, Das CK, Das A, Singh SK, Roy M, Kim H, Sethy N, Kumar A, Sharma RK, Singh SK, Philip D, Das M (2014) Seed treatment with iron pyrite (FeS_2) nanoparticles increases the production of spinach. RSC Adv 4(102):58495–58504. https://doi.org/10.1039/C4RA06861K

Tang Y, He R, Zhao J, Nie G, Xu L, Xing B (2016) Oxidative stressinduced toxicity of CuO nanoparticles and related toxicogenomic responses in *Arabidopsis thaliana*. Environ Pollut 212:605–614

Taran N, Storozhenko V, Svietlova N, Batsmanova L, Shvartau V, Kovalenko M (2017) Effect of zinc and copper nanoparticles on drought resistance of wheat seedlings. Nanoscale Res Lett 12(1):60

Worrall EA, Hamid A, Mody AT, Mitter N, Pappu HR (2018) Nanotechnology for plant disease management. Agronomy 8:285. https://doi.org/10.3390/agronomy8120285

Zhang R, Meng Z, Abid MA, Zhao X (2018) Novel pollen magnetofection system for transformation of cotton plant with magnetic nanoparticles as gene carriers. In: Zhang B (ed) Transgenic cotton. Methods in molecular biology, vol 1902. Humana Press, New York

Zhao X, Meng Z, Wang Y, Chen W, Sun C, Cui B, Cui J, Yu M, Zeng Z, Guo S, Luo D, Cheng JQ, Zhang R, Cui H (2018) Pollen magnetofection for genetic modification with magnetic nanoparticles as gene carriers. Nat Plants 3:956–964

Large-Scale Production/Biosynthesis of Biogenic Nanoparticles

5

Reza Mohammadinejad and G. Ali Mansoori

Abstract

Biogenic nanoparticles are either produced inside the biological entity or outside of it. While this has been the interest of many groups of investigators and a more widely explored area in nanotechnology, there are very few such processes having the potential for large-scale production mostly due to: difficulty to control the process, lack of availability of large-scale industrially abundant raw materials, lack of their environmental friendliness, and raw material expenses. The present chapter defines the process involved in the production of biogenic nanoparticles.

Keywords

Biogenic process · Nanoparticles · Large-scale production · Raw materials · Environmental friendliness · Biosynthesis

5.1 Introduction

Nanoparticles which are produced through biological processes using fungus, microorganisms, other living things, vegetation, etc., are known as "biogenic nanoparticles" (BN), also known as biomineralization (Ingale and Chaudhari 2013; Akhtar et al. 2013). Provided the process is performed with such biological entities

R. Mohammadinejad
Pharmaceutics Research Center, Institute of Neuropharmacology, Kerman University of Medical Sciences, Kerman, Iran

G. A. Mansoori (✉)
Departments of Bio & Chemical Engineering, University of Illinois at Chicago (M/C 063), Chicago, IL, USA
e-mail: mansoori@uic.edu

M. Ghorbanpour et al. (eds.), *Biogenic Nano-Particles and their Use in Agro-ecosystems*, https://doi.org/10.1007/978-981-15-2985-6_5

and solvents, which are environmentally friendly it will be in the category of "Green Nanotechnology Synthesis." Of course there is a need to produce nanoparticles with close to uniform size and shape (Mansoori et al. 2007). To achieve this, it is necessary to choose the process-appropriate conditions, including temperature, pH, salinity, electrostatics, and possibly environmental pressure. Presently such appropriate conditions may be defined by extensive laboratory tests, known as trial-and-error tests (Mansoori 2005).

For practical applications of biogenic nanoparticles it is necessary to design the strategy for their large-scale production (Vahabi et al. 2011). They are officially named *extracellular* or *intracellular* biogenic processes (Akhtar et al. 2013; Ingale and Chaudhari 2013). In processes involving extracellular production, the biological entity used for biogenic nanomaterials may be recyclable. However, in the case of intracellular production, separation of the produced nanoparticles will be a bit more difficult to achieve due to the necessity of disruption/killing the biological entity by breaking its living structure. Such intracellular cases may not be appropriate for large-scale production due to the lack of reuse of biological entity after destruction and production of too much waste.

The known biogenic nanoparticles are metallic nanoparticles, and mostly silver nanoparticles (Vahabi et al. 2011; Mohammadinejad et al. 2013). Metal ions are generally toxic to biological entities. To get rid of the toxicity of the metal ions, biological entities go through certain biological detoxification stages causing the metal ions to convert to insoluble solid metals through chemical reduction and/or precipitation.

5.2 Detoxification Principles

Detoxification is a physiological process to remove toxic substances from a living organism. The main reason for organisms (plants, single-celled life forms, etc.) to synthesize nanoparticles is their inherent detoxification ability. The main route the cell(s) protect themselves from the toxicity of certain soluble ionic species, especially metal ions, is to convert them into insoluble solid metallic forms through a natural reduction process which will end in their precipitation (Fig. 5.1).

While liver in humans and other large animals has the complex multiprocess role of detoxification of their bodies, detoxification in organisms may be due to one of the following five basic processes (Srivastava 2019):

(i) Sequestration in biological processes is the action by organism accumulating a compound or tissue. Detoxification through sequestration is referred to as the conversion of a toxic species (like metal ions) to a nontoxic form (solid metal) and its accumulation within (intracellular sequestration) or outside (extracellular sequestration) the cell.
(ii) Enzyme (catalytic) reduction of the toxic species (like metal ions) to a less- or nontoxic form (solid metal).

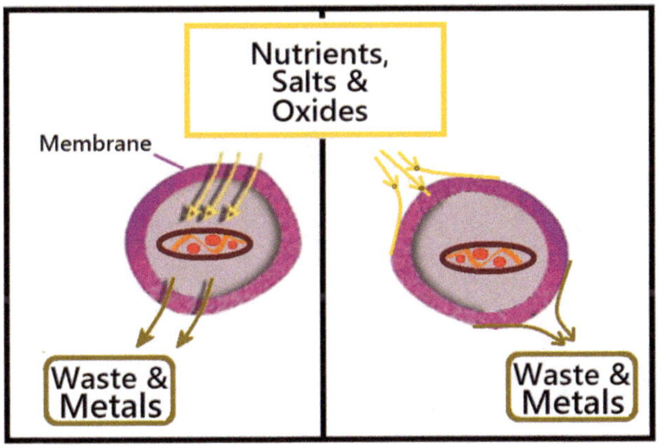

Fig. 5.1 Intracellular and extracellular detoxifications at the cellular level

(iii) Ability of the cellular transport mechanisms to restrict entry of the toxic ions (like metal ions) into the cell and its reduction to nontoxic element (solid metal) upon contact with the outer cell membrane.
(iv) Activation of energy-dependent uptake and efflux pathways in cells to eliminate the toxic species.
(v) In certain other cases, nanoparticles production happens in order to meet a cell's requirements for a functional component. Such nanoparticles production occurs through oxidation and condensation-association.

The more wide application of biosynthesis of biogenic nanoparticles is for the production of silver and gold nanoparticles (AgNP and AuNP). They have found widespread applications in many fields, such as medicine and science and technology (Prabhu and Poulose 2012; Mansoori et al. 2010).

5.3 Biosynthesis of Biogenic Nanoparticles

In Table 5.1 we report a collection of biosynthesis of biogenic nanoparticle (BBN) processes using vegetation/agricultural products/plants which are reported as of now. Vegetation and plants have been shown to be effective in synthesizing silver and gold nanoparticles and the amount of reports available in this topic is quite large. In 2011, we undertook a research project to synthesize silver nanoparticles using *Silybum marianum* seed extract and we published the results of our research in 2013 (Mohammadinejad et al. 2013). We also kept studying biogenic nanoparticles production processes using vegetation/agricultural products/plants. We have produced Table 5.1 comparing characteristics and large-scale production potentials

Table 5.1 Comparison of various biogenic agents and the nature of their applications for production of biogenic nanoparticles

Biogenic agent (Category)	Method: Ex: Extracellular In: Intracellular	Biogenic nanoparticles	Size range [nm]	LSPP	References
Calotropis procera latex extract	Ex	CuNP	15 ± 1.7		Harne et al. (2012)
Terminalia arjuna bark extract	Ex	CuNP	~23		Yallappa et al. (2013)
Silybum marianum seed extract	Ex	AgNP	1–25	✓	Mohammadinejad et al. (2013)
Tabernaemontana divaricate leaf extract	Ex	CuONP	48 ± 4		Sivaraj et al. (2014)
Vitis vinifera extract	Ex	SeNP	3–18	✓	Sharma et al. (2014)
Calendula officinalis seed extract	Ex & In	AgNP	5–10	✓	Baghizadeh et al. (2015)
Ginkgo biloba leaf extract	Ex	CuNP	15–20		Nasrollahzadeh and Sajadi (2015)
Ocimum sanctum leaf extract	Ex	CuNP	60–70	✓	Sadanand et al. (2016)
Artemisia capillaris extract	Ex	AuNP	16.88 ± 5.47~29.93 ± 9.80	✓	Lim et al. (2016)
Gardenia jasminoides extract	Ex	AuNP	20	✓	Chae et al. (2016)
Citrullus colocynthis seed extract	Ex	AgNP	23 ± 2		Azizi et al. (2017)
Sphaeranthus indicus extract	Ex	AuNP	25		Balalakshmi et al. (2017)
Lantana camara root extract	Ex	AuNP	11–32	✓	Ramkumar et al. (2017)
Cibotium barometz root extract	Ex	AuNP	6–23		Wang et al. (2017)

Musa paradisiaca peel extract	Ex	AuNP	50	✓	Vijayakumar et al. (2017)
Sambucus nigra extract	Ex	AuNP	4–26		Opris et al. (2017)
Allium sativum extract	Ex	SeNP	27.6	✓	Ezhuthupurakkal et al. (2017)
Clausena dentata leaf extract	Ex	SeNP	46.32–78.88		Sowndarya et al. (2017)
Plantago asiatica leaf extract	Ex	CuNP	7–35		Nasrollahzadeh et al. (2017)
Terminalia catappa leaf extract	Ex	CuNP	21–30		Muthulakshmi et al. (2017)
Artemisia vulgaris leaf extract	Ex	AuNP	50–100	✓	Sundararajan and Kumari (2017)
Mussaenda glabrata leaf extract	Ex	AgNP & AuNP	51.32 & 10.59		Francis et al. (2017)
Syzygium cumini seed extract	Ex	AuNP	13–30		Kadiyala et al. (2018)
Artemisia dracunculus extract	Ex	AuNP	35–50, 40–75, & 80–130	✓	Wacławek et al. (2018)
Phoenix dactylifera pollen extract	Ex	AgNP & AuNP	27 & 95		Banu et al. (2018)
Prunus serrulata fruit extract	Ex	AgNP & AuNP	66 nm & 65		Singh et al. (2018b)
Crataegus pinnatifida fruit extract	Ex	AgClNP & AuNP	24.80 nm & 14.20	✓	Kang et al. (2018)
Euphrasia officinalis leaf extract	Ex	AgNP & AuNP	40.37 ± 1.8 and 49.72 ± 1.2		Singh et al. (2018a)
Terminalia arjuna bark extract	Ex	AuNP	20–50	✓	Suganthy et al. (2018)

(continued)

Table 5.1 (continued)

Biogenic agent (Category)	Method: Ex: Extracellular In: Intracellular	Biogenic nanoparticles	Size range [nm]	LSPP	References
Indigofera tinctoria leaf extract	Ex	AuNP	9–26	✓	Vijayan et al. (2018)
Chaenomeles sinensis fruit extract	Ex	AgNP & AuNP	5–20 and 20–40		Oh et al. (2018)
Tilia extract	Ex	CuNP	4.7–17.4		Hassanien et al. (2018)
Artemisia haussknechtii leaf extract	Ex	AgNP, CuNP & TiO_2NP	10.69 ± 5.55, 35.36 ± 44.4 & 92.58 ± 56.98		Alavi and Karimi 2018)
Quisqualis indica extract	Ex	CuNP	39.3 ± 5.45		Mukhopadhyay et al. (2018)
Olea europaea leaf extract	Ex	CuNP	20–50	✓	Sulaiman et al. (2018)
Hawthorn fruit extract	Ex	SeNP	113	✓	Cui et al. (2018)
Tea extract	Ex	SeNP	83–160	✓	Zhang et al. (2018)
Arctium lappa burdock root extract	Ex	AgNP & AuNP	21.3 & 24.7	✓	Nguyen et al. (2018)
Punica granatum peel extract	Ex	AgNP	20–40	✓	Devanesan et al. (2018)
Amanita muscaria extract	Ex	AgNP & USIONP	5–25 & 2.2–2.5		Ivashchenko et al. (2018)
Caesalpinia sappan extract	Ex	AgNP	411.8 ± 20.4, 245.3 ± 34.2, 183.7 ± 15.8, 160.8 ± 4.0, & 233.3 ± 42.4		Suwan et al. (2018)
Durio Zibethinus seed extract	Ex	AgNP	20–75		Sumitha et al. (2018)
Matricaria chamomilla extract	Ex	AgNP	45.12	✓	Dadashpour et al. (2018)

Nigella arvensis leaf extract	Ex	AuNP	3–37		Chahardoli et al. (2018)
Bauhinia acuminate flower extract	Ex	AgNP	17		Hu et al. (2019)
Salvia miltiorrhiza leaf extract	Ex	AgNP	100		Zhang et al. (2019)
Dysosma pleiantha rhizome extract	Ex	AgNP	76		Karuppaiya et al. (2019)
Mangifera indica flower extract	Ex	AgNP	10–20		Ameen et al. (2019)
Piper betle leaf extract	Ex	AgNP	6–14	✓	Khan et al. (2019)
Artemisia scoporia extract	Ex	AgNP	10–80		Moulavi et al. (2019)
Carpesium cernuum extract	Ex	AgNP	13.0 ± 0.2		Ahn et al. (2019)
Scoparia dulcis leaf extract	Ex	AgNP	3–18		Parvataneni 2019)
Albizia procera leaf extract	Ex	AgNP	6.18		Rafique et al. (2019)
Selaginella bryopteris extract	Ex	AgNP	5–10		Dakshayani et al. (2019)
Manilkara zapota leaf extract	Ex	AgNP	24		Shaniba et al. (2019)
Arisaema flavum tuber extract	Ex	AgNP	12–20		Rahman et al. (2019)
Tomato leaf extract	Ex	AgNP	< 87	✓	Santiago et al. (2019)
Pogostemon cablin leaf extract	Ex	AgNP	25	✓	Qing et al. (2019)
Sasa borealis leaf extract	Ex	AuNP	10–30		Patil et al. (2018)
Camellia sinensis leaf extract	Ex	AgNP	30–80	✓	Kumar et al. (2019)
Allium sativum extract	Ex	AgNP	26 ± 7	✓	Robles-Martínez et al. (2019)
Tropaeolum majus leaf extract	Ex	AgNP	35–55	✓	Valsalam et al. (2019)

(continued)

Table 5.1 (continued)

Biogenic agent (Category)	Method: Ex: Extracellular In: Intracellular	Biogenic nanoparticles	Size range [nm]	LSPP	References
Cornus Mas fruit extract	Ex	AgNP & AuNP	16 & 19		Filip et al. (2019)
Passiflora edulis leaf extract	Ex	AgNP	3–7	✓	Thomas et al. (2019)
Prosopis juliflora leaf extract	Ex	AgNP	10–20		Arya et al. (2019)
Berberis vulgaris leaf and root extract	Ex	AgNP	30–70	✓	Behravan et al. (2019)
Eucalyptus critriodora leaf extract	Ex	AgNP	8–15	✓	Wintachai et al. (2019)
Bauhinia purpurea leaf extract	Ex	AgNP	< 50		Vijayan et al. (2019)
Elaeis guineensis extract	Ex	AuNP	26.41 ± 7.95		Ahmad et al. (2019)
Mulberry extract	Ex	AuNP	35	✓	Xu et al. (2019)
Ephedra sinica extract	Ex	AuNP	35.04 ± 4.02		Park et al. (2019b)
Musa balbisiana extract	Ex	AuNP	33.83 ± 3.39		Maji et al. (2019)
Dictyopteris divaricata extract	Ex	AuNP	28.01 ± 2.03		Park et al. (2019a)
Green tea extract	Ex	AuNP, Au nanostar & Au nanorod	8.7 ± 1.7, 99.0 ± 47.0, & length 60.4, width 16.4	✓	Lee et al. (2019)
Alternanthera sessilis extract	Ex	AuNP	20–40		Qian et al. (2019)
Anacardium occidentale extract	Ex	AuNP	10–30	✓	Sunderam et al. (2019)
Coleus aromaticus extract	Ex	AuNP	16–18	✓	Boomi et al. (2019)

Abies spectabilis extract	Ex	AuNP	20–200		Wu et al. (2019)
Chenopodium formosanum shell extract	Ex	AuNP	8 ± 6		Chen et al. (2019)
Dracocephalum kotschyi extract	Ex	AuNP	5–21		Chahardoli et al. (2019)
Panax ginseng extract	Ex	AuNP	10–20		Simu et al. (2019)
Cordyceps militaris extract	Ex	AuNP	15–20		Ji et al. (2019)
Emblica officinalis fruit extract	Ex	SeNP	15–40		Lokanadhan et al. (2019)
Psidium guajava leaf extract	Ex	CuONP	2–6	✓	Singh et al. (2019)
Falcaria vulgaris leaf extract	Ex	CuNP	20–25		Zangeneh et al. (2019)
Cissus arnotiana extract	Ex	CuNP	60–90		Rajeshkumar et al. (2019)
Eucalyptus globulus leaf extract	Ex	CuONP	16.78–22.5	✓	Ali et al. (2019)

LSPP in this table stands for "Large-Scale Production Potential." Those methods which are certain for LSPP, in the judgment of the authors, are indicated by a "✓"

This table is produced by Dr. R. Mohammadinejad (r.mohammadinejad@kmu.ac.ir; r.mohammadinejad87@gmail.com), to whom questions need to be addressed to

Abbreviations: *AgNP* Silver nanoparticle, *AgClNP* Silver chloride nanoparticle, *AuNP* Gold nanoparticle, *CuNP* Cupper nanoparticle, *CuONP* Copper oxide nanoparticle, *USIONP* Ultrasmall iron oxide nanoparticle, *SeNP* Selenium nanoparticle, *TiO2NP* Titanium dioxide nanoparticle

(LSPP) of such processes since 2011. Some of the plant species that have been employed to synthesize nanoparticles are listed in Table 5.1. This table covers the processes introduced since 2012, it is quite large and it highlights the extensive amount of research already carried out in this area in the past 7 years.

5.4 Industrial-Scale Production/Biosynthesis of Biogenic Nanoparticles

There exists a great deal of processes in the literature for biogenic nanoparticles production. However, a limited number of such production schemes are suitable for large-scale (industrial-scale) production. Generally, for a process to be suitable for scale-up to industrial scale it needs to meet the following criteria:

(a) The raw materials for the process need to be producible, industrially abundant, and available in large scale.
(b) The raw materials, products, by-products, and wastes must be environmentally friendly.
(c) While the conditions for the process need to be well defined, they need to be wide enough to be able to control the process automatically through available process-control facilities.
(d) Economics of the production process must be appropriate for investment in its production.

The basic stages of a large-scale production scheme for biogenic nanoparticles are shown in Fig. 5.2.

In order to scale up a research laboratory of a process to industrial/large-scale production scheme, the principles of process scale-up need to be understood and followed.

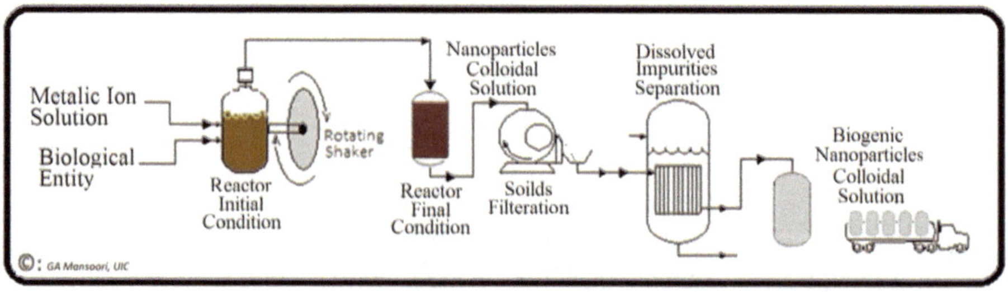

Fig. 5.2 Basic stages of large-scale biogenic nanoparticles production

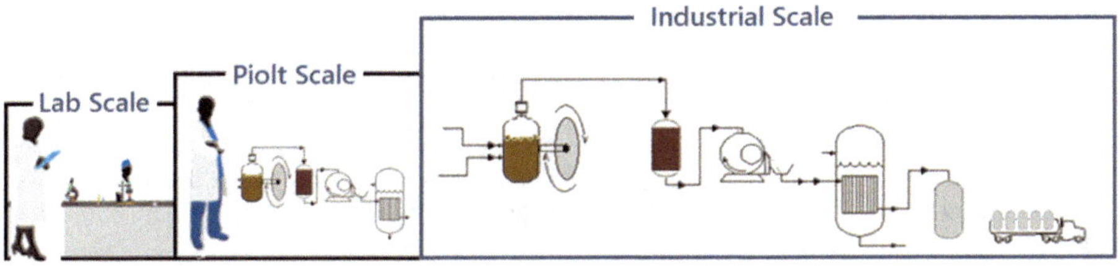

Fig. 5.3 Process scale-up principles

5.5 Process Scale-Up Principles

The goal of scale-up principles is to identify and develop a process that will successfully produce a desired product when manufactured at a commercial scale. To successfully move from the laboratory scale to the large industrial scale, one must understand how size changes impact a number of physical and chemical phenomena. For this purpose, it is customary to build a pilot-plant of the process for which, in principles includes all the stages of the industrial scale, but at a fraction of its production rate (Fig. 5.3). However, scale-up of biosynthesis of biogenic nanomaterials can be challenging with respect to their economics, productivity, and reliability.

For fast reactions, the biosynthesis of biogenic nanomaterials (BBN) is usually reaction-rate limited and production rate is strongly dependent on reactor size. For most, if not all, BBN schemes, only large-sale batch processes are suitable. Thus, scale-up parameters need to include proper design of uniform and proper temperature, pH, and salinity control as well as agitation/mixing of solid-liquid mixture in the batch system. Detailed reaction rate data for most of the BBN processes are unavailable, which is a challenge for developing correlations for process control applications. Unit reliability also needs to be addressed in the design stage of BBN large-scale batch reactors.

References

Ahmad T, Bustam MA, Irfan M, Moniruzzaman M, Asghar HMA, Bhattacharjee S (2019) Mechanistic investigation of phytochemicals involved in green synthesis of gold nanoparticles using aqueous Elaeis guineensis leaves extract: role of phenolic compounds and flavonoids. Biotechnol Appl Biochem 66(4):698–708

Ahn E-Y, Jin H, Park Y (2019) Green synthesis and biological activities of silver nanoparticles prepared by Carpesium cernuum extract. Arch Pharm Res 42(10):1–9

Akhtar MS, Panwar J, Yun Y-S (2013) Biogenic synthesis of metallic nanoparticles by plant extracts. ACS Sustain Chem Eng 1(6):591–602

Alavi M, Karimi N (2018) Characterization, antibacterial, total antioxidant, scavenging, reducing power and ion chelating activities of green synthesized silver, copper and titanium dioxide nanoparticles using Artemisia haussknechtii leaf extract. Artif Cells Nanomedicine Biotechnol 46(8):2066–2081

Ali K, Ahmed B, Ansari SM, Saquib Q, Al-Khedhairy AA, Dwivedi S, Alshaeri M, Khan MS, Musarrat J (2019) Comparative in situ ROS mediated killing of bacteria with bulk analogue, Eucalyptus leaf extract (ELE)-capped and bare surface copper oxide nanoparticles. Mater Sci Eng C 100:747–758

Ameen F, Srinivasan P, Selvankumar T, Kamala-Kannan S, Al Nadhari S, Almansob A, Dawoud T, Govarthanan M (2019) Phytosynthesis of silver nanoparticles using Mangifera indica flower extract as bioreductant and their broad-spectrum antibacterial activity. Bioorg Chem 88:102970

Arya G, Kumari RM, Sharma N, Gupta N, Kumar A, Chatterjee S, Nimesh S (2019) Catalytic, antibacterial and antibiofilm efficacy of biosynthesised silver nanoparticles using Prosopis juliflora leaf extract along with their wound healing potential. J Photochem Photobiol B Biol 190:50–58

Azizi S, Mohamad R, Rahim RA, Mohammadinejad R, Ariff AB (2017) Hydrogel beads bio-nanocomposite based on Kappa-Carrageenan and green synthesized silver nanoparticles for biomedical applications. Int J Biol Macromol 104:423–431

Baghizadeh A, Ranjbar S, Gupta VK, Asif M, Pourseyedi S, Karimi MJ, Mohammadinejad R (2015) Green synthesis of silver nanoparticles using seed extract of Calendula officinalis in liquid phase. J Mol Liq 207:159–163

Balalakshmi C, Gopinath K, Govindarajan M, Lokesh R, Arumugam A, Alharbi NS, Kadaikunnan S, Khaled JM, Benelli G (2017) Green synthesis of gold nanoparticles using a cheap Sphaeranthus indicus extract: impact on plant cells and the aquatic crustacean Artemia nauplii. J Photochem Photobiol B Biol 173:598–605

Banu H, Renuka N, Faheem S, Ismail R, Singh V, Saadatmand Z, Khan SS, Narayanan K, Raheem A, Premkumar K (2018) Gold and silver nanoparticles biomimetically synthesized using date palm pollen extract-induce apoptosis and regulate p53 and Bcl-2 expression in human breast adenocarcinoma cells. Biol Trace Elem Res 186(1):122–134

Behravan M, Panahi AH, Naghizadeh A, Ziaee M, Mahdavi R, Mirzapour A (2019) Facile green synthesis of silver nanoparticles using Berberis vulgaris leaf and root aqueous extract and its antibacterial activity. Int J Biol Macromol 124:148–154

Boomi P, Ganesan R, Poorani G, Prabu HG, Ravikumar S, Jeyakanthan J (2019) Biological synergy of greener gold nanoparticles by using Coleus aromaticus leaf extract. Mater Sci Eng C 99:202–210

Chae SY, Park SY, Park JO, Lee KJ, Park G (2016) Gardenia jasminoides extract-capped gold nanoparticles reverse hydrogen peroxide-induced premature senescence. J Photochem Photobiol B Biol 164:204–211

Chahardoli A, Karimi N, Sadeghi F, Fattahi A (2018) Green approach for synthesis of gold nanoparticles from Nigella arvensis leaf extract and evaluation of their antibacterial, antioxidant, cytotoxicity and catalytic activities. Artif Cells Nanomedicine Biotechnol 46(3):579–588

Chahardoli A, Karimi N, Fattahi A, Salimikia I (2019) Biological applications of phytosynthesized gold nanoparticles using leaf extract of Dracocephalum kotschyi. J Biomed Mater Res A 107(3):621–630

Chen MN, Chan CF, Huang SL, Lin YS (2019) Green biosynthesis of gold nanoparticles using Chenopodium formosanum shell extract and analysis of the particles' antibacterial properties. J Sci Food Agric 99(7):3693–3702

Cui D, Liang T, Sun L, Meng L, Yang C, Wang L, Liang T, Li Q (2018) Green synthesis of selenium nanoparticles with extract of hawthorn fruit induced HepG2 cells apoptosis. Pharm Biol 56(1):528–534

Dadashpour M, Firouzi-Amandi A, Pourhassan-Moghaddam M, Maleki MJ, Soozangar N, Jeddi F, Nouri M, Zarghami N, Pilehvar-Soltanahmadi Y (2018) Biomimetic synthesis of silver nanoparticles using Matricaria chamomilla extract and their potential anticancer activity against human lung cancer cells. Mater Sci Eng C 92:902–912

Dakshayani S, Marulasiddeshwara M, Kumar S, Golla R, Devaraja S, Hosamani R (2019) Antimicrobial, anticoagulant and antiplatelet activities of green synthesized silver nanoparticles using Selaginella (Sanjeevini) plant extract. Int J Biol Macromol 131:787–797

Devanesan S, AlSalhi MS, Balaji RV, Ranjitsingh AJA, Ahamed A, Alfuraydi AA, AlQahtani FY, Aleanizy FS, Othman AH (2018) Antimicrobial and cytotoxicity effects of synthesized silver nanoparticles from Punica granatum peel extract. Nanoscale Res Lett 13(1):315

Ezhuthupurakkal PB, Polaki LR, Suyavaran A, Subastri A, Sujatha V, Thirunavukkarasu C (2017) Selenium nanoparticles synthesized in aqueous extract of Allium sativum perturbs the structural integrity of calf thymus DNA through intercalation and groove binding. Mater Sci Eng C 74:597–608

Filip GA, Moldovan B, Baldea I, Olteanu D, Suharoschi R, Decea N, Cismaru CM, Gal E, Cenariu M, Clichici S (2019) UV-light mediated green synthesis of silver and gold nanoparticles using Cornelian cherry fruit extract and their comparative effects in experimental inflammation. J Photochem Photobiol B Biol 191:26–37

Francis S, Joseph S, Koshy EP, Mathew B (2017) Green synthesis and characterization of gold and silver nanoparticles using Mussaenda glabrata leaf extract and their environmental applications to dye degradation. Environ Sci Pollut Res 24(21):17347–17357

Harne S, Sharma A, Dhaygude M, Joglekar S, Kodam K, Hudlikar M (2012) Novel route for rapid biosynthesis of copper nanoparticles using aqueous extract of Calotropis procera L. latex and their cytotoxicity on tumor cells. Colloids Surf B Biointerfaces 95:284–288

Hassanien R, Husein DZ, Al-Hakkani MF (2018) Biosynthesis of copper nanoparticles using aqueous Tilia extract: antimicrobial and anticancer activities. Heliyon 4(12):e01077

Hu D, Gu X, Si W, Qin W, Jiao J, Hao Y (2019) Biosynthesis of silver nanoparticles using Bauhinia acuminate flower extract and their effect to promote osteogenesis of MSCs and improve meniscus injury healing. J Photochem Photobiol B Biol 197:111536

Ingale AG, Chaudhari A (2013) Biogenic synthesis of nanoparticles and potential applications: an eco-friendly approach. J Nanomedicine Nanotechnol 4(165):1–7

Ivashchenko O, Przysiecka Ł, Peplińska B, Jarek M, Coy E, Jurga S (2018) Gel with silver and ultrasmall iron oxide nanoparticles produced with Amanita muscaria extract: physicochemical characterization, microstructure analysis and anticancer properties. Sci Rep 8(1):13260

Ji Y, Cao Y, Song Y (2019) Green synthesis of gold nanoparticles using a Cordyceps militaris extract and their antiproliferative effect in liver cancer cells (HepG2). Artif Cells Nanomedicine Biotechnol 47(1):2737–2745

Kadiyala NK, Mandal BK, Ranjan S, Dasgupta N (2018) Bioinspired gold nanoparticles decorated reduced graphene oxide nanocomposite using Syzygium cumini seed extract: evaluation of its biological applications. Mater Sci Eng C 93:191–205

Kang JP, Kim YJ, Singh P, Huo Y, Soshnikova V, Markus J, Ahn S, Chokkalingam M, Lee HA, Yang DC (2018) Biosynthesis of gold and silver chloride nanoparticles mediated by Crataegus pinnatifida fruit extract: in vitro study of anti-inflammatory activities. Artif Cells Nanomedicine Biotechnol 46(8):1530–1540

Karuppaiya P, Satheeshkumar E, Tsay HS (2019) Biogenic synthesis of silver nanoparticles using rhizome extract of Dysosma pleiantha and its antiproliferative effect against breast and human gastric cancer cells. Mol Biol Rep 46(5):1–10

Khan S, Singh S, Gaikwad S, Nawani N, Junnarkar M, Pawar SV (2019) Optimization of process parameters for the synthesis of silver nanoparticles from Piper betle leaf aqueous extract, and evaluation of their antiphytofungal activity. Environ Sci Pollut Res:1–13

Kumar V, Wadhwa R, Kumar N, Maurya PK (2019) A comparative study of chemically synthesized and Camellia sinensis leaf extract-mediated silver nanoparticles. 3 Biotech 9(1):7

Lee YJ, Ahn E-Y, Park Y (2019) Shape-dependent cytotoxicity and cellular uptake of gold nanoparticles synthesized using green tea extract. Nanoscale Res Lett 14(1):129

Lim SH, Ahn E-Y, Park Y (2016) Green synthesis and catalytic activity of gold nanoparticles synthesized by Artemisia capillaris water extract. Nanoscale Res Lett 11(1):474

Lokanadhan G, Dass RS, Kalagatur NK (2019) Phytofabrication of Selenium Nanoparticles from Emblica officinalis fruit extract and exploring its biopotential applications: antioxidant, antimicrobial, and biocompatibility. Front Microbiol 10:931

Maji A, Beg M, Das S, Sahoo NK, Jha PK, Islam MM, Hossain M (2019) Binding interaction study on human serum albumin with bactericidal gold nanoparticles synthesized from a leaf extract of Musa balbisiana: a multispectroscopic approach. Luminescence 34(6):563–575

Mansoori GA (2005) Principles of nanotechnology: molecular-based study of condensed matter in small systems. World Scientific, Hackensack

Mansoori GA, George TF, Zhang G, Assoufid L (2007) Molecular building blocks for nanotechnology. In: Molecular building blocks for nanotechnology: from diamondoids to nanoscale materials and applications, vol 111. Springer, New York

Mansoori GA, Brandenburg KS, Shakeri-Zadeh A (2010) A comparative study of two folate-conjugated gold nanoparticles for cancer nanotechnology applications. Cancers 2(4):1911–1928

Mohammadinejad R, Pourseyedi S, Baghizadeh A, Ranjbar S, Mansoori G (2013) Synthesis of silver nanoparticles using Silybum marianum seed extract. Int J Nanosci Nanotechnol 9(4):221–226

Moulavi P, Noorbazargan H, Dolatabadi A, Foroohimanjili F, Tavakoli Z, Mirzazadeh S, Hashemi M, Ashrafi F (2019) Antibiofilm effect of green engineered silver nanoparticles fabricated from Artemisia scoporia extract on the expression of icaA and icaR genes against multidrug-resistant Staphylococcus aureus. J Basic Microbiol 59(7):701–712

Mukhopadhyay R, Kazi J, Debnath MC (2018) Synthesis and characterization of copper nanoparticles stabilized with Quisqualis indica extract: evaluation of its cytotoxicity and apoptosis in B16F10 melanoma cells. Biomed Pharmacother 97:1373–1385

Muthulakshmi L, Rajini N, Nellaiah H, Kathiresan T, Jawaid M, Rajulu AV (2017) Preparation and properties of cellulose nanocomposite films with in situ generated copper nanoparticles using Terminalia catappa leaf extract. Int J Biol Macromol 95:1064–1071

Nasrollahzadeh M, Sajadi SM (2015) Green synthesis of copper nanoparticles using Ginkgo biloba L. leaf extract and their catalytic activity for the Huisgen [3+ 2] cycloaddition of azides and alkynes at room temperature. J Colloid Interface Sci 457:141–147

Nasrollahzadeh M, Momeni SS, Sajadi SM (2017) Green synthesis of copper nanoparticles using Plantago asiatica leaf extract and their application for the cyanation of aldehydes using K4Fe (CN) 6. J Colloid Interface Sci 506:471–477

Nguyen TT-N, Vo T-T, Nguyen BN-H, Nguyen D-T, Dang V-S, Dang C-H, Nguyen T-D (2018) Silver and gold nanoparticles biosynthesized by aqueous extract of burdock root, Arctium lappa as antimicrobial agent and catalyst for degradation of pollutants. Environ Sci Pollut Res 25(34):34247–34261

Oh KH, Soshnikova V, Markus J, Kim YJ, Lee SC, Singh P, Castro-Aceituno V, Ahn S, Kim DH, Shim YJ (2018) Biosynthesized gold and silver nanoparticles by aqueous fruit extract of Chaenomeles sinensis and screening of their biomedical activities. Artif Cells Nanomedicine Biotechnol 46(3):599–606

Opris R, Tatomir C, Olteanu D, Moldovan R, Moldovan B, David L, Nagy A, Decea N, Kiss ML, Filip GA (2017) The effect of Sambucus nigra L. extract and phytosinthesized gold nanoparticles on diabetic rats. Colloids Surf B Biointerfaces 150:192–200

Park SY, Kim YJ, Park G, Kim H-H (2019a) Neuroprotective effect of Dictyopteris divaricata extract-capped gold nanoparticles against oxygen and glucose deprivation/reoxygenation. Colloids Surf B Biointerfaces 179:421–428

Park SY, Yi EH, Kim Y, Park G (2019b) Anti-neuroinflammatory effects of Ephedra sinica Stapf extract-capped gold nanoparticles in microglia. Int J Nanomedicine 14:2861

Parvataneni RJD (2019) Biogenic synthesis and characterization of silver nanoparticles using aqueous leaf extract of Scoparia dulcis L. and assessment of their antimicrobial property. Drug Chem Toxicol:1–15

Patil MP, Jin X, Simeon NC, Palma J, Kim D, Ngabire D, Kim N-H, Tarte NH, Kim G-D (2018) Anticancer activity of Sasa borealis leaf extract-mediated gold nanoparticles. Artif Cells Nanomedicine Biotechnol 46(1):82–88

Prabhu S, Poulose EK (2012) Silver nanoparticles: mechanism of antimicrobial action, synthesis, medical applications, and toxicity effects. Int Nano Lett 2(1):32

Qian L, Su W, Wang Y, Dang M, Zhang W, Wang C (2019) Synthesis and characterization of gold nanoparticles from aqueous leaf extract of Alternanthera sessilis and its anticancer activity on cervical cancer cells (HeLa). Artif Cells Nanomedicine Biotechnol 47(1):1173–1180

Qing S, Shoutian Q, Hongyan G, Ming Y, Swamy MK, Sinniah UR, Umar A, Akhtar MS (2019) Biosynthesis, characterization and biological activities of silver nanoparticles from Pogostemon cablin Benth. Methanolic leaf extract. J Nanosci Nanotechnol 19(7):4109–4115

Rafique M, Sadaf I, Tahir MB, Rafique MS, Nabi G, Iqbal T, Sughra K (2019) Novel and facile synthesis of silver nanoparticles using Albizia procera leaf extract for dye degradation and antibacterial applications. Mater Sci Eng C 99:1313–1324

Rahman AU, Khan AU, Yuan Q, Wei Y, Ahmad A, Ullah S, Khan ZUH, Shams S, Tariq M, Ahmad W (2019) Tuber extract of Arisaema flavum eco-benignly and effectively synthesize silver nanoparticles: photocatalytic and antibacterial response against multidrug resistant engineered E. coli QH4. J Photochem Photobiol B Biol 193:31–38

Rajeshkumar S, Menon S, Kumar SV, Tambuwala MM, Bakshi HA, Mehta M, Satija S, Gupta G, Chellappan DK, Lakshmi T (2019) Antibacterial and antioxidant potential of biosynthesized copper nanoparticles mediated through Cissus arnotiana plant extract. J Photochem Photobiol B Biol 197:111531

Ramkumar R, Balasubramani G, Raja RK, Raja M, Govindan R, Girija EK, Perumal P (2017) Lantana camara Linn root extract-mediated gold nanoparticles and their in vitro antioxidant and cytotoxic potentials. Artif Cells Nanomedicine Biotechnol 45(4):748–757

Robles-Martínez M, González JFC, Pérez-Vázquez FJ, Montejano-Carrizales JM, Pérez E, Patiño-Herrera R (2019) Antimycotic activity potentiation of Allium sativum extract and silver nanoparticles against Trichophyton rubrum. Chem Biodivers 16(4):e1800525

Sadanand V, Rajini N, Rajulu AV, Satyanarayana B (2016) Preparation of cellulose composites with in situ generated copper nanoparticles using leaf extract and their properties. Carbohydr Polym 150:32–39

Santiago TR, Bonatto CC, Rossato M, Lopes CA, Lopes CA, G Mizubuti ES, Silva LP (2019) Green synthesis of silver nanoparticles using tomato leaf extract and their entrapment in chitosan nanoparticles to control bacterial wilt. J Sci Food Agric 99(9):4248–4259

Shaniba V, Aziz AA, Jayasree P, Kumar PM (2019) Manilkara zapota (L.) P. Royen leaf extract derived silver nanoparticles induce apoptosis in human colorectal carcinoma cells without affecting human lymphocytes or erythrocytes. Biol Trace Elem Res 192(2):1–15

Sharma G, Sharma A, Bhavesh R, Park J, Ganbold B, Nam J-S, Lee S-S (2014) Biomolecule-mediated synthesis of selenium nanoparticles using dried Vitis vinifera (raisin) extract. Molecules 19(3):2761–2770

Simu SY, Ahn S, Castro-Aceituno V, Singh P, Mathiyalagan R, Jiménez-Pérez ZE, Hurh J, Oi LZ, Hun NJ, Kim Y-J (2019) Gold nanoparticles synthesized with fresh Panax ginseng leaf extract suppress adipogenesis by downregulating PPARγ/CEBPα signaling in 3T3-L1 mature adipocytes. J Nanosci Nanotechnol 19(2):701–708

Singh H, Du J, Singh P, Yi TH (2018a) Ecofriendly synthesis of silver and gold nanoparticles by Euphrasia officinalis leaf extract and its biomedical applications. Artif Cells Nanomedicine Biotechnol 46(6):1163–1170

Singh P, Ahn S, Kang J-P, Veronika S, Huo Y, Singh H, Chokkaligam M, El-Agamy Farh M, Aceituno VC, Kim YJ (2018b) In vitro anti-inflammatory activity of spherical silver nanoparticles and monodisperse hexagonal gold nanoparticles by fruit extract of Prunus serrulata: a green synthetic approach. Artif Cells Nanomedicine Biotechnol 46(8):2022–2032

Singh J, Kumar V, Kim K-H, Rawat M (2019) Biogenic synthesis of copper oxide nanoparticles using plant extract and its prodigious potential for photocatalytic degradation of dyes. Environ Res 177:108569

Sivaraj R, Rahman PK, Rajiv P, Salam HA, Venckatesh R (2014) Biogenic copper oxide nanoparticles synthesis using Tabernaemontana divaricate leaf extract and its antibacterial activity against urinary tract pathogen. Spectrochim Acta A Mol Biomol Spectrosc 133:178–181

Sowndarya P, Ramkumar G, Shivakumar M (2017) Green synthesis of selenium nanoparticles conjugated Clausena dentata plant leaf extract and their insecticidal potential against mosquito vectors. Artif Cells Nanomedicine Biotechnol 45(8):1490–1495

Srivastava S (2019) Plant-metal interactions. Springer

Suganthy N, Ramkumar VS, Pugazhendhi A, Benelli G, Archunan G (2018) Biogenic synthesis of gold nanoparticles from Terminalia arjuna bark extract: assessment of safety aspects and neuroprotective potential via antioxidant, anticholinesterase, and antiamyloidogenic effects. Environ Sci Pollut Res 25(11):10418–10433

Sulaiman GM, Tawfeeq AT, Jaaffer MD (2018) Biogenic synthesis of copper oxide nanoparticles using olea europaea leaf extract and evaluation of their toxicity activities: an in vivo and in vitro study. Biotechnol Prog 34(1):218–230

Sumitha S, Vasanthi S, Shalini S, Chinni SV, Gopinath SC, Anbu P, Bahari MB, Harish R, Kathiresan S, Ravichandran V (2018) Phyto-mediated photo catalysed green synthesis of silver nanoparticles using Durio Zibethinus seed extract: antimicrobial and cytotoxic activity and photocatalytic applications. Molecules 23(12):3311

Sundararajan B, Kumari BR (2017) Novel synthesis of gold nanoparticles using Artemisia vulgaris L. leaf extract and their efficacy of larvicidal activity against dengue fever vector Aedes aegypti L. J Trace Elem Med Biol 43:187–196

Sunderam V, Thiyagarajan D, Lawrence AV, Mohammed SSS, Selvaraj A (2019) In-vitro antimicrobial and anticancer properties of green synthesized gold nanoparticles using Anacardium occidentale leaves extract. Saudi J Biol Sci 26(3):455–459

Suwan T, Wanachantararak P, Khongkhunthian S, Okonogi S (2018) Antioxidant activity and potential of Caesalpinia sappan aqueous extract on synthesis of silver nanoparticles. Drug Discov Ther 12(5):259–266

Thomas B, Vithiya B, Prasad T, Mohamed S, Magdalane CM, Kaviyarasu K, Maaza M (2019) Antioxidant and photocatalytic activity of aqueous leaf extract mediated green synthesis of silver nanoparticles using passiflora edulis f. flavicarpa. J Nanosci Nanotechnol 19(5):2640–2648

Vahabi K, Mansoori GA, Karimi S (2011) Biosynthesis of silver nanoparticles by fungus Trichoderma reesei (a route for large-scale production of AgNPs). Insciences J 1(1):65–79

Valsalam S, Agastian P, Arasu MV, Al-Dhabi NA, Ghilan A-KM, Kaviyarasu K, Ravindran B, Chang SW, Arokiyaraj S (2019) Rapid biosynthesis and characterization of silver nanoparticles from the leaf extract of Tropaeolum majus L. and its enhanced in-vitro antibacterial, antifungal, antioxidant and anticancer properties. J Photochem Photobiol B Biol 191:65–74

Vijayakumar S, Vaseeharan B, Malaikozhundan B, Gopi N, Ekambaram P, Pachaiappan R, Velusamy P, Murugan K, Benelli G, Kumar RS (2017) Therapeutic effects of gold nanoparticles synthesized using Musa paradisiaca peel extract against multiple antibiotic resistant Enterococcus faecalis biofilms and human lung cancer cells (A549). Microb Pathog 102:173–183

Vijayan R, Joseph S, Mathew B (2018) Indigofera tinctoria leaf extract mediated green synthesis of silver and gold nanoparticles and assessment of their anticancer, antimicrobial, antioxidant and catalytic properties. Artif Cells Nanomedicine Biotechnol 46(4):861–871

Vijayan R, Joseph S, Mathew B (2019) Anticancer, antimicrobial, antioxidant, and catalytic activities of green-synthesized silver and gold nanoparticles using Bauhinia purpurea leaf extract. Bioprocess Biosyst Eng 42(2):305–319

Wacławek S, Gončuková Z, Adach K, Fijałkowski M, Černík M (2018) Green synthesis of gold nanoparticles using Artemisia dracunculus extract: control of the shape and size by varying synthesis conditions. Environ Sci Pollut Res 25(24):24210–24219

Wang D, Markus J, Wang C, Kim Y-J, Mathiyalagan R, Aceituno VC, Ahn S, Yang DC (2017) Green synthesis of gold and silver nanoparticles using aqueous extract of Cibotium barometz root. Artif Cells Nanomedicine Biotechnol 45(8):1548–1555

Wintachai P, Paosen S, Yupanqui CT, Voravuthikunchai SP (2019) Silver nanoparticles synthesized with Eucalyptus critriodora ethanol leaf extract stimulate antibacterial activity against clinically multidrug-resistant Acinetobacter baumannii isolated from pneumonia patients. Microb Pathog 126:245–257

Wu T, Duan X, Hu C, Wu C, Chen X, Huang J, Liu J, Cui S (2019) Synthesis and characterization of gold nanoparticles from Abies spectabilis extract and its anticancer activity on bladder cancer T24 cells. Artif Cells Nanomedicine Biotechnol 47(1):512–523

Xu L, Li W, Shi Q, Li H, Yang Z, Liao D, Li L, Yang X, Zhang J (2019) Synthesis of mulberry leaf extract mediated gold nanoparticles and their ameliorative effect on Aluminium intoxicated and diabetic retinopathy in rats during perinatal life. J Photochem Photobiol B Biol 196:111502

Yallappa S, Manjanna J, Sindhe M, Satyanarayan N, Pramod S, Nagaraja K (2013) Microwave assisted rapid synthesis and biological evaluation of stable copper nanoparticles using T. arjuna bark extract. Spectrochim Acta A Mol Biomol Spectrosc 110:108–115

Zangeneh MM, Ghaneialvar H, Akbaribazm M, Ghanimatdan M, Abbasi N, Goorani S, Pirabbasi E, Zangeneh A (2019) Novel synthesis of Falcaria vulgaris leaf extract conjugated copper nanoparticles with potent cytotoxicity, antioxidant, antifungal, antibacterial, and cutaneous wound healing activities under in vitro and in vivo condition. J Photochem Photobiol B Biol 197:111556

Zhang W, Zhang J, Ding D, Zhang L, Muehlmann LA, Deng S-e, Wang X, Li W, Zhang W (2018) Synthesis and antioxidant properties of Lycium barbarum polysaccharides capped selenium nanoparticles using tea extract. Artif Cells Nanomedicine Biotechnol 46(7):1463–1470

Zhang K, Liu X, Samuel Ravi SOA, Ramachandran A, Aziz Ibrahim IA, Nassir A M, Yao J (2019) Synthesis of silver nanoparticles (AgNPs) from leaf extract of Salvia miltiorrhiza and its anticancer potential in human prostate cancer LNCaP cell lines. Artif Cells Nanomedicine Biotechnol 47(1):2846–2854

Role of Nanotechnology in the Management of Agricultural Pests

6

Nidhi Shukla, Keshav Meghvanshi, and Jayendra Nath Shukla

Abstract

The continuous use of pesticide-mediated insect control has led to the rise in insecticide resistance cases. It has now become a global problem and a matter of serious concern. Therefore, more reliable and advanced methods are urgently required for the control of insect/pests. Nanotechnology, an interdisciplinary field, has revolutionized different sectors of science and technology by introducing nanoparticles. Nanoparticles can be utilized for enhancing the efficacy of insecticides and pesticides in reduced doses. The use of nanotechnology in agriculture is less frequent compared to sectors like medicine and pharmacy. In this chapter, we give a gist of traditional insect/pest control strategies and discuss the potential of nanotechnology as a new tool for insect control.

Keywords

Nanoparticles · Nanotechnology · Pesticides · Nanopesticides · Pests

6.1 Introduction

Biotic and abiotic stresses are the two major components influencing crop losses. Drought, temperature, humidity, and salinity are abiotic stresses which influence the spread of biotic factors like insects/pests, pathogens, and weeds (Pandey et al.

N. Shukla
Department of Biotechnology and Bioinformatics, Birla Institute of Scientific Research, Jaipur, Rajasthan, India

K. Meghvanshi · J. N. Shukla (✉)
Department of Biotechnology, School of Life Sciences, Central University of Rajasthan, Ajmer, India
e-mail: jayendrashukla@curaj.ac.in

M. Ghorbanpour et al. (eds.), *Biogenic Nano-Particles and their Use in Agro-ecosystems*, https://doi.org/10.1007/978-981-15-2985-6_6

2017). Among biotic factors, insects contribute significantly to crop loss by competing with humans and plants for their co-existence in nature. Insects can affect the crop at multiple stages of plant growth, which depends upon the life cycle of the insect. Starting from seedling to standing/mature crop and further during storage of the grains, pests-mediated losses are huge. Farmers spend a huge amount of money in plant protection programs to minimize crop losses by pests. Even today, the use of pesticides in different combinations is considered to be the primary option for pest control. Use of pesticides has several advantages, including high availability, fast action, and reliability, but the side effects of pesticides cannot be ignored (Damalas 2009). Development of insecticide resistance is an ever-increasing, complicated worldwide problem (Hawkins et al. 2019). Additionally, harmful effects of pesticides on nontarget insects and other organisms is a matter of grave concern. Transgenic crops expressing insecticidal proteins are another option to control crop losses but ironically, this strategy is not very efficient against many species of insect/pest (Tabashnik and Carrière 2017). Development of alternative strategies for the control of insect pests is the need of the hour. Pesticides based on nanotechnology could be the promising tool for developing target-specific insect pest management methods (Fig. 6.1). The most promising effect of nanotechnology has been seen in the case of drug delivery: targeted drug delivery and prolonged/sustained delivery of drugs (Patra et al. 2018). A new era has emerged with the discovery of nanomedicines where nanoparticles are engineered to deliver pharmaceutical ingredients (Mu et al. 2018).

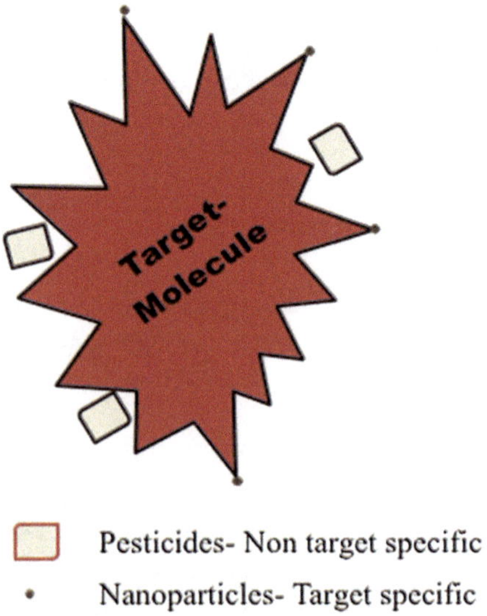

Fig. 6.1 Schematic representation showing the specificity of nanoparticles

6.2 Methods for the Management of Insects/Pests

Different methods for the management of insects/pests can be used based on the type of insects in the field and the extent of their infestation. Following are different methods adopted for the control of insects/pests in the field:

(a) **Cultural control:** Cultural control method is a preventive measure for the establishment and outbreak of pests (Dhaliwal et al. 2004). This includes a shift in agricultural practices (altering the environment, condition of the host or site) from the routine, which negatively impacts the life cycle of pests. Cultivation of resistant plant varieties, change in the timing of planting, harvesting, ploughing, irrigation, and application of recommended fertilizers are some of the practices known to reduce the burden of pests in the field. Further, crop rotation and planting of trap crops have also been shown to be effective methods of pest management (Hokkanen 1991), (Weisz et al. 1994).

(b) **Mechanical and physical control:** This strategy uses mechanical or physical devices for reducing the burden of the pest population in a given locality by preventing the entry of pest in the area, their trapping and removal from the area, directly killing them, or making the environment unsuitable for their propagation. Heat and steam sterilization of soil is used in greenhouses to kill any existing stages of pest in the soil. Physical method also includes the proper sweeping of the floors and benches, use of screens, barriers, fences, and nets, as well as light trapping for the removal of insects (Vincent et al. 2003).

(c) **Biological control:** The principle of the biological control method is the utilization of various natural enemies (predators, parasites, pathogens, and competitors) of the pests for keeping a check on their population (Jonsson et al. 2008). Biological control is often targeted against pests that are not native to the geographical area. Nonnative pests can be a big threat to crops as no natural enemies are available in the area to control or suppress their population. To control their population, natural enemies of the pest population are explored in their native place and released in the targeted area after ensuring them to be harmless, otherwise. This is a highly regulated process with complete surveillance on the introduced biological control agents, as sometimes such agents themselves become pests. Several different types of biological control methods have been used from time to time and at different places, which has shown to be an effective insect control strategy (Johnson et al. 1988) (Divya and Sankar 2009).

Another type of biological control method is based on manipulating the reproductive system of insects, for example, Sterile Insect Technique (SIT), which ultimately leads to reduction in pest population (Hendrichs and Robinson 2009; Dunn and Follett 2017). Mating of sterile males of targeted pest population with wild-type females results in no progeny by the virtue of sterility in the released males. Continuous release of sterilized males ultimately leads to decline in the targeted insect/pest population. SIT has been successfully used in the control of many insects (Scott et al. 2017).

A recent modification of the traditional sterile insect technique is in the trial phase, which uses molecular tools rather than ionizing radiation to render the males sterile (Lees et al. 2015; Wilke and Marrelli 2012). Another biological control method is based on an endosymbiont bacterium, *Wolbachia*, which interferes with the normal reproduction of its host insect. The infection of *Wolbachia* causes feminization, parthenogenesis, male killing, and cytoplasmic incompatibility (CI). CI is also called incompatible insect technique (IIT) and has been effectively used for the control of several insect pests and disease vectors (Nikolouli et al. 2018; Zhang et al. 2015).

(d) **Chemical control:** Pesticides are formulated chemicals to control or completely destroy insects, fungi, weeds, rodents, and microbes. A large number of artificial and natural pesticides are available in the market. Pesticides are very effective in killing insects/pests but continuous use of pesticides leads to the development of resistance in the pest population besides having negative effect on soil structure, fertility, mineral cycles, and soil micro-flora (García-García et al. 2016). Pesticides adversely affect nontarget insect/pests and even human beings (Le Goff and Giraudo 2019). Pesticides are known to cause several deadly diseases in humans, including endocrine disruption, breast cancer, cytotoxicity, and reproductive toxicity (Gangemi et al. 2016; Nicolopoulou-Stamati et al. 2016). It is advisable to use chemical control methods for the management of pest population as a last option, only when the pests cannot be controlled by natural methods.

6.3 Nanoparticles as a New Tool for Pest Management

Nanotechnology, an interdisciplinary field, has enormous potential in different areas such as medicine, pharmaceutical, and agriculture. The word *nano* is derived from the Greek, meaning "dwarf," which justifies the word nanotechnology, which evolved due to the use of particles with size ranging in nanometers (size of 1–100 nm) (Bhattacharyya et al. 2010). Nanotechnology is one of the most reliable and expanding technologies of the twenty-first century. It has enormous advantages, especially in the field of agriculture, which include insect pest management through the formulations of nanomaterials-based pesticides and insecticides, development of insect pest-resistant crop varieties, and increase in agricultural productivity using bio-conjugated nanoparticles (Dimetry and Hussein 2016). Nanoparticles have distinct physical, biological, and chemical properties associated with extraordinary strength and more chemical reactivity.

6.4 Role of Naturally Occurring Nanoparticles in Insects

There are a lot of naturally occurring nanostructures which meet certain specifications, but until now they have been neglected (Ehrlich et al. 2008). A very good example is ordered hexagonal array of structures in cicada wings, e.g., *Psaltoda claripennis* Ashton and termite, family Rhinotermitidae (Zhang et al. 2006). Many studies have shown that the size of nanoparticles varies from 200 to 1000 nm. These wing structures have a round shape at the top and there is a protrusion of about 150–350 nm on the outer side. These features increase the aerodynamic efficiency of the insect. Similar nanostructures are also present in compound eyes of some insects. Interestingly, the bright colored components present in butterfly wings are nothing but nanoparticles. These naturally occurring nanoparticles have been reported to be responsible for plant-insect interactions (Gorb and Gorb 2009). Some ferromagnetic material is found in the different body parts of some insects which are magnetic nanoparticles and act as geomagnetic sensors specifically in social insects (Esquivel 2007). Yang et al. have shown the insecticidal properties of polyethylene-glycol-coated nanoparticles which is responsible for almost 80% mortality in *Tribolium castaneum* due to continuous release of dynamic components nanoparticles.

6.5 Nanoparticles Used in Biopesticides Controlled Release Formulations

The different types of nanomaterials used are:

(i) Nanospheres: Active compound is dispersed into the polymeric matrix.
(ii) Nanocapsules: Active compound is lined by matrix polymer and centered in the core.
(iii) Nanogels: These are cross-linked polymers which are hydrophilic in nature and can absorb large amounts of water.
(iv) Micelles: These are aqueous collection formed by hydrophilic and hydrophobic molecules.

6.6 Classes of Nanoparticles

There are many ways to classify nanoparticles based on their chemical composition and their properties. They can also be classified based on their morphology, structure, dimension, and composition. The five major classes of nanoparticles are carbon-based nanoparticles, metal-containing nanoparticles (including metal oxides), quantum dots, zero-valent metals, and dendrimers, but recent studies have mainly focused on carbon-based nanoparticles (carbon nanotubes and fullerenes) and metal or metal-oxide nanoparticles (Ag NPs, CuO NPs, TiO2).

(i) **Carbon-based nanoparticles:** These materials have gained significant interest in different areas because of their unique structure as well as excellent mechanical, chemical, and optical properties. There are two major categories of carbon-based nanoparticles: carbon nanotubes (CNTs) and fullerenes (Khan et al. 2017). Carbon nanotubes are graphene sheets which are rolled into a tube, which provide structural support as they are almost 100 times stronger than steel. These nanoparticles are very unique as they are thermally conductive along the length and nonconductive across the tube. Fullerenes are carbon allotropes having sixty or more carbon atoms arranged in a pentagonal and hexagonal array in a hollow cage structure. These are commercially very useful because of their high strength and electrical and structural properties.

(ii) **Metal-containing nanoparticles:** These constitute the largest group of nanoparticles, which include oxides such as zinc oxide (ZnO), titanium dioxide (TiO_2), cerium dioxide (CeO_2), copper oxide (CuO), chromium dioxide (CrO_2), molybdenum trioxide (MoO_3), bismuth trioxide (Bi_2O_3), and lithium cobalt dioxide ($LiCoO_2$) (Buzea et al. 2007). These nanoparticles have gained immense popularity over the last few years due to their extensive use in food, chemical, and biological areas.

6.7 Nanopesticides

Nanopesticides are biologically derived complexes with nanoparticles which are designed to be used as pesticides. Nanopesticides consist of organic ingredients such as polymers and inorganic ingredients such as metal oxides (Kookana et al. 2014). Nanoformulation has a significant role in the development of nanopesticides by increasing the apparent solubility of poorly soluble active ingredients and releasing the active ingredient in a targeted manner and protecting the active ingredient against premature degradation (Kah et al. 2013). There are many nanopesticides which are engineered with useful pesticidal properties and it has been shown to be effective against a variety of insects. The salient features which make nanoparticles-based pesticide formulation important include increased stability of formulation which prevents early degradation, more elimination of toxic substances compared to conventional pesticides, improved mobility and higher insecticidal activity due to smaller particle size, and larger surface area, which increases their longevity (Shah and Wani 2016). The benefits of nanoparticles include better and accurate delivery of products as they are designed to transfer particular molecules to a cell or tissue according to requirement. These nanoparticles enter into the plant cell by binding to some carrier protein or through ion channels (Singh and Lillard 2009). Permeability of plant cell wall also plays an important role in this uptake. In spite of its advantageous effect on plants, these nanoparticles are being tested for its phytotoxicity. Different studies have reported different effects of these nanoparticles on seed germination and plant development. For example, it has been shown that nano-ZNO particles promote seedling growth of mung and gram beans at a certain concentration whereas treating castor seeds with silver nanoparticles did not show any effect on seed germination nor on growth of lepidopteran insects on the seeds (Dhoke et al. 2011). Localization of nanoparticles in nucleolus or mitochondria, by scanning electron microscopy confirmed their penetration into cell organelles, which suggest their use for targeted delivery of pesticides.

6.8 Methods to Develop Nanoparticles for Pest Control

Pesticide encapsulation is essential for its controlled release as well as to minimize its toxic properties. There are many chemical and physical methods to develop nanoparticles loaded with pesticides. Some of the uses of nanoparticles for pest management are described below:

(i) **Pesticide-loaded nanoparticles:** There are some nanoparticles which have the property of loading different kinds of pesticides such as insecticides, herbicides, fungicides, and nematicides. Different studies have used different techniques to formulate nanoparticles. Crooks et al. have made aqueous suspension of nanoparticles which contain an organic active ingredient such as pesticides. There have been many insecticides, herbicides, fungicides, aphicides, and miticides which are formulated with this technique. In another study, Ishaque et al. developed pesticide-coated metal oxide nanoparticles comprising a UV photo protective filter. This formulation is really helpful for controlling harmful insects.
(ii) **Insecticide-loaded nanoparticles:** Imidacloprid is a novel insecticide encapsulated with chitosan and sodium alginate and presents in a low amount in soybean plants (Guan et al. 2008). Deltamethrin, a silver nanoparticle conjugated to the pyrethroid pesticide, was shown to be effective against arthropod vectors such as mosquitoes (Sooresh et al. 2011). Another nanoparticle, Pyrifluquinazon, which has a controlled release feature, was found to be effective against *Myzus persicae* (Kang et al. 2012). Recent studies have characterized natural insecticides such as polyethylene-glycol-coated nanoparticles which were more effective against insect pests compared to essential oils (Yang et al. 2009). Nanogels loaded with cumin and ajwain essential oil showed more efficacy against insect pests compared to only oil-loaded nanogels.
(iii) **Fungicide-loaded nanoparticles:** These nanoparticles were prepared using polyvinylpyridine (PVPy) and polyvinylpyridine-co-styrene and were loaded with tebuconazole or chlorothalonil fungicides as aqueous dispersions. This formulation was shown to be effective against many pathogenic fungi (Liu et al. 2003).
(iv) **Nematicide-loaded nanoparticles:** Yin et al. reported the formulation and efficacy of Lansiumamide B, a nematicide-loaded insecticide which was prepared using the microemulsion polymerization method (Yin et al. 2012).
(v) **Herbicide-loaded nanoparticles:** These nanoparticles have very low phytotoxicity on crops and are designed for slow and controlled release of active

substances. A study reported by Silva et al. has shown that Paraquat-loaded alginate/chitosan nanoparticles can change the release profile of a herbicide and its interaction with the soil (Silva et al. 2010). Eventually, herbicide-loaded nanoparticles can be applied to reduce the negative impacts caused by herbicides. Glyphosate isopropylamine, a water-soluble herbicide, was prepared using a green nanoemulsion system, and it was shown to be effective against the weeds creeping foxglove, slender button weed, and buffalo grass (Lim et al. 2013). Another nanoparticle loaded with chloroacetanilide herbicide (alachlor) was shown to reduce the degradation of the herbicide (Thompson et al. 2010).

6.9 Mechanisms of Action of Nanoparticles

Though there is vast data available on the toxicity of nanoparticles, the precise mechanisms of action of nanoparticles against insects are not well understood (Benelli et al. 2018). Understanding the mechanisms is really important to predict the toxicological effects in using these nanoparticles as pesticides. Most of the toxicity studies have been done for silver nanoparticles considering the fact that toxicity is highly influenced by size, shape, and charge of nanoparticles (Foldbjerg et al. 2015). Also, these studies have been largely done on bacterial models or by in vitro cytotoxicity assays; very few studies have been done on insects, which opens up new avenues for further research (L. Santo-Orihuela et al. 2016). One of the most accepted theories about the mechanism of action of nanoparticles is that they penetrate exoskeleton and bind to DNA and proteins, which lead to rapid denaturation of organelles and enzymes. Ultimately, disturbance in proton motive force and decrease in membrane permeability results in loss of cellular functions, leading to cell death. However, the studies done on the mechanism of action of nanoparticles on insects/pests are very limited as even today a PubMed search for the mechanism of action of nanoparticles against insect/pests retrieves very few articles. Some of the research articles showing the possible mechanism of action of nanoparticles are summarized in Table 6.1. Hence, there is a need for further research on this aspect of nanoparticles.

6.10 Conclusion: Nanotechnology Risks and Regulation

Traditional strategies are insufficient for the control of pests in very large areas. Chemical pesticides are very effective but associated with huge toxic effects on other animals as well as human beings. Besides, there is always a threat of emergence of pesticide resistance among insect/pests. Nanotechnology and RNAi-based pesticides, a relatively recent field, can be an alternative approach for the management of insect pests in agriculture without negatively affecting nature. Furthermore, combining pesticides with nanoparticles will have more specific and targeted effects on pest populations. This will also reduce the burden of pesticides as combining

Table 6.1 Current knowledge about the mode of action of nanoparticles against insects

S.No	Tested nanomaterials	Target insect/pest	Mode of action	References
1	Nanodiamonds coated with Neb-colloostatin	*Tenebrio molitor*	Apoptosis of hemocytes and inhibited cellular and humoral immune responses	
2	Titanium dioxide nanoparticles (TDNPs)	*Helicoverpa armigera*	Reduction of detoxifying enzymes such as β-glucosidase and carboxylesterase and increase of glutathione S-transferase	Chinnaperumal et al. (2018)
3.	Ag nanoparticles prepared using *Cassia fistula* extract	*Aedes albopictus* and *Culex pipiens pallens*	4th instar larvae showed a decrease in total protein levels; nanoAg also reduced acetylcholinesterase and α- and ß-carboxylesterase activities	(Fouad et al. 2018)
4.	Ag nanoparticles fabricated using salicylic acid and 3,5-dinitrosalicylic acid	*Aedes albopictus*	4th instar larvae showed a decrease in total proteins, esterase, acetylcholine esterase, and phosphatase enzymes	Ga'al et al. (2018)
5.	Ag nanoparticles	*Drosophila melanogaster*	Accumulation of reactive oxygen species (ROS) leading to ROS-mediated apoptosis, DNA damage, and autophagy; activation of the Nrf2-dependent antioxidant pathway	Mao et al. (2018)
6.	Nanostructured Al_2O_3	*Sitophilus oryzae*	Bind to the beetle cuticle due to triboelectric, results in insect dehydration forces, sorbing its wax layer by surface area phenomena, resulting in insect dehydration	Stadler et al. (2017)
7.	Au nanoparticles fabricated using latex of *Jatropha curcas*	*Aedes aegypti*, beetles, and mealybugs	Triggered trypsin inhibition	Patil et al. (2016)

(continued)

Table 6.1 (continued)

S.No	Tested nanomaterials	Target insect/pest	Mode of action	References
8.	Graphene oxide nanoparticles	*Acheta domesticus*	Increase in enzymatic activity of catalase and glutathione peroxidases, as well as heat shock protein (HSP 70) and increase in total antioxidant capacity levels	Dziewiecka et al. (2016)
9.	TiO2 nanoparticles	*Bombyx mori*	Upregulation of PI3K and P70S6K [mTOR pathways]; 4 cytochrome P450 genes were upregulated; 20-hydroxyecdysone biosynthesis was stimulated; reduced development and molting duration were noted	Li et al. (2014)
10.	Ag nanoparticles	Drosophila melanogaster	Loss of melanin cuticular pigments, reduced activity of Cu-dependent enzymes (tyrosinase and Cu-Zn superoxide dismutase)	Armstrong et al. (2013)
11.	SiO2 nanoparticles	Bombus terrestris	Leads to midgut epithelial injury in intoxicated workers	Mommaerts et al. (2012)
12.	Ag and TiO_2 nanoparticles	Drosophila melanogaster	Loss of progeny and a decrease in developmental success	Philbrook et al. (2011)
13.	Polystyrene nanoparticles	Insect cells (BACULOSOMES®)	Inhibited the enzymatic activity of CYP450 isoenzymes in BACULOSOMES®	Fröhlich et al. (2010)
14.	Carbon black and multiwalled nanotubes	Drosophila melanogaster	Nanomaterials adhere strongly to fly body parts which leads to impaired motor functions, resulting in insect mortality	Liu et al. (2009)
15.	SiO2 nanoparticles	Different species, with special reference to stored product pests	Physio-sorbed by the insect cuticular lipids, causing major damages, followed by insect death	Barik et al. (2008), Debnath et al. (2011) and Athanassiou et al. (2018)

Fig. 6.2 Agricultural field with the use of nanoparticles for the pest control

even a small amount with nanoparticles will have a significantly stronger effect. Though the toxic effects of nanoparticles have been studied on some aquatic organisms, not much data is available on the toxicity of nanoparticles on beneficial insects. This gives an opportunity to the scientific community for further research and validation with the aim of completely harnessing the potential of nanoparticles and nanotechnology in the agricultural field (Fig. 6.2).

References

Armstrong N, Ramamoorthy M, Lyon D, Jones K, Duttaroy A (2013) Mechanism of silver nanoparticles action on insect pigmentation reveals intervention of copper homeostasis. PLoS One 8(1)

Athanassiou CG, Kavallieratos NG, Benelli G, Losic D, Usha Rani P, Desneux N (2018) Nanoparticles for pest control: current status and future perspectives. J Pest Sci 91(1):1–15

Barik TK, Sahu B, Swain V (2008) Nanosilica – From medicine to pest control. Parasitol Res 103(2):253–258

Benelli G, Maggi F, Pavela R, Murugan K, Govindarajan M, Vaseeharan B, Petrelli R, Cappellacci L, Kumar S, Hofer A et al (2018) Mosquito control with green nanopesticides: towards the one health approach? A review of non-target effects. Environ Sci Pollut Res 25(11):10184–10206

Bhattacharyya A, Bhaumik A, Rani PU, Mandal S, Epidi TT (2010) Nano-particles - A recent approach to insect pest control. Afr J Biotechnol 9(24):3489–3493

Buzea C, Pacheco II, Robbie K (2007) Nanomaterials and nanoparticles: sources and toxicity. Biointerphases 2(4):MR17–MR71

Chinnaperumal K, Govindasamy B, Paramasivam D, Dilipkumar A, Dhayalan A, Vadivel A, Sengodan K, Pachiappan P (2018) Bio-pesticidal effects of Trichoderma viride formulated titanium dioxide nanoparticle and their physiological and biochemical changes on Helicoverpa armigera (Hub.). Pestic Biochem Physiol 149:26–36

Damalas CA (2009) Understanding benefits and risks of pesticide use. Sci Res Essays 4(10):945–949

Debnath N, Das S, Seth D, Chandra R, Bhattacharya SC, Goswami A (2011) Entomotoxic effect of silica nanoparticles against Sitophilus oryzae (L.). J Pest Sci 84(1):99–105

Dhaliwal GS, Koul O, Arora R (2004) Integrated pest management: retrospect and prospect. Integrated Pest Management: Potential, Constraints Challenges

Dhoke SK, Mahajan P, Khanna AS (2011) Effect of nano-ZnO particle suspension on growth of mung (Vigna radiata) and gram (Cicer arietinum) seedlings using plant agar method. J Nanotechnol 2011:1–7

Dimetry NZ, Hussein HM (2016) Role of nanotechnology in agriculture with special reference to pest control. Int J Pharm Tech Res 9(10):121–144

Divya K, Sankar M (2009) Entomopathogenic nematodes in pest management. Indian J Sci Technol 2(7):53–60

Dunn DW, Follett PA (2017) The sterile insect technique (SIT) – an introduction. Entomol Exp Appl 163(3):151–154

Dziewiecka M, Karpeta-Kaczmarek J, Augustyniak M, Majchrzycki Ł, Augustyniak-Jabłokow MA (2016) Evaluation of in vivo graphene oxide toxicity for Acheta domesticus in relation to nanomaterial purity and time passed from the exposure. J Hazard Mater 305:30–40

Ehrlich H, Janussen D, Simon P, Bazhenov VV, Shapkin NP, Erler C, Mertig M, Born R, Heinemann S, Hanke T et al (2008) Nanostructural organization of naturally occurring composites-part II: silica-chitin-based biocomposites. J Nanomater 2008:1–8

Esquivel DMS (2007) Magnetic nanoparticles in social insects: are they the geomagnetic sensors? Entomological Society of America. Annual Meeting. Monday, December 10, 2007 NO-0574

Foldbjerg R, Jiang X, Micləuş T, Chen C, Autrup H, Beer C (2015) Silver nanoparticles – wolves in sheep's clothing? Toxicol Res (Camb) 4(3):563–575

Fouad H, Hongjie L, Hosni D, Wei J, Abbas G, Ga'al H, Jianchu M (2018) Controlling Aedes albopictus and Culex pipiens pallens using silver nanoparticles synthesized from aqueous extract of Cassia fistula fruit pulp and its mode of action. Artif Cells Nanomed Biotechnol 46(3):558–567

Fröhlich E, Kueznik T, Samberger C, Roblegg E, Wrighton C, Pieber TR (2010) Size-dependent effects of nanoparticles on the activity of cytochrome P450 isoenzymes. Toxicol Appl Pharmacol 242(3):326–332

Ga'al H, Fouad H, Tian J, Hu Y, Abbas G, Mo J (2018) Synthesis, characterization and efficacy of silver nanoparticles against Aedes albopictus larvae and pupae. Pestic Biochem Physiol 144:49–56

Gangemi S, Miozzi E, Teodoro M, Briguglio G, De Luca A, Alibrando C, Polito I, Libra M (2016) Occupational exposure to pesticides as a possible risk factor for the development of chronic diseases in humans (review). Mol Med Rep 14(5):4475–4488

García-García CR, Parrón T, Requena M, Alarcón R, Tsatsakis AM, Hernández AF (2016) Occupational pesticide exposure and adverse health effects at the clinical, hematological and biochemical level. Life Sci 145:274–283

Gorb EV, Gorb SN (2009) Contact mechanics at the insect-plant interface: how do insects stick and how do plants prevent this? Borodich FM (ed.), IUTAM Symposium on scaling in solid mechanics, IUTAM Bookseries, Springer Science and Business Media B.V. pp 243

Guan H, Chi D, Yu J, Li X (2008) A novel photodegradable insecticide: preparation, characterization and properties evaluation of nano-Imidacloprid. Pestic Biochem Physiol 92(2):83–91

Hawkins NJ, Bass C, Dixon A, Neve P (2019) The evolutionary origins of pesticide resistance. Biol Rev 94(1):135–155

Hendrichs J, Robinson A (2009) Chapter 243 – Sterile Insect Technique A2 - Resh, Vincent H. In Encyclopedia of Insects (2nd Ed)

Hokkanen H (1991) Trap cropping in Pest management. Annu Rev Entomol 36(1):119–138
Johnson DL, Huang HC, Harper AM (1988) Mortality of grasshoppers (Orthoptera: Acrididae) inoculated with a Canadian isolate of the fungus Verticillium lecanii. J Invertebr Pathol 52(2):335–342
Jonsson M, Wratten SD, Landis DA, Gurr GM (2008) Recent advances in conservation biological control of arthropods by arthropods. Biol Control 45(2):172–175
Kah M, Beulke S, Tiede K, Hofmann T (2013) Nanopesticides: state of knowledge, environmental fate, and exposure modeling. Crit Rev Environ Sci Technol 43(16):1823–1867
Kang MA, Seo MJ, Hwang IC, Jang C, Park HJ, Yu YM, Youn YN (2012) Insecticidal activity and feeding behavior of the green peach aphid, Myzus persicae, after treatment with nano types of pyrifluquinazon. J Asia Pac Entomol 15(4):533–541
Khan I, Saeed K, Khan I (2017) Nanoparticles: properties, applications and toxicities. Arab J Chem 12(7):908–931
Kookana RS, Boxall ABA, Reeves PT, Ashauer R, Beulke S, Chaudhry Q, Cornelis G, Fernandes TF, Gan J, Kah M et al (2014) Nanopesticides: guiding principles for regulatory evaluation of environmental risks. J Agric Food Chem 62(19):4227–4240
Le Goff G, Giraudo M (2019) Effects of pesticides on the environment and insecticide resistance. In Olfactory concepts of insect control – alternative to insecticides
Lees RS, Gilles JR, Hendrichs J, Vreysen MJ, Bourtzis K (2015) Back to the future: the sterile insect technique against mosquito disease vectors. Curr Opin Insect Sci 10:156–162
Li F, Gu Z, Wang B, Xie Y, Ma L, Xu K, Ni M, Zhang H, Shen W, Li B (2014) Effects of the biosynthesis and signaling pathway of ecdysterone on Silkworm (Bombyx mori) following exposure to titanium dioxide nanoparticles. J Chem Ecol 40(8):913–922
Lim CJ, Basri M, Omar D, Abdul Rahman MB, Salleh AB, Raja Abdul Rahman RNZ (2013) Green nanoemulsion-laden glyphosate isopropylamine formulation in suppressing creeping foxglove (A. gangetica), slender button weed (D. ocimifolia) and buffalo grass (P. conjugatum). Pest Manag Sci 69(1):104–111
Liu Y, Laks P, Heiden P (2003) Nanoparticles for the controlled release of fungicides in wood: soil jar studies using G. Trabeum and T. Versicolor wood decay fungi. Holzforschung 57(2):135–139
Liu X, Vinson D, Abt D, Hurt RH, Rand DM (2009) Differential toxicity of carbon nanomaterials in Drosophila: larval dietary uptake is benign, but adult exposure causes locomotor impairment and mortality. Environ Sci Technol 43(16):6357
Mao BH, Chen ZY, Wang YJ, Yan SJ (2018) Silver nanoparticles have lethal and sublethal adverse effects on development and longevity by inducing ROS-mediated stress responses. Sci Rep 8(1):1–16
Mommaerts V, Jodko K, Thomassen LCJ, Martens JA, Kirsch-Volders M, Smagghe G (2012) Assessment of side-effects by Ludox TMA silica nanoparticles following a dietary exposure on the bumblebee Bombus terrestris. Nanotoxicology 6(5):554–561
Mu Q, Yu J, McConnachie LA, Kraft JC, Gao Y, Gulati GK, Ho RJY (2018) Translation of combination nanodrugs into nanomedicines: lessons learned and future outlook. J Drug Target 26(5–6):435–447
Nicolopoulou-Stamati P, Maipas S, Kotampasi C, Stamatis P, Hens L (2016) Chemical pesticides and human health: the urgent need for a new concept in agriculture. Front Public Heal 4
Nikolouli K, Colinet H, Renault D, Enriquez T, Mouton L, Gibert P, Sassu F, Cáceres C, Stauffer C, Pereira R et al (2018) Sterile insect technique and Wolbachia symbiosis as potential tools for the control of the invasive species Drosophila suzukii. J Pest Sci 91(2):489–503
Pandey P, Irulappan V, Bagavathiannan MV, Senthil-Kumar M (2017) Impact of combined abiotic and biotic stresses on plant growth and avenues for crop improvement by exploiting physio-morphological traits. Front Plant Sci 8:537
Patil CD, Borase HP, Suryawanshi RK, Patil SV (2016) Trypsin inactivation by latex fabricated gold nanoparticles: a new strategy towards insect control. Enzym Microb Technol 92:18–25
Patra JK, Das G, Fraceto LF, Campos EVR, Rodriguez-Torres M d P, Acosta-Torres LS, Diaz-Torres LA, Grillo R, Swamy MK, Sharma S et al (2018) Nano based drug delivery systems: recent developments and future prospects. J Nanobiotechnol 16(1):1–33

Philbrook NA, Winn LM, Afrooz ARMN, Saleh NB, Walker VK (2011) The effect of TiO_2 and Ag nanoparticles on reproduction and development of Drosophila melanogaster and CD-1 mice. Toxicol Appl Pharmacol 257(3):429–436

Santo-Orihuela PL, Foglia ML, Targovnik AM, Miranda MV, Desimone MF (2016) Nanotoxicological effects of SiO 2 nanoparticles on Spodoptera frugiperda Sf9 cells. Curr Pharm Biotechnol 17(5):465–470

Scott MJ, Concha C, Welch JB, Phillips PL, Skoda SR (2017) Review of research advances in the screwworm eradication program over the past 25 years. Entomol Exp Appl 164(3):226–236

Shah MA, Wani SH (2016) Nanotechnology and insecticidal formulations. J Food Bioeng Nanoprocessing 1(3):285–310

Silva MDS, Cocenza DS, De Melo NFS, Grillo R, Rosa AH, Fraceto LF (2010) Alginate nanoparticles as a controlled release system for clomazone herbicide. Quim Nova 33(9)

Singh R, Lillard JW (2009) Nanoparticle-based targeted drug delivery. Exp Mol Pathol 86(3):215–223

Sooresh A, Kwon H, Taylor R, Pietrantonio P, Pine M, Sayes CM (2011) Surface functionalization of silver nanoparticles: novel applications for insect vector control. ACS Appl Mater Interfaces 3(10):3779–3787

Stadler T, López García GP, Gitto JG, Buteler M (2017) Nanostructured alumina: biocidal properties and mechanism of action of a novel insecticide powder. Bull Insectol 70(1):17–26

Tabashnik BE, Carrière Y (2017) Surge in insect resistance to transgenic crops and prospects for sustainability. Nat Biotechnol 35(10):926–935

Thompson JM, Chisholm BJ, Bezbaruah AN (2010) Reductive dechlorination of chloroacetanilide herbicide (Alachlor) using zero-valent iron nanoparticles. Environ Eng Sci 27(3):227–232

Vincent C, Hallman G, Panneton B, Fleurat-Lessard F (2003) Management of agricultural insects with physical control methods. Annu Rev Entomol 48:261–281

Weisz R, Smilowitz Z, Christ B (1994) Distance, rotation, and border crops affect Colorado potato beetle (Coleoptera: Chrysomelidae) colonization and population density and early blight (Alternaria solani) severity in rotated potato fields. J Econ Entomol 87(3):723–729

Wilke ABB, Marrelli MT (2012) Genetic control of mosquitoes: population suppression strategies. Rev Inst Med Trop Sao Paulo 54(5):287–292

Yang FL, Li XG, Zhu F, Lei CL (2009) Structural characterization of nanoparticles loaded with garlic essential oil and their insecticidal activity against *Tribolium castaneum* (Herbst) (Coleoptera: Tenebrionidae). J AgricFood Chem 11 57(21):10156–10162

Yin Y-h, Guo Q-m, Han Y, Wang L-j, Wan S-q (2012) Preparation, characterization and Nematicidal activity of Lansiumamide B Nano-capsules. J Integr Agric 11(7):1151–1158

Zhang G, Zhang J, Xie G, Liu Z, Shao H (2006) Cicada wings: a stamp from nature for nanoimprint lithography. Small 2(12):1440–1443

Zhang D, Zheng X, Xi Z, Bourtzis K, Gilles JRL (2015) Combining the sterile insect technique with the incompatible insect technique: I-impact of Wolbachia infection on the fitness of triple- and double-infected strains of Aedes albopictus. PLoS One 10(4):e0121126

7 Biogenic Nanomaterials: Synthesis and Its Applications for Sustainable Development

Nitin Kumar, Abarna Balamurugan, Purushothaman Balakrishnan, Kanchan Vishwakarma, and Kumaran Shanmugam

Abstract

Nanotechnology is a promising technology in sensing and preventing pollution due to its nanosized materials and augmenting agricultural production by detecting microbes, humidity, and toxic pollutants. Photocatalysis is an incredible process in nanotechnology to degrade organic pesticides and industrial pollutants into nontoxic and beneficiary product. Nanotechnology jumps into the agricultural fields from the lab, achieving the milestone continuously in different ways. In the present chapter, focus has been given on nanoparticle synthesis and its deployment for sustainable development.

Keywords

Biogenic nanoparticles · Sustainable developement · Green synthesis · Agriculture

7.1 Introduction

The field of nanotechnology, an interdisciplinary area among biology, physics, and chemistry with vast variety of technical features, covers fabrication, characterization, and handling of nanoscale structures and materials. Research related to nanotechnology draws increasing attention due to its good as well as bad impacts on numerous areas like environment, technology, agriculture, and medicine (Seabra and Duran 2010; Zhang et al. 2011; Duran and Seabra 2012a; Kumar et al. 2017a,

N. Kumar · A. Balamurugan · P. Balakrishnan · K. Shanmugam (✉)
Department of Biotechnology, Periyar Maniammai Institute of Science & Technology, Thanjavur, Tamil Nadu, India
e-mail: kumarans@pmu.edu

K. Vishwakarma
Amity Institute of Microbial Technology, Amity University, Noida, Uttar Pradesh, India

M. Ghorbanpour et al. (eds.), *Biogenic Nano-Particles and their Use in Agro-ecosystems*, https://doi.org/10.1007/978-981-15-2985-6_7

b, 2018; Vishwakarma et al. 2017). An eminent emerging field in nanotechnology is nanobiotechnology which implicates various research fields of proficiency like biology, chemistry, physics, engineering, medicine, and material science (de Lima et al. 2012; Duran and Seabra 2012b).

Agriculture is an essential profession in every part of the world. There are a number of professional farmers doing the agriculture in their old traditional methods to satisfy the human and animal needs. On the other hand, the growing population and their need for food are increasing day by day, but the limited arable land to produce larger amount of foods is difficult, and the pressure on the natural source will increase (Wheeler and von Braun 2013). Besides, a number of threats will arise while farming which includes insufficient minerals and nutrients, harmful pathogens, natural disaster, etc. Similarly, the innovative technologies also increase parallelly to support the farming by reducing their risk by giving latest machines, vehicles, nutrition supplement, and pesticides. In this ladder, nanotechnology plays the incredible part to enhance the productivity of the foods drastically because of its robust applications (Nair et al. 2010, Ghormade et al. 2011). Currently, the biogenic nanoparticles have given us a number of new ways to develop agricultural product in the eco-friendly manner to limit the plant diseases (Navrotsky 2000, Hu et al. 2006, Moonjung et al. 2010).

One of the noteworthy activities of the nanotechnology is the antimicrobial activity. Harmful plant pathogens are majorly responsible for the agricultural loss, (Pimentel 2009) in that $45 billion loss universally is due to fungi. This is because of the ability of the microbe to damage the plant at any parts it adheres with (Fernández-Acero et al. 2007). In order to avoid the adherence of the microbes, there should be the requirement of its resistance. For that nanoformulation is pioneer agent to genetically modified plants to act against the pathogens and pests (Ragaei and Sabry 2014). This is because the nanosized particles improve the enhancement and attachment with the plant surface and reduce the spillage (Chen and Yada, 2011).

Universally, various organic and inorganic contaminants from industrialization could play a major role in water pollution, and pollution and scarcity of water affect the livelihood of aquatic organisms (Anjum et al. 2016). In addition to that, it is too difficult to identify the need of purification process in water which was elevated around the world (Zhang et al. 2016). Detection, monitoring, and remediation of pollutants could be influenced through nanotechnology. These applications can be achieved by the production of engineered nanoparticles (NPs) (<100 nm). Chemical reaction will convert the harmful pollutants into harmless chemicals. For example, a dangerous pollutant which is commonly seen in industrial wastewater, namely, trichloroethene, can be catalyzed and treated by nanoparticles (Nutt et al. 2005). In addition to that, nanotech for hazardous waste cleanups is done with nanoscale material which shows enormous difference. Nanomaterials can easily penetrate them by means of their size, and engineered coating lets them suspend in groundwater.

Furthermore, we could treat wastewater in an exact and accurate manner by means of nanomaterials which have the capacity to deliver clean and affordable water treatment technologies. At present, removal and degradation of water pollutants through nanoscale adsorbents and catalysts is an emerging one. Application of biogenic nanomaterials leads wastewater treatment as an escalating area of research.

Presently, there is an augmenting need to evolve sustainable, reliable, and eco-friendly procedures to fabricate nanoparticles with broad scope. Biogenic nanoparticles (NPs) synthesized via nanobiotechnology-related processes have the potential to carry hygienic manufacturing technologies. This latest technology can remarkably decrease the contaminants in the environment and danger to humans due to the usage of poisonous solvents and chemicals. A wide range of bioactive compounds, primary and secondary metabolites which enables the plants and microbes to act as bio factories to fabricate NPs.

7.2 Nanoparticles and Its Classification

The NPs have less than 100 nm size in at least one dimension [1D] with its specific chemical (Parashar et al. 2009), physical, optical (Li and Du, 2003), biological (Schlorf et al. 2011), photoelectrochemical (Mohanpuria et al. 2008), mechanical, electronic, electrical (Krumov et al. 2009), and magnetic (Osterloh et al. 2005) properties which differ from their bulk materials because of the distinctive and noteworthy characteristics that are determined by its monodispersity, size, and shape (Fig. 7.1).

Deducing the size of NPs results in one-dimensional, two-dimensional, or three-dimensional confinements which determine the physical properties of NPs by accelerating the surface energy and surface atoms in numbers and decreasing the imperfections in fabricated particles (Fig. 7.1). Surface-enhanced Raman scattering

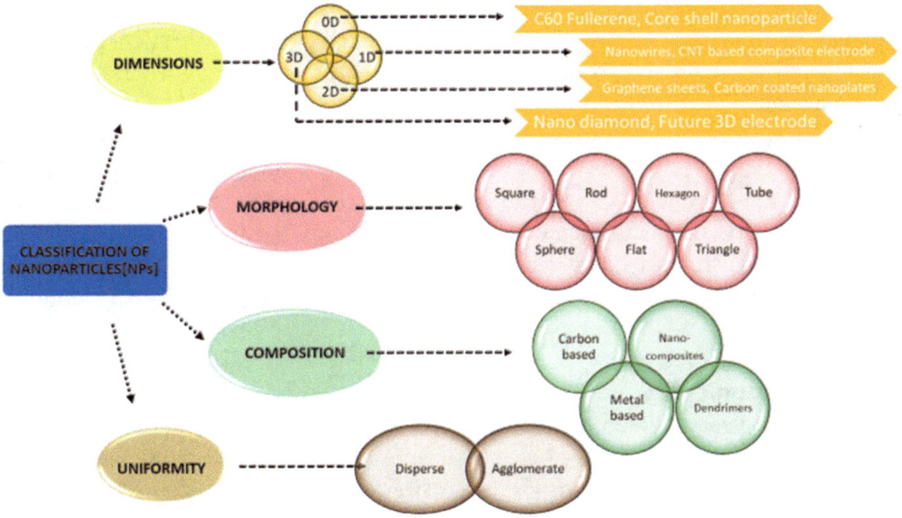

Fig. 7.1 Classification of nanoparticles (NPs)

(SERS), surface plasmon resonance (SPR), and quantum size result in semiconductor particles, and also the outstanding paramagnetic characteristics of magnetic NPs disclose that these NPs have been ideal for the construction of next-generation industrial and scientific instruments (Akbarzadeh et al. 2009; Narayanan and Sakthivel, 2010; Rong et al. 2010). The NP synthesis can have superior control over the particle distribution, purity, size, surface morphology, quantity, and quantity by following ecological friendly processes which have been assessed as a challenge (Fig. 7.1). In addition, special focus is granted to monodispersing and steady particle formation. Due to the size distribution and shape of the NPs, it is having its own property. Moreover, nanomaterials (NMs) act as a link among molecular structures and bulk scale materials (Daniel and Astruc, 2004).

Typical shapes can be developed such as rods, spheres, hexagons, pentagons, particles, triangles, cubes, and wires (Fig. 7.1).

7.3 Different Methods of Synthesis of NPs

7.3.1 Green Chemistry in Fabrication of Nanomaterials/Nanostructures

There are several naturally available bionanostructures or biotemplates like DNA, RNA, microtubules, flagella and S-layer of bacteria, viral capsid, proteins, and amino acids that showed many advantages when compared with chemically consistent carbon nanotubes or other solvents in the nanowire synthesis because of the presence of different functional groups on the surface intensively. Moreover, there is a probability to manipulate their chemistry of exposure with the help of genetic engineering (Zheng et al. 2004; Hinds et al. 2006; Wang et al. 2012). Following by millions of years of evolution, biological bodies have attained optimum development for the biogenic synthesis of NMs and conduct ideal biochemical pathways and also act as templates to make nanosized structures/materials when compared with man-made systems.

The nanostructures and NMs can be procured according to our need by biological heterogeneous techniques. They can be applied in a broad range of applications from removal of dye in wastewater to drug delivery (Sadighi et al. 2012). As it is not a fully inspected research zone of nanobiotechnology, the futuristic viewpoint of functional nanostructures and nanomaterials production will be mainly based on biotemplate-mediated systems, valuable and environmentally benign (Marchiol, 2012).

7.3.1.1 Biogenic NPs

In the current years, "green nanotechnology" has captivated much attention due to its probability to mitigate or eradicate toxic substances. Biogenic nanomaterial syntheses like nanoparticles, nanorods, nanowires, and nanotubes are assessed economically as a valuable substitute to produce nanomaterials for various applications. It proves brand new potentials and beneficial functionalities for the biological entities with immense applications in diverse field of science and technology (Krumov et al. 2009; Yang et al. 2009).

The green synthesis involves biological resources such as plant's extract, biodegradable waste, and microorganisms like bacteria, algae, fungi, virus, and yeast as a reducing agent, and it is referred to as "bionanofactories," "biological factories," "nanofactories," or "green factories." These are eco-friendly, cost-effective, effectual, structured exclusively, and naked, having metal uptake with high potential (Sarkar et al. 2012; Cai et al. 2011; Honary et al. 2012).

By coating biomolecules on the NP surface, the NP becomes biocompatible as compared to the NPs synthesized by physical and chemical methods (Mukherjee et al. 2001; Hakim et al. 2005) (Fig. 7.2).

Our main motive is to highlight the biogenic synthesis of NPs, as this approach is advantageous in term of cost effectiveness, alternative for toxic solvents/chemicals, eco-friendly, can manufacture NPs in a relatively less time period, sustainable, biocompatible, and can provide excellent control on characteristics and its immense application. Biogenic NPs can be synthesized by using bacteria, yeast, fungi, plants, etc. (Fig. 7.3). Furthermore, unicellular/multicellular organisms can produce the NPs intracellularly and extracellularly (Figs. 7.4 and 7.5). Several microbes have shown to manufacture NPs biologically with the help of NADPH-dependent reductase enzymes, which reduce the metallic salt solution into NPs via electron shuttle enzymatic metal reduction process (Bhawana and Fulekar, 2012).

Numerous microbes like bacteria (Stephen and Macnaughtont, 1999; Shivaji et al. 2011), yeast (Dameron et al. 1989; Kowshik et al. 2003), fungi (Syed et al. 2013), actinomycetes (Hassan et al. 2018), plant extracts (Akhtar et al. 2013), and biodegradable waste materials (Kanchi et al. 2014) acted as ideal progenitors to fabricate NPs with broad applications.

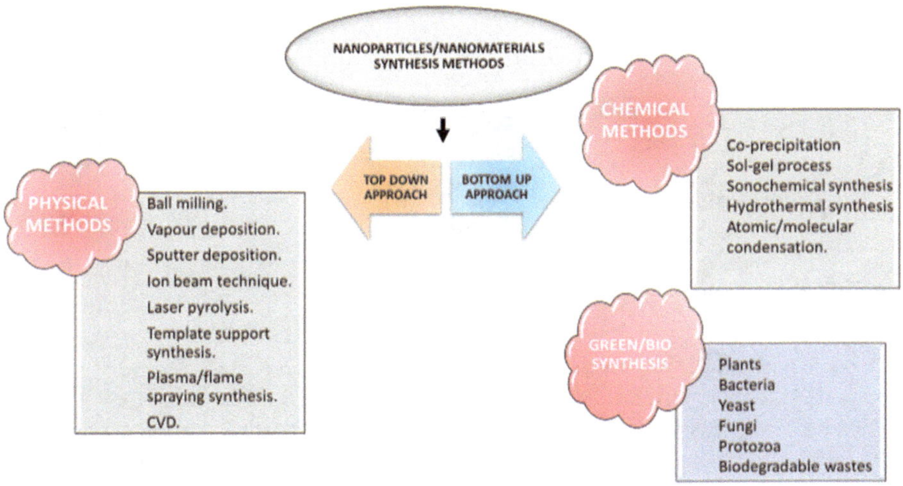

Fig. 7.2 The types of NP synthesis methods

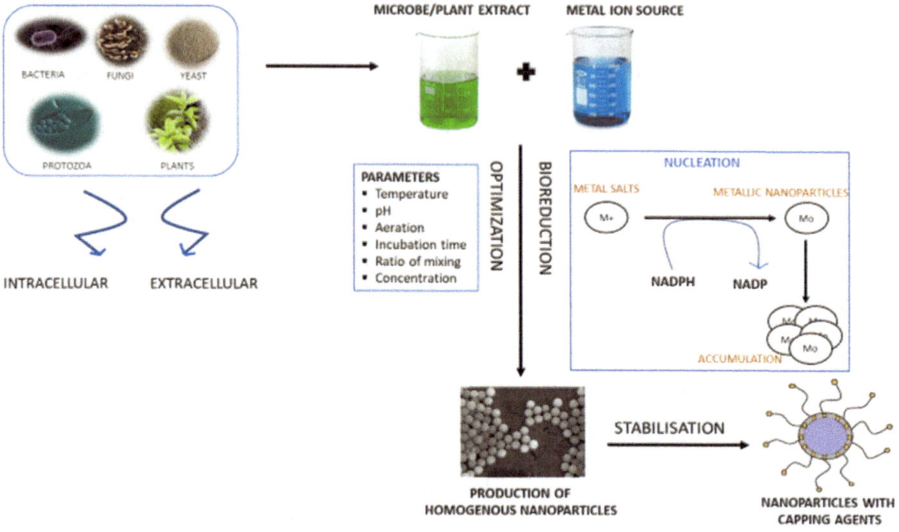

Fig. 7.3 The mechanism of biogenic NP synthesis

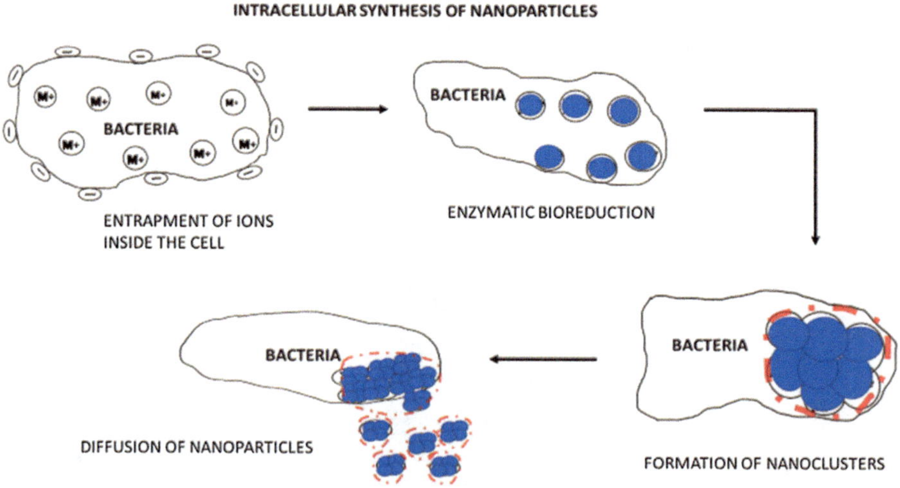

Fig. 7.4 The mechanism intracellular synthesis of NPs

7.3.1.1.1 Microbial Synthesis of Nanoparticles

To synthesize nanoparticles, microorganisms have been used because they are easy to handle, need less safety maintance, can grow in low cost medium such as cellulosic wastes or wastelands and can synthesize nanoparticles by adsorbing the metallic ions and reducing them to nanoparticles with the help of enzymes (Kumar et al.

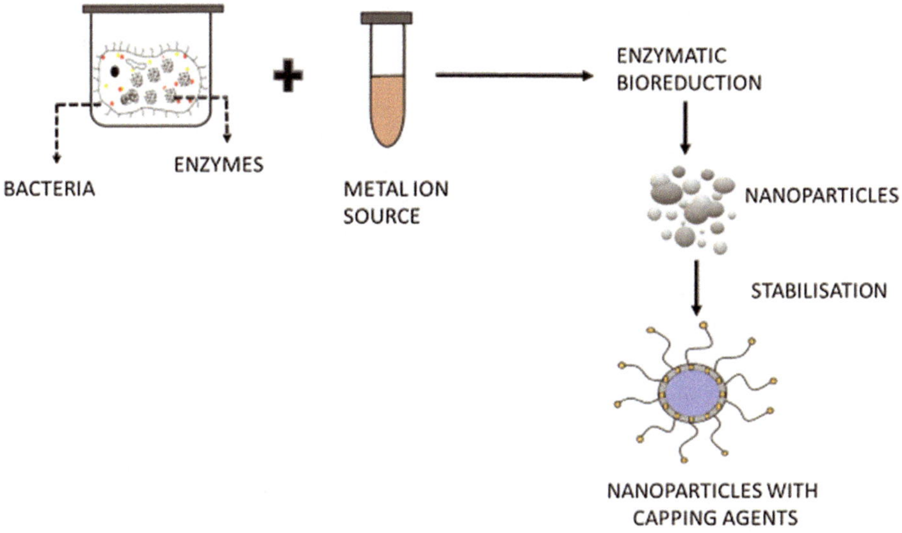

Fig. 7.5 The mechanism extracellular synthesis of NPs

2014; Luo et al. 2014). Depending on the locus, nanoparticle synthesis by microbes can be intracellular and extracellular. Mechanism behind intracellular microbial synthesis of NPs is electrostatic attraction (i.e., movement of certain ions into negatively charged cell wall in company of positively charged metals and get dispersed into cell wall whereas the toxic metals convert into nontoxic metal NPs by the enzymes in the cell wall) (Fig. 7.4). Mechanism of extracellular microbial synthesis of NPs entails enzymes produced by various prokaryotes or fungus such as hydroquinone or nitrate reductase which transforms the metallic ions to metallic NPs (Fig. 7.5). A related mechanism was identified in *Rhodomonas capsulate* for gold NP synthesis (Pacioni et al. 2015; Khandel and Shahi, 2016). While in case of transforming the metals into volatile matters, the microbes perform certain detoxification mechanisms like volatilization or metal binding and vacuole compartmentalization.

Under stress condition of metals, the microbes execute different mechanisms to remove the toxic heavy metals. Toxic metallic ions are reduced into nonmetallic ions by entailing active efflux reaction through cell membranes, and the nontoxic metal ions get cumulated inside the cells. Heavy metals such as silver, lead, gold, nickel, etc. are intruded through ion channels, ion pumps, endocytosis, lipid permeation, or transport using carrier (Issazade et al. 2013). Small ionic binding molecules like siderophores (chelating agent) chelate heavy metals by conducting absorption and aid in transport from the microbial cell (Kumar et al. 2017a, b). Molecules such as metallothioneins (MTs), a low molecular weight protein rich in cysteine (Gomathy and Sabarinathan, 2010), or glutathione, a derived peptide heavy metals (i.e., phytochelatin) (Xu et al. 2014), are extracted from *E. coli*, *Synechococcus* sp., *Cyanobacterium*, and *Pseudomonas putida* which perform metal detoxification.

7.3.1.2 Bacteria-Based NP Synthesis

In recent years, NP production from bacteria has extended comprehensively because of its vast applications. According to the studies, the reports have shown that many bacteria can effectively uptake and reduce the metallic ions as well as facilitate either reduction, biosorption, or oxidation of metal ions. The NP synthesis may be of intracellular or extracellular which depends on the bacterial species (Fig. 7.4 and 7.5). Adversely, there is a limit for accumulation of NPs in the bacteria. While crossing the limit, the NP load becomes toxic to the bacteria (Deepak et al. 2011). This was the first evidence of NP fabrication from bacteria. Pyramidal and hexagonal silver nanoparticles (AgNPs) up to 200 nm were produced by *Pseudomonas stutzeri* AG 259 isolated from silver mine (Table 7.1) (Klaus et al. 1999).

AgNPs were fabricated using many bacteria which include *E. coli*, *Staphylococcus aureus*, *Klebsiella pneumoniae*, *Bacillus subtilis*, *Candida albicans*, *Lactobacillus acidophilus*, and *Enterobacter cloacae* by using silver nitrate at the concentration of 10^{-3} M as a reducing agent. *K. pneumoniae*, *E. coli*, and *E. cloacae* were detected as effective among other organisms for AgNP production extracellularly (Fig. 7.5) (Minaeian et al. 2008).

Klebsiella pneumoniae produced selenium nanoparticles (SeNPs) with average size of 245 nm using selenium chloride, and the SeNPs were procured from the bacteria by sterilizing it for 20 min at 17 psi or 121 °C (Table 7.1) (Fesharaki et al. 2010).

Sunkar and Nachiyar (2012) reported about the ability of *Bacillus* species to lessen silver and fabricated NPs extensively in the range of 10–20 nm in size.

As compared to the bulk gold, gold nanoparticles (AuNPs) show high chemical reactivity. Correa-Lianten et al. (2013) fabricated AuNPs by *Geobacillus* sp. strain ID17, a thermophilic bacterium and exposed to Au^{3+} ions. It was identified and isolated from Deception Island, Antarctica. This reaction was mediated by enzymes and NADP as cofactor (Table 7.1). Cadmium sulfide nanoparticles (CdSNPs) in the range of 12 nm formed after the reduction of cadmium sulfate solution by *Serratia nematodiphila* which was collected and isolated from the effluent of a chemical company. The CdSNPs exhibited antibacterial property against *Klebsiella planticola* and *Bacillus subtilis* (Table 7.1) (Malarkodi et al. 2013).

7.3.1.3 Fungi-Based NP Synthesis

Filamentous fungi have exclusive competency over other microbes like algae and bacteria, as they are highly resistant to metals and have bioaccumulation proficiency. These are favorable in the viability count, biomass handling, and scale-up and downstream processing. They have discharged the enzymes extracellularly making them easily suitable for large-scale production (Fig. 7.5). The shape, size distribution, and biochemical composition of the NPs are regulated by the bioactive compounds.

Itajahia species, a *basidiomycetous* fungus, acts as a stabilizing polymer. Rod-shaped ferrous sulfide nanoparticles (FeS NPs) in the size of 200 nm were synthesized using a polymer of *Itajahia* species in addition to $FeSO_4$ solution and exposed to stirring under nitrogen atmosphere. Surprisingly, it devalued lindane

Table 7.1 Different source-based biogenic nanoparticles

Type of nanoparticle	Source	Morphology	Uses/property	References
Bacteria				
AgNPs	*Pseudomonas stutzeri* AG 259	Pyramidal and hexagonal crystals/up to 200 nm	Thin-film and surface coating technology	Klaus et al. (1999)
SeNPs	*Klebsiella pneumoniae*	Average size of ~ 245 nm	High biological activity and low toxicity	Fesharaki et al. (2010)
AuNPs	*Geobacillus* sp. strain ID17	Quasi-hexagonal/10–20 nm	Drug delivery, gene transfer, bioprobes	Correa-Lianten et al. (2013)
CdSNPs	*Serratia nematodiphila*	Spherical/12 nm	Antibacterial property against *Klebsiella planticola* and *Bacillus subtilis*	Malarkodi et al. (2013)
Fungi				
FeSNPs	*Itajahia* species	Rod/200 nm	Lindane (γ-hexachlorocyclohexane), an organochlorine pesticide	Paknikar et al. (2005)
Algae				
AuNPs	*Spirulina platensis*	Stable spherical/average size of ~ 5 nm	Design and development of nanomedicine	Suganya et al. (2015)
Au NPs	*Ecklonia cava*	Spherical and triangular/30 ± 0.25 nm	Biocompatible with the human keratinocyte cell lines	Venkatesan et al. (2014)
AgNPs	*Ulva armoricana* sp.	–	Advanced material to prepare antimicrobial compounds in the field of cosmetics and biomedicine	Massironi et al. (2019)
Yeast				
AgNPs	Yeast strain MKY3	2–5 nm	Silver-tolerant	Kowshik et al. (2003)
Ag/ AgCl-NPs	Yeasts were obtained from the gut of termites from the species *Cornitermes cumulans*	Circular/range of 2–10 nm	High antimicrobial activity against Gram-positive *Staphylococcus aureus* and Gram-negative *Klebsiella pneumoniae*	Eugenio et al. (2016)

(continued)

Table 7.1 (continued)

Type of nanoparticle	Source	Morphology	Uses/property	References
Plants				
AuNPs	*Croton Caudatus Geisel*	Spherical/range of 20–50 nm	High free radical scavenging property	Kumar et al. (2019)
AgNPs	Extracts of neem, onion, and tomato	–	Efficient drug material against *Staphylococcus aureus*	Chand et al. (2019)
Biodegradable waste				
AuNPs	Mango peel	Small and monodispersed/range of 6.03 ± 2.77 to 18.01 ± 3.67 nm	No biological cytotoxicity on normal cells even at high concentration, drug delivery systems	Yang et al. (2014)
AgNPs	Grape wastes	Spherical/average size of 25 to 35 nm	Antibacterial activity against Gram-positive and Gram-negative bacteria	Xu et al. (2015)
AuNPs	Eggshell membrane [ESM]	Variable shapes and sizes	Applicable for catalysis and sensor development	Devi et al. (2012)
AgNPs	Peel extract of *Annona squamosa* [custard apple]	Irregular spherical/ average size – 35 ± 5 nm	–	Kumar et al. (2012)

(γ-hexachlorocyclohexane), an organochlorine pesticide and a persistent organic pollutant (Table 1) (Paknikar et al. 2005).

7.3.1.4 Algae-Based NP Synthesis

Algae, a photosynthetic eukaryote and an oxygenic aquatic microorganism, have the property of bioaccumulation of heavy metals inward and outward. This character manifests algae as an inexhaustible raw source for the synthesis of NPs (Castro et al. 2013). Numerous studies have opened the secret that biological molecules in the cell wall of various seaweeds act as catalyst in which the precursor metal salt source gets reduced into nucleated NPs (Mahdavi et al. 2013; Kumar et al. 2013), whereas huge amphiphilic biomolecules direct and govern the NPs growth (Stalin Dhas et al. 2012).

Spirulina platensis, a blue-green algae, produced AuNPs in which they have absorbed Au ions and manufactured AuNPs intracellularly especially assembled in the vacuoles by involving some bioactive molecules in the metabolism of fungal cells such as 3-glucan binding proteins, glyceraldehyde-3-phosphate dehydrogenase, and ATPases (Table 7.1) (Fig. 7.4) (Suganya et al. 2015).

Fucoidan, a polysaccharide secreted from the cell wall of marine brown algae like Fucales, Dictyotales, Laminariales, Desmarestiales, and Chordariales, possessed whitening agent, antiaging compound (Fitton et al. 2007), and few pharmacological activities like antitumor (Li et al. 2008), antiviral, anti-peptic (Shibata et al. 2003; Yang et al. 2006), anti-proliferation (Teruya et al. 2007), anti-inflammation, and anticoagulant (Cumashi et al. 2007). Therefore, this has proven to be a valuable alternative to the hazardous old physical and chemical methods (Soisuwan et al. 2010).

Ecklonia cava, a marine brown alga, was exploited to produce spherical- and triangular-shaped pure AuNPs in the range of 30 ± 0.25 nm at 80 °C within 1 minute and were employed with biomolecules that contain hydroxyl group, primary amine group, and few stabilizing functional groups. Notably, they were biocompatible with the human keratinocyte cell lines (Table 7.1) (Venkatesan et al. 2014).

A recent study reported that *Ulva armoricana* sp. green algae having ulvan which is a sulfated polysaccharide was investigated for the first time as it is recognized as a stabilizing and reducing agent for the fabrication of AgNPs. Ulvan would be a promising alternative for citric acid as it's a natural stabilizer for AgNPs. Besides, it will serve as an advanced material to prepare antimicrobial compounds in the field of cosmetics and biomedicine (Table 7.1) (Massironi et al. 2019).

7.3.1.5 Yeast-Based NP Synthesis

Yeast, a single-celled microorganism, possesses many advantages over bacteria due to their mass fabrication of NPs. It is very easy to handle in laboratory circumstances, and they synthesize various enzymes because of their rapid growth by consuming simple nutrients (Dameron et al. 1989). During log phase of its growth cycle, a yeast strain MKY3 resistant to silver in which the culture tolerated up to 0.8 mM of silver and synthesized AgNPs (silver chloride nanoparticles) in the range of 2–5 nm extracellularly in high quantity without any loss in the cell viability and recovered using uncomplicated downstream processing (Kowshik et al. 2003). Also, the fabrication of circular-shaped Ag/AgCl-NPs (silver/silver chloride nanoparticles) in the range of 2–10 nm assisted by yeast strains has been shown which has high resistivity against Gram-positive *Staphylococcus aureus* and Gram-negative *Klebsiella pneumoniae*. Hence, it is highly significant in biomedical field (Table 7.1) (Eugenio et al. 2016).

Both yeast magnetic bionanocomposite (YB-MNP) and magnetite nanoparticles (MNP) synthesized by using the biomass of yeast which procured from the ethanol industry. These magnetic materials can detach mixed used motor oil (MUMO), new motor oil (NMO), and Petroleum 28 °API (P28API) from water applying the ASTM F726–12 method (Debs et al. 2019).

7.3.1.6 Plant-Based NP Synthesis

It renders a one-step biofabrication process. Plants have high preferences to synthesize NPs as it is renewable, biocompatible, and free from toxicity, and it supplies natural stabilizing agents to the NPs readily.

Contrary to classical material science procedures which need nonaqueous solvents, high pH, and temperature, the biological approaches typically permit the reactions to begin at ambient pressure, temperature, and extremely low pH values. Studies have shown the important role of hydroxyl and carboxyl groups of proteins like functional proteins (Mms6 protein in magneto tactic bacteria), enzymes, and few intracellular polypeptides (photoheating), beside polysaccharides (starch) in the transition of some metal ions into their NPs (Fig. 7.4) (Faramarzi and Sadighi 2013). Extracts of plants and its parts like leaf, stem, latex, root, and seed have been utilized to fabricate NPs as it acts as reducing/capping agents (Yallappa et al. 2015).

Using single-pot green method, the spherical-shaped gold nanoparticles (AuNPs) in the range of 20–50 nm were fabricated in the first time using *Croton caudatus* Geisel extract by deducing chloroauric acid ($HAuCl_4$). The polyphenols, a phytochemical which is present in the extract, act as an ideal reducing agent for gold, and the AuNPs show high free radical scavenging property (Table 1) (Kumar et al. 2019).

Silver nanoparticles (AgNPs) were synthesized using the neem extract (NE) and also the combinational extracts of Neem, Onion, and Tomato (NOT) at pH of 5, 7, and 9. The phytochemicals such as flavonoid and terpenoid were as reducing and capping agents which were proven by the FTIR results. Also, it has greater potential for drug delivery, cosmetic products, cryogenics, food storage, biosensor, and dental materials. The AgNPs synthesized using NOT extract employs as an efficient drug material against *Staphylococcus aureus*, a Gram-positive bacterium (Table 7.1) (Chand et al. 2019).

7.3.1.7 Biological Particle-Based NP Synthesis

Bioparticles like proteins, vitamins, virus, enzymes, and peptides could act as a source to fabricate NPs. Cowpea mosaic virus and cowpea chlorotic mottle virus have been exerted to mineralize the inorganic materials (Douglas and Young, 1998;

Douglas et al. 2002). Crystalline and sulfide nanowires have been mineralized by tobacco mosaic virus (TMV) (Shenton et al. 1999). Peptides are efficient in nucleating the nanocrystal growth, whereas it was authenticated from the integrated screen and identified on the M13 bacteriophage surface (Mao et al. 2003).

7.3.1.8 Biodegradable Waste-Based NP Synthesis

Our environment acts as a treasure of waste substances especially food wastes. However, it can be exploited for the different NP production, as its cell wall has various organic compounds like polysaccharides, dietary fibers, carotenoids, phenols, flavonoids, vitamins, and essential oils (Heim et al. 2002; Kim et al. 2012) which function as templates.

In a study, extract of mango peel has produced monodispersed AuNPs in the range of 6.03 ± 2.77 to 18.01 ± 3.67 nm in size. These were examined for toxicity on human's fetal lung fibroblast cells and on African green monkey's kidney cells by treating it for 24 h at various concentrations including 0, 20, 40, 80, and 160 μg mL^{-1} and also showed no toxicity even at high concentration (160 μg ml^{-1}) of mango peel extract (Yang et al. 2014).

Wine industries generated huge amount of grape waste which is actually a raw source of biologically organic molecules and reduced the metals into spherical-shaped AgNPs with average size of 25 to 35 nm and exhibited antibacterial activity for Gram-positive and Gram-negative bacteria (Xu et al. 2015) (Table 7.1).

The eggshell membrane (ESM) of chicken is one of the nature's bonus, which is not edible and used for the fabrication of AuNPs with fluorescence via one-step process under optimum conditions (Devi et al. 2012). It has amino acids which include alanine, glycine, and uronic acid and glycoproteins like collagen (Arias et al. 1991) which employed as a template for the production of AuNPs (Devi et al. 2019).

Custard apple or *Annona squamosa* belongs to Annonaceae family which has anticancer, antimicrobial, medicinal, and insecticidal properties (Dwivedi and Gopal, 2010; Madhumitha et al. 2012). It is edible, and the peels were considered as waste; however, the peels consist of hydrophilic hydroxyl and ketone groups which actually account for the transition of silver ions into AgNPs, and these groups offer stability by establishing thin layer on the surface of NPs (Kumar et al. 2012) (Table 7.1).

7.4 Factors that Impact the Synthesis of Nanoparticles

The synthesis of nanoparticles is influenced by various physiochemical parameters. Factors that have been reported for green synthesis of nanoparticles were size, shape, and reaction rate. Control of shape and size are the major pivotal challenges for the biosynthesis of nanoparticles.

7.4.1 Impact of pH

The solution's pH is one of the deciding factors for nanoparticle formation of diverse size, shape, and bio-reduction rate (Armendariz et al. 2004). Even in various plant extracts taken from different parts of the same plant may have various pH values. Hence, optimization is indispensable to obtain the ideal NP synthesis. The effect of pH in the range of 1–11 has been shown in the AgNP synthesis from the bark extract and powder of *Cinnamomum zeylanicum* (Sathishkumar et al. 2009). At basic/high pH, high dispersed tiny spherical nanoparticles were found in large quantity because the Ag (I) complexes in high numbers can bind with plant extract and nucleate individually from many NPs with small diameter, whereas at acidic/low pH, big ellipsoidal nanoparticles were found in small quantity.

7.4.2 Impact of Temperature

Temperature also plays a prominent role that determines the size, rate of reaction, and shape. AgNPs were synthesized from the bark extract of *Pinus eldarica* and observed the decline in size, whereas the sharpness increased in absorption peak with the increased temperature which includes 25 °C, 50 °C, 100 °C, and 150 °C (Iravani and Zolfaghari, 2013). In a study, the synthesis of NPs gets enhanced with the increase in temperature and reaction rate (Dwivedi and Gopal, 2010; Philip 2009).

In a comparative study, synthesis of bimetallic Au-Ag NPs involved the reduction of gold and silver ions by the leaf extract of *Anacardium occidentale* from lower to higher temperature to attain the optimum conditions for the synthesis. More extract was needed at lower temperature for the stable NP synthesis. At 100 °C, 0.6 ml of extract was needed for the synthesis, whereas at 27 °C, 2.5 ml of extract was needed for the synthesis. Reduction takes place at high temperatures which lead to the synthesis of large-sized stable NPs (Sheny et al. 2011).

7.4.3 Impact of Plant Extract Concentration

Depending on the concentration of plant extract, the rate of reaction, the shape, and the size of NPs can be determined. The tansy fruit extract (*Tanacetum vulgare*) with different concentrations such as 0.5 mL, 1.0 mL, 1.8 mL, 2.8 mL, 3.8 mL, and 4.8 mL as well as *Chenopodium album* leaf extract was used for the AgNP and AuNP synthesis and observed the rise in absorption peaks with decreased particle size while increasing the extract concentration (Dwivedi and Gopal, 2010; Dubey et al. 2010). In the synthesis of AgNPs by using the leaf extract of *Coleus amboinicus* with multiple concentrations like 0.5 mL, 2 mL, and 4 mL, the particle size and the surface plasmon resonance (SPR) bands were declined with the increment in extract concentration, and it was proved by HR-TEM analysis (Narayanan and Sakthivel, 2011).

7.4.4 Impact of Metal Ion Concentration

To determine the size, shape, and reduction rate of NPs, metal ion concentration also plays a pivotal role. AgNPs and AuNPs were synthesized using the extract of tansy fruit with diverse concentrations such as 1–3 mM. They reported that the absorption peak was broadened while increasing the concentration of silver ions from 1 mM to 2 mM. Larger size of AgNPs was found at higher concentration of metal ion. By comparing AuNPs and AgNPs at high metal ion concentration, the AuNPs had larger size shown by TEM images (Dubey et al. 2010). Similarly, Dwivedi and Gopal in 2010 synthesized AgNPs and AuNPs from the leaves of *Chenopodium album*. They have noticed that the particle size and absorbance peak have increased with increase in metal ion concentration, whereas at low metal ion concentration, the synthesis rate as well as particle size and absorbance peak was observed to be slow.

7.4.5 Impact of Reaction Time

The virtue of biogenic nanoparticles synthesis also relies on the reaction time/incubation period which greatly holds the yield, size, shape, stability, and optimum synthesis of NPs. AgNPs were synthesized with 5 h of incubation period from the leaf extract of *Capsicum annuum* which were in spherical, 10 ± 2 nm, and the diffraction ring evinced that the NPs were polycrystalline whereas the size of the NPs got increased to 25 ± 3 nm and 40 ± 5 nm at 9 h and 13 h of incubation time, respectively (Li et al. 2007). In 2010, Dubey et al. synthesized AgNPs and AuNPs from the extract of tansy fruit by giving starting incubation time of 10 min and then observed the increased sharpness in the absorption peak while increasing the incubation period for both AgNPs and AuNPs. To fulfil the nucleation process and to stabilize the NPs, Veerasamy et al. (2011) have reported that 60 min is the ideal incubation period for the extract of mango leaf to synthesis AgNPs.

7.4.6 Capping Agents

Capping agents has a salient role in the synthesis of NPs. Both Gittins et al. (2000) and Liu et al. (2005) reported that it was used to stabilize and functionalize the NPs. Moreover, it imparts functional properties by governing the size and morphology and prevents aggregation by protecting the surface. Because aggregation reduces the interfacial free energy, particle reactivity, and specific surface area, hence, it is extremely essential to strengthen the stability of the NPs for its overall life span, storage, and transportation.

To alter the NPs with desired size and shape, several surfactants were used. The commercial surfactants were onerous to remove and degrade, altering the NPs physically and chemically, and occupy vital mass fraction (> 50%) of nanoparticle system as well as dangerous to the environment (He and Zhao 2007; Stubbs and Gilman

2007). While looking at the limitations of these chemicals, there is an immediate need to employ biocompatible (i.e., hydrophilic, nontoxic, and non-immunogenic), eco-friendly stabilizing agents/green capping agents such as vitamins (B, C, D, K), enzymes, polysaccharides, citric acid, silica, and phytochemicals including polyphenols and biodegradable polymers like chitosan, cellulose, polyethylene glycol (PEG), and polylactides (PLA). In this chapter, some of them have been discussed with their potent role.

7.4.6.1 Green Capping Agents

It has predominant importance with properties such as renewable, efficient, effective, nontoxic, and inexpensive and can be applicable for diverse NPs. For instance, amino acids can act as stabilizing agents as well as reducing agents. Currently, it is grabbing an increased attention amidst the scientific community.

7.4.6.2 Polysaccharides

Polysaccharides belong to the class of linear/branched polymeric carbohydrates whose molecules consist of repetition units of either monosaccharide (glucose, fructose, galactose) or disaccharide (sucrose, lactose) connected by glycosidic bonds. As it is cheap, stable, harmless, hydrophilic, and biodegradable, it is one of the best capping agents for NPs. Toxic solvents can be eliminated by using water as a solvent to synthesis NPs (Akhlaghi et al. 2013; Duan et al. 2015). In addition, it speeds up the rate of sol-gel process due to their catalytic reaction (Boury and Plumejeau, 2015). AuNPs which are spherical, ~15 nm in size, were synthesized using water as solvent and organic honey as reducing/stabilizing agents. In honey, the proteins act as capping agents, whereas the fructose acts as reducing agent (Philip, 2009).

Dextran is a polycationic soluble carbohydrate which consists of more glucose molecules with varied lengths of chains. It is nontoxic, hydrophilic, efficient, and utilized as capping agent for several metal NPs (Virkutyte and Varma, 2011).

7.4.6.3 Carboxymethyl Cellulose (CMC)

It is one of the most available polysaccharide polymers, biocompatible and renewable, and can be altered facile into sodium carboxymethyl cellulose (CMC) by exchanging the native group (CH_2OH) with carboxymethyl group ($-CH_2-COOH$) in the glucose unit. Nanomaterials stabilized with CMC are persistently acquiring significance due to their distinctive properties (e.g., highly sensitive, low toxicity) in many technologies like water and land remediation and biological and medical applications such as biolabeling, antimicrobial coatings, biomedical devices, and food packaging (He et al. 2007, 2009). CMC is not capable to synthesize PdNPs, PtNPs, and AuNPs because at room temperature, noble metals do not chelate with carboxyl groups (Nadagouda and Varma, 2007).

7.4.6.4 Polyethylene Glycol (PEG)

It has gained much recognition as it is nontoxic, hydrophilic, and inexpensive. Its character can be changed by tuning the molecular weight (Ma et al. 2008; Chen et al. 2005). In 2008, Nadagouda and Varma synthesized metal NPs like Pt, Ag, and

Fe and stabilized simultaneously with PEG by using water as solvent. At room temperature, they mixed suitable aqueous metal salt solution with PEG by assisted microwave method at 280 psi for 1 h at 100 °C. Such technique does not need reducing agents/surfactants, so that it did not create any hazardous wastes. This can be suited for diverse applications.

7.5 Applications

7.5.1 Agriculture

Nowadays a number of researchers use nanotechnology as a tool to eradicate the harmful pathogens and pest from the agriculture. In the scenario, silver nanoparticles synthesized by Abdelmalek and Salaheldin (2016) act against *Penicillium digitatum*, *Alternaria citri*, and *Alternaria alternata* which are the citrus phytopathogenic fungi. Zinc oxide nanoparticles were observed to act against *Fusarium oxysporum* and *Aspergillus niger* (Patra et al. 2012). In Central India, *F. oxysporum*, *F. equiseti*, and *F. culmorum*, which were the most harmful fungi for crops, were eradicated by the nanoparticles from copper (Bramhanwade et al. 2016). In addition to that, 91% of *Psilocybe cubensis* and 69% of *Pseudomonas syringae* pv. were reduced by TiO_2 (Cui et al. 2009). Furthermore, the other significant work by Mishra S and team was the silver nanoparticles from *Serratia* sp. BHU-S4, an agro-supporting bacteria, controlled the most pioneer disease in wheat called spot blotch disease caused by the *B. sorokiniana* (Mishra et al. 2014).

7.5.1.1 Significance of Nanotechnology in Pest Management

Nanotechnology supports the plants to not only act against the pathogens but also to give the strong fight to the pests by one of the products called "nanocapsules." Nanocapsules insert the active substance via cuticles and tissues and release them in a regular manner so it is also called as a "magic bullets" (Perez-de-Luque and Rubiales 2009). On the other hand, 3 nm mesoporous silica nanoparticle is used as a carrier by Torney and team to transport DNA and chemicals into plant cells and then trigger it to fix the particles via cell walls without any deadly effects. This method is successfully established in corn and tobacco plants (Torney et al. 2007).

7.5.1.2 Stress Reduction

7.5.1.2.1 Application of Silicon Nanoparticles to Eradicate Drought and Salinity Stress in Plants

Both salinity and drought had negative effect on the growth of plants by reducing its growth of roots (Gupta and Huang, 2014). By expanding the area of the root, the absorption of the diffusible ions can be facilitated (Barber 1995), and this expansion of the root could be achieved by silicon NP (SiNP) which improves root growth, biomass of shoot, and augmented water uptake even in drought condition and improves root activity (Hattori et al. 2008; Lee et al. 2010; Ahmed et al. 2011; Chen and Yada 2011; Hameed et al. 2013; Kim et al. 2014). This incredible property of

SiNP is achieved by adjusting the levels of solutes such as carbohydrates (Ming et al. 2012), free amino acids and total soluble sugars (Sonobe et al. 2010; Hajiboland et al. 2016) polyols, phenolics (Hashemi et al. 2010), glycine, betaine (Torabi et al. 2015), and proline (Lee et al. 2010; Yin et al. 2013); all these things reduced the osmotic shock formed by NaCl ionic toxicity.

7.5.1.2.2 Application of Si Nanoparticles to Eradicate Disease in Plants

The amorphous silica precipitation is one of the mechanical barriers to the plants (Fauteux et al. 2005). Potential physical obstruction postulation depends upon the Si type deposited in leaves (Fawe et al. 2001). Universally it is accepted that the plant cell wall and apoplast contain polymerized Si which prevents the penetration of pathogens (Table 7.2) (Fleck et al. 2010) and hence prevents plants from harmful infections.

Table 7.2 Biogenic nanoparticles and its applications

Type of nanoparticle	Applications	References
Polymerized Si NPs	Prevents the penetration of pathogens	Fleck et al. (2010)
Pd-on-Au bimetallic nanoparticle	Used as catalyst to treat water pollutant 'trichloroethene'	Nutt et al. (2005)
Rod-shaped Fe2O3 NPs using *Aloe vera* leaf extract	Used for arsenic(V) remediation and having high sorption capacity of arsenic(V)	Mukherjee et al. (2016)
Magnetic iron oxide (Fe_3O_4) nanoparticles from tea waste	Eradication of As [arsenic].	Lunge et al. (2014)
Iron-oxide nanowires	New adsorbent for the removal of arsenic from water	Andjelkovic et al. (2017)
Fe_3O_4 nanocomposite-modified cells of *Yarrowia lipolytica* (NCIM 3589 and NCIM 3590)	Removal of hexavalent chromium [Cr(VI)]	Rao et al. (2013)
Zero-valent iron nanoparticles (ZVNI) from the leaf extracts of *Eucalyptus globules*	High adsorption capacity of hexavalent chromium (Cr(VI))	Madhavi et al. (2013)
Magnetic inverse spinel iron oxide nanoparticles (MISFNPs)	Capable of recycling and removal of heavy metals without loss of its stability	Lingamdinne et al. (2017)
CdSNPs from *P. aeruginosa* JP-11	Used for Cd removal from simulated wastewater	Raj et al. (2016)
Iron oxide nanoparticles	Adsorbent for cadmium ion removal	Ehrampoush et al. (2015)
Pd NPs (palladium) and MnOx NPs (manganese oxides)	Used for continuous elimination of pollutants from pharmaceutical industry	Forrez et al. (2011)
MnOx NPs	Degradation of ciprofloxacin	Tu et al. (2014)
Pt nanocatalyst	Biocatalysts for removal of ciprofloxacin, sulfamethoxazole, ibuprofen, and 17β-estradiol	Martins et al. (2017)

7.5.2 Wastewater Treatment

7.5.2.1 Biogenic NPs for Heavy Metal Removal

For the last few decades, heavy metals have been used enormously which results in damaging terrestrial and aquatic environments, drainages, and industrial and domestic effluents and mostly affects the agricultural cultivable lands, though heavy metals are even at low concentration because of their high bioaccumulation. Heavy metals include chromium, cadmium, mercury, nickel, lead, and arsenic (Chipasa, 2003).

Comparatively, arsenic (As) is considered to be an extremely hazardous pollutant. This heavy metal could be eradicated with the help of biologically synthesized rod-shaped Fe2O3 nanoparticles from *Aloe vera* leaf extract which has polysaccharides and anthraquinones and enables aqueous phase sequestration. At 20 °C, the maximum adsorption capacity was calculated to be 38.47 mg/g (Table 7.2) (Mukherjee et al. 2016). For As(V) and As(III), it is 153.8 and 188.69 mg/g, respectively (Lunge et al. 2014). Biologically synthesized nanoparticles were encapsulated with chitosan to prepare magnetic organic-nano-Fe hybrid with a ratio of 50:50. Recycling study also showed that prepared nanoparticles could be recycled up to four times (Table 7.2) (Andjelkovic et al. 2017) which extracted iron oxide nanowires from chemoautotrophic bacteria *M. ferrooxydans* on a bacterial biofilm and applied it for the removal of trivalent and pentavalent As. As removal is pH sensitive with As(III) having a high degree of removal at pH 6–10 and As(V) have high degree of removal at acidic pH 3. The maximum adsorption capacities for As(V) and As(III) were estimated to be 48.06 and 104.53 mg/g, respectively.

In a study by Rao et al. (2013), *P. granatum* extract fabricated a biocompatible FeNPs and modified by yeast *Y. lipolytica* and then used it for the eradication of Cr(VI) in aqueous solutions. With respect to contact time, pH, and temperature, the extent of removal was decreased with increasing temperature, and maximum removal was estimated at 60 min of contact time at pH 2. The maximum adsorption capacities for NCIM 3590 and NCIM 3589 were found to be 156.3 and 125 mg/1, respectively (Table 7.2). Madhavi et al. (2013) synthesized zero-valent iron nanoparticles of size 50–80 nm from the leaf extract of *Eucalyptus globulus* which results in the successful separation of Cr(VI) up to 98.1% in 30 min contact time (Table 7.2).

Xiao et al. (2016) prepared iron nanoparticles from 15 different plant extracts to remove Cr(VI) with a removal capacity of 0.5 g/g and average particle size of 13.7 ± 5.0 nm. *S. jambos* was also found to be potential in separating Cr(VI) with removal capacity of 0.69 g/g. Lingamdinne et al. (2017) performed biogenic reductive preparation of magnetic iron nanoparticles from *Cnidium monnieri* seed extract for the removal of Pb(II) and Cr(III) from aqueous matrices with maximum adsorption capacity of 102.3 and 105.6 mg/g for Cr(III) and Pb(II), respectively (Table 7.2).

Raj et al. (2016) prepared biogenic CdS nanoparticles from *P. aeruginosa* JP-11 and used them for Cd removal from simulated wastewater of particle sizes ranging from 20 to 40 nm in size. It reveals that these NPs removed 88.66% of Cd in 48 h contact time (Table 7.2). Ehrampoush et al. (2015) synthesized iron nanoparticles from tangerine peel extract by aqueous phase and estimated to remove Cd(II) up to 91% removal at pH 4.0 and 90 min contact time at equilibrium state (Table 7.2).

Jain et al. (2015) produced biogenic selenite nanoparticles for Zn(II) removal by incubating Se with anaerobic granular sludge which attains this state at 4 h contact time. Interestingly, the SeNPs effectively sequestrated Zn (II). Kandasamy (2017) synthesized the FeNPs from *S. thermolineatus* and employed them to separate Cu ions from the effluents of pigment industries with and efficiency of 85%. Wang et al. (2018) has fabricated selenite nanoparticles from bacterial PS for the removal of Hg from groundwater which shows 81.2% efficacy.

Srivastava et al. (2015) synthesized flower-shaped MgONPs using acacia gum and applied to remove several divalent heavy metals such as Pb(II), Co(II), Ni(II), Cu(II), Cd(II), Zn(II), and Mn(II) with varying concentrations of acacia gum and named as AGM-20 and AGM-10. The removal trend for AGM-20 and AGM-10 was estimated to be Zn(II) < Mn(II) < Pb(II) < Co(II) < Cu(II) < Cd(II) < Ni(II) and Mn(II) < Co(II) < Zn(II) < Cu(II) < Cd(II) < Pb(II) < Ni(II), respectively. Zhou et al. (2015) synthesized the biogenic manganese oxide NPs from *P. putida* MnB1 and applied for the removal of divalent heavy metals such as Cd(II), Pb(II), and Zn(II). The removal of Pb(II) was higher when compared to Zn(II) and Cd(II).

7.5.2.2 Organic Pollutant Removal

An organic pollutant is the one which is comprised of nitrogen, sulfur, halogens, phosphorous, insecticides, synthetic dyes, pharmaceuticals, pesticides, aromatic hydrocarbons, phenols, and halogenated hydrocarbons (Lapworth et al. 2012). Moreover, the persistent organic pollutants are highly toxic to aquatic environment when discharged from industrial effluent to agricultural fields (Yadav et al. 2015). There are many more products such as nanofilms, nanocatalysts, and nanoadsorbents which have been produced by nano-enabled technology (Gautam et al. 2015), and some of them create carcinogenic and allergic effects. Khan et al. (2016) synthesized silver nanoparticles using aqueous extract of *C. japonicum* and performed fast removal of bromophenol blue through their catalytic reduction with size 8–10 nm.

Huang et al. (2014) fabricated iron nanoparticles from tea extract and subjected to remove carcinogenic dye malachite green, which resulted in 74.5% removal. Weng et al. (2013) synthesized iron nanoparticles from aqueous green tea extract and determined the catalytic reactivity with malachite green which resulted in 91% removal with 9 h of contact time. Devi et al. (2016) synthesized silver nanoparticles from *A. agallocha* leaves by aqueous phase to remove Victoria blue which resulted in in 99.46% removal in 1–2 h treatment time.

Shahwan et al. (2011) synthesized biogenic FeNPs using green tea and employed them for the removal of methyl orange and methylene blue from wastewater. Smuleac et al. (2011) synthesized iron and iron/palladium bimetallic nanoparticles of sizes 20 to 30 nm from tea extract and applied them for the acute elimination of more toxic pollutants such as trichloroethane (TCE) from water. Zaheer et al. (2016) synthesized silver nanocatalysts from mint leaves and applied them for the removal of Congo red which was then capped and stabilized by SDS and CTAB. Vanaamudan et al. (2016) synthesized biogenic silver nanoparticles from palm shell extract and capped with H_2O_2 for the removal of a Cu phthalocyanine-based dye and xanthene dye.

Bogireddy et al. (2016) synthesized AgNPs from *S. acuminata* fruit aqueous extract for the elimination of some organic dyes, viz., methylene blue, 4-nitrophenol, Direct blue 24 methyl orange, and phenol red from consumable water. Nadaf and Kanase (2016) fabricated biogenic gold nanoparticles from *B. marisflavi* to remove Congo red and methylene blue from wastewaters. Qu et al. (2017) developed biocompatible gold NPs from the supernatant of *Aspergillus* sp. and employed it to remove aromatic pollutants such as 3-nitrophenol, 4-nitrophenol, and 2-nitrophenol. Shen et al. (2017) synthesized Au-NPs from *Aspergillus* sp. extract and used it for the removal of 4-nitrophenol. Wang et al. (2017) produced bioinspired manganese oxides nano-biocomposite using *Desmodesmus* sp. for the complete removal of bisphenol from wastewater. Zinatloo-Ajabshir et al. (2018) synthesized biomodulated Dy2Ce2O7 nanostructures from *Vitis vinifera* juice and employed them for the catalytic removal of rhodamine B, methyl orange, and B naphthol. Beddow et al. (2014) demonstrated that silver NPs could reduce the rate of nitrification potential of *Nitrosococcus* sp. and *Nitrosomonas europaea*.

7.5.2.3 Radioactive and Inorganic Pollutants Removal

Radioactivity possessing pollutants such as radioisotopes and radionuclide might cause chronic and acute toxicity to the humankinds when exposed to aquatic environments from mines and effluents exposing from nuclear reactors (Gawande and Jenkins-Smith, 2001).

In the above perspective, Handley-Sidhu et al. developed a biogenic nanometered hydroxyapatite material (Bio-HAP) using NCIMB 40259 strain of *Serratia* sp. to separate strontium (Sr^{2+}) from simulated groundwater. Size and adsorption capacity of Bio-HAP were calculated to be 25 nm and 5 mg/g, respectively. In another work, they fabricated Bio-HAP nanomaterials from *Serratia* sp. and used it to remove the biohazardous radionuclides such as cobalt, uranium, strontium, and europium from artificial groundwater (Handley-Sidhu et al. 2014). At 400 °C, the Bio-HAP nanomaterials have the largest surface area of $115 m^2/g$.

They also synthesized Bio-HAP nanomaterials of sizes between 20 and 90 nm for the sequestration of Sr^{2+} and Co^{2+} from the aqueous phase (Handley-Sidhu et al. 2011). Choi et al. 2017 fabricated biogenic gold nanoparticles using *D. radiodurans* and used for the removal of 125I from effluent of nuclear reactor which size ranges from 40–60 nm. Wang et al. (2014) synthesized two different types of Fe nanoadsorbents from aqueous leaf extracts of green tea and eucalyptus and further studied their ability to remove nitrate which showed 59% and 41.4% removal of nitrate, respectively.

Paixão et al. (2017) synthesized Cu nanoparticles from *H. sabdariffa* flower extract and impregnated onto the surface of activated carbon. At 15 °C, it was calculated that the maximum adsorption capacity was 45.01 mg/g. Katata-Seru et al. (2017) synthesized two different iron nanoparticles from *M. oleifera* seed and leaf extracts and applied them for the nitrate removal from water samples which resulted in 85% and 26% removal for seed and leaf-based Fe nanoparticles, respectively.

Yong et al. (2004) formulated a nanopowder from *Serratia* sp. and used it for phosphate removal in water which resulted in decrease in the phosphate concentration from ~24 ppm to ~3 ppm. Cao et al. (2016) synthesized nanoparticles from *Eucalyptus* leaf extract and coated with a cationic surfactant and applied for phosphate removal which increased from 71.0% to 97.3% after coating.

7.5.2.4 Metallic and Nonorganic Pollutant Removal

Zhang (2003) demonstrated that the laboratory-synthesized nanoscale metallic iron has strong destroying activity against chlorinated and brominated methanes, trihalomethanes, chlorinated ethenes and benzenes, pesticides, and dyes. Nanoscale zero-valent iron could reduce both organic contaminants and inorganic contaminants such as anion nitrate which is reduced to NH_3 and perchlorate/chlorite which is reduced to arsenite, chloride, arsenate, and chromate powder and also removes some metals such as Pb, Ni, and Li.

Macaskie et al. (1997) isolated *A. Serratia* from a metal-contaminated site and provided with organic phosphate and inorganic phosphate as heavy metals. The cells also immobilized as a biofilm to the removal of uranyl ion from U-mine wastes.

7.5.2.5 Pharmaceutical Pollutant Removal

Nowadays, the effluent from the pharmaceutical industry produces a new area of micropollutants containing antibiotics, endocrine-disrupting compounds (EDCs), steroids, and hormones which cause adverse effects in living things and enter in to the water system (Kim et al. 2007). To eradicate this problem, there are a number of techniques used, in that nano-based techniques showed promising results (Malik et al. 2017). One research team has formulated Bio-Pd (bio-palladium) and BioMnOx (biogenic manganese oxides) nanoparticles for continuous elimination of pollutants from pharmaceutical industry using membrane bioreactors (MBR) at lab scale (Table 7.2) (Forrez et al. 2011).

In this contest, another unique nanoparticle named BioMnOx-MBR (via oxidative degradation) has strongly eliminated 14 out of 29 various types of micropollutants from the effluent, they were removed more than 90% of ibuprofen, naproxen, diuron, codeine (b), N-acetyl-sulfamethoxazole, exceed 80% of chlorophene, diclofenac, mecoprop, triclosan, go beyond 70% of clarithromycin, iohexol, more than 60% of iopromide, iomeprol and 52% of sulfamethoxazole. On the other hand, using bio-Pd through catalytic reduction, some other pollutants were removed, and the maximum up to 97% elimination was found from iomeprol, iohexol, and iopromide and 90% from diatrizoate. Likewise, BioMnOx synthesis from *P. putida* eliminates the large amount of micropollutants in marine environment (Furgal et al. 2015). Similarly, BioMnOx obtained from G7 species of *Pseudomonas* (using oxidation process) successfully done degradation of ciprofloxacin in aqueous phase (Table 7.2) (Tu et al. 2014). Recently, the wide range of pharmaceutical hazards were degraded using bioelectrochemical method, which was done by the combination process of biogenic Pd NPs, with chemically synthesized Fe_3O_4 and MnO_2 nanoparticles (Xu et al. 2018).

Further, about 94.85% elimination was observed for sulfamethoxazole and 17β-estradiol, whereas 70% was observed for ciprofloxacin using synthesized NPs. Significant reduction in diatrizoate by improving electrochemical reduction was achieved by the bio-Pd NPs which were entrenched to cathode in MEC (microbial electrolysis cell) and observed that dehalogenation reaction was achieved by elevated voltage given to the MEC and at – 0.8volt diatrizoate eradication was achieved in 2 h (Gusseme et al. 2012). Furthermore, ciprofloxacin, sulfamethoxazole, and 17β-estradiol were recently removed using the bio-Pt nanocatalyst synthesized from *D. vulgaris* (Martins et al. 2017). In addition to that, the combination of zero-valent Au NPs with bio-Pd NPs was used to treat diclofenac treated water, and in 24 h, a larger amount elimination of diclofenac was observed. However, another noteworthy work was done first by De Corte et al. (2011) and got succeeded. In this work, *S. oneidensis* synthesized Pd/Au bimetallic nanocatalyst was resulted in decontamination of diclofenac and trichloroethylene in water. Interestingly, both of the nanocatalysts such as bio-Au and monometallic bio-Pd individually were not working, but the combination showed remarkable activities and achieved 78% elimination of pollutants at 24 h. Henceforth, the crystalline nanostructure of Au-Pd was confirmed by the results of TEM and XRD.

7.5.3 Growth Promotion

Iron nanoparticles enhance the leaf and pod dry weight and improve yield of soybean, wheat, and peanut seedlings through FeNPs (Sheykhbaglou et al. 2010). CuNPs have caused the shoot and root growth of mung bean, whereas CsNPs increase the pod weight of peanuts (Prasad et al. 2012). By showing increasing sugar content and super oxide dismutase activity followed by proteomic analysis, it was confirmed that the uptake of copper and iron nanoparticles improved the yield in wheat through physiological impacts (Yasmeen et al. 2017).

Besides, the photosynthetic activity of cucumber and germination of spinach seeds were effectively increased due to TiO_2 NPs and positive effect on plants. Interestingly, chlorophyll formation in spinach was improved by 28% by rutile and 19% by TiO_2 (Zheng et al. 2005). On the other side, growth of spinach was enhanced by the processes of photoreduction of atmospheric N_2 to NH_3 by TiO_2 nanoparticles (Yang et al. 2007).

Carbon nanotube, a precious member of nanotechnology, plays a major role by increased uptake of water to enhance the plant growth and seed germination. In addition to that, it has pioneer place as nano-fertilizers to advance the agriculture sector. Furthermore, it increases the bioavailability of hazardous metals in agricultural crops, by inducing apoptosis through reactive oxygen species (ROS) generation, and drastically reduces the diversity in microbes used in agricultural fields (Vithanage et al. 2017).

7.5.4 Present Challenges in Agriculture and Nanotechnology-Based Solutions

Nowadays, NPs, liposome, and fluorescent dendrimers are used as a pesticide nanocarrier to control pest in the agricultural crops. Nanoscience delivers a number of innovative methods for disease management at molecular level which increases the yield of plants via nanofertilizers/nanoinsecticides, delivers efficient use of natural resource, improves the uptake of nutrient (Peters et al. 2016), maintains soil health, and reduces the amount of growth promoters (Baker et al. 2017).

The biosynthesized Ag nanoparticles at concentration 5 μg/mL could have the ability to fight against blight disease in tomato without disturbance of the soil health. Prevention of the infection establishment, increasing the host plant resistance by decreasing the biotic stress, and increasing the chlorophyll content and antioxidant levels could be done with the pre-treatment of Ag nanoparticles on tomato leaves.

Though cucumber plants are water crucial and adequate irrigation has to be performed in order to prevent the fall of fruits and flowers from the plant (Hashem et al. 2011; Sahin et al. 2015), the cucumber plants grown in a greenhouse condition with amorphous silica nanoparticles (SiNPs) showed the normal growth without any symptoms of irrigating water shortage up to 85%. The highest yield of fruit per plant was obtained at 85%. Silver nanoparticles with a rate of 200 mg/kg have the higher positive effect on yield characteristics and vegetation of the plant in comparison with the lower and higher concentrations. This results that silver nanoparticles play a major role in balancing the nutrient uptake by the plants by increased uptake of potassium, nitrogen, and silicon and decreased uptake of sodium in various parts of the plant.

$Mg(OH)_2$NP-treated *Z. mays* seeds showed the significant improvement in their germination rate at 500 ppm concentration (Jayarambabu et al. 2016). In addition to that, these biogenic $Mg(OH)_2$ NPs synthesized from the fungus showed the remarkable activity in the enhancement of percentage and for breaking the seed dormancy. Furthermore, nanoparticles inoculated with MS medium showed an enhanced growth in *Z. mays* seedlings in comparison with the other medium. Similarly, $Mg(OH)_2$NPs have the potential to enhance the growth of *Z. mays* plant in fields.

The chlorophyll fluorescence study also established that the nanoparticle-treated plant has high-performance index and low energy losses when compared to normal/untreated plants. Also, the roots and leaves of NP-treated plants showed high content of Mg in comparison with the salt-treated plants. These overall outcomes conclude that the treatment of $Mg(OH)_2$ NPs could be easily penetrated and translocated in various plant parts which result in increased seed germination percentage and plant growth promotion.

7.6 Conclusion

The approach of synthesizing nanoparticles by green nanotechnology possesses the high potential to become a strong leader among the industries. As the biogenesis of nanoparticles achieves more commercialization, the synthesis from bacteria, fungi, algae, biodegradable wastes, and plants becomes more prominent leading to generation of safe, secure, environment-friendly nanomaterials. However, several studies in pharmaceutical development still involve challenges with respect to the optimization of various factors such as optimum temperature, pH, stabilizers, etc. Knowing the applications of variety of nanomaterials in agriculture, pest management, stress reduction, wastewater treatment, cosmetics, etc., they can be considered to improve the quality of life of human beings and environment as well as promote ethical values in nanotechnology.

References

Abdelmalek GAM, Salaheldin TA (2016) Silver nanoparticles as a potent fungicide for Citrus phytopathogenic fungi. J Nanomed Res 3(5):00065

Ahmed M, Qadeer U, Aslam MA (2011) Silicon application and drought tolerance mechanism of sorghum. Afr J Agric Res 6:594–607

Akbarzadeh A, Zare D, Farhangi A, Mehrabi MR, Norouzian D (2009) Synthesis and characterization of gold nanoparticles by Tryptophane. Am J Appl Sci 6:691–695

Akhlaghi SP, Peng B, Yao Z, Tam KC (2013) Sustainable nanomaterials derived from polysaccharides and amphiphilic compounds. Soft Matter 9(33):7905–7918

Akhtar MS, Panwar J, Yun YS (2013) Biogenic synthesis of metallic nanoparticles by plant extracts. ACS Sustain Chem Eng 1(6):591–602

Andjelkovic I, Azari S, Erkelens M, Forward P, Lambert MF, Losic D (2017) Bacterial iron-oxide nanowires from biofilm waste as a new adsorbent for the removal of arsenic from water. RSC Adv 7:3941–3948

Anjum M, Miandad R, Waqas M, Gehany F, Barakat MA (2016) Remediation of wastewater using various nano-materials. Arab J Chem:1–23

Arias JL, Fernandez MS, Dennis JE, Caplan AI (1991) Collagens of the chicken eggshell membranes. Connect Tissue Res 26(1–2):37–45

Armendariz V, Herrera I, Jose-Yacaman M, Troiani H, Santiago P, Gardea-Torresdey JL (2004) Size controlled gold nanoparticle formation by Avena sativa biomass: use of plants in Nanobiotechnology. J Nanopart Res 6(4):377–382

Baker S, Volova T, Prudnikova SV, Satish S, Prasad MNN (2017) Nanoagroparticles emerging trends and future prospect in modern agriculture system. Environ Toxicol Pharmacol 53:10–17

Barber SA (1995) Soil nutrient bioavailability: a mechanistic approach. John Wiley & Sons

Beddow J, Stolpe B, Cole P, Lead JR, Sapp M, Lyons BP, Colbeck I, Whitby C (2014) Effects of engineered silver nanoparticles on the growth and activity of ecologically important microbes. Environ Microbiol Rep 6(5):448–458

Bhawana P, Fulekar M (2012) Nanotechnology: remediation technologies to clean up the environmental pollutants. Res J Chem Sci ISSN 2231:606X

Bogireddy NKR, Kumar Kiran HA, Mandal BK (2016) Biofabricated silver nanoparticles as green catalyst in the degradation of different textile dyes. J Environ Chem Eng 4:56–64

Boury B, Plumejeau S (2015) Metal oxides and polysaccharides: an efficient hybrid association for materials chemistry. Green Chem 17(1):72–88

Bramhanwade K, Shende S, Bonde S, Gade A, Rai M (2016) Fungicidal activity of cu nanoparticles against Fusarium causing crop diseases. Environ Chem Lett 14:229–235

Cai F, Li J, Suna J, Jia Y (2011) Biosynthesis of gold nanoparticles by biosorption using *Magnetospirillum gryphiswaldense* MSR-1. Chem Eng J 175:70–75

Cao D, Jin X, Gan L, Wang T, Chen Z (2016) Removal of phosphate using iron oxide nanoparticles synthesized by eucalyptus leaf extract in the presence of CTAB surfactant. Chemosphere 159:23–31

Castro L, Blázquez ML, Muñoz JA, González F, Ballester A (2013) Biological synthesis of metallic nanoparticles using algae. IET Nanobiotechnol 7(3):109–116

Chand K, Abro MI, Aftab U, Shah AH, Lakhan MN, Cao D, Mohamed AMA (2019) Green synthesis characterization and antimicrobial activity against Staphylococcus aureus of silver nanoparticles using extracts of neem, onion and tomato. RSC Adv 9(30):17002–17015

Chen H, Yada R (2011) Nanotechnologies in agriculture: new tools for sustainable development. Trends Food Sci Technol 22(11):585–594

Chen J, Spear SK, Huddleston JG, Rogers RD (2005) Polyethylene glycol and solutions of polyethylene glycol as green reaction media. Green Chem 7(2):64–82

Chipasa KB (2003) Accumulation and fate of selected heavy metals in a biological wastewater treatment system. Waste Manag 23:135–143

Choi MH, Jeong SW, Shim HE, Yun SJ, Mushtaq S, Choi DS, Jang BS, Yang JE, Choi YJ, Jeon J (2017) Efficient bioremediation of radioactive iodine using biogenic gold nanomaterial-containing radiation-resistant bacterium, *Deinococcus radiodurans* R1. Chem Commun 53:3937–3940

Correa-Llantén DN, Muñoz-Ibacache SA, Castro ME, Muñoz PA, Blamey JM (2013) Gold nanoparticles synthesized by *Geobacillus* sp strain ID17 a thermophilic bacterium isolated from Deception Island. Antarctica Microbial Cell Fact 12(1):75

Cui H, Zhang P, Gu W, Jiang J (2009) Application of anatase TiO 2 sol derived from peroxotitannic acid in crop diseases control and growth regulation. Nanotech Conf Expo:286–289

Cumashi A, Ushakova NA, Preobrazhenskaya ME, D'Incecco A, Piccoli A et al (2007) A comparative study of the anti-inflammatory, anticoagulant, antiangiogenic and antiadhesive activities of nine different fucoidans from brown seaweeds. Glycobiology 17:541–552

Dameron CT, Reese RN, Mehra RK, Kortan AR, Carroll PJ, Steigerwald ML, Brus LE, Winge DR (1989) Biosynthesis of cadmium sulphide quantum semiconductor crystallites. Nature 338:596–597

Daniel M-C, Astruc D (2004) Gold nanoparticles: assembly, supramolecular chemistry, quantum-size-related properties, and applications toward biology, catalysis, and nanotechnology. Chem Rev 104:293–346

De Corte S, Hennebel T, Fitts JP, Sabbe T, Bliznuk V, Verschuere S, van der Lelie D, Verstraete W, Boon N (2011) Biosupported bimetallic Pd–au nanocatalysts for dechlorination of environmental contaminants. Environ Sci Technol 45:8506–8513

de Lima R, Seabra AB, Duran N (2012) Silver nanoparticles: a brief review of cytotoxicity and genotoxicity of chemically and biogenically synthesized nanoparticles. J Appl Toxicol 32(11):867–879

Debs KB, Cardona DS, da Silva HD, Nassar NN, Carrilho EN, Haddad PS, Labuto G (2019) Oil spill cleanup employing magnetite nanoparticles and yeast-based magnetic bionanocomposite. J Environ Manag 230:405–412

Deepak V, Kalishwaralal K, Pandian SRK, Gurunathan S (2011) An insight into the bacterial biogenesis of silver nanoparticles, industrial production and scale-up. In: Rai M, Duran N (eds) Metal nanoparticles in microbiology. Springer, Berlin/Heidelberg, pp 17–35

Devi PS, Banerjee S, Chowdhury SR, Kumar GS (2012) Eggshell membrane: a natural bio template to synthesize fluorescent gold nanoparticles. RSC Adv 2(30):11578–11585

Devi TB, Begum S, Ahmaruzzaman M (2016) Photo-catalytic activity of Plasmonic ag@ AgCl nanoparticles (synthesized via a green route) for the effective degradation of Victoria blue B from aqueous phase. J Photochem Photobiol B Biol 160:260–270

Devi GK, Suruthi P, Veerakumar R, Vinoth S, Subbaiya R, Chozhavendhan S (2019) A review on metallic gold and silver nanoparticles. Res J Pharm Technol 12(2):935–943

Douglas T, Young M (1998) Host-guest encapsulation of materials by assembled virus protein cages. Nature 393(6681):152–155

Douglas T, Strable E, Willits D, Aitouchen A, Libera M et al (2002) Protein engineering of a viral cage for constrained nanomaterials synthesis. Adv Mater 14(6):415–418

Duan H, Wang D, Li Y (2015) Green chemistry for nanoparticle synthesis. Chem Soc Rev 44(16):5778–5792

Dubey SP, Lahtinen M, Sillanpää M (2010) Tansy fruit mediated greener synthesis of silver and gold nanoparticles. Process Biochem 45(7):1065–1071

Durán N, Seabra AB (2012a) Microbial syntheses of metallic sulphide nanoparticles: an overview. Curr Biotechnol 1:287–296

Durán N, Seabra AB (2012b) Metallic oxide nanoparticles: state of the art in biogenic syntheses and their mechanisms. Appl Microbiol Biotechnol 95:275–288

Dwivedi AD, Gopal K (2010) Biosynthesis of silver and gold nanoparticles using Chenopodium album leaf extract. Colloids Surf A Physicochem Eng Asp 369(1–3):27–33

Ehrampoush MH, Miria M, Salmani MH, Mahvi AH (2015) Cadmium removal from aqueous solution by green synthesis iron oxide nanoparticles with tangerine peel extract. J Environ Health Sci Eng 13:84b

Eugenio M, Müller N, Frasés S, Almeida-Paes R, Lima LMT, Lemgruber L, Sant'Anna C (2016) Yeast-derived biosynthesis of silver/silver chloride nanoparticles and their antiproliferative activity against bacteria. RSC Adv 6(12):9893–9904

Faramarzi MA, Sadighi A (2013) Insights into biogenic and chemical production of inorganic nanomaterials and nanostructures. Adv Colloid Interf Sci:189:1–189:18920

Fauteux F, Rémus-Borel W, Menzies JG, Bélanger RR (2005) Silicon and plant disease resistance against pathogenic fungi. FEMS Microbiol Lett 249:1–6

Fawe A, Menzies JG, Chérif M, Bélanger RR (2001) Silicon and disease resistance in dicotyledons Stud. Plant Sci 8:159–169

Fernández-Acero FJ, Jorge I, Calvo E, Vallejo I, Carbú M, Camafeita E, Garrido C, López JA, Jorrin J, Cantoral JM (2007) Proteomic analysis of phytopathogenic fungus Botrytis cinerea as a potential tool for identifying pathogenicity factors, therapeutic targets and for basic research. Arch Microbiol 187:207–215

Fesharaki PJ, Nazari P, Shakibaie M, Rezaie S, Banoee M, Abdollahi M, Shahverdi AR (2010) Biosynthesis of selenium nanoparticles using *Klebsiella pneumoniae* and their recovery by a simple sterilization process. Braz J Microbiol 41(2):461–466

Fitton JH, Irhimeh M, Falk N (2007) Macroalgal fucoidan extracts: a new opportunity for marine cosmetics. Cosmet Toiletr 122:55–64

Fleck AT, Nye T, Repenning C, Stahl F, Zahn M, Schenk MK (2010) Silicon enhances suberization and lignification in roots of rice (Oryza sativa). J Exp Bot 62(6):2001–2011

Forrez I, Carballa M, Fink G, Wick A, Hennebel T, Vanhaecke L, Ternes T, Boon N, Verstraete W (2011) Biogenic metals for the oxidative and reductive removal of pharmaceuticals, biocides and iodinated contrast media in a polishing membrane bioreactor. Water Res 45:1763–1773

Furgal KM, Meyer RL, Bester K (2015) Removing selected steroid hormones, biocides and pharmaceuticals from water by means of biogenic manganese oxide nanoparticles in situ at ppb levels. Chemosphere 136:321–326

Gautam PK, Gautam RK, Banerjee S, Lofrano G, Sanroman MA, Chattopadhyaya MC, Pandey JD (2015) Preparation of activated carbon from Alligator weed(Alternanthera philoxeroides) and its application for tartrazine removal: isotherm, kinetics and spectroscopic analysis. J Environ Chem Eng 3:2560–2568

Gawande K, Jenkins-Smith H (2001) Nuclear waste transport and residential property values: estimating the effects of perceived risks. J Environ Econ Manag 42:207–233

Ghormade V, Deshpande MV, Paknikar KM (2011) Perspectives for nanobiotechnology enabled protection and nutrition of plants. Biotechnol Adv 29(6):792–803

Gittins DI, Bethell D, Schiffrin DJ, Nichols RJ (2000) A nanometre-scale electronic switch consisting of a metal cluster and redox-addressable groups. Nature 408(6808):67–69

Gomathy M, Sabarinathan KG (2010) Microbial mechanisms of heavy metal tolerance –a review. Agric Rev 31(2):133–138

Gupta B, Huang B (2014) Mechanism of salinity tolerance in plants: physiological, biochemical, and molecular characterization. Int J Genom 2014

Gusseme BD, Soetaert M, Hennebel T, Vanhaecke L, Boon N, Verstraete W (2012) Catalytic dechlorination of diclofenac by biogenic palladium in a microbial electrolysis cell. Microbial Biotechnol 5:396–402

Hajiboland R, Cherghvareh L, Dashtebani F (2016) Effect of silicon supplementation on wheat plants under salt stress. J Plant Proc Func 5:1–12

Hakim LF, Portman JL, Casper MD, Weimer AW (2005) Aggregation behavior of nanoparticles in fluidized beds. Powder Technol 160(3):149–160

Hameed A, Sheikh MA, Jamil A, Basra SMA (2013) Seed priming with sodium silicate enhances seed germination and seedling growth in wheat (*Triticum aestivum L*) under water deficit stress induced by polyethylene glycol. Pak J Life Soc Sci 11:19–24

Handley-Sidhu S, Renshaw J, Moriyama S, Stolpe B, Mennan C, Bagheriasl S, Yong P, Stamboulis A, Paterson-Beedle M, Sasaki K (2011) Uptake of Sr2+ and Co2+ into biogenic hydroxyapatite: implications for biomineral ion exchange synthesis. Environ Sci Technol 45:6985–6990

Handley-Sidhu S, Hriljac J, Cuthbert M, Renshaw J, Pattrick R, Charnock J, Stolpe B, Lead J, Baker S, Macaskie L (2014) Bacterially produced calcium phosphate nanobiominerals: sorption capacity, site preferences, and stability of captured radionuclides. Environ Sci Technol 48:6891–6898

Hashem FA, Medany MA, Abd El-Moniem EM, Abdallah MMF (2011) Influence of greenhouse cover on potential evapotranspiration and cucumber water requirements. Ann Agric Sci 56:49–55

Hashemi A, Abdolzadeh A, Sadeghipour HR (2010) Beneficial effects of silicon nutrition in alleviating salinity stress in hydroponically grown canola, *Brassica napus L*, plants. Soil Sci Plant Nutr 56:244–253

Hassan SED, Salem SS, Fouda A, Awad MA, El-Gamal MS, Abdo AM (2018) New approach for antimicrobial activity and bio-control of various pathogens by biosynthesized copper nanoparticles using endophytic actinomycetes. J Radiat Res Appl Sc 11(3):262–270

Hattori T, Sonobe K, Araki H, Inanaga S, An P, Morita S (2008) Silicon application by sorghum through the alleviation of stress-induced increase in hydraulic resistance. J Plant Nutr 31(8):1482–1495

He F, Zhao D (2007) Manipulating the size and dispensability of zerovalent iron nanoparticles by use of carboxymethyl cellulose stabilizers. Environ Sci Technol 41:6216–6221

He F, Zhao D, Liu J, Roberts CB (2007) Stabilization of FePd nanoparticles with sodium carboxymethyl cellulose for enhanced transport and dechlorination of trichloroethylene in soil and groundwater. Ind Eng Chem Res 46(1):29–34

He F, Liu J, Roberts CB, Zhao D (2009) One-step "green" synthesis of Pd nanoparticles of controlled size and their catalytic activity for trichloroethene hydro dechlorination. Ind Eng Chem Res 48(14):6550–6557

Heim KE, Tagliaferro AR, Bobilya DJ (2002) Flavonoid antioxidants: chemistry, metabolism and structure-activity relationships. J Nutr Biochem 13(10):572–584

Hinds S, Taft BJ, Levina L, Sukhovatkin V, Dooley CJ, Roy MD et al (2006) Nucleotide directed growth of semiconductor nanocrystals. J Am Chem Soc 128(1):64

Honary S, Gharaei-Fathabad E, Paji ZK, Eslamifar M (2012) A novel biological synthesis of gold nanoparticle by Enterobacteriaceae family. Trop J Pharm Res 11(6):887–891

Hu C, Lan YQ, Qu JH, Hu XX, Wang AM (2006) Ag/AgBr/TiO2 visible light photocatalyst for destruction of azo dyes and bacteria. J Phys Chem B 110(9):4066–4072

Huang L, Weng X, Chen Z, Megharaj M, Naidu R (2014) Synthesis of iron-based nanoparticles using oolong tea extract for the degradation of malachite green. Spectrochim Acta A Mol Biomol Spectrosc 117:801–804

Iravani S, Zolfaghari B (2013) Green synthesis of silver nanoparticles using Pinus eldarica bark extract. Biomed Res Int 2013

Issazade K, Jahanpour N, Pourghorbanali F, Raeisi G, Faekhondeh J (2013) Heavy metals resistance by bacterial strains. Ann Biol Res 4(2):60–63

Jain R, Jordan N, Schild D, Van Hullebusch ED, Weiss S, Franzen C, Farges F, Hübner R, Lens PN (2015) Adsorption of zinc by biogenic elemental selenium nanoparticles. Chem Eng J 260:855–863

Jayarambabu N, Kumari SB, Rao VK, Prabhu YT (2016) Enhancement of growth in maize by biogenic- synthesized MgO nanoparticles. Int J Pure Appl Zool 4(3):262–270

Kanchi S, Kumar G, Lo AY, Tseng CM, Chen SK, Lin CY, Chin T-S (2014) Exploitation of de-oiled jatropha waste for gold nanoparticles synthesis: a green approach. Arab J Chem 11(2):247–255

Kandasamy R (2017) A novel single step synthesis and surface functionalization of iron-oxide magnetic nanoparticles and thereof for the copper removal from pigment industry effluent. Separ Purif Technol 188:458–467
Katata-Seru L, Moremedi T, Aremu OS, Bahadur I (2017) Green synthesis of iron nanoparticles using Moringa oleifera extracts and their applications: removal of nitrate from water and antibacterial activity against Escherichia coli. J Mol Liq 256:296–304
Khan ZUH, Khan A, Shah A, Wan P, Chen Y, Khan GM, Khan AU, Tahir K, Muhammad N, Khan HU (2016) Enhanced photocatalytic and electrocatalytic applications of green synthesized silver nanoparticles. J Mol Liq 220:248–257
Khandel P, Shahi SK (2016) Microbes mediated synthesis of metal nanoparticles: current status and future prospects. Int J Nanomater Biostruct 6(1):1–24
Kim SD, Cho J, Kim IS, Vanderford BJ, Snyder SA (2007) Occurrence and removal of pharmaceuticals and endocrine disruptors in south Korean surface, drinking, and waste waters. Water Res 41:1013–1021
Kim H, Mosaddik A, Gyawali R, Ahn KS, Cho SK (2012) Induction of apoptosis by ethanolic extract of mango peel and comparative analysis of the chemical constitutes of mango peel and flesh. Food Chem 133(2):416–422
Kim YH, Khan AL, Waqas M, Shim JK, Kim DH, Lee KY (2014) Lee IJ silicon application to rice root zone influenced the phytohormonal and antioxidant responses under salinity stress. J Plant Growth Regul 33(2):137–149
Klaus T, Joerger R, Olsson E, Granqvist CG (1999) Silver-based crystalline nanoparticles, microbially fabricated. PNAS 96(24):13611–13614
Kowshik M, Ashtaputre S, Kharrazi S, Vogel W, Urban J et al (2003) Extracellular synthesis of silver nanoparticles by a silver-tolerant yeast strain MKY3. Nanotechnology 14(1):95–100
Krumov N, Perner-Nochta I, Oder S, Gotcheva V, Angelov A, Posten C (2009) Production of inorganic nanoparticles by microorganisms. Chem Eng Technol 32:1026–1035
Kumar R, Roopan SM, Prabhakarn A, Khanna VG, Chakroborty S (2012) Agricultural waste Annona squamosa peel extract: biosynthesis of silver nanoparticles. Spectrochim Acta A Mol Biomol Spectrosc 90:173–176
Kumar P, Selvi SS, Govindaraju M (2013) Seaweed-mediated biosynthesis of silver nanoparticles using Gracilaria corticata for its antifungal activity against Candida spp. Appl Nanosci 3(6):495–500
Kumar KS, Kumar G, Prokhorov E, Sanchez IC (2014) Exploitation of anaerobic enriched mixed bacteria (AEMB) for the silver and gold nanoparticles synthesis. Colloids Surf A Physicochem Eng Asp 462:264–270
Kumar N, Tripathi P, Nara S (2017a) Gold nanomaterials to plants: impact of bioavailability, particle size and surface coating. In: Nanomaterials in plants, algae and micro-organism: concepts and controversies. Elsevier Press, London, pp 195–220
Kumar VS, Menon S, Agarwal H, Gopalakrishnan D (2017b) Characterization and optimization of bacterium isolated from soil samples for the production of siderophores. Resour Effic Technol: 1–6
Kumar N, Sharma S, Nara S (2018) Dual gold nanostructure-based electrochemical immunosensor for CA125 detection. Appl Nanosci 8(7):1843–1853
Kumar PV, Kala SMJ, Prakash KS (2019) Green synthesis of gold nanoparticles using croton caudatus geisel leaf extract and their biological studies. Mater Lett 236:19–22
Lapworth D, Baran N, Stuart M, Ward R (2012) Emerging organic contaminants in groundwater: a review of sources and fate and occurrence. Environ Pollut 163:287–303
Lee S, Sohn E, Hamayun M, Yoon J, Lee I (2010) Effect of silicon on growth and salinity stress of soybean plant grown under hydroponic system. Agro for Syst 80(3):333–340
Li Z, Du Y (2003) Biomimic synthesis of CdS nanoparticles with enhanced luminescence. Mater Lett 57(16–17):2480–2484
Li S, Shen Y, Xie A, Yu X, Qiu L, Zhang L et al (2007) Green synthesis of silver nanoparticles using Capsicum annuum L extract. Green Chem 9(8):852–858
Li B, Lu F, Wei X, Zhao R (2008) Fucoidan: structure and bioactivity. Molecules 13:1671–1695

Lingamdinne LP, Chang YY, Yang JK, Singh J, Choi EH, Shiratani Koduru JR, Attri P (2017) Biogenic reductive preparation of magnetic inverse spinel iron oxide nanoparticles for the adsorption removal of heavy metals. Chem Eng J 307(74–84)

Liu FK, Ko FH, Huang PW, Wu CH, Chu TC (2005) Studying the size/shape separation and optical properties of silver nanoparticles by capillary electrophoresis. J Chromatogr A1062(1):139–145

Lunge S, Singh S, Sinha A (2014) Magnetic iron oxide (Fe3O4) nanoparticles from tea waste for arsenic removal. J Magn Magn Mater 356:21–31

Luo P, Liu Y, Xia Y, Xu H, Xie G (2014) Aptamer biosensor for sensitive detection of toxin a of Clostridium difficile using gold nanoparticles synthesized by Bacillus stearothermophilus. Biosens Bioelectron 54:217–221

Ma X, Jiang T, Han B, Zhang J, Miao S, Ding K, An G, Xie Y, Zhou Y, Zhu A (2008) Palladium nanoparticles in polyethylene glycols: efficient and recyclable catalyst system for hydrogenation of olefins. Catal Commun 9(1):70–74

Macaskie LE, Yong P, Doyle TC, Roig MG, Diaz M, Manzano T (1997) Bioremediation of uranium-bearing wastewater: biochemical and chemical factors influencing bioprocess application. Biotechnol Bioeng 53:100–109

Madhumitha G, Rajakumar G, Roopan SM, Rahuman AA, Priya KM, Saral AM, Kamaraj C (2012) Acaricidal, insecticidal, and larvicidal efficacy of fruit peel aqueous extract of Annona squamosa and its compounds against blood-feeding parasites. Parasitol Res 111(5):2189–2199

Mahdavi M, Namvar F, Ahmad M, Mohamad R (2013) Green biosynthesis and characterization of magnetic iron oxide (Fe3O4) nanoparticles using seaweed (Sargassum muticum) aqueous extract. Molecules 18(5):5954–5964

Malarkodi C, Rajeshkumar S, Paulkumar K, Jobitha GG, Vanaja M (2013) Biosynthesis of semiconductor nanoparticles by using sulfur reducing bacteria *Serratia nematodiphila*. Adv Nano Res 1(2):83–91

Malik B, Pirzadah TB, Kumar M, Rehman RU (2017) Biosynthesis of nanoparticles and their application in pharmaceutical industry. In: Metabolic engineering for bioactive compounds. Springer, Singapore, pp 331–349

Mao C, Flynn CE, Hayhurst A, Sweeney R, Qi J et al (2003) Viral assembly of oriented quantum dot nanowires. Proc Natl Acad Sci U S A 100:6946–6951

Marchiol L (2012) Synthesis of metal nanoparticles in living plants. Ital J Agron 7(3):274–282

Martins M, Mourato C, Sanches S, Noronha JP, Crespo MB, Pereira IA (2017) Biogenic platinum and palladium nanoparticles as new catalysts for the removal of pharmaceutical compounds. Water Res 108:160–168

Massironi A, Morelli A, Grassi L, Puppi D, Braccini S, Maisetta G, Chiellini F (2019) Ulvan as novel reducing and stabilizing agent from renewable algal biomass: application to green synthesis of silver nanoparticles. Carbohydr Polym 203:310–321

Minaeian S, Shahverdi AR, Nohi AS, Shahverdi HR (2008) Extracellular biosynthesis of silver nanoparticles by some bacteria. J Sci I A U 17(66):1–4

Ming DF, Pei ZF, Naeem MS, Gong HJ, Zhou WJ (2012) Silicon alleviates PEG-induced water-deficit stress in upland rice seedlings by enhancing osmotic adjustment. J Agron Crop Sci 198:14–26

Mishra S, Singh BR, Singh A, Keswani C, Naqvi AH, Singh HB (2014) Bio fabricated silver nanoparticles act as a strong fungicide against *Bipolaris sorokiniana* causing spot blotch disease in wheat. PLoS One 9:e97881

Mohanpuria P, Rana NK, Yadav SK (2008) Biosynthesis of nanoparticles: technological concepts and future applications. J Nanopart Res 10(3):507–517

Moonjung C, Kyoung-Hwan S, Jyongsik J (2010) Plasmonic photocatalytic system using silver chloride/silver nanostructures under visible light. J Colloid Interface Sci 341:83–87

Mukherjee P, Ahmad A, Mandal D, Senapati S, Sainkar SR, Khan MI, Parishcha R, Ajaykumar PV, Alam M, Kumar R, Sastry M (2001) Fungus-mediated synthesis of silver nanoparticles and their immobilization in the mycelial matrix: a novel biological approach to nanoparticle synthesis. Nano Lett 1(10):515–519

Mukherjee D, Ghosh S, Majumdar S, Annapurna K (2016) Green synthesis of α-Fe2O3 nanoparticles for arsenic (V) remediation with a novel aspect for sludge management. J Environ Chem Eng 4:639–650

Nadaf NY, Kanase SS (2016) Biosynthesis of gold nanoparticles by Bacillus marisflavi and its potential in catalytic dye degradation. Arab J Chem. https://doi.org/10.1016/jarabjc201609020

Nadagouda MN, Varma RS (2007) Synthesis of thermally stable carboxymethyl cellulose/metal biodegradable nanocomposites for potential biological applications. Biomacromolecules 8(9):2762–2767

Nair R, Varghese SH, Nair BG, Maekawa T, Yoshida Y, Kumar DS (2010) Nanoparticulate material delivery to plants. Plant Sci 179(3):154–163

Narayanan KB, Sakthivel N (2010) Biological synthesis of metal nanoparticles by microbes. Adv Colloid Interf Sci 156:1–13

Narayanan KB, Sakthivel N (2011) Extracellular synthesis of silver nanoparticles using the leaf extract of Coleus amboinicus. Lour Mater Res Bull 46(10):1708–1713

Navrotsky (2000) A nanomaterials in the environment, agriculture, and technology (NEAT). J Nanopart 2:321–323

Nutt MO, Hughes JB, Wong MS (2005) Designing Pd-on-au bimetallic nanoparticle catalysts for trichloroethene hydrodechlorination. Environ Sci Technol 39:1346–1353

Osterloh FE, Hiramatsu H, Dumas RK, Liu K (2005) Fe3O4-LiMo3Se3 nanoparticle clusters as superparamagnetic nanocompasses. Langmuir 21(21):9709–9713

Pacioni NL, Borsarelli CD, Rey V, Veglia AV (2015) Synthetic routes for the preparation of silver nanoparticles. Springer, Cham, pp 13–46

Paixão RM, Reck IM, Bergamasco R, Vieira MF, Vieira AMS (2017) Activated carbon of Babassu coconut impregnated with copper nanoparticles by green synthesis for the removal of nitrate in aqueous solution. Environ Technol 3:1–10

Paknikar KM, Nagpal V, Pethkar AV et al (2005) Degradation of lindane from aqueous solutions using iron sulfide nanoparticles stabilized by biopolymers. Sci Technol Adv Mater 6:370–374

Parashar UK, Saxena PS, Srivastava A (2009) Bioinspired synthesis of silver nanoparticles. Dig J Nanomater Bios 4:159–166

Patra P, Mitra S, Debnath N, Goswami A (2012) Biochemical-, biophysical-, and microarray-based antifungal evaluation of the buffer-mediated synthesized nano zinc oxide: an in vivo and in vitro toxicity study. Langmuir 28(49):16966–16978

Pérez-de-Luque A, Rubiales D (2009) Nanotechnology for parasitic plant control. Pest Manag Sci 65:540–545

Peters RJ, Bouwmeester H, Gottardo S, Amenta V, Arena M, Brandhoff P, Marvin HJ, Mech A, Moniz FB, Pesudo LQ (2016) Nanomaterials for products and application in agriculture, feed and food. Trends Food Sci Technol 54:155–164

Philip D (2009) Biosynthesis of au, ag and au–ag nanoparticles using edible mushroom extract. Spectrochim Acta A Mol Biomol Spectrosc 73(2):374–381

Pimentel D (2009) Pesticide and pest control. In: Peshin P, Dhawan AK (eds) Integrated pest management: innovation-development process. Springer, Dordrecht, Netherlands, pp 83–87

Prasad T, Sudhakar P, Sreenivasulu Y, Latha P, Munaswamy V, Reddy KR, Sreeprasad T, Sajanlal P, Pradeep T (2012) Effect of nanoscale zinc oxide particles on the germination, growth and yield of peanut. J Plant Nutr 35:905–927

Qu Y, Pei X, Shen W, Zhang X, Wang J, Zhang Z, Li S, You S, Ma F, Zhou J (2017) Biosynthesis of gold nanoparticles by aspergillum sp WL-au for degradation of aromatic pollutants. Phys E Low-dimens Syst Nanostruct 88:133–141

Ragaei M, Sabry AH (2014) Nanotechnology for insect pest control. Int J Sci Environ Technol 3:528–545

Raj R, Dalei K, Chakraborty J, Das S (2016) Extracellular polymeric substances of a marine bacterium mediated synthesis of CdS nanoparticles for removal of cadmium from aqueous solution. J Colloid Interface Sci 462:166–175

Rao A, Bankar A, Kumar AR, Gosavi S, Zinjarde S (2013) Removal of hexavalent chromium ions by Yarrowia lipolytica cells modified with phyto-inspired Fe0/Fe3O4 nanoparticles. J Contam Hydrol 146:63–73

Rong C, Jiliang W, Hui L, Gang C, Zhong L, Chi-Ming C (2010) Fabrication of gold nanoparticles with different morphologies in HEPES buffer. Rare Metals 29:180–186

Sadighi A, Ostad SN, Rezayat SM, Foroutan M, Faramarzi MA, Dorkoosh FA (2012) Mathematical modelling of the transport of hydroxypropyl-β-cyclodextrin inclusion complexes of ranitidine hydrochloride and furosemide loaded chitosan nanoparticles across a Caco-2 cell monolayer. Int J Pharm 422:479–488

Sahin U, Kuslu Y, Kiziloglu FM (2015) Response of cucumber to different irrigation regimes applied through drip-irrigation system. J Anim Plant Sci 25(1):198–205

Sarkar J, Ray S, Chattopadhyay D, Laskar A, Acharya K (2012) Mycogenesis of gold nanoparticles using a phytopathogen *Alternaria alternate*. Bioprocess Biosyst Eng 35(4):637–643

Sathishkumar M, Sneha K, Won S, Cho CW, Kim S, Yun YS (2009) Cinnamon zeylanicum bark extract and powder mediated green synthesis of Nano-crystalline silver particles and its bactericidal activity. Colloids Surf B Biointerfaces 73:332–338

Schlorf T, Meincke M, Kossel E, Glüer CC, Jansen O, Mentlein R (2011) Biological properties of iron oxide nanoparticles for cellular and molecular magnetic resonance imaging. Int J Mol Sci 12:12–23

Seabra AB, Duran N (2010) Nitric oxide-releasing vehicles for biomedical applications. J Mater Chem 20:1624–1637

Shahwan T, Sirriah SA, Nairat M, Boyacı E, Eroğlu AE, Scott TB, Hallam KR (2011) Green synthesis of iron nanoparticles and their application as a Fenton-like catalyst for the degradation of aqueous cationic and anionic dyes. Chem Eng J 172:258–266

Shen W, Qu Y, Pei X, Li S, You S, Wang J, Zhang Z, Zhou J (2017) Catalytic reduction of 4-nitrophenol using gold nanoparticles biosynthesized by cell-free extracts of Aspergillus sp WL-au. J Hazard Mater 321:299–306

Shenton W, Douglas T, Young M, Stubbs G, Mann S (1999) Inorganic-organic nanotube composites from template mineralization of tobacco mosaic virus. Adv Mater 11:253–256

Sheny D, Mathew J, Philip D (2011) Phytosynthesis of au, ag and au–ag bimetallic nanoparticles using aqueous extract and dried leaf of Anacardium occidentale. Spectrochim Acta A Mol Biomol Spectrosc 79(1):254–262

Sheykhbaglou R, Sedghi M, Shishevan MT, Sharifi RS (2010) Effects of nano-iron oxide particles on agronomic traits of soybean. Not Sci Biol 2:112

Shibata H, Iimuro M, Uchiya N, Kawamori T, Nagaoka M et al (2003) Preventive effects of Cladosiphon fucoidan against helicobacter pylori infection in Mongolian gerbils. Helicobactor 8:59–65

Shivaji S, Madhu S, Singh S (2011) Extracellular synthesis of antibacterial silver nanoparticles using psychrophilic bacteria. Process Biochem 46(9):1800–1807

Smuleac V, Varma R, Sikdar S, Bhattacharyya D (2011) Green synthesis of Fe and Fe/Pd bimetallic nanoparticles in membranes for reductive degradation of chlorinated organic. J Membr Sci 379:131–137

Soisuwan S, Warisnoicharoen W, Lirdprapamongkol K, Svasti J (2010) Eco-friendly synthesis of fucoidan-stabilized gold nanoparticles. Am J Appl Sci 7(8):1038–1048

Sonobe K, Hattori T, An P, Tsuji W, Eneji AE, Kobayashi S, Kawamura Y, Tanaka K, Inanaga S (2010) Effect of silicon application on sorghum root responses to water stress. J Plant Nutr 34(1):71–82

Srivastava V, Sharma Y, Sillanpää M (2015) Green synthesis of magnesium oxide nanoflower and its application for the removal of divalent metallic species from synthetic wastewater. Ceram Int 41:6702–6709

Stalin Dhas T, Ganesh Kumar V, Abraham LS, Karthick V, Govindaraju K (2012) *Sargassum myriocystum* mediated biosynthesis of gold nanoparticles. Spectrochim Acta A Mol Biomol Spectrosc 99:97–101

Stephen JR, Macnaughtont SJ (1999) Developments in terrestrial bacterial remediation of metals. Curr Opin Biotech 10(3):230–233

Stubbs D, Gilman P (2007) Nanotechnology applications in environmental health: big plans for little particles. Oak Ridge Centre for Advanced Studies

Suganya KSU, Govindaraju K, Kumar VG (2015) Blue green alga mediated synthesis of gold nanoparticles and its antibacterial efficacy against Gram positive organisms. Mater Sci Eng C47:351–356

Sunkar S, Nachiyar CV (2012) Microbial synthesis and characterization of silver nanoparticles using the endophytic bacterium *Bacillus cereus* a novel source in the benign synthesis. Glob J Med Res 12:43–49

Syed A, Saraswati S, Kundu GC, Ahmad A (2013) Biological synthesis of silver nanoparticles using the fungus Humicola sp and evaluation of their cytoxicity using normal and cancer cell lines. Spectrochim Acta A Mol Biomol Spectrosc 114:144–147

Teruya T, Konishi T, Uechi S, Tamaki H, Tako M (2007) Anti-proliferative activity of over sulfated fucoidan from commercially cultured *Cladosiphon okamuranus* TOKIDA in U937 cells. Inter J Biol Macromol 41:221–226

Torabi F, Majd A, Enteshari S (2015) The effect of silicon on alleviation of salt stress in borage (Borago officinalis L) soil. Sci Plant Nutr 61:788–798

Torney F, Trewyn BG, Lin VS, Wang K (2007) Mesoporous silica nanoparticles deliver DNA and chemicals into plants. Nat Nanotechnol 2:295–300

Tu J, Yang Z, Hu C, Qu J (2014) Characterization and reactivity of biogenic manganese oxides for ciprofloxacin oxidation. J Environ Sci 26:1154–1161

Vanaamudan A, Soni H, Sudhakar PP (2016) Palm shell extract capped silver nanoparticles—as efficient catalysts for degradation of dyes and as SERS substrates. J Mol Liq 215:787–794

Veerasamy R, Xin TZ, Gunasagaran S, Xiang TFW, Yang EFC, Jeyakumar N et al (2011) Biosynthesis of silver nanoparticles using mangosteen leaf extract and evaluation of their antimicrobial activities. J Saudi Chem Soc 15(2):113–120

Venkatesan J, Manivasagan P, Kim SK, Kirthi AV, Marimuthu S, Rahuman AA (2014) Marine algae-mediated synthesis of gold nanoparticles using a novel Ecklonia cava. Bioprocess Biosyst Eng 37(8):1591–1597

Virkutyte J, Varma RS (2011) Green synthesis of metal nanoparticles: biodegradable polymers and enzymes in stabilization and surface functionalization. Chem Sci 2(5):837–846

Vishwakarma K, Upadhyay N, Kumar N, Tripathi DK, Sharma S (2017) Potential applications and avenues of nanotechnology in sustainable agriculture. In: Nanomaterials in plants, algae and microorganism: concepts and controversies. Elsevier Press, London, pp 473–500

Vithanage M, Seneviratne M, Ahmad M, Sarkar B, Ok YS (2017) Contrasting effects of engineered carbon nanotubes on plants: a review. Environ Geochem Health 39:1421–1439

Wang Z, Tang L, Tan LH, Li J, Lu Y (2012) Discovery of the DNA "genetic code" for a biological gold nanoparticle morphologies. Angew Chem Int Ed 51:9078–9082

Wang T, Lin J, Chen Z, Megharaj M, Naidu R (2014) Green synthesized iron nanoparticles by green tea and eucalyptus leaves extracts used for removal of nitrate in aqueous solution. J Clean Prod 83:413–419

Wang R, Wang S, Tai Y, Tao R, Dai Y, Guo J, Yang Y, Duan S (2017) Biogenic manganese oxides generated by green algae Desmodesmus sp WR1 to improve bisphenol a removal. J Hazard Mater 339:310–319

Wang X, Zhang D, Qian H, Liang Y, Pan X, Gadd GM (2018) Interactions between biogenic selenium nanoparticles and goethite colloids and consequence for remediation of elemental mercury contaminated ground water. Sci Total Environ 613:672–678

Weng X, Huang L, Chen Z, Megharaj M, Naidu R (2013) Synthesis of iron-based nanoparticles by green tea extract and their degradation of malachite. Ind Crop Prod 51:342–347

Wheeler T, Von Braun J (2013) Climate change impacts on global food security. Science 341:508–513

Xiao Z, Yuan M, Yang B, Liu Z, Huang J, Sun D (2016) Plant-mediated synthesis of highly active iron nanoparticles for Cr (VI) removal investigation of the leading biomolecules. Chemosphere 150:357–364

Xu P, Liu L, Zeng G, Huang D, Lai C, Zhao M, Zhang C (2014) Heavy metal-induced glutathione accumulation and its role in heavy metal detoxification in Phanerochaete chrysosporium. Appl Microbiol Biotechnol 98(14):6409–6418

Xu H, Wang L, Su H, Gu L, Han T, Meng F, Liu C (2015) Making good use of food wastes: green synthesis of highly stabilized silver nanoparticles from grape seed extract and their antimicrobial activity. Food Biophys 10:12–18

Xu H, Quan X, Xiao Z, Chen L (2018) Effect of anodes decoration with metal and metal oxides nanoparticles on pharmaceutically active compounds removal and power generation in microbial fuel cells. Chem Eng J 335:539–547

Yadav IC, Devi NL, Syed JH, Cheng Z, Li J, Zhang G, Jones KC (2015) Current status of persistent organic pesticides residues in air, water, and soil, and their possible effect on neighbouring countries: a comprehensive review of India. Sci Total Environ 511:123–137

Yallappa S, Manjanna J, Dhananjaya BL (2015) Phyto synthesis of stable au, ag and au-ag alloy nanoparticles using J sambac leaves extract, and their enhanced antimicrobial activity in presence of organic antimicrobials. Spectrochim Acta A Mol Biomol Spectrosc 137:236–243

Yang JW, Yoon SY, Oh SJ, Kim SK, Kang KW (2006) Bifunctional effects of fucoidan on the expression of inducible nitric oxide synthase. Biochem Biophys Res Commun 346:345–350

Yang F, Liu C, Gao F, Su M, Wu X, Zheng L, Hong F, Yang P (2007) The improvement of spinach growth by nano-anatase TiO2 treatment is related to nitrogen photoreduction. Biol Trace Elem Res 119:77–88

Yang P, Lipowsky R, Dimova R (2009) Nanoparticle formation in giant vesicles: synthesis in biomimetic compartments. Small 5:2033

Yang N, WeiHong L, Hao L (2014) Biosynthesis of au nanoparticles using agricultural waste mango peel extract and its in vitro cytotoxic effect on two normal cells. Mater 134:67–70

Yasmeen F, Raja NI, Razzaq A, Komatsu S (2017) Proteomic and physiological analyses of wheat seeds exposed to copper and iron nanoparticles. Biochim Biophys Acta Protein Proteomics 1865:28–42

Yin L, Wang S, Li J, Tanaka K, Oka M (2013) Application of silicon improves salt tolerance through ameliorating osmotic and ionic stresses in the seedling of Sorghum bicolor. Acta Physiol Plant 35:3099–3107

Yong P, Macaskie L, Sammons R, Marquis P (2004) Synthesis of nanophase hydroxyapatite by a Serratia sp from waste-water containing inorganic phosphate. Biotechnol Lett 26:1723–1730

Zaheer Z, Aazam ES, Kosa SA (2016) Effects of cationic and anionic micelles on the morphology of biogenic silver nanoparticles, and their catalytic activity for Congo red. J Mol Liq 220:364–369

Zhang WX (2003) Nanoscale iron particles for environmental remediation: an overview. J Nanopart Res 5:323–332

Zhang X, Yan S, Tyagu RD, Surampalli RY (2011) Synthesis of nanoparticles by microorganisms and their application in enhancing microbiological reaction rates. Chemosphere 82:489–494

Zhang Y, Wu B, Xu H, Liu H, Wang M, He Y, Pan B (2016) Nanomaterials-enabled water and wastewater treatment. NanoImpact 3:22–39

Zheng L, Brody JP, Burke PJ (2004) Electronic manipulation of DNA, proteins, and nanoparticles for potential circuit assembly. Biosens Bioelectron 20:606–619

Zheng L, Hong F, Lu S, Liu C (2005) Effect of nano-TiO2 on strength of naturally aged seeds and growth of spinach. Biol Trace Elem Res 104:83–91

Zhou D, Kim DG, Ko SO (2015) Heavy metal adsorption with biogenic manganese oxide: generated by Pseudomonas putida strain MnB1. J Ind Eng Chem 24:132–139

Zinatloo-Ajabshir S, Salehi Z, Salavati-Niasari M (2018) Green synthesis and characterization of Dy2Ce2O7 ceramic nanostructures with good photocatalytic properties under visible light for removal of organic dyes in water. J Clean Prod 192:678–687

An Overview on the Effect of Soil Physicochemical Properties on the Immobilization of Biogenic Nanoparticles

8

Kamal Khalkhal, Behnam Asgari Lajayer, and Mansour Ghorbanpour

Abstract

There are various methods available today for the synthesis of biogenic nanoparticles (by plants, algae, yeast, bacteria, fungi and waste material), which, due to their cheapness and environmental effects, are superior to chemical synthesis. Soil microorganisms by the secretion of various substances (such as a variety of enzymes, proteins, amino acids, etc.) play a critical role in the synthesis and bioavailability of biogenic nanoparticles in the soil. Synthesized biogenic nanoparticles, due to their nature as well as their surface properties, can lead to the bioremediation of inorganic and organic contaminants in the soil. Among the physicochemical properties of the soil, pH, organic matter content and clay seem to have the most influence on the distribution and immobilization of soil nanoparticles. These nanoparticles may be accumulated in the soil for various reasons or may be transported to groundwater and lead to contamination in humans, animals, plants and microorganisms. Therefore, it is necessary to understanding the behaviour of nanoparticles in the soil, which influence soil physicochemical properties, and to assessing possibility hazards.

Keywords

Aggregation · Immobilization · Soil science · Nanoparticles

K. Khalkhal · B. Asgari Lajayer
Department of Soil Science, Faculty of Agriculture, University of Tabriz, Tabriz, Iran

M. Ghorbanpour (✉)
Department of Medicinal Plants, Faculty of Agriculture and Natural Resources, Arak University, Arak, Iran
e-mail: m-ghorbanpour@araku.ac.ir

M. Ghorbanpour et al. (eds.), *Biogenic Nano-Particles and their Use in Agro-ecosystems*, https://doi.org/10.1007/978-981-15-2985-6_8

8.1 Introduction

The idea of nanosize and nanoscale (10^{-9} m) has been known since the nineteenth century. The term "nanoparticles" (NPs), nevertheless, has been public for about 30 years, these particles were formerly called "ultrafine", "submicron" or "fine-grained." The prefix "nano" (such as nanotechnology, nanoparticles, nano-fertilizers, etc.) is nowadays a part of scientific literature, especially in soil science. For instance, refers to rock weathering as nanosize decomposition and to soil as "a more complex nanoparticle–mixed inorganic, organic, biological, and various sizes" where "most of the transport of heavy metals, nutrients, pollutants, organics happen at the nanosize" (Navrotsky 2003; Maghsoodi et al. 2019). Recently, biogenic nanoparticles (BioNPs) production has given a charming alternative to chemical methods due to their cheapness and reduced environmental effects. There are various methods available today for the synthesis of biogenic nanoparticles using a variety biosystems containing plants, algae, fungi, yeast or bacteria (Narayanan and Sakthivel 2010, 2011a, b; Sinha et al. 2009; Mandal et al. 2006; Rai et al. 2009; Mohanpuria et al. 2008; Durán et al. 2010, 2011; Varshney et al. 2012; Durán and Marcato 2012; Rubilar et al. 2013).

Industrial nanoparticles synthesized by humans eventually enter the soil through sewage and dust, resulting in the pollution of soil and water (Maroufpour et al. 2019). Soil is the main sink of engineered nanoparticles relative to air and water (Gottschalk et al. 2009; Rajput et al. 2017; Keller et al. 2013). Soil physicochemical properties, such as organic matter (OM), pH, soil texture, soil structure, compactness, ionic strength, microorganisms and redox potential (Eh) play major roles influencing the immobilization, bioavailability and distribution of NPs (Pachapur et al. 2016; Fierer and Jackson 2006; Jiménez-Lamana and Slaveykova 2016).These nanoparticles are adsorbed into soil for various reasons and cause contamination in humans, animals, plants and microorganisms. Therefore, it is necessary to understanding the behaviour of NPs in the soil, which influence soil physicochemical properties, and to assessing possibility hazards (Shrestha et al. 2013).

8.2 Synthesis of Biogenic Nanoparticles

According to Fig. 8.1, the synthesis mechanism of BioNPs mostly include bioprecipitation and bioreduction using amino acids, carbohydrates, peptides, polysaccharides, polyphenols, flavonoids, biopolymers, vitamins, alkaloids attained from living organisms in nature (Gautam et al. 2019; Park et al. 2011). Various macro- and microorganisms, such as bacteria (Stephen and Macnaughtont 1999; Shivaji et al. 2011), algae (Castro et al. 2013; Shankar et al. 2016), fungi (Chan and Mat Don 2013; Syed et al. 2013), actinomycetes (Fayaz et al. 2011), yeast (Kowshik et al. 2002), and waste material such as fruit waste (Kanchi et al. 2018) and plants (Alshehri et al. 2017; Makarov et al. 2014) have operated as precursors for the synthesis of BioNPs. As shown in Table 8.1. various synthesis of BioNPs result in the production of NPs with varied sizes and morphologies. The benefits of nanoparticle biosynthesis over chemical synthesis include green technology, biological process,

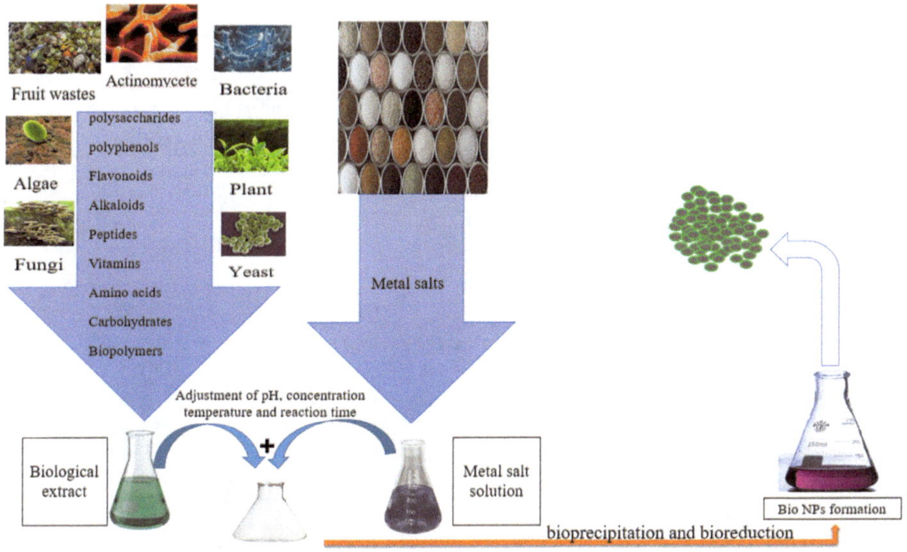

Fig. 8.1 Schematic representation for biosynthesis of NPs

reliable, economical, biocompatible with environment, higher catalytic activity, energy saving, lack of high temperature and no harmful chemicals in the synthesis process (Tripp et al. 2002; Bhattacharya and Gupta 2005; Mukherjee et al. 2001; Hakim et al. 2005). Iravani (2011) reported that plants probably are the appropriate choice for synthesis of BioNPs due to their easy availability, abundance and diversity, as well as lesser time compared to microbial synthesis. As result, the choice of different plant species for extraction is a significant factor (Parsons et al. 2007). Vanaja and Annadurai (2013) reported different plant species (terrestrial to aquatic) have been successfully used in the biosynthesis of NPs. Some researchers report that fresh/biomass can be used to synthesize NPs instead of powdered/dried biomass (Dauthal and Mukhopadhyay 2016).On the other, Patel et al. 2015 reported that different forms of algae can be considered as model systems for nanoparticle synthesis because they have significant ability for metal bioremediation.

8.3 Applications of Biogenic Synthesis of NPs

The synthesis of BioNPs with different crystalline nature, shape and particle size has been one of the important goals in chemistry that could be used for different purposes (Table 8.2), such as agriculture (bioremediation), industry (biosensors and catalysts) and medicine (biomedical) (Staniland 2007; Frances et al. 2009; Antonyraj et al. 2013. The characteristics of these BioNPs vary as per the NPs produced through chemical and other common methods because no surfactants/detrimental materials are involved. Therefore, synthesis of NPs by bacteria, fungi, algae, yeast, waste materials and plant represents a wide range of applications.

Table 8.1 Synthesis of different biogenic NPs

BioNPs	Species	Shape	Size (nm)	References
	Bacteria strain			
Au	*Pseudomonas fluorescens* 417	Spherical	5–50	Husseiny et al. (2007)
Au	*Rhodopseudomonas capsulata*	Spherical	10–20	Syed et al. (2016)
Au	*Shewanella algae*	Spherical	9.6	Mishra et al. (2011)
Au	*Escherichia coli* DH5a	Spherical	20	Suganya et al. (2015)
Au	*Escherichia coli*	Spherical	5–20	Gopal et al. (2013)
Au	*Klebsiella pneumonia*	Spherical	5–65	Kumar et al. (2014)
Au	*Bacillus stearothermophilus*	Triangular	5–30	Luo et al. (2014)
Au	*Stenotrophomonas maltophilia*	Spherical	~40	Srinath and Rai (2015)
Au	*Geobacillusstearo thermophilus*	Spherical	12–14	Gomathy and Sabarinathan (2010)
Au	*Magnetospirillum Gryphiswaldense* MSR-1	Spherical	10–40	Cai et al. (2011)
Au	*Shewanella oneidensis*	Spherical	2–50	Suresh et al. (2011)
Au	*Sporosarcina koreensis* DC4	Spherical	30–50	Singh et al. (2016)
Au	*Staphylococcus epidermidis*	Spherical	20–25	Ogi et al. (2010)
Fe_3O_4	*Geobacter sulfurreducens*	–	10–50	Byrne et al. (2011)
CuO	*Serratia* sp.	–	10–30	Saif Hasan et al. (2008)
Cu_2O	*Lactobacillus* sp. and *S. cerevisiae*	–	10–20	Prasad et al. (2010)
CuO	*E. coli*	Variable shapes	Varied	V Singh et al. (2010)
CuO	Biomass of *Pseudomonas stutzeri*	Spherical	8–15	Varshney et al. (2010)
CuO	*Pseudomonas stutzeri* bacterial strain from electroplating wastewater	–	50–150	Varshney et al. (2011)
Se^0	*Citrobacter freundii* Y9	Amorphous	71 ± 16	Wang et al. (2017)
	Plants			
Au, Ag, Au-Ag	Leaf extracts of *Jasminum sambac*	–	–	Yallappa et al. (2015)

(continued)

Table 8.1 (continued)

BioNPs	Species	Shape	Size (nm)	References
Ag	*J. curcas*	Cubical	10–20	Bar et al. (2009)
Ag°	Commercial green tea extract (*Camellia sinensis*)	Spherical	–	Rolim et al. (2019)
CuO	*Aloe vera* extract	–	15–30	Sangeetha et al. (2012)
Cu	Lemongrass tea	Spherical	2.9 ± 0.64	Brumbaugh et al. (2014)
CuO	Leaf extracts of Magnolia	–	40–100	Lee et al. (2011)
Fe	*A. spinosus* leaf	Spherical	–	Muthukumar and Matheswaran (2015)
	Fungi strain			
Ag^+	*Verticillium*	–	25 ± 12	Mukherjee et al. (2001)
Ag	*Verticillium*	Spherical	25	Sharma et al. (2015)
Ag	*Trichoderma harzianum*	Spherical	58 ± 4 20 and 30	Guilger et al. (2017)
Au	*Aureobasidium pullulans*	Spherical	35–23	Suganya et al. (2015)
Au	*Alternaria alternata*	Spherical and triangular	2–30	Sarkar et al. (2012)
Au	*Botrytis cinerea*	Hexagonal, spherical, pyramidal, Triangularand decahedral	1–100	Vijayaraghavan et al. (2011)
Au	*Penicillium crustosum*	Spherical	100	Roy and Das (2016)
Au	*Penicillium rugulosum*	Spherical, triangular and hexagonal	30–70 and 20–80	Mishra et al. (2012)
Au	*Phanerochaete chrysosporium*	Spherical	10–100	Sheikhloo and Salouti (2011)
Au	*Rhizopus oryzae*	Spherical	5–65	Sanghi et al. (2011)
Au	*Neurospora crassa*	Spherical	3–100	Castro et al. (2014)
Au	*Fusarium semitectum*	Spherical	10–35	Barabadi et al. (2014)
Au	*Fusarium solani*	Spherical	20–50	Castro-Longoria et al. (2011)
Au	*Penicillium chrysogenum*	Spherical, triangle and rod	5–100	Sawle et al. (2008)
Au	*Sclerotium rolfsii*	Spherical	25	Das et al. (2012)

(continued)

Table 8.1 (continued)

BioNPs	Species	Shape	Size (nm)	References
Au	*Trichodermaviride hypocrealixii*	Spherical	20–30	Narayanan and Sakthivel (2011a, b)
Au	*Aspergillus foetidus*	Spherical	30–50	Singaravelu et al. (2007)
Copper oxides	*Penicillium aurantiogriseum*,	Spherical	91–119	Rubilar et al. (2013)
Copper oxides	*Penicillium citrinum*	Spherical	91–133	Rubilar et al. (2013)
Copper oxides	*Penicillium waksmanii*	Spherical	80–88	Rubilar et al. (2013)
	Algal strain			
Ag	*C. racemose*, *Colpmenia sinusa*, *Jania rubins*, capillacae, *Ulva faciata*, *Pterocladia*	Spherical	30	Edison et al. (2016)
Ag	*Cystophora moniliformis*	Spherical	50–100	Prasad et al. (2013)
Au	Brown, *Ecklonia cava*	Triangular and spherical	30 ± 0.25	Kathiraven et al. (2015)
Ag	*Chlamydomonas reinhardtii*	Rectangular or round	5–35	Barwal et al. (2011)
Au	*Padina gymnospora*	Spherical	53–67	Singh et al. (2013)
Au	*Tetraselmis kochinensis*	Triangular and spherical	5–35	Venkatesan et al. (2014)
Au	Brown, *Sargassum muticum*	Spherical	5.4 ± 1.2	Namvar et al. (2015)
Au	*Fucus vesiculosus*	Spherical	Variable	Mata et al. (2009)
Au	*Chlorella vulgaris*	–	2–10	Annamalai and Nallamuthu (2015)
Au	Natural honey (fructose)	–	15	Philip (2009)
CdS	*Phaeodactylum tricornutum*	–	–	Scarano and Morelli (2003)
	Waste material			
Ag	Industrial milk waste	Nanorods	–	Sivakumar et al. (2013)
Ag	*Satsumamandarin* (Citrus unshiu) peel extract	Spherical	5–20	Basavegowda and Rok Lee (2013)
Au	Rice bran	Spherical	50–100	Malhotra et al. (2014)
Au	Grape seeds, stalk and skin	Quasi-spherical	20–25	Krishnaswamy et al. (2014)

(continued)

Table 8.1 (continued)

BioNPs	Species	Shape	Size (nm)	References
Fe	Citrine juices	Irregular, cylindrical and spherical	3–300	Machado et al. (2014)
N-CNTs	Chicken feather	–	–	Gao et al. (2014)
Pb	Watermelon rind	Spherical	96	Lakshmipathy et al. (2015)
Silicon carbide	Electronic compact discs char	Spherical	40–90	Rajarao et al. (2014)
Cellulose	Cotton fibres	Spherical	40–90	Fattahi Meyabadi et al. (2014)
	Actinomycete strain			
Au	*Thermomonospora*	Spherical	8	Fayaz et al. (2011)
Au	*Streptomyces hygroscopicus*	Spherical	20	Husseiny et al. (2007)
Au	*Gordoniaamarae*	Spherical	15–40	Montes et al. (2011)
Au	*Streptomyces fulvissimus*	Spherical	20–50	Balagurunathan et al. (2011)
Au	*Streptomyces* sp. VITDDK3	Hexagonal, cubical, brick and irregular	90	Soltani Nejad et al. (2015)
Au	*Thermomonospora*	Spherical	8	Fayaz et al. (2011)
Au	*Streptomyces viridogens* (HM10)	Spherical and rod	18–20	Arumugam and Berchmans (2011)
Au	*Streptomyces hygroscopicus*	Spherical	20	Khan et al. (2016)
Au	*Gordoniaamarae*	Spherical	15–40	Elavazhagan and Arunachalam (2011)

8.4 Biogenic Nanoparticles in Soils

Soils contain many kinds of inorganic and organic particles with at least one dimension in the nanoscale or colloidal range (<100 nm). Natural NPs in soil include oxides (Al, Fe and Mn), humic substances, clay minerals, mobile colloids, enzymes and viruses (Kretzschmar and Schäfer 2005). Organic NPs are in contact with inorganic NPs or occur as coatings on inorganic NPs surfaces (Chorover et al. 2007; Oades 1989). Due to surface properties and their nature, NPs in environments participate in vital ecological services, ranging from regulating element cycling and water remediation, via transport or adsorption of biological and chemical pollutants, to serving as a sink/source of plant nutrients and organic carbon. The clay fraction of the soil usually represents particles less than 2 μm in diameter. So, mineralogists and soil scientists have confirmed that this clay fraction (i.e. colloidal particles) contains nanoscale particles (<100 nm) (Theng and Yuan 2008).

Table 8.2 Applications of biogenic synthesized NPs

BioNPs	Roles	Effective	References
Fe	Contaminants immobilization in soils and waters	The removal of As (V) and As (III) from water and soil	Shipley et al. (2011)
Fe_3O_4	Bioremediation	Reducing toxic Cr(VI) to less toxic Cr(III)	Byrne et al. (2011)
FeS	Remediation and adsorbent	Removal of Cr(VI) (convert Cr(VI) to Cr(III)) in chromium-contaminated soil and water	Watson et al. (1995) and Mullet et al. (2004)
Se^0	Bioremediationof contaminated soil	Immobilization Hg^0 in soil under both aerobic and anaerobic conditions	Wang et al. (2017)
Se	Removal from wastewater	Adsorption of Zn^{2+} ions onto Se BioNPs	Jain et al. (2015)
Se	Metal immobilization in soils and waters	Remediation of Hg^0 contamination	Wang et al. (2017, 2018a, 2019a)
Ag	Antifungal and antibacterial	Against bacterial pathogens such as *Staphylococcus aureus*, *E. coli*, *Vibrio parahaemolyticus*, *Shewanella putrefaciens* Against fungal pathogen of *Candida albicans*	Nayak et al. (2018)
Ag	Biocompatible agent and antimicrobial	Against *E.coli*	Eckhardt et al. (2013) and Jung et al. (2008)
Ag	Antimicrobial	Against the phytopathogenic fungus *Sclerotinia sclerotiorum* (responsible for white mould disease)	Guilger et al. (2017)
Ag°	Antimicrobial	Against gram-positive *Staphylococcus*, gram-negative *Pseudomonas aeruginosa*, *Klebsiella pneumoniae*, *Salmonella enterica* and *E. coli*	Rolim et al. (2019)
Au	Antibacterial	Against *S. aureus* and *B.* subtilis.	Uma Suganya et al. (2015)
Ag	Antimicrobial	Against pathogenic bacteria (*Staphylococcus aureus*, *Bacillus subtilis*, *E. coli*, *Pseudomonas aeruginosa*) and against yeast (*Candida albicans*)	Ibrahim (2015)
Cu_2O	Antibacterial	Against *E. coli*	Gopalakrishnan et al. (2012)
Au	Biomedical applications	Emerging alternative for DNA modeling and life-threatening diseases	Khan et al. (2013) and Gupta and Gupta (2005)
Iron and iron oxide	Biomedical applications	Tissue repair, cell labeling, magnetic resonance imaging and drug delivery	Pankhurst et al. (2003) and Catherine and Adam (2003)

(continued)

Table 8.2 (continued)

BioNPs	Roles	Effective	References
Pd	Biocatalysts	Reductive removal of contaminants, namely azo dyes, trichloroethylene, polychlorobifenyls and Cr(VI)	Quan et al. (2015) and Hennebel et al. (2012)
Pd	Biocatalysts	Removal of 4 pharmaceutical products (PhP): 17b-estradiol, ibuprofen, sulfamethoxazole and ciprofloxacin	Martins et al. (2017)
Pt	Biocatalysts	Removal of 4 pharmaceutical products (PhP): 17b-estradiol, ibuprofen, sulfamethoxazole and ciprofloxacin	Martins et al. (2017)
Ag	Catalytic applications	Reduction of 4-nitrophenol	Liang et al. (2014)
Fe_3O_4	Catalytic applications	Removal of contaminants	Magnacca et al. (2014)
Pd	Catalytic applications	Degradation of azo dyes	Petla et al. (2012)
Pd	Catalytic applications	Catalyst for Mizoroki–Heck cross-coupling reaction	Eroglu et al. (2013)
Pd	Catalytic applications	The reaction of methyl acrylate and iodobenzene	Parker et al. (2015)
Au	Biosensor applications	Detection of hormone (HCG) in pregnant women	Kuppusamy et al. (2014)
Au-Ag	Biosensor applications	Detecting cancer	Zhang et al. (2012)
Au or PEG-coated Au	Biosensor applications	Cancer nanotechnology, anticancer	Visaria et al. (2006), Van Horssen et al. (2006) and Cai et al. (2008)
Pt	Biosensor applications	Determination of adrenaline	Brondani et al. (2009)

The transformation and formation of BioNPs in soil may be accomplished via either a biological or an abiotic pathway, or both. For instance, humic substances are formed through biological pathways and aluminosilicates are formed through non-biological pathways, while some nano-oxides Fe and Mn are formed through a combination of biological and non-biological pathways. (Theng and Yuan 2008). On the other hand, soil is the final sink of engineered nanoparticles (ENPs) from the industry. The increasing influx of ENPs (photocatalysts, fullerenes, carbon nanotubes semiconductors, black carbon, soot, vehicle exhaust emissions and zero-valent iron/nickel) into soil has increased concerns about their harmful effects on human, animal and plant health (Nowack and Bucheli 2007).

Since most of the organic colloids are coated with metal oxides in the soil or in some way related to mineral surfaces, the characteristics of the soil minerals are greatly affected by organic colloids. For instance, the formation of complexes between allophane and humic acid results in a significant increase in copper and cadmium adsorption (Yuan et al. 2002). Most of the research in this field is related to synthesized laboratory NPs. Even if NPs reactivity studies are carried out in the laboratory, it would be difficult to interpret the results due to some variables, such as loading of contaminant, ionic strength and pH, that affect NP pollutant interactions (Theng and Yuan 2008; Chorover et al. 2007).

Microorganisms also play a key role in the production of mineral BioNPs, such as phyllosilicates. Microorganisms such as bacteria are suitable for mediating the formation of mineral NPs due to their high surface area and negative electrical charge on the cell wall. This negative charge on the cell wall causes the adsorption and accumulation of cations and then the anions of the environment to produce various types of biogenic NPs. On the other hand, these bacteria play as oxidizing/and or reducing bacteria in the soil, thereby increasing the precipitation or solubility of NPs (Bargar et al. 2008). Although inadequate, Information is available on the microbially mediated alteration and formation of minerals in soil, except for such environments as acid sulphate soils and acid mine drainage systems (Banfield and Zhang 2001; Douglas and Beveridge 1998).

Most soils include diverse oxides/hydroxides (Al, Fe and Mn), formed using microbial process or through weathering of primary and secondary silicate minerals. These nanominerals (in PZC ˃pH = 8) play a key role in the retention and adsorption of nutrient anions (sulfate, phosphate, molybdate, borate) by ligand exchange and electrostatic interactions, and they actively aggregate soil stabilization and promote clay flocculation (Schwertmann 2008; McBride 1994). Unlike other (hydro)oxides, most oxides and hydroxides of Mn have a low PZC (pH <4). Since manganese oxides and hydroxides are in the pH range of most of the soils having a negative charge, these oxides are efficient scavengers and sorbents of heavy metals (Tebo et al. 2004; McKenzie 1989).

8.5 Effect of Soil Physicochemical Properties on Immobilization of Biogenic Nanoparticles

In soil, the successful performance of remediation by BioNPs is nearly associated with the environmental fate of NPs. Soil physicochemical properties, such as pH, salinity, ionic strength, clay minerals, texture, structure, compactness, colloids, organic matter and soil microbial community play important roles influencing the dispersion, aggregation, immobilization, bioavailability, stability and transport of NPs (Fierer and Jackson 2006; French et al. 2009; Van Hoecke et al. 2011; Hua et al. 2015; Pachapur et al. 2016; Jiménez-Lamana and Slaveykova 2016; Ellis et al. 2016; Maghsoodi et al. 2019). Soil can contain large quantities of organic matter, such as dissolved organic matter (DOM), natural organic matter (NOM) and particulate organic matter (POM), that affect the mobility of NPs. DOM is a complex mixture of various functional groups, molecular weights and different structures such as protein, fulvic acid and humic acid (Chen et al. 2013). The heterogeneous molecular weight of DOM affects the binding behaviour between NPs, metals and DOM (Wu et al. 2012), whose effects on the fate of nanoparticles (such as Ag, Au and fullerenes) are observed (Shen et al. 2015; Louie et al. 2013). DOM-coated NPs influence their stability and can cause declined aggregation via steric stabilization and surface charge effects (Li et al. 2016; Erhayem and Sohn 2014). However, bridging mechanisms and charge neutralization between DOM and NPs can also lead to enhanced aggregation (Buffle et al. 1998). Biochemical processes, source of materials and hydrology may cause different composition as well as different effects of DOM (Aiken et al. 2011). The application of DOM increased the stability of SeNPs and thus favoured the immobilization of Hg^0. Upon interaction with intermediate molecular weights, DOM demonstrated the biggest effect (Wang et al. 2019a).

Natural organic matter (NOM) plays an important role in NPs aggregation and is one of the major constituents of colloidal natural materials. Fulvic acid and humic acid are considered to be POM that have been studied. NOM can stabilize colloids and NPs due to the coating effect on the surface of the NPs via a steric stabilization and charge mechanism. Giasuddin et al. (2007) has reported that interactions between NOM and NPs affect the stability of colloidal NPs and chemical reactions. Apart from DOM, there is another important and dynamic fraction of the NOM called particulate organic matter (POM) that effect on the fate of Se NPs and mercury immobilization in the soil, which has not been fully investigated (Wang et al. 2019c).The POM consists of organic debris, rhizoplan and fine-grained fragments that are between 2–53 mm in size and can affect the fate and performance of NPs in the remediation process (Leifeld 2006; Cambardella and Elliott 1992; Wang et al. 2019c). Wang et al. (2019c) reported that low concentrations of more negatively charged POM hindered heteroaggregation and homoaggregation with SeNPs which had a lower negative charge via electrostatic repulsion. In high concentrations of POM, SeNPs were formed to bind to POM with further Hg^0 presence in the POM due to a higher content of NPs, which result in successful collisions. Although, mercury immobilization capability by SeNPs was not remarkably affected using the application of POM (Wang et al. 2019c).

Soil microorganisms produce BioNPs in the soil, but on the other hand, these microorganisms are affected by the toxicity of the NPs in the soil. In other words, the physicochemical properties of the soil affect microorganisms by affecting the toxicity of NPs. The toxicity of NPs cannot be limited to a single soil property, and features such as soil texture, pH and OM at high concentrations are among the most important. Comparison of AgNP toxicity with soil physical and chemical properties showed that the toxicity decreased with increase in clay percentage and increasing pH. But, AgNPs toxicity did not seem to be influenced by soil organic matter (Schlich and Hund-Rinke 2015).

Soil biological systems reduce the number of Cu ions to reduce the toxicity of copper, or, like the rhizospheres of *Iris pseudoacorus* and *Phragmites australis*, convert copper particles into biogenic NPs and prevent its toxicity (Manceau et al. 2008a, b). Biomolecular responses to Cu stress by plant roots in oxygenated environments have resulted in its transformation and immobilization (Manceau et al. 2008a, b).CuBioNPs may be formed in phytoremediated soils around peat bogs or swamps. For example, CuNPs have been formed by plant rhizospheres in swamp peats around New Brunswick, Canada. The physicochemical properties of the rhizosphere result in the immobilization of CuNPs and prevent their entry into plants and food chains that can infect humans and animals (Shotyk 1988). The addition of peats that either act as templating substrates or include CuNPs could raise the efficacy of using swamp plants for immobilization.

The retention and immobilization of AgNPs in soils rely on many physicochemical factors, for example redox potential (Eh), soil organic matter, clay, various microorganisms (Li et al. 2017; Hoppe et al. 2014) and exposure period (Whitley et al. 2013). For instance, positively charged clay edges and Al and Fe oxides/hydroxides controlled AgNPs immobilization (Cornelis et al. 2010). In addition, Hedberg et al. (2015) reported that greater retention of AgNPs occurred in acid and clay soils than in natural and sandy soils, respectively. Wang et al. (2018b) reported the results of multiple regression analyses indicated that the retention capacity of Ag^+ ions on soils was largely correlated to soil OM content, while retention capacity of AgNPs was positively correlated with the Fe oxide content. Highest retention capacity of Ag^+ ion and AgNPs in soils is diverse with soil physical and chemical attributes, and especially, soil organic matter contents and iron oxide are two pivotal factors controlling Q_{max} (Wang et al. 2018b).

Another important and influential feature of soil on the immobilization and bioremediation of NPs is salinity and ionic strength. Wang et al. (2019b) reported that on aggregation, Hg^0 immobilization was suppressed due to reduced surface area and decreased availability sites of the Se BioNPs, and the inhibitory effect was more pronounced with increased salinity in soil rather than in soil solutions. It is known that Se BioNPs have a higher Hg sorption capacity than chemical Se ENPs (Johnson et al. 2008). According to the following reaction, Se BioNPs are a successful bioremediation method for mercury immobilization in water and soil:

$$\mathrm{Hg}^0 + \mathrm{Se}^0 \rightarrow \mathrm{HgSe}\left(\Delta G^0 = -38.1\ \mathrm{kjmol}^{-1}\right)(\text{Wang et al. 2017, 2018c})$$

Joo et al. (2013) reported that the reduced electrical double layer with increasing salinity led to an increase in the hydrodynamic diameter of AgNPs. Generally, higher salinity can lead to less toxicity of NPs as they may be released from metal nano-oxides due to lower metal ions concentration (Yung et al. 2017). On the contrary, the toxicity of AgNPs coated with polyvinylpyrrolidone significantly increased with enhancing salinity although the toxicity of Ag^+ decreased (Macken et al. 2012).

Microorganisms such as algae have been properly recognized due to their phytochemical action to immobilize, stabilize and synthesize the NPs (Jena et al. 2013). Fungi, on the other hand, are capable of immobilizing, stabilization, and synthesizing of metallic NPs by secreting different enzymes, polypeptides and proteins, etc. and act as nanocatalysts and nanoadsorbents (Yadav et al. 2015). Fungi such as wasp nest soil fungi (Nayak et al. 2018) are more advanced than other microorganisms in the synthesis of NPs, since most of the fungi possess high wall-binding capacity to immobilize, are easy to handle, have intracellular metal uptake capabilities and require simple nutrients to grow (Bhat et al. 2015). Coutris et al. (2012) evaluated the behaviour of three types of Ag, namely, uncoated AgNPs (19 nm), silver nitrate and citrate stabilized AgNPs (5 nm), in two soils with various OM content, and to follow changes in binding strength over time. Their findings showed that silver nitrate is quickly immobilized in soil, while uncoated Ag NPs can act as a persistent source of bioaccessible Ag.

MnNPs whose high retention potential for toxic trace metals, especially Pb, is well established (Villalobos et al. 2005). Many researchers have investigated the mechanisms of trace metal retention on biogenic and synthetic Mn^{4+}NPs (O'Reilly and Hochella 2003; Tebo et al. 2004; Nelson and Lion 2003). These researchers noted the key role played by negatively charged sites in the (hydro)oxides and by inner-sphere surface complexation of the metals (i.e. direct chemical binding of the metals to the mineral surface). Tebo et al. (2004) and Manceau et al. (2003) have also reported high sorption affinity of Mn oxides in sediments and natural soils for other bivalent trace metal cations, such as Co, Cd, Zn and Ni. Biogenic Mn oxides therefore have two strong metal cation retention mechanisms: the binding of species at the external edges of crystallites and the structural binding of species at interlayer sites (Villalobos et al. 2005). These mechanisms demonstrate the observed high affinity of birnessite NPs for lead. It has been clear that the biogenic magnetite NPs serve as efficient adsorbents for metal contaminants cations such as Mn^{2+}, Co^{2+}, Ni^{2+} and Zn^{2+} (Iwahori et al. 2014).

8.6 Influence of Soil Properties on the Toxicity of Nanoparticles

Soil is a dynamic, heterogeneous, porous, complex and live system composed of three phases: solid (organic and inorganic), liquid and gas. Agricultural soils are exposed to metal oxide NPs due to the use of sewage sludge as fertilizers and different pesticides (Larue et al. 2014; Suppan 2013). Soil physicochemical properties such as pH, soil texture and OM contents substantially affect the bioavailability and behaviour of conventional contaminants like ENPs, polycyclic aromatic hydrocarbons, heavy metals and pesticides (Giller et al. 1998; Martins and Mermoud 1998; Labud et al. 2007; Ranjard et al. 2000). Metal NPs interact with OM and clay minerals due to their small size (2–53 and < 2 nm) (Antisari et al. (2013). Since most of the soil bacteria are in the fine soil (less than 20 nm), the metal NPs in the clay fraction are in direct contact with microbial communities (Ranjard and Richaume 2001; Antisari et al. 2013). Among the various soil parameters, soil microbial communities have been identified as sensitive ecological indicators of soil response to environmental stress due to soil microbial impact on ecosystem function (Doran and Zeiss 2000; Schloter et al. 2003; Lejon et al. 2008; Baptist et al. 2008) and may be good models to examinw TiO_2NPs impact on soil quality and functioning (Holden et al. 2014). Some researchers reported that metal NPs can alter the bacterial community structure and reduce microbial activity in soils (Ge et al. 2011; Du et al. 2011; Nogueira et al. 2012). Simonin et al. (2015) evaluated the influence of TiO_2NPs on soil microbial communities in six agricultural soils displaying different OM contents and textures in microcosms. TiO_2NPs only in silty-clay soil (due to high OM and carbon mineralization) significantly decreased the abundance and activity of microbial communities. In this study, OM and pH were the main parameters affecting the toxicity of NPs in soil, and soil texture showed no significant effect on the toxicity of TiO_2NPs.

8.7 Toxicity of Nanoparticles in Soil

In recent years, the synthesis and application of NPs in agricultural and plant sciences has grown exponentially and many advances have been made (Ghorbanpour and Hadian 2015, 2017; Hatami et al. 2013, 2014, 2016, 2017, 2019; Ghorbanpour et al. 2015, 2018; Ghorbanpour and Hatami 2014, 2015; Ghorbanpour 2015; Ghorbanpour and Fahimirad 2017; Hatami and Ghorbanpour 2013, 2014; Baiazidi-Aghdam et al. 2016; Hatami 2017; Chegini et al. 2017; Mohammadi et al. 2018; Fahimirad et al. 2019; Tian et al. 2018; Ahmadi et al. 2018), but these advances have been along with the diffusion of different NPs into the environment and increased pollution (Aitken et al. 2006; Maghsoodi et al. 2019; Khadem Moghadam et al. 2019). In the case of metal-based nanoparticles such as Ag, ZnO, CeO and TiO_2, the toxicity has largely been due to the emission and release of free metal ions into the environment (Auffan et al. 2009), and as a result of high specificity and reactivity, its toxicity effects are expected to increase (Van Gestel et al. 2010). The NPs can be divided into two classes of organic and inorganic nanoparticles. Inorganic NPs mainly include, metal oxides (Fe_3O_4, Al_2O_3, Cu_2O, CuO, CeO_2, La_2O_3, In_2O_3, NiO, MgO, SnO_2, TiO_2, ZrO_2, ZnO), metals (Ag, Al, Au, Bi, Cu, Co, Mo, In, Ni Fe, Sn, Zn, Ti, W) and quantum dots, while carbon nanotubes and fullerenes are organic NPs (Rajput et al. 2018a, b). Among inorganic NPs, metal NPs are more broadly used because of their dangerous effects (as biocides to suppress microorganisms) and have many effects on the diversity, abundance, activity and habitat of living organisms (Rajput et al. 2018a, b). Thus, these NPs should be examined for their toxic fate and effects in soil. NPs released into the environment may be adsorbed into soil, create leaching or drain to groundwater or degradation by abiotic and biotic processes (Boxall et al. 2007; Khadem Moghadam et al. 2019). Therefore, it is necessary to assess the risks associated with nanoparticle toxicity in soil. Contaminant NPs mobility via the soil profile rely on the kind of soil minerals, size, shape, charge and soil physicochemical and biological properties (Petosa et al. 2010). For example, zinc oxide NPs are adsorbed in the surface of soil colloids. These ZnONPs have low mobility depending on the ionic strength (Zhao et al. 2012), and are more strongly adsorbed than the Zn^{2+}ion. Adsorption of $Zn^{2+}$and ZnONPs increases with increasing pH (Rajput et al. 2018a, b).

Biodiversity, microbial biomass and abundance are important biological properties of soil affected by NPs (Bondarenko et al. 2013). Silver, copper and silicon NPs in arctic soils significantly reduced microbial biomass (Kumar et al. 2011). ZnO and CuONPs had significant toxic effects on the soil microbial community and caused enzymatic activities, such as fluorescein diacetate hydrolysis, urease and catalase (Chai et al. 2015; Dinesh et al. 2012). The toxic effects of some metal and metal oxide NPs on microorganisms have been reported more than carbon nanotubes and fullerenes (Simonin and Richaume 2015; Frenk et al. 2013). Ge et al. (2011) investigated the influence of ZnO and TiO_2 on soil microbial communities and indicated that both NPs decrease diversity and microbial mass. The effects of silver and copper oxide NPs on leaf microbial decomposition led to a reduction in leaf decomposition rate (Pradhan et al. 2011). Fe_3O_4 and CuONPs did not affect some of the macroscopic as well as organic matter, but in contrast to the microscopic properties, the bacterial community and humic substances in DOM were changed, although the effects of CuO NPs were greater than that of Fe_3O_4NPs. The effect depended on NPs concentration, NP types and soil types (Ben-Moshe et al. 2013). The effects of CuONPs, ZnONPs and TiO_2NPs on different soils resulted in a significant reduction in the diversity of microbial community, soil microbial activity and enzymatic activity (Xu et al. 2015; Rousk et al. 2012). The toxicity of CuONPs was more severe than TiO_2NPs, because CuONPs (heavy metal) altered nutrient bioavailability, producing free radicals and high dissolution of soil microbes, while the effect of TiO_2 NPs were mainly caused by titanium particles themselves (Gajjar et al. 2009; Xu et al. 2015). Burke et al. (2014) reported that the variation in arbuscular mycorrhiza fungal community in the TiO_2NP amended soil, in contrast, a notable enhancement in soil richness was reported by Shah et al. (2014). This difference of soil biological response to NPs can be due to the coactions of some parameters, such as exposure time and dose, microbiological species, inherent toxicity differences among NPs soil property (Eh, pH, iron strength, water content, OM, etc.) or other experimental conditions.

8.8 Conclusion and Future Research

As the bulk of industrial nanoparticles eventually enter the soil and lead to contamination of soil, groundwater and surface water, future research should focus on the synthesis and use of biogenic nanoparticles instead of engineered nanoparticles, and further research on soil bioremediation and immobilization of various types of contaminants (organic and inorganic) are needed in soil using biogenic nanoparticles. Soil physicochemical properties (pH, EC, OM, Eh, ionic strength, clay minerals, texture, structure, compactness, colloids and soil microorganisms) play important roles influencing the aggregation, immobilization, bioavailability, stability and transport of NPs, however, clay content, organic matter content, and pH have received the most attention. Studies on some physicochemical properties of the soil such as calcium carbonate (CCE), active calcium carbonate equivalent (ACCE), cations exchange capacity (CEC), soil water content and clay type (2: 1 and 2: 1) have not mentioned the behaviour of biogenic nanoparticles and this should be included in future research.

References

Ahmadi SZ, Ghorbanpour M, Hadian J, Salehi-Arjmand H (2018) Impact of foliar spray of spherical nano-carbon and salicylic acid on physiological traits and parthenolide content in two feverfew cultivars (*Tanacetum parthenium* Linn. cv. Pharmasaat and Jelitto). J Med Plants 17(4):82–98

Aiken GR, Hsu-Kim H, Ryan JN (2011) Influence of dissolved organic matter on the environmental fate of metals, nanoparticles, and colloids. Environ Sci Technol 45:3196–3201

Aitken RJ, Chaudhry MQ, Boxall ABA, Hull M (2006) Manufacture and use of nanomaterials: current status in the UK and global trends. Occup Med Oxford 56:300–306

Alshehri A, Malik MA, Khan Z, Al-Thabaiti SA, Hasan N (2017) Biofabrication of Fe nanoparticles in aqueous extract of Hibiscus sabdariffa with enhanced photocatalytic activities. RSC Adv 7(40):25149–25159

Annamalai J, Nallamuthu T (2015) Characterization of biosynthesized gold nanoparticles from aqueous extract of Chlorella vulgaris and their anti-pathogenic properties. Appl Nanosci 5(5):603–607

Antisari LV, Carbone S, Gatti A, Vianello G, Nannipieri P (2013) Toxicity of metal oxide (CeO2, Fe3O4, SnO2) engineered nanoparticles on soil microbial biomass and their distribution in soil. Soil Biol Biochem 60:87–94

Antonyraj CA, Jeong J, Kim B, Shin S, Kim S, Lee KY, Cho JK (2013) Selective oxidation of HMF to DFF using Ru/γ-alumina catalyst in moderate boiling solvents toward industrial production. J Ind Eng Chem 19(3):1056–1059

Arumugam P, Berchmans S (2011) Synthesis of gold nanoparticles: an ecofriendly approach using Hansenula anomala. ACS Appl Mater Interfaces 3(5):1418–1425

Auffan M, Rose J, Wiesner MR, Bottero J-Y (2009) Chemical stability of metallic nanoparticles: a parameter controlling their potential cellular toxicity in vitro. Environ Pollut 157:1127–1133

Baiazidi-Aghdam MT, Mohammadi H, Ghorbanpour M (2016) Effects of nanoparticulate anatase titanium dioxide on physiological and biochemical performance of *Linum usitatissimum* (Linaceae) under well watered and drought stress conditions. Braz J Bot 39:139–146

Balagurunathan R, Radhakrishnan M, Rajendran RB, Velmurugan D (2011) Biosynthesis of gold nanoparticles by actinomycete Streptomyces viridogens strain HM10. Indian J Biochem Biophys 48(5):331–335

Banfield JF, Zhang H (2001) Nanoparticles in the environment. In: Banfield JF, Navrotsky A (eds) Nanoparticles and the environment, Reviews in Mineralogy & Geochemistry, vol 44. Mineralogical Society of America, Chantilly, pp 1–58

Baptist F, Zinger L, Clement JC, Gallet C, Guillemin R, Martins JMF et al (2008) Tannin impacts on microbial diversity and the functioning of alpine soils: a multidisciplinary approach. Environ Microbiol 10(3):799–809

Bar H, Bhui DK, Sahoo GP, Sarkar P, De SP, Misra A (2009) Green synthesis of silver nanoparticles using latex of Jatropha curcas. Colloids Surf A Physicochem Eng Asp 339(1–3):134–139

Barabadi H, Honary S, Ebrahimi P, Mohammadi MA, Alizadeh A, Naghibi F (2014) Microbial mediated preparation, characterization and optimization of gold nanoparticles. Braz J Microbiol 45(4):1493–1501

Bargar JR, Bernier-Latmani R, Giammar DE, Tebo BM (2008) Biogenic uraninite nanoparticles and their importance for uranium remediation. Elements 4:407–412

Barwal I, Ranjan P, Kateriya S, Yadav S (2011) Cellular oxido-reductive proteins of Chlamydomonas reinhardtii control the biosynthesis of silver nanoparticles. J Nanobiotechnol 9(1):56

Basavegowda N, Rok Lee Y (2013) Synthesis of silver nanoparticles using Satsuma mandarin (Citrus unshiu) peel extract: a novel approach towards waste utilization. Mater Lett 109:31–33

Ben-Moshe T, Frenk S, Dror I, Minz D, Berkowitz B (2013) Effects of metal oxide nanoparticles on soil properties. Chemosphere 90(2):640–646

Bhat MA, Nayak BK, Nanda A (2015) Evaluation of bactericidal activity of biologically synthesised silver nanoparticles from Candida albicans in combination with ciprofloxacin. Mater Today Proc 2(9):4395–4401

Bhattacharya D, Gupta RK (2005) Nanotechnology and potential of microorganisms. Crit Rev Biotechnol 25(4):199–204

Bondarenko O, Juganson K, Ivask A, Kasemets K, Mortimer M, Kahru A (2013) Toxicity of Ag, CuO and ZnO nanoparticles to selected environmentally relevant test organisms and mammalian cells in vitro: a critical review. Arch Toxicol 87(7):1181–1200

Boxall A, Chaudhry Q, Cinclair C, Jones A, Aitken R, Jefferson B, Watts C (2007) Current and future predicted environmental exposure to engineered nanoparticles. Central Science Laboratory, Sand Hutton

Brondani D, Scheeren CW, Dupont J, Vieira IC (2009) Biosensor based on platinum nanoparticles dispersed in ionic liquid and laccase for determination of adrenaline. Sensors Actuators B Chem 140(1):252–259

Brumbaugh AD, Cohen KA, St. Angelo SK (2014) Ultrasmall copper nanoparticles synthesized with a plant tea reducing agent. ACS Sustain Chem Eng 2(8):1933–1939

Buffle J, Wilkinson KJ, Stoll S, Filella M, Zhang J (1998) A generalized description of aquatic colloidal interactions: the three-colloidal component approach. Environ Sci Technol 32(19):2887–2899

Burke DJ, Zhu S, Pablico-Lansigan MP, Hewins CR, Samia ACS (2014) Titanium oxide nanoparticle effects on composition of soil microbial communities and plant performance. Biol Fertil Soils 50:1169–1173

Byrne JM, Telling ND, Coker VS, Pattrick RAD, Van Der Laan G, Arenholz E et al (2011) Control of nanoparticle size, reactivity and magnetic properties during the bioproduction of magnetite by Geobacter sulfurreducens. Nanotechnology 22(45):455709

Cai W, Gao T, Hong H, Sun J (2008) Applications of gold nanoparticles in cancer nanotechnology. Nanotechnol Sci Appl 1:17–32

Cai F, Li J, Sun J, Ji Y (2011) Biosynthesis of gold nanoparticles by biosorption using Magnetospirillum gryphiswaldense MSR-1. Chem Eng J 175:70–75

Cambardella CA, Elliott ET (1992) Particulate soil organic-matter changes across a grassland cultivation sequence. Soil Sci Soc Am J 56(3):777–783

Castro L, Blázquez ML, Muñoz JA, González F, Ballester A (2013) Biological synthesis of metallic nanoparticles using algae. IET Nanobiotechnol 7:109–116

Castro ME, Cottet L, Castillo A (2014) Biosynthesis of gold nanoparticles by extracellular molecules produced by the phytopathogenic fungus Botrytis cinerea. Mater Lett 115:42–44

Castro-Longoria E, Vilchis-Nestor AR, Avalos-Borja M (2011) Biosynthesis of silver, gold and bimetallic nanoparticles using the filamentous fungus Neurospora crassa. Colloids Surf B: Biointerfaces 83(1):42–48

Catherine CB, Adam SGC (2003) Functionalisation of magnetic nanoparticles for applications in biomedicine. J Phys D Appl Phys 36(13):R198

Chai H, Yao J, Sun J, Zhang C, Liu W, Zhu M, Ceccanti B (2015) The effect of metal oxide nanoparticles on functional bacteria and metabolic profiles in agricultural soil. Bull Environ Contam Toxicol 94(4):490–495

Chan YS, Mat Don M (2013) Biosynthesis and structural characterization of Ag nanoparticles from white rot fungi. Mater Sci Eng C 33(1):282–288

Chegini E, Ghorbanpour M, Hatam M, Taghizadeh M (2017) Effect of multi-walled carbon nanotubes on physiological traits, phenolic contents and antioxidant capacity of *Salvia mirzayanii* Rech. f. & Esfand. under drought stress. J Med Plants 16(2):191–207

Chen W, Smith DS, Guéguen C (2013) Influence of water chemistry and dissolved organic matter (DOM) molecular size on copper and mercury binding determined by multiresponse fluorescence quenching. Chemosphere 92:351–359

Chorover J, Kretzschmar R, Garcia-Pichel F, Sparks DL (2007) Soil biogeochemical processes within the critical zone. Elements 3:321–326

Cornelis G, Kirby JK, Beak D, Chittleborough D, McLaughlin MJ (2010) A method for determination of retention of silver and cerium oxide manufactured nanoparticles in soils. Environ Chem 7(3):298–308

Coutris C, Joner EJ, Oughton DH (2012) Aging and soil organic matter content affect the fate of silver nanoparticles in soil. Sci Total Environ 420:327–333

Das SK, Dickinson C, Lafir F, Brougham DF, Marsili E (2012) Synthesis, characterization and catalytic activity of gold nanoparticles biosynthesized with Rhizopus oryzae protein extract. Green Chem 14(5):1322–1334

Dauthal P, Mukhopadhyay M (2016) Noble metal nanoparticles: plant-mediated synthesis, mechanistic aspects of synthesis, and applications. Ind Eng Chem Res 55(36):9557–9577

Dinesh R, Anandaraj M, Srinivasan V, Hamza S (2012) Engineered nanoparticles in the soil and their potential implications to microbial activity. Geoderma 173–174:19–27

Doran JW, Zeiss MR (2000) Soil health and sustainability: managing the biotic component of soil quality. Appl Soil Ecol 15:3–11

Douglas S, Beveridge TJ (1998) Mineral formation by bacteria in natural microbial communities. FEMS Microbiol Ecol 26:79–88

Du W, Sun Y, Ji R, Zhu J, Wu J, Guo H (2011) TiO 2 and ZnO nanoparticles negatively affect wheat growth and soil enzyme activities in agricultural soil. J Environ Monit 13(4):822–828

Durán N, Marcato PD (2012) Biotechnological routes to metallic nanoparticles production: mechanistics aspects, antimicrobial activity, toxicity and industrial applications. In: Rai M, Cioffi N (eds) Nano-antimicrobials: progress and prospects, vol part 3. Springer, Berlin, pp 337–374

Durán N, Marcato PD, Conti RD, Alves OL, Costa F, Brocchi M (2010) Potential use of silver nanoparticles on pathogenic bacteria, their toxicity and possible mechanisms of action. J Braz Chem Soc 21(6):949–959

Durán N, Marcato PD, Durán M, Yadav A, Gade A, Rai M (2011) Mechanistic aspects in the biogenic synthesis of extracellular metal nanoparticles by peptides, bacteria, fungi and plants. Appl Microbiol Biotechnol 90:1609–1624

Eckhardt S, Brunetto PS, Gagnon J, Priebe M, Giese B, Fromm KM (2013) Nanobio silver: its interactions with peptides and bacteria, and its uses in medicine. Chem Rev 113(7):4708–4754

Edison TNJI, Atchudan R, Kamal C, Lee YR (2016) Caulerpa racemosa: a marine green alga for eco-friendly synthesis of silver nanoparticles and its catalytic degradation of methylene blue. Bioprocess Biosyst Eng 39(9):1401–1408

Elavazhagan T, Arunachalam KD (2011) Memecylon edule leaf extract mediated green synthesis of silver and gold nanoparticles. Int J Nanomedicine 6:1265

Ellis LJA, Valsami-Jones E, Lead JR, Baalousha M (2016) Impact of surface coating and environmental conditions on the fate and transport of silver nanoparticles in the aquatic environment. Sci Total Environ 568:95–106

Erhayem M, Sohn M (2014) Stability studies for titanium dioxide nanoparticles upon adsorption of Suwannee River humic and fulvic acids and natural organic matter. Sci Total Environ 468:249–257

Eroglu E, Chen X, Bradshaw M, Agarwal V, Zou J, Stewart SG, Duan X, Lamb RN, Smith SM, Raston CL, Iyer KS (2013) Biogenic production of palladium nanocrystals using microalgae and their immobilization on chitosan nanofibers for catalytic applications. RSC Adv 3(4):1009–1012

Fahimirad S, Ajalloueian F, Ghorbanpour M (2019) Synthesis and therapeutic potential of silver nanomaterials derived from plant extracts. Ecotoxicol Environ Saf 168:260–278

Fattahi Meyabadi T, Dadashian F, Mir Mohamad Sadeghi G, Ebrahimi Zanjani Asl H (2014) Spherical cellulose nanoparticles preparation from waste cotton using a green method. Powder Technol 261:232–240

Fayaz AM, Girilal M, Rahman M, Venkatesan R, Kalaichelvan PT (2011) Biosynthesis of silver and gold nanoparticles using thermophilic bacterium Geobacillus stearothermophilus. Process Biochem 46(10):1958–1962

Fierer N, Jackson RB (2006) The diversity and biogeography of soil bacterial communities. Proc Natl Acad Sci 103(3):626–631

Frances N, Nikolay AP, Michael JFB, Tim G, Paul AM (2009) Novel one-pot synthesis and characterization of bioactive thiol-silicate nanoparticles for biocatalytic and biosensor applications. Nanotechnology 20(5):055612

French RA, Jacobson AR, Kim B, Isley SL, Penn RL, Baveye PC (2009) Influence of ionic strength, pH, and cation valence on aggregation kinetics of titanium dioxide nanoparticles. Environ Sci Technol 43(5):1354–1359

Frenk S, Ben-Moshe T, Dror I, Berkowitz B, Minz D (2013) Effect of metal oxide nanoparticles on microbial community structure and function in two different soiltypes. PLoS One 8(12):e84441

Gajjar P, Pettee B, Britt DW, Huang W, Johnson WP, Anderson AJ (2009) Antimicrobial activities of commercial nanoparticles against an environmental soil microbe, Pseudomonas putida KT2440. J Biol Eng 3:1–13

Gao L, Li R, Sui X, Li R, Chen C, Chen Q (2014) Conversion of chicken feather waste to N-doped carbon nanotubes for the catalytic reduction of 4-nitrophenol. Environ Sci Technol 48(17):10191–10197

Gautam PK, Singh A, Misra K, Sahoo AK, Samanta SK (2019) Synthesis and applications of biogenic nanomaterials in drinking and wastewater treatment. J Environ Manag 231:734–748

Ge Y, Schimel JP, Holden PA (2011) Evidence for negative effects of TiO2 and ZnO nanoparticles on soil bacterial communities. Environ Sci Technol 45(4):1659–1664

Ghorbanpour M (2015) Major essential oil constituents, total phenolics and flavonoids content and antioxidant activity of *Salvia officinalis* plant in response to nano-titanium dioxide. Indian J Plant Physiol 20(3):249–256

Ghorbanpour M, Fahimirad SH (2017) Plant nanobionics a novel approach to overcome the environmental challenges. In: Ghorbanpour M, Varma A (eds) Medicinal plants and environmental challenges. Springer, Cham. https://doi.org/10.1007/978-3-319-68717-9_14

Ghorbanpour M, Hadian J (2015) Multi-walled carbon nanotubes stimulate callus induction, secondary metabolites biosynthesis and antioxidant capacity in medicinal plant *Saturejakhuzestanica* grown *in vitro*. Carbon 94:749–759

Ghorbanpour M, Hadian J (2017) Engineered nanomaterials and their interactions with plant cells: injury indices and detoxification pathways. In: Ghorbanpour M et al (eds) Nanoscience and plant–soil systems, Soil biology 48. Springer, Cham. https://doi.org/10.1007/978-3-319-46835-8_13

Ghorbanpour M, Hatami M (2014) Spray treatment with silver nanoparticles plus thidiazuron increases anti-oxidant enzyme activities and reduces petal and leaf abscission in four cultivars of geranium (*Pelargonium zonale*) during storage in the dark. J Hortic Sci Biotechnol 89(6):712–718

Ghorbanpour M, Hatami H (2015) Changes in growth, antioxidant defense system and major essential oils constituents of *Pelargonium graveolens* plant exposed to nano-scale silver and thidiazuron. Indian J Plant Physiol 20(2):116–123

Ghorbanpour M, Hatami M, Hatami M (2015) Activating antioxidant enzymes, hyoscyamine and scopolamine biosynthesis of *Hyoscyamusniger* L. plants with nano-sized titanium dioxide and bulk application. Acta Agric Slov 105:23–32

Ghorbanpour M, Khaltabadi Farahani AH, Hadian J (2018) Potential toxicity of nano-graphene oxide on callus cell of *Plantago major* L. under polyethylene glycol-induced dehydration. Ecotoxicol Environ Saf 148:910–922

Giasuddin AB, Kanel SR, Choi H (2007) Adsorption of humic acid onto nanoscale zerovalent iron and its effect on arsenic removal. Environ Sci Technol 41(6):2022–2027

Giller KE, Witter E, Mcgrath SP (1998) Toxicity of heavy metals to microorganisms and microbial processes in agricultural soils: a review. Soil Biol Biochem 30(10–11):1389–1414

Gomathy M, Sabarinathan KG (2010) Microbial mechanisms of heavy metal tolerance-a review. Agric Rev 31(2):133–138

Gopal JV, Thenmozhi M, Kannabiran K, Rajakumar G, Velayutham K, Rahuman AA (2013) Actinobacteria mediated synthesis of gold nanoparticles using Streptomyces sp. VITDDK3 and its antifungal activity. Mater Lett 93:360–362

Gopalakrishnan K, Ramesh C, Ragunathan V, Thamilselvan M (2012) Antibacterial activity of Cu2O nanoparticles on E. coli synthesized from Tridax procumbens leaf extract and surface coating with polyaniline. Dig J Nanomater Biostruct 7(2):833–839

Gottschalk F, Sonderer T, Scholz RW, Nowack B (2009) Modeled environmental concentrations of engineered nanomaterials (TiO2, ZnO, Ag, CNT, fullerenes) for different regions. Environ Sci Technol 43:9216–9222

Guilger M, Pasquoto-Stigliani T, Bilesky-Jose N, Grillo R, Abhilash PC, Fraceto LF, De Lima R (2017) Biogenic silver nanoparticles based on trichoderma harzianum: synthesis, characterization, toxicity evaluation and biological activity. Sci Rep 7:44421

Gupta AK, Gupta M (2005) Synthesis and surface engineering of iron oxide nanoparticles for biomedical applications. Biomaterials 26(18):3995–4021

Hakim LF, Portman JL, Casper MD, Weimer AW (2005) Aggregation behavior of nanoparticles in fluidized beds. Powder Technol 160(3):149–160

Hatami M (2017) Stimulatory and inhibitory effects of nanoparticulates on seed germination and seedling vigor indices. In: Ghorbanpour M et al (eds) Nanoscience and plant–soil systems, Soil biology 48. Springer, Cham. https://doi.org/10.1007/978-3-319-46835-8_13

Hatami M, Ghorbanpour M (2013) Effect of nanosilver on physiological performance of *Pelargonium* plants exposed to dark storage. J Hortic Res 21(1):15–20

Hatami M, Ghorbanpour M (2014) Defense enzymes activity and biochemical variations of *Pelargonium zonale*in response to nanosilver particles and dark storage. Turk J Biol 38:130–139

Hatami M, Hatamzadeh A, Ghasemnezhad M, Ghorbanpour M (2013) The comparison of antimicrobial effects of silver nanoparticles (SNP) and silver nitrate($AgNo_3$) to extend the vase life of 'red ribbon' cut rose flowers. Trakia J Sci 2:144–151

Hatami M, Ghorbanpour M, Salehiarjomand H (2014) Nano-anatase TiO_2 modulates the germination behavior and seedling vigority of the five commercially important medicinal and aromatic plants. J Biol Environ Sci 8(22):53–59

Hatami M, Kariman K, Ghorbanpour M (2016) Engineered nanomaterial-mediated changes in the metabolism of terrestrial plants. Sci Total Environ 571:275–291

Hatami M, Hadian J, Ghorbanpour M (2017) Mechanisms underlying toxicity and stimulatory role of single-walled carbon nanotubes in *Hyoscyamus niger* during drought stress simulated by polyethylene glycol. J Hazard Mater 324:306–320

Hatami M, Hosseini SM, Ghorbanpour M, Kariman K (2019) Physiological and antioxidative responses to GO/PANI nanocomposite in intact and demucilaged seeds and young seedlings of Salvia mirzayanii. Chemosphere 233:920–935

Hedberg J, Oromieh AG, Kleja DB, Wallinder IO (2015) Sorption and dissolution of bare and coated silver nanoparticles in soil suspensions—influence of soil and particle characteristics. J Environ Sci Health A 50(9):891–900

Hennebel T, De Corte S, Verstraete W, Boon N (2012) Microbial production and environmental applications of Pd nanoparticles for treatment of halogenated compounds. Curr Opin Biotechnol 23(4):555–561

Holden PA, Schimel JP, Godwin HA (2014) Five reasons to use bacteria when assessing manufactured nanomaterial environmental hazards and fates. Curr Opin Biotechnol 27:73–78

Hoppe M, Mikutta R, Utermann J, Duijnisveld W, Guggenberger G (2014) Retention of sterically and electrosterically stabilized silver nanoparticles in soils. Environ Sci Technol 48(21):12628–12635

Hua Z, Tang Z, Bai X, Zhang J, Yu L, Cheng H (2015) Aggregation and resuspension of graphene oxide in simulated natural surface aquatic environments. Environ Pollut 205:161–169

Husseiny MI, El-Aziz MA, Badr Y, Mahmoud MA (2007) Biosynthesis of gold nanoparticles using Pseudomonas aeruginosa. Spectrochim Acta A Mol Biomol Spectrosc 67(3–4):1003–1006

Ibrahim HMM (2015) Green synthesis and characterization of silver nanoparticles using banana peel extract and their antimicrobial activity against representative microorganisms. J Radiat Res Appl Sci 8(3):265–275

Iravani S (2011) Green synthesis of metal nanoparticles using plants. Green Chem 13(10):2638–2650

Iwahori K, Watanabe JI, Tani Y, Seyama H, Miyata N (2014) Removal of heavy metal cations by biogenic magnetite nanoparticles produced in Fe (III)-reducing microbial enrichment cultures. J Biosci Bioeng 117(3):333–335

Jain R, Jordan N, Schild D, Van Hullebusch ED, Weiss S, Franzen C et al (2015) Adsorption of zinc by biogenic elemental selenium nanoparticles. Chem Eng J 260:855–863

Jena J, Pradhan N, Dash BP, Sukla LB, Panda PK (2013) Biosynthesis and characterization of silver nanoparticles using microalga Chlorococcum humicola and its antibacterial activity. Int J Nanomater Biostruct 3(1):1–8

Jiménez-Lamana J, Slaveykova VI (2016) Silver nanoparticle behaviour in lake water depends on their surface coating. Sci Total Environ 573:946–953

Johnson NC, Manchester S, Sarin L, Gao Y, Kulaots I, Hurt RH (2008) Mercury vapor release from broken compact fluorescent lamps and in situ capture by new nanomaterial sorbents. Environ Sci Technol 42(15):5772–5778

Joo HS, Kalbassi MR, Yu IJ, Lee JH, Johari SA (2013) Bioaccumulation of silver nanoparticles in rainbow trout (Oncorhynchus mykiss): influence of concentration and salinity. Aquat Toxicol 140:398–406

Jung WK, Koo HC, Kim KW, Shin S, Kim SH, Park YH (2008) Antibacterial activity and mechanism of action of the silver ion in Staphylococcus aureus and Escherichia coli. Appl Environ Microbiol 74(7):2171–2178

Kanchi S, Kumar G, Lo AY, Tseng CM, Chen SK, Lin CY, Chin TS (2018) Exploitation of de-oiled jatropha waste for gold nanoparticles synthesis: a green approach. Arab J Chem 11(2):247–255

Kathiraven T, Sundaramanickam A, Shanmugam N, Balasubramanian T (2015) Green synthesis of silver nanoparticles using marine algae Caulerpa racemosa and their antibacterial activity against some human pathogens. Appl Nanosci 5(4):499–504

Keller AA, McFerran S, Lazareva A, Suh S (2013) Global life cycle releases of engineered nanomaterials. J Nanopart Res 15(6):1692

Khadem Moghadam N, Hatami M, Rezaei S, Bayat M, Asgari Lajayer B (2019) Induction of plant defense machinery against nanomaterials exposure. In: Ghorbanpour M, Wani SH (eds) Advances in phytonanotechnology: from synthesis to application. Elsevier Inc, London

Khan MS, Vishakante GD, Siddaramaiah H (2013) Gold nanoparticles: a paradigm shift in biomedical applications. Adv Colloid Interf Sci 199–200:44–58

Khan ME, Khan MM, Cho MH (2016) Fabrication of WO 3 nanorods on graphene nanosheets for improved visible light-induced photocapacitive and photocatalytic performance. RSC Adv 6(25):20824–20833

Kowshik M, Ashtaputre S, Kharrazi S, Vogel W, Urban J, Kulkarni SK, Paknikar KM (2002) Extracellular synthesis of silver nanoparticles by a silver-tolerant yeast strain MKY3. Nanotechnology 14(1):95

Kretzschmar R, Schäfer T (2005) Metal retention and transport on colloidal particles in the environment. Elements 1:205–210

Krishnaswamy K, Vali H, Orsat V (2014) Value-adding to grape waste: green synthesis of gold nanoparticles. J Food Eng 142:210–220

Kumar N, Shah V, Walker VK (2011) Perturbation of an arctic soil microbial community by metal nanoparticles. J Hazard Mater 190(1–3):816–822

Kumar DA, Palanichamy V, Roopan SM (2014) Green synthesis of silver nanoparticles using Alternanthera dentata leaf extract at room temperature and their antimicrobial activity. Spectrochim Acta A Mol Biomol Spectrosc 127:168–171

Kuppusamy P, Mashitah MY, Maniam GP, Govindan N (2014) Biosynthesized gold nanoparticle developed as a tool for detection of HCG hormone in pregnant women urine sample. Asian Pac J Trop Dis 4(3):237

Labud V, Garcia C, Hernandez T (2007) Effect of hydrocarbon pollution on the microbial properties of a sandy and a clay soil. Chemosphere 66(10):1863–1871

Lakshmipathy R, Palakshi Reddy B, Sarada NC, Chidambaram K, Khadeer Pasha S (2015) Watermelon rind-mediated green synthesis of noble palladium nanoparticles: catalytic application. Appl Nanosci 5(2):223–228

Larue C, Castillo-Michel H, Sobanska S, Cecillon L, Bureau S, Barthès V, Ouerdane L, Carriere M, Sarret G (2014) Foliar exposure of the cropLactucasativa to silver nanoparticles: evidence for internalization and changes in Agspeciation. J Hazard Mater 264:98–106

Lee HJ, Lee G, Jang NR, Yun JH, Song JY, Kim BS (2011) Biological synthesis of copper nanoparticles using plant extract. Nanotechnology 1(1):371–374

Leifeld J (2006) Application of diffuse reflectance FT-IR spectroscopy and partial least-squares regression to predict NMR properties of soil organic matter. Eur J Soil Sci 57(6):846–857

Lejon DP, Martins JM, Lévêque J, Spadini L, Pascault N, Landry D et al (2008) Copper dynamics and impact on microbial communities in soils of variable organic status. Environ Sci Technol 42(8):2819–2825

Li S, Ma H, Wallis LK, Etterson MA, Riley B, Hoff DJ, Diamond SA (2016) Impact of natural organic matter on particle behavior and phototoxicity of titanium dioxide nanoparticles. Sci Total Environ 542:324–333

Li M, Wang P, Dang F, Zhou DM (2017) The transformation and fate of silver nanoparticles in paddy soil: effects of soil organic matter and redox conditions. Environ Sci Nano 4(4):919–928

Liang M, Su R, Huang R, Qi W, Yu Y, Wang L, He Z (2014) Facile in situ synthesis of silver nanoparticles on procyanidin-grafted eggshell membrane and their catalytic properties. ACS Appl Mater Interfaces 6(7):4638–4649

Louie SM, Tilton RD, Lowry GV (2013) Effects of molecular weight distribution and chemical properties of natural organic matter on gold nanoparticle aggregation. Environ Sci Technol 47(9):4245–4254

Luo P, Liu Y, Xia Y, Xu H, Xie G (2014) Aptamer biosensor for sensitive detection of toxin A of Clostridium difficile using gold nanoparticles synthesized by Bacillus stearothermophilus. Biosens Bioelectron 54:217–221

Machado S, Grosso JP, Nouws HPA, Albergaria JT, Delerue-Matos C (2014) Utilization of food industry wastes for the production of zero-valent iron nanoparticles. Sci Total Environ 496:233–240

Macken A, Byrne HJ, Thomas KV (2012) Effects of salinity on the toxicity of ionic silver and Ag-PVP nanoparticles to Tisbe battagliai and Ceramium tenuicorne. Ecotoxicol Environ Saf 86:101–110

Maghsoodi MR, Asgari Lajayer B, Hatami M (2019) Challenges and opportunities of nanotechnology in plants-soil mediated systems: beneficial role, phytotoxicity and phytoextraction. In: Ghorbanpour M, Wani SH (eds) Advances in phytonanotechnology: from synthesis to application. Elsevier Inc, London

Magnacca G, Allera A, Montoneri E, Celi L, Benito DE, Gagliardi LG, Gonzalez MC, Mártire DO, Carlos L (2014) Novel magnetite nanoparticles coated with waste sourced bio based substances as sustainable and renewable adsorbing materials. ACS Sustain Chem Eng 2(6):1518–1524

Makarov VV, Makarova SS, Love AJ, Sinitsyna OV, Dudnik AO, Yaminsky IV et al (2014) Biosynthesis of stable iron oxide nanoparticles in aqueous extracts of Hordeum vulgare and Rumex acetosa plants. Langmuir 30(20):5982–5988

Malhotra A, Sharma N, Navdezda, Kumar N, Dolma K, Sharma D, Nandanwar HS, Choudhury AR (2014) Multi-analytical approach to understand biomineralization of gold using rice bran: a novel and economical route. RSC Adv 4(74):39484–39490

Manceau A, Tamura N, Celestre RS, MacDowell AA, Geoffroy N, Sposito G, Padmore HA (2003) Molecular-scale speciation of Zn and Ni in soil ferromanganese nodules from loess soils of the Mississippi Basin. Environ Sci Technol 37:75–80

Manceau A, Nagy KL, Marcus MA, Lanson M, Geoffroy N, Jacquet T, Kirpichtchikova T (2008a) Formation of metallic copper nanoparticles at the soil-root interface. Environ Sci Technol 42:1766–1777

Manceau A, Nagy KL, Marcus MA, Lanson M, Geoffroy N, Jacquet T, Kirpichtchikova T (2008b) Formation of metallic copper nanoparticles at the soil-root interface. Environ Sci Technol 42(5):1766–1772

Mandal D, Bolander ME, Mukhopadhyay D, Sarkar G, Mukherjee P (2006) The use of microorganisms for the formation of metal nanoparticles and their application. Appl Microbiol Biotechnol 69(5):485–492
Maroufpour N, Mousavi M, Hatami M, Rasoulnia A, Asgari Lajayer B (2019) Mechanisms involved in stimulatory and toxicity effects of nanomaterials on seed germination and early seedling growth. In: Ghorbanpour M, Wani SH (eds) Advances in phytonanotechnology: from synthesis to application. Elsevier Inc, London
Martins JM, Mermoud A (1998) Sorption and degradation of four nitroaromatic herbicides in mono and multi-solute saturated/unsaturated soil batch systems. J Contam Hydrol 33(1–2):187–210
Martins M, Mourato C, Sanches S, Noronha JP, Crespo MB, Pereira IA (2017) Biogenic platinum and palladium nanoparticles as new catalysts for the removal of pharmaceutical compounds. Water Res 108:160–168
Mata YN, Torres E, Blazquez ML, Ballester A, González FMJA, Munoz JA (2009) Gold (III) biosorption and bioreduction with the brown alga Fucus vesiculosus. J Hazard Mater 166(2–3):612–618
McBride MB (1994) Environmental chemistry of soils. Oxford University Press, New York. 416 pp
McKenzie RM (1989) Manganese oxides and hydroxides. In: Dixon JB, Weed SB (eds) Minerals in soil environments, 2nd edn. Soil Science Society of America, Madison, pp 439–465
Mishra A, Tripathy SK, Yun SI (2011) Bio-synthesis of gold and silver nanoparticles from Candida guilliermondii and their antimicrobial effect against pathogenic bacteria. J Nanosci Nanotechnol 11(1):243–248
Mishra A, Tripathy SK, Yun S-I (2012) Fungus mediated synthesis of gold nanoparticles and their conjugation with genomic DNA isolated from Escherichia coli and Staphylococcus aureus. Process Biochem 47(5):701–711
Mohammadi M, Hatami M, Feghezadeh K, Ghorbanpour M (2018) Mitigating effect of nano-zerovalent iron, iron sulfate and EDTA against oxidative stress induced by chromium in Helianthus annuus L. Acta Physiol Plant 40:69
Mohanpuria P, Rana NK, Yadav SK (2008) Biosynthesis of nanoparticles: technological concepts and future applications. J Nanopart Res 10:507–517
Montes MO, Mayoral A, Deepak FL, Parsons JG, Jose-Yacamán M, Peralta-Videa JR, Gardea-Torresdey JL (2011) Anisotropic gold nanoparticles and gold plates biosynthesis using alfalfa extracts. J Nanopart Res 13(8):3113–3121
Mukherjee P, Ahmad A, Mandal D, Senapati S, Sainkar SR, Khan MI, Parishcha R, Ajaykumar PV, Alam M, Kumar R, Sastry M (2001) Fungus-mediated synthesis of silver nanoparticles and their immobilization in the mycelial matrix: a novel biological approach to nanoparticle synthesis. Nano Lett 1(10):515–519
Mullet M, Boursiquot S, Ehrhardt JJ (2004) Removal of hexavalent chromium from solutions by mackinawite, tetragonal FeS. Colloids Surf A Physicochem Eng Asp 244(1–3):77–85
Muthukumar H, Matheswaran M (2015) Amaranthus spinosus leaf extract mediated FeO nanoparticles: physicochemical traits, photocatalytic and antioxidant activity. ACS Sustain Chem Eng 3(12):3149–3156
Namvar F, Azizi S, Ahmad M, Shameli K, Mohamad R, Mahdavi M, Tahir P (2015) Green synthesis and characterization of gold nanoparticles using the marine macroalgae Sargassum muticum. Res Chem Intermed 41(8):5723–5730
Narayanan KB, Sakthivel N (2010) Biological synthesis of metal nanoparticles by microbes. Adv Colloid Interf Sci 156(1–2):1–13
Narayanan KB, Sakthivel N (2011a) Facile green synthesis of gold nanostructures by NADPH-dependent enzyme from the extract of Sclerotium rolfsii. Colloids Surf A Physicochem Eng Asp 380(1–3):156–161
Narayanan KB, Sakthivel N (2011b) Green synthesis of biogenic metal nanoparticles by terrestrial and aquatic phototrophic and heterotrophic eukaryotes and biocompatible agents. Adv Colloid Interf Sci 169:59–79
Navrotsky A (2003) Materials and nanotechnology. J Frankl Inst 340:263–268

Nayak BK, Nanda A, Prabhakar V (2018) Biogenic synthesis of silver nanoparticle from wasp nest soil fungus, Penicillium italicum and its analysis against multi drug resistance pathogens. Biocatal Agric Biotechnol 16:412–418
Nelson YM, Lion LW (2003) Formation of biogenic manganese oxides and their influence on the scavenging of toxic trace metals. In: Selim HM, Kingerly WL (eds) Geochemical and hydrological reactivity of heavy metals in soils. CRC Press, Boca Raton, pp 169–186
Nogueira V, Lopes I, Rocha-Santos T, Santos AL, Rasteiro GM, Antunes F et al (2012) Impact of organic and inorganic nanomaterials in the soil microbial community structure. Sci Total Environ 424:344–350
Nowack B, Bucheli TD (2007) Occurrence, behavior and effects of nanoparticles in the environment. Environ Pollut 150(1):5–22
O'Reilly SE, Hochella MF (2003) Lead sorption efficiencies of natural and synthetic Mn and Fe-oxides. Geochim Cosmochim Acta 67:4471–4487
Oades JM (1989) An introduction to organic matter in mineral soils. In: Dixon JB, Weed SB (eds) Minerals in soil environments, 2nd edn. Soil Science Society of America, Madison, pp 89–159
Ogi T, Saitoh N, Nomura T, Konishi Y (2010) Room-temperature synthesis of gold nanoparticles and nanoplates using Shewanella algae cell extract. J Nanopart Res 12(7):2531–2539
Pachapur VL, Larios AD, Cledon M, Brar SK, Verma M, Surampalli RY (2016) Behavior and characterization of titanium dioxide and silver nanoparticles in soils. Sci Total Environ 563:933–943
Pankhurst QA, Connolly J, Jones SK, Dobson J (2003) Applications of magnetic nanoparticles in biomedicine. J Phys D Appl Phys 36(13):R167
Park Y, Hong Y, Weyers A, Kim Y, Linhardt R (2011) Polysaccharides and phytochemicals: a natural reservoir for the green synthesis of gold and silver nanoparticles. IET Nanobiotechnol 5:69–78
Parker HL, Dodson JR, Budarin VL, Clark JH, Hunt AJ (2015) Direct synthesis of Pd nanoparticles on alginic acid and seaweed supports. Green Chem 17(4):2200–2207
Parsons JG, Peralta-Videa JR, Gardea-Torresdey JL (2007) Use of plants in biotechnology: synthesis of metal nanoparticles by inactivated plant tissues, plant extracts, and living plants. Dev Environ Sci 5:463–485
Patel V, Berthold D, Puranik P, Gantar M (2015) Screening of cyanobacteria and microalgae for their ability to synthesize silver nanoparticles with antibacterial activity. Biotechnol Rep 5:112–119
Petla RK, Vivekanandhan S, Misra M, Mohanty AK, Satyanarayana N (2012) Soybean (Glycine Max) leaf extract based green synthesis of palladium nanoparticles. J Biomater Nanobiotechnol 3(1):14–19
Petosa AR, Jaisi DP, Quevedo IR, Elimelech M, Tufenkji N (2010) Aggregation and deposition of engineered nanomaterials in aquatic environments: role of physicochemical interactions. Environ Sci Technol 44(17):6532–6549
Philip D (2009) Biosynthesis of Au, Ag and Au–Ag nanoparticles using edible mushroom extract. Spectrochim Acta A Mol Biomol Spectrosc 73(2):374–381
Pradhan A, Seena S, Pascoal C, Cássio F (2011) Can metal nanoparticles be a threat to microbial decomposers of plant litter in streams? Microb Ecol 62(1):58–68
Prasad K, Jha AK, Prasad K, Kulkarni AR (2010) Can microbes mediate nano-transformation? Indian J Phys 84(10):1355–1360
Prasad TVKV, Kambala V, Naidu R (2013) Phyconanotechnology: synthesis of silver nanoparticles using brown marine algae Cystophora moniliformis and their characterisation. J Appl Phycol 25(1):177–182
Quan X, Zhang X, Xu H (2015) In-situ formation and immobilization of biogenic nanopalladium into anaerobic granular sludge enhances azo dyes degradation. Water Res 78:74–83
Rai M, Yadav A, Gade A (2009) Current trends in phytosynthesis of metal nanoparticles (vol 28, pg 277, 2008). Crit Rev Biotechnol 29(1):78–78
Rajarao R, Ferreira R, Sadi SHF, Khanna R, Sahajwalla V (2014) Synthesis of silicon carbide nanoparticles by using electronic waste as a carbon source. Mater Lett 120:65–68

Rajput P, Anjum MH, Gupta T (2017) One year record of bioaerosols and particles concentration in Indo-Gangetic Plain: implications of biomass burning emissions to high-level of endotoxin exposure. Environ Pollut. 224:98–106. https://doi.org/10.1016/j.envpol.2017.01.045
Rajput VD, Minkina TM, Behal A, Sushkova SN, Mandzhieva S, Singh R et al (2018a) Effects of zinc-oxide nanoparticles on soil, plants, animals and soil organisms: a review. Environ Nanotechnol Monit Manag 9:76–84
Rajput VD, Minkina T, Sushkova S, Tsitsuashvili V, Mandzhieva S, Gorovtsov A, Gromakova N (2018b) Effect of nanoparticles on crops and soil microbial communities. J Soils Sediments 18(6):2179–2187
Ranjard L, Richaume A (2001) Quantitative and qualitative microscale distribution of bacteria in soil. Res Microbiol 152(8):707–716
Ranjard L, Nazaret S, Gourbière F, Thioulouse J, Linet P, Richaume A (2000) A soil microscale study to reveal the heterogeneity of Hg (II) impact on indigenous bacteria by quantification of adapted phenotypes and analysis of community DNA fingerprints. FEMS Microbiol Ecol 31(2):107–115
Rolim WR, Pelegrino MT, de Araújo Lima B, Ferraz LS, Costa FN, Bernardes JS et al (2019) Green tea extract mediated biogenic synthesis of silver nanoparticles: characterization, cytotoxicity evaluation and antibacterial activity. Appl Surf Sci 463:66–74
Rousk J, Ackermann K, Curling SF, Jones DL (2012) Comparative toxicity of nanoparticulate CuO and ZnO to soil bacterial communities. PLoS One 7:e34197
Roy S, Das TK (2016) Effect of biosynthesized silver nanoparticles on the growth and some biochemical parameters of Aspergillus foetidus. J Environ Chem Eng 4(2):1574–1583
Rubilar O, Rai M, Tortella G, Diez MC, Seabra AB, Durán N (2013) Biogenic nanoparticles: copper, copper oxides, copper sulphides, complex copper nanostructures and their applications. Biotechnol Lett 35(9):1365–1375
Saif Hasan S, Singh S, Parikh RY, Dharne MS, Patole MS, Prasad BLV, Shouche YS (2008) Bacterial synthesis of copper/copper oxide nanoparticles. J Nanosci Nanotechnol 8(6):3191–3196
Sangeetha G, Rajeshwari S, Rajendran V (2012) Aloe barbadensis Miller mediated green synthesis of mono-disperse copper oxide nanoparticles: optical properties. Spectrochim Acta A 97:1140–1144
Sanghi R, Verma P, Puri S (2011) Enzymatic formation of gold nanoparticles using Phanerochaete chrysosporium. Adv Chem Eng Sci 1(03):154
Sarkar J, Ray S, Chattopadhyay D, Laskar A, Acharya K (2012) Mycogenesis of gold nanoparticles using a phytopathogen Alternaria alternata. Bioprocess Biosyst Eng 35(4):637–643
Sawle BD, Salimath B, Deshpande R, Bedre MD, Prabhakar BK, Venkataraman A (2008) Biosynthesis and stabilization of Au and Au–Ag alloy nanoparticles by fungus, Fusarium semitectum. Sci Technol Adv Mater 9(3):035012
Scarano G, Morelli E (2003) Properties of phytochelatin-coated CdS nanocrystallites formed in a marine phytoplanktonic alga (Phaeodactylum tricornutum, Bohlin) in response to Cd. Plant Sci 165(4):803–810
Schlich K, Hund-Rinke K (2015) Influence of soil properties on the effect of silver nanomaterials on microbial activity in five soils. Environ Pollut 196:321–330
Schloter M, Dilly O, Munch JC (2003) Indicators for evaluating soil quality. Agric Ecosyst Environ 98(1–3):255–262
Schwertmann U (2008) Iron oxides. In: Chesworth W (ed) Encyclopedia of soil science. Springer, Dordrecht, pp 363–369
Shah V, Jones J, Dickman J, Greenman S (2014) Response of soil bacterial community to metal nanoparticles in biosolids. J Hazard Mater 274:399–403
Shankar PD, Shobana S, Karuppusamy I, Pugazhendhi A, Ramkumar VS, Arvindnarayan S, Kumar G (2016) A review on the biosynthesis of metallic nanoparticles (gold and silver) using bio-components of microalgae: formation mechanism and applications. Enzym Microb Technol 95:28–44
Sharma D, Kanchi S, Bisetty K (2015) Biogenic synthesis of nanoparticles: a review. Arab J Chem

Sheikhloo Z, Salouti M (2011) Intracellular biosynthesis of gold nanoparticles by the fungus Penicillium chrysogenum. Int J Nanosci Nanotechnol 7(2):102–105

Shen MH, Yin YG, Booth A, Liu JF (2015) Effects of molecular weight-dependent physicochemical heterogeneity of natural organic matter on the aggregation of fullerene nanoparticles in mono-and di-valent electrolyte solutions. Water Res 71:11–20

Shipley HJ, Engates KE, Guettner AM (2011) Study of iron oxide nanoparticles in soil for remediation of arsenic. J Nanopart Res 13(6):2387–2397

Shivaji S, Madhu S, Singh S (2011) Extracellular synthesis of antibacterial silver nanoparticles using psychrophilic bacteria. Process Biochem 46(9):1800–1807

Shotyk W (1988) Review of the inorganic geochemistry of peats and peatland waters. Earth Sci Rev 25(2):95–176

Shrestha B, Acosta-Martinez V, Cox SB, Green MJ, Li S, Cañas-Carrell JE (2013) An evaluation of the impact of multiwalled carbon nanotubes on soil microbial community structure and functioning. J Hazard Mater 261:188–197

Simonin M, Richaume A (2015) Impact of engineered nanoparticles on the activity, abundance, and diversity of soil microbial communities: a review. Environ Sci Pollut Res 22(18):13710–13723

Simonin M, Guyonnet JP, Martins JM, Ginot M, Richaume A (2015) Influence of soil properties on the toxicity of TiO2 nanoparticles on carbon mineralization and bacterial abundance. J Hazard Mater 283:529–535

Singaravelu G, Arockiamary JS, Kumar VG, Govindaraju K (2007) A novel extracellular synthesis of monodisperse gold nanoparticles using marine alga, Sargassum wightii Greville. Colloids Surf B: Biointerfaces 57(1):97–101

Singh AV, Patil R, Anand A, Milani P, Gade WN (2010) Biological synthesis of copper oxide nano particles using Escherichia coli. Curr Nanosci 6(4):365–369

Singh M, Kalaivani R, Manikandan S, Sangeetha N, Kumaraguru AK (2013) Facile green synthesis of variable metallic gold nanoparticle using Padina gymnospora, a brown marine macroalga. Appl Nanosci 3(2):145–151

Singh P, Singh H, Kim YJ, Mathiyalagan R, Wang C, Yang DC (2016) Extracellular synthesis of silver and gold nanoparticles by Sporosarcina koreensis DC4 and their biological applications. Enzym Microb Technol 86:75–83

Sinha S, Pan L, Chanda P, Sen SK (2009) Nanoparticles fabrication using ambient biological resources. J Appl Biosci 19:1113–1130

Sivakumar P, Sivakumar P, Anbarasu K, Pandian K, Renganathan S (2013) Synthesis of silver nanorods from food industrial waste and their application in improving the keeping quality of milk. Ind Eng Chem Res 52(49):17676–17681

Soltani Nejad M, Shahidi Bonjar GH, Khaleghi N (2015) Biosynthesis of gold nanoparticles using streptomyces fulvissimus isolate. Nanomed J 2(2):153–159

Srinath BS, Rai VR (2015) Rapid biosynthesis of gold nanoparticles by Staphylococcus epidermidis: its characterisation and catalytic activity. Mater Lett 146:23–25

Staniland SS (2007) Magnetosomes: bacterial biosynthesis of magnetic nanoparticles and potential biomedical applications. In: Nanotechnologies for the life sciences. WileyVCH Verlag GmbH & Co. KGaA

Stephen JR, Macnaughtont SJ (1999) Developments in terrestrial bacterial remediation of metals. Curr Opin Biotechnol 10(3):230–233

Suganya KU, Govindaraju K, Kumar VG, Dhas TS, Karthick V, Singaravelu G, Elanchezhiyan M (2015) Blue green alga mediated synthesis of gold nanoparticles and its antibacterial efficacy against gram positive organisms. Mater Sci Eng C 47:351–356

Suppan S (2013) Nanomaterials in soil. Institute for Agriculture and Trade Policy

Suresh AK, Pelletier DA, Wang W, Broich ML, Moon JW, Gu B et al (2011) Biofabrication of discrete spherical gold nanoparticles using the metal-reducing bacterium Shewanella oneidensis. Acta Biomater 7(5):2148–2152

Syed A, Saraswati S, Kundu GC, Ahmad A (2013) Biological synthesis of silver nanoparticles using the fungus Humicola sp. and evaluation of their cytoxicity using normal and cancer cell lines. Spectrochim Acta A Mol Biomol Spectrosc 114:144–147

Syed B, Prasad NM, Satish S (2016) Endogenic mediated synthesis of gold nanoparticles bearing bactericidal activity. J Microsc Ultrastruct 4(3):162–166

Tebo BM, Bargar JR, Clement BG, Dick GJ, Murray KJ, Parker D, Verity R, Webb SM (2004) Biogenic manganese oxides: properties and mechanisms of formation. Annu Rev Earth Planet Sci 32:287–328

Theng BK, Yuan G (2008) Nanoparticles in the soil environment. Elements 4(6):395–399

Tian H, Ghorbanpour M, Kariman K (2018) Manganese oxide nanoparticle-induced changes in growth, redox reactions and elicitation of antioxidant metabolites in deadly nightshade (*Atropa belladonna* L.). Ind Crop Prod 126:403–414

Tripp SL, Pusztay SV, Ribbe AE, Wei A (2002) Self-assembly of cobalt nanoparticle rings. J Am Chem Soc 124(27):7914–7915

Van Gestel CA, Kool PL, Diez Ortiz M (2010) Metal-based nanoparticles in soil: new research themes should not ignore old rules and theories. Comments on the paper by Hu et al. 2010 "Toxicological effects of TiO2 and ZnO nanoparticles in soil on earthworm Eisenia fetida". Soil Biol Biochem 42:1892–1893

Van Hoecke K, De Schamphelaere KA, Van der Meeren P, Smagghe G, Janssen CR (2011) Aggregation and ecotoxicity of CeO2 nanoparticles in synthetic and natural waters with variable pH, organic matter concentration and ionic strength. Environ Pollut 159(4):970–976

Van Horssen R, Ten Hagen TL, Eggermont AM (2006) TNF-alpha in cancer treatment: molecular insights, antitumor effects, and clinical utility. Oncologist 11(4):397–408

Vanaja M, Annadurai G (2013) Coleus aromaticus leaf extract mediated synthesis of silver nanoparticles and its bactericidal activity. Appl Nanosci 3(3):217–223

Varshney R, Bhadauria S, Gaur MS, Pasricha R (2010) Characterization of copper nanoparticles synthesized by a novel microbiological method. JOM-J Miner Met Mater Soc 62:102–104

Varshney R, Bhadauria S, Gaur MS, Pasricha R (2011) Copper nanoparticles synthesis from electroplating industry effluent. Nano Biomed Eng 3(2):115–119

Varshney R, Bhadauria S, Gaur MS (2012) A review: biological synthesis of silver and copper nanoparticles. Nano Biomed Eng 4(2):99–106

Venkatesan J, Manivasagan P, Kim SK, Kirthi AV, Marimuthu S, Rahuman AA (2014) Marine algae-mediated synthesis of gold nanoparticles using a novel Ecklonia cava. Bioprocess Biosyst Eng 37(8):1591–1597

Vijayaraghavan K, Mahadevan A, Sathishkumar M, Pavagadhi S, Balasubramanian R (2011) Biosynthesis of Au (0) from Au (III) via biosorption and bioreduction using brown marine alga Turbinaria conoides. Chem Eng J 167(1):223–227

Villalobos M, Bargar J, Sposito G (2005) Trace metal retention on biogenic manganese oxide nanoparticles. Elements 1(4):223–226

Visaria RK, Griffin RJ, Williams BW, Ebbini ES, Paciotti GF, Song CW, Bischof JC (2006) Enhancement of tumor thermal therapy using gold nanoparticle–assisted tumor necrosis factor-α delivery. Mol Cancer Ther 5(4):1014–1020

Wang X, Zhang D, Pan X, Lee DJ, Al-Misned FA, Golam Mortuza M, Gadd GM (2017) Aerobic and anaerobic biosynthesis of nano-selenium for remediation of mercury contaminated soil. Chemosphere 170:266–273

Wang X, Zhang D, Qian H, Liang Y, Pan X, Gadd GM (2018a) Interactions between biogenic selenium nanoparticles and goethite colloids and consequence for remediation of elemental mercury contaminated groundwater. Sci Total Environ 613:672–678

Wang R, Du H, Wang Y, Wang D, Sun Q, Zhou D (2018b) Retention of silver nanoparticles and silver ion to natural soils: effects of soil physicochemical properties. J Soils Sediments 18(7):2491–2499

Wang X, Song W, Qian H, Zhang D, Pan X, Gadd GM (2018c) Stabilizing interaction of exopolymers with nano-Se and impact on mercury immobilization in soil and groundwater. Environ Sci Nano 5:456–466

Wang X, Pan X, Gadd GM (2019a) Soil dissolved organic matter affects mercury immobilization by biogenic selenium nanoparticles. Sci Total Environ 658:8–15

Wang X, Pan X, Gadd GM (2019b) Immobilization of elemental mercury by biogenic Se nanoparticles in soils of varying salinity. Sci Total Environ 668:303–309

Wang X, Wang S, Pan X, Gadd GM (2019c) Heteroaggregation of soil particulate organic matter and biogenic selenium nanoparticles for remediation of elemental mercury contamination. Chemosphere 221:486–492

Watson JHP, Ellwood DC, Deng Q, Mikhalovsky S, Hayter CE, Evans J (1995) Heavy metal adsorption on bacterially produced FeS. Miner Eng 8(10):1097–1108

Whitley AR, Levard C, Oostveen E, Bertsch PM, Matocha CJ, von der Kammer F, Unrine JM (2013) Behavior of Ag nanoparticles in soil: effects of particle surface coating, aging and sewage sludge amendment. Environ Pollut 182:141–149

Wu J, Zhang H, Shao L, He P (2012) Fluorescent characteristics and metal binding properties of individual molecular weight fractions in municipal solid waste leachate. Environ Pollut 162:63–71

Xu C, Peng C, Sun L, Zhang S, Huang H, Chen Y, Shi J (2015) Distinctive effects of TiO2 and CuO nanoparticles on soil microbes and their community structures in flooded paddy soil. Soil Biol Biochem 86:24–33

Yadav A, Kon K, Kratosova G, Duran N, Ingle AP, Rai M (2015) Fungi as an efficient mycosystem for the synthesis of metal nanoparticles: progress and key aspects of research. Biotechnol Lett 37(11):2099–2120

Yallappa S, Manjanna J, Dhananjaya BL (2015) Phytosynthesis of stable Au, Ag and Au-Ag alloy nanoparticles using J. sambac leaves extract, and their enhanced antimicrobial activity in presence of organic antimicrobials. Spectrochim Acta A Mol Biomol Spectrosc 137:236–243

Yuan G, Percival HJ, Theng BKG, Parfitt RL (2002) Sorption of copper and cadmium by allophane-humic complexes. In: Violante A, Huang PM, Bollag J-M, Gianfreda L (eds) Developments in soil science, vol 28A, pp 37–47

Yung MM, Kwok KW, Djurišić AB, Giesy JP, Leung KM (2017) Influences of temperature and salinity on physicochemical properties and toxicity of zinc oxide nanoparticles to the marine diatom Thalassiosira pseudonana. Sci Rep 7(1):3662

Zhang Y, Gao G, Qian Q, Cui D (2012) Chloroplasts-mediated biosynthesis of nanoscale Au-Ag alloy for 2-butanone assay based on electrochemical sensor. Nanoscale Res Lett 7(1):475

Zhao L, Peralta-Videa JR, Ren M, Varela-Ramirez A, Li C, Hernandez-Viezcas JA et al (2012) Transport of Zn in a sandy loam soil treated with ZnO NPs and uptake by corn plants: electron microprobe and confocal microscopy studies. Chem Eng J 184:1–8

Biogenic Nanoparticles as Novel Sustainable Approach for Plant Protection

9

Nariman Maroufpour, Mahdieh Mousavi, Mahsa Abbasi, and Mansour Ghorbanpour

Abstract

Around the world due to plant diseases, the amount of crop decreases annually. Different natural and synthetic approaches to manage and reduce damage diseases are used, out of which using nanoparticles is one of them. In the last decades, interest in research on using nanoparticles has increased because of the global concern about environmental pollution. A variety of traditional physical and chemical processes are used to produce nanoscale materials, but nowadays environment-friendly green chemistry-based techniques are available to biologically synthesize materials. Recently, nanotechnology and biology have convergence to create a new field called nanobiotechnology which incorporates the use of biological entities such as actinomycetes, algae, bacteria, fungi, viruses, yeasts, and plants in a number of processes, either biochemical or biophysical. Nanobiotechnology processes have a significant potential to boost nanoparticle production and reduce the use of harsh, toxic, and expensive chemicals that are commonly used in the conventional physical and chemical processes of production.

N. Maroufpour
Department of Plant Protection, Faculty of Agriculture, University of Tabriz, Tabriz, Iran

M. Mousavi
Department of Plant Protection, Faculty of Agriculture, Urmia University, Urmia, Iran

M. Abbasi
Department of Horticultural Science, Gorgan University of Agricultural Sciences and Natural Resources, Gorgan, Iran

M. Ghorbanpour (✉)
Department of Medicinal Plants, Faculty of Agriculture and Natural Resources, Arak University, Arak, Iran
e-mail: m-ghorbanpour@araku.ac.ir

M. Ghorbanpour et al. (eds.), *Biogenic Nano-Particles and their Use in Agro-ecosystems*, https://doi.org/10.1007/978-981-15-2985-6_9

Keywords
Biogenic · Nanoparticles · Nanotechnology · Plant diseases

9.1 Introduction

The most challenging aspect of the farmer for crop production is control of plant diseases. The primary food source for humans and almost other organisms on the earth are plants, among which six plant species of sweet potato, wheat, corn, cassava, rice, and potato provide 80% of calories consumed by humans worldwide (FAO 2012; Goudie and Cuff 2001). As respects crop casualties in resources, because of diseases, are usually overstated and not founded on a comprehensive strategy considering all possible production limitations. A recent investigation handling with all production limitation (containing diseases) in 13 Asian and African agriculture systems for six main crops (rice, wheat, cassava, cowpea, chickpea, and sorghum) revealed that casualties created by diseases ranged from 3% to 14%, while yield casualties to all crop production restrictions ranged from 36% to 65% and yield casualties because of all biotic stresses ranged from 16% to 37% (Waddington et al. 2010). To control plant diseases, different kinds of synthetic and natural methods are used, out of which the application of chemical pesticides is highly prevalent. Recently, environmental risk of chemical pesticides has caused scientists to search for an alternative approach with the least environmental hazards (Jo et al. 2009). In order to control plant diseases, agricultural scientists and experts seek alternative eco-friendly and less capital-intensive methods (Parthiban et al. 2019). As a replacement to chemical pesticides, usage of nanoparticles as antimicrobial factors has become highly prevalent as technological progression, making their production affordable (Malandrakis et al. 2019; Sahadan et al. 2019). A new area of utilization of NPs in agriculture is seen in their usage as a fertilizer or a plant protection product (Gogos et al. 2012; Khot et al. 2012; Rai and Ingle 2012) (Fig. 9.1).

The increasing demand for nanomaterials should be accompanied by "green" synthesis methods in an effort to reduce generated hazardous waste from this industry. Green chemistry would help minimize the use of unsafe products and maximize the efficiency of chemical processes (Parthiban et al. 2019). An advantage of biogenic synthesis, over conventional chemical synthesis, is the safer and easier handling of microbial cultures and the simpler downstream processing of biomass as compared to synthetic methods (Rai et al. 2011). Hence, biogenic NP synthesis represents a very interesting greener and more environmentally friendly manufacturing alternative, due to the use of chemicals of lower toxicity and to the use of lower ambient temperatures and lower pressures in the synthesis (Chhipa 2019).

In an attempt to overcome the limitations of these conventional methods and the growing demand to come up with synthesis methods of nanomaterials with the desired size and shape that are fast and eco-friendly, researchers have developed biogenic synthesis principles for nanomaterials with the use of biological resources such as plants and microorganisms or their products (Rónavári et al. 2018; Schröfel et al. 2011).

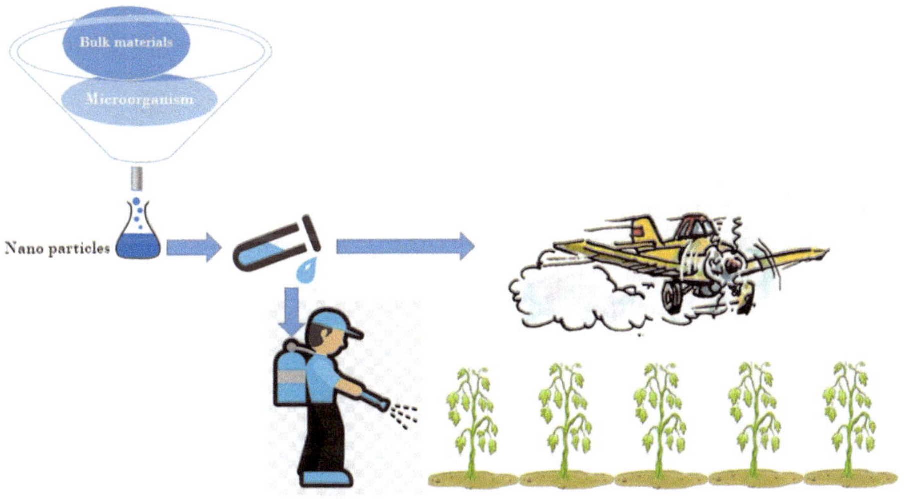

Fig. 9.1 Application of nanoparticle in plant protection

9.2 Biogenic Nanoparticle

Diverse microorganisms of both prokaryotic and eukaryotic origins are used for biosynthesis of metallic nanoparticles such as silver, gold, platinum, zirconium, palladium, iron, and cadmium and metal oxides such as titanium oxide, zinc oxide, etc. as a green and eco-friendly technology (Hasan 2015; Luo et al. 2018). Bacteria, actinomycetes, fungi, and algae are part of these microorganisms. Based on the location of nanoparticles, the synthesis process can be intracellular or extracellular (Hulkoti and Taranath 2014; Mann 2001).

The most extensively researched natural resource for the synthesis of metallic nanoparticles are prokaryotic bacteria (Luo et al. 2018), which the relative ease to manipulate them for nanoparticles synthesis is one of the reasons for the "bacterial preference" (Slawson et al. 1992). Microbial synthesis is one of such processes, a green chemistry approach that interlinks nanotechnology and microbial biotechnology (Luo et al. 2018). The major bacterial species used for the synthesis of metallic nanoparticles include *Acinetobacter* spp., *Escherichia coli*, *Klebsiella pneumoniae*, *Lactobacillus* spp., *Bacillus cereus*, *Corynebacterium* sp., and *Pseudomonas* sp. (Iravani 2014; Mohanpuria et al. 2008; NVKV Prasad et al. 2011).

Synthesis of metal nanoparticles in microbes takes place in the intracellular or extracellular environment (Ghandehari et al. 2018), the former of which requires additional steps such as ultrasound treatment or reactions with suitable detergents in order to isolate the synthesized nanoparticles (Kalimuthu et al. 2008). Meanwhile, the latter is cheaper and requires simpler downstream processing. The extracellular form is in favor of large-scale production of silver nanoparticles to further explore its potential applications. In regard to this, a lot of researchers have focused their work on extracellular methods of metal nanoparticle synthesis (Durán et al. 2005).

Considering the fast rate of microbe reproduction, it makes them a perfect model to exploit their use in various aspects of biological sciences. Microbiology is the result of biotechnology joining hands and emerging as an initiative to study microbes and their various characteristics.

Microorganisms, called nanofactories because of their small size of 10–6 nm, generate nanoparticles. Because of their presence in nature, they are also called biofactories (Deepak et al. 2011). Bacteria synthesize metal nanoparticles as a defense mechanism (resistance mechanism), since the resistance of bacterial cell on metal ions in the environment is the cause for nanoparticle synthesis, and cell wall (negatively charged) interacts electrostatically with the metal ions (positively charged) (Luo et al. 2018).

The enzymes present within cell wall bioreduce the metal ions to nanoparticles, and finally the smaller-sized nanoparticles get diffused of through the cell wall, and the nanoparticles are produced (Mukherjee et al. 2001). The first synthesis of Ag nanoparticles by bacteria was reported in 2000. Joerger et al. (2000) used *Pseudomonas stutzeri* AG259 to synthesize Ag nanoparticles with a size of less than 200 nm.

The rapid synthesis of silver nanoparticles using the reduction of aqueous Ag^+ ion in culture supernatants of *Klebsiella pneumoniae*, *Escherichia coli*, and *Enterobacter cloacae* (Enterobacteriaceae) has been reported by Shahverdi et al. (2007) and (Lee 1996). *Lactobacillus*, a common bacterial strain present in the buttermilk, synthesizes both Au and AgNPs under standard conditions (Nair and Pradeep 2002).

Also, the synthesis of spherical Cu nanoparticles using nonpathogenic *Pseudomonas stutzeri* has been observed in a rapid biological synthesis technique performed by Varshney et al. (2011). In a recent study, an innovative approach used *Pseudomonas stutzeri* bacterial strain to synthesize copper nanoparticles from electroplating waste water (Varshney et al. 2011).

In relation to other plant material and microbes for NP synthesis, fungi have a number of advantages. Several fungal strains such as *Aspergillus*, *Fusarium*, *Penicillium*, and *Verticillium* have been used as promising resources for nanoparticle fabrication. The fungi because they are simpler to handle in the laboratory and produce large quantities of enzymes have the potential to use the synthesis of NPs (Mandal et al. 2006; Mitra et al. 2019; Mohanpuria et al. 2008). Many studies have been reported the synthesis of NPs by fungi. The synthesis of nanoparticles by fungi may be intracellular or extracellular. The authors reported that *Fusarium oxysporum* synthesized the AgNPs by intra- and extracellular method (Birla et al. 2013; Khosravi and Shojaosadati 2009; Mitra et al. 2019; Zhang et al. 2019). In another study, synthesis of magnetite NPs by *F. oxysporum* and *Verticillium* sp. was demonstrated (Durán et al. 2010). Also, biosynthesis of NPs by *Penicillium* sp. was reported by other researchers (Kathiresan et al. 2009; Maliszewska et al. 2014; Nayak et al. 2011; Shaligram et al. 2009; Singh et al. 2014). The investigations proved that biosynthesis of metal NPs is affected by culture conditions (Hamad 2019; Ingle et al. 2008; Saravanan and Nanda 2010), and some factors such as the biomass concentration of the fungal species, incubation time and conditions, pH, temperature, nature of the metal species or parent compound, and colloidal interaction conditions (Jain et al. 2011; Kumar et al. 2008; Prakash and Soni 2011; Saravanan and Nanda 2010; Sunkar and Nachiyar 2013).

9.3 Nanoparticle and Pathogen Control

In agriculture, many NPs have been studied for their potential application such as nano-zinc oxide (ZnO), nano-silica (SiO_2), nanosilver (Ag), nano-titanium dioxide (TiO_2), nano-copper (Cu), carbon nanotubes (CNT), and nano-aluminum (Al). With approximately half of the NP constituents being designed as the active ingredient and the other half as the additive, plant protection is the principal purpose (75%). Furthermore, fertilization and UV protection are other purposes (Gogos et al. 2012). As physicochemical properties of nano-forms considerably differ from bulk forms, they become significant to test the impact of NPs on microorganisms in order to restrain the benefit of this technology in the plant protection exclusively against phytopathogens. Nanoparticles may affect activity of microorganisms through too small size, even smaller than a virus bit and high responsiveness (Khan and Rizvi 2014). Various types of nanomaterials have also been tested like copper, zinc, titanium (Gu et al. 2003), magnesium, gold, alginate (Ahmad et al. 2006), and silver, but AgNPs have shown to be the most efficient ones as they have excellent antimicrobial efficiency against bacteria, viruses, fungi, and other eukaryotic microorganisms (Alvarez-Puebla et al. 2004; Gong et al. 2007; Lead and Wilkinson 2006).

Based on the possible reactions between nanoparticles and macromolecules of living organisms, numerous investigations have been conducted. Due to different charges with microorganisms, a nanoparticle acts as an electromagnetic absorber between the microbe and the nanoparticle and helps the nanoparticle to stick to the cell surface, leading to cell death. Many of these contacts cause oxidation of surface molecules of microbes, eventually leading to death (Lin and Xing 2007).

Bacterial development and stress resistance, plant susceptibility to bacterial infection, and mechanisms of interaction among plant and related bacteria are influenced by some NPs (Degrassi et al. 2012). Many researchers consider the antibacterial characteristics of nanoparticles so that many articles are daily published concerning the wonderful property of nanoparticles to overcome pathogens resistant to current antibiotics (Elechiguerra et al. 2005). A number of nanomaterials such as carbon nanotubes (Kang et al. 2007; Liu et al. 2009), iron-based nanoparticles (Hu et al. 2010), silver (Sondi and Salopek-Sondi 2004), graphene-based nanomaterials (Hu et al. 2010), zinc, copper, and titanium oxide nanoparticles (Ge et al. 2011; Kasemets et al. 2009) have been noted to have a toxic influence on pure cultures of bacteria (He et al. 2011) and toxic impact on both harmful and useful rhizosphere microorganisms (Gajjar et al. 2009; Gunawan et al. 2011; Jones et al. 2008) containing *Pseudomonas chlororaphis* (Dimkpa et al. 2012), *Pseudomonas putida* (Gajjar et al. 2009), *Escherichia coli*, *Bacillus subtilis*, *Streptococcus aureus* (Baek and An 2011), and *Campylobacter jejuni* (Xie et al. 2011). Nevertheless, the reports are incompatible about the effect of nanoparticles on secondary metabolites of microbes. The investigation of the antibacterial action of zinc NPs has demonstrated that the morphology and oxidative stress play significant roles in the antibacterial activity (Dwivedi et al. 2014; Raghupathi et al. 2011). Several elements such as ZnO, SiO_2, and inorganic TiO_2 have been reported to exert a toxic impact on bacteria that toxicity of these elements considerably intensifies in attendance of light (Adams et al. 2006).

The most sensitive microbial process of the nitrogen cycle may be nitrification, albeit moderate stimulatory impacts can also consequence on exposure to a narrow range of sublethal AgNP concentrations. Reduction in nitrification activity occurs before other nitrogen cycling operations are affected if AgNP concentrations reach the inhibitory levels (Yang et al. 2013). Mechanisms of AgNPs against bacteria involved alteration of cell wall and cytoplasm (Łysakowska et al. 2015), alteration of membrane permeability and respiration (Manjumeena et al. 2014), alteration of membrane (Naraginti and Sivakumar 2014), morphological alterations, disjuncture of the cytoplasmic membrane from the cell wall, plasmolysis (Tamayo et al. 2014), change of membrane with inhibition respiratory activity (Wang et al. 2014), irremediable injury on bacterial cells, change of membrane penetrance and respiration (Morones et al. 2005), irreversible damage on bacterial cells (Jain et al. 2009), deterrence of bacterial DNA replication, bacterial cytoplasm membrane injury, and alteration of intracellular ATP levels (Shameli et al. 2012). For sustaining life, bacteria are believed to use an enzyme to metabolize oxygen, but silver ions disable the enzyme and prevent the metabolization of oxygen suffocating the bacteria, resulting in death (Alghuthaymi et al. 2015). The deterrent impact of nanoparticles against bacterial types can be due to injury to the bacterial enzymes or plasma membrane. Leakage of the cytoplasmic content to the surroundings impaired metabolic pathways, leading to the death of bacterial cells (Li et al. 2006).

Some researchers have reported the antifungal impact of silver nanoparticles on some pathogenic fungi (Kim et al. 2009). The considerable decrease in mycelial development and incubated spore germination is observed with silver nanoparticles (Morones et al. 2005). In the nanoparticles, the inbitory impact can be owing to the release of extracellular enzymes and metabolites working as a factor for their own survival when exposed to stress from poisonous elements and temperature alterations as confirmed in the case of fungus *Trichoderma reesei* and other fungi (Pérez-de-Luque and Rubiales 2009).

A number of silver nanoparticles have antifungal property on some fungi such as *Fusarium* species, wood rotting fungi, and other phytopathogenic fungi. The investigation has shown that the antifungal activity against fungal pathogens can be due to repression of enzymes and toxins utilized by the fungal pathogens for pathogenesis (Bhainsa and D'souza 2006; Khabat et al. 2011). Min et al. (2009) determined the antifungal activity of silver nanoparticles against sclerotium-forming phytopathogens, *S. sclerotiorum*, *S. minor*, and *R. solani*. According to the obtained information, the nanoparticles strongly inhibited the sclerotial germination growth and fungal development. In addition, it is suggested that nanometer-sized silvers have various attributes, which might be the result of physiological, morphological, and structural changes (Baek and An 2011). The data collected from the microscope demonstrated that silver nanoparticle-treated hyphae were extremely damaged on walls causing the plasmolysis of hyphae (Min et al. 2009). A recent study indicated that transport systems, including ion efflux, were disrupted by silver nanoparticles (Morones et al. 2005). The result of dysfunction of ion efflux may be a quick accumulation of silver ions, preventing cellular activities at their lower concentrations such as metabolism and respiration by responding to molecules. Furthermore, silver ions through reaction with oxygen can generate reactive oxygen species (ROS), being harmful to cells and leading to injury to nucleic acids, lipids, and proteins (Hwang et al. 2008).

At 100 ppm concentration of silver nanoparticles, most fungi have shown high deterrence efficacy. Generally, with the increase in the concentration of AgNPs, inhibition increased. This increase occurs owing to the above density at which the solution could impregnate and stick to fungal hyphae, inactivating plant pathogenic fungi. The DNA loses its capability to replicate after treatment with Ag finishing in the deactivated expression of ribosomal subunit proteins. Moreover, some other cellular proteins and enzymes are necessary to the ATP production. For Ag^+, it is assumed that Ag^+ primarily impacts the function of membrane-bound enzymes such as those in the respiratory chain (Kim et al. 2012). Nowadays, the application of the antimicrobial properties of nano-sized silver particles has become more accepted as technological progress made their manufacturing more economical. In order to control spore-producing fungal plant pathogens, the antifungal activity of nanoparticle silver or ionic has high potential. Studies on the antifungal activity of different forms of silver ions and nanoparticles have been conducted on the two plant pathogenic fungi, namely, *Bipolaris sorokiniana* and *Magnaporthe grisea* (Jo et al. 2009). According to the results of investigations, the effect of silver is highly influenced by utilization time and preventive applications of silver nanoparticles work better before fungal isolates penetrate and colonize within the plant tissue (Kim et al. 2012).

9.4 Conclusions

Biogenic methods for synthesis of nanoparticles are still in the development stages, and furthermore, biogenic synthesized NPs in comparison with chemically synthesized ones are more polydisperse. Optimization of important parameters controls the growth condition of organisms, cellular activities, and enzymatic processes which in turn can control the properties of NPs. Since the mechanism of production has not been described clearly, more investigations are needed to understand the exact mechanisms of biogenic methods and identify the enzymes and proteins involved in nanoparticle biosynthesis. Given that in many fields, these nanomaterials can be used, but environmental risks about them exist, so we need more study in this area. Top of this matter, it's true that the nanoparticles have a good potential in control and nanoparticles are a new weapon of human for the fight with plant diseases.

References

Adams LK, Lyon DY, Alvarez PJ (2006) Comparative eco-toxicity of nanoscale TiO_2, SiO_2, and ZnO water suspensions. Water Res 40:3527–3532

Ahmad Z, Pandey R, Sharma S, Khuller G (2006) Alginate nanoparticles as antituberculosis drug carriers: formulation development, pharmacokinetics and therapeutic potential. Indian J Chest Dis Allied Sci 48:171–176

Alghuthaymi MA, Almoammar H, Rai M, Said-Galiev E, Abd-Elsalam KA (2015) Myconanoparticles: synthesis and their role in phytopathogens management. Biotechnol Biotechnol Equip 29:221–236

Alvarez-Puebla R, Dos Santos D Jr, Aroca R (2004) Surface-enhanced Raman scattering for ultrasensitive chemical analysis of 1 and 2-naphthalenethiols. Analyst 129:1251–1256

Baek Y-W, An Y-J (2011) Microbial toxicity of metal oxide nanoparticles (CuO, NiO, ZnO, and Sb_2O_3) to *Escherichia coli*, *Bacillus subtilis*, and Streptococcus aureus. Sci Total Environ 409:1603–1608

Bhainsa KC, D'souza S (2006) Extracellular biosynthesis of silver nanoparticles using the fungus *Aspergillus fumigatus*. Colloids Surf B: Biointerfaces 47:160–164

Birla SS, Gaikwad SC, Gade AK, Rai MK (2013) Rapid synthesis of silver nanoparticles from Fusarium oxysporum by optimizing physicocultural conditions. Sci World J 2013:18

Chhipa H (2019) Mycosynthesis of nanoparticles for smart agricultural practice: a green and eco-friendly approach. In: Green synthesis, characterization and applications of nanoparticles. Elsevier, Amsterdam, pp 87–109

Deepak V, Kalishwaralal K, Pandian SRK, Gurunathan S (2011) An insight into the bacterial biogenesis of silver nanoparticles, industrial production and scale-up. In: Metal nanoparticles in microbiology. Springer, Berlin, pp 17–35

Degrassi G, Bertani I, Devescovi G, Fabrizi A, Gatti A, Venturi V (2012) Response of plant-bacteria interaction models to nanoparticles EQA-international. J Environ Qual 8:39–50

Dimkpa CO, Zeng J, McLean JE, Britt DW, Zhan J, Anderson AJ (2012) Production of indole-3-acetic acid via the indole-3-acetamide pathway in the plant-beneficial bacterium *Pseudomonas chlororaphis* O_6 is inhibited by ZnO nanoparticles but enhanced by CuO nanoparticles. Appl Environ Microbiol 78:1404–1410

Durán N, Marcato PD, Alves OL, De Souza GI, Esposito E (2005) Mechanistic aspects of biosynthesis of silver nanoparticles by several Fusarium oxysporum strains. J Nanobiotechnol 3:8

Durán N, Marcato PD, Alves OL, Da Silva JP, De Souza GI, Rodrigues FA, Esposito E (2010) Ecosystem protection by effluent bioremediation: silver nanoparticles impregnation in a textile fabrics process. J Nanopart Res 12:285–292

Dwivedi S, Wahab R, Khan F, Mishra YK, Musarrat J, Al-Khedhairy AA (2014) Reactive oxygen species mediated bacterial biofilm inhibition via zinc oxide nanoparticles and their statistical determination. PLoS One 9:e111289

Elechiguerra JL, Burt JL, Morones JR, Camacho-Bragado A, Gao X, Lara HH, Yacaman MJ (2005) Interaction of silver nanoparticles with HIV-1. J Nanobiotechnol 3:6

FAO F (2012) Agriculture Organization of the United Nations, FAOSTAT database

Gajjar P, Pettee B, Britt DW, Huang W, Johnson WP, Anderson AJ (2009) Antimicrobial activities of commercial nanoparticles against an environmental soil microbe, *Pseudomonas putida* KT2440. J Biol Eng 3:9

Ge Y, Schimel JP, Holden PA (2011) Evidence for negative effects of TiO_2 and ZnO nanoparticles on soil bacterial communities. Environ Sci Technol 45:1659–1664

Ghandehari F, Fani M, Rezaee M (2018) Biosynthesis of iron oxide nanoparticles by cytoplasmic extract of bacteria *Lactobacillus Fermentum*. J Med Chem Sci 1:28–30

Gogos A, Knauer K, Bucheli TD (2012) Nanomaterials in plant protection and fertilization: current state, foreseen applications, and research priorities. J Agric Food Chem 60:9781–9792

Gong P et al (2007) Preparation and antibacterial activity of Fe_3O_4@Ag nanoparticles. Nanotechnology 18:285604

Goudie AS, Cuff DJ (2001) Encyclopedia of global change: environmental change and human society, vol 1. Oxford University Press, Oxford

Gu H, Ho P, Tong E, Wang L, Xu B (2003) Presenting vancomycin on nanoparticles to enhance antimicrobial activities. Nano Lett 3:1261–1263

Gunawan C, Teoh WY, Marquis CP, Amal R (2011) Cytotoxic origin of copper (II) oxide nanoparticles: comparative studies with micron-sized particles, leachate, and metal salts. ACS Nano 5:7214–7225

Hamad M (2019) Biosynthesis of silver nanoparticles by fungi and their antibacterial activity. Int J Environ Sci Technol 16:1015–1024

Hasan S (2015) A review on nanoparticles: their synthesis and types. Res J Recent Sci 2277:9–11

He S, Feng Y, Ren H, Zhang Y, Gu N, Lin X (2011) The impact of iron oxide magnetic nanoparticles on the soil bacterial community. J Soils Sediments 11:1408–1417

Hu W et al (2010) Graphene-based antibacterial paper. ACS Nano 4:4317–4323

Hulkoti NI, Taranath T (2014) Biosynthesis of nanoparticles using microbes—a review. Colloids Surf B: Biointerfaces 121:474–483

Hwang ET, Lee JH, Chae YJ, Kim YS, Kim BC, Sang BI, Gu MB (2008) Analysis of the toxic mode of action of silver nanoparticles using stress-specific bioluminescent bacteria. Small 4:746–750

Ingle A, Gade A, Pierrat S, Sonnichsen C, Rai M (2008) Mycosynthesis of silver nanoparticles using the fungus *Fusarium acuminatum* and its activity against some human pathogenic bacteria. Curr Nanosci 4:141–144

Iravani S (2014) Bacteria in nanoparticle synthesis: current status and future prospects. Int Sch Res Not 2014:18

Jain J, Arora S, Rajwade JM, Omray P, Khandelwal S, Paknikar KM (2009) Silver nanoparticles in therapeutics: development of an antimicrobial gel formulation for topical use. Mol Pharm 6:1388–1401

Jain N, Bhargava A, Majumdar S, Tarafdar J, Panwar J (2011) Extracellular biosynthesis and characterization of silver nanoparticles using *Aspergillus flavus* NJP08: a mechanism perspective. Nanoscale 3:635–641

Jo Y-K, Kim BH, Jung G (2009) Antifungal activity of silver ions and nanoparticles on phytopathogenic fungi. Plant Dis 93:1037–1043

Joerger R, Klaus T, Granqvist CG (2000) Biologically produced silver–carbon composite materials for optically functional thin-film coatings. Adv Mater Res 12:407–409

Jones N, Ray B, Ranjit KT, Manna AC (2008) Antibacterial activity of ZnO nanoparticle suspensions on a broad spectrum of microorganisms. FEMS Microbiol Lett 279:71–76

Kalimuthu K, Babu RS, Venkataraman D, Bilal M, Gurunathan S (2008) Biosynthesis of silver nanocrystals by *Bacillus licheniformis*. Colloids Surf B: Biointerfaces 65:150–153

Kang S, Pinault M, Pfefferle LD, Elimelech M (2007) Single-walled carbon nanotubes exhibit strong antimicrobial activity. Langmuir 23:8670–8673

Kasemets K, Ivask A, Dubourguier H-C, Kahru A (2009) Toxicity of nanoparticles of ZnO, CuO and TiO_2 to yeast Saccharomyces cerevisiae. Toxicol In Vitro 23:1116–1122

Kathiresan K, Manivannan S, Nabeel M, Dhivya B (2009) Studies on silver nanoparticles synthesized by a marine fungus, *Penicillium fellutanum* isolated from coastal mangrove sediment. Colloids Surf B: Biointerfaces 71:133–137

Khabat V, Mansoori GA, Karimi S (2011) Biosynthesis of silver nanoparticles by fungus *Trichoderma Reesei*. Insciences J 1:65–79

Khan MR, Rizvi TF (2014) Nanotechnology: scope and application in plant disease management. Plant Pathol J 13:214–231

Khosravi A, Shojaosadati S (2009) Evaluation of silver nanoparticles produced by fungus *Fusarium oxysporum*. Int J Nanotechnol 6:973–983

Khot LR, Sankaran S, Maja JM, Ehsani R, Schuster EW (2012) Applications of nanomaterials in agricultural production and crop protection: a review. Crop Prot 35:64–70

Kim SW et al (2009) An in vitro study of the antifungal effect of silver nanoparticles on oak wilt pathogen *Raffaelea* sp. J Microbiol Biotechnol 19:760–764

Kim SW, Jung JH, Lamsal K, Kim YS, Min JS, Lee YS (2012) Antifungal effects of silver nanoparticles (AgNPs) against various plant pathogenic fungi. Mycobiology 40:53–58
Kumar R, Liu D, Zhang L (2008) Advances in proteinous biomaterials. J Biobased Mater Bioenergy 2:1–24
Lead JR, Wilkinson KJ (2006) Aquatic colloids and nanoparticles: current knowledge and future trends. Environ Chem 3:159–171
Lee SY (1996) High cell-density culture of Escherichia coli. Trends Biotechnol 14:98–105
Li Y, Leung P, Yao L, Song Q, Newton E (2006) Antimicrobial effect of surgical masks coated with nanoparticles. J Hosp Infect 62:58–63
Lin D, Xing B (2007) Phytotoxicity of nanoparticles: inhibition of seed germination and root growth. Environ Pollut 150:243–250
Liu S et al (2009) Sharper and faster "nano darts" kill more bacteria: a study of antibacterial activity of individually dispersed pristine single-walled carbon nanotube. ACS Nano 3:3891–3902
Luo K, Jung S, Park K-H, Kim Y-R (2018) Microbial biosynthesis of silver nanoparticles in different culture media. J Agric Food Chem 66:957–962
Łysakowska ME, Ciebiada-Adamiec A, Klimek L, Sienkiewicz M (2015) The activity of silver nanoparticles (Axonnite) on clinical and environmental strains of Acinetobacter spp. Burns 41:364–371
Malandrakis AA, Kavroulakis N, Chrysikopoulos CV (2019) Use of copper, silver and zinc nanoparticles against foliar and soil-borne plant pathogens. Sci Total Environ 670:292–299
Maliszewska I, Juraszek A, Bielska K (2014) Green synthesis and characterization of silver nanoparticles using ascomycota fungi *Penicillium nalgiovense* AJ12. J Clust Sci 25:989–1004
Mandal D, Bolander ME, Mukhopadhyay D, Sarkar G, Mukherjee P (2006) The use of microorganisms for the formation of metal nanoparticles and their application. Appl Microbiol Biotechnol 69:485–492
Manjumeena R, Duraibabu D, Sudha J, Kalaichelvan P (2014) Biogenic nanosilver incorporated reverse osmosis membrane for antibacterial and antifungal activities against selected pathogenic strains: an enhanced eco-friendly water disinfection approach. J Environ Sci Health A 49:1125–1133
Mann S (2001) Biomineralization: principles and concepts in bioinorganic materials chemistry, vol 5. Oxford University Press, New York. on Demand
Min J-S et al (2009) Effects of colloidal silver nanoparticles on sclerotium-forming phytopathogenic fungi. Plant Pathol J 25:376–380
Mitra C, Gummadidala PM, Merrifield R, Omebeyinje MH, Jesmin R, Lead JR, Chanda A (2019) Size and coating of engineered silver nanoparticles determine their ability to growth-independently inhibit aflatoxin biosynthesis in *Aspergillus parasiticus*. Appl Microbiol Biotechnol 103:4623–4632
Mohanpuria P, Rana NK, Yadav SK (2008) Biosynthesis of nanoparticles: technological concepts and future applications. J Nanopart Res 10:507–517
Morones JR, Elechiguerra JL, Camacho A, Holt K, Kouri JB, Ramírez JT, Yacaman MJ (2005) The bactericidal effect of silver nanoparticles. Nanotechnology 16:2346–2353
Mukherjee P et al (2001) Bioreduction of AuCl4− ions by the fungus, *Verticillium* sp. and surface trapping of the gold nanoparticles formed. Angew Chem Int Ed 40:3585–3588
Nair B, Pradeep T (2002) Coalescence of nanoclusters and formation of submicron crystallites assisted by *Lactobacillus strains*. Cryst Growth Des 2:293–298
Naraginti S, Sivakumar A (2014) Eco-friendly synthesis of silver and gold nanoparticles with enhanced bactericidal activity and study of silver catalyzed reduction of 4-nitrophenol. Spectrochim Acta A Mol Biomol Spectrosc 128:357–362
Nayak RR, Pradhan N, Behera D, Pradhan KM, Mishra S, Sukla LB, Mishra BK (2011) Green synthesis of silver nanoparticle by *Penicillium purpurogenum* NPMF: the process and optimization. J Nanopart Res 13:3129–3137
NVKV Prasad T, Subba Rao Kambala V, Naidu R (2011) A critical review on biogenic silver nanoparticles and their antimicrobial activity. Curr Nanosci 7:531–544

Parthiban E, Manivannan N, Ramanibai R, Mathivanan N (2019) Green synthesis of silver-nanoparticles from Annona reticulata leaves aqueous extract and its mosquito larvicidal and anti-microbial activity on human pathogens. Biotechnol Rep 21:e00297
Pérez-de-Luque A, Rubiales D (2009) Nanotechnology for parasitic plant control. Pest Manag Sci 65:540–545
Prakash N, Soni N (2011) Factors affecting the geometry of silver nanoparticles synthesis in *Chrysosporium tropicum* and *Fusarium oxysporum*. Am J Biochem Biotechnol 2:112–121
Raghupathi KR, Koodali RT, Manna AC (2011) Size-dependent bacterial growth inhibition and mechanism of antibacterial activity of zinc oxide nanoparticles. Langmuir 27:4020–4028
Rai M, Ingle A (2012) Role of nanotechnology in agriculture with special reference to management of insect pests. Appl Microbiol Biotechnol 94:287–293
Rai M, Gade A, Yadav A (2011) Biogenic nanoparticles: an introduction to what they are, how they are synthesized and their applications. In: Metal nanoparticles in microbiology. Springer, Berlin, pp 1–14
Rónavári A et al (2018) Biosynthesized silver and gold nanoparticles are potent antimycotics against opportunistic pathogenic yeasts and dermatophytes. Int J Nanomedicine 13:695
Sahadan MYB et al (2019) Phomopsidione nanoparticles coated contact lenses reduce microbial keratitis causing pathogens. Exp Eye Res 178:10–14
Saravanan M, Nanda A (2010) Extracellular synthesis of silver bionanoparticles from Aspergillus clavatus and its antimicrobial activity against MRSA and MRSE. Colloids Surf B: Biointerfaces 77:214–218
Schröfel A, Kratošová G, Bohunická M, Dobročka E, Vávra I (2011) Biosynthesis of gold nanoparticles using diatoms—silica-gold and EPS-gold bionanocomposite formation. J Nanopart Res 13:3207–3216
Shahverdi AR, Minaeian S, Shahverdi HR, Jamalifar H, Nohi A-A (2007) Rapid synthesis of silver nanoparticles using culture supernatants of Enterobacteria: a novel biological approach. Process Biochem 42:919–923
Shaligram NS, Bule M, Bhambure R, Singhal RS, Singh SK, Szakacs G, Pandey A (2009) Biosynthesis of silver nanoparticles using aqueous extract from the compactin producing fungal strain. Process Biochem 44:939–943
Shameli K, Ahmad MB, Jazayeri SD, Shabanzadeh P, Sangpour P, Jahangirian H, Gharayebi Y (2012) Investigation of antibacterial properties silver nanoparticles prepared via green method. Chem Cent J 6:73
Singh D, Rathod V, Ninganagouda S, Hiremath J, Singh AK, Mathew J (2014) Optimization and characterization of silver nanoparticle by endophytic fungi Penicillium sp. isolated from Curcuma longa (turmeric) and application studies against MDR *E. coli* and *S. aureus*. Bioinorg Chem Appl 2014:1–8
Slawson RM, Van Dyke MI, Lee H, Trevors JT (1992) Germanium and silver resistance, accumulation, and toxicity in microorganisms. Plasmid 27:72–79
Sondi I, Salopek-Sondi B (2004) Silver nanoparticles as antimicrobial agent: a case study on *E. coli* as a model for gram-negative bacteria. J Colloid Interface Sci 275:177–182
Sunkar S, Nachiyar CV (2013) Endophytic fungi mediated extracellular silver nanoparticles as effective antibacterial agents. Int J Pharm Pharm Sci 5:95–100
Tamayo L et al (2014) Release of silver and copper nanoparticles from polyethylene nanocomposites and their penetration into Listeria monocytogenes. Mater Sci Eng C 40:24–31
Varshney R, Bhadauria S, Gaur M, Pasricha R (2011) Copper nanoparticles synthesis from electroplating industry effluent. Nano Biomed Eng 3:115–119
Waddington SR, Li X, Dixon J, Hyman G, De Vicente MC (2010) Getting the focus right: production constraints for six major food crops in Asian and African farming systems. Food Sec 2:27–48
Wang C, Huang X, Deng W, Chang C, Hang R, Tang B (2014) A nano-silver composite based on the ion-exchange response for the intelligent antibacterial applications. Mater Sci Eng C 41:134–141

Xie Y, He Y, Irwin PL, Jin T, Shi X (2011) Antibacterial activity and mechanism of action of zinc oxide nanoparticles against *Campylobacter jejuni*. Appl Environ Microbiol 77:2325–2331
Yang Y, Wang J, Xiu Z, Alvarez PJ (2013) Impacts of silver nanoparticles on cellular and transcriptional activity of nitrogen-cycling bacteria. Environ Toxicol Chem 32:1488–1494
Zhang H, Zhou H, Bai J, Li Y, Yang J, Ma Q, Yuanyuan Q (2019) Biosynthesis of selenium nanoparticles mediated by fungus *Mariannaea* sp. HJ and their characterization. Colloids Surf A Physicochem Eng Asp 571:9–16

Biogenic Nanoparticles in the Insect World: Challenges and Constraints

10

Nariman Maroufpour, Mahdieh Mousavi,
Behnam Asgari Lajayer, and Mansour Ghorbanpour

Abstract

Insects are found in a variety of environmental conditions and occupy little more than two-thirds of the known species of animals in the world. Traditional pest management tactics used in plant protection are insufficient, and synthetic pesticides are costly and have adverse effects on human and environment. A brilliant approach to pest control is using nanoparticles to help reduce the application of synthetic pesticides and environmental pollution, therefore providing green and efficient alternative approaches for pest control in plant protection by the help of nanotechnology without harming the environment. Nowadays, biosynthesis of nanoparticles by microorganisms and plants is being efficiently used in plant protection.

Keywords

Insects · Plant protection · Nanoparticles · Nanotechnology

N. Maroufpour
Department of Plant Protection, Faculty of Agriculture, University of Tabriz, Tabriz, Iran

M. Mousavi
Department of Plant Protection, Faculty of Agriculture, Urmia University, Urmia, Iran

B. Asgari Lajayer
Department of Soil Science, Faculty of Agriculture, University of Tabriz, Tabriz, Iran

M. Ghorbanpour (✉)
Department of Medicinal Plants, Faculty of Agriculture and Natural Resources, Arak University, Arak, Iran
e-mail: m-ghorbanpour@araku.ac.ir

M. Ghorbanpour et al. (eds.), *Biogenic Nano-Particles and their Use in Agro-ecosystems*, https://doi.org/10.1007/978-981-15-2985-6_10

10.1 Introduction

Insects are one of the largest animal populations that the first insects living in the Devonian layers have been seen. Insects have existed for more than 500 million years in all geological periods on Earth. Because of successful development, insects have been scattered in all possible environments throughout the world. Features like high reproductive potential, desiccation-resistant eggs, metamorphosis, malleable exoskeleton, and wings are just a few of these evolutionary successes. According to scientific estimates, about 4/5 of the known animal species of the world are insects. On the other hand, the number of insect species to date has been reported by over one million species, which is only 1/4 up to 1/3 of the total insects on Earth. Insects in human societies cause damage to health and the economy issues through vectors of many diseases and agricultural pests. These insects can also be useful species, such as honey bee or parasitoids and predators for controlling agricultural pests (Ragaei and Sabry 2014). We are faced daily with nanoscience in our environment from butterflies with iridescent colors to geckos that walk upside down on a ceiling, apparently against gravity. In nature, we encounter some outstanding solutions to complex problems in the form of good nanostructures with which accurate functions are concerned.

10.2 Biogenic Nanoparticles and Insects

By natural nanomaterials mean to us that these are materials belonging to nature (animal and mineral), without manipulating and changing humans, as well as having interesting properties due to their intrinsic structure (Filipponi et al. 2010). Natural nanomaterials have led to that the human integrates nanoscience in their environmental issues and uses it in its own affairs. Living organisms rely on nanometer-shaped protein machines to be able to do anything, whether as small as whipping bacterial flagella or flexing of muscles (Huck 2008). Although natural nanostructures are usually ignored, they are in fact sources of special properties (Watson and Watson 2004). Natural nanomaterials are not only intended to figure out the unusual properties of biological materials but also to design and fabricate new materials with more useful properties (Filipponi et al. 2010).

The industries that are emerging on the basis of nanotechnology in the world have not used very much of free technology available in nature (Ehrlich et al. 2008). The presence of natural nanoparticles is seen throughout nature, including the world of insects. Nanoparticles are seen from different parts of the body, including the compound eyes, antenna, wings, and abdomen, that isolated nanoparticles of abdomen and antennae have diameters of about 12 and 11 nm, respectively. A good example for the insect wings would be cicada (*Psaltoda claripennis* Ashton) and termite (Rhinotermitidae) (Zhang et al. 2006).

Researches have indicated that these wing nanoparticles help in the aerodynamic efficiency of the insect and consist of a round-shaped apex that protrudes some 150–350 nm out from the surface plane (Ragaei and Sabry 2014). Butterfly wings with bright color elements have nanoparticles. The wings of butterflies often display amazing colors which are a result of the wing's surface and its interplay with light. The wings also display iridescence, which is the change in color of an object when seen at various angles. The interaction of light with the physical structure of the surface causes an iridescence, which is a physical color. To interact with visible light, these structures need to be within the range of the nanoscale that this interaction causes constructive or destructive interference (Ragaei and Sabry 2014). Factors like the refractive index of the substrate, thickness, frequency of the incident light, and incident angle can affect the intensity, angles, and color of iridescence. The natural iridescence happened in materials such as opals because of packed silica spheres in the nanometer range uniform in size and arranged in layers. This presents suitable conditions for intervention. But the interference in these insects (butterflies and moths) for the iridescence is produced by a different method. In studies that researchers conducted on the structure of the wings of *Morpho rhetenor*, they found that each scale is about 70 × 200 μm and these are formed of rows of scales arranged such as tiles in a roof. Also, they found that scales have a smaller structure on its surface, a very intricate and highly ordered nanometer organization of ridges that every ridge has about 800 nm wide (Ragaei and Sabry 2014). In this structure, the spaces between scales lead to the formation of natural photonic crystal that can generate constructive and destructive interference. The image of the cross section of the ridges on the wings using the SEM analysis shows a more intricate structure that looks like fir trees, called setae, and these are about 400 nm long. These setae are responsible for producing constructive interference in the blue wavelengths which create a strong blue color (Filipponi et al. 2010). Furthermore, a shell is arisen by a layer of cells that first lays down a coating of protein that these proteins work same as nano-assembly mechanism to control the growth of carbon carbonate crystals. The honeycomb-like matrix of protein and chitin remains around each crystal that this flexible envelope is necessary for the mechanical characteristics of the shell and mitigates cracking (Filipponi et al. 2010).

The fine structure of opals and butterflies directly connected to their colors. Actually, this structure packed nanostructures that work as a diffraction grid and persuade iridescence. In the case of butterflies, this color can be due to pigments that absorb specific colors in the wings or in some species, such as *Morpho rhetenor*, due to the presence of nanostructures which are photonic crystals. Meanwhile, in the opals, this is because of packed silica spheres in the nanometer range, uniform in size and arranged in layers (Filipponi et al. 2010).

Another example of the presence of biogenic in insects is spider silk. Silk is the material known with about five times higher strength that of steel of the same weight. The remarkable characteristics of spider silk are because of the proteins that build the silk (mainly fibroin) and its supramolecular organization which is at the nanoscale level (Filipponi et al. 2010).

10.3 Engineering Biogenic Nanoparticle in Control of Insects

Many pest control strategies such as integrated pest management have been used in agriculture that are insufficient, therefore needing to use the chemical pesticides. Considering the many problems for human health and the environment caused by pesticides (Benelli et al. 2017b; Naqqash et al. 2016), some of the more important ones include poorly water-soluble molecules (e.g., oil-in-water (O/W) emulsions and emulsifiable concentrates, ECs) and formulations like ECs that consist of organic, toxic, expensive, and flammable solvents or a blend of surfactant emulsifiers used in the water and spray tanks (Knowles 2009). O/W emulsions do not have such problems, but the process of emulsification requires high-energy input which is a major drawback (Kah et al. 2013; Kah and Hofmann 2014).

In the world of pest control and pesticides, nanotechnology has been embraced in the recent decade (Mukunthan et al. 2011). Researches have shown that nanoparticles (NPs) have insecticidal properties and can be used as nano-carriers and also bioinsecticides (Barik et al. 2008; Elango et al. 2016). Because nanoparticles have novel properties (size, shape, specific surface area, and chemical composition), they are thought to be potentially toxic (Sharifi et al. 2012).

Nanoparticles can be used in the production of new compounds (pesticides and insect repellants) for pest control (Owolade et al. 2008). Many natural materials with different shapes and compositions were used or synthesized to make nanoparticles, including carbon, silicates, polymers, ceramics, proteins, emulsions, dendrimers, lipids, semiconductor quantum dots (QDs), metal oxides, and metal (Niemeyer 2001; Oskam 2006). Different methods were extended for the synthesis of these nanoparticles like hydrothermal synthesis, thermal decomposition, chemical reduction, ultrasonic technique, sol-gel, microemulsion, precipitation method, and electrochemical and microwave-assisted process (Lin et al. 2013a; Singhal et al. 2011). Disadvantages of these approaches are hazardous chemicals, the difficulty of scaling up the process, high energy requirements, and low material conversion. Given these disadvantages, the development of nanoparticle production methods without the use of hazardous materials is very necessary.

The concerns of consumers about the environmental use of synthetic materials propelled scientists to find alternatives for the production of nanomaterials, the so-called green synthesis (Benelli and Lukehart 2017).

The idea of this method is based on the fact that various organisms have the capacity to generate nonorganic materials (Simkiss and Wilbur 2012). Synthesis of nanoparticles by microorganisms such as bacteria, viruses, fungi, yeasts, actinomycetes, and plant extracts is suggested as possible eco-friendly alternatives to physical and chemical approaches (Dubey et al. 2009; Mohanpuria et al. 2008; Narayanan and Sakthivel 2010).

Compared to other methods, green synthesis is regarded as cost-effective, safe, sustainable, and environment-friendly. The use of plants can be advantageous over other biological processes due to being easily available, low cost, eco-friendly, and safe to handle. Also, it possesses a broad variability of metabolites that may aid in the reduction and a single-step method for biosynthesis process (Govindarajan and

Benelli 2016; Govindarajan et al. 2016; Huang et al. 2007; Shankar et al. 2004; Teimouri et al. 2018). Many investigations on different metal nanomaterials are performed to improve the potential tools for alternative and effective control of agricultural or stored product pests especially of pests related to humans' and animals' health. A lot of researches have been done about the toxicity of nanoparticles toward various arthropod species of economic value, and more than half of them were done using the so-called green synthesis method (Athanassiou et al. 2018; Banumathi et al. 2017; Benelli 2018; Benelli et al. 2017b; Rajan et al. 2015). Recent investigations indicate the potential of the green synthesis for metal NPs, mainly AgNPs, in controlling a wide range of pests in the field and laboratory.

The efficacy of a number of plant-synthesized NPs has been tested against arthropod pests with economic importance such as lice (Jayaseelan et al. 2011), hard ticks (Zahir et al. 2012), louse flies (Jayaseelan et al. 2012), beetles (Zahir et al. 2012), moths (Roni et al. 2015), and mosquitoes (Benelli 2016a, b). Also, a lot of researches associated with public health issues concerning other common pests have been done.

Furthermore, plant-synthesized NPs have shown promising value as ovicides and adulticides for many pest species, for example, all larval instars and pupae of the cotton bollworm, *Helicoverpa armigera* (Hübner) (Lepidoptera: Noctuidae), where they exhibited susceptibility to AgNPs synthesized by leaf aqueous extract of *Euphorbia hirta* L. (Malpighiales: Euphorbiaceae) (Devi et al. 2014). Similar results were obtained with larvae of the mosquitoes *C. quinquefasciatus* and *A. subpictus* exposed to AgNPs synthesized from the leaf aqueous extract of *Mimosa pudica* L. (Fabales: Fabaceae) (Marimuthu et al. 2011).

In the same study, larvae of the tick *Rhipicephalus microplus* Canestrini (Acari: Ixodidae) were proven very susceptible to AgNPs (Marimuthu et al. 2011). In Arjunan et al. (2012) study, AgNPs synthesized by leaf aqueous extract of *Annona squamosa* L. (Magnoliales: Annonaceae) showed very good mortality in pupa and all larvae instars of *C. quinquefasciatus* and *Anopheles stephensi* Liston (Diptera: Culicidae). Similar results were obtained in the reports of Suresh et al. (2014) who observed AgNPs triggered 100% mortality of second-instar larvae of *A. aegypti* L. (Diptera: Culicidae). They biosynthesized AgNPs using the root extract of *Delphinium denudatum* Wall (Ranunculales: Ranunculaceae) (Suresh et al. 2014). Field experiments of Dinesh et al. (2015) reported that AgNPs synthesized by leaf aqueous extract of *Aloe vera* (L.) Burm. f. (Asparagales: Xanthorrhoeaceae) were toxic for all larvae instars of *A. stephensi*. Suresh et al. (2015) also reported that *A. aegypti* larvae after application of AgNPs synthesized by leaf aqueous extract of *Phyllanthus niruri* L. (Malpighiales: Phyllanthaceae) resulted in good mortality. Similar results were reported by Subramaniam et al. (2015) who observed the susceptibility of *A. stephensi* and *A. albopictus* to AgNPs biosynthesized using *Mimusops elengi* L. (Ericales: Sapotaceae).

AgNPs not only indicated highly promising toxicity against mosquitoes and ticks but also adulticidal, ovicidal, and oviposition deterrence effects have been observed (Benelli 2018; Benelli et al. 2017a; Rajaganesh et al. 2016; Suresh et al. 2015). For example, the housefly, *Musca domestica* L. (Diptera: Muscidae), was completely suppressed at the adult stage after 4 h of exposure with AgNPs synthesized by leaf aqueous extract of *Manilkara zapota* (L.) (Ericales: Sapotaceae) (Kamaraj et al. 2012). About adulticidal toxicity of nanoparticles, only a few studies are available. Jayaseelan et al. (2011) reported that the head louse, *P. humanus capitis* De Geer (Phthiraptera: Pediculidae), can be controlled when adult exposed with AgNPs is synthesized by leaf aqueous extract of *Tinospora cordifolia* (Thunb.) (Ranunculales: Menispermaceae). Similarly, Veerakumar and Govindarajan (2014) indicated that adults of *A. stephensi*, *A. aegypti*, and *C. quinquefasciatus* controlled by AgNPs are synthesized using *Feronia elephantum* Corrêa (Sapindales: Rutaceae) leaf extract. In the same study, adults of *A. stephensi*, *A. aegypti*, and *C. quinquefasciatus* showed a good mortality after applications of AgNPs synthesized by leaf extract of *Heliotropium indicum* L. (Eudicotidae: Boraginaceae) (Veerakumar and Govindarajan 2014). AgNPs biosynthesized using *Phyllanthus niruri* were very toxic to *A. aegypti* adults (Suresh et al. 2015). Also, suppressed female fecundity, egg hatchability, and reduced longevity in both sexes by biosynthesized nanoparticles against mosquitoes were observed in some investigations (Madhiyazhagan et al. 2015; Roni et al. 2015). Meanwhile, among the different tested mosquitoes' species, the most resistant to the toxic activity of plant-synthesized NPs are reported from *Culex quinquefasciatus* larvae and pupae (Benelli 2016b). Marine plants, apart from terrestrial plants, have been used for the biosynthesis of metal NPs as a novel biological control strategy against insect pest species that impact agriculture and public health. Marine plants like *Sargassum muticum* (Yendo) Fensholt (Fucales: Sargassaceae), *Caulerpa scalpelliformis* (R. Brown ex Turner), and *C. agardh* (Bryopsidales: Caulerpaceae) were used for fabricating nanoparticles (Moorthi et al. 2015; Murugan et al. 2015b).

The biosynthesis of metal NPs by fungus also indicated interesting prospects for the management of insect pest species (Amerasan et al. 2016). The scientists reported that the nanoparticles synthesized by some fungi such as *Chrysosporium tropicum* J. W. Carmich. (Onygenales: Onygenaceae), *Trichoderma harzianum* Rifai (Hypocreales: Hypocreaceae), *Fusarium oxysporum*, and *Cochliobolus lunatus* R. R. Nelson and Haasis (Pleosporales: Pleosporaceae) controlled very well mosquitoes' species (Salunkhe et al. 2011; Soni and Prakash 2012, 2013; Sundaravadivelan and Padmanabhan 2014).

10.3.1 Mode of Action of Nanoparticles Against Insect Pests

We have strictly inadequate knowledge about the possible mode of action of nanoparticles against insect pests which is important information to forecast the toxicological outcomes arising from the real-world application of nanoparticles as pesticides.

The cytochrome (CYP) P450 monooxygenases (p450s) include a group of enzymes found in most living organisms (Werck-Reichhart and Feyereisen 2000). The CYP genes (CYP2, CYP3, CYP4 and mitochondrial) are involved in the first and second step of proceeding drug metabolism, in detoxification of numerous xenobiotics, endogenous substances, and detoxification (Afrin and Wait 2018; Fröhlich et al. 2010; Martignoni et al. 2006; Pelkonen et al. 1998). Meanwhile, resistance in the insects to insecticides is often linked with one or more detoxifying genes such as p450s, esterases, and glutathione S-transferases (Martínez-Paz et al. 2012; Niu et al. 2011). Nevertheless, the important roles in metabolism and inactivation of xenobiotic compounds like pesticides and insecticides are with the p450 genes (Iga and Kataoka 2012; Lin et al. 2013b).

There are only a few investigations carried out about express CYP p450 gene in arthropod when they exposed to toxic concentrations of AgNPs (Fröhlich et al. 2010; Lamb et al. 2010; Warisnoicharoen et al. 2011). Furthermore, an important role has been demonstrated for CYP6BG1 in metabolizing toxic nanoparticles in the insects (Afrin and Wait 2018).

Also, about the toxicity of several types of nanoparticles, a theory is widely accepted that they achieve toxicity triggering oxidative stress in arthropod tissues (Foldbjerg et al. 2015; Mao et al. 2018; Nair and Choi 2011, 2012). The penetration of nanoparticles into the cell organelles and localization of the NPs at mitochondria or nucleolus in insect tissues were demonstrated by transmission electron microscopy (TEM). With this information, scientists suggest that mitochondria or nucleolus in insect tissues can be used for targeted delivery of pesticides (Yasur and Rani 2013, 2015). Moreover, nanoparticle in the intracellular space binds to phosphorus from DNA or to sulfur from proteins, and the consequence of this linking is organelles' and enzymes' rapid denaturation. Subsequently, these changes can cause a disturbance in proton motive force, decrease in membrane permeability, and finally loss of cellular function and cell death (Benelli 2016a; Jiang et al. 2015). Sundararajan and Kumari (2017) reported that AuNPs biosynthesized by *Artemisia vulgaris* L. leaf extract cause accumulation of this nanoparticle in the midgut region and damage of an epithelial cell, cortex, and midgut in third and fourth instars of *A. aegypti*. The same result for AuNPs against *A. aegypti* was recorded (Suganya et al. 2017). In another research, toxicity of grapheme oxide nanoparticles on *Acheta domesticus* (L.) was investigated. The authors observed that this nanoparticle causes increased enzymatic activity of glutathione peroxidases, heat shock protein (HSP 70), catalase, and total antioxidant capacity levels (Dziewięcka et al. 2016).

Between different nanoparticles, many investigations have been conducted about the AgNP mechanism. Fouad et al. (2018) produce AgNPs by the *Cassia fistula* L. fruit pulp and then was used against fourth-instar larvae of *Culex pipiens pallens* (Coquilett) and *Aedes albopictus* (Skuse). The authors found that nanoparticles reduced α- and ß-carboxylesterase, acetylcholinesterase activities, and total protein levels (Fouad et al. 2018). Also, Ga'al et al. (2018) reported that AgNPs on the fourth-instar larvae of *Ae. albopictus* cause decrease of total proteins, acetylcholinesterase, esterase, and phosphatase enzymes. In a study done by Kalimuthu et al. (2017), *A. aegypti* were exposed to biosynthesized AgNPs by *Hedychium coronarium* J. Koenig. In this mosquito pest, AgNPs cause partial lyses of the midgut epithelial cells, vesicles, and damaged membranes at the apical side of epithelial cells. Meanwhile, some evidence demonstrated that particle size or surface area is mainly responsible for AgNP toxicity. Also, after oxidation of AgNPs, Ag^+ released from that could enter into the body and interact with biological molecules (Lin et al. 2010; Moore 2006; Park et al. 2010). Nair and Choi (2011) evaluated the impact of the AgNPs on the expression of glutathione S-transferase (GST) genes, which are linked with the occurrence of oxidative stress. They reported that all GST genes showed up- or downregulation to varying levels based upon the tested concentration and duration of exposure to the contaminant, with the highest mRNA expression in Delta3, Sigma4, and Epsilon1 GST class (Nair and Choi 2011). In another study, Nair et al. (2011) showed that AgNPs in *Chironomus riparius* cause downregulation of the ribosomal protein gene (CrL15) regulating ribosomal assembly, thus the synthesis of proteins. In addition, upregulation of the gonadotropin-releasing hormone gene (CrGnRH1) and the Balbiani ring protein gene (CrBR2.2) can indicate the activation of gonadotropin-releasing hormone mediated signal transduction pathways and reproductive failure and the organism protection mechanism against the AgNPs, respectively. In the same study, Nair et al. (2013) reported that the transcript levels of catalase, phospholipid hydroperoxide glutathione peroxidase 1, and thioredoxin reductase 1 were upregulated after exposure to AgNPs. Yasur and Rani (2015) also reported AgNPs in *Spodoptera litura* (Fabricius), and *Achaea janata* (L.) induced oxidative stress in moth larval guts, as proved by the enhanced antioxidant enzyme levels. Moreover, Bharani and Namasivayam (2017) demonstrated that AgNPs biosynthesized by the *Punica granatum* L. in third-instar larvae of *S. litura* affected lipase, protease, invertase, and amylase activities. Furthermore, they observed the extracellular enzyme production, gut microflora, total heterotrophic bacterial population, pH, and weight decreased in the larvae (Bharani and Namasivayam 2017). Armstrong et al. (2013) indicated that AgNPs in *Drosophila melanogaster* Meigen (wild type) cause loss of the melanin cuticular pigments, achieving a lighter body color and strongly decreasing the activity of Cu-dependent enzymes (i.e., tyrosinase and Cu-Zn superoxide dismutase). In further research, DNA damage, ROS-mediated apoptosis, and autophagy due to the accumulation of reactive oxygen species (ROS) in the fly tissues in *D. melanogaster* have been indicated as well (Mao et al. 2018).

10.4 Conclusions

Nowadays, management of insect pests is focused on the modern approaches of nanotechnology using nanomaterials instead of traditional strategies based on chemical insecticides (Ragaei and Sabry 2014). Nanotechnology decreases the usage of crop protection chemicals which makes agriculture eco-friendly and profitable (Athanassiou et al. 2018). The efficacy of nanoparticle biosynthesis is really promising and inspires many study groups worldwide to manage vectors and pests with new methods. There are a number of publications and successfully explored examples with considerable confidence that biosynthesis of nanoparticles has a bright future and potential for developing eco-friendly method for the control of agricultural pests related to humans' and animals' health. Natural pesticides in comparison to the conventional pesticide have advantages. For instance, insects are less likely to evolve resistance to the toxins. Many studies have been conducted to understand the mechanism of nanoparticles, but the actual mechanism of nanoparticles inside the insect's body requires more detailed studies and our knowledge is still limited about this field. The challenge in nanoparticle biosynthesis technology is to characterize the nanoparticles by controlling their shape and size, so the impact of nano-size, shape, and charge on the various potential mechanism(s) of action will be elucidated. Moreover, further works are required in the field of plant protection to validate the proposed nano-pesticides in field conditions to monitoring their stability and sublethal effects on nontarget organisms (Baun et al. 2008; Foldbjerg et al. 2015; Fruijtier-Pölloth 2012). Nevertheless, since there are concerns about the potential toxicity of nanomaterials, which are neither well explored nor well understood or standardized yet, this technology will likely go through strong scrutiny by international and national safety regulators that request far more investigation on human and environmental impacts of these materials (Athanassiou et al. 2018; Benelli 2015; Murugan et al. 2015a, b; Fahimirad et al. 2019).

References

Afrin T, Wait A (2018) Effects of engineered carbon and silver nanoparticles on gene expression in *Plutella xylostella* to assess toxicity. J Genet Genet Eng 2:9–17

Amerasan D, Nataraj T, Murugan K, Panneerselvam C, Madhiyazhagan P, Nicoletti M, Benelli G (2016) Myco-synthesis of silver nanoparticles using *Metarhizium anisopliae* against the rural malaria vector *Anopheles culicifacies* Giles (Diptera: Culicidae). J Pest Sci 89:249–256

Arjunan NK, Murugan K, Rejeeth C, Madhiyazhagan P, Barnard DR (2012) Green synthesis of silver nanoparticles for the control of mosquito vectors of malaria, filariasis, and dengue. Vector Borne Zoonotic Dis 12:262–268

Armstrong N, Ramamoorthy M, Lyon D, Jones K, Duttaroy A (2013) Mechanism of silver nanoparticles action on insect pigmentation reveals intervention of copper homeostasis. PLoS One 8:e53186

Athanassiou C, Kavallieratos N, Benelli G, Losic D, Rani PU, Desneux N (2018) Nanoparticles for pest control: current status and future perspectives. J Pest Sci 91:1–15

Banumathi B et al (2017) Exploitation of chemical, herbal and nanoformulated acaricides to control the cattle tick, *Rhipicephalus* (Boophilus) *microplus*—a review. Vet Parasitol 244:102–110

Barik T, Sahu B, Swain V (2008) Nanosilica—from medicine to pest control. Parasitol Res 103:253–258

Baun A, Hartmann NB, Grieger K, Kusk KO (2008) Ecotoxicity of engineered nanoparticles to aquatic invertebrates: a brief review and recommendations for future toxicity testing. Ecotoxicology 17:387–395

Benelli G (2015) Research in mosquito control: current challenges for a brighter future. Parasitol Res 114:2801–2805

Benelli G (2016a) Green synthesized nanoparticles in the fight against mosquito-borne diseases and cancer—a brief review. Enzym Microb Technol 95:58–68

Benelli G (2016b) Plant-mediated biosynthesis of nanoparticles as an emerging tool against mosquitoes of medical and veterinary importance: a review. Parasitol Res 115:23–34

Benelli G (2018) Gold nanoparticles–against parasites and insect vectors. Acta Trop 178:73–80

Benelli G, Lukehart CM (2017) Applications of green-synthesized nanoparticles in pharmacology, parasitology and entomology. J Clust Sci 28:1–2

Benelli G et al (2017a) Nanoparticles as effective acaricides against ticks—a review. Ticks Tick Borne Dis 8:821–826

Benelli G, Pavela R, Maggi F, Petrelli R, Nicoletti M (2017b) Commentary: making green pesticides greener? The potential of plant products for nanosynthesis and pest control. J Clust Sci 28:3–10

Bharani RA, Namasivayam SKR (2017) Biogenic silver nanoparticles mediated stress on developmental period and gut physiology of major lepidopteran pest *Spodoptera litura* (Fab.) (Lepidoptera: Noctuidae)—an eco-friendly approach of insect pest control. J Environ Chem Eng 5:453–467

Devi GD, Murugan K, Selvam CP (2014) Green synthesis of silver nanoparticles using *Euphorbia hirta* (Euphorbiaceae) leaf extract against crop pest of cotton bollworm, *Helicoverpa armigera* (Lepidoptera: Noctuidae). J Biopest 7:54–66

Dinesh D et al (2015) Mosquitocidal and antibacterial activity of green-synthesized silver nanoparticles from Aloe vera extracts: towards an effective tool against the malaria vector Anopheles stephensi? Parasitol Res 114:1519–1529

Dubey M, Bhadauria S, Kushwah B (2009) Green synthesis of nanosilver particles from extract of *Eucalyptus hybrida* (safeda) leaf. Dig J Nanomater Biostruct 4:537–543

Dziewięcka M, Karpeta-Kaczmarek J, Augustyniak M, Majchrzycki Ł, Augustyniak-Jabłokow MA (2016) Evaluation of in vivo graphene oxide toxicity for *Acheta domesticus* in relation to nanomaterial purity and time passed from the exposure. J Hazard Mater 305:30–40

Ehrlich H et al (2008) Nanostructural organization of naturally occurring composites-part II: silica-chitin-based biocomposites. J Nanomater 3:1–8

Elango G, Roopan SM, Dhamodaran KI, Elumalai K, Al-Dhabi NA, Arasu MV (2016) Spectroscopic investigation of biosynthesized nickel nanoparticles and its larvicidal, pesticidal activities. J Photochem Photobiol B Biol 162:162–167

Fahimirad S, Ajalloueian F, Ghorbanpour M (2019) Synthesis and therapeutic potential of silver nanomaterials derived from plant extracts. Ecotoxicol Environ Saf 168:260–278

Filipponi L, Sutherland D, Center IN (2010) Introduction to nanoscience and nanotechnologies Interdisciplinary Nanoscience Center (iNano): Aarhus University, NANOYOU Teachers Training in Nanoscience and Nanotechnologies, pp 2–29

Foldbjerg R, Jiang X, Miclăuş T, Chen C, Autrup H, Beer C (2015) Silver nanoparticles–wolves in sheep's clothing? Toxicol Res 4:563–575

Fouad H, Hongjie L, Hosni D, Wei J, Abbas G, Ga'al H, Jianchu M (2018) Controlling Aedes albopictus and *Culex pipiens* pallens using silver nanoparticles synthesized from aqueous extract of *Cassia fistula* fruit pulp and its mode of action. Artif Cells Nanomed Biotechnol 46:558–567

Fröhlich E, Kueznik T, Samberger C, Roblegg E, Wrighton C, Pieber TR (2010) Size-dependent effects of nanoparticles on the activity of cytochrome P450 isoenzymes. Toxicol Appl Pharmacol 242:326–332

Fruijtier-Pölloth C (2012) The toxicological mode of action and the safety of synthetic amorphous silica—A. nanostructured material. Toxicology 294:61–79

Ga'al H, Fouad H, Tian J, Hu Y, Abbas G, Mo J (2018) Synthesis, characterization and efficacy of silver nanoparticles against *Aedes albopictus* larvae and pupae. Pestic Biochem Physiol 144:49–56

Govindarajan M, Benelli G (2016) One-pot green synthesis of silver nanocrystals using *Hymenodictyon orixense*: a cheap and effective tool against malaria, chikungunya and Japanese encephalitis mosquito vectors? RSC Adv 6:59021–59029

Govindarajan M, Khater HF, Panneerselvam C, Benelli G (2016) One-pot fabrication of silver nanocrystals using *Nicandra physalodes*: a novel route for mosquito vector control with moderate toxicity on non-target water bugs. Res Vet Sci 107:95–101

Huang J et al (2007) Biosynthesis of silver and gold nanoparticles by novel sundried *Cinnamomum camphora* leaf. Nanotechnology 18:105104

Huck WT (2008) Responsive polymers for nanoscale actuation. Mater Today 11:24–32

Iga M, Kataoka H (2012) Recent studies on insect hormone metabolic pathways mediated by cytochrome P450 enzymes. Biol Pharm Bull 35:838–843

Jayaseelan C et al (2011) Synthesis of pediculocidal and larvicidal silver nanoparticles by leaf extract from heartleaf moonseed plant, *Tinospora cordifolia* Miers. Parasitol Res 109:185–194

Jayaseelan C et al (2012) Efficacy of plant-mediated synthesized silver nanoparticles against hematophagous parasites. Parasitol Res 111:921–933

Jiang X et al (2015) Fast intracellular dissolution and persistent cellular uptake of silver nanoparticles in CHO-K1 cells: implication for cytotoxicity. Nanotoxicology 9:181–189

Kah M, Hofmann T (2014) Nanopesticide research: current trends and future priorities. Environ Int 63:224–235

Kah M, Beulke S, Tiede K, Hofmann T (2013) Nanopesticides: state of knowledge, environmental fate, and exposure modeling. Crit Rev Environ Sci Technol 43:1823–1867

Kalimuthu K et al (2017) Control of dengue and Zika virus vector *Aedes aegypti* using the predatory copepod *Megacyclops formosanus*: synergy with *Hedychium coronarium*-synthesized silver nanoparticles and related histological changes in targeted mosquitoes. Process Saf Environ Prot 109:82–96

Kamaraj C, Rajakumar G, Rahuman AA, Velayutham K, Bagavan A, Zahir AA, Elango G (2012) Feeding deterrent activity of synthesized silver nanoparticles using *Manilkara zapota* leaf extract against the house fly, *Musca domestica* (Diptera: Muscidae). Parasitol Res 111:2439–2448

Knowles A (2009) Global trends in pesticide formulation technology: the development of safer formulations in China. Outlook Pest Manag 20:165–170

Lamb JG, Hathaway LB, Munger MA, Raucy JL, Franklin MR (2010) Nanosilver particle effects on drug metabolism in vitro. Drug Metab Dispos 38:2246–2251

Lin J, Zhang H, Chen Z, Zheng Y (2010) Penetration of lipid membranes by gold nanoparticles: insights into cellular uptake, cytotoxicity, and their relationship. ACS Nano 4:5421–5429

Lin PC, Lin HJ, Liao YY, Guo HR, Chen KT (2013a) Acute poisoning with neonicotinoid insecticides: a case report and literature review. Basic Clin Pharmacol Toxicol 112:282–286

Lin Q et al (2013b) Transcriptome analysis of chlorantraniliprole resistance development in the diamondback moth *Plutella xylostella*. PLoS One 8:e72314

Madhiyazhagan P et al (2015) Sargassum muticum-synthesized silver nanoparticles: an effective control tool against mosquito vectors and bacterial pathogens. Parasitol Res 114:4305–4317

Mao B-H, Chen Z-Y, Wang Y-J, Yan S-J (2018) Silver nanoparticles have lethal and sublethal adverse effects on development and longevity by inducing ROS-mediated stress responses. Sci Rep 8:2445

Marimuthu S et al (2011) Evaluation of green synthesized silver nanoparticles against parasites. Parasitol Res 108:1541–1549

Martignoni M, Groothuis GM, de Kanter R (2006) Species differences between mouse, rat, dog, monkey and human CYP-mediated drug metabolism, inhibition and induction. Expert Opin Drug Metab Toxicol 2:875–894

Martínez-Paz P, Morales M, Martínez-Guitarte JL, Morcillo G (2012) Characterization of a cytochrome P450 gene (CYP4G) and modulation under different exposures to xenobiotics (tributyltin, nonylphenol, bisphenol A) in *Chironomus riparius* aquatic larvae. Comp Biochem Physiol Part C: Toxicol Pharmacol 155:333–343

Mohanpuria P, Rana NK, Yadav SK (2008) Biosynthesis of nanoparticles: technological concepts and future applications. J Nanopart Res 10:507–517

Moore M (2006) Do nanoparticles present ecotoxicological risks for the health of the aquatic environment? Environ Int 32:967–976

Moorthi PV, Balasubramanian C, Mohan S (2015) An improved insecticidal activity of silver nanoparticle synthesized by using *Sargassum muticum*. Appl Biochem Biotechnol 175:135–140

Mukunthan K, Elumalai E, Patel TN, Murty VR (2011) Catharanthus roseus: a natural source for the synthesis of silver nanoparticles. Asian Pac J Trop Biomed 1:270–274

Murugan K et al (2015a) Cymbopogon citratus-synthesized gold nanoparticles boost the predation efficiency of copepod *Mesocyclops aspericornis* against malaria and dengue mosquitoes. Exp Parasitol 153:129–138

Murugan K et al (2015b) Datura metel-synthesized silver nanoparticles magnify predation of dragonfly nymphs against the malaria vector *Anopheles stephensi*. Parasitol Res 114:4645–4654

Nair PMG, Choi J (2011) Identification, characterization and expression profiles of *Chironomus riparius* glutathione S-transferase (GST) genes in response to cadmium and silver nanoparticles exposure. Aquat Toxicol 101:550–560

Nair PMG, Choi J (2012) Modulation in the mRNA expression of ecdysone receptor gene in aquatic midge, *Chironomus riparius* upon exposure to nonylphenol and silver nanoparticles. Environ Toxicol Pharmacol 33:98–106

Nair PMG, Park SY, Lee S-W, Choi J (2011) Differential expression of ribosomal protein gene, gonadotropin releasing hormone gene and Balbiani ring protein gene in silver nanoparticles exposed *Chironomus riparius*. Aquat Toxicol 101:31–37

Nair PMG, Park SY, Choi J (2013) Evaluation of the effect of silver nanoparticles and silver ions using stress responsive gene expression in *Chironomus riparius*. Chemosphere 92:592–599

Naqqash MN, Gökçe A, Bakhsh A, Salim M (2016) Insecticide resistance and its molecular basis in urban insect pests. Parasitol Res 115:1363–1373

Narayanan KB, Sakthivel N (2010) Biological synthesis of metal nanoparticles by microbes. Adv Colloid Interf Sci 156:1–13

Niemeyer CM (2001) Nanoparticles, proteins, and nucleic acids: biotechnology meets materials science. Angew Chem Int Ed 40:4128–4158

Niu G, Rupasinghe SG, Zangerl AR, Siegel JP, Schuler MA, Berenbaum MR (2011) A substrate-specific cytochrome P450 monooxygenase, CYP6AB11, from the polyphagous navel orangeworm (*Amyelois transitella*). Insect Biochem Mol Biol 41:244–253

Oskam G (2006) Metal oxide nanoparticles: synthesis, characterization and application. J Sol-Gel Sci Technol 37:161–164

Owolade O, Ogunleti D, Adenekan M (2008) Titanium dioxide affects disease development and yield of edible cowpea. J Agric Food Chem 7:2942–2947

Park E-J, Yi J, Kim Y, Choi K, Park K (2010) Silver nanoparticles induce cytotoxicity by a Trojan-horse type mechanism. Toxicol in Vitro 24:872–878

Pelkonen O, Mäeenpäeä J, Taavitsainen P, Rautio A, Raunio H (1998) Inhibition and induction of human cytochrome P450 (CYP) enzymes. Xenobiotica 28:1203–1253

Ragaei M, Sabry A-kH (2014) Nanotechnology for insect pest control. Int J Sci Environ Technol 3:528–545

Rajaganesh R et al (2016) Fern-synthesized silver nanocrystals: towards a new class of mosquito oviposition deterrents? Res Vet Sci 109:40–51

Rajan R, Chandran K, Harper SL, Yun S-I, Kalaichelvan PT (2015) Plant extract synthesized silver nanoparticles: an ongoing source of novel biocompatible materials. Ind Crop Prod 70:356–373

Roni M et al (2015) Characterization and biotoxicity of *Hypnea musciformis*-synthesized silver nanoparticles as potential eco-friendly control tool against *Aedes aegypti* and *Plutella xylostella*. Ecotoxicol Environ Saf 121:31–38

Salunkhe RB, Patil SV, Patil CD, Salunke BK (2011) Larvicidal potential of silver nanoparticles synthesized using fungus *Cochliobolus lunatus* against *Aedes aegypti* (Linnaeus, 1762) and *Anopheles stephensi* Liston (Diptera; Culicidae). Parasitol Res 109:823–831

Shankar SS, Rai A, Ankamwar B, Singh A, Ahmad A, Sastry M (2004) Biological synthesis of triangular gold nanoprisms. Nat Mater 3:482–488

Sharifi S, Behzadi S, Laurent S, Forrest ML, Stroeve P, Mahmoudi M (2012) Toxicity of nanomaterials. Chem Soc Rev 41:2323–2343

Simkiss K, Wilbur KM (2012) Biomineralization. Academic/Elsevier, San Diego, p 337

Singhal G, Bhavesh R, Kasariya K, Sharma AR, Singh RP (2011) Biosynthesis of silver nanoparticles using *Ocimum sanctum* (Tulsi) leaf extract and screening its antimicrobial activity. J Nanopart Res 13:2981–2988

Soni N, Prakash S (2012) Efficacy of fungus mediated silver and gold nanoparticles against *Aedes aegypti* larvae. Parasitol Res 110:175–184

Soni N, Prakash S (2013) Possible mosquito control by silver nanoparticles synthesized by soil fungus (Aspergillus niger 2587). Adv Nanopart 2:125–132

Subramaniam J et al (2015) Eco-friendly control of malaria and arbovirus vectors using the mosquitofish *Gambusia affinis* and ultra-low dosages of Mimusops elengi-synthesized silver nanoparticles: towards an integrative approach? Environ Sci Pollut Res 22:20067–20083

Suganya P et al (2017) Biopolymer zein-coated gold nanoparticles: synthesis, antibacterial potential, toxicity and histopathological effects against the zika virus vector *Aedes aegypti*. J Photochem Photobiol B Biol 173:404–411

Sundararajan B, Kumari BR (2017) Novel synthesis of gold nanoparticles using *Artemisia vulgaris* L. leaf extract and their efficacy of larvicidal activity against dengue fever vector *Aedes aegypti* L. J Trace Elem Med Biol 43:187–196

Sundaravadivelan C, Padmanabhan MN (2014) Effect of mycosynthesized silver nanoparticles from filtrate of *Trichoderma harzianum* against larvae and pupa of dengue vector *Aedes aegypti* L. Environ Sci Pollut Res 21:4624–4633

Suresh G et al (2014) Green synthesis of silver nanoparticles using *Delphinium denudatum* root extract exhibits antibacterial and mosquito larvicidal activities. Spectrochim Acta A Mol Biomol Spectrosc 127:61–66

Suresh U et al (2015) Tackling the growing threat of dengue: *Phyllanthus niruri*-mediated synthesis of silver nanoparticles and their mosquitocidal properties against the dengue vector *Aedes aegypti* (Diptera: Culicidae). Parasitol Res 114:1551–1562

Teimouri M, Khosravi-Nejad F, Attar F, Saboury AA, Kostova I, Benelli G, Falahati M (2018) Gold nanoparticles fabrication by plant extracts: synthesis, characterization, degradation of 4-nitrophenol from industrial wastewater, and insecticidal activity—a review. J Clean Prod 184:740–753

Veerakumar K, Govindarajan M (2014) Adulticidal properties of synthesized silver nanoparticles using leaf extracts of *Feronia elephantum* (Rutaceae) against filariasis, malaria, and dengue vector mosquitoes. Parasitol Res 113:4085–4096

Warisnoicharoen W, Hongpiticharoen P, Lawanprasert S (2011) Alteration in enzymatic function of human cytochrome P450 by silver nanoparticles. Res J Environ Toxicol 5:58–64

Watson GS, Watson JA (2004) Natural nano-structures on insects—possible functions of ordered arrays characterized by atomic force microscopy. Appl Surf Sci 235:139–144

Werck-Reichhart D, Feyereisen R (2000) Cytochromes P450: a success story. Genome Biol 1:Reviews3003

Yasur J, Rani PU (2013) Environmental effects of nanosilver: impact on castor seed germination, seedling growth, and plant physiology. Environ Sci Pollut Res 20:8636–8648

Yasur J, Rani PU (2015) Lepidopteran insect susceptibility to silver nanoparticles and measurement of changes in their growth, development and physiology. Chemosphere 124:92–102

Zahir AA, Bagavan A, Kamaraj C, Elango G, Rahuman AA (2012) Efficacy of plant-mediated synthesized silver nanoparticles against *Sitophilus oryzae*. J Biopest 11:921–933

Zhang G, Zhang J, Xie G, Liu Z, Shao H (2006) Cicada wings: a stamp from nature for nanoimprint lithography. Small 2:1440–1443

Biogenic Synthesis of Gold Nanoparticles and Their Potential Application in Agriculture

11

Fatemeh Graily-Moradi, Ayda Maadani Mallak, and Mansour Ghorbanpour

Abstract

Nanotechnology is a new approach for the production of particles with unique features at the nanoscale dimensions. Among the various routes available for the synthesis of these nanoparticles, biogenic synthesis is a simple, low-cost, and eco-friendly method. The biosynthesis of gold nanoparticles is provided by various natural sources including plants, fungi, bacteria, actinomycetes, yeasts, and algae. Gold nanoparticles of various shapes and sizes are synthesized using biomass and/or extract of the organism. Enzymes secreted by microorganisms and metabolites of plants act as reducing, stabilizing, and capping agents for the production of the nanoparticles. The gold nanoparticles have antibacterial/antifungal properties that can be used to protect plants against pathogens. In addition, they can be applied for pesticide identification and water purification. This chapter focuses on the biosynthesis of gold nanoparticles, their characterization, and application in agriculture.

Keywords

Agriculture · Biogenic synthesis · Biosynthesis · Extracellular · Gold nanoparticle · Intracellular · Nanotechnology

F. Graily-Moradi
Department of Plant Protection, Faculty of Agriculture, University of Tabriz, Tabriz, Iran

A. Maadani Mallak
Department of Soil Science, Faculty of Agriculture, Ferdowsi University of Mashhad, Mashhad, Iran

M. Ghorbanpour (✉)
Department of Medicinal Plants, Faculty of Agriculture and Natural Resources, Arak University, Arak, Iran
e-mail: m-ghorbanpour@araku.ac.ir

M. Ghorbanpour et al. (eds.), *Biogenic Nano-Particles and their Use in Agro-ecosystems*, https://doi.org/10.1007/978-981-15-2985-6_11

11.1 Introduction

In recent years, nanotechnology has development as an effective field in biology and material science. Nanotechnology is a technology at nanometer scale (1–100 nm) that controls the shape and size of particles. The nanoparticles have unique properties that are related to the very small size of particles and the increase of the surface to volume ratio (Ochekpe et al. 2009; Khadem Moghadam et al. 2019; Maghsoodi et al. 2019).

Nanoparticles are produced by different approaches, including physical, chemical, and biological methods. Biogenic synthesis of nanoparticles is a process that utilizes the biological agents such as plants, bacteria, fungi, etc. to produce nanoparticles (Fig. 11.1). The biological synthesis of nanoparticles is important because of its environment-friendly approach.

In recent years, the biosynthesis of noble metal nanoparticles (gold, silver, palladium, and platinum) has been considered due to the development of eco-friendly technologies in material synthesis (Chandran et al. 2006; Aromal and Philip 2012; Jia et al. 2009; Song et al. 2010). Among metal nanoparticles, gold is a very popular element due to being chemically inert and non-toxic (Connor et al. 2005). The gold nanoparticles are most stable and resistant to oxidation (Daniel and Astruc 2004). They are used in a variety of fields, including catalysis, gene expression, nonlinear optics, and delivery systems. The biosynthesis of metal nanoparticles is carried out using the "bottom-up" approach of nanotechnology (Golinska et al. 2014). In this method, the nanoparticles are formed through the growth or assembly of atoms or molecules that are the building units.

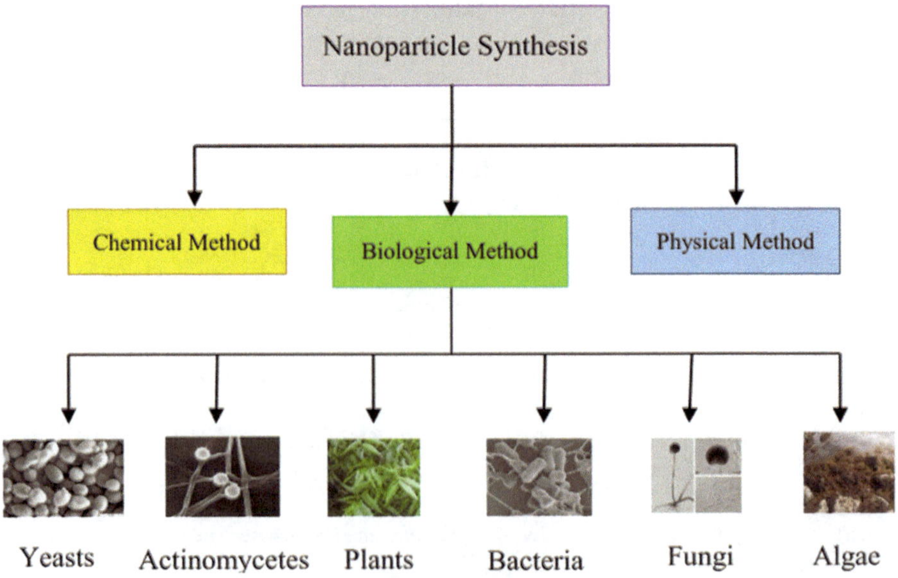

Fig. 11.1 Synthesis of nanoparticles using biological method

Biosynthesis of nanoparticles is formed through reduction/oxidation reactions of metal. In biogenic synthesis of metal nanoparticles, enzymes secreted by microbial agents and metabolites of plants are responsible for the occurrence of these reactions (Prabhu and Poulose 2012).

11.2 Biosynthesis of Nanoparticles

11.2.1 Synthesis of Gold Nanoparticles Using Plant

The use of plant extracts is preferred to produce metal nanoparticles compared with the use of microorganisms. The synthesis rate of nanoparticles using plants is faster than microbial agents, and nanoparticles obtained are more stable (Iravani 2011). In addition, plants are known as an important source of various metabolites, and they have the potential for synthesis of metal nanoparticles in large scale (Jha et al. 2009). However, reaction time required for biogenic synthesis methods is longer than chemical methods for the production of nanoparticles (Song and Kim 2009). Biomass and extract of different parts of plants such as leaf, root, flower, seed, stem, and fruit are used for the biosynthesis of gold nanoparticles. Extract of plants can act as a stabilizing, reducing, and capping agent for the synthesis of nanoparticles (Sharma et al. 2015).

11.2.1.1 Plant Biomass

The presence of metal elements, especially in drinking water, is a serious concern for global health. The use of plant biomass for the removal of heavy metals from aqueous solutions can be valuable as an eco-friendly method and also because of their potential application in removing contaminants from industrial wastewater in the future.

For this reason, many researchers have studied the role of plants in the absorption and accumulation of metal nanoparticles. The formation of gold nanoparticles from living plants was first reported by Gardea-Torresdey et al. (2002). The gold nanoparticles are synthesized inside live alfalfa plants (*Medicago sativa*) by gold ion uptake from the $AuCl_4$-rich agar solid media. The absorption and formation of gold nanoparticles within the plant were confirmed by X-ray absorption spectroscopy (XAS) and transmission electron microscopy (TEM). TEM images showed that gold nanoparticles were in crystalline state, but also twinned crystal structures and icosahedral nanoparticles were found.

In another study, Armendariz et al. (2004a) reported the synthesis of gold nanoparticles using oat (*Avena sativa*) biomass. The binding trend of Au(III) to oat and the possible formation of gold nanoparticles were studied at different pH values (pH 2–6). The size of the nanoparticles produced by oat biomass was dependent on the pH of the solution, while the shape of the nanoparticles was not significantly affected by the different pH values. Similar results have been reported for gold nanoparticles formed by wheat biomass (Armendariz et al. 2004b).

11.2.1.2 Plant Extracts

The synthesis of gold nanoparticles using plant leaf extracts has been demonstrated by many researchers (Table 11.1). Dubey et al. (2010a) reported the rapid synthesis of gold nanoparticles using leaf extract of *Rosa rugosa* within 10 min. In addition, they evaluated the effect of leaf extract quantity and concentration of metal solution (auric acid) in order to optimize the synthesis route of the metal nanoparticles. The formation and stability of the biosynthesized gold nanoparticles was confirmed using spectroscopic characterizations of UV-Vis, TEM, FTIR (Fourier transform infrared spectroscopy), and zeta potential. The sharpness, shape, size, and rate of formation of gold nanoparticles depend on the concentrations of leaf extract and metal ion. Sharp and symmetrical nanoparticles were formed at higher concentrations of leaf extract. Comparatively larger size of gold nanoparticles (50–250 nm) was found at higher gold ion concentration, and the rate of formation of the nanoparticles was slower at lowest concentration.

Gold nanoparticles were formed when the leaves of *Pelargonium graveolens* were exposed to aqueous chloroaurate ions. The rapid bioreduction of metal ions led to the formation of stable gold nanoparticles of different sizes. The size of the nanoparticles was in the range of 20–40 nm, and their shape was mainly decahedral and icosahedral (Shankar et al. 2003).

Shankar et al. (2004) reported the synthesis of pure metallic silver and gold nanoparticles and bimetallic Au core-Ag shell nanoparticles using the broth of neem leaves (*Azadirachta indica*). They proposed that the presence of reducing sugars and/or terpenoids in the broth can possibly facilitate the reduction of metal ions. The time of reduction of Au^+ ion (2 h) by neem leaf extract was faster than that observed for Ag^+ ion (4 h).

The biological synthesis of gold nanoparticles using olive leaf extracts has been reported (Khalil et al. 2012). The characterization of gold nanoparticles exhibited that the morphology of the gold nanoparticles depends on the extract concentration and the solution pH. The nanoparticles formed at lower concentrations of leaf broth were mainly triangular in shape, while spherical shaped nanoparticles were obtained at higher concentrations of leaf broth. The increase of pH also results in the production of smaller nanoparticles.

Green synthesis of gold nanoparticles using fruit extracts has been demonstrated by some researchers. For instance, Ankamwar et al. (2005) used *Emblica officinalis* (amla) fruit extract to produce gold nanoparticles. Chloroauric acid solution was treated with amla fruit extract (as the reducing agent), which results in the formation of highly stable gold nanoparticles. The size of the nanoparticles produced was in the range of 15–25 nm.

The effect of pH on the morphology of gold nanoparticles prepared from pear fruit extract has been investigated (Ghodake et al. 2010). According to the results of the investigation, gold nanostructures produced in an alkaline condition were very efficient and provide an optimal quantity of pure nanomaterial (Fig. 11.2b). The triangular and hexagonal nanoplates were formed in the range of 200–500 nm in size, depending on the shape (Fig. 11.2c, d). It was suggested that the mechanism of induction of these nanostructures is alkaline-responsive phytochemicals, such as

Table 11.1 Biological synthesis of gold nanoparticles using plants

No.	Name of the plants	Biomass/plant extract	Morphology	Size	References
01	*Medicago sativa*	Biomass	Twinned crystal structures and icosahedral nanoparticles	4 nm, 6–10 nm	Gardea-Torresdey et al. (2002)
02	*Avena sativa*	Biomass	Tetrahedral, decahedral, hexagonal, icosahedral multitwinned, irregular, and rod shaped	5–20 nm (pH 3 & 4), 25–85 nm (pH 2)	Armendariz et al. (2004a)
03	*Triticum aestivum*	Biomass	Tetrahedral, decahedral, hexagonal, icosahedral multitwinned, irregular, and rod shaped	10–30 nm	Armendariz et al. (2004b)
04	*Pelargonium graveolens*	Leaf extract	Decahedral and icosahedral	20–40 nm	Shankar et al. (2003)
05	*Rosa rugosa*	Leaf extract	Triangular and hexagonal	11 nm	Dubey et al. (2010a)
06	*Azadirachta indica*	Leaf extract	Spherical, triangular, hexagonal	50–100 nm (au/ag)	Shankar et al. (2004)
07	Olive leaf broth	Leaf extract	Triangular, hexagonal, and spherical	50–100 nm	Khalil et al. (2012)
08	*Emblica officinalis*	Fruit extract	Polyhedron, extracellular	15–25 nm	Ankamwar et al. (2005)
09	Pear fruit extract	Fruit extract	Triangular, hexagonal	200–500 nm	Ghodake et al. (2010)
10	*Tanacetum vulgare*	Fruit extract	Spherical, triangular	11 nm	Dubey et al. (2010b)
11	*Prunus domestica*	Fruit extract	Spherical	20 ± 6 nm	Dauthal and Mukhopadhyay (2012)
12	*Nyctanthes arbor-tristis*	Flower extract	Triangular, pentagonal, rod shaped, and spherical	19.8 ± 5.0 nm	Das et al. (2011)
13	*Cuminum cyminum*	Seed powder extract	Triangular, octahedral, and spherical	5–400 nm	Sneha et al. (2011)
14	*Morinda citrifolia*	Root extract	Cubic	12.17–38.26 nm	Suman et al. (2014)
15	*Eucalyptus globulus*	Bark extract	Spherical	20.1–100.9 nm	Pinto et al. (2017)

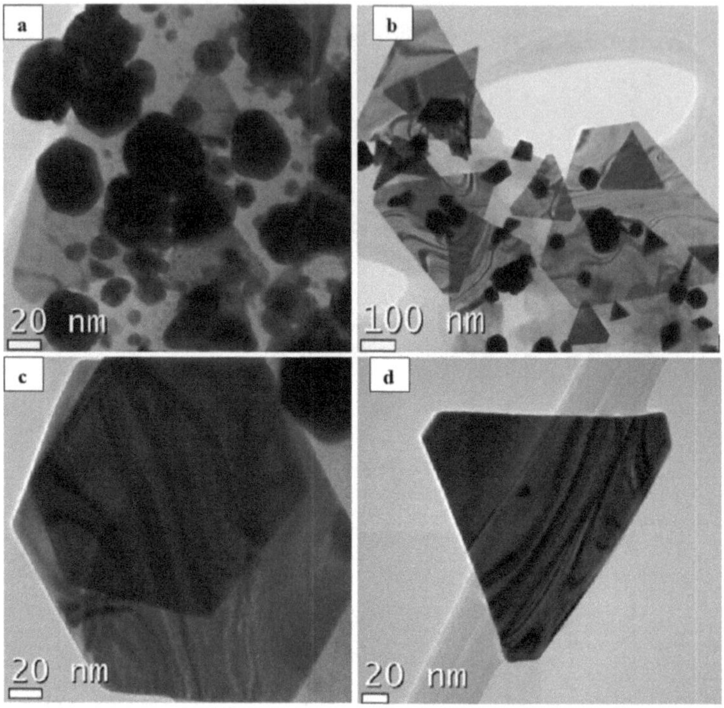

Fig. 11.2 HR-TEM micrographs of the gold nanoparticles formed from pear fruit extract under the normal (**a**) and alkaline (**b**) conditions and HR-TEM micrographs of a gold nanohexagon and nanotriangle formed under alkaline conditions (**c** and **d**) (Ghodake et al. 2010)

organic acids, amino acids, peptides, and/or proteins. Gold nanoparticles obtained under the normal conditions showed plate-like morphologies with low production efficiency (Fig. 11.2a).

Biosynthesis of gold nanoparticles using tansy fruit extract (*Tanacetum vulgare*) has also been reported (Dubey et al. 2010b). Zeta potential is an index of surface charge of the nanoparticles that is used to predict the stability of colloidal particles (Heurtault et al. 2003). The effect of pH on zeta potential of the nanoparticles produced by tansy fruit extract indicates that the zeta potential value of nanoparticles depends on the pH of the solution. The zeta potential value of gold nanoparticles in alkaline pH was slightly higher than that of acidic pH. Furthermore, size of the particles produced was increased by decreasing the pH.

In another study, gold nanoparticles have been fabricated by treatment of the $HAuCl_4$ solution with *Prunus domestica* (plum) fruit extracts (Dauthal and Mukhopadhyay 2012). The catalytic activity of gold nanoparticles dispersed in the fruit extract was studied for 4-nitrophenol reduction to 4-aminophenol. FTIR analysis suggested that the water-soluble polyols like flavanols, glycosides, and phenols were responsible for the reduction of Au^{3+} ions. Biosynthesized gold nanoparticles showed dose-dependent catalytic activity for 4-NP reduction. The catalytic activity of 4-nitrophenol increased with increasing dosage of colloidal gold nanoparticles.

Flower extract of the plant *Nyctanthes arbor-tristis* has been used as the reducing and capping agent for the synthesis of gold nanoparticles (Das et al. 2011). TEM images of the nanoparticles showed a mixture of different shapes (triangular, pentagonal, rod shaped, and spherical) with an average size of 19.8 ± 5.0 nm. Sneha et al. (2011) also reported the formation of gold nanoparticles using cumin seeds (*Cuminum cyminum*). They stated that the particles were predominately monodispersed at higher pH and polydispersed particles formed at lower pH. Table 11.1 summarizes the important examples of gold nanoparticles synthesized by plants.

11.2.2 Synthesis of Gold Nanoparticles Using Bacteria

Many microorganisms can produce various biomolecules either intracellularly or extracellularly. In synthesis of nanoparticles outside the cell, extracellularly, the enzymes secreted by microorganism play an important role in the bioreduction of metal ions. In synthesis of nanoparticles inside the cell, intracellularly, the enzymes present in the cell wall of the microorganisms involve in the reduction of metal ions to metal nanoparticles (Hulkoti and Taranath 2014). The nanoparticles produced inside the organism can have a smaller size than extracellularly formed nanoparticles (Narayanan and Sakthivel 2010).

Beveridge and Murray (1980) synthesized the gold nanoparticles using the cell wall of *Bacillus subtilis*. They were chemically modified amine and carboxyl groups of the cell wall of *B. subtilis* to determine their contribution to the metal uptake values. Their results indicated that chemical modifications of amine functions did not decrease the metal uptake values, whereas alteration of carboxyl groups was severely restricted metal deposition of most of the metals tested.

Deplanche and Macaskie (2008) demonstrated microbial reduction of gold using *Escherichia coli* and *Desulfovibrio desulfuricans* and determined the location and size of the formed gold particles. According to their report, hydrogenases are responsible in the bacteria-mediated reduction of the gold ions. The size and shape of the gold nanoparticles produced depend on the solution pH and the location of the formation of the nanoparticles. The nanoparticles ranged from 5 to 50 nm and located in the periplasmic space and on the cell surface as well as intracellularly.

The extracellular synthesis of gold nanoparticles using the gram-negative soil bacterium *Pseudomonas fluorescens* has been proven (Rajasree and Suman 2012). In a recent study, a human pathogenic bacterium *Salmonella enterica* subsp. *enterica* serovar Typhi isolated from blood and stool specimens of patients provided the biogenic synthesis of gold nanoparticles (Mortazavi et al. 2017). Characterizations of some gold nanoparticles synthesized by bacteria are enlisted in Table 11.2.

Table 11.2 Biological synthesis of gold nanoparticles using bacteria, actinomycetes, algae, and yeast

No.	Name of microorganism	Extracellular/ intracellular	Morphology	Size	References
Bacteria					
01	*Escherichia coli*, *Desulfovibrio desulfuricans*	Intracellular	Spherical, triangles, hexagons, and rods	5–50 nm	Deplanche and Macaskie (2008)
02	*Pseudomonas fluorescens*	Extracellular	Spherical	50–70 nm	Rajasree and Suman (2012)
03	*Salmonella enterica* subsp. *enterica* serovar Typhi	Extracellular	–	42 ± 2 nm	Mortazavi et al. (2017)
Actinomycetes					
01	*Thermomonospora* sp.	Extracellular	Spherical	8 nm	Ahmad et al. (2003a)
02	*Rhodococcus* sp.	Intracellular	Spherical	5–15 nm	Ahmad et al. (2003b)
03	*Streptomyces fulvissimus*	Extracellular	Spherical	20–50 nm	Soltani Nejad et al. (2015)
Algae					
01	*Plectonema boryanum*	Extracellular/ intracellular	Cubic	< 10–25	Lengke et al. (2006)
02	*Sargassum wightii* Greville	Extracellular	Thin planar structures	8–12 nm	Singaravelu et al. (2007)
03	*Galaxaura elongata*	–	Spherical	3.85–77.13 nm	Abdel-Raouf et al. (2017)
Yeast					
01	*Yarrowia lipolytica NCIM 3589*	Intracellular	Hexagonal, triangular	15 nm	Agnihotri et al. (2009)
02	*Magnusiomyces ingens LH-F1*	–	Sphere, triangle, and hexagon	80.1 ± 9.8 nm	Zhang et al. (2016)

11.2.3 Synthesis of Gold Nanoparticles Using Actinomycetes, Algae and Yeast

Actinomycetes are a group of gram-positive bacteria that have the characteristics of fungi. They produce various biomolecules including proteins, enzymes, antibiotics, and vitamins. Actinomycetes can be used as stabilizing and capping agents for the synthesis of metal nanoparticles, including gold. Among the bioactive agents secreted by actinomycetes, proteins play an important role in the synthesis of the nanoparticles. It is proven that free amine groups or cysteine residues in the proteins can bind to gold nanoparticles (Gole et al. 2001).

The alkalothermophilic actinomycete *Thermomonospora* sp. has been explored for extracellular synthesis of stable gold nanoparticles (Ahmad et al. 2003a).

Intracellular synthesis of gold nanoparticles has been provided by *Rhodococcus* sp. actinomycetes (Ahmad et al. 2003b). Spherical nanoparticles with size range of 5–15 nm were achieved.

Algae are photosynthetic eukaryotic microorganisms used for synthesis of gold nanoparticles. Cell walls of algae contain biomolecules, including polysaccharides, proteins, and enzymes which act as reducing agents for the reduction of gold ions (Sharma et al. 2015).

The powder and the ethanolic extract of marine red alga *Galaxaura elongate* have been used for synthesis of gold nanoparticles. The formation of gold nanoparticles by powder (3 h) was faster than alcoholic extract (2–5 min) (Abdel-Raouf et al. 2017).

Yeasts are single-cell eukaryotic microorganisms that are classified in the fungus kingdom. The biogenic synthesis of gold nanoparticles by the non-conventional yeasts *Yarrowia lipolytica* and *Magnusiomyces ingens* is described (Agnihotri et al. 2009; Zhang et al. 2016). X-ray diffraction (XRD) data and TEM images of *Y. lipolytica* showed that the nanoparticles are synthesized with a size of 15 nm and located on the wall of the cells (Agnihotri et al. 2009). TEM images and dynamic light scattering (DLS) data of *M. ingens* indicated that the average size of gold nanoparticles was 80.1 ± 9.8 and 137.8 ± 4.6 nm, respectively. According to the results of the investigation, some biomolecules were absorbed on the surface of the nanoparticles, which can act as organic ligands in the formation of gold nanoparticles (Zhang et al. 2016). Important examples of biosynthesis of gold nanoparticles by actinomycetes, algae, and yeasts are summarized in Table 11.2.

11.2.4 Synthesis of Gold Nanoparticles Using Fungi

The biosynthesis of gold nanoparticles using fungi has been demonstrated (Table 11.3). Among the microorganisms used for the synthesis of metal nanoparticles, fungi are a suitable candidate for the production of different enzymes, which have high growth capacity and are easy to handle.

The exact mechanism of synthesis of nanoparticles using biological agents is still unknown, but it has been demonstrated that different biomolecules have a significant role in the synthesis of nanoparticles. It has been revealed that the enzyme nitrate reductase is involved in biosynthesis of nanoparticles by fungi (Kumar et al. 2007a, b). The mechanisms for intracellular and extracellular synthesis of nanoparticles are different. Moreover, the shape and size of nanoparticles produced can be affected by enzymes and mechanisms involved in the synthesis of nanoparticles.

Mukherjee et al. (2001) reported intracellular synthesis of gold nanoparticles by bioreduction of aqueous $AuCl_4^-$ ions using the fungus *Verticillium* sp. TEM image of a single cell showed that the gold nanoparticles were formed on both the cell wall (outer boundary) and the cytoplasmic membrane (inner boundary). The number of gold nanoparticles on the cytoplasmic membrane was more than on the cell wall. The shape of gold nanoparticles was mostly spherical, although a few triangular and hexagonal particles were observed.

Table 11.3 Biological synthesis of gold nanoparticles using fungi

No.	Name of the fungi	Extracellular/ intracellular	Morphology	Size	References
01	*Verticillium* sp.	Intracellular	Spherical, triangular, and hexagonal	20 ± 8 nm	Mukherjee et al. (2001)
02	*Fusarium oxysporum*	Extracellular	Spherical, triangular	20–40 nm	Mukherjee et al. (2002)
03	*Rhizopus oryzae*	–	–	10 nm	Das et al. (2009)
04	*F. oxysporum* f. sp. *cubense*	Extracellular	–	22 nm	Thakker et al., (2013)
05	*Penicillium aurantiogriseum*, *P. citrinum*, and *P. waksmanii*	–	Spherical	153.3, 172, and 160.1 nm (90 = 300 nm)	Honary et al. (2013)
06	*Aspergillus sydowii*	Extracellular/ intracellular	Spherical	8.7–15.6 nm	Vala (2015)
07	*Alternaria* sp.	Extracellular	Quasi-spherical, spherical, and anisotropic nanoparticles	7–13, 15–18, and 69–93 nm	Dhanasekar et al. (2015)
08	*Pleurotus ostreatus*	Extracellular	–	22–39 nm	El-Batal et al. (2015)
09	*Penicillium aculeatum*	Extracellular	Spherical	60 nm	Barabadi et al. (2017)
10	*Aspergillus* sp. WL-Au	Extracellular	Spherical	2.5–6.7 nm	Shen et al. (2017)
11	Thermophilic fungi	Intracellular/ extracellular	Spherical, hexagonal, and amorphous	6–40 nm	Molnar et al. (2018)

In addition, they reported extracellular synthesis of gold nanoparticles by a eukaryotic system such as fungi for the first time (Mukherjee et al. 2002). The nanoparticles were synthesized by treatment of the fungus *Fusarium oxysporum* with $AuCl_4^-$ solution. TEM pictures showed that gold particles have spherical and triangular morphology with a size range of 20–40 nm. Indeed, the gold nanoparticle formed by reaction of gold ions with extracellular secreted enzymes by the fungus. Thakker et al. (2013) synthesized gold nanoparticles using a plant pathogenic fungus *F. oxysporum* f. sp. *cubense* and reported their antibacterial activity against *Pseudomonas* sp.

The microbial synthesis of gold nanoparticles has been investigated using the fungus *Rhizopus oryzae* to remove different organophosphorus pesticides (model) from water along with some microorganisms (Das et al. 2009). The gold nanoparticles were formed on the surface of *R. oryzae* and were stable even up to 6 months. FTIR spectra after treatment of *R. oryzae* with $HAuCl_4^-$ revealed the presence of amide I, II, and III groups and the disappearance of carboxyl groups present in mycelia. Based on the FTIR results, it was suggested that polypeptides/proteins are involved in the reduction of gold ions. Indeed, the gold nanoparticles are formed by surface-bound protein molecules that act as both reducing and stabilizing agents.

In another study, the use of a marine-derived fungus *Aspergillus sydowii* resulted in the formation of spherical gold nanoparticles with an average size of 10 nm. The fungus could synthesize gold nanoparticles extra-/intracellularly depending on the applied gold ion concentration (Vala 2015). Fungus-mediated synthesis of gold nanoparticles by *Penicillium aurantiogriseum*, *P. citrinum*, and *P. waksmanii* has been demonstrated (Honary et al. 2013).

11.3 Characterization of Gold Nanoparticles

The characteristics of gold nanoparticles are determined using various techniques such as scanning electron microscopy (SEM), atomic force microscopy (AFM), TEM, DLS, FTIR, XRD, and UV-Vis spectroscopy. The shape and size of nanoparticles are determined by TEM, SEM, and AFM. DLS is also used for determination of size, dispersity, and zeta potential of nanoparticles. Furthermore, FTIR and XRD are applied for the determination of structural characteristics and crystallinity of formed particles.

In the biogenic synthesis of gold nanoparticles, the change of color of the reaction mixture from pale yellow to dark purple/deep red reflects the formation of gold particles. The different colors of gold nanoparticle solution are due to their surface plasmon resonance properties (He et al. 2007). Generally, UV-visible spectroscopy is utilized to confirm formation of metal nanoparticles including gold. The UV-visible spectrum of the reaction mixtures (organism-gold ions) represents the formation of a gold surface plasmon resonance (detection of gold nanoparticles) that ranged from 500 to 600 nm (Deplanche and Macaskie 2008).

11.4 Applications of Gold Nanoparticles in Agriculture

In recent years, the use of nanotechnology in various fields, including pharmaceuticals, engineering, and agriculture, has been developed. The application of nanotechnology in the agricultural sector has improved, especially in the area of food industry and plant protection. Gold nanoparticles have many potential applications in agriculture due to their antimicrobial activity and unique optical property.

Gold nanoparticles can be applied as a sensor in a series of colorimetric assays. In this assay, the interaction between the analyte and the gold nanoparticles can induce the aggregation of gold nanoparticles and consequently solution color changes from red to purple. This feature can be used to identify different molecules, including pesticides. For instance, Bai et al. (2010) have studied gold nanoparticles as colorimetric probes for screening insecticide pymetrozine. It has been demonstrated that compounds containing nitrogen heterocycles and amine groups can be bound to the surface of metal nanoparticles and induce the accumulation of the nanoparticles (Gittins and Caruso 2001; Ai et al. 2009). Chemical structure of pymetrozine contains multiple binding sites including one exocyclic secondary amine and four-nitrogen hybrid ring. Indeed, the color change and aggregation of gold nanoparticles can be attributed to the specific interactions between the functional groups of gold nanoparticle and pymetrozine.

In addition, gold nanoparticle-based sensors can be utilized to determine the residue of different pesticides in plants and food products. For example, Bai and his colleagues (2010) determined the concentration of pymetrozine with the low detection limit (1×10^{-6} M) and reported the high sensitivity of this method for pymetrozine compared with other 11 pesticides. The detection sensitivity of this system could be increased by adding salt and reducing the pH. The use of bacterial-derived gold nanoparticles to detect organophosphorus pesticide residues in fruits and vegetables has been proven (Malarkodi et al. 2017).

Gold nanoparticles can also be useful in water purification. For instance, Zhang et al. (2014) fabricated imidazole ionic liquid functionalized gold nanoparticles for the recognition of imidacloprid. The researchers suggested that the detection system could be used to determine and remove imidacloprid in different water samples based on the aggregation phenomena of gold nanoparticles. The application of fungus-mediated synthesized gold nanoparticles to remove pesticides and pathogens from water has been reported (Das et al. 2009).

Gold nanoparticles have antibacterial and antifungal properties that can be used in plant disease management, food safety, and medical applications. Jayaseelan et al. (2013) synthesized gold nanoparticles using seed aqueous extract of *Abelmoschus esculentus* and posed antifungal activity of the nanoparticles against *Candida albicans*, *Aspergillus niger*, *Aspergillus flavus*, and *Puccinia graminis tritci*. The antibacterial activity of the biosynthesized gold nanoparticles against *Klebsiella pneumoniae*, *Bacillus subtilis*, *Pseudomonas aeruginosa*, and *Escherichia coli* has been reported (Annamalai et al. 2013; Muthuvel et al. 2014).

The mycelial growth inhibition of *Phomopsis theae* by *Trichoderma atroviride*-mediated biosynthesized gold nanoparticles has been demonstrated (Ponmurugan

2016). Field experiments conducted with soil application and wound dressing of the nanoparticles confirmed the efficacy of the nanoparticles for control of *Phomopsis* canker disease in tea plants.

It has been revealed that biosynthesized gold nanoparticles can be useful to control pests in agriculture and public health (Thakur et al. 2018; Sundararajan and Kumari 2017). Thakur et al. (2018) studied the effect of biosynthesized gold nanoparticles on root-knot nematodes (*Meloidogyne incognita*) in tomato crop. The nanoparticles showed suitable nematicidal effect and had no negative impact on plant growth. All articles on the insecticidal activity of gold nanoparticles focused on mosquito species of medical and veterinary importance.

Many researchers have demonstrated that gold nanoparticles induce cell division, seed and pollen germination, and plant growth (Arora et al. 2012; Gopinath et al. 2013; Mahakham et al. 2016; Thakur et al. 2018). Therefore, application of gold nanoparticles in agriculture and plant sciences could be beneficial to increase the plant growth and crop yield like several types of engineered nanomaterials (Baiazidi-Aghdam et al. 2016; Ghorbanpour and Hadian 2015; Hatami et al. 2013, 2016, 2017, 2019; Ghorbanpour et al. 2015, 2018; Ghorbanpour and Hatami 2014, 2015; Ghorbanpour 2015; Ghorbanpour and Hadian 2017; Ghorbanpour and Fahimirad 2017; Hatami et al. 2014; Hatami and Ghorbanpour 2013, 2014; Hatami 2017; Chegini et al. 2017; Mohammadi et al. 2018; Tian et al. 2018; Ahmadi et al. 2018).

11.5 Conclusions

The main goal of most nanotechnology research is to design and produce nanoparticles with new features. Compared with the chemical method, the biological synthesis of gold nanoparticles by organisms is an environmentally friendly and reliable method. The gold nanoparticles of a variety of shapes and sizes can be easily synthesized from different types of plants and microbes. The synthesis of gold nanoparticles depends on various factors including the concentration of plant extract/biomass and metal salt, pH of the solution, temperature, reaction time, and the location of nanoparticle formation (extracellular/intracellular). Applications of such eco-friendly nanoparticles in agriculture to purify rivers and lakes from pesticides can reduce the harmful impacts on nontarget organisms. Biosynthesized gold nanoparticles can be effective to protect the various crop plants against plant pathogens and can be a suitable alternative to chemical pesticides that are toxic to human and the environment.

References

Abdel-Raouf N, Al-Enazi NM, Ibraheem IB (2017) Green biosynthesis of gold nanoparticles using *Galaxaura elongata* and characterization of their antibacterial activity. Arab J Chem 10:S3029–S3039

Agnihotri M, Joshi S, Kumar AR, Zinjarde S, Kulkarni S (2009) Biosynthesis of gold nanoparticles by the tropical marine yeast *Yarrowia lipolytica* NCIM 3589. Mater Lett 63:1231–1234

Ahmad A, Senapati S, Khan MI, Kumar R, Sastry M (2003a) Extracellular biosynthesis of monodisperse gold nanoparticles by a novel extremophilic actinomycete, *Thermomonospora* sp. Langmuir 19:3550–3553

Ahmad A, Senapati S, Khan MI, Kumar R, Ramani R, Srinivas V, Sastry M (2003b) Intracellular synthesis of gold nanoparticles by a novel alkalo tolerant actinomycete, *Rhodococcus* species. Nanotechnology 14:824

Ahmadi SZ, Ghorbanpour M, Hadian J, Salehi-Arjmand H (2018) Impact of foliar spray of spherical Nano-carbon and Salicylic acid on physiological traits and Parthenolide content in two feverfew cultivars (*Tanacetum parthenium* Linn. cv. Pharmasaat and Jelitto). J Med Plant 17(4):82–98

Ai K, Liu Y, Lu L (2009) Hydrogen-bonding recognition-induced color change of gold nanoparticles for visual detection of melamine in raw milk and infant formula. J Am Chem Soc 131:9496–9497

Ankamwar B, Damle C, Ahmad A, Sastry M (2005) Biosynthesis of gold and silver nanoparticles using *Emblica officinalis* fruit extract, their phase transfer and transmetallation in an organic solution. J Nanosci Nanotechnol 5:1665–1671

Annamalai A, Christina VLP, Sudha D, Kalpana M, Lakshmi PTV (2013) Green synthesis, characterization and antimicrobial activity of Au NPs using *Euphorbia hirta* L. leaf extract. Colloids Surf B Biointerfaces 108:60–65

Armendariz V, Herrera I, Jose-yacaman M, Troiani H, Santiago P, Gardea-Torresdey JL (2004a) Size controlled gold nanoparticle formation by *Avena sativa* biomass: use of plants in nanobiotechnology. J Nanopart Res 6:377–382

Armendariz V, Jose-Yacaman M, Duarte Moller A, Peralta-Videa JR, Troiani H, Herrera I, Gardea-Torresdey JL (2004b) HRTEM characterization of gold nanoparticles produced by wheat biomass. Revista Mexicana de Fisica Supplement 50:7–11

Aromal SA, Philip D (2012) Green synthesis of gold nanoparticles using *Trigonellafoenum*-graecum and its size-dependent catalytic activity. Spectrochim Acta A Mol Biomol Spectrosc 97:1–5

Arora S, Sharma P, Kumar S, Nayan R, Khanna PK, Zaidi MGH (2012) Gold-nanoparticle induced enhancement in growth and seed yield of *Brassica juncea*. Plant Growth Regul 66:303–310

Bai LY, Zhang YP, Chen J, Zhou XM, Hu LF (2010) Rapid, sensitive and selective detection of pymetrozine using gold nanoparticles as colorimetric probes. Micro Nano Lett 5:304–308

Baiazidi-Aghdam MT, Mohammadi H, Ghorbanpour M (2016) Effects of nanoparticulate anatase titanium dioxide on physiological and biochemical performance of *Linum usitatissimum* (Linaceae) under well watered and drought stress conditions. Braz J Bot 39:139–146

Barabadi H, Honary S, Mohammadi MA, Ahmadpour E, Rahimi MT, Alizadeh A, Naghibi F, Saravanan M (2017) Green chemical synthesis of gold nanoparticles by using *Penicillium aculeatum* and their scolicidal activity against hydatid cyst protoscolices of *Echinococcus granulosus*. Environ Sci Pollut Res 24:5800–5810

Beveridge TJ, Murray RG (1980) Sites of metal deposition in the cell wall of *Bacillus subtilis*. J Bacteriol 141:876–887

Chandran SP, Chaudhary M, Pasricha R, Ahmad A, Sastry M (2006) Synthesis of gold nanotriangles and silver nanoparticles using *Aloe vera* plant extract. Biotechnol Prog 22:577–583

Chegini E, Ghorbanpour M, Hatam M, Taghizadeh M (2017) Effect of multi-walled carbon nanotubes on physiological traits, phenolic contents and antioxidant capacity of *Salvia mirzayanii* Rech. f & Esfandunder drought stress. J Med Plant 16(2):191–207

Connor EE, Mwamuka J, Gole A, Murphy CJ, Wyatt MD (2005) Gold nanoparticles are taken up by human cells but do not cause acute cytotoxicity. Small 1:325–327

Daniel MC, Astruc D (2004) Gold nanoparticles: assembly, supramolecular chemistry, quantum-size-related properties, and applications toward biology, catalysis, and nanotechnology. Chem Rev 104:293–346

Das SK, Das AR, Guha AK (2009) Gold nanoparticles: microbial synthesis and application in water hygiene management. Langmuir 25:8192–8199

Das RK, Gogoi N, Bora U (2011) Green synthesis of gold nanoparticles using *Nyctanthesarbortristis* flower extract. Bioprocess Biosyst Eng 34:615–619

Dauthal P, Mukhopadhyay M (2012) *Prunus domestica* fruit extract-mediated synthesis of gold nanoparticles and its catalytic activity for 4-nitrophenol reduction. Ind Eng Chem Res 51:13014–13020

Deplanche K, Macaskie LE (2008) Biorecovery of gold by *Escherichia coli* and *Desulfovibrio desulfuricans*. Biotechnol Bioeng 99:1055–1064

Dhanasekar NN, Rahul GR, Narayanan KB, Raman G, Sakthivel N (2015) Green chemistry approach for the synthesis of gold nanoparticles using the fungus *Alternaria* sp. J Microbiol Biotechnol 25:1129–1135

Dubey SP, Lahtinen M, Sillanpää M (2010a) Green synthesis and characterizations of silver and gold nanoparticles using leaf extract of *Rosa rugosa*. Colloids Surf A Physicochem Eng Asp 364:34–41

Dubey SP, Lahtinen M, Sillanpää M (2010b) Tansy fruit mediated greener synthesis of silver and gold nanoparticles. Process Biochem 45:1065–1071

El-Batal AI, ElKenawy NM, Yassin AS, Amin MA (2015) Laccase production by *Pleurotus ostreatus* and its application in synthesis of gold nanoparticles. Biotechnol Reps 5:31–39

Gardea-Torresdey JL, Parsons JG, Gomez E, Peralta-Videa J, Troiani HE, Santiago P, Yacaman MJ (2002) Formation and growth of Au nanoparticles inside live alfalfa plants. Nano Lett 2:397–401

Ghodake GS, Deshpande NG, Lee YP, Jin ES (2010) Pear fruit extract-assisted room-temperature biosynthesis of gold nanoplates. Colloids Surf B Biointerfaces 75:584–589

Ghorbanpour M (2015) Major essential oil constituents, total phenolics and flavonoids content and antioxidant activity of *Salvia officinalis* plant in response to nano-titanium dioxide. Ind J Plant Physiol 20(3):249–256

Ghorbanpour M, Fahimirad SH (2017) Plant nanobionics a novel approach to overcome the environmental challenges. In: Ghorbanpour M, Varma A (eds) Medicinal plants and environmental. Springer. https://doi.org/10.1007/978-3-319-68717-9_14

Ghorbanpour M, Hadian J (2015) Multi-walled carbon nanotubes stimulate callus induction, secondary metabolites biosynthesis and antioxidant capacity in medicinal plant *Satureja khuzestanica* grown *in vitro*. Carbon 94:749–759

Ghorbanpour M, Hadian (2017) Engineered nanomaterials and their interactions with plant cells: injury indices and detoxification pathways. In: Ghorbanpour M et al (eds) Nanoscience and plant–soil systems, Soil biology 48. https://doi.org/10.1007/978-3-319-46835-8_13

Ghorbanpour M, Hatami M (2014) Spray treatment with silver nanoparticles plus thidiazuron increases anti-oxidant enzyme activities and reduces petal and leaf abscission in four cultivars of geranium (*Pelargonium zonale*) during storage in the dark. J Hortic Sci Biotechnol 89(6):712–718

Ghorbanpour M, Hatami H (2015) Changes in growth, antioxidant defense system and major essential oils constituents of *Pelargonium graveolens* plant exposed to nano-scale silver and thidiazuron. Ind J Plant Physiol 20(2):116–123

Ghorbanpour M, Hatami M, Hatami M (2015) Activating antioxidant enzymes, hyoscyamine and scopolamine biosynthesis of *Hyoscyamus niger* L. plants with nano-sized titanium dioxide and bulk application. Acta Agric Slov 105:23–32

Ghorbanpour M, Khaltabadi Farahani AH, Hadian J (2018) Potential toxicity of nano-graphene oxide on callus cell of *Plantago major* L. under polyethylene glycol-induced dehydration. Ecotoxicol Environ Saf 148:910–922

Gittins DI, Caruso F (2001) Spontaneous phase transfer of nanoparticulate metals from organic to aqueous media. Angew Chem Int Ed 40:3001–3004

Gole A, Dash C, Ramakrishnan V, Sainkar SR, Mandale AB, Rao M, Sastry M (2001) Pepsin-gold colloid conjugates: preparation, characterization, and enzymatic activity. Langmuir 17:1674–1679

Golinska P, Wypij M, Ingle AP, Gupta I, Dahm H, Rai M (2014) Biogenic synthesis of metal nanoparticles from actinomycetes: biomedical applications and cytotoxicity. Appl Microbiol Biotechnol 98:8083–8097

Gopinath K, Venkatesh KS, Ilangovan R, Sankaranarayanan K, Arumugam A (2013) Green synthesis of gold nanoparticles from leaf extract of *Terminalia arjuna*, for the enhanced mitotic cell division and pollen germination activity. Ind Crop Prod 50:737–742

Hatami M (2017) Stimulatory and inhibitory effects of Nanoparticulates on seed germination and seedling vigor indices. In: Ghorbanpour M et al (eds) Nanoscience and plant–soil systems, Soil biology 48. https://doi.org/10.1007/978-3-319-46835-8_13

Hatami M, Ghorbanpour M (2013) Effect of nanosilver on physiological performance of *Pelargonium* plants exposed to dark storage. J Hortic Res 21(1):15–20

Hatami M, Ghorbanpour M (2014) Defense enzymes activity and biochemical variations of *Pelargonium zonale* in response to nanosilver particles and dark storage. Turk J Biol 38:130–139

Hatami M, Hatamzadeh A, Ghasemnezhad M, Ghorbanpour M (2013) The comparison of antimicrobial effects of silver nanoparticles (SNP) and silver nitrate($AgNo_3$) to extend the vase life of 'red ribbon' cut rose flowers. Trakia J Sci 2:144–151

Hatami M, Ghorbanpour M, Salehiarjomand H (2014) Nano-anatase TiO_2 modulates the germination behavior and seedling vigority of the five commercially important medicinal and aromatic plants. J Biol Environ Sci 8(22):53–59

Hatami M, Kariman K, Ghorbanpour M (2016) Engineered nanomaterial-mediated changes in the metabolism of terrestrial plants. Sci Total Environ 571:275–291

Hatami M, Hadian J, Ghorbanpour M (2017) Mechanisms underlying toxicity and stimulatory role of single-walled carbon nanotubes in *Hyoscyamus niger* during drought stress simulated by polyethylene glycol. J Hazard Mater 324:306–320

Hatami M, Hosseini SM, Ghorbanpour M, Kariman K (2019) Physiological and antioxidative responses to GO/PANI nanocomposite in intact and demucilaged seeds and young seedlings of Salvia mirzayanii. Chemosphere 233:920–935

He S, Guo Z, Zhang Y, Zhang S, Wang J, Gu N (2007) Biosynthesis of gold nanoparticles using the bacteria *Rhodopseudomonas capsulata*. Mater Lett 61:3984–3987

Heurtault B, Saulnier P, Pech B, Proust JE, Benoit JP (2003) Physico-chemical stability of colloidal lipid particles. Biomaterials 24:4283–4300

Honary S, Gharaei-Fathabad E, Barabadi H, Naghibi F (2013) Fungus-mediated synthesis of gold nanoparticles: a novel biological approach to nanoparticle synthesis. J Nanosci Nanotechnol 13:1427–1430

Hulkoti NI, Taranath TC (2014) Biosynthesis of nanoparticles using microbes-a review. Colloids Surf B Biointerfaces 121:474–483

Iravani S (2011) Green synthesis of metal nanoparticles using plants. Green Chem 13:2638–2650

Jayaseelan C, Ramkumar R, Rahuman AA, Perumal P (2013) Green synthesis of gold nanoparticles using seed aqueous extract of *Abelmoschus esculentus* and its antifungal activity. Ind Crop Prod 45:423–429

Jha AK, Prasad K, Prasad K, Kulkarni AR (2009) Plant system: nature's nanofactory. Colloids Surf B Biointerfaces 73:219–223

Jia L, Zhang Q, Li Q, Song H (2009) The biosynthesis of palladium nanoparticles by antioxidants in Gardenia jasminoides Ellis: long lifetime nanocatalysts for p-nitrotoluene hydrogenation. Nanotechnology 20:385601

Khadem Moghadam N, Hatami M, Rezaei S, Bayat M, Asgari Lajayer B (2019) Induction of plant defense machinery against nanomaterials exposure. In: Ghorbanpour M, Wani SH (eds) Advances in phytonanotechnology: from synthesis to application. Elsevier, London

Khalil MH, Ismail EH, El-Magdoub F (2012) Biosynthesis of Au nanoparticles using olive leaf extract. Arab J Chem 5(4):431–437

Kumar SA, Abyaneh MK, Gosavi SW, Kulkarni SK, Pasricha R, Ahmad A, Khan MI (2007a) Nitrate reductase-mediated synthesis of silver nanoparticles from AgNO3. Biotechnol Lett 29:439–445

Kumar SA, Ansary AA, Ahmad A, Khan MI (2007b) Extracellular biosynthesis of CdSe quantum dots by the fungus, *Fusarium oxysporum*. J Biomed Nanotechnol 3:190–194

Lengke MF, Fleet ME, Southam G (2006) Morphology of gold nanoparticles synthesized by filamentous cyanobacteria from gold (I)– thiosulfate and gold (III)– chloride complexes. Langmuir 22:2780–2787

Maghsoodi MR, AsgariLajayer B, Hatami M, Mirjalili MH (2019) Challenges and opportunities of nanotechnology in plants-soil mediated systems: beneficial role, phytotoxicity and phytoextraction. In: Ghorbanpour M, Wani SH (eds) Advances in phytonanotechnology: from synthesis to application, Elsevier, London

Mahakham W, Theerakulpisut P, Maensiri S, Phumying S, Sarmah AK (2016) Environmentally benign synthesis of phytochemicals-capped gold nanoparticles as nanopriming agent for promoting maize seed germination. Sci Total Environ 573:1089–1102

Malarkodi C, Rajeshkumar S, Annadurai G (2017) Detection of environmentally hazardous pesticide in fruit and vegetable samples using gold nanoparticles. Food Control 80:11–18

Mohammadi M, Hatami M, Feghezadeh K, Ghorbanpour M (2018) Mitigating effect of nano-zerovalent iron, iron sulfate and EDTA against oxidative stress induced by chromium in Helianthus annuus L. Acta Physiol Plant 40:69

Molnár Z, Bódai V, Szakacs G, Erdélyi B, Fogarassy Z, Sáfrán G, Varga T, Kónya Z, Tóth-Szeles E, Szűcs R, Lagzi I (2018) Green synthesis of gold nanoparticles by thermophilic filamentous fungi. Sci Rep 8:3943

Mortazavi SM, Khatami M, Sharifi I, Heli H, Kaykavousi K, Poor MHS, Kharazi S, Nobre MAL (2017) Bacterial biosynthesis of gold nanoparticles using *Salmonella enterica* subsp. enterica serovar Typhi isolated from blood and stool specimens of patients. J Clust Sci 28:2997–3007

Mukherjee P, Ahmad A, Mandal D, Senapati S, Sainkar SR, Khan MI, Ramani R, Parischa R, Ajayakumar PV, Alam M, Sastry M, Kumar R (2001) Bioreduction of AuCl4– ions by the fungus, *Verticillium* sp. and surface trapping of the gold nanoparticles formed. Angew Chem Int Ed 40:3585–3588

Mukherjee P, Senapati S, Mandal D, Ahmad A, Khan MI, Kumar R, Sastry M (2002) Extracellular synthesis of gold nanoparticles by the fungus *Fusarium oxysporum*. Chembiochem 3:461–463

Muthuvel A, Adavallan K, Balamurugan K, Krishnakumar N (2014) Biosynthesis of gold nanoparticles using *Solanum nigrum* leaf extract and screening their free radical scavenging and antibacterial properties. Biomed Prev Nutr 4:325–332

Narayanan KB, Sakthivel N (2010) Biological synthesis of metal nanoparticles by microbes. Adv Colloid Interf Sci 156:1–13

Ochekpe NA, Olorunfemi PO, Ngwuluka NC (2009) Nanotechnology and drug delivery part 2: nanostructures for drug delivery. Trop J Pharm Res 8(3):275–287

Pinto RJ, Lucas JM, Morais MP, Santos SA, Silvestre AJ, Marques PA, Freire CS (2017) Demystifying the morphology and size control on the biosynthesis of gold nanoparticles using *Eucalyptus globulus* bark extract. Ind Crop Prod 105:83–92

Ponmurugan P (2016) Biosynthesis of silver and gold nanoparticles using *Trichoderma atroviride* for the biological control of *Phomopsis* canker disease in tea plants. IET Nanobiotechnol 11:261–267

Prabhu S, Poulose EK (2012) Silver nanoparticles: mechanism of antimicrobial action, synthesis, medical applications, and toxicity effects. Int Nano Lett 2:1–10

Rajasree SR, Suman TY (2012) Extracellular biosynthesis of gold nanoparticles using a gram negative bacterium *Pseudomonas fluorescens*. Asian Pac J Trop Dis 2:S796–S799

Shankar SS, Ahmad A, Pasricha R, Sastry M (2003) Bioreduction of chloroaurate ions by geranium leaves and its endophytic fungus yields gold nanoparticles of different shapes. J Mater Chem 13:1822–1826

Shankar SS, Rai A, Ahmad A, Sastry M (2004) Rapid synthesis of Au, Ag, and bimetallic Au core-Ag shell nanoparticles using neem (*Azadirachta indica*) leaf broth. J Colloid Interface Sci 275:496–502

Sharma D, Kanchi S, Bisetty K (2015) Biogenic synthesis of nanoparticles: a review. Arab J Chem. https://doi.org/10.1016/j.arabjc.2015.11.002

Shen W, Qu Y, Pei X, Li S, You S, Wang J, Zhang Z, Zhou J (2017) Catalytic reduction of 4-nitrophenol using gold nanoparticles biosynthesized by cell-free extracts of *Aspergillus* sp. WL-Au. J Hazard Mater 321:299–306

Singaravelu G, Arockiamary JS, Kumar VG, Govindaraju K (2007) A novel extracellular synthesis of monodisperse gold nanoparticles using marine alga, *Sargassum wightii* Greville. Colloids Surf B Biointerfaces 57:97–101

Sneha K, Sathishkumar M, Lee SY, Bae MA, Yun YS (2011) Biosynthesis of Au nanoparticles using cumin seed powder extract. J Nanosci Nanotechnol 11:1811–1814

Soltani Nejad M, Shahidi Bonjar GH, Khaleghi N (2015) Biosynthesis of gold nanoparticles using *Streptomyces fulvissimus* isolate. Nanomed J 2:153–159

Song JY, Kim BS (2009) Rapid biological synthesis of silver nanoparticles using plant leaf extracts. Bioprocess Biosyst Eng 32:79–84

Song JY, Kwon EY, Kim BS (2010) Biological synthesis of platinum nanoparticles using *Diopyros kaki* leaf extract. Bioprocess Biosyst Eng 33:159–164

Suman TY, Rajasree SR, Ramkumar R, Rajthilak C, Perumal P (2014) The green synthesis of gold nanoparticles using an aqueous root extract of *Morinda citrifolia* L. Spectrochim Acta A Mol Biomol Spectrosc 118:11–16

Sundararajan B, Kumari BR (2017) Novel synthesis of gold nanoparticles using Artemisia vulgaris L. leaf extract and their efficacy of larvicidal activity against dengue fever vector *Aedes aegypti* L. J Trace Elem Med Biol 43:187–196

Thakker JN, Dalwadi P, Dhandhukia PC (2013) Biosynthesis of gold nanoparticles using *Fusarium oxysporum* f. sp. cubense JT1, a plant pathogenic fungus. ISRN Biotechnol 2013:1–5

Thakur RK, Shirkot P, Dhirta B (2018) Studies on effect of gold nanoparticles on *Meloidogyne incognita* and tomato plants growth and development. Bio Rxiv 428144. https://doi.org/10.1101/428144

Tian H, Ghorbanpour M, Kariman K (2018) Manganese oxide nanoparticle-induced changes in growth, redox reactions and elicitation of antioxidant metabolites in deadly nightshade (*Atropa belladonna* L.). Ind Crop Prod 126:403–414

Vala AK (2015) Exploration on green synthesis of gold nanoparticles by a marine-derived fungus *Aspergillus sydowii*. Environm Prog Sustain Energy 34:194–197

Zhang X, Sun Z, Cui Z, Li H (2014) Ionic liquid functionalized gold nanoparticles: synthesis, rapid colorimetric detection of imidacloprid. Sensors Actuators B Chem 191:313–319

Zhang X, Qu Y, Shen W, Wang J, Li H, Zhang Z, Li S, Zhou J (2016) Biogenic synthesis of gold nanoparticles by yeast *Magnusiomyces ingens* LH-F1 for catalytic reduction of nitrophenols. Colloids Surf A Physicochem Eng Asp 497:280–285

Application of Biogenic and Non-biogenic Synthesized Metal Nanoparticles on Longevity of Agricultural Crops

12

Mousa Solgi

Abstract

Agricultural crops includes horticultural (vegetables, fruits and ornamental plants), agronomic and aromatic medicinal herbs. Human population is growing fast, and consequently providing enough and healthy food is becoming a very significant problem in the near future. Nowadays, decreasing postharvest waste through using the findings of innovative technical studies like nanotechnology and nanobiotechnology in crops could be planned as one of the best resolutions to this problem. Progressing in time proved development in technology that showed the ability of metals of nanoscale to perform specific utilities better than the bulk form of metals. Nanotechnology by means of specific characters of nanoparticles can be an identical valuable knowledge in various industry and science divisions. Therefore, the current chapter especially focuses on the uses of biological or biogenic and non-biological (biogenic) on the shelf life of agricultural crops.

Keywords

Biogenic nanoparticles · Non-biogenic nanoparticles · Shelf life · Postharvest · Green nanoparticles

12.1 Introduction

The effects of metal nanoparticle on postharvest process of agricultural crops, particularly horticultural crops, have been widely studied. Solgi et al. (2009) first reported that 1 or 2 mgL^{-1} of silver nanoparticles (SNP) along with 6% sucrose can

M. Solgi (✉)
Department of Horticultural Science, Faculty of Agriculture and Natural Resources, Arak University, Arak, Iran
e-mail: M-solgi@araku.ac.ir

M. Ghorbanpour et al. (eds.), *Biogenic Nano-Particles and their Use in Agro-ecosystems*, https://doi.org/10.1007/978-981-15-2985-6_12

increase the vase life of gerbera flowers (Dune) from 8.3 (control) to 16 days. This study aimed at evaluating the effects of SNP and essential oils acting as new antimicrobial agents to extend the vase life of gerbera (*Gerbera jamesonii* cv. 'Dune') cut flowers. The vase life of flowers kept in a solution including 5 mgL^{-1} of SNP and 6% sucrose showed significantly more than those treated with 8-hydroxyquinoline citrate (8-HQC) or distilled water (control). The flowers kept in SNP solutions indicated significantly more relative fresh weight (RFW) compared to the control group. The vase life of gerbera flowers was also increased using 1 or 2 mgL^{-1} of SNP as well as 50 or 100 mgL^{-1} of carvacrol from 8.3 to 16 days. Moreover, RFW and solution uptake of gerbera flowers were extended adding 1 or 2 mgL^{-1} of SNP and 100 mg L^{-1} of essential oils than the control groups. Based on the results, SNP or essential oils can be used as new alternatives to usual agents, like silver nitrate and 8-HQC for gerbera flowers.

Based on Liu et al. (2009), 24 h pulsing by 5 mgL^{-1} of nano-silver solution along with holding in deionized water (DI) retained water uptake, increased vase life and prevented bacterial growth on cut gerbera 'Ruikou' flowers. Solgi et al. (2011) indicated that vase life of gerbera can be ended via stem bending, breaking or petal wilting. In addition, the impacts of vase solutions including 1 mgL^{-1} of silver nanoparticles using or without 6% sucrose on gerbera 'Deep Purple' cut flowers were also measured. In this regard, they measured the following parameters: the vase life, RFW, relative solution uptake, stem bending/breaking, stem end discolouration and ethylene synthesis via petals, stem ends and stem necks. According to their findings, the combined SNP and 6% sucrose extended vase life to 8 days than that of the control group treated with deionized water. They also reduced petal wilting and stem breaking in comparison to the control group. The combined SNP and 6% sucrose treatment increased ethylene synthesis by petals, stem ends and stem necks through vase life.

However, no significant adverse effect was found in ethylene enhancement on vase life factors. In general, SNP (1 mgL^{-1}) with 6% sucrose showed commercial benefits to use as vase solution for cut gerbera flowers (Table 12.1 and Figs. 12.1 and 12.2).

Table 12.1 Effects of deionized water (control), 6% sucrose, 1 mgL^{-1} SNP and 1 mgL^{-1} SNP plus 6% sucrose on gerbera stem breakage and consequent ethylene production by stem necks (Solgi et al. 2011)

	Breakage[a]		No breakage[a]	
Treatments	No. of stems	Ethylene production[b]	No. of stems	Ethylene production[b]
Deionized water	6	0.417 ± 0.106	18	0.079 ± 0.019
6% sucrose	0	0.147 ± 0.001	24	0.102 ± 0.016
1 mgL^{-1} SNP	4	0.63 ± 0.221	20	0.146 ± 0.046
1 mgL^{-1} SNP + 6% sucrose	0	0.364 ± 0.001	24	0.216 ± 0.028

[a]The total number of stems in the experiment was 24

[b]Ethylene production: Log_{10} (1 + μL kg^{-1} fresh $weight^{-1}$)

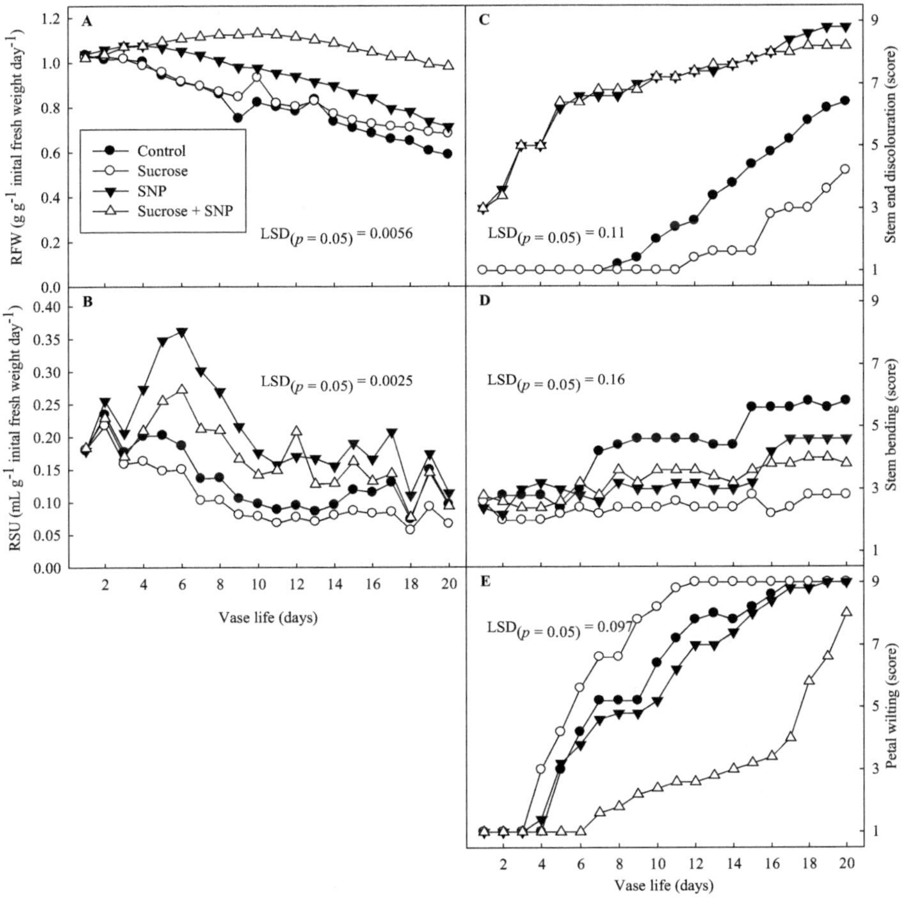

Fig. 12.1 Effects of deionized water (●), 6% sucrose (○) and 1 mgL^{-1} SNP with (Δ) or without (▼) 6% sucrose on gerbera flower relative fresh weight (RFW; **a**), relative solution uptake (RSU; **b**), stem end discolouration (**c**), stem bending (**d**) and petal wilting (**e**) during gerbera vase period. LSD ($P = 0.05$) values are present for comparison of means ($n = 10$) (Solgi et al. 2011)

The imbalanced vase solution uptake and water loss in cut flowers can result in petal wilting and stem breaking as well. Microbial population in vase solutions caused by xylem blockage has been shown as the major reason for negative water balance and terminated vase life of cut flowers (van Meeteren 1978; van Doorn and De Witte 1994; Liu et al. 2009).

It is essential to develop postharvest treatments for reducing vase life difficulties for ornamental plants. Cut flowers like roses, carnations and gerbera usually needed vase solutions consisting of carbohydrate supply plus an antimicrobial agent for maximizing vase life. Sucrose is commonly applied for maintaining metabolic activity and antimicrobial agents, including silver nitrate and 8-HQC employed for

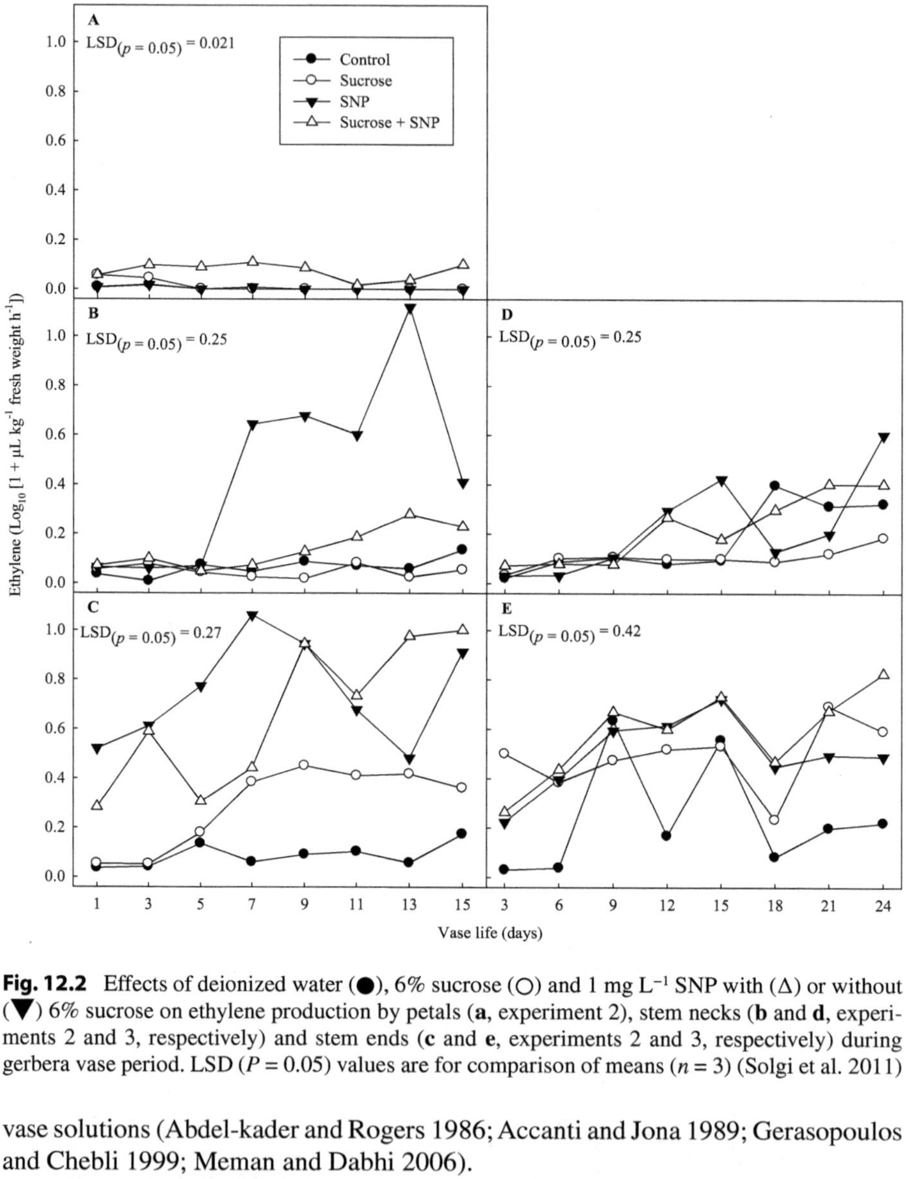

Fig. 12.2 Effects of deionized water (●), 6% sucrose (○) and 1 mg L^{-1} SNP with (Δ) or without (▼) 6% sucrose on ethylene production by petals (**a**, experiment 2), stem necks (**b** and **d**, experiments 2 and 3, respectively) and stem ends (**c** and **e**, experiments 2 and 3, respectively) during gerbera vase period. LSD ($P = 0.05$) values are for comparison of means ($n = 3$) (Solgi et al. 2011)

vase solutions (Abdel-kader and Rogers 1986; Accanti and Jona 1989; Gerasopoulos and Chebli 1999; Meman and Dabhi 2006).

In cut flowers, stem bending generally happened about 10 cm under their head (capitulum) (Ferrante et al. 2007). The synthesis of ethylene can be seen in bent or breaking stems than the straight parts (Mencarelli et al. 1995; Balestra et al. 2005). According to Ferrante et al. (2007), the prevalence of stem bending is lower after treatment with ethylene, which is linked to the increased phenylalanine ammonia-lyase (PAL) activity in several cut flowers, such as gerbera.

The silver ion is the main cause of increasing vase life of cut flower due to the efficacy of silver nitrate and silver thiosulphate (STS). Antimicrobial efficacy and antiviral and algicidal effects of Ag^+ have been indicated (Sondi and Salopek-Sondi 2004; Blaser et al. 2008; Navarro et al. 2008). Ag^+ is able to interact with thiol groups (–SH) of enzymes and other proteins essential for metabolic and maintenance processes, such as respiration and substance transportation through cell-bounding membranes and also through the cells (Maneerung et al. 2008).

Using silver nanoparticles has been recently increased for several purposes, including textile industries, electronics, medicine, cosmetics and environmental measures (Navarro et al. 2008). Their size is lower than 100 nm in at least one dimension. SNP has shown great antimicrobial properties than bulk silver metal, due to the large specific surface area leading to elevated fraction of surface atoms. The interaction of SNP with bacterial membranes leads to structural modifications, proton motive force dissipation and free radical manufacturing resulting in loss of DNA replication ability and consequently cell death (Feng et al. 2000; Sondi and Salopek-Sondi 2004; Danilczuk et al. 2006).

In contrast, many investigations have indicated the association between vascular blockage and further senescence of cut flowers and bacterial plugging of stem factors (Zagory and Reid 1986). Microbes in vase solution of cut flowers are the most usual factor for stem blockage and reduced longevity, which can lead to physical plugging of the cut stem, releasing toxic agents and enzymes or ethylene synthesis (Rodney and Hill 1993). Accordingly, several antimicrobial materials, including 8-hydroxyquinoline sulphate, silver nitrate and aluminium sulphate, have been applied in vase solution for cut flowers for increasing vase life (longevity) through increasing water uptake. The vital role of silver nanoparticles as antibacterial materials has been widely shown (Lu et al. 2010).

It has widely shown that the genera, including *Bacillus*, *Pseudomonas* and *Acinetobacter*, are the most type of bacteria found in vase solution of some cut flowers (Balestra et al. 2005; Van Doorn and De Witte 1994; Jowkar et al. 2012; Zagory and Reid 1986; Solgi 2014).

The human food chain is affected by several microbial, physical and chemical health problems due to this main safety issue. Vegetables and fruits are usually connected with soil, insects, humans and animals not only through the growth and harvesting periods but also in processing. Therefore, natural contamination can influence vegetable and fruit surface. Several fresh products have shown to have 104–106 microorganism cells per 1 g (Ukuku 2004), of which *Salmonella, E. coli, Listeria monocytogenes* and *Staphylococcus aureus* have been reported as the certain pathogenic microorganisms (Trias et al. 2008).

In contrast, grey mould disease induced by *Botrytis cinerea* has been shown as an important postharvest pathogen worldwide, which causes decay on numerous economically important fruits and vegetables through the growing season and also postharvest storage. It has been also found as a main challenge for long-distance transportation and storage. The leaves, stems, flowers and fruits are infected by *Botrytis cinerea* via direct penetration or wounds through cultivation practices. Grey mould control is essentially crucial in storage period, since it grows in low-temperature environments and spreads promptly between fruits and vegetables. General knowledge about the adverse effect of synthetic fungicide residues on human health and environment has prompted deregulation of main chemical fungicides (Yadollahi et al. 2010).

Nanotechnology has been used for developing antifungal agents in several types of fruits and vegetables. A technology has been recently introduced for producing zinc oxide nanoparticles by microbial strategy. Some nanoparticles have been applied as antifungal in vitro and also for postharvest storage of banana, carrot, tomato, onion, etc. (Yadollahi et al. 2010).

Storing fruits and vegetables is linked to their microbiological deterioration usually due to adverse effects of bacteria, moulds and yeasts. It can be inhibited using nano-silver solutions. Silver nanoparticles exert antibacterial activity, which is able to break down the bacteria, like *Escherichia coli* and *Staphylococcus aureus*. Ag^+ is interacted electro statistically with bacterial cells that are negatively loaded. Silver is mainly active in cellular structural proteins and enzymatic proteins, which are essential for the accurate function of microorganisms (Atiyeh et al. 2007; Solgi et al. 2009). The impact of metal nanoparticles on longevity of agricultural crops, such as fruits, vegetables, cut flowers and ornamental plants, has been indicated (Table 12.2).

12.2 Application of Biological (Biogenic/Green) Synthesized Nanoparticles to Improve Longevity of Agricultural Crops During Postharvest

However, using physicochemical production of metal nanoparticles on the postharvest quality of several agricultural crops has been extensively reported. Limited information is available on using biological or green or biogenic synthesized metal nanoparticles on postharvest quality of crops.

Saffron is largely produced in Iran with 230 tons per year, which accounts for 93.7% of the world saffron production (Ghorbani 2008). Saffron petals are by-products of saffron processing and are throwing out following harvesting. Petals are produced more than 10,000 tons annually. Currently, dye extraction is the main and only application of saffron petals, which has not increased yet (Abbasi-Alikamar et al. 2007). Solgi (2014) aimed at investigating the saffron (*Crocus sativus*) petal extract-mediated biosynthesis of silver nanoparticles and their influence on some bacteria genera involved in reducing flowers' longevity for finding the possible use of huge amounts of petal residues as green biosynthesis. Green synthesis of silver nanoparticles via petal extract of saffron petals as a reducing agent from 5 mM $AgNO_3$ has been evaluated in this research. Several ratios of petal extracts (1:20, 1:10, 1:5, 1:1 and 2:1 saffron petal extract/silver nitrate V/V) as well as reaction times (zero time, 30 min and 2 h following reaction) were employed to produce SNPs. The obtained SNPs were characterized using UV-Vis spectra at the wavelength of 200–700 nm, Philips X'Pert 1 X-ray diffractometer and FTIR methods (spectral range of 500–4000 cm^{-1} and resolution of 4 cm^{-1}). According to the findings, silver nanoparticles were produced quickly in 30 min of incubation, and the resulted SNPs indicated an absorption peak at 380–400 nm in the UV-Vis spectrum.

Table 12.2 The application of different types of metal nanoparticles on postharvest of different agricultural crops

Agricultural crop name	Metal nanoparticle type	Concentration of metal nanoparticle	Year	References
Rose 'Avalanche' and 'Fiesta' cut flowers	Silver	0, 50, 100 and 200 mgL^{-1}	2013	Rafi and Ramezanian
Rose 'Cherry Brandy' cut flower	Silver	1, 2.5 or 5%	2013	Jowkar
Strawberry fruits	Nano-silver based on polyethylene, nano-silver based on polypropylene, nano-silicate based on polyethylene, nano-silicate based on polypropylene	–	2013	Zandi et al.
Tuberose cut flower (*Polianthes tuberosa*)	Silver	0, 15, 30 and 45 mgL^{-1}	2014	Bahrehmand et al.
Gerbera cut flower	Silver	0, 5, 10 mgL^{-1}	2015	Jafarpour et al.
Fresh-cut melon fruit	Silver	150, 250 and 500 ppm (w/v)	2015	Danza et al.
Tomato and apple	Silver	0, 2.5 and 50 ppm	2015	
Gladiolus (*Gladiolus hybridus*) cut flower	Silver	10, 25 and 50 mgL^{-1}	2017	Li et al.
Eustoma grandiflorum 'Echo' cut flower	Silver and silicon	0, 10, 20 and 40 mgL^{-1}	2017	Kamiab et al.
Anthurium cut flower	Silver	5, 10 and 15 mgL^{-1}	2017	Amin
Tomato and cabbage	Silver (AgNPs impregnated cellulosic packets)	10% AgNPs colloidal suspension	2017	Singh and Sahareen
Tomato vegetable	Silver	0%, 1%, 5% and 10%)	2018	Zhang et al.
Fig (*Ficus carica*) fruit	Zinc oxide	25, 50, 75, 100, 125, 150, 175 and 200 ppm	2018	Lakshmi et al.
Tomato vegetable	Copper	50, 125, 250 and 500 mg L^{-1}	2018	López-Vargas et al.

Metallic silver synthesis was confirmed by XRD spectrum. Green-synthesized silver nanoparticles were applied as antimicrobial material for *Bacillus*, *Pseudomonas* and *Acinetobacter* contaminated thc preservative solution of cut flowers as well. The biosynthesized SNPs by saffron petals successfully controlled these bacteria leading to consider them as promising options for production of new generation of antimicrobial agents. This method is quick, cost-effective and simple, and no hazardous agents are produced for the synthesized SNPs. The antibacterial activities of various amounts of synthesized SNPs were significant than the pure saffron petal extracts (Fig. 12.3). Synthesized SNPs, especially with the ratios of 1:20 and 1:5, indicated antibacterial activity for both tested gram-negative and gram-positive bacteria. However, pure petal extract was not found to exert antibacterial activity (Figs. 12.4 and 12.5). Saffron petal extract includes phenolic compounds, including flavonoids (kaempferol) and anthocyanins (anthocyanidin, delphinidin and pelargonidin) (Isao and Ikuyo 1999; Hadizadeh et al. 2003). The hydroxyl groups of such materials can bind to silver ions and play a role in the biosynthesis of SNPs and also act as reducing agent for the reduction of silver ions (Ag^{+}) to silver nanoparticles (Ag^{0}) (Dubey et al. 2010; Bankar et al. 2010; Jain et al. 2009; Solgi and Taghizadeh 2012).

The plant-mediated SNPs have shown to have antibacterial activities. Bankar et al. (2010) have indicated the antibacterial effect of synthesized SNPs by banana peel extracts against *E. coli*, *E. aerogenes*, *Klebsiella* sp. and *Shigella* sp. Kaviya et al. (2011) reported the higher antibacterial effect of synthesized SNPs with *Citrus sinensis* peel extracts for *E. coli* and *P. aeruginosa* (gram negative) compared to *S. aureus* (gram positive) as well.

Carnation has been regarded as one of the most crucial and used cut flowers throughout the world. Solgi (2018) has recently examined the impacts of eco-friendly agents, such as green silver nanoparticles (25 and 50 mgL^{-1}), thymol (25

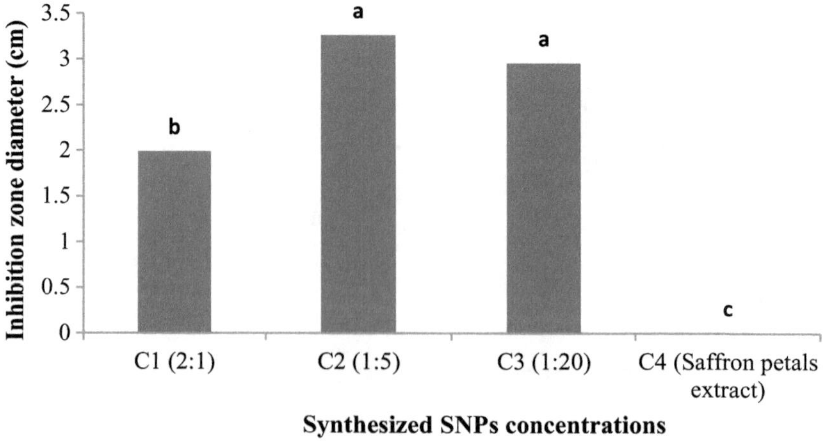

Fig. 12.3 Antibacterial activity of different concentrations of synthesized SNPs by saffron petal extract in comparison to pure saffron petals (Solgi 2014)

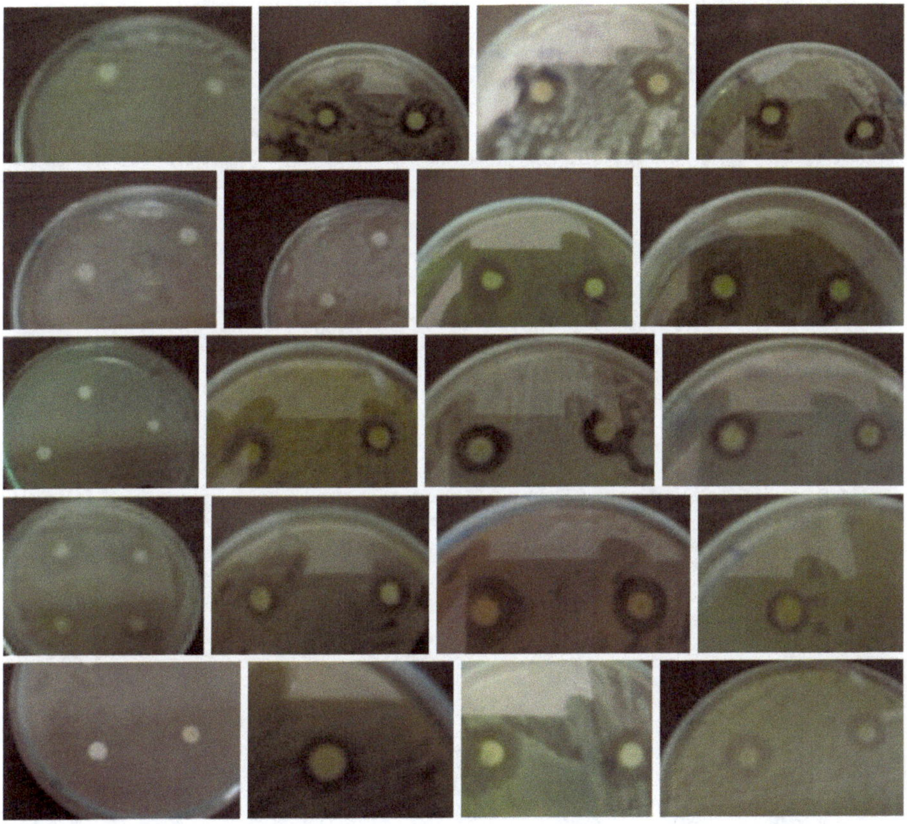

Fig. 12.4 Antibacterial activity of synthesized SNPs by saffron petal extract (from left to right; saffron extracts, 2:1, 1:5 and 1:20) on different gram-negative and gram-positive bacteria (from top row to bottom; *Bacillus cereus*, *Pseudomonas aeruginosa*, *Pseudomonas fluorescens*, *Acinetobacter* and *Bacillus subtilis*) (Solgi 2014)

and 50 mgL^{-1}) and chitosan (25 and 50 mgL^{-1}), than the control group (distilled water) and 2% sucrose on postharvest quality of cut carnation 'Tabor'. Studies with completely randomized design (CRD) in three replications have been performed, in which vase life, RFW, relative solution uptake, chlorophyll amount and ion leakage of leaves' and petals' tissues were the evaluated characteristics.

Biosynthesis of green silver nanoparticles was performed by saffron petals (*Crocus sativus*) (Solgi 2014). In brief, saffron petal powder (2 g) was placed in a 250 mL beaker including 100 mL of distilled water and located on a boiling steam bath for within 15 min at 65–70 °C until colour changes. They were cooled at the room temperature, gradually pressed and filtered via Whatman No. 40 filter paper. Silver nitrate (purity 99.9%; Sigma) solution (5 mM) was prepared and reduced by saffron petal extract at room temperature. Green silver nanoparticles were produced quickly in 30 min of incubation and UV-Vis spectrum or XRD spectrum approved their production.

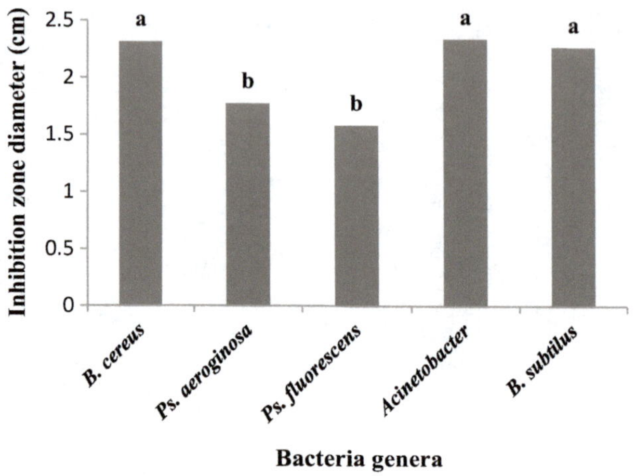

Fig. 12.5 Antibacterial activity of synthesized SNPs by saffron petal extract on different gram-negative and gram-positive bacteria (Solgi 2014)

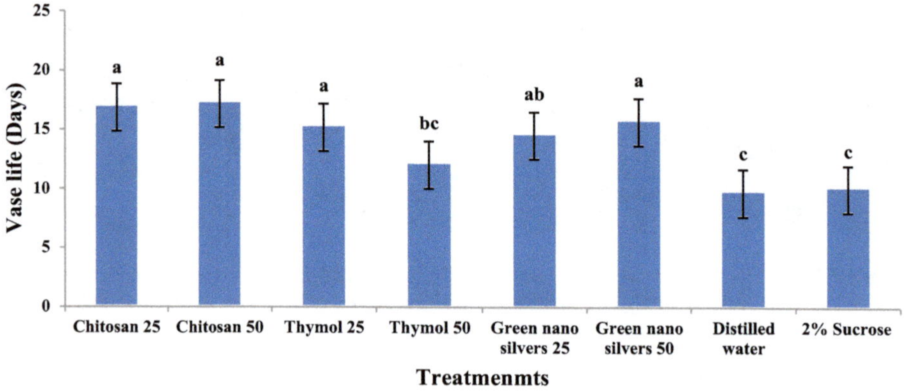

Fig. 12.6 Effects of different compounds on vase life of carnation cut flowers 'Tabor'

Based on the findings, the vase life of cut carnations was increased by adding 25 or 50 mgL^{-1} chitosan for 17 days or 50 mgL^{-1} green silver nanoparticles for 15.7 days than the control group (10 days) (Fig. 12.6). The influences of various treatments on RFW (except for the days 3 and 4) and relative solution uptake were significant in the studied days (Figs. 12.7 and 12.8). Although chlorophyll amount did not change significantly, all agents significantly ($P < 0.05$) sustained the ion leakage of petals (Fig. 12.9).

In addition, Solgi and Taghizadeh (2017) examined the impacts of green silver nanoparticles (25, 50 and 100 mg L^{-1}), thymol (25, 50 and 100 mg L^{-1}) and chitosan (25, 50 and 100 mg L^{-1}) for silver nitrate (25, 50 and 100 mg L^{-1}) and control (2%

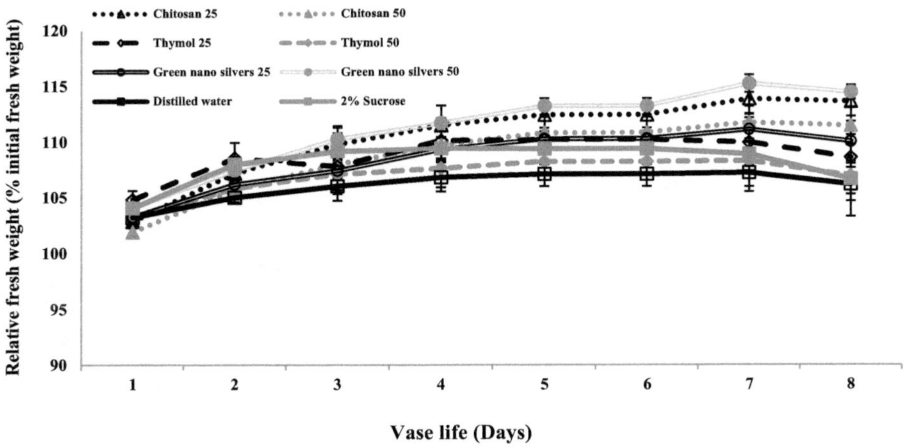

Fig. 12.7 Effect of different compounds on relative fresh weight during 8 days of carnation cut flowers' vase life. Vertical bars show standard errors of means (n = 6)

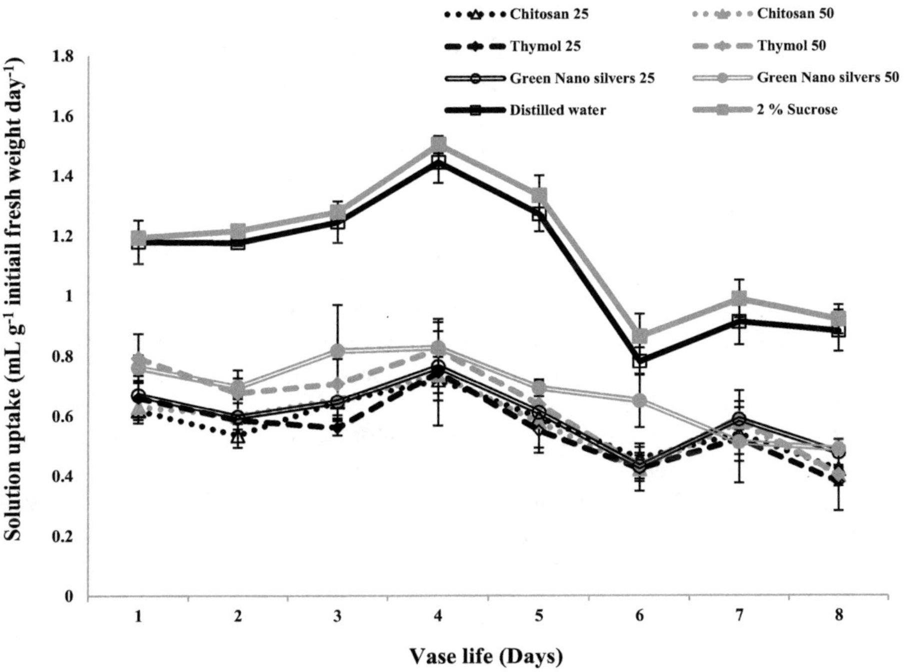

Fig. 12.8 Effect of different compounds on relative solution uptake during 8 days of carnation cut flowers' vase life. Vertical bars show standard errors of means (n = 6)

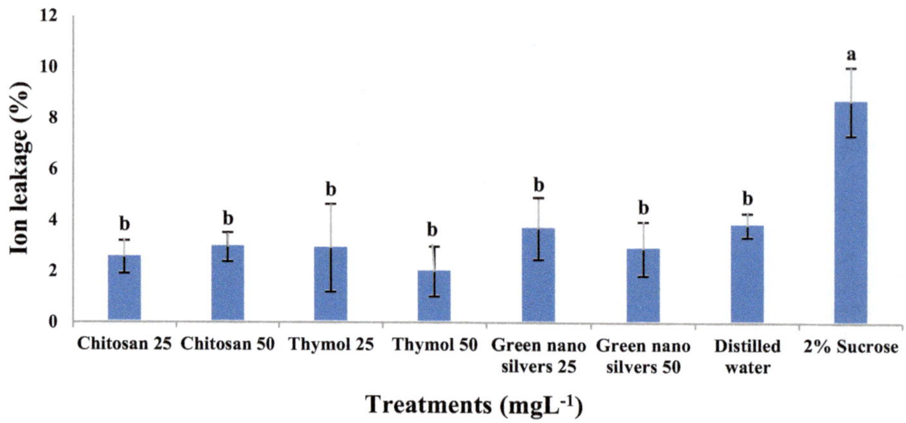

Fig. 12.9 Effects of different compounds on petals' ion leakage of carnation cut flowers 'Tabor' in day 9

sucrose) on vase life of carnation cut flowers 'White Liberty'. Two per cent sucrose was existed in all treatments. Cut flowers were treated for 24 h, followed by replacing with distilled water. This CRD research was done in three replications. Vase life, RFW (on days 2, 3, 4, 7, 9 and 10), solution uptake (on days 1, 2, 3, 4, 7, 8, 9 and 10), ion leakage of petals (on day 11) and chlorophyll contents (on days 2, 4 and 8) were assessed. We found that different treatments were significantly effective on vase life, and thymol 25 (13 days), green silver nanoparticles and chitosan 25 (12.5 days) showed the highest vase life than the control group (Fig. 12.10). The treatments were significantly effective on RFW and solution uptake in the studied days. However, the chlorophyll content of leaves and ion leakage of petals were not significant.

We indicated that pulse treatment by chitosan, thymol and green silver nanoparticles (25 mg L^{-1}) significantly enhanced the vase life of cut carnations cv. 'White Liberty'. Two sections of sucrose as nutrition source and antimicrobial factors are possibly essential for carnation cut flowers. Several studies have displayed their antimicrobial effects.

We also observed that the minimum longevity by control treatments approved this idea. Different bacteria in holding solution can result in vascular occlusion and interrupt the water balance leading to flower wilting (Van Meeteren 1978; Van Doorn and De Witte 1994; Balestra et al. 2005; Solgi et al. 2009, 2011; Hassan et al. 2014). Accordingly, two sections of sucrose as nutrition source in respiration and antimicrobial agents are probably effective for many cut flowers (Halevy and Mayak 1979; Solgi et al. 2009; Jowkar et al. 2012). Silver nanoparticles have been widely shown as effective antimicrobials (Sondi and Salopek-Sondi 2004; Navarro et al. 2008; Bankar et al. 2010; Ahmad et al. 2010; Solgi 2014).

The chemically produced silver nanoparticle in vase solutions of numerous cut flowers, such as gerbera (Solgi et al. 2009; Liu et al. 2009; Solgi et al. 2011) and rose (Lu et al. 2010; Rafi and Ramezanian 2013), has been applied. However, using green silver nanoparticles as antimicrobial agents in preservation solution is still a

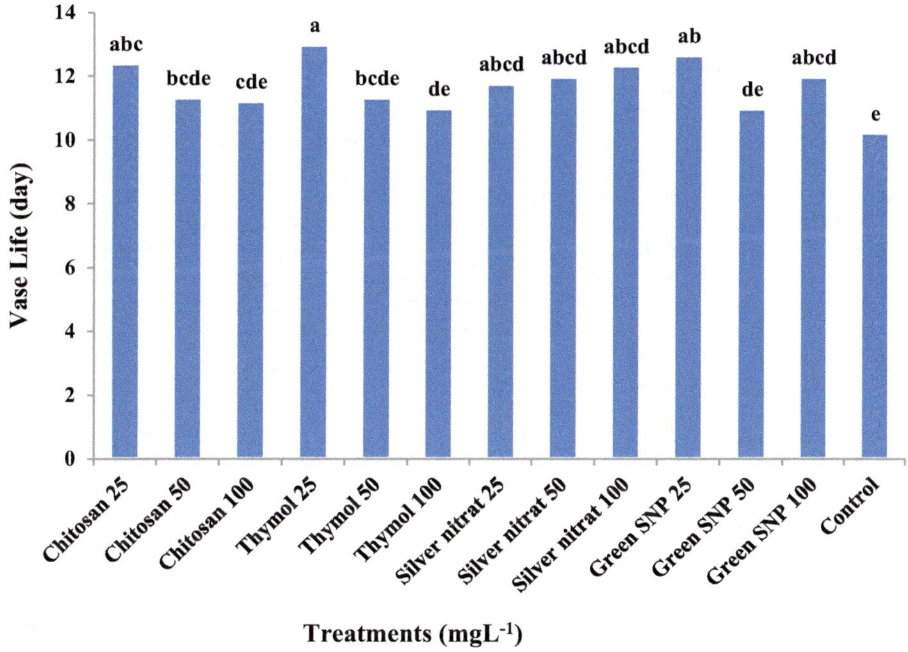

Fig. 12.10 Effects of different treatments on vase life of cut carnation 'White Liberty' (Solgi and Taghizadeh 2017)

developing method, which was first reported by Hassan et al. (2014). They evaluated the impacts of 25, 50 and 100 mg L^{-1} biologically synthesized silver nanoparticles on postharvest quality of rose cut flowers 'First Red' and indicated that all levels of biologically synthesized silver nanoparticles significantly increased the vase life than the control group. The microbial growth was prevented in vase solution, whereas RFW, relative water content, chlorophyll amount and membrane stability index were constant. Moreover, stomatal conductance, ethylene, malondialdehyde and hydrogen peroxide (H_2O_2) synthesis were reduced, while antioxidant enzyme activities were enhanced.

References

Abbasi-Alikamar R, Eskandari M, Tatari M (2007) The effect of water extract of Saffron's petals on germination and seedling growth of wheat (cultivar: Azar2). Second international symposium on saffron biology and technology. Acta Hortic 739

Abdel-kader H, Roger MN (1986) Postharvest treatment of *Gerbera jamesonii*. Acta Horic. Abs. 181

Accanti EG, Jona R (1989) Parameters influencing gerbera cut flower longevity. Acta Hortic 261:63–68

Ahmad N, Sharma S, Alama MK, Singh VN, Shamsi SF, Mehta BR, Fatmae A (2010) Rapid synthesis of silver nanoparticles using dried medicinal plant of basil. Colloids Surf B: Biointerfaces 81:81–86

Amin OA (2017) Influence of Nanosilver and Stevia extract on cut Anthurium inflorescences. Middle East Journal of Applied Sciences 7(2):299–313

Atiyeh BS, Costagliola M, Hayek SN, Dibo SA (2007) Effect of silver on burn wound infection control and healing: review of the literature. Burns 33:139–148

Bahrehmand S, Razmjoo J, Farahmand H (2014) Effects of nano-silver and sucrose applications on cut flower longevity and quality of tuberose (Polianthus tuberosa). Int J Hortic Sci Technol 1:67–77

Balestra GM, Agostini R, Bellincontro A, Mencarelli F, Varvaro L (2005) Bacterial populations related to gerbera (*Gerbera jamesonii* L.) stem break. Phytopathol Mediterr 44:291–299

Bankar A, Joshi B, Kumar AR, Zinjarde S (2010) Banana peel extract mediated novel rout for the synthesis of silver nanoparticles. Colloid Surf A Physicochem Eng Asp 368:58–63

Blaser SA, Scheringer M, Mac Leod M, Hungerbuhler K (2008) Estimation of cumulative aquatic exposure and risk due to silver: contribution of nano-functionalized plastics and textiles. Sci Total Environ 390:396–409

Danilczuk M, Lund A, Saldo J, Yamada H, Michali KJ (2006) Conduction electron spin resonance of small silver particles. Spectrochimaca Acta A 63:189–191

Danza A, Conte A, Mastromatteo M, Del Nobile MA (2015) A new example of nanotechnology applied to minimally processed fruit: the case of fresh-cut melon. J Food Process Technol 6(4):1–4

Dubey SP, Lahtineb M, Sillanpaa M (2010) S green synthesis and characterization of silver and gold nanoparticles using leaf extract of *Rosa rugosa*. Colloids Surf A Physiochem Eng Asp 364:34–41

Feng QL, Wu J, Chen GQ, Cui FZ, Kim JO (2000) A mechanistic study of the antibacterial effect of silver ions on *Escherichia coli* and *Staphylococcus aureus*. J Biomed Mater Res 52:662–668

Ferrante A, Alberici A, Antonacci S (2007) Effects of promoter and inhibitors of phenylalanine ammonia-lyase enzymes on stem bending of cut gerbera flowers. Acta Horic 755:471–476

Gerasopoulos D, Chebli B (1999) Effects of pre- and postharvest calcium applications on the vase-life of cut gerberas. J Hort Sci & Biotech 74:78–81

Ghorbani M (2008) The efficiency of Saffron's marketing channel in Iran. World Appl Sci J 4(4):523–527

Ghorbani HR (2013) Biosynthesis of silver nanoparticles using Salmonella typhimurium. J Nanostruct Chem 3:29

Hadizadeh F, Khalili N, Hosseinzadeh H, Khair-Aldine R (2003) Kaempferol from saffron petals. Iran J Pharm Res 2:251–252

Hassan FAS, Ali EF, El-Deebc B (2014) Improvement of postharvest quality of cut rose cv. 'First red' by biologically synthesized silver nanoparticles. Sci Hortic 179:340–348

Halevy AH, Mayak S (1979) Senescence and post-harvest physiology of cut flowers: part I. Hortic Rev 1:204–236

Isao K, Ikuyo KH (1999) Flavonols from saffron flowers, tyrosinase activity and inhibition mechanism. J Agric Food Chem 47:4121–4125

Jafarpour M, Golparvar AR, Askarikhorasgani O, Amini S (2015) Improving postharvest vase-life and quality of cut gerbera flowers using natural and chemical preservatives. J Cent Eur Agric 16(2):199–211

Jain D, Kumar Daima H, Kachhwaha S, Kothari SL (2009) Synthesis of plant-mediated silver nanoparticles using papaya fruit extract and evaluation of their antimicrobial activities. Dig Nanomater Biostruct 4(3):557–563

Jowkar MM, Kafi M, Khalighi A, Hasanzadeh N (2012) Postharvest physiological and microbial impact of hydroxy quinoline citrate as 'cherry brandy' rose vase solution biocide. Ann Biol Res 3(5):2238–2247

Jowkar MM, Khalighi A, Kafi M, Hassanzadeh N (2013) Nano silver application impact as vase solution biocide on postharvest microbial and physiological properties of 'cherry brandy' rose. Journal of Food, Agriculture & Environment 11(1):1045–1050

Kamiab F, Shahmoradzadeh Fahreji S, Zamani Bahramabadi E (2017) Antimicrobial and physiological effects of silver and silicon nanoparticles on vase life of Lisianthus (Eustoma grandiflora cv. Echo) flowers. International journal of horticultural. Sci Technol 4(1):135–144

Kaviya S, Santhanalakshmi J, Viswanathan B, Muthumary J, Srinivasan K (2011) Biosynthesis of silver nanoparticles using citrus sinensis peel extract and its antibacterial activity. Spectrochim Acta A 79:594–598

Lakshmi SJ, Bai RSR, Sharanagouda H, Ramachandra CT, Nadagouda S, Nidoni U (2018) Effect of biosynthesized zinc oxide nanoparticles coating on quality parameters of fig (Ficus carica L.) fruit. J Pharmacogn Phytochem 7(3):10–14

Li H, Lia H, Liua J, Luoa Z, Joyce D, He S (2017) Nano-silver treatments reduced bacterial colonization and bio film formation at the stem-ends of cut gladiolus 'Eerde' spikes. Postharvest Biol Technol 123:102–111

Liu J, He S, Zhang Z, Cao J, LV P, Joyce DC (2009) Nano-silver pulse treatments inhibit stem-end bacteria on cut gerbera cv. Ruikou flowers. Postharvest Biol Technol 54:59–62

López-Vargas ER, Ortega-Ortíz H, Cadenas-Pliego G, Romenus KA, Fuente MC, Mendoza AB, Juárez-Maldonado A (2018) Foliar application of copper nanoparticles increases the fruit quality and the content of bioactive compounds in tomatoes. Appl Sci 8:1020. https://doi.org/10.3390/app8071020

Lu P, He S, Li H, Cao J, Xu H (2010) Effects of nano-silver treatment on vase life of cut rose cv. Movie star flowers. J Food Agric Environ 8(2):1118–1122

Maneerung T, Tokura S, Rujiravant R (2008) Impregnation of silver nanoparticles into bacterial cellulose for antimicrobial wound dressing. Carbohydr Polym 72:43–51

Meman MA, Dabhi KM (2006) Effects of different stalk lengths and certain chemical substances on vase life of gerbera (*Gerbera jamesonii* hook.) cv. 'Savana red'. J Appl Hortic 8:147–150

Mencarelli F, Agostini R, Botondi R, Massantini R (1995) Ethylene production, ACC content, PAL and POD activity in excised sections of straight and bent gerbera scapes. J Hortic Sci 70:409–416

Navarro E, Baun A, Behra R, Hartman NB, Filser J, Miao AJ, Quiagg A, Santachi PH, Sigg L (2008) Environmental behavior and ecotoxicity of engineered nanoparticles to algae, plants, and fungi. Ecotoxicology 17:372–386

Rafi ZN, Ramezanian A (2013) Vase life of cut rose cultivars 'avalanche' and 'fiesta' as affected by Nano-silver and S-carvone treatments. S Afr J Bot 86:68–72

Rodney BJ, Hill M (1993) The effect of germicides on the longevity of cut flowers. J Am Soc Hortic Sci 118(3):350–354

Singh M, Sahareen T (2017) Investigation of cellulosic packets impregnated with silver nanoparticles for enhancing shelf-life of vegetables. LWT Food Sci Technol 86:116–122

Solgi M (2014) Evaluation of plant-mediated silver nanoparticles synthesis and its application in postharvest physiology of cut flowers. Physiol Mol Biol Plants 20(30):279–285

Solgi M (2018) The application of new environmentally friendly compounds on postharvest characteristics of cut carnation (*Dianthus caryophyllus* L.). Rev Bras Bot 41:515–522

Solgi M, Taghizadeh M (2012) Silver nanoparticles ecofriendly synthesis by two medicinal plants. Int J Nanomater Biostruct 2(4):60–64

Solgi M, Taghizadeh M (2017) The effects of silver nitrate, thymol, green silver nanoparticles and chitosan on vase life of carnation cut flowers cv. White liberty. Plant Prod 40(2):1–13. (In Persian)

Solgi M, Kafi M, Taghavi TS, Naderi R (2009) Essential oils and silver nanoparticles (SNP) as novel agents to extend vase life of gerbera (*Gerbera jameso*nii cv. 'Dune'). Postharvest Biol Technol 53:155–158

Solgi M, Kafi M, Taghavi TS, Naderi R, Eyre J, Joyce DC (2011) Effects of silver nanoparticles (SNP) on *Gerbera jamesonii* cut flowers. Int J Postharvest Innov 2(3):274–285

Sondi I, Salopek-sondi B (2004) Silver nanoparticles as an antimicrobial agent: a case study on *E. coli* as a model for gram-negative bacteria. J Colloid Interf Sci 27:177–182

Trias R, Badosa E, Montesinos E, Baneras L (2008) Bioprotective Leuconostoc strains against Listeria monocytogenes in fresh fruits and vegetables. Int J Food Microbiol 127:91–98

Ukuku DO (2004) Effect of hydrogen peroxide treatment on microbial quality and appearance of whole and fresh-cut melons contaminated with Salmonella spp. Int J Food Microbiol 95:137–146

Van Doorn WG, De Witte Y (1994) Effect of bacteria on scape bending in cut *Gerbera jamesonii* flowers. J Am Soc Hortic Sci 119:568–571

Van Meeteren U (1978) Water relations and keeping-quality of cut gerbera flowers. I. The cause of stem break. Sci Hortic 8:65–74

Yadollahi A, Arzani K, Khoshghal H (2010) The role of nanotechnology in horticultural crops postharvest management. Acta Hortic (875):49–56

Zagory D, Reid MS (1986) Role of vase solution microorganisms in the life of cut-flowers. J Am Soc Hortic Sci 111(1):154–158

Zandi K, Weisany W, Ahmadi H, Bazargan I, Naseri L (2013) Effect of nanocomposite-based packaging on postharvest quality of strawberry during storage. Bull Environ Pharmacol Life Sci 2(5):28–36

Zhang C, Li W, Zhu B, Chen H, Chi H, Li L, Qin Y, Xue J (2018) The quality evaluation of postharvest strawberries stored in Nano-Ag packages at refrigeration temperature. Polymers 10:894. https://doi.org/10.3390/polym10080894

Nanoparticles and Their Fate in Soil Ecosystem

13

Mujeebur Rhaman Khan and Mohammad Akram

Abstract

The nanoparticles (NPs) can be synthesized by different methods, importantly the chemical and biological methods. At the present state of technology, chemical methods are widely used to synthesize large quantities of NPs for various commercial applications. However, biological methods which use plants or microorganisms to synthesize nanoparticles, are low cost and eco-friendly. The phytosynthesis method in particular offers a handy and easy operation and has great potential for future use. Soil is considered as an ultimate sink for NPs introduced in our ecosystem. Since NPs are quite resistant to degradation, their accumulation and persistence in soil may profusely influence the soil microbial community and plant roots. The impact of NPs on microorganisms and plant may vary with the microbe/plant species, soil type and the NP species. In the presented chapter, a concise note on various methods of NP synthesis is presented, followed by a detailed description on the fate of NPs in soil with regard to their impact on growth of plants and microbes in different prospective.

Keywords

Nanoparticles · Biosynthesis · Antimicrobial effects · Phytotoxicity

Nanotechnology is one of the recently explored technologies that possess a wide application in different sectors including agriculture. The nanomaterials have potential scope for use in the production and protection of crops and livestock as well as in the natural resource conservation (Khan and Rizvi 2017). Nanotechnology may also provide solutions to the challenges in the areas of environmental contamination and water purification. Nanotechnology has two major aspects; the first aspect is

M. R. Khan (✉) · M. Akram
Department of Plant Protection, Aligarh Muslim University, Aligarh, India

M. Ghorbanpour et al. (eds.), *Biogenic Nano-Particles and their Use in Agro-ecosystems*, https://doi.org/10.1007/978-981-15-2985-6_13

synthesis of nanoparticles in the particle range size of 1–100 nm, and the second is application of the nanomaterials for a desired use. The synthesis of nanoparticles requires specific skill and facilities depending on the method to be used (Khan and Rizvi 2014). Once the synthesis is achieved, characterization of nanoparticles is done to ensure that the required particle size of the material has been achieved. In this chapter we shall present an overlook on the methods used to synthesize the nanoparticles, followed by the fate of nanoparticles in the soil and their interaction with the soil microorganisms and plant roots.

13.1 Synthesis of Nanoparticles

The nanoparticles (NPs) can be synthesized using mainly the chemical and biological methods depending on the need, as efficiency of synthesis varies with the method (Fig. 13.1). At the present state of technology, chemical methods to synthesize nanoparticles are considered to be the most efficient and used to produce large quantities of nanoparticles for various applications.

13.1.1 Chemical Methods of NP Synthesis

The chemical methods are commonly used to produce large quantities of nanoparticles for commercial use. Since there are different chemical methods for NP synthesis, choice of the methods may depend on the type of material and amount of NPs required. Some of the important chemical methods of NP synthesis are reduction method, sonochemical method, solvothermal decomposition method, etc. which are briefly summarized below.

13.1.1.1 Chemical Reduction Method

The chemical reduction of metal ions is simple and is in common use to synthesize metallic nanoparticles. Song et al. (2004) reported that the reduction method offers

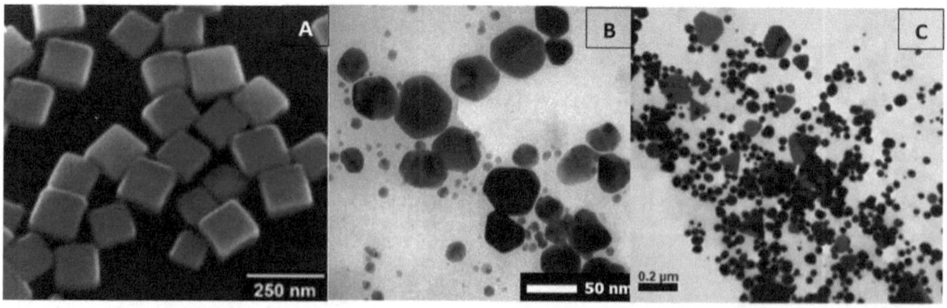

Fig. 13.1 SEM images of silver nanoparticle synthesized by chemical method (A; Im et al. 2005), plant material, *Lantana camara* (B; Fatimah and Indriani 2018) and microbe, *Aspergillus oryzae* var. *viridis* (C; Binupriya et al. 2010)

the production of nanoparticles of uniform shape and size. In the chemical reduction techniques, a metal salt, for example, copper sulphate, is reduced by a reducing agent such as sodium borohydride (Aslam et al. 2002), ascorbate (Wang et al. 2006), polyol (Park et al. 2007), isopropyl alcohol with cetyl trimethyl ammonium bromide (CTAB) (Athawale et al. 2005) or ascorbic acid (Umer et al. 2012).

13.1.1.2 Electrochemical Method

In this method, the NP synthesis is achieved by passing an electric current between two electrodes separated by an electrolyte of the metal salt (Gupta et al. 2013). The NP synthesis takes place at the electrode-electrolyte interface. To synthesize Cu NPs, the electrolytic solution of copper salt and sulphuric acid are usually used. The electrochemical technique is low cost, simple, handy, with high flexibility and eco-friendly (Kajbafvala et al. 2013). The copper nanoparticles of 40–60 nm size were formed in 30 min when 4V/5A current was passed through electrodes in copper sulphate and sulphuric acid as the electrolytic solution (Raja et al. 2008).

13.1.1.3 Solvothermal Decomposition

In this method the chemical reaction takes place in a sealed vessel where solvents are heated at a temperature much higher than the boiling points. Byrappa and Yoshimura (2001) reported that in a hydrothermal process (using water as a solvent), heterogeneous chemical reaction is carried out in a closed system, in the presence of aqueous or non-aqueous solvent above the room temperature and at a pressure > 1 atm at 647.15°K and 221 bar, water is said to be a supercritical fluid, which act both as a liquid and gas. A supercritical fluid decreases the surface tension at the interference of solid and dissolves the chemical compounds which are very difficult to be dissolved at NTP. This technique is used to drastically enhance solubility and reactivity of metal. The use of supercritical condition is becoming very popular for the synthesis of copper, silver, gold, platinum, germanium and silicon nanoparticles because of the rapid reaction rates and particle fast growth (Reverchon and Adami 2006; Chen et al. 2010). Yang et al. (2001) reported that the reaction initiated under ambient conditions before the mixture was transferred to an autoclave and heated to 100 °C. By varying the solvent and reaction temperatures, the NPs of varying sizes and morphologies were produced. Amendola et al. (2006) observed controlled production of Cu NPs with regard to shape and structure by regulating the reaction temperature and quantity of sodium dodecylbenzenesulfonate.

13.1.1.4 Microemulsion/Colloidal Method

In this method, two immiscible fluids such as water in oil (W/O) or oil in water (O/W) or water in supercritical carbon dioxide become a thermodynamically stable dispersion with the aid of a surfactant. Chen et al. (2006) reported that a typical emulsion is a single phase of three components, water, oil and a surfactant. Kitchens and Roberts (2004) stated that oil and water are immiscible, but with the addition of a surfactant, the oil and water become miscible because the surfactant is able to bridge the interfacial tension between the two fluids. The metallic nanoparticles

(Cu, Ag, Co, Al), oxides (TiO_2, SiO_2), metal sulphides (CdS, ZnS) and various other nanomaterials can be prepared using this microemulsion technique (Cason et al. 2001). Pileni and Lisiecki (1993) demonstrated that copper and silver nanoparticles can be synthesized using reverse micelle system.

13.1.1.5 Sonochemical Method

This method uses ultrasound irradiations of 10–20 kHz to enhance the chemical reaction of molecules (Suslick et al. 1991). Suslick et al. (1996) used this method to synthesize iron NPs. The sonochemical technique has also been found quite efficient to synthesize other metals and metal oxides (Pol et al. 2003). The major advantage of this method is its simplicity, operating conditions and good control of NP size (Vijayakumar et al. 2000). The sound waves at a fixed frequency are allowed to pass through a slurry or solution of carefully selected metal complex precursors. In the solution at a vapour pressure of a certain threshold, the alternating waves of expansion and compression induce the formation of cavities, which grow and implode. Sonochemical reactions of volatile organometallics have been exploited as a general approach to synthesize various nanophase materials by changing the reaction medium (Pol et al. 2003).

13.1.1.6 Microwave Method

The microwave irradiation is another simple, easy and efficient method of the NP synthesis (Komarneni 2003). Microwaves, a form of electromagnetic energy at a frequency of 300 MHz to 300 GHz, commonly 2.456 GHz, can induce synthesis of crystalline particles with radius ranging from 90 to 260 nm (Blosi et al. 2011). The microwaves induce fast and homogeneous reaction conditions during the synthesis of Cu NPs (Blosi et al. 2011). The efficiency of Cu NP synthesis may be increased further by using copper sulphate as a precursor and sodium hypophosphite as the reducing agent in ethyl glycol under microwave irradiation (Zhu et al. 2004).

13.1.2 Biological Methods of NP Synthesis

Biological methods or green synthesis methods involve use of plants or microorganisms for the synthesis of nanoparticles (Fig. 13.1; Khan and Rizvi 2017). Plants (Khan et al. 2019b) and soil microorganisms are important biological agents and have immense potential to mediate the synthesis of nanoparticles (Khan et al. 2019a). In view of the risk of soil and water contamination due to chemical processes, exploitation and development of efficient and eco-friendly processes for synthesis of nanoparticles shall be crucial in popularizing the use of nanotechnology in agriculture.

13.1.2.1 Phytosynthesis of Nanoparticles

Using plants or their extracts to bring about synthesis of nanoparticles is a low-cost, handy and environment-friendly process. Plant extracts contain a range of biochemical molecules such as proteins, sugars, phenol, alkaloids, flavonoids, hormones, terpenoids, anthraquinones, etc. which bring about the synthesis of nanoparticles through reduction of ions in an aqueous solution of a metal salt (Khan et al. 2019b). The plant extracts from *Centella asiatica*, *Citrus sinensis*, *Ocimum tenuiflorum*, *Solanum trilobatum* and *Syzygium cumini* are well documented to induce reduction of Ag ions to Ag NPs in the silver nitrate solution. Patil et al. (2012) reported that *O. tenuiflorum* leaf extract brought about the reduction and stabilization of Ag ions to Ag NPs of 25–40 nm size. The leaf extract of *Piper longum* also acted as reducing and capping agent and yielded Ag NPs in 1 mM $AgNO_3$ solution. The leaf extracts of *Artemisia nilagirica* (Asteraceae) produced 70–90 nm Ag NPs (Vijayakumar et al. 2013).

In addition to above, leaf extracts of several other plants such as *Avena sativa* (Armendariz et al. 2004), *Azadirachta indica* (Shankar et al. 2004a, b), *Aloe vera* (Chandran et al. 2006), *Acanthella elongate* (Inbakandan et al. 2010), *Cinnamomum camphora* (Huh and Kwon 2011), *Coriandrum sativum* (Badrinarayanan and Sakthivel 2008), *Carica papaya* (Mude et al. 2009), *Emblica officinalis* (Ankamwar et al. 2005), *Medicago sativa* (Gardea-Torresdey et al. 2003; Schabes-Retchkiman et al. 2006; Lukman et al. 2011), *Pelargonium graveolens* (Shankar et al. 2003a, b), *Parthenium hysterophorus* (Parashar et al. 2009), *Sesuvium portulacastrum* (Nabikhan et al. 2010), *Tamarindus indica* (Shanker et al. 2004a, b) and *Tritium vulgare* (Armendariz et al. 2009) have also been reported to induce reduction of metal ions to metal NPs. As the above citations reveal, generally the leaf extracts are used to synthesize the NPs. This is probably for the soft and succulent tissue with easy and bulk availability of leaves. Further, higher contents of reducing and stabilizing biochemicals such as isoflavonoids, flavonoids, proteins, alkaloids, terpenoids, carbohydrates and anthraquinones occur in the leaf tissue (Suman et al. 2013). Huang et al. (2007) reported that dry leaf power of plants, for example, *Cinnamon camphora*, may also induce synthesis of nanoparticles such as silver and gold.

13.1.3 Microbial Synthesis of Nanoparticles

Soil microorganisms constitute one of the vast and diversified natural biotic resources that can be exploited in NP synthesis (Fariq et al. 2017; González-Garcinuño et al. 2019). A wide range of microorganisms have been found to catalyse the formation of nanoparticles (Fig. 13.1; El-Rafie et al. 2012, Khan and Rizvi 2014). The microbial metabolites generally act as reducing and capping agents for conversion of ions into nanoparticles. Gurunathan et al. (2009) reported that culture supernatant of *Escherichia coli* induced the reduction of Ag + ions to Ag NPs. The extracellular biosynthesis of Ag NPs using *E. coli* has also been achieved (Manonmani and Juliet 2011). Law et al. (2008) successfully achieved the reduction of Ag (I) as insoluble AgCl or Ag (+) ions by using *Geobacter sulfurreducens*. The Ag NPs were also synthesized through the mediation by a gram-positive bacteria *Morganella morganii* in the culture medium (Abd et al. 2013)

Besides Ag NPs, gold nanoparticles have also been synthesized by using bacteria. Ahmad et al. (2003a) reported that extremophilic bacteria, *Thermomonospora* sp., catalysed the synthesis of Ag NPs in the medium. The Au NPs were synthesized with the help of culture of mesophilic bacterium, *Shewanella* (Konishi et al. 2004). He et al. 2007 noticed that culture of *Rhodopseudomonas capsulata* facilitated the production of 10–20 nm Au NPs. Similarly, Husseiny et al. (2007) used the cell supernatants of *Pseudomonas aeruginosa* to achieve reduction of Au ions to Au NPs. Ahmad et al. (2003b) successfully obtained the extracellular synthesis of spherical 8 nm gold nanoparticles using the actinomycete, *Thermomonospora* sp. The extremophilic yeast has been found effective in mediating the synthesis of Ag and Au NPs (Mourato et al. 2011). Kalimuthu et al. (2008) reported successful synthesis of Ag NPs (20 nm) and Au NPs (30–100 nm) using the yeast biomass. They used *Bacillus licheniformis* to catalyse synthesis of 50 nm Ag/Au NPs.

Woolfolk and Whiteley (1962) reported that cell-free extracts of *Micrococcus lactilyticus* reduced the uranium (vi) to uranium (iv). Similarly, Lovley et al. (1989) reported that *Alteromonas putrefaciens* culture in the presence of hydrogen as electron donor and U (vi) as electron acceptor reduced U (vi) to U (iv). Further, reduction of soluble U (vi) to insoluble U (iv) was achieved when *G. metallireducens* GS-15 was grown anaerobically in the presence of acetate and U (vi) as electron donor and electron acceptor (Lovley et al. 1991). Reduction of selenite to elemental selenium and accumulation of selenium granules in the cytoplasm or extracellular space was achieved by using *Stenotrophomonas maltophilia* SELTE02 (Di Gregorio et al. 2005). Using different groups of bacteria such as facultative anaerobes, *E. cloacae* (Losi and Frankenberger 1997), nonsulphur bacterium *Rhodospirillum rubrum* and anaerobic *Desulfovibrio desulfuricans* (Tomei et al. 1995), reduction of selenite to selenium was achieved successfully. Silverberg et al. (1976) reported that *E. coli* mediated the synthesis of Se NPs. Similarly, aerobic reduction of selenite to elemental selenium was achieved with the help of *P. stutzeri* (Lortie et al. 1992). Another phosphate-solubilizing bacteria, *P. aeruginosa*, reduced Se ions to spherical Se NPs (Yadav et al. 2008).

Microbial synthesis of different nanoparticles has also been reported using a wide range of fungi such as *Aspergillus fumigatus* (Bhainsa and D'Souza 2006), *A. niger* (Gade et al. 2008), *Colletotrichum* sp. (Shankar et al. 2003a, b), *Cladosporium cladosporioides* (Balaji et al. 2009), *Coriolus versicolor* (Sanghi and Verma 2009), *Fusarium oxysporum* (Bansal et al. 2004), *F. solani* (Ingle et al. 2009), *F. semitectum* (Basavaraja et al. 2008), *Phanerochaete chrysosporium* (Vigneshwaran et al. 2006), *Phoma glomerata* (Birla et al. 2009), *Penicillium brevicompactum* (Shaligram et al. 2009), *Penicillium fellutanum* (Kathiresan et al. 2009), *Trichothecium* sp., *Trichoderma asperellum*, *T. viride* (Khandel and Shahi 2018) and *Volvariella volvacea* (Philip 2009) which have also been found quite effective in catalysing the formation of nanoparticles. El-Rafie et al. (2012) examined the effect of some factors such as biomass concentration of *F. udum*, pH of the reaction medium, $AgNO_3$ concentration and the ratio of $AgNO_3$ to biomass concentration to synthesize Ag NPs. Optimum synthesis of Ag NPs was attained 10 g using biomass of *F. solani* /100 ml, 0.078 g $AgNO_3$/100 ml concentration pH 12; 25 °C temperature and 24 h duration. This set of conditions yielded stabilized concentration of 2000 ppm Ag NPs of 8–15 nm size. Gade et al. (2008) reported extracellular synthesis of silver nanoparticles could be achieved by *Aspergillus niger*. The cyanobacteria, *Anabaena*,

Calothrix and *Leptolyngbya*, have been found to produce Au, Ag, Pd and Pt nanoparticles extracellularly. The NPs produced in the culture medium were stabilized by the algal polysaccharides (Brayner et al. 2007). Dameron et al. (1989) demonstrated that yeast *Candida glabrata* can intracellularly produce spherical and CdS quantum dots measuring 2 nm. The yeast *Schizosaccharomyces pombe* also produced wurtzite-typed hexagonal lattice-structured CdS nanoparticles of 1–1.5 nm size (Kowshik et al. 2002). Another yeast, *Pichia jadinii*, produced spherical, triangular and hexagonal gold nanoparticles of 100 nm size (Gericke and Pinches 2006). Gold nanoparticles in the size range of 2–10 nm were produced by *S. cerevisiae* (Jha et al. 2009). Microbial synthesis of other metals has also been well documented. Prasad et al. (2007) reported that the extracellular culture filtrate of *Lactobacillus* sp. synthesized spherical 40–60 nm titanium NPs. Konishi et al. (2007) achieved the microbial synthesis of Pt (ii) organics and metallic platinum nanoparticles of 30–300 nm size extracellularly using gram-negative cyanobacterium *P. boryanum* similarly. Yong et al. (2002) reported that anaerobic sulphate-reducing bacterium *Desulfovibrio desulfuricans* also reduced palladium (2+) ions to palladium NPs extracellularly. An iron-reducing bacterium, *S. oneidensis*, reduced Pd (ii) to Pd (0) nanoparticles in the presence of lactate as electron donor inter- and extracellularly (De Windt et al. 2005). In the same study, the Au NPs of 5–15 nm were formed using alkalotolerant actinomycete, *Rhodococcus* sp.

13.2 Fate of Nanoparticles in Soil

Soil is supposed to be the largest recipient of NPs. When nanoparticles or nanoformulations are applied on the plants, the material finally reaches to the soil. Hence, it becomes very important to understand the fate and behaviour of NPs in the soil, as soil is the ultimate sink for nanoparticles. Upon reaching in soil, the nanomaterial is exposed to a variety of factors or agents especially plant roots and microorganisms and develops different relationship with them, most commonly synergistic or antagonistic relationship. Soil is also considered as an externally important environmental matrix, richest in natural NPs both as primary particles and aggregates. Anthropogenic introduction of NPs into the soil might have significant impact on the soil biota. The protection of soil microbial biomass and diversity as well as root growth are the major issues for sustainable use of NPs in crop production (Torsvik and Øvreås 2002; Khan and Rizvi 2014). Hence, we discuss under the interactions between nanoparticles and plants/microorganisms with regard to NP effects.

13.2.1 Effect of Nanoparticles and Materials on Plants

The nanoparticles or nanomaterials when applied to plants through foliar spray on soil application are ultimately absorbed by the plant foliage and/or roots and cause a complexity of effects. The effects, however, may vary with the material or NP species and the amount. Since nanoparticles are very small in size, they have potential to enter, translocate and penetrate physiological barriers to travel within the plant tissues with fair degree of free movement. However, long-term NP effects in plant systems are not fully understood. Some researchers suggest that the NPs may cause morphological, physiological and genetic modifications leading to quantitative and/or qualitative changes in the plant growth. It is well understood that nanoparticles enter plant tissues either via root tips, rhizodermis and lateral root junctions when present in the soil or via cuticles, trichomes, stomata, etc. when applied on the foliage. The effects of nanoparticles on plants may be growth suppressive or promontory depending on the NP species and its concentration.

13.2.1.1 Suppressive Effects on Plants

The nanoparticles may easily penetrate the root and may cause toxic effects which may limit to a group of cells to expression of toxic symptoms and suppression of plant growth. Toxic effects of nanoparticles on plants may appear in the form of seed germination, root/shoot growth, biomass production, transpiration photosynthesis, biomolecule synthesis, etc. (Ghosh et al. 2016). Nanoparticles may cause decline in the plant growth, degrade its quality and lower the nutritive value of the produce (Le Van et al. 2016a, b; Rajput et al. 2018a, b). From various studies it is now well recognized that NPs adversely affect crop development and yield and accumulate in the plant tissues, fruits, grains, etc. It has been found that NPs may inhibit seed germination, reduce plant growth and decrease in photosynthetic rate and pigments of a number of crops such as cucumber (Moon et al. 2014), coriander (Zuverza-Mena et al. 2015), lettuce (Trujillo-Reyes et al. 2014; Hong et al. 2015), onion (Cvjetko et al. 2017), rice (Shaw and Hossain 2013), mung bean (Nair et al. 2014), radish (Corral-Diaz et al. 2014), spinach (Singh and Kumar 2016), soybean (Nair and Chung 2014, Yang et al. 2015, Priester et al. 2017), tobacco (Frazier et al. 2014), wheat (Vannini et al. 2014) and mustard (Nair and Chung 2015). Toxic effects of NP on plants have been summarized in Table 13.1.

13.2.1.2 Promontory Effects on Plants

Nanoparticles of some chemicals especially those which are useful for plant growth such as nitrogen, phosphorus, zinc, etc. stimulate the plant growth and enhance the biomass production and yield. Nitrogen is a vital nutrient for plant growth and development, and 50–70% of the applied N dose is lost to the soil due to leaching and other factors. It has been found that application of nano-nitrogen shall be far more effective than even polymer-coated conventional slow-release N fertilizers due to the high surface area to volume ratio (Hossain et al. 2008; De Rosa et al. 2010). Similarly, phosphorus fertilizers are also lost due to eutrophication problem in surface waters (Correll 1998; Carpenter 2005; Conley et al. 2009). The commercially available P fertilizers are water-soluble phosphate salts, which easily get dissolved in the soil solution and available for plant uptake (Fageria 2009). However, the phosphates soluble in water are highly mobile in the soil, and large portions are lost due to surface run-off or seepage. On the other hand, solid forms of P such as naturally occurring phosphate rocks and apatites have also been attempted as P fertilizers where the phosphate is locked in a solid form and is less easily available

Table 13.1 Phytotoxic effects of some important nanoparticles

NPs, size and concentration	Plants/species	Effects	References
Ag 1–10 mg/L	*Triticum aestivum*	Decreased seedling growth and modified root tip cells	Vannini et al. (2014)
Ag 1000–3000 mM	*Pisum sativum*	Decreased photosynthetic pigments	Tripathi et al. (2017)
Ag 5–25 nm; 40 mg/L	*Phaseolus radiatus*	Ag accumulation in roots and shoots. Growth inhibited	Lee et al. (2012)
Ag 5–20 μg/L	*Allium cepa*	Chromosomal aberrations in both mitotic and meiotic cells	Saha and Gupta (2017)
Ag 0.6–2 nm; 0–100 mg/L	Flax, ryegrass, barley	Reduced seed germination and shoot length	El-Temsah and Joner (2012)
Ag 1000 mg/L	*Oryza sativa*	Phytotoxic effect on plant	Mazumdar and Ahmed (2011)
Al 2000 mg/L	Corn, lettuce	Decline in root length	Lin and Xing (2007)
Al 20 mg/L	Ryegrass, corn	Reduced germination and root length	Lin and Xing (2007)
Al_2O_3 13 nm	Carrots, cabbage, cucumber, maize	Decline in root growth	Yang and Watts (2005)
Au 25–100 mg L	*Arabidopsis thaliana*, *Medicago sativa*	Up to 75% reduction root length	Taylor et al. (2014)
CeO_2 0–500 mg kg	*R. sativus*	Retarded germination	Corral-Diaz et al. (2014)
CeO_2 7 nm; 0–4000 mg/L	Alfalfa, cucumber, maize, soybean, tomato	Reduced germination, biomass, shoot and root growth	Lopez-Moreno et al. (2010)
Cu 1000 mg/L	Mung bean, wheat	Reduced seedling and shoot growth	Lee et al. (2008)
CuO 0–80 mg/kg	*Coriandrum sativum*	Reduced germination and shoot elongation	Zuverza-Mena et al. (2015)
CuO 10–1000 mg/L	Cotton	Shoot and root length decreased and senescence	Le Van et al. (2016b)
CuO 200–500 mg L	*Vigna radiata* L.	Affected root length, biomass, total chlorophyll content, primary and lateral root growth	Nair et al. (2014)
CuO 0–20 mg/L	*Lactuca sativa*, *Medicago sativa*	Reduced the root length; increased Cu, P and S in alfalfa shoots; and decreased P and Fe in lettuce	Hong et al. (2015)
CuO 50–500 mg/L	*Glycine max*	Significantly reduced the shoot growth, weight, total chlorophyll content and root length	Nair and Chung (2014)
CuO 0–500 mg/L	*Brassica juncea*	Affected shoot and root growth	Nair and Chung (2015)

(continued)

Table 13.1 (continued)

NPs, size and concentration	Plants/species	Effects	References
CuO 0–50 mg/L	Oak	Disturbed shape, plastoglobules and starch contents of leaves' cells	Olchowik et al. (2017)
CuO 1–1000 mg/L	Carrot	Biomass decreased significantly	Ebbs et al. (2016)
CuO 10–1000 mg/L	Cotton	Reduced the uptake of minerals (B, Mo, Mn, Mg, Zn, Fe)	Le Van et al. (2016a, b)
CuO 0–80 mg/L	Onion	Reduced/stopped root growth and caused showed deformation on root cap surface and meristematic zone	Deng et al. (2016)
Fe NPs 50–100 mg/L	Sunflower	Reduced uptake and translocation of nutrients and root hydraulic conductivity	Martinez-Fernandez et al. (2016)
Fe NPs 50–100 mg/L	Tomato	Inhibited root hydraulic conductivity	Martinez-Fernandez and Komarek (2016)
Fe NPs 1–50 mg/L	Lettuce	Decreased root elongation	Liu et al. (2016)
Fe NPs 2–200 mg/L	Rice	Inhibited phytohormones	Gui et al. (2015)
ZnO 500–1500 mg/L	Mustard	Reduced seed germination and seedling growth	Zafar et al. (2017)
ZnO 0.05–0.5 mg/L	Soybean	Leaf chlorosis, necrosis and affected photo system II quantum efficiency	Priester et al. (2017)
ZnO 1000 mg/L	Spinach	Reduced root and shoot length, total weight, chlorophyll and carotenoid content	Singh and Kumar (2016)
ZnO 100–500 mg/L	Bean	Inhibited growth, imbalanced nutrient in shoots, Na increased, Fe, Mn, Zn and Ca decreased	Dimkpa et al. (2015)
ZnO 2000 mg/L	Maize, rice	Inhibited root elongation	Yang et al. (2015)
ZnO	Rice	Induced phytotoxicity, reduced ROS production and lipid peroxidation	Chen et al. (2015)
ZnO 500 mg/L	Soybean	Reduced root and shoot growth, seed formation	Yoon et al. (2014)
ZnO 2000 mg/L	Radish, rapeseed, ryegrass, lettuce, corn, cucumber	Reduced root growth and elongation	Lin and Xing (2007)

to plant roots and also less easily transportable by run-off or soil erosion (Fageria 2009). However, these solids are less effective in providing nutrient P to plants (Fageria 2009). In addition, application of solid phosphates in agriculture is hindered by the large size of the particles. However, nano-sized apatite particles could be as effective in providing the nutrient P as the commonly used soluble P fertilizers, but shall greatly minimize the eutrophication.

In a greenhouse test, application of nano-sized hydroxyapatite increased the height of soybean plants by 30% over regular P fertilizer treatment. The treatment also enhanced the growth, dry biomass and seed yield of soybean greater than regular P applied soybean plants (Liu and Lal 2014).

Around 50% of the arable soils of the world are Zn deficient (Sillanpaa 1990). The effectiveness of Zn fertilizers for providing plants with Zn in Zn-deficient soils largely depends on the solubility of the Zn in soil. The inorganic ZnO is a most commonly used Zn fertilizer (Martens and Westermann 1991). Use of ZnO NPs in place of conventional ZnO appears a most promising approach where a very small dose would fulfil the Zn requirement of the plant (Khan and Rizvi 2017). Application of ZnO NPs may improve the efficiency of the fertilizer and Zn availability to plants by enhancing the rate and extent of Zn dissolution (Panwar 2012). Pot trials with foliar spray have demonstrated that plants sprayed with 20 mg ml^{-1} ZnO NPs solution showed improved growth and biomass production over conventional Zn fertilizer (De la Rosa et al. 2013). The treatment with TiO_2 nanoparticles promoted the plant growth of maize; however, the effect of TiO_2 bulk treatment was negligible. In another experiment, a compound of SiO_2 and TiO_2 nanoparticles increased the activity of nitrate reductase in soybeans and intensified plant absorption capacity, making an efficient use of water and fertilizer (Lu et al. 2002).

13.2.2 Effect of Nanoparticles on Microorganisms

Nanoparticles are generally found toxic to microorganisms (Khan et al. 2019a). Accumulation of NPs in the soil may adversely affect the microbial community and diversity. The effects on the microbial communities as well as on the individual microorganisms may vary with NP species and are described below.

13.2.2.1 Effects of Nanoparticles on Soil Microbial Community

At the present state of knowledge, it is not properly understood how nanoparticles can affect the microbial communities in the soil (Table 13.2). The nanoparticles may directly affect the bioavailability of other toxins and nutrients or indirectly via interactions with natural organic compounds (Haris and Ahmad 2017). Nanoparticles and their ions can produce free radicals which can damage microbial cell membrane, DNA and mitochondria resulting in death of the microbes (Rajput et al. 2018a, b). Soil respiration and enzymatic activities are important determinants to assess the impact of nanoparticles on microbial activity (Simonin and Richaume 2015). Metallic nanoparticles may influence enzymatic activities and bacterial abundance in soil; soil types may also play an important role in the NP toxicity (You et al. 2018). The

Table 13.2 Effects of nanoparticles on microbes and their communities

Nanoparticles	Microorganisms	Effects	References
Ag (0.14 mg/kg)	Bacterial community	Modified the community	Colman et al. (2013)
Ag (1.25 μg–30 mg/kg)	Soil microbial community	Decreased the enzymatic activities	Peyrot et al. (2014)
Ag NPs	*Aspergillus niger*, *Staphylococcus* sp., *Bacillus* sp. and *E. coli*	100% inhibition in the colonization	Rajkishore et al. (2013)
Ag NPs	*Escherichia coli*	33–45% damage to the cells	Beddow et al. (2014)
Ag NPs	*Chlamydomonas reinhardtii*	Toxicity	Navarro et al. (2008a, b)
Ag	*Escherichia coli*	MIC 1 g/ml	Vertelov et al. (2008)
Ag	Ammonia bacteria oxidizing *Nitrosomonas europaea*, *Nitrosospira multiformis* and *Nitrosococcus oceani*	Significant inhibition to the nitrification potential rates	Beddow et al. (2014)
Ag0	*Escherichia coli*	MIC 100 g/ml	Chudasama et al. (2010)
Ag, Al_2O_3, SiO_2 (50 mg/kg)	Soil bacteria	Reduced dehydrogenase and urease activity	McGee et al. (2017)
Au (0.1–100 mg/kg)	Soil bacterial community	Decreased the enzyme activities, community and nutrient cycling	Asadishad et al. (2017)
CuO	*Bacillus subtilis*	Inhibited colonization	Baek and An (2011)
CuO	*Klebsiella pneumoniae*, *P. aeruginosa*, *Salmonella paratyphi* and *Shigella*	Antibacterial activity inhibited the growth	Rajkishore et al. (2013)
CuO (10 mg/kg soil)	Soil microbial community	Inhibiting	Ben-Moshe et al. (2013)
CuO (upto 1000 mg/kg soil)	Soil microbial community	Decreased the microbial biomass, enzymatic activities, disturbed community structures	Xu et al. (2015)
Cu-doped TiO_2	*Shewanella oneidensis*	Inhibited colonization	Wu et al. (2011)
CdSe QDs (quantum dot)	*Escherichia coli*		Pokhrel et al. (2012)
CuO, Fe_3O_4 (10 g/kg	Soil microbial community	Affected the microbial community	Ben-Moshe et al. (2013)
CdSe/ZnS QDs	*Pseudomonas aeruginosa*	Inhibited colonization	Yang et al. (2012a, b)
CeO_2	*Escherichia coli*	Inhibited colonization	Pelletier et al. (2010)

(continued)

Table 13.2 (continued)

Nanoparticles	Microorganisms	Effects	References
Fe (550 mg/kg soil)	Soil microbial community	Adverse effect on bacterial groups	Shah et al. (2014)
Fe_3O_4 (0.1–10.0 mg/kg soil)	Soil bacterial community	Decreased the bacteria count in soil	Cao et al. (2016)
SiO_2	*Escherichia coli*	Toxic effect	Li et al. (2012)
TiO_2	*Vibrio fischeri*	Toxic effect	Heinlaan et al. (2008)
TiO_2	*Pseudomonas aeruginosa*	Toxic effect	Hessler et al. (2012)
TiO_2 (20 g/kg)	Bacterial community	Decreased the diversity	Ge et al. (2012)
ZnO (0.1 mg/L)	*Escherichia coli*	Inhibited colonization	Applerot et al. (2009)
ZnO (20 mg/L)	*Escherichia coli*	100% mortality	Jiang et al. (2009)
ZnO (960 mg/L)	Escherichia coli O157:H7	100% mortality	Liu et al. (2009)
ZnO (1000 mg/kg soil)	*Azotobacter*, P-solubilizing and K-solubilizing bacteria	Inhibited enzymatic activities	Chai et al. (2015)
ZnO (500–2000 mg/kg soil)		Altered soil bacterial community structure and decreased bacterial diversity	Ge et al. (2011)
ZnS QDs	*Nitrosomonas europaea*	Toxic effect	Yang et al. (2012a, b)

introduction of NPs into the natural environments may pose threat to beneficial microbial communities also. The TiO_2 and CuO NPs decreased soil microbial biomass and enzymatic activities and affected their community structures in flooded paddy soil (Xu et al. 2015). You et al. (2018) reported similar effects of ZnO, TiO_2, CeO_2 and Fe_3O_4 NPs on the microbial enzymes such as invertase, urease, catalase, phosphatase and bacterial communities in two different soil types. High concentration of Fe_3O_4 NPs significantly decreased the soil of bacteria count (Cao et al. 2016). Chai et al. (2015) observed decrease in the plate counts of *Azotobacter*, P-solubilizing and K-solubilizing bacteria and enzymatic activities due to ZnO and CeO_2 NPs. Similarly titanium dioxide NPs reduced the enzymatic activity and abundance and diversity of rhizoplane bacteria (Buzea et al. 2007; Solanki et al. 2008). The CuO NPs were found very toxic to native soil bacteria, causing cavities, holes, membrane degradation and lysis in the microbial cells (Concha-Guerrero et al. 2014). Xu et al. (2015) reported drastic decrease in the microbial biomass and enzymatic activity of the microbial community in soil due to 1000 mg Cu NPs/kg soil. However, Baek and An (2011) reported inhibitory effect of Cu NPs on soil microbial community at a much lower concentration, i.e. 10 mg/kg soil.

13.2.2.2 NP Effects on Individual Microbes

13.2.2.2.1 Silver Nanoparticles

The silver ions and Ag NPs were tested for their effect on *Bipolaris sorokiniana* and *Magnaporthe grisea* (Jo et al. 2009). The NP treatments significantly suppressed the colonization of both the fungi, and the EC_{50} of Ag NPs was much lower than the ionic Ag. A significant reduction in the colonization of two fungi on perennial rye-grass was recorded due to Ag NP treatment (Kasprowicz et al. 2010). Safavi et al. (2011) reported that nanosilver had a good potential for removing the bacterial contaminants during tissue culturing of tobacco plant. The antibacterial activity of the Ag NPs/polyvinylpyrrolidone (PVP) against three different groups of bacteria, *Staphylococcus aureus* (gram-positive bacteria), *E. coli* (gram-negative bacteria), *P. aeruginosa* (nonfermenting gram-negative bacteria), as well as against spores of *Bacillus subtilis* has been reported (Bryaskova et al. 2011). Guzman et al. (2009) found that Ag NPs showed high bactericidal activity against gram-positive bacteria such as *E. coli*, *P. aeruginosa* and *S. aureus*. The Ag NPs/PVP was also tested for fungicidal activity against different yeasts and moulds such as *Candida albicans*, *C. krusei*, *C. tropicalis*, *C. glabrata* and *Aspergillus brasiliensis*. The hybrid materials showed strong antifungal effects against the tested microbes (Bryaskova et al. 2011).

The silver nanoparticles at 15 mg/l greatly inhibited the growth of *Alternaria alternata*, *Sclerotinia sclerotiorum*, *Macrophomina phaseolina*, *Rhizoctonia solani*, *B. cinerea* and *Curvularia lunata* (Krishnaraj et al. 2012). Application of different concentrations of Ag NPs before and after disease outbreak on cucurbits under field conditions showed that 100 ppm Ag NPs provided better disease control when applied both before and after the outbreak of the disease (Lamsa et al. 2011). Silver nanoparticles have a high surface area and fraction of surface atoms; as a result, they have high antimicrobial effect tested compared to the bulk silver (Singh et al. 2015).

13.2.2.2.2 Zinc and Sulphur Nanoparticles

Like Ag NPs, the zinc oxide nanoparticles have also been found more antimicrobial than the conventional ZnO powder. Tayel et al. (2011) tested efficacy of ZnO NPs against nine bacterial strains in comparison to conventional ZnO. The ZnO NPs were found more suppressive to bacteria than powder. Gram-positive bacteria were generally more sensitive to ZnO than gram negatives. The ZnO NP treatment to *Salmonella typhimurium* and *Staphylococcus aureus* caused 100% cell death within 8 and 4 h of application, respectively (Tayel et al. 2011). Antibacterial activity of ZnO NPs against *P. aeruginosa* has been reported by Jayaseelan et al. (2012). The maximum zone of inhibition in the colonization of the bacteria (22 ± 1.8 mm) was recorded at 25 ng/mL ZnO NPs. Antifungal efficiency of sulphur NPs (S NPs) against *Fusarium solani* and *Venturia inaequalis* was investigated in vitro (Rao and Paria 2013). The S NPs (35 nm) were found more effective than the bigger particles in suppressing the fungal colonization. Microscopic study revealed that the accumulation of S NPs on the surface of the spores and hyphae and subsequent damage were the main cause for the fungal suppression (Rao and Paria 2013). Study on the

ZnO NPs against *Botrytis cinerea* and *Penicillium expansum* showed that ZnO NPs at 3 mmol/l significantly inhibited the growth of *B. cinerea* and *P. expansum*; the latter fungus was found more sensitive to the treatments. Jayaseelan et al. (2012) reported that ZnO NPs at 25 mg/ml suppressed the colonization of *A. flavus*.

13.2.2.2.3 Copper and Silica Nanoparticles

Cioffi et al. (2004) reported that Cu NPs suppressed the colonies of *Botrytis cinerea* greater than the copper salt powder. The Cu NP antifungal activities of polymer-based copper nanocomposites against pathogenic fungi (Cioffi et al. 2004) and silica-silver nanoparticles against *Botrytis cinerea*, *R. solani* and *Colletotrichum gloeosporioides* have been reported (Park et al. 2006). Copper nanoparticles in soda lime glass powder showed efficient antimicrobial activity against gram-positive and gram-negative bacteria and fungi (Esteban-Tejeda et al. 2009). The antibacterial activities of CuO nanoparticles have also been reported against *S. aureus*, *Bacillus subtilis*, *P. aeruginosa* and *E. coli* (Azam et al. 2012).

13.3 Conclusion

The critical analysis of the relevant literature has revealed that there are various methods of synthesis of nanoparticles and nanomaterials, but for simplicity and eco-friendly nature, green synthesis methods, especially phytosynthesis techniques, appear to have substantial potential for future exploitation and use. Since soil is an ultimate sink for final accumulation/deposition of nanomaterials, the plant roots and soil inhibiting microorganisms are exposed and affected directly to nanoparticles. The information available on this aspect shows that the nanoparticles especially the nano-form of inorganic fertilizers like N, P, Zn, etc. can greatly enhance their efficacy and can substantially reduce the dose of application. However, the NPs especially of the toxic metals may suppress the growth of plant and soil microbes. The nanoparticles generally prove antimicrobial at a concentration not causing phytotoxic effects. Hence, the antimicrobial characteristics of nanoparticles offer a diversified application of nanotechnology against plant pathogens. However, antimicrobial nature of NPs may also lead to the suppression of beneficial soil microorganisms. Before commercial use of nanoparticles and nanomaterials in crop production, the biosafety and environment safety aspects associated with NP applications as well as their effects on plants and beneficial microbes are required to be ascertained in a fool-proof manner.

References

Abd FG, Al-Kawaz AJAH, Al-Dahmoshi HOM (2013) Phenotypic and genotypic investigation on silver nanoparticles of *Morganella morganii* recovered from (cauti) Iraq. Int J Med Pharm Sci 3:29–38

Ahmad A, Mukherjee P, Senapati S, Mandal D, Khan MI, Kumar R, Sastry M (2003a) Extracellular biosynthesis of silver nanoparticles using the fungus *Fusarium oxysporum*. Colloids Surf B Biointerfaces 28:313–318

Ahmad A, Senapati S, Khan MI, Kumar R, Sastry M (2003b) Extracellular biosynthesis of monodisperse gold nanoparticles by a novel extremophilic actinomycete, *Thermomonospora* sp. Langmuir 19:3550–3553

Amendola V, Polizzi S, Meneghetti M (2006) Laser ablation synthesis of gold nanoparticles in organic solvents. J Phys Chem B 110:7232–7237

Ankamwar B, Damle C, Ahmad A, Sastry M (2005) Biosynthesis of gold and silver nanoparticles using *Emblica officinalis* fruit extract, their phase transfer and transmetallation in an organic solution. J Nanosci Nanotechnol 5:1665–1671

Applerot G, Lipovsky A, Dror R, Perkas N, Nitzan Y, Lubart R, Gedanken A (2009) Enhanced antibacterial activity of nanocrystalline ZnO due to increased ROS-mediated cell injury. Adv Funct Mater 19:842–852

Armendariz V, Herrera I, Jose-yacaman M, Troiani H, Santiago P, Gardea-Torresdey JL (2004) Size controlled gold nanoparticle formation by *Avena sativa* biomass: use of plants in nanobiotechnology. J Nanopart Res 6:377–382

Armendariz V, Parsons JG, Lopez ML, Peralta-Videa JR, Jose-Yacaman M, Gardea-Torresdey JL (2009) The extraction of gold nanoparticles from oat and wheat biomasses using sodium citrate and cetyltrimethylammonium bromide, studied by X-ray absorption spectroscopy, high-resolution transmission electron microscopy, and UV–visible spectroscopy. Nanotechnology 20:105607

Asadishad B, Chahal S, Cianciarelli V, Zhou K, Tufenkji N (2017) Effect of gold nanoparticles on extracellular nutrient-cycling enzyme activity and bacterial community in soil slurries: role of nanoparticle size and surface coating. Environ Sci Technol 4:907–918

Aslam M, Gopakumar G, Shoba TL, Mulla IS, Vijayamohanan K (2002) Formation of Cu and Cu2O nanoparticles by variation of the surface ligand: preparation, structure and insulating-to-metallic transition. J Colloid Interf Sci 255:79–90

Athawale AA, Katre PP, Kumar M, Majumdar MB (2005) Synthesis of CTAB-IPA reduced copper nanoparticles. Mater Chem Phys 91:507–512

Azam A, Ahmed AS, Oves M, Khan MS, Memic A (2012) Size-dependent antimicrobial properties of CuO nanoparticles against gram-positive and-negative bacterial strains. Int J Nanomedicine 7:3527–3535

BadriNarayanan K, Sakthivel N (2008) Coriander leaf mediated biosynthesis of gold nanoparticles. Mater Lett 62:4588–4590

Baek YW, An YJ (2011) Microbial toxicity of metal oxide nanoparticles (CuO, NiO, ZnO, and Sb2O3) to *Escherichia coli, Bacillus subtilis,* and Streptococcus aureus. Sci Total Environ 409:1603–1608

Balaji DS, Basavaraja S, Deshpande R, Mahesh DB, Prabhakar BK, Venkataraman A (2009) Extracellular biosynthesis of functionalized silver nanoparticles by strains of *Cladosporium cladosporioides* fungus. Colloids Surf B Biointerfaces 68:88–92

Bansal V, Rautaray D, Ahmad A, Sastry M (2004) Biosynthesis of zirconia nanoparticles using the fungus *Fusarium oxysporum*. J Mater Chem 14:3303–3305

Basavaraja S, Balaji SD, Lagashetty A, Rajasab AH, Venkataraman A (2008) Extracellular biosynthesis of silver nanoparticles using the fungus *Fusarium semitectum*. Mater Res Bull 43:1164–1170

Beddow J, Stolpe B, Cole P, Lead JR, Sapp M, Lyons BP, Colbeck I, Whitby C (2014) Effects of engineered silver nanoparticles on the growth and activity of ecologically important microbes. Environ Microbiol Rep 6:448–458

Ben-Moshe T, Frenk S, Dror I, Minz D, Berkowitz B (2013) Effects of metal oxide nanoparticles on soil properties. Chemosphere 90:640–646

Bhainsa KC, D'Souza SF (2006) Extracellular biosynthesis of silver nanoparticles using the fungus *Aspergillus fumigatus*. Colloids Surf B Biointerfaces 47:160–164

Binupriya AR, Sathishkumar M, Vijayaraghavan K, Yun SI (2010) Bioreduction of trivalent aurum to nano-crystalline gold particles by active and inactive cells and cell-free extract of *Aspergillus oryzae var. viridis*. J Hazard Mater 177:539–545

Birla SS, Tiwari VV, Gade AK, Ingle AP, Yadav AP, Rai MK (2009) Fabrication of silver nanoparticles by *Phoma glomerata* and its combined effect against *Escherichia coli, Pseudomonas aeruginosa* and *Staphylococcus aureus*. Lett Appl Microbiol 48:173–179

Blosi M, Albonetti S, Dondi M, Martelli C, Baldi G (2011) Microwave-assisted polyol synthesis of Cu nanoparticles. J Nanopart Res 13:127–138

Brayner R, Barberousse H, Hemadi M, Djedjat C, Yepremian C (2007) Cyanobacteria as bioreactors for the synthesis of Au, Ag, Pd and Pt nanoparticles via an enzyme-mediated route. J Nanosci Nanotechnol 7:2696–2708

Bryaskova R, Pencheva D, Nikolov S, Kantardjiev T (2011) Synthesis and comparative study on the antimicrobial activity of hybrid materials based on silver nanoparticles (AgNps) stabilized by polyvinylpyrrolidone (PVP). J Chem Biol 4:185–191

Buzea C, Pacheco II, Robbie K (2007) Nanomaterials and nanoparticles: sources and toxicity. Biointerphases 2:MR17–MR71

Byrappa K, Yoshimura M (2001) Hydrothermal technology – principles and applications. In: Handbook of hydrothermal technology, pp 1–52

Cao J, Feng Y, Lin X, Wang J (2016) Arbuscular mycorrhizal fungi alleviate the negative effects of iron oxide nanoparticles on bacterial community in rhizospheric soils. Front Environ Sci 4(10)

Carpenter SR (2005) Eutrophication of aquatic ecosystems: bistability and soil phosphorus. Proc Natl Acad Sci 102(29):10002–10005

Cason JP, Miller ME, Thompson JB, Roberts CB (2001) Solvent effects on copper nanoparticle growth behavior in AOT reverse micelle systems. J Phys Chem B 105:2297–2302

Chai H, Yao J, Sun J, Zhang C, Liu W, Zhu M, Ceccanti B (2015) The effect of metal oxide nanoparticles on functional bacteria and metabolic profiles in agricultural soil. Bull Environ Contam Toxicol 94:490–495

Chandran SP, Chaudhary M, Pasricha R, Ahmad A, Sastry M (2006) Synthesis of gold nanotriangles and silver nanoparticles using Aloevera plant extract. Biotechnol Prog 22:577–583

Chen L, Zhang D, Chen J, Zhou H, Wan H (2006) The use of CTAB to control the size of copper nanoparticles and the concentration of alkylthiols on their surfaces. Mater Sci Eng A 415:156–161

Chen H, Lee JH, Kim YH, Shin DW, Park SC, Meng X, Yoo JB (2010) Metallic copper nanostructures synthesized by a facile hydrothermal method. J Nanosci Nanotechnol 10:629–636

Chen J, Liu X, Wang C, Yin SS, Li XL, Hu WJ, Simon M, Shen ZJ, Xiao Q, Chu CC, Peng XX (2015) Nitric oxide ameliorates zinc oxide nanoparticles-induced phytotoxicity in rice seedlings. J Hazard Mater 297:173–182

Chudasama B, Vala AK, Andhariya N, Mehta RV, Upadhyay RV (2010) Highly bacterial resistant silver nanoparticles: synthesis and antibacterial activities. J Nanopart Res 12:1677–1685

Cioffi N, Torsi L, Ditaranto N, Sabbatini L, Zambonin PG, Tantillo G, Ghibelli L, D'Alessio M, Bleve-Zacheo T, Traversa E (2004) Antifungal activity of polymer-based copper nanocomposite coatings. Appl Phys Lett 85:2417–2419

Colman BP, Arnaout CL, Anciaux S, Gunsch CK, Hochella MF Jr, Kim B, Lowry GV, McGill BM, Reinsch BC, Richardson CJ, Unrine JM (2013) Low concentrations of silver nanoparticles in biosolids cause adverse ecosystem responses under realistic field scenario. PLoS One 8:57189

Concha-Guerrero SI, Brito EMS, Piñón-Castillo HA, Tarango-Rivero SH, Caretta CA, Luna-Velasco A, Duran R, Orrantia-Borunda E (2014) Effect of CuO nanoparticles over isolated bacterial strains from agricultural soil. J Nanomater 2014:206

Conley DJ, Paerl HW, Howarth RW, Boesch DF, Seitzinger SP, Havens KE, Lancelot C, Likens GE (2009) Controlling eutrophication: nitrogen and phosphorus. Science 323:1014–1015

Corral-Diaz B, Peralta-Videa JR, Alvarez-Parrilla E, Rodrigo-García J, Morales MI, Osuna-Avila P, Niu G, Hernandez-Viezcas JA, Gardea-Torresdey JL (2014) Cerium oxide nanoparticles alter the antioxidant capacity but do not impact tuber ionome in *Raphanus sativus* (L). Plant Physiol Biochem 84:277–285

Correll DL (1998) The role of phosphorus in the eutrophication of receiving waters: a review. J Environ Qual 27:261–266

Cvjetko P, Milošić A, Domijan AM, Vrček IV, Tolić S, Štefanić PP, Letofsky-Papst I, Tkalec M, Balen B (2017) Toxicity of silver ions and differently coated silver nanoparticles in *Allium cepa* roots. Ecotoxicol Environ Saf 137:18–28

Dameron CT, Reese RN, Mehra RK, Kortan AR, Carroll PJ (1989) Biosynthesis of cadmium sulphide quantum semiconductor crystallites. Nature 338:596–597

De la Rosa G, López-Moreno ML, de Haro D, Botez CE, Peralta-Videa JR, Gardea-Torresdey JL (2013) Effects of ZnO nanoparticles in alfalfa, tomato, and cucumber at the germination stage: root development and X-ray absorption spectroscopy studies. Pure Appl Chem 85:2161–2174

De Rosa MC, Monreal C, Schnitzer M, Walsh R, Sultan Y (2010) Nanotechnology in fertilizers. Nat Nanotechnol 5:91

De Windt W, Aelterman P, Verstraete W (2005) Bioreductive deposition of palladium (0) nanoparticles on Shewanella oneidensis with catalytic activity towards reductive dechlorination of polychlorinated biphenyls. Environ Microbiol 7:314–325

Deng F, Wang S, Xin H (2016) Toxicity of CuO nanoparticles to structure and metabolic activity of *Allium cepa* root tips. Bull Environ Contam Toxicol 97:702–708

Di Gregorio S, Lampis S, Vallini G (2005) Selenite precipitation by a rhizospheric strain of Stenotrophomonas sp. isolated from the root system of *Astragalus bisulcatus*: a biotechnological perspective. Environ Int 31:233–241

Dimkpa CO, McLean JE, Britt DW, Anderson AJ (2015) Nano-CuO and interaction with nano-ZnO or soil bacterium provide evidence for the interference of nanoparticles in metal nutrition of plants. Ecotoxicology:119–129

Ebbs SD, Bradfield SJ, Kumar P, White JC, Musante C, Ma X (2016) Accumulation of zinc, copper, or cerium in carrot (*Daucus carota*) exposed to metal oxide nanoparticles and metal ions. Environ Sci Nano 3:114–126

El-Rafie MH, Shaheen TI, Mohamed AA, Hebeish A (2012) Bio-synthesis and applications of silver nanoparticles onto cotton fabrics. Carbohydr Polym 90:915–920

El-Temsah YS, Joner EJ (2012) Impact of Fe and Ag nanoparticles on seed germination and differences in bioavailability during exposure in aqueous suspension and soil. Environ Toxicol 27:42–49

Esteban-Tejeda L, Malpartida F, Esteban-Cubillo A, Pecharromán C, Moya JS (2009) Antibacterial and antifungal activity of a soda-lime glass containing copper nanoparticles. Nanotechnology 20:505701

Fageria NK (2009) The use of nutrıents in crop plants. CRC Press, Boca Raton

Fariq A, Khan T, Yasmin A (2017) Microbial synthesis of nanoparticles and their potential applications in biomedicine. J Appl Biomed 15:241–248

Fatimah I, Indriani N (2018) Silver nanoparticles synthesized using *Lantana camara* flower extract by reflux, microwave and ultrasound methods. Chem J Moldova 13:95–102

Frazier TP, Burklew CE, Zhang B (2014) Titanium dioxide nanoparticles affect the growth and microRNA expression of tobacco (*Nicotiana tabacum*). Funct Integr Genomics 14:75–83

Gade AK, Bonde P, Ingle AP, Marcato PD, Duran N, Rai MK (2008) Exploitation of *Aspergillus niger* for synthesis of silver nanoparticles. J Biobaased Mater Bioenergy 2:243–247

Gardea-Torresdey JL, Gomez E, Peralta-Videa JR, Parsons JG, Troiani H, Jose-Yacaman M (2003) Alfalfa sprouts: a natural source for the synthesis of silver nanoparticles. Langmuir 19:1357–1361

Ge Y, Schimel JP, Holden PA (2011) Evidence for negative effects of TiO2 and ZnO nanoparticles on soil bacterial communities. Environ Sci Technol 45:1659–1664

Ge Y, Schimel JP, Holden PA (2012) Identification of soil bacteria susceptible to TiO2 and ZnO nanoparticles. Appl Environ Microbiol 78:6749–6758

Gericke M, Pinches A (2006) Biological synthesis of metal nanoparticles. Hydrometallurgy 83:132–140

Ghosh M, Jana A, Sinha S, Jothiramajayam M, Nag A, Chakraborty A, Mukherjee A, Mukherjee A (2016) Effects of ZnO nanoparticles in plants: cytotoxicity, genotoxicity, deregulation of antioxidant defenses, and cell-cycle arrest. Mutat Res Genet Toxicol Environ Mutagen 807:25–32

González-Garcinuño Á, Tabernero A, Marcelo G, Sebastián V, Arruebo M, Santamaría J, del Valle EM (2019) Differences in Levan nanoparticles depending on their synthesis route: microbial vs cell-free systems. Int J Biol Macromol 137:62–68

Gui X, Deng Y, Rui Y, Gao B, Luo W, Chen S, Li X, Liu S, Han Y, Liu L, Xing B (2015) Response difference of transgenic and conventional rice (*Oryza sativa*) to nanoparticles (γFe 2 O 3). Environ Sci Pollut Res 22:17716–17723

Gupta V, Gupta AR, Kant V (2013) Synthesis, characterization and biomedical application of nanoparticles. Sci Int 1:167–174

Gurunathan S, Kalishwaralal K, Vaidyanathan R, Venkataraman D, Pandian SRK (2009) Biosynthesis, purification and characterization of silver nanoparticles using Escherichia coli. Colloids Surf B Biointerfaces 74:328–335

Guzman MG, Dille J, Godet S (2009) Synthesis of silver nanoparticles by chemical reduction method and their antibacterial activity. Int J Chem Biol Eng 2:104–111

Haris Z, Ahmad I (2017) Impact of metal oxide nanoparticles on beneficial soil microorganisms and their secondary metabolites. Int J Life-Sci Sci Res 3:1020–1030

He SY, Guo ZR, Zhang Y, Zhang S, Wang J, Gu N (2007) Biosynthesis of gold nanoparticles using the bacteria *Rhodopseudomonas capsulata*. Mater Lett 61:3984–3987

Heinlaan M, Ivask A, Blinova I, Dubourguier HC, Kahru A (2008) Toxicity of nanosized and bulk ZnO, CuO and TiO2 to bacteria Vibrio fischeri and crustaceans Daphnia magna and Thamnocephalus platyurus. Chemosphere 71:1308–1316

Hessler CM, Wu MY, Xue Z, Choi H, Seo Y (2012) The influence of capsular extracellular polymeric substances on the interaction between TiO2 nanoparticles and planktonic bacteria. Water Res 46:4687–4696

Hong J, Rico CM, Zhao L, Adeleye AS, Keller AA, Peralta-Videa JR, Gardea-Torresdey JL (2015) Toxic effects of copper-based nanoparticles or compounds to lettuce (*Lactuca sativa*) and alfalfa (*Medicago sativa*). Environ Sci Process Impacts 17:177–185

Hossain KZ, Monreal CM, Sayari A (2008) Adsorption of urease on PE-MCM-41 and its catalytic effect on hydrolysis of urea. Colloids Surf B Biointerfaces 62:42–50

Huang J, Li Q, Sun D, Lu Y, Su Y, Yang X, Wang H, Wang Y, Shao W, He N, Hong J (2007) Biosynthesis of silver and gold nanoparticles by novel sundried *Cinnamomum camphora* leaf. Nanotechnology 18:105104

Huh AJ, Kwon YJ (2011) "Nanoantibiotics": a new paradigm for treating infectious diseases using nanomaterials in the antibiotics resistant era. J Control Release 156:128–145

Husseiny MI, El-Aziz MA, Badr Y, Mahmoud MA (2007) Biosynthesis of gold nanoparticles using *Pseudomonas aeruginosa*. Spectrochim Acta A Mol Biomol Spectrosc 67:1003–1006

Im SH, Lee YT, Wiley B, Xia Y (2005) Large-scale synthesis of silver nanocubes: the role of hcl in promoting cube perfection and monodispersity. Angew Chem Int Ed 44:2154–2157

Inbakandan D, Venkatesan R, Khan SA (2010) Biosynthesis of gold nanoparticles utilizing marine sponge *Acanthella elongata* (Dendy, 1905). Colloids Surf B Biointerfaces 81:634–639

Ingle A, Rai M, Gade A, Bawaskar M (2009) *Fusarium solani*: a novel biological agent for the extracellular synthesis of silver nanoparticles. J Nanopart Res 11:2079–2085

Jayaseelan C, Rahuman AA, Kirthi AV, Marimuthu S, Santhoshkumar T, Bagavan A, Gaurav K, Karthik L, Rao KB (2012) Novel microbial route to synthesize ZnO nanoparticles using *Aeromonas hydrophila* and their activity against pathogenic bacteria and fungi. Spectrochim Acta A Mol Biomol Spectrosc 90:78–84

Jha AK, Prasad K, Prasad K (2009) A green low-cost biosynthesis of Sb2O 3 nanoparticles. Biochem Eng J 43:303–306

Jiang W, Mashayekhi H, Xing B (2009) Bacterial toxicity comparison between nano-and micro-scaled oxide particles. Environ Pollut 157:1619–1625

Jo YK, Kim BH, Jung G (2009) Antifungal activity of silver ions and nanoparticles on phytopathogenic fungi. Plant Dis 93:1037–1043

Kajbafvala A, Li M, Bahmanpour H, Maneshian MH, Kauffmann A (2013) Nano/microstructured materials: rapid, low-cost, and eco-friendly synthesis methods. J Nanopart 2013

Kalimuthu K, Babu RS, Venkataraman D, Bilal M, Gurunathan S (2008) Biosynthesis of silver nanocrystals by *Bacillus licheniformis*. Colloids Surf B Biointerfaces 65:150–153

Kasprowicz MJ, Kozioł M, Gorczyca A (2010) The effect of silver nanoparticles on phytopathogenic spores of *Fusarium culmorum*. Can J Microbiol 56:247–253

Kathiresan K, Manivannan S, Nabeel MA, Dhivya B (2009) Studies on silver nanoparticles synthesized by a marine fungus, *Penicillium fellutanum* isolated from coastal mangrove sediment. Colloids Surf B Biointerfaces 71:133–137

Khan MR, Rizvi TF (2014) Nanotechnology: scope and application in plant disease management. Plant Pathol J 13:214–231

Khan MR, Rizvi TF (2017) Application of nanofertilizer and nanopesticides for improvements in crop production and protection. In: Nanoscience and plant–soil systems. Springer, Cham, pp 405–427

Khan MR, Adam V, Rizvi TF, Zhang B, Ahamad F, Jośko I, Zhu Y, Yang M, Mao C (2019a) Nanoparticle–plant interactions: two-way traffic. Small 15:1901794

Khan MR, Ahamad F, Rizvi TF, (2019b) Effect of nanoparticles on plant pathogens. In: Advances in phytonanotechnology. Academic Press, London, pp 215–240

Khandel P, Shahi SK (2018) Mycogenic nanoparticles and their bio-prospective applications: current status and future challenges. J Nanostruct Chem 8:369–391

Kitchens CL, Roberts CB (2004) Copper nanoparticle synthesis in compressed liquid and supercritical fluid reverse micelle systems. Ind Eng Chem Res 43:6070–6081

Komarneni S (2003) Nanophase materials by hydrothermal, microwave hydrothermal and microwave-solvothermal methods. Curr Sci 85:1730–1734

Konishi Y, Ohno K, Saitoh N, Namura T, Nagmine S (2004) Microbial synthesis of gold nanoparticles by metal reducing bacterium. Trans Mater Res Soc Jpn 29:2341–2343

Konishi Y, Ohno K, Saitoh N, Nomura T, Nagamine S (2007) Bioreductive deposition of platinum nanoparticles on the bacterium Shewanella algae. J Biotechnol 128:648–653

Kowshik M, Deshmukh N, Vogel W, Urban J, Kulkarni SK, Paknikar KM (2002) Microbial synthesis of semiconductor CdS nanoparticles, their characterization and their use in the fabrication of an ideal diode. Biotechnol Bioeng 78:583–588

Krishnaraj C, Ramachandran R, Mohan K, Kalaichelvan PT (2012) Optimization for rapid synthesis of silver nanoparticles and its effect on phytopathogenic fungi. Spectrochim Acta A Mol Biomol Spectrosc 93:95–99

Lamsa K, Kim SW, Jung JH, Kim YS, Kim KS, Lee YS (2011) Inhibition effects of silver nanoparticles against powdery mildews on cucumber and pumpkin. Mycobiology 39:26–32

Law N, Ansari S, Livens FR, Renshaw JC, Lloyd JR (2008) Formation of nanoscale elemental silver particles via enzymatic reduction by *Geobacter sulfurreducens*. Appl Environ Microbiol 74:7090–7093

Le Van N, Ma C, Shang J, Rui Y, Liu S, Xing B (2016a) Effects of CuO nanoparticles on insecticidal activity and phytotoxicity in conventional and transgenic cotton. Chemosphere 144:661–670

Le Van N, Rui Y, Cao W, Shang J, Liu S, Nguyen Quang T, Liu L (2016b) Toxicity and bio-effects of CuO nanoparticles on transgenic Ipt-cotton. J Plant Interact 11:108–116

Lee WM, An YJ, Yoon H, Kweon HS (2008) Toxicity and bioavailability of copper nanoparticles to the terrestrial plants mung bean (*Phaseolus radiatus*) and wheat (*Triticum aestivum*): plant agar test for water -insoluble nanoparticles. Environ Toxicol Chem 27:1915–1921

Lee WM, Kwak JI, An YJ (2012) Effect of silver nanoparticles in crop plants *Phaseolus radiatus* and *Sorghum bicolor*: media effect on phytotoxicity. Chemosphere 86:491–499

Li Y, Zhang W, Niu J, Chen Y (2012) Mechanism of photogenerated reactive oxygen species and correlation with the antibacterial properties of engineered metal-oxide nanoparticles. ACS Nano 6:5164–5173

Lin D, Xing B (2007) Phytotoxicity of nanoparticles: inhibition of seed germination and root growth. Environ Pollut 150:243–250

Liu R, Lal R (2014) Synthetic apatite nanoparticles as a phosphorus fertilizer for soybean (*Glycine max*). Sci Rep 4:5686

Liu Y, He L, Mustapha A, Li H, Hu ZQ, Lin M (2009) Antibacterial activities of zinc oxide nanoparticles against *Escherichia coli* O157: H7. J Appl Microbiol 107:1193–1201

Liu R, Zhang H, Lal R (2016) Effects of stabilized nanoparticles of copper, zinc, manganese, and iron oxides in low concentrations on lettuce (*Lactuca sativa*) seed germination: nanotoxicants or nanonutrients. Water Air Soil Pollut 227:42

López-Moreno ML, De la Rosa G, Hernández-Viezcas JA, Peralta-Videa JR, Gardea-Torresdey JL (2010) XAS corroboration of the uptake and storage of CeO2 nanoparticles and assessment of their differential toxicity in four edible plant species. J Agric Food Chem 58:3689

Lortie L, Gould WD, Rajan S, Meeready RGL, Cheng KJ (1992) Reduction of selenate and selenite to elemental selenium by a *Pseudomonas stutzeri* isolate. Appl Environ Microbiol 58:4042–4044

Losi ME, Frankenberger WT (1997) Reduction of selenium oxyanions by *Enterobacter cloacae* SLD1a-1: isolation and growth of the bacterium and its expulsion of selenium particles. Appl Environ Microbiol 63:3079–3084

Lovley DR, Phillips EJP, Lonergan DJ (1989) Hydrogen and formate oxidation coupled to dissimilatory reduction of iron or manganese by *Alteromonas putrefaciens*. Applied. Environ Microbiol 55:700–706

Lovley DR, Phillips EJP, Gorby YA, Landa ER (1991) Microbial reduction of uranium. Nature 350:413–416

Lu CM, Zhang CY, Wen JQ, Wu GR (2002) Effects of nano material on germination and growth of soybean. Soybean Sci 21:168–171

Lukman AI, Gong B, Marjo CE, Roessner U, Harris AT (2011) Facile synthesis, stabilization, and anti-bacterial performance of discrete Ag nanoparticles using *Medicago sativa* seed exudates. J Colloid Interface Sci 353:433–444

Manonmani V, Juliet V (2011) Biosynthesis of Ag nanoparticles for the detection of pathogenic bacteria. In: Proceedings of the 2nd International conference on innovation, management and service September 16–18, 2011, Singapore, pp 307–311

Martens DC, Westermann DT (1991) Fertilizer applications for correcting micronutrient deficiencies. In: Mortvedt JJ, Cox FR, Shuman LM, Welch RM (eds) Micronutrients in agriculture. Madison, Soil Science Society of America, pp 549–592

Martínez-Fernández D, Komárek M (2016) Comparative effects of nanoscale zero-valent iron (nZVI) and Fe2O3 nanoparticles on root hydraulic conductivity of *Solanum lycopersicum* L. Environ Exp Bot 131:128–136

Martínez-Fernández D, Barroso D, Komárek M (2016) Root water transport of *Helianthus annuus* L. under iron oxide nanoparticle exposure. Environ Sci Pollut Res 23:1732–1741

Mazumdar H, Ahmed GU (2011) Phytotoxicity effect of silver nanoparticles on *Oryza sativa*. Int J ChemTech Res 3:1494–1500

McGee CF, Storey S, Clipson N, Doyle E (2017) Soil microbial community responses to contamination with silver, aluminium oxide and silicon dioxide nanoparticles. Ecotoxicology 26:449–458

Moon YS, Park ES, Kim TO, Lee HS, Lee SE (2014) SELDI-TOF MS-based discovery of a biomarker in *Cucumis sativus* seeds exposed to CuO nanoparticles. Environ Toxicol Pharmacol 38:922–931

Mourato A, Gadanho M, Lino AR, Tenreiro R (2011) Biosynthesis of crystalline silver and gold nanoparticles by extremophilic yeasts. Bioinorg Chem Appl 8

Mude N, Ingle A, Gade A, Rai M (2009) Synthesis of silver nanoparticles using callus extract of *Carica papaya* – a first report. J Plant Biochem Biotechnol 18:83–86

Nabikhan A, Kandasamy K, Raj A, Alikunhi NM (2010) Synthesis of antimicrobial silver nanoparticles by callus and leaf extracts from saltmarsh plant, *Sesuvium portulacastrum* L. Colloids Surf B 79:488–493

Nair PMG, Chung IM (2014) A mechanistic study on the toxic effect of copper oxide nanoparticles in soybean (*Glycine max* L.) root development and lignification of root cells. Biol Trace Elem Res 162:342–352

Nair PMG, Chung IM (2015) Study on the correlation between copper oxide nanoparticles induced growth suppression and enhanced lignification in Indian mustard (*Brassica juncea* L.). Ecotoxicol Environ Saf 113:302–313

Nair PMG, Kim SH, Chung IM (2014) Copper oxide nanoparticle toxicity in mung bean (*Vigna radiata* L.) seedlings: physiological and molecular level responses of in vitro grown plants. Acta Physiol Plant 36:2947–2958

Navarro E, Baun A, Behra R, Hartmann NB, Filser J, Miao AJ, Quigg A, Santschi PH, Sigg L (2008a) Environmental behavior and ecotoxicity of engineered nanoparticles to algae, plants, and fungi. Ecotoxicology 17:372–386

Navarro E, Piccapietra F, Wagner B, Marconi F, Kaegi R, Odzak N, Sigg L, Behra R (2008b) Toxicity of silver nanoparticles to *Chlamydomonas reinhardtii*. Environ Sci Technol 42:8959–8964

Olchowik J, Bzdyk R, Studnicki M, Bederska-Błaszczyk M, Urban A, Aleksandrowicz-Trzcińska M (2017) The effect of silver and copper nanoparticles on the condition of english oak (*Quercus robur* L.) seedlings in a container nursery experiment. Forests 8:310

Panwar J (2012) Positive effect of zinc oxide nanoparticles on tomato plants: A step towards developing nano-fertilizers 01/2012. In: Proceeding of 3rd International conference on environmental research and technology (ICERT), May 30–June 1 2012, Penang

Parashar V, Parashar R, Sharma B, Pandey AC (2009) Parthenium leaf extract mediated synthesis of silver nanoparticles: a novel approach towards weed utilization. Dig J Nanomater Biostruct 4:1

Park HJ, Kim SH, Kim HJ, Choi SH (2006) A new composition of nanosized silica-silver for control of various plant diseases. Plant Pathol J 22:295–302

Park BK, Jeong S, Kim D, Moon J, Lim S, Kim JS (2007) Synthesis and size control of monodisperse copper nanoparticles by polyol method. J Colloid Interface Sci 311:417–424

Patil RS, Kokate MR, Kolekar SS (2012) Bioinspired synthesis of highly stabilized silver nanoparticles using *Ocimum tenuiflorum* leaf extract and their antibacterial activity. Spectrochim Acta A Mol Biomol Spectrosc 91:234–238

Pelletier DA, Suresh AK, Holton GA, McKeown CK, Wang W, Gu B, Mortensen NP, Allison DP, Joy DC, Allison MR, Brown SD (2010) Effects of engineered cerium oxide nanoparticles on bacterial growth and viability. Appl Environ Microbiol 76:7981–7989

Peyrot C, Wilkinson KJ, Desrosiers M, Sauvé S (2014) Effects of silver nanoparticles on soil enzyme activities with and without added organic matter. Environ Toxicol Chem 33:115–125

Philip D (2009) Biosynthesis of Au, Ag and Au-Ag nanoparticles using edible mushroom extract. Spectrochim Acta A Mol Biomol Spectrosc 73:374–374

Pileni MP, Lisiecki I (1993) Anometer metallic copper particle synthesis in reverse micelles. Colloids Surf A Physicochem Eng Asp 80:63–68

Pokhrel LR, Silva T, Dubey B, El Badawy AM, Tolaymat TM, Scheuerman PR (2012) Rapid screening of aquatic toxicity of several metal-based nanoparticles using the MetPLATE™ bioassay. Sci Total Environ 426:414–422

Pol VG, Motiei M, Gedanken A, Calderon-Moreno J, Mastai Y (2003) Sonochemical deposition of air-stable iron nanoparticles on monodispersed carbon spherules. Chem Mater 15:1378–1384

Prasad K, Jha AK, Kulkarni AR (2007) Lactobacillus assisted synthesis of titanium nanoparticles. Nanoscale Res Lett 2:248

Priester JH, Moritz SC, Espinosa K, Ge Y, Wang Y, Nisbet RM, Schimel JP, Goggi AS, Gardea-Torresdey JL, Holden PA (2017) Damage assessment for soybean cultivated in soil with either CeO2 or ZnO manufactured nanomaterials. Sci Total Environ 579:1756–1768

Raja M, Subha J, Ali FB, Ryu SH (2008) Synthesis of copper nanoparticles by electroreduction process. Mater Manuf Process 23:782–785

Rajkishore SK, Subramanian KS, Natarajan N, Gunasekaran K (2013) Nanotoxicity at various trophic levels: a review. Bioscan 8:975–982
Rajput VD, Minkina T, Sushkova S, Tsitsuashvili V, Mandzhieva S, Gorovtsov A, Nevidomskyaya D, Gromakova N (2018a) Effect of nanoparticles on crops and soil microbial communities. J Soils Sediments 18:2179–2187
Rajput VD, Minkina T, Suskova S, Mandzhieva S, Tsitsuashvili V, Chapligin V, Fedorenko A (2018b) Effects of copper nanoparticles (CuO NPs) on crop plants: a mini review. BioNanoScience 8:36–42
Rao KJ, Paria S (2013) Use of sulfur nanoparticles as a green pesticide on *Fusarium solani* and *Venturia inaequalis* phytopathogens. RSC Adv 3:10471–10478
Reverchon E, Adami R (2006) Nanomaterials and supercritical fluids. J Supercrit Fluids 37:1–22
Safavi K, Mortazaeinezahad F, Esfahanizadeh M, Javad Asgari M (2011) In vitro antibacterial activity of nanomaterial for using in tobacco plants tissue culture. World Acad Sci Eng Technol 79:372–373
Saha N, Gupta SD (2017) Low-dose toxicity of biogenic silver nanoparticles fabricated by *Swertia chirata* on root tips and flower buds of *Allium cepa*. J Hazard Mater 330:18–28
Sanghi R, Verma P (2009) Biomimetic synthesis and characterisation of protein capped silver nanoparticles. Bioresour Technol 100:501–504
Schabes-Retchkiman PS, Canizal G, Herrera-Becerra R, Zorrilla C, Liu HB, Ascencio JA (2006) Biosynthesis and characterization of Ti/Ni bimetallic nanoparticles. Opt Mater 29:95–99
Shah V, Collins D, Walker VK, Shah S (2014) The impact of engineered cobalt, iron, nickel and silver nanoparticles on soil bacterial diversity under field conditions. Environ Res Lett 9:024001
Shaligram SN, Bule M, Bhambure R, Singhal SR, Singh SK, Szakacs G, Pandey A (2009) Biosynthesis of silver nanoparticles using aqueous extract from the compactin producing fungal strain. Process Biochem 44:939–943
Shankar SS, Ahmad A, Pasricha R, Sastry M (2003a) Bioreduction of chloroaurate ions by geranium leaves and its endophytic fungus yields gold nanoparticles of different shapes. J Mater Chem 13:1822–1826
Shankar SS, Ahmad A, Sastry M (2003b) Geranium leaf assisted biosynthesis of silver nanoparticles. Biotechnol Prog 19:1627–1631
Shankar SS, Rai A, Ahmad A, Sastry M (2004a) Rapid synthesis of Au, Ag, and bimetallic Au core–Ag shell nanoparticles using Neem (*Azadirachta indica*) leaf broth. J Colloid Interface Sci 275:496–502
Shankar SS, Rai A, Ankamwar B, Singh A, Ahmad A, Sastry M (2004b) Biological synthesis of triangular gold nanoprisms. Nat Mater 3:482
Shaw AK, Hossain Z (2013) Impact of nano-CuO stress on rice (*Oryza sativa* L.) seedlings. Chemosphere 93:906–915
Sillanpaa M (1990) Micronutrient assessment at the country level: an international study. FAO, Rome, p 208
Silverberg PA, Wong PTS, Chau YK (1976) Localization of selenium in bacterial cells using TEM and energy dispersive X-ray analysis. Arch Microbiol 107:1–6
Simonin M, Richaume A (2015) Impact of engineered nanoparticles on the activity, abundance, and diversity of soil microbial communities: a review. Environ Sci Pollut Res 22:13710–13723
Singh D, Kumar A (2016) Impact of irrigation using water containing CuO and ZnO nanoparticles on *Spinach oleracea* grown in soil media. Bull Environ Contam Toxicol 97:548–553
Singh S, Singh BK, Yadav SM, Gupta AK (2015) Applications of nanotechnology in agricultural and their role in disease management. Res J Nanosci Nanotechnol 5:1–5
Solanki A, Kim JD, Lee KB (2008) Nanotechnology for regenerative medicine: nanomaterials for stem cell imaging. Nanomedicine 3:567–578
Song X, Zhang W, Yin Z (2004) A method for the synthesis of spherical copper nanoparticles in the organic phase. J Colloid Interface Sci 273:463
Suman TY, Rajasree SR, Kanchana A, Elizabeth SB (2013) Biosynthesis, characterization and cytotoxic effect of plant mediated silver nanoparticles using *Morinda citrifolia* root extract. Colloids Surf B Biointerfaces 106:74–78

Suslick KS, Choe SB, Cichowlas AA, Grinstaff MW (1991) Sonochemical synthesis of amorphous iron. Nature 353:414–416
Suslick KS, Hyeon T, Fang M, Cichowlas AA (1996) Sonochemical preparation of nanostructured catalyst. In: Moser WR (ed) Advanced catalysts and nanostructured materials: modern synthetic methods. Academic Press, New York, pp 197–212. ISBN-13: 978-0125084604
Tayel AA, EL-TRAS WF, Moussa S, EL-BAZ AF, Mahrous H, Salem MF, Brimer L (2011) Antibacterial action of zinc oxide nanoparticles against foodborne pathogens. J Food Saf 31:211–218
Taylor AF, Rylott EL, Anderson CW, Bruce NC (2014) Investigating the toxicity, uptake, nanoparticle formation and genetic response of plants to gold. PLoS One 9:93793
Tomei FA, Barton LL, Lemanski CL, Zocco TG, Fink NH, Sillerud LO (1995) Transformation of selenate and selenite to elemental selenium by *Desulfovibrio desulfuricans*. J Ind Microbiol 14:329–336
Torsvik V, Øvreås L (2002) Microbial diversity and function in soil: from genes to ecosystems. Curr Opin Microbiol 5:240–245
Tripathi DK, Singh S, Singh S, Srivastava PK, Singh VP, Singh S, Prasad SM, Singh PK, Dubey NK, Pandey AC, Chauhan DK (2017) Nitric oxide alleviates silver nanoparticles (AgNps)-induced phytotoxicity in *Pisum sativum* seedlings. Plant Physiol Biochem 110:167–177
Trujillo-Reyes J, Majumdar S, Botez CE, Peralta-Videa JR, Gardea-Torresdey JL (2014) Exposure studies of core–shell Fe/Fe3O4 and Cu/CuO NPs to lettuce (*Lactuca sativa*) plants: are they a potential physiological and nutritional hazard. J Hazard Mater 267:255–263
Umer A, Naveed S, Ramzan N, Rafique MS (2012) Selection of a suitable method for the synthesis of copper nanoparticles. Brief Rep Rev 7:1–18
Vannini C, Domingo G, Onelli E, De Mattia F, Bruni I, Marsoni M, Bracale M (2014) Phytotoxic and genotoxic effects of silver nanoparticles exposure on germinating wheat seedlings. J Plant Physiol 171:1142–1148
Vertelov GK, Krutyakov YA, Efremenkova OV, Olenin AY, Lisichkin GV (2008) A versatile synthesis of highly bactericidal Myramistin® stabilized silver nanoparticles. Nanotechnology 19:355707
Vigneshwaran N, Kathe AA, Varadarajan PV, Nachane RP, Balasubramanya RH (2006) Biomimetics of silver nanoparticles by white rot fungus, *Phaenerochaete chrysosporium*. Colloids Surf B 53:55–59
Vijayakumar R, Koltypin Y, Felner I, Gedanken A (2000) Sonochemical synthesis and characterization of pure nanometer-sized Fe3O4 particles. Mater Sci Eng A 286:101–105
Vijayakumar M, Priya K, Nancy FT, Noorlidah A, Ahmed ABA (2013) Biosynthesis, characterisation and anti-bacterial effect of plant-mediated silver nanoparticles using *Artemisia nilagirica*. Ind Crop Prod 41:235–240
Wang YH, Chen PL, Liu MH (2006) Synthesis of well-defined copper nanocubes by a one-pot solution process. Nanotechnology 17:6000–6006
Woolfolk CA, Whiteley HR (1962) Reduction of inorganic compounds with molecular hydrogen by *Micrococcus lactilyticus*. I. Stoichiometry with compounds of arsenic, selenium, tellurium, transition and other elements. J Bacteriol 84:647–658
Wu B, Zhuang WQ, Sahu M, Biswas P, Tang YJ (2011) Cu-doped TiO2 nanoparticles enhance survival of *Shewanella oneidensis* MR-1 under Ultraviolet Light (UV) exposure. Sci Total Environ 409:4635–4639
Xu C, Peng C, Sun L, Zhang S, Huang H, Chen Y, Shi J (2015) Distinctive effects of TiO2 and CuO nanoparticles on soil microbes and their community structures in flooded paddy soil. Soil Biol Biochem 86:24–33
Yadav V, Sharma N, Prakash R, Raina KK, Bharadwaj LM, Prakash NT (2008) Generation of selenium containing nano-structures by soil bacterium, *Pseudomonas aeruginosa*. Biotechnology 7:299–304
Yang L, Watts DJ (2005) Particle surface characteristics may play an important role in phytotoxicity of alumina nanoparticles. Toxicol Lett 158:22–132

Yang J, Cheng GH, Zeng JH, Yu SH, Liu XM, Qian YT (2001) Shape control and characterization of transition metal diselenides MSe2 (M = Ni, co, Fe) prepared by a solvothermal-reduction process. Chem Mater 13:848–853

Yang Y, Mathieu JM, Chattopadhyay S, Miller JT, Wu T, Shibata T, Guo W, Alvarez PJ (2012a) Defense mechanisms of *Pseudomonas aeruginosa* PAO1 against quantum dots and their released heavy metals. ACS Nano 6:6091–6098

Yang Y, Wang J, Zhu H, Colvin VL, Alvarez PJ (2012b) Relative susceptibility and transcriptional response of nitrogen cycling bacteria to quantum dots. Environ Sci Technol 46:3433–3441

Yang Z, Chen J, Dou R, Gao X, Mao C, Wang L (2015) Assessment of the phytotoxicity of metal oxide nanoparticles on two crop plants, maize (*Zea mays* L.) and rice (*Oryza sativa* L.). Int J Environ Res Public Health 12:15100–15109

Yong P, Rowson NA, Farr JPG, Harris IR, Macaskie LE (2002) Bioaccumulation of palladium by *Desulfovibrio desulfuricans*. J Chem Technol Biotechnol 77:593–601

Yoon SJ, Kwak JI, Lee WM, Holden PA, An YJ (2014) Zinc oxide nanoparticles delay soybean development: a standard soil microcosm study. Ecotoxicol Environ Saf 100:131–137

You T, Liu D, Chen J, Yang Z, Dou R, Gao X, Wang L (2018) Effects of metal oxide nanoparticles on soil enzyme activities and bacterial communities in two different soil types. J Soils Sediments 18:211–221

Zafar H, Ali A, Zia M (2017) CuO nanoparticles inhibited root growth from *Brassica nigra* seedlings but induced root from stem and leaf explants. Appl Biochem Biotechnol 181:365–378

Zhu HT, Zhang CY, Yin YS (2004) Rapid synthesis of copper nanoparticles by sodium hypophosphite reduction in ethylene glycol under microwave irradiation. J Cryst Growth 270:722–728

Zuverza-Mena N, Medina-Velo IA, Barrios AC, Tan W, Peralta-Videa JR, Gardea-Torresdey JL (2015) Copper nanoparticles/compounds impact agronomic and physiological parameters in cilantro (*Coriandrum sativum*). Environ Sci Process Impacts 17:1783–1793

Impact of Nanoparticles on PGPR and Soil Nutrient Contents

14

Ashish, Deepti Singh, and Neeraj Gupta

Abstract

Rhizosphere is a site of high microbial diversity that makes soil more physiologically active. High diversity of bacteria in the rhizosphere depends on physico-chemical composition of soil, its pH, partial pressure of oxygen (pO_2) and water potential. The continuous increase of population and availability of low-production land causes high demand of crop production with a significant decrease of synthetic chemical fertilizers and pesticide use which is a huge challenge nowadays. Plant growth-promoting rhizobacteria are free-living, soilborne rhizobacteria that play a significant role in the sustainable agriculture. Soil diversity of microbes shows key role in maintaining soil fertility, functions and crop productivity. The use of metal nanoparticles to increase the soil microbial diversity has been reported recently. Although metal nanoparticle could meaningfully produce ecotoxicity and kill phytostimulatory soil bacteria, so engineered nanoparticles (ENPs) using gold, silver and aluminium should be added to test as a possible ecofriendly agent, and others on the PGPR.

Keywords

PGPR · Sustainable agriculture · Nanotechnology

Ashish
Department of Bioengineering, Integral University, Lucknow, Uttar Pradesh, India

D. Singh · N. Gupta (✉)
Institute of Bioscience and Technology Shri Ramswaroop Memorial University, Barabanki, Uttar Pradesh, India
e-mail: neeraj.bio@srmu.ac.in

M. Ghorbanpour et al. (eds.), *Biogenic Nano-Particles and their Use in Agro-ecosystems*, https://doi.org/10.1007/978-981-15-2985-6_14

14.1 Introduction

As the worldwide population grows exponentially day by day, so simultaneously the demand of food production also increases; this may soon cause insufficiency to feed all of the world's population. To overcome from this problem, developing such agricultural techniques and strategies that can enhance the yield to feed all of these individuals is an important challenge in the twenty-first century (Ladeiro 2012; Glick 2014; Goswami et al. 2016).

Major factors that play an important role for agriculture productivity are water quality, soil composition, climate change, organic matter of soil and soil environment. Soil content is regulated by a number of factors, such as organic carbon content, moisture, nitrogen, phosphorous and potassium content and other biotic and abiotic factors. However, increase in use of chemical fertilizers has led to substantial soil pollution by altering pH and making these nutrients unavailable to crops; this will result in loss of productivity (Gupta et al. 2015). One possible way to overcome this problem is the use of soil microorganisms, such as bacteria, fungi and algae (Vejan et al. 2016). Association between microbes and leguminous plants will result in biomineralization and synergistically improve soil quality and fertility (Herrera Paredes and Lebeis 2016; Agler et al. 2016; Gouda et al. 2018).

Kloepper and Schroth in 1978 introduced the term 'rhizobacteria' that form symbiotic relationships between plant root and colonizing bacteria and stimulate the plant growth, thereby reducing the incidence of plant diseases. These beneficial rhizobacteria (free-living soil bacteria) are termed as plant growth-promoting rhizobacteria (PGPR). PGPR play a vital role in converting poor-quality soil into fertile soil and enhance plant growth and health by suppressing plant pathogens and making different nutrients available to plants. The term PGPR was coined by Kloepper around the 1970s; PGPR can also be termed as plant health-promoting rhizobacteria (PHPR) or nodule-promoting rhizobacteria (NPR) and are attached with the rhizosphere that is an important ecological environment of soil for plant-microbe interactions (Tariq et al. 2017; Gouda et al. 2018).

Various species that have been reported in the group belong to PGPR, and some of them have been commercialized, such as *Acinetobacter*, *Agrobacterium*, *Arthrobacter*, *Azotobacter*, *Azospirillum*, *Burkholderia*, *Bradyrhizobium*, *Rhizobium*, *Frankia*, *Serratia*, *Thiobacillus*, *Pseudomonas*, *Bacillus*, *Enterobacter*, *Klebsiella*, *Variovorax*, *Serratia*, *Alcaligenes*, *Flavobacterium*, *Mesorhizobium*, *Rhodococcus*, *Streptomyces*, etc. (Tariq et al. 2017). Generally, it was observed that about 2–5% of rhizosphere bacteria are PGPR (Goswami et al. 2016; Vejan et al. 2016; Tariq et al. 2017; Hayat et al. 2010).

There are many factors that may affect the microbial flora present in soil and affect the PGPR efficacy by altering the nutrient value of soil, pH, chemical composition, etc. These factors could be soil contaminants such as hydrocarbon, nanoparticles and heavy metals and change in geographical area, etc. Mostly all nanoparticles are widely used, and its use increases day by day due to wide range of consumer and industrial products such as textiles, food packaging, wound dressings, drug delivery, medical devices, paints, water treatment processes, electronics, defence and biomedical, etc. Nanoparticles are generally synthesized from metal oxides such as Al_2O_3, MgO, ZrO_2, CeO_2, TiO_2, ZnO, Fe_2O_3, SnO, etc. Increased uses of metal oxide nanoparticle led to their accumulation and release in the environment, which may cause toxic effects. Most of metal oxide nanoparticles exhibit excellent antimicrobial activity against Gram-positive and Gram-negative bacteria and show undesirable toxic effects on both deleterious and beneficial soil microorganism. Soil acts as a substantial sink for nanoparticles so it is directly affected by toxicity which deteriorates the population of agriculturally important microorganism. Deposition of metal oxide nanoparticle in agricultural field could negatively affect PGPR such as mineral solubilization, secondary metabolite production, phytohormone production, nitrogen fixation, siderophore production, etc. (Karunakaran et al. 2014; Haris and Ahmad 2017).

14.2 PGPR

Plants have always been in a symbiotic relationship with soil microbes (bacteria and fungus) during their growth and development. The symbiotic free-living soil microorganisms inhabiting the rhizosphere of many plant species and having diverse beneficial effects on the host plant through different mechanisms such as nitrogen fixation and nodulation are generally referred to as plant growth-promoting rhizobacteria (PGPR). PGPR and their interactions with plants are exploited commercially and have scientific applications for sustainable agriculture growth. Applications of these associations have been investigated in oat, canola, soy, potato, maize, peas, tomato, lentil, barley, wheat, radicchio and cucumber (Ladeiro 2012; Gouda et al. 2018; Gray and Smith 2005).

14.2.1 Role of PGPR as a Plant Growth Enhancer

Plant growth and health improvement by PGPR are regulated in two different ways, indirectly or directly. PGPR may affect plant growth and development by using any one, or more, of these mechanisms as shown in Fig. 14.1.

14.2.1.1 Direct Mechanism

Direct promotion of plant growth by PGPR directly enhances plant growth either by providing nutrients or by producing growth regulators/metabolites to the plant with a compound that is synthesized by the bacterium, for example, phytohormones. In short, direct mechanism includes facilitating the uptake of certain nutrients from the environment and enhanced availability of nutrients. This is directly responsible to plant growth so termed as ‘direct mechanism of PGPR’ (Glick 2012; Gouda et al. 2018).

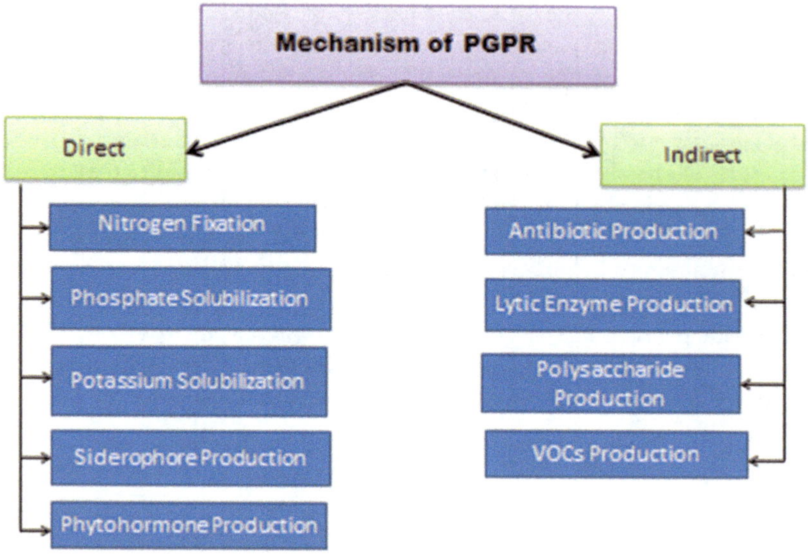

Fig. 14.1 Mechanism of plant growth promotion by PGPR

14.2.1.2 Indirect Mechanism

Indirect mechanisms of growth promotion by PGPR induced systemic resistance, antibiotic protection against pathogens, reduction of iron availability by sequestration with siderophores and synthesis of antifungal enzymes or lytic enzymes (Burdman et al. 2000; Ashraf et al. 2013). In indirect mechanism, plant gets benefited by reducing the impact of diseases, which include antibiosis, induction of systemic resistance and competition for nutrients and niches (Egamberdieva and Lugtenberg 2014). Indirect mechanism helps plant to grow healthily under environmental stresses (abiotic stress) and induction of systemic resistance to plant pathogens (biotic stress) (Goswami et al. 2016).

14.3 Beneficial Aspects of PGPR

Rapid industrialization and increase in population result in the increase of demand for crop production with a significant reduction of synthetic chemical fertilizers and pesticide use which is a big challenge nowadays. The use of PGPR is steadily increased in agriculture and offers an attractive way to replace chemical fertilizers, pesticides and supplements (Dasgupta et al. 2015). The use of PGPR has been proven to be an environmentally sound way of increasing crop yields by facilitating plant growth. PGPR applications and its mode of action were briefly shown in Table 14.1.

Table 14.1 PGPR application and its mode of action to promote plant health and growth

Microbe	Uses	PGPR mode of action	References
Pseudomonas and *Acinetobacter*, *Microbacterium arabinogalactanolyticum*	Phytoremediation of heavy metals	Enhance the phytoavailability of metal by their direct effect on plant growth dynamics or, indirectly, by acidification, chelation, precipitation or immobilization of heavy metals in the rhizosphere	Tak et al. (2013)
Kocuria Turkanensis 2M4, *Arthrobacter*, *Bacillus*, *Beijerinckia*, *Burkholderia*, *Enterobacter*, *Microbacterium*, *Pseudomonas*, *Erwinia*, *Rhizobium*, *Mesorhizobium*, *Flavobacterium*, *Rhodococcus* and *Serratia*	Phosphate solubilization	Solubilizing unavailable complexed phosphate into available inorganic phosphate ion in soil	Jha and Saraf (2015) and Gouda et al. (2018)
Rhizobium sp., *Azoarcus* sp., *Beijerinckia* sp., *Pantoea agglomerans* and *K. pneumoniae*	Nitrogen fixation	Improves soil quality and enhances nodule formation	Gouda et al. (2018), Ahemad and Kibret (2014), and Damam et al. (2016)
Acidithiobacillus sp., *Bacillus edaphicus*, *Ferrooxidans* sp., *Bacillus mucilaginosus*, *Pseudomonas* sp., *Burkholderia* sp. and *Paenibacillus* sp.	Potassium solubilization	Solubilize potassium rock by producing and secreting organic acids, potassium-solubilizing PGPR used as biofertilizer	Pawar et al. (2013), Liu et al. (2012), and Setiawati and Mutmainnah (2016)
Pseudomonas putida, *Enterobacter asburiae*, *Pseudomonas aeruginosa*, *Paenibacillus polymyxa*, *Stenotrophomonas maltophilia*, *Mesorhizobium ciceri*, *Klebsiella oxytoca*, *Azotobacter chroococcum* and *Rhizobium leguminosarum*	Phytohormone (gibberellins, cytokinins, abscisic acid, ethylene, brassino steroids and auxins) production	PGPR provide such external environmental signals to hormone-mediated pathways that enhance the production of phytohormones	Damam et al. (2016), Sureshbabu et al. (2016), and Ahemad and Kibret (2014)

(continued)

Table 14.1 (continued)

Microbe	Uses	PGPR mode of action	References
Pseudomonas sp.	Siderophore production	PGPR can produce siderophores to increase iron uptake capacity of plants	Saha et al. (2016), Beneduzi et al. (2012), and Jha and Saraf (2015)
Rhizobium leguminosarum, *Azotobacter vinelandii*, *Bacillus drentensis*, *Enterobacter cloacae*, *Agrobacterium* sp., *Xanthomonas* sp. and *Rhizobium* sp.	Exopolysaccharide production	Maintaining water potential, aggregating soil particles, ensuring obligate contact between plant roots and rhizobacteria, sustaining the host under conditions of stress or pathogenesis	Jha and Saraf (2015), Ahemad and Kibret (2014), Sanalibaba and Çakmak (2016), and Pawar et al. (2013)
Pseudomonas, *Bacillus*, *Arthrobacter*, *Stenotrophomonas* and *Serratia*	Production of volatile organic compounds (VOCs)	Increased disease resistance, abiotic stress tolerance and plant biomass	Santoro et al. (2016), Sharifi and Ryu (2016), and Kanchiswamy et al. (2015)

14.4 Disadvantages of PGPR

Besides the various advantages of PGPR in agriculture industries, there are certain disadvantages of PGPR when we compare them with chemical compounds used in agro-industries. These disadvantages are as follows (Labuschagne et al. 2010):

- Due to the use of live microorganisms, they are more sensitive to environmental conditions such as temperatures, soil conditions, desiccation, etc. So microorganisms cannot be able to produce desirable metabolites until they are in their optimal conditions; this may limit the use of PGPR in wide area or condition.
- Shelf life of commercial PGPR in general is shorter than that of the chemical pesticides or fungicides.
- The major important disadvantage is the fact that the efficacy of PGPR in general has been inconsistent under field conditions. Many of researches and scientific reports on PGPR were done under controlled environmental conditions or in green house, but it was found that once they were implemented in field condition, the efficacy of PGPR was reduced.

These were considered as major disadvantages of PGPR. However, this does not detract from the large number of beneficial aspects of PGPR.

14.5 Nanotechnology in PGPR

In current scenario, agriculture as a food source is becoming more and more important in the world of diminishing resources and demographic growth (Brennan 2012). To fulfil the sufficient demand of the global population, advanced technologies are required in agriculture and food science, such as nanotechnology. In the area of agriculture, many researchers face a major challenge, to feed the constantly growing global population without degrading the soil health and agroecosystems.

The use of nanotechnology approaches to deal with the problem of increasing demand for food production and obtaining a significant role in agriculture areas like plant protection, monitoring plant development, detection of plants and animals' relevant diseases, enhancement of the quality of foodstuff and minimization of renewable strengthening of spoil are the areas of focus in nanotechnology.

Agriculture production has growing tremendously using chemical fertilizers to fulfill the raising demand of peoples worldwide. As long as the massive usage of fertilizers and pesticides in agricultural lands has great impact on degradation in soil quality and fertility, thus the agricultural land progression with rich soil is nearly impractical; therefore, some scientific personnel focus on improving the safety and fertile techniques of agricultural activities.

Plant growth-promoting rhizobacteria (PGPR) has been proven to be co-evolution among plants and microorganisms revealing conflicting and symbiotic interactions with microbes and the soil. Microbial rejuvenation through plant growth stimulators is being obtained by direct and indirect methods such as bio-fertilization, stimulating root development, phytoremediation, resistance to disease, etc. Despite that there are large varieties of PGPR and their associates have major role, application for ecofriendly agriculture remains as doubtful and limited.

The instability in their functioning of PGPR might be caused by a lot of ecological conditions that can influence their progress and proliferation in the plants. These constraints might be tackled by innovative attempts and techniques like nanotechnology. Its advances belong to the fast-growing industry offering novel approaches and future prospects with enormous possibilities leading to economically and ecologically significant implications. The field of nanotechnology becomes known as a technological innovation which can transform the agricultural field-related areas by introducing the latest resources for the molecular discipline (Chen and Yada 2011; Kah and Hofmann 2014; Khot et al. 2012).

In recent time nanotechnology is focused on the biotic and abiotic stresses, quick disease identification and management and improvement of the capacity of plants that consume nutrients or plant protection agents. Researchers have also been investigating the impact of nanoparticles towards enhancement of plant growth and development that found positive response (Karunakaran et al. 2013a, b, 2014).

14.5.1 Nanoparticles

The relations of PGPR and plants is very auspicious in the field of agriculture and to maintain an environment-friendly sustainable environment. More significantly, introducing the nanoparticles into the environment could have considerable influence as they might be highly resistant to degradation and have the ability to accumulate in bodies of water or in soil, although nanoparticles are capable to operate on living cells at the nano-grade, consequently in biological positive effect.

Nanosensors or nanotechnology is an emerging technology that are currently being used in agricultural industries and upgrading the crop production capability (Handford et al. 2014; Parisi et al. 2015).

Nanotechnology-based devices like nanotubes, nanowires, fullerene derivatives and quantum dots are also being explored and received a huge interest in the area of plant breeding and genetic transformation (Torney et al. 2007; Prasad et al. 2014). Similarly between these accomplishments, dynamic encapsulate components such as fertilizers, herbicides, fungicides, insecticides and micronutrients in sustained-release matrix classified among the highly promising and appropriate choices to resolve the constraints in agriculture field for sustainable development and food abundance in global climate change also reduce the toxicity and environmental contaminants (Cota-Arriola et al. 2013).

This particularly applies via formulations of encapsulated nanoparticle-based fertilizers or pesticides as nanoemulsions or nanocapsules could improve their targeted delivery and productivity.

Overall plant growth is stimulated by commercially based fertilizers, although majority of fertilizers have toxic nature towards humans, animals and ecological system. Similarly, demand of fertilizer also confronts challenges like hydrolysis, disintegration and seepage process. Therefore, nanoparticle-based and nano-encapsulated fertilizers are used instead of conventional fertilizers and found to be efficient in improving plant growth by encouraging certain release of nutrients to the plant (DeRosa et al. 2010; Nair et al. 2010). Nano-fertilizers also have tendency to promote the fertility of soil and eventually come to the aid of elimination of waterlogging contaminants (Giraldo et al. 2014; Galbraith 2007; Torney et al. 2007; Lahiani et al. 2013; Siddiqui and Al-Whaibi 2014).

14.5.1.1 Gold Nanoparticles

The use of microorganisms to synthesize functional nanoparticles has been of great interest recently (Philip 2009). Gold nanoparticles (GNPs) recently considered as many biological applications. GNPs is synthesized chemically and its impact was studies on PGPR. GNPs can be exploited as nano-biofertilizers for selected PGPR microorganisms. The chemical inertness makes gold as a 'crucial substance' that demand is for making nano-scale based tools and devices (Shankar et al. 2004).

14.5.1.2 Silver Nanoparticles

Silver nanoparticles (SNPs) showed stronge bactericidal characteristics, even it present in less concentration. In situ studies have shown that silver, even in larger particle form, prevents microbial growth below concentrations of other heavy metals. Toxicity of nanosilver has been reported in heterotrophic (ammonifying/nitrogen fixing/PGPR) and chemolithotrophic soil formation bacteria (Throbäck et al. 2007).

References

Agler MT, Ruhe J, Kroll S, Morhenn C, Kim ST, Weigel D, Kemen EM (2016) Microbial hub taxa link host and abiotic factors to plant microbiome variation. PLoS Biol 14(1):e1002352

Ahemad M, Kibret M (2014) Mechanisms and applications of plant growth promoting rhizobacteria: current perspective. J King Saud Univ Sci 26(1):1–20

Ashraf MA, Asif M, Zaheer A, Malik A, Ali Q, Rasool M (2013) Plant growth promoting rhizobacteria and sustainable agriculture: a review. Afr J Microbiol Res 7(9):704–709

Beneduzi A, Ambrosini A, Passaglia LM (2012) Plant growth-promoting rhizobacteria (PGPR): their potential as antagonists and biocontrol agents. Genet Mol Biol 35(4):1044–1051

Brennan B (2012) Nanobiotechnology in agriculture. Menlo Park, Strategic Business Insights

Burdman S, Jurkevitch E, Okon Y (2000) Recent advances in the use of plant growth promoting rhizobacteria (PGPR) in agriculture. In: Microbial interactions in agriculture and forestry, vol II. Science Publishers, Enfield, pp 229–250

Chen H, Yada R (2011) Nanotechnologies in agriculture: new tools for sustainable development. Trends Food Sci Technol 22(11):585–594

Cota-Arriola O, Onofre Cortez-Rocha M, Burgos-Hernández A, Marina Ezquerra-Brauer J, Plascencia-Jatomea M (2013) Controlled release matrices and micro/nanoparticles of chitosan with antimicrobial potential: development of new strategies for microbial control in agriculture. J Sci Food Agric 93(7):1525–1536

Damam M, Kaloori K, Gaddam B, Kausar R (2016) Plant growth promoting substances (phytohormones) produced by rhizobacterial strains isolated from the rhizosphere of medicinal plants. Int J Pharm Sci Rev Res 37(1):130–136

Dasgupta D, Ghati A, Sarkar A, Sengupta C, Paul G (2015) Application of plant growth promoting rhizobacteria (PGPR) isolated from the rhizosphere of Sesbania bispinosa on the growth of chickpea (Cicer arietinum L.). Int J Curr Microbiol App Sci 4(5):1033–1042

DeRosa MC, Monreal C, Schnitzer M, Walsh R, Sultan Y (2010) Nanotechnology in fertilizers. Nat Nanotechnol 5(2):91

Egamberdieva D, Lugtenberg B (2014) Use of plant growth-promoting rhizobacteria to alleviate salinity stress in plants. In: Use of microbes for the alleviation of soil stresses, vol 1. Springer, New York, pp 73–96

Galbraith DW (2007) Nanobiotechnology: silica breaks through in plants. Nat Nanotechnol 2(5):272

Giraldo JP, Landry MP, Faltermeier SM, McNicholas TP, Iverson NM, Boghossian AA, Reuel NF, Hilmer AJ, Sen F, Brew JA, Strano MS (2014) Plant nanobionics approach to augment photosynthesis and biochemical sensing. Nat Mater 13(4):400

Glick BR (2012) Plant growth-promoting bacteria: mechanisms and applications. Scientifica 2012:1–15

Glick BR (2014) Bacteria with ACC deaminase can promote plant growth and help to feed the world. Microbiol Res 169(1):30–39

Goswami D, Thakker JN, Dhandhukia PC (2016) Portraying mechanics of plant growth promoting rhizobacteria (PGPR): a review. Cogent Food Agric 2(1):1127500

Gouda S, Kerry RG, Das G, Paramithiotis S, Shin HS, Patra JK (2018) Revitalization of plant growth promoting rhizobacteria for sustainable development in agriculture. Microbiol Res 206:131–140

Gray EJ, Smith DL (2005) Intracellular and extracellular PGPR: commonalities and distinctions in the plant–bacterium signaling processes. Soil Biol Biochem 37(3):395–412
Gupta G, Parihar SS, Ahirwar NK, Snehi SK, Singh V (2015) Plant growth promoting rhizobacteria (PGPR): current and future prospects for development of sustainable agriculture. J Microb Biochem Technol 7(2):096–102
Handford CE, Dean M, Henchion M, Spence M, Elliott CT, Campbell K (2014) Implications of nanotechnology for the agri-food industry: opportunities, benefits and risks. Trends Food Sci Technol 40(2):226–241
Haris Z, Ahmad I (2017) Impact of metal oxide nanoparticles on beneficial soil microorganisms and their secondary metabolites. Int J Life Sci Sci Res 3(3):1020–1030
Hayat R, Ali S, Amara U, Khalid R, Ahmed I (2010) Soil beneficial bacteria and their role in plant growth promotion: a review. Ann Microbiol 60(4):579–598
Herrera Paredes S, Lebeis SL (2016) Giving back to the community: microbial mechanisms of plant–soil interactions. Funct Ecol 30(7):1043–1052
Jha CK, Saraf M (2015) Plant growth promoting rhizobacteria (PGPR): a review. J Agric Res Dev 5(2):108–119
Kah M, Hofmann T (2014) Nanopesticide research: current trends and future priorities. Environ Int 63:224–235
Kanchiswamy CN, Malnoy M, Maffei ME (2015) Chemical diversity of microbial volatiles and their potential for plant growth and productivity. Front Plant Sci 6:151
Karunakaran G, Suriyaprabha R, Manivasakan P, Yuvakkumar R, Rajendran V, Kannan N (2013a) Impact of nano and bulk ZrO_2, TiO_2 particles on soil nutrient contents and PGPR. J Nanosci Nanotechnol 13(1):678–685
Karunakaran G, Suriyaprabha R, Manivasakan P, Yuvakkumar R, Rajendran V, Prabu P, Kannan N (2013b) Effect of nanosilica and silicon sources on plant growth promoting rhizobacteria, soil nutrients and maize seed germination. IET Nanobiotechnol 7(3):70–77
Karunakaran G, Suriyaprabha R, Manivasakan P, Rajendran V, Kannan N (2014) Influence of nano and bulk SiO_2 and Al_2O_3 particles on PGPR and soil nutrient contents. Curr Nanosci 10(4):604–612
Khot LR, Sankaran S, Maja JM, Ehsani R, Schuster EW (2012) Applications of nanomaterials in agricultural production and crop protection: a review. Crop Prot 35:64–70
Labuschagne N, Pretorius T, Idris AH (2010) Plant growth promoting rhizobacteria as biocontrol agents against soil-borne plant diseases. In: Plant growth and health promoting bacteria. Springer, Berlin/Heidelberg, pp 211–230
Ladeiro B (2012) Saline agriculture in the 21st century: using salt contaminated resources to cope food requirements. J Bot 2012:1–7
Lahiani MH, Dervishi E, Chen J, Nima Z, Gaume A, Biris AS, Khodakovskaya MV (2013) Impact of carbon nanotube exposure to seeds of valuable crops. ACS Appl Mater Interfaces 5(16):7965–7973
Liu D, Lian B, Dong H (2012) Isolation of Paenibacillus sp. and assessment of its potential for enhancing mineral weathering. Geomicrobiol J 29(5):413–421
Nair R, Varghese SH, Nair BG, Maekawa T, Yoshida Y, Kumar DS (2010) Nanoparticulate material delivery to plants. Plant Sci 179(3):154–163
Parisi C, Vigani M, Rodríguez-Cerezo E (2015) Agricultural nanotechnologies: what are the current possibilities? Nano Today 10(2):124–127
Pawar ST, Bhosale AA, Gawade TB, Nale TR (2013) Isolation, screening and optimization of exopolysaccharide producing bacterium from saline soil. J Microbiol Biotechnol Res 3(3):24–31
Philip D (2009) Biosynthesis of Au, Ag and Au–Ag nanoparticles using edible mushroom extract. Spectrochim Acta A Mol Biomol Spectrosc 73(2):374–381
Prasad R, Kumar V, Prasad KS (2014) Nanotechnology in sustainable agriculture: present concerns and future aspects. Afr J Biotechnol 13(6):705–713
Saha M, Sarkar S, Sarkar B, Sharma BK, Bhattacharjee S, Tribedi P (2016) Microbial siderophores and their potential applications: a review. Environ Sci Pollut Res 23(5):3984–3999

Sanalibaba P, Çakmak GA (2016) Exopolysaccharides production by lactic acid bacteria. Appl Microbiol Open Access 2:115
Santoro MV, Bogino PC, Nocelli N, Cappellari LD, Giordano WF, Banchio E (2016) Analysis of plant growth-promoting effects of fluorescent Pseudomonas strains isolated from Mentha piperita rhizosphere and effects of their volatile organic compounds on essential oil composition. Front Microbiol 7:1085
Setiawati TC, Mutmainnah L (2016) Solubilization of potassium containing mineral by microorganisms from sugarcane rhizosphere. Agric Agric Sci Proc 9:108–117
Shankar SS, Rai A, Ankamwar B, Singh A, Ahmad A, Sastry M (2004) Biological synthesis of triangular gold nanoprisms. Nat Mater 3(7):482
Sharifi R, Ryu CM (2016) Are bacterial volatile compounds poisonous odors to a fungal pathogen Botrytis cinerea, alarm signals to Arabidopsis seedlings for eliciting induced resistance, or both? Front Microbiol 7:196
Siddiqui MH, Al-Whaibi MH (2014) Role of nano-SiO_2 in germination of tomato (*Lycopersicum esculentum* seeds Mill.). Saudi J Biol Sci 21(1):13–17
Sureshbabu K, Amaresan N, Kumar K (2016) Amazing multiple function properties of plant growth promoting rhizobacteria in the rhizosphere soil. Int J Curr Microbiol App Sci 5(2):661–683
Tak HI, Ahmad F, Babalola OO (2013) Advances in the application of plant growth-promoting rhizobacteria in phytoremediation of heavy metals. In: Reviews of environmental contamination and toxicology, vol 223. Springer, New York, pp 33–52
Tariq M, Noman M, Ahmed T, Hameed A, Manzoor N, Zafar M (2017) Antagonistic features displayed by plant growth promoting rhizobacteria (PGPR): a review. J Plant Sci Phytopathol 1:38–43
Throbäck IN, Johansson M, Rosenquist M, Pell M, Hansson M, Hallin S (2007) Silver (Ag^+) reduces denitrification and induces enrichment of novel nirK genotypes in soil. FEMS Microbiol Lett 270(2):189–194
Torney F, Trewyn BG, Lin VS, Wang K (2007) Mesoporous silica nanoparticles deliver DNA and chemicals into plants. Nat Nanotechnol 2(5):295
Vejan P, Abdullah R, Khadiran T, Ismail S, Nasrulhaq Boyce A (2016) Role of plant growth promoting rhizobacteria in agricultural sustainability – a review. Molecules 21(5):573

New Insights into Application of Nanoparticles for Plant Growth Promotion: Present and Future Prospects

15

Anmol Gupta, Ambreen Bano, Smita Rai, Neelam Pathak, and Swati Sharma

Abstract

Nanotechnology has opened up new avenues in precision and sustainable agriculture by offering more efficient fertilizers and pesticides. The effects of use of these nanomaterials include increased seed germination, length of root-shoot, and biomass of the seedlings along with enhancement of the physiological parameters that enhance nitrogen metabolism and photosynthetic activity in many crop plants. They also provide many other benefits as reducing the amount of chemical used and increasing the absorption of nutrients from the soil, hence reducing the agricultural inputs. Nanotechnology holds the promises controlled release of agrochemicals as well as targeted delivery of several macromolecules. This technology may be used to make nanoscale sensors for monitoring the soil quality as well as nutritional status of agricultural field. Precise and on-demand application of nanopesticides or nanofertilizers can enhance the productivity and prove protection against several pests without harming the environment.

Keywords

Nanofetilizer · Nanoparticles · Carrier · Nanofomulation · Nanobiosensor · Nanopeticides

A. Gupta · A. Bano · S. Rai · S. Sharma
Department of Biosciences, Integral University, Lucknow, Uttar Pradesh, India

N. Pathak (✉)
Department of Biochemistry, Dr. Rammanohar Lohia Avadh University, Ayodhya, Uttar Pradesh, India
e-mail: neelampathak@rmlau.ac.in

M. Ghorbanpour et al. (eds.), *Biogenic Nano-Particles and their Use in Agro-ecosystems*, https://doi.org/10.1007/978-981-15-2985-6_15

15.1 Introduction

In present-day scenario, nanotechnology is getting huge attention of scientists, and the field of nanotechnology is also being explored as a new source for key improvements in various areas as industries, medicine, as well as agriculture. Agriculture occupies second place in the list of uses of nanotechnology. The term "nanotechnology" was first defined in 1974 by Norio Taniguchi of Tokyo Science of University. Nanobiotechnology is the multidisciplinary integration of biotechnology, nanotechnology, chemical processing, material science, and system engineering into biochips, molecular motors, nanocrystals, and nanobiomaterials (Huang et al. 2007). Nanotechnology can be used to create many new materials and devices having vast range of applications in Nano-Bio farming as development of new tools for the treatment of plant diseases, rapid detection of pathogens using nano-based kits, as well as improving the ability of plants to absorb nutrients. Nanomaterials find applications in precision agriculture, smart plant protection, nutrition, and management practices in farms due to small size, high surface to volume ratio, and unique optical properties (Ghormade et al. 2011). There are many types of nanomaterials which can be used in agriculture as carbon based including single-walled and multi-walled carbon nanotubes (SWCNT/MWCNT); metal- and metal oxide-based dendrimers (nano-sized polymers) as aluminum (Al), copper (Cu), gold (Au), silver (Ag), silica (Si), and zinc (Zn) nanoparticles, magnetized iron (Fe) nanoparticles, zinc oxide (ZnO), titanium dioxide (TiO_2), and cerium oxide (Ce_2O_3); or biocomposite nanomaterials (EPA 2007; Nair et al. 2010) or nanoparticles like ceramics, semiconductor, quantum dots, polymers (synthetic or natural), dendrimers, and emulsions (Puoci et al. 2008).

Nanoparticles have high reactivity, enhanced bioavailability and bioactivity, adherence effects, and surface effects; hence, use of nanoparticles can support sustainable agriculture and decrease environmental challenges by reduced use of pesticide and chemical fertilizers and plant disease control by using the nanoparticles in an environmentally friendly way along with improving the efficiency of pesticides with a lower dose (Singh et al. 2014; Bhattacharyya et al. 2016) (Gutiérrez et al. 2011). Apart from direct agricultural benefits, nanotechnology also has various additional benefits such as environmental remediation, wastewater treatment, food processing and packaging, and the development of smart sensors (Khot et al. 2012).

Currently, the key focus areas for nanotechnology agricultural research are:

- Controlled release nanofertilizers and nanocomplexes
- Nanopesticides and nanoherbicides
- Agricultural diagnostics, drug delivery, and nanobiosensors
- Nanogenetic manipulation of agricultural crops

15.2 Nanofertilizers

A nanofertilizer is any product that is made using nanotechnology to improve nutrient efficiency or with nanoparticles.

There are basically three classes of nanofertilizers which are as follows:

1. Nanoscale fertilizer (nanoparticles which contain nutrients)
2. Nanoscale additives (traditional fertilizers with nanoscale additives)
3. Nanoscale coating (traditional fertilizers coated or loaded with nanoparticles)

"Nanofertilizers are modified nanoforms of traditional fertilizers or fertilizer bulk materials to improve the quality, soil fertility, and productivity. Due to their small particle size and high surface area, they are highly reactive and hence facilitate diverse metabolic process in the plant system, resulting in increased rate of synthesis, ultimately increasing the yield. Since nanofertilizers enhance nutrient availability for crop plants, they also increase the quality parameters of the plant by including carbohydrate, protein, and oil content in the plant system. Iron and zinc nanoparticles have been reported to enhance the carbohydrates (starch), protein, chlorophyll, and IAA content in the grain (Rajaie and Ziaeyan 2009). Similarly, nano-Fe_2O_3 has been observed to promote the photosynthesis and growth of the peanut plant (Liu et al. 2005).

15.2.1 Effects of Nanofertilizers on Seed Germination and Growth Parameters of the Plant

Nanoparticles, in general, have both negative and positive effects on the plant growth (Nadi et al. 2013). Nanofertilizers can directly influence the seed germination and alter the seed vigor which increases the root-shoot length. They increase the photosynthesis rate, chlorophyll formation, and dry matter production which result in the overall growth of plant (Salama 2012); Kannan et al. 2012; Mahajan et al. 2013; Suriyaprabha et al. 2012). Application of titanium dioxide (TiO_2) has shown excellent reduction in diseases as bacterial leaf blight disease, the *Curvularia* leaf spot of potatoes, and rice blast as well as reported to promote plant growth, increase the photosynthetic rate, and enhance yield by 30% in rice, wheat, and soybean crops. The application of TiO_2 significantly reduced the incidence of rice blast and tomato spray mold with a correspondent 20% increase in grain weight due to the growth-promoting effect of TiO_2 nanoparticles as reported by Mahmoodzadeh et al. (2000). Interestingly, some nanoparticles may also show negative/inhibitory effects on the seedling growth and germination of the plant, in case the concentration is more than the optimum.

15.2.2 Effects of Foliar Application of Nanofertilizers

Their foliar application of encapsulated nanofertilizers was also found to increase the nutrient uptake/availability and yield of the crops (Tarafdar et al. 2012b). Joseph and Morrison (2006) observed that zeolite-based nanofertilizers can increase availability of nutrient to the crop though out the growth period by preventing nutrient loss from leaching, denitrification, fixation in the soil especially NO_3-N and NH_4-N, as well as volatilization which release the nutrient slowly to the crop plant enhancing the growth. ZnO nanoparticles were also found to be beneficial to plant under stressed environmental conditions (Tarafdar et al. 2012b).

Currently, the plant growth promontory effect of many nanoparticles has been studied; some studies are mentioned in Table 15.1.

15.2.2.1 Advantages of Nanofertilizers

Nanofertilizers increase the quality parameters of the crop by increasing the nutrient efficacy and increase the yield of soil over conventional fertilizers. Nanofertilizers also improve the quality of the taste and the nutritional content of crops, protect the plants from various diseases, and improve stability of the plants by deeper rooting of crops and anti-bending, hence enhancing the plant growth. They are less harmful and nontoxic to humans and environment. By the use of nanofertilizers, one can increase the profit by minimizing the cost of environment protection (Naderi and Abedi 2012). Tarafdar et al. (2012a) proposed that through nanotechnology, balanced fertilization to the crop plant may also be achieved for precision agriculture.

15.2.2.2 Controlled Release of Nanofertilizers and Nanocomplexes

Nanotech fertilizers are being engineered for slow and efficient dosage release (Singh 2012). Nanomaterials used in recommended doses may sometimes fail to exert the desired effects as concentration of these materials is much below the minimum effective concentration required of the chemicals that reaches the target site of crops due to obstacles such as leaching of chemicals, degradation by photolysis/ microbes, and hydrolysis (Agrawal and Rathore 2014).

15.3 Nanopesticides

Plant pests and pathogens cause significant reductions in crop production of up to 20%–40% per year in global crop production (Flood 2010). The current pest management relies severely on application of chemical pesticides including fungicides, herbicides, and insecticides. Pesticides also have harmful effects toward non-target organisms and can cause resistance development in pest population (Stephenson 2003). Furthermore, 90% of applied pesticides are lost during or after application as per the estimate (Stephenson 2003; Ghormade et al. 2011) which results in high-performing cost of pesticides.

Currently, nanotechnology has shown immense potential in managing the problems in agriculture such as nanosensors, nanobarcoding, plant hormone delivery,

Table 15.1 Nanoparticles used as nanofertilizers on various crops

Nanoparticles and their concentration	Plants/organisms/herbs/weeds	References
ZnO and CuO (NPs at 2000–4000 ppm)	Buckwheat (*Fagopyrum esculentum*) seedlings	Lee et al. (2013)
ZnO (NPs at 1000 ppm)	*Schoenoplectus tabernaemontani*	Zhang et al. (2015)
ZnO NPs (NPs at 1 ~ 20 ppm)	Mung bean (*Vigna radiata*) and chickpea (*Cicer arietinum*)	Mahajan et al. (2011)
ZnO NPs, MnOx NPs, and FeOx	Stimulate lettuce seedlings by 12–54%	Liu et al. (2016)
Silver nanoparticles-chitosan encapsulated paraquate	*Eichhornia crassipes*	Namasivayam et al. (2014)
Ag, Cu, Fe, Zn, Mn	*Allium cepa* (L.)	Konotop et al. (2014)
TiO2	*Anabaena variabilis*, *Triticum aestivum* L. var. *Pishtaz*	Cherchi and Gu (2010) and Feizi et al. (2012)
Al	*Lolium perenne*	Lin and Xing (2007)
Al2O3	*Zea mays*, *Cucumis sativus*, *Brassica oleracea*, *Daucus carota*	Yang and Watts (2005)
CeO_2	*Lycopersicon esculentum* and *Zea mays*	López-Moreno et al. (2010)
SiO2	*Lycopersicon esculentum Mill*	Siddiqui and Al-Whaibi (2014)
Graphene oxide (NPs at 400 and 800 mg/L)	*Vicia faba* L.	Anjum et al. (2014)
CNTs (NPs at 40 μg/mL)	*Lycopersicon esculentum*	Morla et al. (2011)
SWCNTs (NPs at 9, 56, 315, and 1750 mg/L)	*Allium cepa*, *Cucumis sativus*	Cañas et al. (2008)
MWCNTs (NPs at 25–100 μg/mL)	*Hordeum vulgare* L., *Glycine max*, *Zea mays*	Lahiani et al. (2013)
ZnO NPs (NPs at 400 mg/kg)	*Cucumis sativus* fruit	Zhao et al. (2014)
GNPs (NPs at 10 and 80 μg/ml)	*Arabidopsis thaliana*	Kumar et al. (2013)
Ag NPs (NPs at 10–30 μg/mL)	*Boswellia ovalifoliolata*	Savithramma et al. (2012)
Sulfur NPs (NPs at 500, 1000, 2000, and 4000 ppm)	*Vigna radiata*	Patra et al. (2013)
SiO2 NPs (NPs at 15 kg/ha)	*Zea mays* L.	Yuvakkumar et al. (2011) and Suriyaprabha et al. (2012)
TiO2 NPs (NPs at 400 mg/L)	*Arabidopsis thaliana*	Lee et al. (2010)
Aluminum oxide NPs (NPs at 400–4000 mg/L)	*Arabidopsis thaliana*	Lee et al. (2010)
CeO_2 (NPs at 250 ppm)	*Arabidopsis thaliana*	Ma et al. (2013)

transfer of target genes, seed germination, water management, and controlled release of agrochemicals or in developing agricultural products (Hayles et al. 2017).

Material scientists have engineered nanoparticles with desired characteristics like pore size, shape, and surface properties so that they can then be used as protectants or for precise and targeted delivery via conjugation, encapsulation, and/or adsorption of an active ingredient, such as a pesticide (Khandelwal et al. 2016). Nanoparticles can protect plants (a) directly as nanoparticles themselves, can provide protection to the crop or (b) nanoparticles as carriers of existing pesticides, and can be applied by spray application or drenching/soaking onto seeds, foliar tissue, or roots. Agriculture nanotechnology can be exploited to create effective formulations for crop plants after conducting short- and long-term field trials toward target pests.

15.4 Nanoparticles That Act as Carriers

Nanoparticles are also commonly used as carriers to encapsulate, absorb, entrap, or attach active molecules to develop effective agricultural formulations, which have been used as carriers for herbicides, insecticides, fungicides, and RNAi-inducing molecules, which are summarized in Table 15.2.

15.4.1 Nanoparticles as Carriers for Insecticides

The loading of insecticides into nanoparticles was first started in the early 2000s. Since then, studies of variety of nanoparticles with conventional insecticides as well as bioactive compounds with insecticidal properties have been shown which reduce

Table 15.2 Antimicrobial action of nanoparticles toward plant pathogens

Nanoparticles	Antimicrobial action toward plant pathogens	References
Ag, Au, and TiO2	*Bacillus cereus*, *Escherichia coli*, *Bacillus subtilis*, *Streptococcus thermophilus*, and *Fusarium oxysporum*	Sunkar and Nachiyar (2012), El-Shanshoury et al. (2011), and Ahmad et al. (2003)
Cu	*Candida* spp., *Aspergillus* spp., *Fusarium* spp., *Klebsiella pneumoniae*, *Pseudomonas aeruginosa*, etc.	Ramyadevi et al. (2012)
Cu-chitosan	*Alternaria alternata*, *Macrophomina phaseolina*, and *Rhizoctonia solani*	Saharan et al. (2013)
Zn and S	*Aspergillus niger* and *Fusarium oxysporum*	Patra et al. (2012) and Choudhury et al. (2011)
MOx NPs (e.g., CuO, ZnO, and Mn2O3 NPs)	*Escherichia coli*	Kaweeteerawat et al. (2015)

the toxicity and increase their solubility. Till now many low-water-soluble insecticides have been effectively loaded into porous silica (Wang et al. 2014) and modified chitosan (Feng and Peng 2012; Zhang et al. 2013; Lu et al. 2013). However, further study needs to be done to examine as well as reduce the environmental toxicity of these insecticides. Essential oils rapidly evaporate due to their chemical instability in the presence of air, high temperatures, moisture, and light but also known for inducing insecticidal effects (Lai et al. 2006). A garlic essential oil was encapsulated into polyethylene glycol (PEG) and applied to harvested rice, then overspread with red flour beetles (*Tribolium castaneum*) (Yang et al. 2009). The toxicity of insecticides is potentially decreased due to slow release of active molecules.

15.4.2 Nanoparticles as Carriers for Fungicides

Since the beginning in 1997, early studies on nanofungicides were conducted on incorporating fungicides into solid wood (Liu et al. 2001; Liu et al. 2002). Since then, studies on wide range of nanoparticles with conventional biocides or antifungal properties and fungicides have been conducted (Table 15.2). The most frequently investigated nanoparticle carriers are polymer mixes, silica, and chitosan. An extensive variety of fungi were studied using nanocarrier containing insecticides and nanofungicide. The efficiency of nanoparticles can be exploited to improve stability while providing a slow sustained release, improve low-water-solubility issues, and decrease volatilization.

15.4.3 Nanoparticles as Carriers for Herbicides

Most of the studies are primarily focused on reducing the environmental impact caused by herbicide nanocarrier (Table 15.3) and their non-target toxicity toward herbicides. Various wide varieties of herbicide-based nanoparticles have been developed for example, nano-sized rice husks (Chidambaram 2016) and amino-activated iron (II, III) oxide magnetic nanoparticles (Viirlaid et al. 2009).

Nanoparticles, as carriers, can provide numerous benefits, like (i) enhancing shelf-life, (ii) ameliorating solubility of poorly water-soluble pesticides, (iii) reducing toxicity, and (iv) boosting site-specific uptake into the target pest (Hayles et al. 2017).

15.5 Pesticide Nanoformulations

Many fungicide nanoformulations are being developed by using nanoscale ingredients or by nanoencapsulating them as a fungicide formulation containing nanoparticles is being developed by Syngenta, for example, Banner MAXX Fungicide having active ingredient propiconazole and Apron MAXX containing active ingredient fludioxonil used for seed treatments (Gogoi et al. 2009), and Primo MAXX

Table 15.3 Summary of use of nanoparticles as nanopesticides toward various plant pathogens

Target pests	Nanopesticides (IRAC MoA classification)	Type of pesticide	Nanoparticle/nanocarriers	Crops	References
Red flour beetle (*Tribolium castaneum*)	Garlic essential oil	Insecticides	PEG	Rice (harvested)	Yang et al. (2009)
Groundnut bruchid (*Caryedon serratus*)	Azadirachtin (UN)	Insecticides	Zinc oxide and chitosan	–	Jenne et al. (2018)
–	Abamectin	Insecticides	LDH (layered double hydroxide)	Mustard leaves (*B. juncea*)	Nguyen et al. (2014)
House fly (M*usca domestica*)	Nicotine	Insecticide	Chitosan/TPP	–	Yang et al. (2018)
Gloeophyllum trabeum	Tebuconazole	Fungicide	PVP and PVP copolymer	Southern pine sapwood	Liu et al. (2001)
Fusarium oxysporum and *Aspergillus parasiticus*	Carbendazim	Fungicide	Chitosan/pectin	Cucumber, maize	Kumar et al. (2017)
Fusarium graminearum	*Cymbopogon martinii* essential oil	Fungicide	Chitosan	Maize grains	Kalagatur et al. (2018)
Maize-mustard (*Brassica* sp.)	Paraquat	Herbicide	Chitosan/tripolyphosphate	–	Grillo et al. (2014)
–	Atrazine	Herbicide	Polycaprolactone coated in chitosan	Maize	Grillo et al. (2014)
Black-jack (*Bidens pilosa*)	Imazapic and imazapyr combined	Herbicide	Alginate/Chitosan/tripolyphosphate	*Allium cepa*	Maruyama et al. (2016)
Echinochloa crus-galli	Diuron	Herbicide	Chitosan	Maize	Yu et al. (2015)
–	Glyphosate	Herbicide	Amino-activated aminopropyl silane coated magnetic iron oxide	–	Viirlaid et al. (2009)

Table 15.4 Nanoformulations and their significance on various crop diseases

Nanoformulations	Active ingredients	Significance	References
Banner MAXX Fungicide	Propiconazole	Role in seed treatments	Gogoi et al. (2009)
Apron MAXX	Fludioxonil	Role in seed treatments	Gogoi et al. (2009)
Primo MAXX	Cyclopropyl derivative of cyclohexenone	Helps the plant in withstanding abiotic as well as biotic stresses including plant pathogens	Banik and Sharma (2011)
Nano-5	–	Controls several plant pathogens and pests	Banik and Sharma (2011)
Nano-Gro	Coded sugar pellets less than $\frac{1}{8}''$ in diameter	Increases the average plant yield and protects from diseases	Banik and Sharma (2011)
Nano Green	Bio-based chemicals	Eliminates blast disease (*Magnaporthe grisea*) from infected rice plant	Banik and Sharma (2011)

containing cyclopropyl derivative of cyclohexenone has been established as plant growth regulator and ameliorating the biotic and abiotic stresses including plant pathogens. "Nano-5" is a natural mucilage organic solution which is used to control many plant pathogens and pests in addition to improving crop yield. "Nano-Gro" (nanotechnology product) has been launched (Agro Nanotechnology Corp., Florida, http://www.agronano.com). Plants treated with "Nano-Gro" showed an increase of 50% for grain yield of sunflower and 10% increase in protein and sugar content enables plants to be protected from various diseases. "Nano-Gro" is certified to be an organic one and harmless to soil and plants. Many bio-based chemicals are mixed to prepare "Nano Green" used to eliminate blast disease (*Magnaporthe grisea*) from infected rice plant. Few important nanoformulations which were found to be effective in various plant diseases are listed in Table 15.4.

15.6 Nanobiosensors and Agriculture

Nanobiosensors are the sensors with an immobilized bioreceptor probes which are selective for target molecules. Their applications in agriculture include detection of analytes like glucose, urea pesticides, herbicides, fungicides, insecticides, and soil pH and moisture. Biosensors are used for monitoring metabolites as well as detection of various pathogens/microorganisms (Rai et al. 2012). Aptamer-based nanosensors proved to be useful in studying the origin as well as reactions of cell metabolites in crop rhizosphere because they have the ability to detect individual chemical species in a specific location. Moreover, nanosensors designed to detect the level of desired as well as undesired bacteria in soil by comparing the amount of oxygen utilized by them during respiration can be used for soil diagnosis (Kaushal and Wani 2017).

15.6.1 Pesticide Nanobiosensors

- The nanobiosensors are developed using photosystem II, known to bind several groups of herbicides, and are isolated from photosynthetic organisms which may have potential of monitoring polluting chemicals, leading to setup of an easy-to-use apparatus, low cost, able to reveal specific herbicides, a wide range of organic compounds present in urban and industrial effluents, sewage sludge, ground water, landfill leak water, and irrigation water (Rai et al. 2012).
- Nanosensors that are linked with zeolite can release minerals, or water retained in it based on any deficiency in soil can be detected by the sensor. Similarly, pesticides that are linked to nanoparticles are being produced which released in a controlled manner based on an environmental trigger; protein and DNA detecting biosensors can prove to be useful for detection of biomarkers and in differentiation of one plant species from another.
- Nanotechnology is applied to two fields of pesticides in agri-food: as a pesticide trace-amount detector and as pesticide delivery vector to attain pesticide management. The nanoparticles can modify/trap pesticides to target the insect pest slowly that will help to arrest pollution of both top soil and ground water, reduce pesticide amounts, and elevate its efficiency (Sinha et al. 2017).
- In addition, whole cell-based biosensors applied in herbicide and pesticide detection include bacterial, fungal, and algal cells; they also assist in the development of rapid, accurate, and cost-effective techniques in decontamination procedures and prevent damage casualties (Gheorghe et al. 2017; Husu et al. 2013).

15.6.2 Nanobiosensors for Heavy Metal Detection

Heavy metals, including Ag^{+}, Cd^{2+}, As^{3+}, Pb^{2+}, Zn^{2+}, and Hg^{2+}, have been classified as chemical contaminants (Verma and Kaur 2016). Heavy metal detecting biosensors are based on a green fluorescent signal amplifier and genetically modified bacterial cells to detect arsenite in the foods (Pola-López et al. 2018). The biosensors with aptamer and DNA-based properties can detect these heavy metals on nanoscale levels and on very large scale that are suitable for screening and monitoring of food safety.

15.6.3 Nanofertilizer Nanobiosensors

Nanosensors can also be used in monitoring plant growth by assessing the cross talk between rhizosphere and roots, leading to the development of a dynamic, precise, and intelligent nanofertilizer delivery platform. The micronutrients can also be delivered according to the spatial requirement in plants (Kaushal and Wani 2017). Nanosensors are portable, small, precise, extra sensitive, and reliable, and also they can be used for real-time monitoring. These properties give them an edge over present sensors.

15.7 Root Exudates (Metabolites) Nanosensors and Pathogen Detection

Various metabolites are readily absorbed by the plants from the rhizosphere which secrete various exudates into the rhizosphere (Walker et al. 2003). As any change in these exudates is an indicative of change in plant health, by using the biosensor, their content can be easily monitored. Carbohydrate exudate includes pentoses, hexoses, disaccharides, and trisaccharides (Jalali and Suryanarayana 1971). Marschner (1996) estimated that around 5–21% of photosynthetically fixed carbon can be released from the roots. Using the reporter-based nanosensors, the concentration of sugars and amino acids can be measured, and the highest concentration of sucrose was found near the tip of *Avena* roots (Farrar et al. 2003; Jaeger III et al. 1999). However, during the pathogenic attack, the exudation appeared to be altered (Walker et al. 2003). Plants used sucrose as a principal sugar for long-distance transport (Lalonde et al. 2004).

15.8 Nanosensors for Detection of Changes in Rhizosphere Microenvironment

The nanosensors can be used to study the parameters of root zone and signal induction when rhizosphere changes. The parameters include soil oxygen, pH, temperature, and moisture.

15.8.1 Soil Oxygen Nanobiosensors

Molecular oxygen is the most essential metabolite for all plants. The external oxygen supply drops down mainly during heavy rainfall when water obstructs up the soil, resulting damage in the root system, which can cause adverse effects on the growth and development of the plant (Bailey-Serres et al. 2012). Moreover, oxygen sensors can be helpful in understanding the consumption of oxygen and their distribution system in plants, as well as maintaining the optimum supply.

An indicator dye capable of sensing oxygen is encapsulated with an oxygen-permeable matrix, in order to make an oxygen detecting nanosensor. Mostly the indicator dyes such as from metalloporphyrins group (e.g., palladium (II) and platinum (II) porphyrins) and transition metal complexes with pyridine derivatives (e.g., ruthenium (II) and iridium (III)) are being used because they are characterized by strong luminescence and longer life span. However, the phosphorescent platinum (II) and palladium (II) porphyrins are widely used as they show high signal intensity and low cytotoxicity. Polymethylmethacrylates, polystyrene, glass-like materials, and fluoropolymers made by sol-gel process are the commonly used matrices (Klimant et al. 1999; Amao 2003; García et al. 2005). The dye-based nanosensors are the best technique which is used to sense the intracellular oxygen which utilizes the endogenous expression of fluorescent proteins, sensitive toward oxygen, and immobilized in a solid membrane, optical fiber, or microplate (solid state sensor). Additionally, it can be linked to polymers which enable it to diffuse through the active soluble sensors. However, the positive features of both of these sensors were combined by the creation of nanoparticle sensors, in which the probes are encapsulated in nanoparticles. This conveys strength and stability to the solid state sensor and prevents it from leaching out in the cell, which is also soluble and minimally invasive and can be injected into gene gun, living system using pico-injection, or liposomal transfer (Clark et al. 1999). An example of this includes the encapsulation of PtPFPP (platinum complex of tetra pentafluorophenyl porphyrins) in microbeads of polystyrene which, even though need to be microinjected into the plant cell, do not affect cell viability significantly (Schmälzlin et al. 2005).

15.8.2 Soil Temperature/Moisture/pH Nanobiosensors

Temperature and moisture are crucial parameters for rhizosphere, as they monitor the exchange of heat energy and water between soil and atmosphere. Nanotechnology-based micro-electromechanical system sensors have been developed that sense temperature as well as moisture of the soil. As stress sensitivity and shear stress of the nano-resistor are used to study the changes on the surface of microcantilever, these changes are affected by changes in the moisture. An on-chip temperature sensor is used to detect changes in temperature of the water vapor (Jackson et al. 2008). A recent research has led to the development of the luminescent nanosensors that are capable of detecting oxygen, temperature, and pH simultaneously at single emission wavelength in the PBS buffer and the cell culture with bovine serum albumin (Wang et al. 2019).

15.8.3 Sensors for Toxin Detection

For development in food safety, electrochemical biosensors are used for assessment and fast detection of food toxins (Patra et al. 2017). Other biosensors like piezoelectric and optic sensing have been applied to chemical detection and toxin in food production (Bahadır and Sezgintürk 2017). Fluorescent nanoparticles are developed to sense toxins in crops and foods including on surface and inside of foods (Burris and Stewart 2012). The key challenges in developing these automatic toxin detectors are the toxin extraction from complex food samples (Moran et al. 2016). Upcoming systems are expected to process, extract, and measure different toxins to determine their harmful effects and levels in food and water samples (Moran et al. 2016; Beck and Villarroel Walker 2013). Recently, advanced separation technique is coupled with surface-enhanced Raman scattering to identify, quantify, and discriminate chemical toxins in food matrices (Liao and Lu 2016).

15.9 Other Applications of Nanotechnology in the Field of Agriculture

Nanotechnology in agriculture has the potential to transform the different sectors of food industry and agricultural with modern techniques for the treatment of rapid disease detection, diseases, amplifying the ability of plants to absorb minerals and nutrients. Nanotechnology can also indirectly protect the environment through the use of alternative or renewable energy supplies and catalysts or filters to clean up and reduce existing pollutants (Tungittiplakorn et al. 2005). Bhattacharyya et al. (2011) reported applications of nanotechnology in different fields such as nano-food, nano-farming, and nano-food packaging as well as emphasized on the nanoparticles and their consequences on ecological balance.

15.10 Challenges and Opportunities

15.10.1 Regulatory Demands in Nanotechnology

Many of nanoparticles have a short life span as they frequently agglomerate or they dissolve in water. Human body has developed several mechanisms for filtering and removing some of these particles (Lin 2007). The engineered nanoparticles can be better to elude the body's defenses because of their size and protective coatings. However, the health and environmental risks raised due to the exposure to engineered nanoparticles need further study. It is essential to study the effect of nanotechnology on human health as well as on the environment, and in recent past, greater emphasis has been given toward nanotoxicology. Furthermore, the potential environmental impacts on exposure of nanomaterials are less understood than human health effects. Therefore, extensive research is required to understand the mechanism of nanomaterials' toxicity and their impacts on the natural environment (Fig. 15.1).

15.10.2 Potential Consumer Safety Issues

Nanoparticles when used in agriculture have a potential to enter food chain. The core route of entry of nano-sized or microparticles into the gut is through the food and drink consumption. The nanotechnology applications are also essentially linked to the physicochemical nature of the nanoparticles for their consumer safety implications in food and the possibility through consumption of nano-foods. Concerns over ingestion of nano-sized ingredients are arising as the free engineered nanoparticles can cross cellular barriers and their exposure from a growing body of scientific evidence indicates to oxidative damage to the cell and enhanced oxy-radical production (Li et al. 2003; Donaldson et al. 2004; Geiser et al. 2005). It is known that the potential effects of nanoparticles may exhibit substantially different physicochemical and biological properties having much larger surface areas, and their

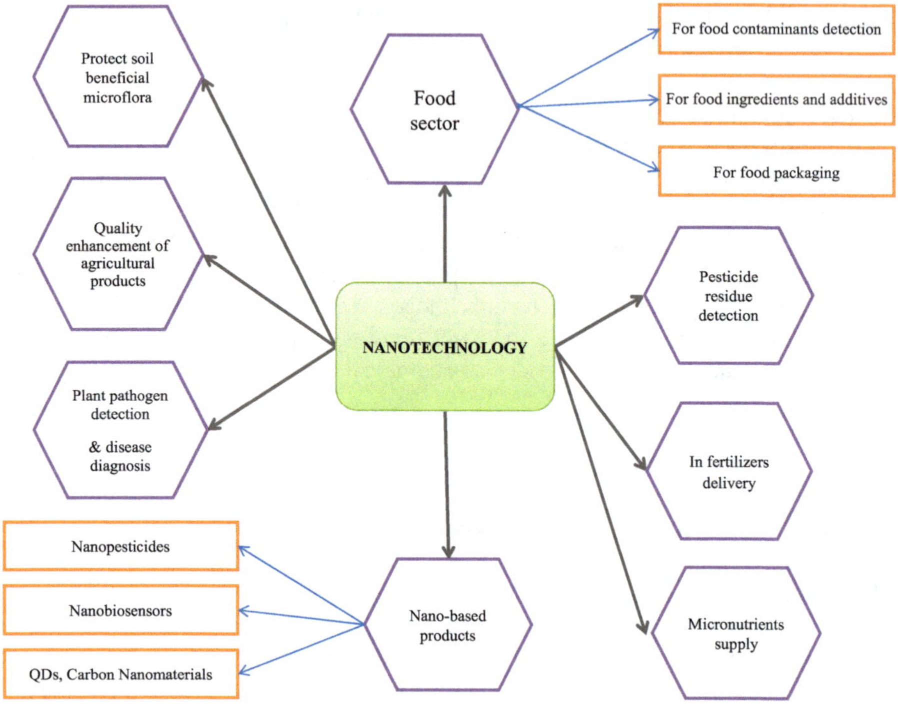

Fig. 15.1 Applications of nanotechnology in various fields

effects through the gastrointestinal (GI) route are largely unknown. It is also possible that the nanoparticles may not persist in a free form in the gut, due to various transformations such as aggregation, adsorption, agglomeration, or reaction with digestive enzymes and acid or binding with other food components and, hence, may not be translocated in the human body. There are various applications of nanomaterials that lead to toxicity in ecosystems and agriculture such as nanomaterial phytotoxicity and potential residue carry-over in foodstuff (Chaudhari and Castle 2011). For example, Kahru and Dubourguier (2010) discussed the health hazards of various nanomaterials, while Bouwmeester et al. (2009) have reviewed the health concerns of nanomaterials related to plant production in detail. Furthermore, there is a necessity to evaluate the "toxicodynamics and toxicokinetics" of nanomaterials (Bouwmeester et al. 2009; Bergeson 2010) used for agricultural production. Nair et al. (2010) highlighted the effect of these nanomaterials which will considerably affect another cropping system on various agricultural crops/plants if not degraded quickly. The summary of FAO/WHO meeting report, 2010, represents the major issues related to the crop protection and nanomaterial application in agricultural production as (a) the precise nanomaterial characterization for in-depth understanding of biological system toxicity, (b) nanomaterial interaction with biology, (c) dose-response considerations, (d) exposure evaluation and characterization studies, (e) product life cycle considerations, (f) background levels of nanomaterials in food and feed matrices, and (g) nanomaterial form and amount in foodstuff due to crop protection and their use in agricultural production. The detailed studies on nanomaterial characteristics such as dosage of nanoparticles in different environments, their physical and chemical characterization, the mechanisms allowing them to pass through cell walls and cellular membranes, toxic effects of nanoparticles with their specific properties, and the mechanism underlying nanoparticle trophic transfers needed to be conducted were reported by Navarro et al. (2008). Many researchers such as Reijnders (2006) and Suh et al. (2009) highlighted the hazard of nanomaterials toward human health. There are various uptake routes of nanomaterials in human body such as inhalation, ingestion, and dermal exposures (Xu et al. 2010). Xu et al. (2010) also highlighted the various factors including chemical composition, surface structure, the size, solubility, and accumulation of the nanomaterials for their associated exposure risk effects to humans. Studies which assess the overall risk are still required in detail.

15.11 Conclusion

Nanotechnology may also be helpful for revolutionizing the agriculture advancement and to find the solutions against many agriculture-related problems and improved the crop varieties. Nanoparticles can be easily synthesized from various biological sources and can be applied in agriculture. They can help in early detection of soil stresses and in alleviating stress effects, disease/toxin detection, disease protection in plants, and slow and on-demand release of nutrients and hence can increase crop production. Delivery of nanoparticles using nanotechnology positively results in enhancing the plant growth due to site-specific delivery of essential nutrients or pesticides. The use of nano-encapsulation improves the effectiveness of pesticides allowing their slow and sustained release. Nanofertilizers can also play an important role in enhancing crop production reducing the fertilizers cost and minimizing the pollution hazard. In the future, nanoscale devices could be used to make agricultural systems smart and to detect the pesticide residues in the field. Smart sensors and smart delivery systems will help the agricultural industry in combating viruses, spores, and other crop pathogens. Hence, the use of nanotechnology can provide green, eco-friendly, and efficient strategy for sustainable agriculture without harming the nature. Though detailed studies are still in pipeline, in near future, the use of nanotechnology could authorize rapid advances in agricultural research.

Acknowledgment The authors are grateful to the Department of Science and Technology (DST FIST, DST INSPIRE-IF160803) and Uttar Pradesh Council of Science and Technology for providing financial support.

References

Agrawal S, Rathore P (2014) Nanotechnology pros and cons to agriculture: a review. Int J Curr Microbiol App Sci 3(3):43–55

Ahmad A, Mukherjee P, Senapati S et al (2003) Extracellular biosynthesis of silver nanoparticles using the fungus Fusarium oxysporum. Colloids Surf B: Biointerfaces 28(4):313–318

Amao Y (2003) Probes and polymers for optical sensing of oxygen. Microchim Acta 143:12

Anjum NA, Singh N, Singh MK et al (2014) Single-bilayer graphene oxide sheet impacts and underlying potential mechanism assessment in germinating faba bean (Vicia faba L.). Sci Total Environ 472:834–841

Bahadır EB, Sezgintürk MK (2017) Biosensor technologies for analyses of food contaminants. In: Nanobiosensors. Elsevier, Amsterdam, pp 289–337. ISBN 978-0-12-804301-1

Bailey-Serres J, Fukao T, Gibbs DJ et al (2012) Making sense of low oxygen sensing. Trends Plant Sci 17:129–138

Banik S, Sharma P (2011) Plant pathology in the era of nanotechnology. Indian Phytopathol 64(2):120–127

Beck MB, Villarroel Walker R (2013) On water security, sustainability, and the water-food-energy-climate nexus. Front Environ Sci Eng 7:626–639

Bergeson LL (2010) Nanosilver: US EPA's pesticide office considers how best to proceed. Environ Qual Manag 19(3):79e85

Bhattacharyya A, Datta PS, Chaudhuri P et al (2011) Nanotechnology: a new frontier for food security in socio economic development. In: Proceeding of disaster, risk and vulnerability conference 2011 held at School of Environmental Sciences, Mahatma Gandhi University, India in association with the Applied Geoinformatics for Society and Environment, Germany, pp 12–14

Bhattacharyya A, Duraisamy P, Govindarajan M et al (2016) Nano-biofungicides: emerging trend in insect pest control. In: Prasad R (ed) Advances and applications through fungal nanobiotechnology. Springer, Cham, pp 307–319

Bouwmeester H, Dekkers S, Noordam MY (2009) Review of health safety aspects of nanotechnologies in food production. Regul Toxicol Pharmacol 53:52e62

Burris KP, Stewart CN (2012) Fluorescent nanoparticles: sensing pathogens and toxins in foods and crops. Trends Food Sci Technol 28:143–152

Cañas JE, Long M, Nations S et al (2008) Effects of functionalized and nonfunctionalized single-walled carbon nanotubes on root elongation of select crop species. Environ Toxicol Chem Int J 27(9):1922–1931

Chaudhari Q, Castle L (2011) Food applications of nanotechnologies: an overview of opportunities and challenges for developing countries. Trends Food Sci Technol 22:595e603

Cherchi C, Gu AZ (2010) Impact of titanium dioxide nanomaterials on nitrogen fixation rate and intracellular nitrogen storage in Anabaena variabilis. Environ Sci Technol 44(21):8302–8307

Chidambaram R (2016) Application of rice husk nanosorbents containing 2, 4-dichlorophenoxyacetic acid herbicide to control weeds and reduce leaching from soil. J Taiwan Inst Chem Eng 63:318–326

Choudhury SR, Ghosh M, Mandal A et al (2011) Surface-modified sulfur nanoparticles: an effective antifungal agent against *Aspergillus niger* and *Fusarium oxysporum*. Appl Microbiol Biotechnol 90(2):733–743

Clark HA, Hoyer M, Philbert MA (1999) Optical nanosensors for chemical analysis inside single living cells. 1. Fabrication, characterization, and methods for intracellular delivery of PEBBLE sensors. Anal Chem 71:4831–4836

Donaldson K, Stone V, Tran CL (2004) Nanotoxicology. Occup Environ Med 619:727–728

El-Shanshoury AERR, ElSilk SE, Ebeid ME (2011) Extracellular biosynthesis of silver nanoparticles using Escherichia coli ATCC 8739, Bacillus subtilis ATCC 6633, and Streptococcus thermophilus ESh1 and their antimicrobial activities. ISRN Nanotechnol 2011:1–7

EPA (2007) Nanotechnology white paper. U.S. Environmental Protection Agency publication, Washington, DC. Available at: http://www.epa.gov/nanoscience/files/epa_nano_wp_2007.pdf. Accessed 22 Jan 11

Farrar J, Hawes M, Jones D et al (2003) How roots control the flux of carbon to the rhizosphere. Ecology 84:827–837

Feizi H, Moghaddam PR, Shahtahmassebi N (2012) Impact of bulk and nanosized titanium dioxide (TiO 2) on wheat seed germination and seedling growth. Biol Trace Elem Res 146(1):101–106

Feng BH, Peng LF (2012) Synthesis and characterization of carboxymethyl chitosan carrying ricinoleic functions as an emulsifier for azadirachtin. Carbohydr Polym 88:576–582

Flood J (2010) The importance of plant health to food security. Food Sec 2(3):215–231

García EA, Fernández RG, Días-García ME (2005) Tris(bipyridine)ruthenium(II) doped sol-gel materials for oxygen recognition in organic solvents. Microporous Mesoporous Mater 77:235–239

Geiser M, Rothen-Rutishauser B, Kapp N et al (2005) Ultrafine particles cross cellular membranes by nonphagocytic mechanisms in lungs and in cultured cells. Environ Health Perspect 113:1555–1560

Gheorghe I, Czobor I, Lazar V et al (2017) Present and perspectives in pesticides biosensors development and contribution of nanotechnology. In: New pesticides and soil sensors. Elsevier, Amsterdam, pp 337–372. ISBN 978-0-12-804299-1

Ghormade V, Deshpande MV, Paknikar KM (2011) Perspectives for nano-biotechnology enabled protection and nutrition of plants. Biotechnol Adv 29(6):792–803

Gogoi R, Dureja P, Singh PK (2009) Nanoformulations- a safer and effective option for agrochemicals. Indian Farm 59(8):7–12

Grillo R, Pereira AE, Nishisaka CS et al (2014) Chitosan/tripolyphosphate nanoparticles loaded with paraquat herbicide: an environmentally safer alternative for weed control. J Hazard Mater 278:163–171

Gutiérrez FJ, Mussons ML, Gatón P et al (2011) Nanotechnology and food industry. Scientific, health and social aspects of the food industry. Tech, Croatia Book Chapter

Hayles J, Johnson L, Worthley C et al (2017) Nanopesticides: a review of current research and perspectives. In: New pesticides and soil sensors. Elsevier, San Diego, pp 193–225

Huang J, Li Q, Sun D et al (2007) Biosynthesis of silver and gold nanoparticles by novel sundried Cinnamomum camphora leaf. Nanotechnology 18:105104

Husu I, Rodio G, Touloupakis E et al (2013) Insights into photo-electrochemical sensing of herbicides driven by Chlamydomonas reinhardtii cells. Sensors Actuators B Chem 185:321–330

Jackson T, Mansfield K, Saafi M et al (2008) Measuring soil temperature and moisture using wireless MEMS sensors. Measurement 41:381–390

Jaeger CH III, Lindow SE, Miller W et al (1999) Mapping of sugar and amino acid availability in soil around roots with bacterial sensors of sucrose and tryptophan. Appl Environ Microbiol 65:2685–2690

Jalali B, Suryanarayana D (1971) Shift in the carbohydrate spectrum of root exudates of wheat in relation to its root-rot disease. Plant Soil 34:261–267

Jenne M, Kambham M, Tollamadugu NP et al (2018) The use of slow releasing nanoparticle encapsulated Azadirachtin formulations for the management of *Caryedon serratus* O. (groundnut bruchid). IET Nanobiotechnol 12(7):963–967

Joseph T, Morrison M (2006) Nanoforum report: nanotechnology in agriculture and food. European Nanotechnology Gateway. Available at: http://ftp.cordis.europa.eu/pub/nanotechnology/docs/nanotechnology_in_agriculture_and_food.pdf. Accessed 18 Apr 14

Kahru A, Dubourguier HL (2010) From ecotoxicology to nanoecotoxicology. Toxicology 269:105e119

Kalagatur NK, Ghosh N, Sivaraman O (2018) Antifungal activity of chitosan nanoparticles encapsulated with Cymbopogon martinii essential oil on plant pathogenic fungi Fusarium graminearum. Front Pharmacol 9:610

Kannan N, Rangaraj S, Gopalu K et al (2012) Curr Nanosci 8:902–908

Kaushal M, Wani SP (2017) Nanosensors: frontiers in precision agriculture. In: Nanotechnology. Springer, Singapore, pp 279–291

Kaweeteerawat C, Ivask A, Liu R et al (2015) Toxicity of metal oxide nanoparticles in Escherichia coli correlates with conduction band and hydration energies. Environ Sci Technol 49(2):1105–1112

Khandelwal N, Barbole RS, Banerjee SS et al (2016) Budding trends in integrated pest management using advanced micro-and nano-materials: challenges and perspectives. J Environ Manag 184:157–169

Khot LR, Sankaran S, Maja JM et al (2012) Applications of nanomaterials in agricultural production and crop protection: a review. Crop Prot 35:64–70

Klimant I, Ruckruh F, Liebsch G (1999) Fast response oxygen micro-optodes based on novel soluble ormosil glasses. Mikrochim Acta 131:35–46

Konotop YO, Kovalenko MS, Ulynets VZ et al (2014) Phytotoxicity of colloidal solutions of metal-containing nanoparticles. Cytol Genet 48(2):99–102

Kumar V, Guleria P, Kumar V et al (2013) Gold nanoparticle exposure induces growth and yield enhancement in Arabidopsis thaliana. Sci Total Environ 461:462–468

Kumar S, Kumar D, Dilbaghi N (2017) Preparation, characterization, and bio-efficacy evaluation of controlled release carbendazim-loaded polymeric nanoparticles. Environ Sci Pollut Res 24(1):926–937

Lahiani MH, Dervishi E, Chen J et al (2013) Impact of carbon nanotube exposure to seeds of valuable crops. ACS Appl Mater Interfaces 5(16):7965–7973

Lai F, Wissing SA, Müller RH et al (2006) Artemisia arborescens L essential oil-loaded solid lipid nanoparticles for potential agricultural application: preparation and characterization. AAPS Pharm Sci Tech 7:E10

Lalonde S, Wipf D, Frommer WB (2004) Transport mechanisms for organic forms of carbon and nitrogen between source and sink. Annu Rev Plant Biol 55:341–372

Lee CW, Mahendra S, Zodrow K et al (2010) Developmental phytotoxicity of metal oxide nanoparticles to Arabidopsis thaliana. Environ Toxicol Chem: Int J 29(3):669–675

Lee S, Chung H, Kim S et al (2013) The genotoxic effect of ZnO and CuO nanoparticles on early growth of buckwheat, *Fagopyrum esculentum*. Water Air Soil Pollut 224:1668. https://doi.org/10.1007/s11270-013-1668-0

Li N, Sioutas C, Cho A et al (2003) Ultrafine particulate pollutants induce oxidative stress and mitochondrial damage. Environ Health Perspect 111:455–460

Liao W, Lu X (2016) Determination of chemical hazards in foods using surface-enhanced Raman spectroscopy coupled with advanced separation techniques. Trends Food Sci Technol 54:103–113

Lin CA (2007) Size matters: regulating nanotechnology. Harv Environ Law Rev 31:350–407

Lin D, Xing B (2007) Phytotoxicity of nanoparticles: inhibition of seed germination and root growth. Environ Pollut 150(2):243–250

Liu Y, Yan L, Heiden P et al (2001) Use of nanoparticles for controlled release of biocides in solid wood. J Appl Polym Sci 79:458–465

Liu Y, Laks P, Heiden P (2002) Controlled release of biocides in solid wood. III. Preparation and characterization of surfactant-free nanoparticles. J Appl Polym Sci 86:615–621

Liu XM, Zhang FD, Zhang SQ et al (2005) Effects of nano-ferric oxide on the growth and nutrients absorption of peanut. Plant Nutr Fert Sci 11:14–18

Liu R, Zhang H, Lal R (2016) Effects of stabilized nanoparticles of copper, zinc, manganese, and iron oxides in low concentrations on lettuce (*Lactuca sativa*) seed germination: nanotoxicants or nanonutrients. Water Air Soil Pollut 227(1):42

López-Moreno ML, de la Rosa G, Hernández-Viezcas JÁ et al (2010) Evidence of the differential biotransformation and genotoxicity of ZnO and CeO_2 nanoparticles on soybean (Glycine max) plants. Environ Sci Technol 44(19):7315–7320

Lu W, Lu ML, Zhang QP et al (2013) Octahydrogenated retinoic acid-conjugated glycol chitosan nanoparticles as a novel carrier of azadirachtin: synthesis, characterization, and in vitro evaluation. J Polym Sci A Polym Chem 51:3932–3940

Ma C, Chhikara S, Xing B et al (2013) Physiological and molecular response of Arabidopsis thaliana (L.) to nanoparticle cerium and indium oxide exposure. ACS Sustain Chem Eng 1(7):768–778

Mahajan P, Dhoke SK, Khanna AS (2011) Effect of nano-ZnO particle suspension on growth of mung (*Vigna radiata*) and gram (*Cicer arietinum*) seedlings using plant agar method. J Nanotechnol 2011:1–7

Mahajan P, Shailesh K, Dhoke RK et al (2013) Nanotechnology 3:4052–4081

Mahmoodzadeh H, Nabavi M, Kashefi H (2000) Effect of nanoscale titanium dioxide particles on the germination and growth of canola Brassica napus. J Ornam Hortic Plants 3:25–32

Marschner H (1996) Mineral nutrition of higher plants. Academic Press/Harcourt Brace & Co, London

Maruyama CR, Guilger M, Pascoli M et al (2016) Nanoparticles based on chitosan as carriers for the combined herbicides imazapic and imazapyr. Sci Rep 6:19768

Moran KLM, Fitzgerald J, McPartlin DA et al (2016) Biosensor-based technologies for the detection of pathogens and toxins. In: Comprehensive analytical chemistry, vol 74. Elsevier, Amsterdam, pp 93–120. ISBN 978-0-444-63579-2

Morla S, Rao CR, Chakrapani R (2011) Factors affecting seed germination and seedling growth of tomato plants cultured in vitro conditions. J Chem Biol Phys Sci (JCBPS) 1(2):328

Naderi MR, Abedi A (2012) J Nanotech 11(1):18–26

Nadi E, Aynehband A, Mojaddam M (2013) Int J Biosci 3:267–272

Nair R, Varghese SH, Nair BG (2010) Nanoparticulate material delivery to plants. Plant Sci 179:154e163

Namasivayam KRS, Aruna A, Gokila (2014) Evaluation of silver nanoparticles-chitosan encapsulated synthetic herbicide paraquate (AgNp-CS-PQ) preparation for the controlled release and improved herbicidal activity against *Eichhornia crassipes*. Res J Biotechnol 9(9):19–27

Navarro E, Baun A, Behra R (2008) Environmental behavior and ecotoxicity of engineered nanoparticles to algae, plants, and fungi. Ecotoxicology 17:372e386

Nguyen TNQ, Le VA, Hua QC, Nguyen TT (2014) Enhancing insecticide activity of anacardic acid by intercalating it into MgAl layered double hydroxides nanoparticles. J Vietnam Environ 6(3):208–211. https://doi.org/10.13141/jve

Patra P, Mitra S, Debnath N et al (2012) Biochemical-, biophysical-, and microarray-based antifungal evaluation of the buffer-mediated synthesized nano zinc oxide: an in vivo and in vitro toxicity study. Langmuir 28(49):16966–16978

Patra P, Choudhury SR, Mandal S et al (2013) Effect sulfur and ZnO nanoparticles on stress physiology and plant (Vignaradiata) nutrition. In: Advanced nanomaterials and nanotechnology. Springer, Berlin/Heidelberg, pp 301–309

Patra S, Roy E, Madhuri R et al (2017) A technique comes to life for security of life: the food contaminant sensors. In: Nanobiosensors. Elsevier, Amsterdam, pp 713–772

Pola-López LA, Camas-Anzueto JL, Martínez-Antonio A et al (2018) Novel arsenic biosensor "POLA" obtained by a genetically modified E. coli bioreporter cell. Sensors Actuators B Chem 254:1061–1068

Puoci F, Iemma F, Picci N (2008) Stimuli-responsive molecularly imprinted polymers for drug delivery: a review. Current Drug Deliv 5(2):85–96

Rai V, Acharya S, Dey N (2012) Implications of nanobiosensors in agriculture. J Biomater Nanobiotechnol 3:315–324

Rajaie M, Ziaeyan AH (2009) Int J Plant Prod 3(3):35–440

Ramyadevi J, Jeyasubramanian K, Marikani A et al (2012) Synthesis and antimicrobial activity of copper nanoparticles. Mater Lett 71:114–116

Reijnders L (2006) Cleaner nanotechnology and hazard reduction of manufactured nanoparticles. J Clean Prod 14:124e133

Salama HMH (2012) Effects of silver nanoparticles in some crop plants, Common bean (Phaseolus vulgaris L.) and corn (Zea mays L.). Int Res J Biotechnol 3:190–197

Saharan V, Mehrotra A, Khatik R et al (2013) Synthesis of chitosan based nanoparticles and their in vitro evaluation against phytopathogenic fungi. Int J Biol Macromol 62:677–683

Savithramma N, Ankanna S, Bhumi G (2012) Effect of nanoparticles on seed germination and seedling growth of Boswellia ovalifoliolata an endemic and endangered medicinal tree taxon. Nano Vis 2(1):2

Schmälzlin E, van Dongen JT, Klimant I (2005) An optical multifrequency phase-modulation method using microbeads for measuring intracellular oxygen concentrations in plants. Biophys J 89:1339–1345

Siddiqui MH, Al-Whaibi MH (2014) Role of nano-SiO2 in germination of tomato (Lycopersicon esculentum seeds Mill.). Saudi J Biol Sci 21(1):13–17

Singh S (2012) Achieving second green revolution through nanotechnology in India. Agric Situat India:545–572

Singh S, Singh BK, Yadav SM (2014) Applications of nanotechnology in agricultural and their role in disease management. Res J Nanosci Nanotechnol 5:1–5. https://doi.org/10.3923/rjnn.2014

Sinha K, Ghosh J, Sil PC (2017) New pesticides: a cutting-edge view of contributions from nanotechnology for the development of sustainable agricultural pest control. In: New pesticides and soil sensors. Elsevier, Amsterdam, pp 47–79. ISBN 978-0-12-804299-1

Stephenson GR (2003) Pesticide use and world food production: risks and benefits. In: Environmental fate and effects of pesticides. American Chemical Society, Washington, DC, pp 261–270

Suh WH, Suslick KS, Stucky GD et al (2009) Nanotechnology, nanotoxicology and neuroscience. Prog Neurobiol 87:133e170

Sunkar S, Nachiyar CV (2012) Biogenesis of antibacterial silver nanoparticles using the endophytic bacterium Bacillus cereus isolated from *Garcinia xanthochymus*. Asian Pac J Trop Biomed 2(12):953–959

Suriyaprabha R, Karunakaran G, Yuvakkumar R et al (2012) J Curr Nanosci 8:902–908

Taniguchi N, Arakawa C, Kobayashi T (1974) On the basic concept of 'nano-technology'. In Proceedings of the international conference on production engineering, 1974–8, vol 2, pp 18–23

Tarafdar JC, Raliya R, Rathore I (2012a) Microbial synthesis of phosphorous nanoparticle from tri-calcium phosphate using Aspergillus tubingensis TFR-5. J Bionanosci 6(2):84–89

Tarafdar JC, Agarwal A, Raliya R et al (2012b) Adv Sci Eng Med 4:1–5

Tungittiplakorn W, Cohen C, Lion LW (2005) Engineered polymeric nanoparticles for bioremediation of hydrophobic contaminants. Environ Sci Technol 39(5):1354–1358

Verma N, Kaur G (2016) Trends on biosensing systems for heavy metal detection. In: Comprehensive analytical chemistry, vol 74. Elsevier, Amsterdam, pp 33–71. ISBN 978-0-444-63579-2

Viirlaid E, Riiberg R, Mäeorg U et al (2009) Glyphosate attachment on aminoactivated carriers for sample stabilization and concentration. Agron Res 13:1152–1159

Walker TS, Bais HP, Grotewold E et al (2003) Root exudation and rhizosphere biology. Plant Physiol 132:44–51

Wang Y, Cui H, Sun C et al (2014) Construction and evaluation of controlled-release delivery system of Abamectin using porous silica nanoparticles as carriers. Nanoscale Res Lett 9:2490

Wang C, Otto S, Dorn M et al (2019) Luminescent TOP nanosensors for simultaneously measuring temperature, oxygen, and pH at a single excitation wavelength. Anal Chem 91(3):2337–2344

Xu L, Liu Y, Bai R et al (2010) Applications and toxicological issues surrounding nanotechnology in the food industry. Pure Appl Chem 82:349e372

Yang L, Watts DJ (2005) Particle surface characteristics may play an important role in phytotoxicity of alumina nanoparticles. Toxicol Lett 158(2):122–132

Yang FL, Li XG, Zhu F et al (2009) Structural characterization of nanoparticles loaded with garlic essential oil and their insecticidal activity against *Tribolium castaneum* (Herbst) (Coleoptera: Tenebrionidae). J Agric Food Chem 57(21):10156–10162

Yang Y, Cheng J, Garamus VM et al (2018) Preparation of an environmentally friendly formulation of the insecticide nicotine hydrochloride through encapsulation in chitosan/tripolyphosphate nanoparticles. J Agric Food Chem 66(5):1067–1074

Yu Z, Sun X, Song H et al (2015) Glutathione-responsive carboxymethyl chitosan nanoparticles for controlled release of herbicides. Mater Sci Appl 6(06):591

Yuvakkumar R, Elango V, Rajendran V et al (2011) Influence of nanosilica powder on the growth of maize crop (Zea mays L.). Int J Green Nanotechnol 3(3):180–190
Zhang J, Li M, Fan T et al (2013) Construction of novel amphiphilic chitosan copolymer nanoparticles for chlorpyrifos delivery. J Polym Res 20:107
Zhang D, Hua T, Xiao F et al (2015) Phytotoxicity and bioaccumulation of ZnO nanoparticles in Schoenoplectus tabernaemontani. Chemosphere 120:211–219
Zhao L, Peralta-Videa JR, Rico CM et al (2014) CeO_2 and ZnO nanoparticles change the nutritional qualities of cucumber (Cucumis sativus). J Agric Food Chem 62(13):2752–2759

Nanomaterials: Emerging Trends and Future Prospects for Economical Agricultural System

16

Nitin Kumar, Abarna Balamurugan, M. Mohiraa Shafreen, Afrin Rahim, Siddharth Vats, and Kanchan Vishwakarma

Abstract

In developing countries, when it comes to national economy, one of the important building blocks is agriculture. The food production rate has risen, which has a substantial role in a country's gross domestic production. The application of pesticides and fertilizers determines the rate of food production. Agricultural growth and food production are very much dependent on parameters like soil health, water availability, climate change, etc. Since the world population is expanding at an alarming rate, the food production needs to be enhanced, and adverse agricultural conditions have to be regulated. Supporting the massive increase in population, the sustainable development of agriculture is required. With latest advancements, new avenues have been opened up by nanotechnology in the field of food processing and crop improvement. The present chapter highlights the role and emergence of nanomaterials in agriculture system.

Keywords

Nanomaterials · Sustainable agriculture · Nanofertilizers · Targeted delivery

N. Kumar · A. Balamurugan · M. Mohiraa Shafreen · A. Rahim
Department of Biotechnology, Periyar Maniammai Institute of Science and Technology, Thanjavur, Tamil Nadu, India

S. Vats
Shri Ram Swaroop Memorial University, Lucknow, Uttar Pradesh, India

K. Vishwakarma (✉)
Amity Institute of Microbial Technology, Amity University, Noida, Uttar Pradesh, India
e-mail: kvishwakarma@amity.edu

M. Ghorbanpour et al. (eds.), *Biogenic Nano-Particles and their Use in Agro-ecosystems*, https://doi.org/10.1007/978-981-15-2985-6_16

16.1 Introduction

Nanoparticles (NPs) have unique optical, electronic, magnetic, and chemical characteristics which make them widely applicable and have a great significance in various fields like agriculture, medical, energy, and environment. The application of nanoparticles in biological field depends upon the mode in which nanomaterials are synthesized and their metallic properties. Different routes such as chemical, biological, and physical methods have been approached for the synthesis of nanomaterials. A few physical methods have found to pose risks to human health and are not environment-friendly, which include pyrolysis, radiation, and arc discharge. Likewise, in chemical synthesis, the use of harmful reductants such as hydrazine and sodium bborohydride generates toxic byproducts (Jain et al. 2011; Rauwel et al. 2015). The nanomaterial toxicity concerns are less when it comes to the synthesis of nanomaterials through biological methods opening the door for biological application (i.e., drug delivery, agriculture, and biosensing) (Netala et al. 2016).

Nanomaterials are designed to be utilized in the field of personal care items, agribusiness, food, medicinal purpose, and biotechnology which are being exposed to the environment and earth. It is perceived that nanomaterials may add up effectively for the betterment of analysis, diagnosis, avoidance of diseases including microbe-based infections or carcinoma, and treatment. It has been reported that nanoparticles have diverse medical and industrial applications along with the presence of nanoparticles in various materials including clothes, cosmetics, etc. which are being used in our day-to-day life (Dubchak et al. 2010; Singh et al. 2019).

As per the alarming rate of increase in human population, the agricultural production has to be tripled in the coming years. High yields must be achieved in the existing production areas since the cultivable lands are already being used (Tilman et al. 2002; Fountain and Wratten 2013; Kumar 2013; Gerland et al. 2014). After the Green Revolution, higher yields were generated, but now the growth has been stabilized in most of the crops (yield stabilization phenomenon). Pests and diseases are the major reasons for this damage (Grassini et al. 2013; Archana Singh 2014).

Applying nanotechnology in the field of agriculture has many benefits when it comes to plant disease management and overall growth. For example, micro/macro nutrients and nanoformulations containing fertilizers, when mixed and applied for crop production, not only increased the crop yield but also have acted as biocontrol agent against various plant pathogens (Keswani et al. 2016; Vishwakarma et al. 2017a). The effectiveness of nanoparticles can be altered and improved by using particulate systems. Eco-friendly alternatives are offered by nanotechnology for the management of plant diseases which in turn plays a vital role in food security, global food production, and food safety. Metallic NPs (such as Ti, Au, Ni, Zn, and Ag) have been used as antimicrobial agents against phytopathogens. There are various advantages of nanopesticide formulations over conventional pesticides as they increase the solubility of poorly soluble active ingredients, the reduced premature degradation of active ingredients, and the target-oriented slow release of active ingredients. Through physical, biological, and chemical methods, various types of nanomaterial products for agricultural practices have been developed which include nanofertilizers, nanopesticides, and nanosensors.

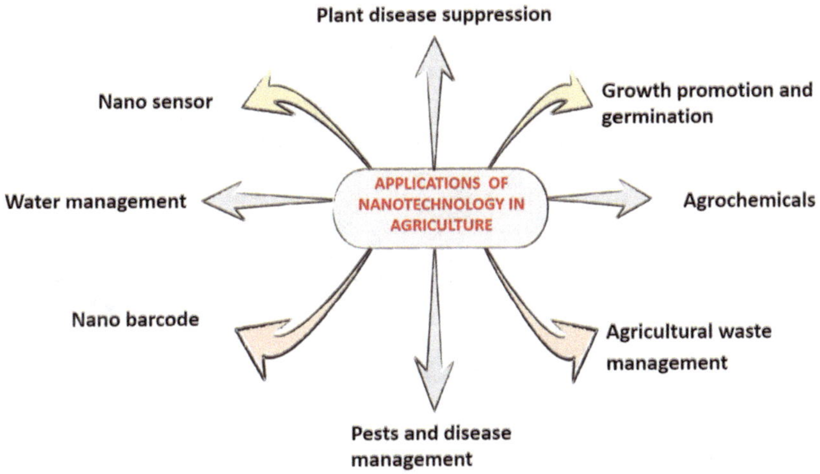

Fig. 16.1 Applications of nanotechnology in agriculture

Nanoparticles between the range of 1 nm and 100 nm are classified as components or aggregates (Ball 2002). Nanoparticles are considered as building blocks of nanotechnology which has high reactivity and a high surface/volume ratio which can cross the plasma membrane and the cell (Stern and McNeil 2008; Farré et al. 2011; Tripathi et al. 2017a; Vishwakarma et al. 2017b), and its surface may attain intrinsic properties (Donaldson et al. 2004). It is important to note that nanoparticles possess these extraordinary properties that make them different from bulk materials. They possess both beneficial and harmful properties depending on their own characteristics and plant host properties (Singh et al. 2017; Tripathi et al. 2017c; Vishwakarma et al. 2019). Agricultural researches are empowered by nanotechnology by a broad range of advancements in transformation of wastes from agricultural and food products to energy, reproductive science and technology, control of diseases, their treatment in plants through using various NPs, and other useful byproducts through nano-enzymatic processing (Fig. 16.1) (Moraru et al. 2003). Major agricultural risks can be possibly solved by nanotechnology. It also helps to understand the biology of various crops which in turn helps to enhance the yield and nutritional value with control over pest incidences and plant diseases (Nair et al. 2010).

16.2 Nanotechnology in Agriculture

16.2.1 Plant Disease Suppression

A wide range of pathogenic bacteria, fungi, nematodes, and viruses can cause plant diseases which in turn result in reduced yield, shelf life, and quality of the product (Patel et al. 2014). Hence, novel systems must be utilized to overcome these kinds

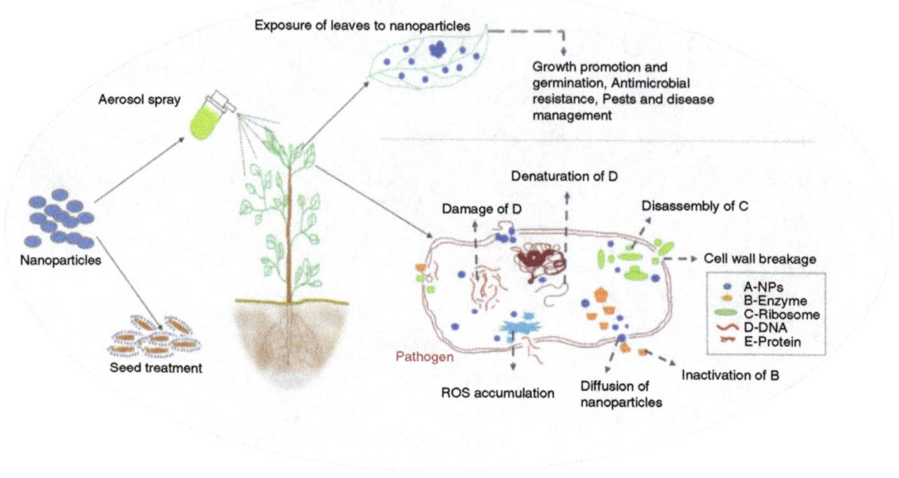

Fig. 16.2 Mode of plant disease suppression

of problems. One such novel system includes nanoparticle utilization for suppressing and preventing plant diseases. Nutrients are retained more efficiently in nanomaterials which have large surface area and functions as long-lasting and stable mineral store for crops (Navarro et al. 2008). Utilization of pesticides and fungicides will be reduced in the coming future if supply of smart nano-supply systems is monitored and activated in remote areas which will give agriculture growers a great assistance (Rai and Ingle 2012). Utilizing metal nanoparticles and metal oxide can have a positive impact on pathogen suppression and crop yield, which is depicted in Fig. 16.2. Some nanoparticles are involved in the suppression of plant stress/diseases directly through antimicrobial/antioxidant/heavy metal uptake activity, for example, zinc oxide, magnesium, titanium oxide, silicon, and silver nanoparticles (Tripathi et al. 2012; Prasad and Prasad 2014; Tripathi et al. 2017b; Rastogi et al. 2019).

The assessment of silver nanoparticles and ions was carried out to examine the effect of antifungal activity on *Bipolaris sorokiniana* and *Magnaporthe grisea* (Woo et al. 2009). When compared with silver nanoparticles, increased soil fertility and lesser toxicity have been observed in the case of zinc oxide nanoparticles. It was observed that the growth of *Fusarium graminearum* was diminished in mung bean by zinc oxide nanoparticles (Dimkpa et al. 2013). For plant growth increase, much attention has been given to carbon nanoparticles, multiwalled carbon nanotubes, and fullerols (Khot et al. 2012; Shweta et al. 2017). A study to check the effect of functionalized as well as nonfunctionalized multiwalled carbon nanotube on plants has shown that it has caused a reduction of chlordane in shoots and roots of plant lettuce (Hamdi et al. 2014).

16.2.2 Plant Growth and Germination

Surrounding environment significantly influences germination and plant growth. Germination and growth may or may not be affected by a chemical's presence (Shojaei et al. 2009). To broaden the application of nanomaterials on germination and growth, several investigations have been carried out. The nanoparticles used for the study are ytterbium oxide (Yb_2O_3), gadolinium (III) oxide (Gd_2O_3), lanthanum (III) oxide (La_2O_3), CeO_2, Zn, Al, ZnO, aluminum oxide (Al_2O_3), palladium (Pd), Si, FITC-labeled silica nanoparticles, Cu, Au, and TiO_2. These NPs are used to check the germination of different plants such as tomato, lettuce, spinach, rice, rape canola, radish, corn, cucumber, cabbage, wheat, and ryegrass. Positive effects have been confirmed on photosynthetic rate, chlorophyll formation, dry weight, and germination rate. The nanoparticle size rate is inversely correlated to seed germination. Germination rate will be higher when the size of nanoparticle is smaller. Enhancement in photogeneration and photo sterilization of active oxygen is the reason for the increase in germination rate (Fig. 16.2).

An important role is played by the microorganisms in decomposition of soil organic matter and biogeochemical cycles. Beneficial relationships are formed by some microorganisms with the plant roots which in turn make the essential nutrients such as phosphorous and nitrogen available to plants (Kumar et al. 2017). Plants that are colonized with fungi have imparted tolerance to heat and drought as well as resistance to plant pathogens and insects. For NP-microbe interaction, NP's concentration plays an important role. Morphological changes are induced in *Rhizobium leguminosarum* by ZnO NPs (250–750 ppm), affecting the root nodulation process which in turn resulted in delayed biological nitrogen fixation (Huang et al. 2014). The nitrogen-fixing ability is declined in *Rhizobium* with an increase in concentration of the AgNPs (0.6–6.6%) (Kumar et al. 2014). However, the type of NPs becomes an important concern when studies are performed at low concentrations (50–100 ppm). At 50 ppm of AgNPs, root nodulation was enhanced in cowpea (Mehta et al. 2016). There was an increased nitrogenase activity in moth bean, cowpea, cluster bean, and green gram when ZnO nanoparticles were used at 1.5 ppm concentration (Kumar et al. 2015).

Results have shown that the concentration of AgNPs may have a great influence in the nitrification and N cycle. Toxicity was exhibited by AgNPs at a concentration > 0.25 ppm, whereas the gene expression of nitrogen-fixing bacteria and denitrifiers at a concentration of 0.025–0.05 ppm was not affected. Using AgNPs at such a low concentration in *N. europaea*, it was observed that the gene expression was upregulated by 2–3 folds (Yang et al. 2013). There was a reduced nodulation frequency in *Sinorhizobium meliloti* on *Medicago truncatula* when Ag, TiO_2, and ZnO were combined at 100, 1400, and 2400 ppm, respectively, whereas the nodulation frequency was not reduced when AgNPs were used alone at a low concentration of 100 ppm (Judy et al. 2016).

Another study showed that the ammonia-oxidizing bacteria's activity was affected by Fe and AgNPs which in turn resulted in reduced nitrification by 90% and 71%, respectively. These nanoparticles get attached to the bacterial surface which in turn resulted in inhibition of bacterial growth (Fig. 16.2) (Michels et al. 2017).

The effect of TiO_2 NPs on biological nitrogen fixation activity and cell growth was studied by Cherchi and Gu (2010) using Ohm's law as the inactivation model in *Anabaena variabilis*. It was found that the growth rate and nitrogen fixation activity were reduced at 0.62 ppm and 0.4 ppm, respectively. At the same time, due to an increase of stress, cyanophycin grana proteins (CGPs) were produced at an increased rate in response to NPs. So the study has concluded that the growth of the plants is greatly influenced by exposure time than the NP toxicity. A decline in nitrogen fixation was observed using an analysis based on pyrosequencing due to the presence of ZnO and TiO_2 in a dose-dependent manner (Ge et al. 2012). Red clover and the symbionts of red clover acquired nutrients effectively with the presence of the titanium oxide (TiO_2) nanoparticles with the concentration range of 10–1000 ppm (Moll et al. 2016). On the contrary, TiO_2 presence in peas has delayed the root nodule development and biological nitrogen fixation (Fan et al. 2014). The presence of CeO_2 NPs at a concentration of 100 ppm in planted soil has altered soil microbial communities when compared to unplanted soil. Plant growth is promoted with the help of micronutrients such as zinc, iron, boron, copper, chlorine, manganese, and molybdenum optimally. Nanoformulations of micronutrients may be used to spray crops for enhanced foliar uptake or can be used as a soil addition for their slow release to promote plant growth and improve soil health (Fig. 16.2) (Peteu et al. 2010). A chitosan-based plant growth hormone is synthesized by Tao et al. (2012) in which 1-naphthylacetic acid is conjugated with chitosan. The release of this nanoformulation depends greatly on temperature and pH, which also enables slow release of the hormones.

16.2.3 Nano-barcodes

Barcodes play a vital role for a successful national and international marketing. It illustrates various factors of electronic information which includes place and date of fabrication and packaging, cost of the product, methods followed, and the chemicals used. An electronic bar reader is necessary to read the barcodes. It is difficult to track and control the products due to the recent increment in the shipment of agricultural commodities (Shrivastava and Dash 2012). Quality control, data monitoring, and tracking of the products in a short time period have become possible through nanotechnology (Nam et al. 2003; Kress et al. 2005). Nano-barcodes were developed using fluorescent NPs doped with scarce earth materials. Optical microscope and UV lamp are mostly employed to read the nano-barcodes. Date and place of manufacture, security and safety of the product, followed methodology, and chemicals used while production will be provided using an identity preservation

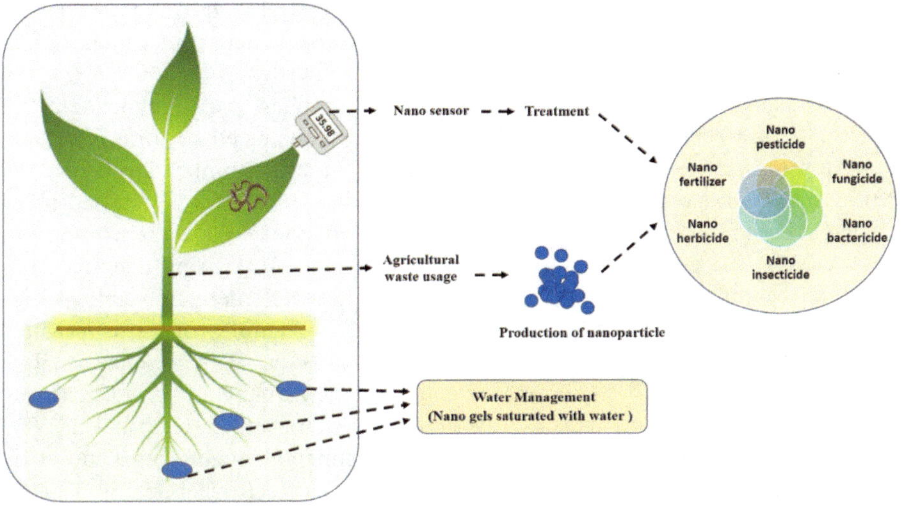

Fig. 16.3 Sustainable approaches in agriculture for growth, waste, and water management/treatment

(IP) system for consumers as well as stakeholders (Meyers et al. 2004). It can transfigure the field by introducing biodegradable sensors for biological and physical characteristics such as texture, stiffness, ripening, flavor, contamination, and color of agricultural products.

16.2.4 Nanobiosensor

A scanning device named millipede has been introduced for agriculture. It has a combined concept of terabit capacity, high data rate, small form factor, and ultrahigh density, which is being introduced in data storage (Fig. 16.3). For monitoring, data collection, and management of plant and soil health, nanosensors have a major contribution (Rai et al. 2012; Khandelwal and Joshi 2018; Kumar et al. 2018). Characteristic properties are possessed by carbon nanotube (CNT) devices for precise sensing, drug delivery, and diagnosis in livestock health management and pest control (Shweta et al. 2017).

To inhibit and detect plant pathogens, smart sensors are used which are driven by nano-based technology. For monitoring real-time crop growth and soil health conditions, nanosensors can use global positioning system (GPS). Crop productivity is enhanced with the use of smart sensors, and also farmers get real information, thereby enabling them to make important decision about the crops. For quantification and detection of contaminants of minute quality such as virus, bacteria, and toxins, bioanalytical nanosensors have been used in food-based systems in agriculture (Seo et al. 2011; Rai et al. 2012).

An antimicrobial chitosan-PVA-based hydrogel is developed by Agnihotri et al. (2012) which has the behavior of a nanoreactor and has a matrix that immobilizes silver nanoparticles (AgNPs) having antibacterial properties.

For organophosphate detection, chitosan-TiO_2-graphene nanocomposite-based nanosensor has been developed. Studies have shown enhanced stability in biosensors by using porous nanocomposites through enzyme immobilization (Cui et al. 2018). In peanut samples, the presence of aflatoxin B1 (AFB1) is detected using graphene-based biosensors which contain gold and graphene oxide nanocomposites (GO/AuNCs) (Li et al. 2018). Biosensors are fabricated by using nanomaterials which are modified morphologically which provide better chemical stability, high surface area to volume ratio, and biocompatibility. Examples of those modified nanostructures are nanoflakes, nanosheets, nanoflowers, etc. The presence of silver (Ag^+) ions and mercury (Hg^{2+}) are detected simultaneously in serum, drinking water, and cell lysate using biosensor based on tungsten disulfide nanosheets (WS_2) which have implications in diagnosis and environmental monitoring (Zuo et al. 2016). Combining electrochemical techniques such as electrochemical impedance spectroscopy (EIS) and cyclic voltammetry (CV) with nanomaterials for detecting pathogen is gaining much popularity in agriculture and biomedical sector.

Crop productivity is limited due to the reoccurring of diseases. Out of many, diseases which are incurred by virus are difficult to control and have most devastating effects. The application of pesticides is of no use once the symptoms start to appear. Hence to eradicate diseases, it has to be predicted earlier. Biosensors have wide applications in this field; using nanobiosensor, diagnosis is much more beneficial as it is more efficient with active surface area, and the detection speed is increased. Nanobiosensing devices of increased sensitivity can be developed by using various elements, nanomaterials, nanosurfaces, etc. Biosensors based on nanoparticles or nanosurfaces could be developed for screening applications which could be used for the detection of the presence of any chemical substance or bacteria which possess fluorescent properties and in addition can give chemical or electrical signal in response when the trace elements are detected (Kahveci et al. 2016). For nutrients to be assimilated better in our body, methods that introduce nanomaterials directly into food are being developed. To enhance efficiency of plants, herbicides are delivered in the form of nanocapsules in slow and controlled manner which get triggered by the target's presence (Khoobdel et al. 2017). This also helps in minimizing plant contamination. Nanobiosensors which can assess the plant's needs such as water and nutrient content will have the most important application in agricultural field. For monitoring continuous physiological processes of plant growth and development and soil conditions, nanosensors are now considered as the smart "eye ". To evaluate and monitor the overall quality of food, the packaging material film can be integrated with the nanobiosensors. Those biosensors have the exclusive property of causing color change when it finds any change in reaction which helps in easy prediction of spoilage and shelf life of food. Due to its compactness and small size, nanobiosensors have an added advantage in chemical and biological analysis (Steinborn et al. 2017). In future prospects, when it comes to developing technologies for precision farming, the main goal has to be focused on increasing crop yield as well as minimizing the chemical input usage.

Smart field system can be done by monitoring soil conditions and crop growth and automatically collecting real-time data using nanostructure sensors by scattering nanosensor network. Fluctuations in environmental conditions which affect the efficiency of the crop can be watched by farmers by collecting real-time data.

16.2.5 Agricultural Waste Management

The lack of skilled personnel and mechanization has led to the production of waste in huge amount. The agricultural waste can either rotten or defiled due to certain limitations in the processing of waste that would affect the crop. The effect of biofuel production using agricultural wastes including vegetable oils, sugarcane, corncobs, cotton stalk, rice husk, coconut shell, cotton, animal fats, and groundnut shell was enhanced by employing nanometallic catalysts through nano-bioengineering (Shrivastava and Dash 2012; Sarkar and Praveen 2017;Bharati and Suresh 2017).

To alleviate, demolish, and modify the dangerous agricultural wastes into nonpoisonous matter, NPs could be utilized as a reactant agent (Ditta 2012). Various approaches involving nanotechnology have been developed to treat wastewater. Photocatalysis is the foremost nano-based wastewater treatment technique and can apply for decomposition, filtration, decontamination, and purification of water and air. Moreover, it can remove the pathogenic or harmful agents with the incorporation of semiconductors. It's a process in which the chemical reaction gets accelerated by the catalyst that decreases the activation energy for the occurrence of primary reaction. In this case, the valence electrons present in the outer layer get excited to develop electron-hole pairs, when a NP subjected to UV light. NPs containing metal oxides and sulfides like ZnS, ZnO, TiO_2, and SnO_2 are employed as catalysts (Bhatkhande et al. 2002; Li and Haneda 2003; Ko et al. 2009; Feigl et al. 2010).

16.2.6 Water Management

In micro-irrigation technology, the consumption of plastics has increased the application for the subsurface irrigation. This system poses burden in terms of weight, nonflexibility and trouble shooting. In the event of any disruption or malfunction, water wastage can increase dramatically in conventional piping systems. Due to these drawbacks, conservative irrigation systems have to be promoted for irrigation and water preservation by employing low-cost, flexible, biodegradable, and efficient goods in the subsurface piping systems, especially for lands that have limited water availability. Nanomaterials can be taken into account as they provide high surface area, strength, and flexibility (Shrivastava and Dash 2012). Recently, utilizing the irrigation systems based on nanogel such as silica for the perennial plant, reduced water utilization was noted. In this event, a saturated nanogel is fixed under each tree, and the water moves from the gel to the root system of a plant through the differences of osmotic pressure (Fig. 16.3) (Vundavalli et al. 2015).

16.3 Agrochemical Applications

16.3.1 Nanopesticides

To alleviate the limitations like enhanced solubility and rate of soil run-off solubility, pesticides are prepared in a secured casing or within a specific shells (Chinnamuthu and Boopathi 2009). *Spodoptera littoralis*, a primary level pest that attacks the plant kingdom mostly by initiating its resistance to all pesticides virtually, was restrained toward tomato plant covered with hydrophobic nanosilica without any obstacle and exhibits positive reaction by ruining the pests at 300–350 ppm (El-bendary and El-Helaly 2013). Multiwalled CNT-induced pesticides like zineb, mancozeb were mixed with citric acid molecules and confined in an aqueous solution that would facilitate the encapsulation of the pesticides to ensure deterioration to the fungi (i.e., *Alternaria alternata*) (Sarlak et al. 2014). Yang and group have examined the impact of NPs coated with polyethylene glycol equipped with essential oils toward *Tribolium castaneum* (Yang et al. 2009). A pesticide valindamycin was later utilized after filling into the hollow porous silica NPs (Liu et al. 2006a, 2006b). Likewise, many other NMs like ZnO NPs, TiO_2 NPs, and AgNPs have been investigated for the insecticidal properties and resistivity against silkworm disease and rice pests (Goswami et al. 2010).

AgNPs have pesticidal impact against harmful fungi, described to have inhibitory outcomes on conidial germination of the genus *Raffaelea*, which leads to mortality in the oak trees (Nair et al. 2010). Moreover, avermectin which is a pesticide that inhibits neurotransmission in the insects has a short lifetime with 6 hours of half-life. The life span has been shown to increase by employing porous AgNPs with shell having pore diameter of 4–5 nm and thickness of 15 nm. AgNPs, which encapsulate avermectin and arrest degradation and its potential under ultraviolet rays, have expanded lifetime till 30 days roughly (Ghormade et al. 2011).

Biogenic AgNPs fabricated using *Tinospora cordifolia* have displayed high pesticide influence on the fourth instar larvae of *Anopheles subpictus*, *Culex quinquefasciatus*, and top louse *Pediculus humanus* (Jayaseelan et al. 2011). AgNPs at a concentration of 100 mg/kg have reduced the growth of mycelium and conidial germination on pumpkins and cucurbits that belong to the family Cucurbitaceae against powdery mildew (Lamsal et al. 2011a, b). Phenolic suspension of hydrophobic alumina-silicate nanoparticles is so vigorous against grasserie disease in the leaves of *Bombyx mori* (Goswami et al. 2010). Nano-encapsulated pesticides adhered on the floor of the plant assist the sustained delivery for a long period when compared with the ordinary pesticides which run off in the rain (Scrinis et al. 2007).

Inorganic nanosized particles like SiO_2, Al_2O_3, TiO_2, or Fe_2O_3 are used as carriers for pesticide to increase the bioactivity. Stable polymeric nanospheres were obtained in the range of 135 nm with encapsulation rate of 3.5%. Despite the minimum amount of active component, the overall yield of this formulation was notable in the bioavailability of the pesticide (RPA 107382) to plants (Boehm et al. 2003).

Lately, a pesticide company established an aqueous dispersion prepared with nanosized biocide (Banner MAXX® from Syngenta) having diverse spectrum of systemic antifungal action. Its active component controls powdery mildew diseases, blights, leaf spots, rusts on several ornamental and horticultural plants (Latin 2006; Ghormade et al. 2011). Silicon nanoparticles (SiNPs) with pore size of 4–5 nm and shell thickness of 15 nm could carry avermectin of nearly 600 g/kg and could control the delivery speed up to 30 days after utilization (Ghormade et al. 2011).

16.3.2 Nanofungicide

Pathogenic fungus is one of the foremost barriers to the crop's growth, which accounts for >70% of diseases and predominantly decreases the yield up to 100% loss (Baker et al. 2017). Various nanoformulations and metal nanoparticles have been inspected to control the pathogenic fungus. In 2013, researchers found that AgNPs declined the *Magnaporthe grisea* growth which develops rice blast disease. AgNPs and fluconazole have showed high antifungal activity toward *Phoma herbarum*, *Fusarium semitectum*, and *Phoma glomerata* (Gajbhiye et al. 2009). Likewise, Ag_2S NPs and AgNPs have displayed antifungal activity toward *Aspergillus niger* and *Fusarium oxysporum*, *Magnaporthe grisea*, *Bipolaris sorokiniana*, *Fusarium culmorum*, and *Colletotrichum* (Jo et al. 2009; Min et al. 2009; Kasprowicz et al. 2010; Musarrat et al. 2010; Aguilar-Méndez et al. 2011).

Ag-silica NPs (nanocomposites) sprayed over the pumpkin leaves also showed antifungal activity toward powdery mildew infection within 3 days of spraying (Park et al. 2006). In Korea, AgNPs were employed as antifungal agents for *Raffaelea* sp., a virulence fungus of oak trees (Kim et al. 2007). Copper nanocomposite with a polymer was fabricated and found its effectual antifungal activity (Cioffi et al. 2005). MgO and ZnO were reported for its controlled growth of *Fusarium oxysporum*, *Alternaria alternata*, *Rhizopus stolonifer*, and *Mucor plumbeus* (Wani and Shah 2012). In fact, ZnO NPs induced the immune system and systemic defense mechanism in plants.

Using foliar spray of ZnO NPs at 1000 ppm, tomato and eggplant have showed optimistic results against *Fusarium* by inducing the defense system of the plant (Elmer and White 2016), and against *Botrytis cinerea*, *Aspergillus niger*, *Penicillium expansum*, and *Aspergillus flavus* via growth reduction (He et al. 2011; Jayaseelan et al. 2012).

For solid wood conservation, polyvinyl pyridine-co-styrene and polyvinyl pyridine nanoparticles were utilized for the controlled delivery of chlorothalonil and tebuconazole fungicides (Dzung et al. 2002). This has given near quantitative integration of active ingredients. After a few years, polymeric nanocapsules were employed as carriers for the pesticides like ivermectin and acetamiprid (Ben-shalom et al. 2003).

16.3.3 Nanobactericide

It is well known that many nanomaterials show antibacterial properties, and hence studies have been carried out for their use in crop protection from harmful bacteria. One such study has shown the use of copper nanoparticles for the effective removal of leaf spot and rice blast pathogens *Xanthomonas campestris* and *Xanthomonas oryzae*. In 2012, Mondal and Mani did studies to find the effectiveness of copper nanoformulation on bacterial pathogens and found that it is very effective than the conventional treatments using copper oxychloride. Studies with green synthesized silver nanoparticles have shown antibacterial properties against *Erwinia cacticida* and *Citrobacter freundii* (Paulkumar et al. 2014). From *Macrophomina phaseolina* (Tassi) Goid, the protein-capped silver nanoparticles synthesized have also shown antibacterial activity (Chowdhury et al. 2014). Another group has found that green synthesized silver nanoparticles have bactericidal property against bacterial pathogens *Xanthomonas axonopodis* and *Ralstonia solanacearum* (Aravinthan et al. 2015). *Erwinia cacticida* and *Citrobacter freundii* are phytopathogenic bacteria, and their growth has been restricted using *Piper nigrum* stem and leaf extract (Paulkumar et al. 2014). *Nephrolepis exaltata* L extract has been used to make biogenic silver nanoparticles, which also have shown significant growth reduction against *Xanthomonas axonopodis pv.* punicae (Bhor et al. 2014). These studies clearly show that nanomaterials, especially silver and copper, have a tremendous application in the eradication of bacterial diseases in crops.

16.3.4 Nanoinsecticide

There is a huge potential in nanotechnology-based products for controlling microbial pathogens in crops. Similarly, studies have shown by using nanoinsecticides there is a significant decrease in the proliferation of various insects in agricultural fields. A study has shown the growth suppression of the pest *Spodoptera littoralis* by using nanosilica, which ultimately decreased the crop yield losses (El-bendary and El-Helaly 2013). These silica nanoparticles have been found to be lethal against pests such as rice weevil, mustard weevil, coconut mite, and white fly. The mechanism of silica nanoparticles having insecticidal activity was because of their physiosorption in cuticle lipids of pests which stimulate killing of the insect (Fig. 16.3) (Barik et al. 2012).

Chitosan-derived nanoparticles increase insecticide rotenone loading by 13,000 times when compared to free rotenone in water and also increase solubility in NOSCS micelles aqueous solution which ultimately paved the way for encapsulation and slow leaching of agrochemicals for plant protection (Lao et al. 2010).

Cashew tree gum and chitosan-composed microspheres were used as transporters for essential oil extracted from *Lippia sidoides* and have shown insecticidal properties (Paula et al. 2011). This result has shown the potential of chitosan as a carrier molecule for insecticides to control the spread of larvae of crop-harming insects. Likewise, microcapsules of chitosan and alginate were synthesized, characterized, and assessed as a carrier molecule for imidacloprid (Guan et al. 2008). When release assays studies have been performed, a significant slow release of insecticide was observed when compared to the free insecticide. Further, the insecticide release can also be modulated by changing the concentrations of chitosan and alginate. A study on ricinoleic acid (RA)-based carboxymethyl chitosan (CM-C) as a carrier for azadirachtin (AZA) biopesticide has shown assistance of this nanoformulation in making lipid-soluble pesticide to become water soluble which is advantageous for agricultural applications (Feng and Peng 2012).

16.3.5 Nanoherbicides

Undesirable weeds have negative impact on the productivity and sustainability of farming. To control and keep a check on the growth of the weeds, many chemical-based herbicides are used, but due to their chemical base, they are hard to be degraded by the nature and have residual toxicity. To overcome such situation, nanotechnology-based herbicides are used and hence termed as nanoherbicides. Nanoherbicides are safe because of their properties. They are stable chemical, have photodecomposition ability and simple solubility, and are easily absorbed by the soil. Nanoherbicides are created by formulating nanoparticles (epsilon-caprolactone) coupled with active ingredients like atrazine. The mobility of the chemicals was reduced, and so is their genotoxicity, by their coupling with nanoparticle-based carriers (Pereira et al. 2014).

Toxicity toward the environment was further reduced by encapsulation. Chitosan and sodium compound-based nanoparticles upon encapsulation have less toxicity compared to sole active components (Grillo et al. 2014). Researchers have prepared the controlled release of herbicides from cross-linked chitosan nanoparticles with disulfide bonds of diuron, based on the glutathione concentration. The results obtained were positive toward the plant's growth and supported the strategies for reduced toxicity (Fig. 16.3) (Yu et al. 2015). In a study, hybrid nanocapsules were formulated from the polymeric polycaprolactones (PCL). These nanocapsules were filled with three different herbicides, namely, atrazine, ametryn, and simazine, and displayed a systemic controlled release (Grillo et al. 2014).

For targeted and systemic delivery of herbicides, nanocapsules are very effective. Nanoparticles inject herbicides in the plants via tissues and cuticles. The releases of active substance through nanocapsules were constant and slow (Duran and Maezrcato 2013).

Toxic chemical-based herbicide, paraquat, controls grass and weeds. Alginate/chitosan-based nanoparticles were used by Silva et al. (2011) as a carrier system for its application as systemic weedicides. The study based on the release profile of the herbicide (paraquat) with alginate/chitosan nanoparticles demonstrated that its association with nanoparticles alters not just its release but also its interaction with the soil. This is an effective system for controlling and minimizing the negative effects of paraquat. There is one more study based on adsorbent conducted by Celis ct al. (2012) involving bio-nanocomposite material made of chitosan and clay (montmorillonite). Herbicide clopyralid present in an aqueous solution or in a mixture of water and soil can be removed by the bio-nanocomposites. Superb herbicide adsorption capacity was shown by bio-nanocomposites at pH levels at which the anionic form of the active principle and the cationic form of chitosan predominated. The efficiency of the adsorbent for the removal of herbicides from aqueous solution was more effective when a higher concentration of chitosan was used in the bio-nanocomposite. Bio-composites can be used to remove herbicides (clopyralid) from the soil at pH below neutral, and the use of this type of formulation reduces the mobility of the pesticides/herbicides of anionic nature.

Microbes like viruses are wonderful example of naturally occurring nanoparticles and behave similar to manmade nanoparticles with core and outer coating. Tobacco mild green mosaic virus (TMGMV) is a perfect example where bioherbicides were released to control a very invasive perennial weed (TSA; *Solanum viarum*) of pastures and other non-cropland areas in the southeastern and mid-southern regions of the United States. This systemic control of weed has been patented. Nanoparticles were prepared involving one adjuvant which is an organosilicon and a carborundum abrasive. Carborundum helps virus to allow its entry. The formulation can be delivered using simple standard application equipment (Charudattan and Hiebert 2007; Ferrell et al. 2008). This TMGMV-based bioherbicide application system is compatible as well as remains infective even in the presence of several herbicides and control TSA and other weeds that often occur together at weed-infested sites (Ferrell et al. 2008).

16.3.6 Nanofertilizers

Nanosized carriers decrease the quantity of chemicals to be employed, environmental complications, and their run-off. For the controlled release of agrochemicals, clay nanotubes have proven to be a good one whereas it increases the connection between the plants and chemicals, thereby decreases the amount and price of the chemicals 70%–80% (Ditta 2012). In a study, anionic nanoclay was employed for the controlled release of agrochemicals and plant growth regulators (Oancea et al. 2009).

Nanoemulsions addition to fertilizers, pesticides, or herbicides micelles can increase the efficacy of fertilizers and pesticides while decilne the volume of agrochemicals utilized (Fig. 16.3). Moreover, it can improve and strengthen the moisturizing ability and dispersion ability of agrochemicals and decrease the unenviable chemical run-off (Bergeson 2010). In addition, these are quickly degraded in soil, but exhibit slow degradation in plants (Khot et al. 2012). For instance, to control soybean pests, a nano-encapsulated imidacloprid was evolved by employing sodium dodecyl sulfate (SDS) modified Ag/TiO_2 (Guan et al. 2010).

An increased plant growth was noticed in the tomato seeds that were grown in a soil comprising CNTs. Due to the penetration of CNTs into the seeds, there was an increment in the water uptake. Furthermore, CNT could be used as a carrier to direct the desirable molecules into the seeds at the germination time. It would protect the plant from the diseases beside its growth-promoting and nontoxic properties (Khodakovskaya et al. 2009).

An optimized composition of gentle delivery of fertilizers and excellent absorbent polymers may not only crucially promote the yield and plant nutrition but it might be a way to alleviate the impact of stressed environment, decrease frequency of irrigation, and decline water loss to evaporation (Davidson and Gu 2012). Compound NPK fertilizer coated with chitosan was evolved with water retention and steady release properties, by coating the chitosan inside and poly(acrylic acid-co acrylamide) [P(AA-co-AM)] outside, which is an absorbent polymer (Wu et al. 2008). Nutrients were delivered in slow and steady manner. At the 30th day, 75% was not reached by the nutrients that have been delivered. Chitosan is a readily available biodegradable matter, while P(AA-co-AM) can be decayed in the soil. These products have a substantial possibility of being environmentally friendly nanofertilizers, especially for drought-prone areas with less availability of water. In a similar case, Corradini et al. (2010) furnished the possibility of consuming the chitosan nanoparticles for slow and steady delivery of NPK fertilizer, while Hussain et al. (2012) observed the steady release of urea from the chitosan microspheres. It may be costly than the simple wide application of fertilizer as well as it can prevent nutrient losses (Fig. 16.3).

16.4 Targeted Genetic Engineering

For agronomic and economic trait enhancement, nanotechnology is being used to modify crop's genetic constitution (Agrawal and Rathore 2014). For releasing genetic materials, chitosan-modified nanoparticles with polyethylene glycol (PEG) have been used as gene carriers. Fluorescent molecules are used for labeling starch nanoparticles which in turn are used for transporting genetic materials across the cell wall by producing aperture channels. For transporting DNA into the nucleus, plasmic DNA is used which is coated with silver nanoparticles. For transporting genes directly into the cell, the gene gun is the most widely used tool. For successful expression and delivery of plasmatic DNA into maize and tobacco, bombardment material gold nanoparticles were used (Nair et al. 2010). Mutation either natural or artificial plays an important role in crop improvement. Instead of using chemical compounds and/or physical mutagens, a remarkable ability has been shown by nanotechnology for inducing mutation. Hence by using nanotechnological approaches from traditional rice, which is purple in color, a new white-grained rice variety has been developed (Shrivastava and Dash 2012).

Plants can be genetically manipulated by combining nanotechnology with other technologies like biotechnology. For example, by using nanocapsules or nanofibers as vectors instead of using conventional viral vectors, genetic expression can be easily triggered and can carry a large number of genes (Miller et al. 2008; Nair et al. 2010). Transgenic plants can be produced effectively by using nanoparticles instead of transporting exogenous DNA into the cells. Better transfection methods are enabled by linking nanoparticles with lipids, ligands, and proteins which break the endosomal barriers and cross the membrane more efficiently.

Using nanomaterials as genetic tool, it has wide application in drug delivery, environmental monitoring, the use of nanofiber array for efficient delivery of genetic material into the cells, and crop engineering (Miller and Kinnear 2007).

Traditional methods for transferring gene into plants such as electroporation, Agrobacterium-mediated gene transfer, particle gun bombardment, PEG-mediated gene transfer, etc. are quite costly and also have several other disadvantages, for example, they have low efficiency around 0.01%–20% (Sivamani et al. 2009). Using nanotechnology for crop improvement has significant advantages when compared to other traditional and conventional techniques. Some of the main advantages are avoidance of transgenic silencing by combining nanoparticles with DNA copies and no requirement of complex carriers during the transformation mediated by nanoparticles. These key features can be taken into consideration for using nanoparticles as gene carriers.

Several works have been carried out that (Torney et al. 2007) used silica nanoparticles to deliver DNA and other chemicals into plant cells without the aid of any specialized equipment. Mesoporous silica nanoparticles were employed (Martin et al. 2012) to co-deliver DNA and protein into plant cells using biolistic methods.

In gene delivery, biodegradable chitosan is used because of its protonating ability in acidic solution, thereby forming a complex with DNA. The complex formation happens because of electrostatic interactions (Duceppe and Tabrizian 2010), but chitosan has low transfection efficiency. Certain factors like pH of the transfecting medium, molecular weight of chitosan, and deacetylation degree influence the transfection efficiency of chitosan (Mao et al. 2010). Due to the cationic nature of chitosan, complex formation between chitosan and small interfering RNA (siRNA) is easier and has grabbed much attention in recent times, and hence for delivery of siRNA, it can be used as a carrier (Katas and Alpar 2006; Malmo et al. 2012; Ragelle et al. 2013). dsRNA has been delivered successfully to mosquito larvae in stabilized form using chitosan nanoparticles (Zhang et al. 2010). For efficient delivery of dsRNA, chitosan nanoparticles could be the best choice as it can bind with RNA efficiently. DNA-coated chitosan nanoparticles have a real potential in the near future to be used as nanocarrier for plant tissues and cell bombardment and gene gun to attain targeted gene transfer.

16.5 Labeling and Imaging

The transition of elements and several molecules in both soils and plants could be investigated through labeling. Labeling enables the researchers to get critical information about effectiveness, significance, and mechanism of action of important chemicals. Traditional labeling involves using organic dyes, but the major disadvantage it has is fast fluorescence degradation which has limited its application. In luminescent labeling field labeling molecules by quantum dots (QD) has grabbed much attention recently. As an alternative to bioanalytical methods, nanosized QD have been considered due to its exclusive optical features such as high fluorescence intensity, narrow emission spectrum, broad tenability, and photostability. QD is more beneficial when compared to conventional fluorescence dyes as it has broad excitation spectrum, high quantum yield, narrow emission, extended fluorescence lifetime, and high degree of stability against photobleaching. For molecular imaging, fluorescent QD has been used commonly nowadays (Zimnitsky et al. 2007; Frasco and Chaniotakis 2009).

16.6 Future Prospects

Despite all the improvement in nanotechnological field, the application of nanoparticle in the field of crop improvement and agriculture has not yet reached to a satisfying level, and more research is needed (Mishra et al. 2014). As per the data given by the Food and Agriculture Organization (FAO) assessment, each year there has been a loss of up to 20–40% in the overall global crop production. This loss has happened due to diseases and pests even though pesticides have been applied (King 2017). Pesticides are severe threat to humans and animals as well as to the environment. In June 2009, a joint venture held by the World Health Organization (WHO) and FAO signified the importance of applying nanotechnology in the field of agriculture and food (Takeuchi et al. 2014). And thus, current trends show that studies are being conducted in the direction of controlled release of pesticides, improved storage, and smart delivery of agrochemicals (Devi et al. 2019). To overcome the threat posed to the environment by pesticides, as a pest control agent, nanocarriers have been applied recently, which also have a major role in drug delivery (Liu et al. 2015). Nanoscience has paved the way for crop improvement by improving the nutrient uptake capability of plants, thereby increasing the yield, and managing the disease at the molecular level (Park et al. 2006). A smart delivery system has been represented by NPs which acts as a "magic bullet" delivering organics, beneficial genes, and herbicides to the targeted plant parts by which productivity is enhanced (Marchiol et al. 2014). In agricultural application area, nanobiopolymers have a great role in dosage reduction of agrochemicals, enhancing the efficiency of nutrient management and in maintenance of soil health. Drug-resistant microbes can be inactivated by bionanohybrid (Baker et al. 2017).

Similarly, researchers are getting motivation from biochar, to make biochar-based eco-friendly fertilizer which helps to improve crop production by controlling the release of fertilizer, thereby minimizing nutrient loss. Due to its porous nature, biochar can be incorporated with essential nutrients and other materials such as phosphates (Xu et al. 2014), nitrates (Hale et al. 2013; Kammann et al. 2015), urea (Khan et al. 2008; Xie et al. 2011), and ammonium salts (Spokas et al. 2012), thus making it available on slow release basis to soil and crops. Besides helping in reducing and managing waste, this kind of approach also helps in preventing pollution by producing wealth from waste.

Future use of nanotechnology in agriculture will cover suppression of plant stress/diseases directly through antimicrobial/antioxidant/heavy metal uptake activity (Tripathi et al. 2017b; Rastogi et al. 2019). Sustainable growth in agriculture should be the prime focus, as enough destruction of nature has been done, and ultimately it is going to harm human beings. So, nanotechnology has shown us the way, and researchers have come with lots of alternatives which are environment-friendly and will also ensure high yield from crops.

References

Agnihotri S, Mukherji S, Mukherji S (2012) Antimicrobial chitosan–PVA hydrogel as a nanoreactor and immobilizing matrix for silver nanoparticles. Appl Nanosci 2(3):179–188

Aguilar-Méndez MA, San Martín-Martínez E, Ortega-Arroyo L, Cobián-Portillo G, Sánchez-Espíndola E (2011) Synthesis and characterization of silver nanoparticles: effect on phytopathogen Colletotrichum gloesporioides. J Nanopart Res 13(6):2525–2532

Aravinthan A, Govarthanan M, Selvam K, Praburaman L, Selvankumar T, Balamurugan R, Kim JH (2015) Sunroot mediated synthesis and characterization of silver nanoparticles and evaluation of its antibacterial and rat splenocyte cytotoxic effects. Int J Nanomedicine 10:1977

Baker S, Volova T, Prudnikova SV, Satish S, Prasad N (2017) Nanoagroparticles emerging trends and future prospect in modern agriculture system. Environ Toxicol Phar 53:10–17

Ball P (2002) Natural strategies for the molecular engineer. Nanotechnology 13(5):R15

Barik TK, Kamaraju R, Gowswami A (2012) Silica nanoparticle: a potential new insecticide for mosquito vector control. Parasitol Res 111(3):1075–1083

Ben-shalom N, Ardi R, Pinto R, Aki C, Fallik E (2003) Controlling gray mould caused by *Botrytis cinerea* in cucumber plants by means of chitosan. Crop Prot 22:285–290

Bergeson LL (2010) Nanosilver: US EPA's pesticide office considers how best to proceed. Environ Qual Manag 19(3):79–85

Bharati S, Suresh A (2017) Review on nano-catalyst from waste for production of biofuel-via-bioenergy. In: Biofuels and bioenergy (BICE2016). Springer, Cham, pp 25–32

Bhatkhande DS, Pangarkar VG, Beenackers AA (2002) Photocatalytic degradation for environmental applications – a review. J Chem Technol Biotechnol 77(1):102–116

Bhor G, Maskare S, Hinge S, Singh L, Nalwade A (2014) Synthesis of silver nanoparticles by using leaflet extract of Nephrolepis exaltata L and evaluation of antibacterial activity against human and plant pathogenic bacteria. Asian J Pharm Technol 02(07):6

Boehm AL, Martinon I, Zerrouk R, Rump E, Fessi H (2003) Nanoprecipitation technique for the encapsulation of agrochemical active ingredients. J Microencapsul 20(4):433–441

Celis R, Adelino MA, Hermosín MC, Cornejo J (2012) Montmorillonite-chitosan bionanocomposites as adsorbents of the herbicide clopyralid in aqueous solution and soil/water suspensions. J Hazard Mater 209–210:67–76

Charudattan R, Hiebert E (2007) A plant virus as a bioherbicide for tropical soda apple, *Solanum viarum*. Outlooks Pest Manag 18(4):167–171

Chinnamuthu CR, Boopathi PM (2009) Nanotechnology and agroecosystem. Madras Agric J 96(1/6):17–31

Chowdhury S, Basu A, Kundu S (2014) Green synthesis of protein capped silver nanoparticles from phytopathogenic fungus Macrophomina phaseolina (Tassi) Goid with antimicrobial properties against multidrug-resistant bacteria. Nanoscale Res Lett 9(1):65

Cioffi N, Ditaranto N, Torsi L, Picca RA, Sabbatini L, Valentini A, Novello G, Tantillo T, Zambonin PG, Bleve-Zacheo (2005) Analytical characterization of bioactive fluoropolymer ultra-thin coatings modified by copper nanoparticles. Anal Bioanal Chem 381(3):607–616

Corradini E, De Moura MR, Mattoso LHC (2010) A preliminary study of the incorporation of NPK fertilizer into chitosan nanoparticles. Express Polym Lett 4(8):509–515

Cui HF, Wu WW, Li MM, Song X, Lv Y, Zhang TT (2018) A highly stable acetylcholinesterase biosensor based on chitosan-TiO2-graphene nanocomposites for detection of organophosphate pesticides. Biosens Bioelectron 99:223–229

Davidson D, Gu FX (2012) Materials for sustained and controlled release of nutrients and molecules to support plant growth. J Agric Food Chem 60(4):870–876

Devi PV, Duraimurugan P, Chandrika KSVP (2019) Bacillus thuringiensis-based nanopesticides for crop protection. In: Nano-biopesticides today and future perspectives. Academic, London, pp 249–260

Dimkpa CO, McLean JE, Britt DW, Anderson AJ (2013) Antifungal activity of ZnO nanoparticles and their interactive effect with a biocontrol bacterium on growth antagonism of the plant pathogen Fusarium graminearum. Biometals 26(6):913–924

Ditta A (2012) How helpful is nanotechnology in agriculture? Adv Nat Sci Nanosci Nanotechnol 3(3):033002

Donaldson K, Stone V, Tran CL, Kreylin W, Borm PJA (2004) Nanotoxicology. Occup Environ Med 61:727–728

Dubchak S, Ogar A, Mietelski JW, Turnau K (2010) Influence of silver and titanium nanoparticles on arbuscular mycorrhiza colonization and accumulation of radiocaesium in *Helianthus annuus*. Span J Agric Res 8(1):103–108

Duceppe N, Tabrizian M (2010) Advances in using chitosan-based nanoparticles for in vitro and in vivo drug and gene delivery. Expert Opin Drug Deliv 7:1191–1207

Duran N, Maezrcato PD (2013) Nanobiotechnology perspectives role of nanotechnology in the food industry: a review. Int J Food Sci Technol 48:1127–1134

Dzung NA, Thang NT, Suchiva VK, Chandrkrachang S, Methacanon P, Peter MG (2002) Effects of oligoglucosamine prepared by enzyme degradation on the growth of soybean. Adv Chitin Sci Bangkok 5:463–467

El-Bendary HM, El-Helaly AA (2013) First record nanotechnology in agricultural: silica nanoparticles a potential new insecticide for pest control. App Sci Rep 4(3):241–246

Elmer WH, White JC (2016) The use of metallic oxide nanoparticles to enhance growth of tomatoes and eggplants in disease infested soil or soilless medium. Environ Sci Nano 3(5):1072–1079

Fan R, Huang YC, Grusak MA, Huang CP, Sherrier DJ (2014) Effects of nano TiO2 on the agronomically-relevant Rhizobium–legume symbiosis. Sci Total Environ 466:503–512

Farré M, Sanchís J, Barceló D (2011) Analysis and assessment of the occurrence, the fate and the behavior of nanomaterials in the environment. Trac-Trend Anal Chem 30(3):517–527

Feigl C, Russo SP, Barnard AS (2010) Safe, stable and effective nanotechnology: phase mapping of ZnS nanoparticles. J Mater Chem 20(24):4971–4980

Feng BH, Peng LF (2012) Synthesis and characterization of carboxymethyl chitosan carrying ricinoleic functions as an emulsifier for azadirachtin. Carbohydr Polym 88(2):576–582

Ferrell J, Carudattan R, Elliott M, Hiebert E (2008) Effects of selected herbicides on the efficacy of tobacco mild green mosaic virus to control tropical soda apple (*Solanum viarum*). Weed Sci 56(1):128–132

Fountain ED, Wratten SD (2013) Conservation biological control and biopesticides in agricultural. In: Reference module in earth systems and environmental sciences. Elsevier, San Diego, pp 377–381

Frasco MF, Chaniotakis N (2009) Semiconductor quantum dots in chemical sensors and biosensors. Sensors 9(9):7266–7286

Gajbhiye M, Kesharwani J, Ingle A, Gade A, Rai M (2009) Fungus-mediated synthesis of silver nanoparticles and their activity against pathogenic fungi in combination with fluconazole. Nanomed-Nanotechnol 5(4):382–386

Ge Y, Schimel JP, Holden PA (2012) Identification of soil bacteria susceptible to TiO2 and ZnO nanoparticles. Appl Environ Microbiol 78(18):6749–6758

Gerland P, Raftery AE, Ševcíková H, Li N, Gu D, Spoorenberg T, Alkema L, Fosdick BK, Chunn J, Lalic N, Bay G, Buettner T, Heilig GK, Wilmoth J (2014) World population stabilization unlikely this century. Science 346(6206):234–237
Ghormade V, Deshpande MV, Paknikar KM (2011) Perspectives for nano-biotechnology enabled protection and nutrition of plants. Biotechnol Adv 29:792–803
Goswami A, Roy I, Sengupta S, Debnath N (2010) Novel applications of solid and liquid formulations of nanoparticles against insect pests and pathogens. Thin Solid Films 519(3):1252–1257
Grassini P, Eskridge KM, Cassman KG (2013) Distinguishing between yield advances and yield plateaus in historical crop production trends. Nat Commun 4:2918
Grillo R, Pereira AES, Nishisaka CS, Lima RD, Oehlke K, Greiner R, Leonardo F, Fraceto LF (2014) Chitosan/tripolyphosphate nanoparticles loaded with paraquat herbicide: an environmentally safer alternative for weed control. J Hazard Mater 278:163–171
Guan H, Chi D, Yu J, Li H (2010) Dynamics of residues from a novel nano-imidacloprid formulation in soyabean fields. Crop Prot 29(9):942–946
Guan H, Chi D, Yu J, Li X (2008) A novel photodegradable insecticide: preparation, characterization and properties evaluation of nano-Imidacloprid. Pestic Biochem Physiol 92:83–91
Hale S, Alling V, Martinsen V, Mulder J, Breedveld G, Cornelissen G (2013) The sorption and desorption of phosphate-P, ammonium-N and nitrate-N in cacao shell and corn cob biochars. Chemosphere 91:1612–1619
Hamdi H, De La Torre-Roche R, Hawthorne J, White JC (2014) Impact of non-functionalized and amino-functionalized multiwall carbon nanotubes on pesticide uptake by lettuce (Lactuca sativa L). Nanotoxicology 9(2):172–180
He L, Liu Y, Mustapha A, Lin M (2011) Antifungal activity of zinc oxide nanoparticles against Botrytis cinerea and Penicillium expansum. Microbiol Res 166(3):207–215
Huang YC, Fan R, Grusak MA, Sherrier JD, Huang CP (2014) Effects of nano ZnO on the agronomically relevant Rhizobium–legume symbiosis. Sci Total Environ 497:78–90
Hussain MR, Devi R, Maji TK (2012) Controlled release of urea from chitosan microspheres prepared by emulsification and cross-linking method. Iran Polym J 21:473–479
Jain N, Bhargava A, Majumdar S, Tarafdar JC, Panwar J (2011) Extracellular biosynthesis and characterization of silver nanoparticles using Aspergillus flavus NJP08: a mechanism perspective. Nanoscale 3(2):635–641
Jayaseelan C, Rahuman AA, Kirthi AV, Marimuthu S, Santhoshkumar T, Bagavan A et al (2012) Novel microbial route to synthesize ZnO nanoparticles using Aeromonas hydrophila and their activity against pathogenic bacteria and fungi. Spectrochim Acta A 90:78–84
Jayaseelan C, Rahuman AA, Rajakumar G, Kirthi AV, Santhoshkumar T, Marimuthu S et al (2011) Synthesis of pediculocidal and larvicidal silver nanoparticles by leaf extract from heartleaf moonseed plant, *Tinospora cordifolia Miers*. Parasitol Res 109(1):185–194
Jo YK, Kim BH, Jung G (2009) Antifungal activity of silver ions and nanoparticles on phytopathogenic fungi. Plant Dis 93(10):1037–1043
Judy JD, Kirby JK, McLaughlin MJ, McNear D, Bertsch PM (2016) Symbiosis between nitrogen-fixing bacteria and Medicago truncatula is not significantly affected by silver and silver sulfide nanomaterials. Environ Pollut 214:731–736
Kahveci Z, Martinez-Tome MJ, Mallavia R, Mateo CR (2016) Fluorescent biosensor for phosphate determination based on immobilized polyfluorene-liposomal nanoparticles coupled with Alkaline Phosphatase. ACS Appl Mater Interface 9(1):136–144
Kammann CI, Schmidt H-P, Messerschmidt N, Linsel S, Steffens D, Müller C, Koyro H-W, Conte P, Stephen J (2015) Plant growth improvement mediated by nitrate capture in co-composted biochar. Sci Rep 5:11080
Kasprowicz MJ, Kozioł M, Gorczyca A (2010) The effect of silver nanoparticles on phytopathogenic spores of Fusarium culmorum. Can J Microbiol 56(3):247–253
Katas H, Alpar HO (2006) Development and characterisation of chitosan nanoparticles for siRNA delivery. J Control Release 115:216–225
Keswani C, Sarma BK, Singh HB (2016) Synthesis of policy support, quality control, and regulatory management of biopesticides in sustainable agriculture. In: Singh HB, Sarma BK,

Keswani C (eds) Agriculturally important microorganisms: commercialization and regulatory requirements in Asia. Springer, Singapore, pp 3–12
Khan MA, Kim K-W, Mingzhi W, Lim B-K, Lee W-H, Lee J-Y (2008) Nutrient-impregnated charcoal: an environmentally friendly slow-release fertilizer. Environmentalist 28:231–235
Khandelwal A, Joshi R (2018) Synthesis of nanoparticles and their application in agriculture. Acta Sci Agric 2(3):10–13
Khodakovskaya MV, Dervishi E, Mahmood M, Xu Y, Li Z, Watanabe F, Biris AS (2009) Carbon nanotubes are able to penetrate plant seed coat and dramatically affect seed germination and plant growth. ACS Nano 3:3221–3227
Khoobdel M, Ahsaei SM, Farzaneh M (2017) Insecticidal activity of polycaprolactone nanocapsules loaded with Rosmarinus officinalis essential oil in Tribolium castaneum (Herbst). Entomol Res 47(3):175–184
Khot LR, Sankaran S, Maja JM, Ehsani R, Schuster EW (2012) Applications of nanomaterials in agricultural production and crop protection: a review. Crop Prot 35:64–70
Kim JS, Kuk E, Yu KN, Kim JH, Park SJ, Lee HJ et al (2007) Antimicrobial effects of silver nanoparticles. Nanomed-Nanotechnol 3(1):95–101
King A (2017) The future of agriculture. Nature 544(7651):S21–S23
Ko YD, Kang JG, Park JG, Lee S, Kim DW (2009) Self-supported SnO2 nanowire electrodes for high-power lithium-ion batteries. Nanotechnology 20(45):455701
Kress WJ et al (2005) Use of DNA barcodes to identify flowering plants. Proc Natl Acad Sci U S A 102(23):8369–8374
Kumar N, Palmer GR, Shah V, Walker VK (2014) The effect of silver nanoparticles on seasonal change in arctic tundra bacterial and fungal assemblages. PLoS One 9:e99953
Kumar N, Sharma S, Nara S (2018) Dual gold nanostructure-based electrochemical immunosensor for CA125 detection. Appl Nanosci 8(7):1843–1853
Kumar N, Tripathi P, Nara S (2017) Gold nanomaterials to plants: impact of bioavailability, particle size and surface coating. In: Nanomaterials in plants, algae and micro-organism: concepts and controversies. Elsevier, London, pp 195–220
Kumar P, Burman U, Santra P (2015) Effect of nano-zinc oxide on nitrogenase activity in legumes: an interplay of concentration and exposure time. Int Nano Lett 5:191–198
Lamsal K, Kim SW, Jung JH, Kim YS, Kim KS, Lee YS (2011a) Application of silver nanoparticles for the control of Colletotrichum species in vitro and pepper anthracnose disease in field. Mycobiology 39:194–199
Lamsal K, Kim SW, Jung JH, Kim YS, Kim KS, Lee YS (2011b) Inhibition effects of silver nanoparticles against powdery mildews on cucumber and pumpkin. Mycobiology 39(1):26–32
Lao SB, Zhang ZX, Xu HH, Jiang GB (2010) Novel amphiphilic chitosan derivatives: synthesis, characterization and micellar solubilization of rotenone. Carbohydr Polym 824:1136–1142
Latin R (2006) Residual efficacy of fungicides for control of dollar spot on creeping bentgrass. Plant Dis 50:571–575
Li D, Haneda H (2003) Morphologies of zinc oxide particles and their effects on photocatalysis. Chemosphere 51(2):129–137
Li Z, Xue N, Ma H, Cheng Z, Miao X (2018) An ultrasensitive and switch-on platform for aflatoxin B 1 detection in peanut based on the fluorescence quenching of graphene oxide-gold nanocomposites. Talanta 181:346–351
Liu F, Wen LX, Li ZZ, Yu W, Sun HY, Chen JF (2006a) Porous hollow silica nanoparticles as controlled delivery system for water-soluble pesticide. Mater Res Bull 41:2268–2275
Liu XM, Feng ZB, Zhang FD, Zhang SQ, He XS (2006b) Preparation and testing of cementing and coating nano-subnanocomposites of slow/controlled-release fertilizer. Agric Sci China 5(9):700–706
Liu X, He B, Xu Z, Yin M, Yang W, Zhang H, Cao J, Shen J (2015) A functionalized fluorescent dendrimer as a pesticide nanocarrier: application in pest control. Nanoscale 7:445–449
Malmo J, Sørgård H, Vårum KM, Strand SP (2012) siRNA delivery with chitosan nanoparticles: molecular properties favoring efficient gene silencing. J Control Release 158:261–268

Mao S, Sun W, Kissel T (2010) Chitosan-based formulations for delivery of DNA and siRNA. Adv Drug Deliv Rev 62:12–27

Marchiol L, Mattiello A, Pošćić F, Giordano C, Musetti R (2014) In vivo synthesis of nanomaterials in plants: location of silver nanoparticles and plant metabolism. Nanoscale Res Lett 9:101

Martin OS, Valenstein JS, Lin VSY, Trewyn BG, Wang K (2012) Gold functionalized mesoporous silica nanoparticle mediated protein and DNA codelivery to plant cells via the biolistic method. Adv Funct Mater 22:3576–3582

Mehta CM, Srivastava R, Arora S, Sharma AK (2016) Impact assessment of silver nanoparticles on plant growth and soil bacterial diversity. Biotech 6:254–263

Meyers BC, Galbraith DW, Nelson T, Agrawal V (2004) Methods for transcriptional profiling in plants. Be fruitful and replicate. Plant Physiol 135(2):637–652

Michels C, Perazzoli S, Soares HM (2017) Inhibition of an enriched culture of ammonia oxidizing bacteria by two different nanoparticles: silver and magnetite. Sci Total Environ 586:995–1002

Miller G, Kinnear S (2007) Nanotechnology the new threat to food. Clean Food Org 4:31–33

Miller G, Lowrey N, Senjen R (2008) Out of the laboratory and on to our plates: nanotechnology in food & agriculture. Friends of the Earth, Melbourne

Min JS, Kim KS, Kim SW, Jung JH, Lamsal K, Kim SB et al (2009) Effects of colloidal silver nanoparticles on sclerotium-forming phytopathogenic fungi. Plant Pathol J 25(4):376–380

Mishra V, Mishra RK, Dikshit A, Pandey AC (2014) Interactions of nanoparticles with plants: an emerging prospective in the agriculture industry. In: Emerging technologies and management of crop stress tolerance. Academic, Burlington, pp 159–180

Moll J, Gogos A, Bucheli TD, Widmer F, Heijden MG (2016) Effect of nanoparticles on red clover and its symbiotic microorganisms. J Nanobiotechnol 14(1):36

Mondal KK, Mani C (2012) Investigation of the antibacterial properties of nanocopper against Xanthomonas axonopodis pv punicae, the incitant of pomegranate bacterial blight. Ann Microbiol 62(2):889–893

Moraru CI, Panchapakesan CP, Huang Q, Takhistov P, Liu S, Kokini JL (2003) Nanotechnology: a new frontier in food science. Food Technol 57:24–29

Musarrat J, Dwivedi S, Singh BR, Al-Khedhairy AA, Azam A, Naqvi A (2010) Production of antimicrobial silver nanoparticles in water extracts of the fungus Amylomyces rouxii strain KSU-09. Bioresour Technol 101(22):8772–8776

Nair R, Varghese SH, Nair BG, Maekawa T, Yoshida Y, Kumar DS (2010) Nanoparticulate material delivery to plants. Plant Sci 179(3):154–163

Nam J-M, Thaxton CS, Mirkin CA (2003) Nanoparticle-based bio-bar codes for the ultrasensitive detection of proteins. Science 301(5641):1884–1886

Navarro E, Baun A, Behra R et al (2008) Environmental behavior and ecotoxicity of engineered nanoparticles to algae, plants, and fungi. Ecotoxicology 17(5):372–386

Netala VR, Kotakadi VS, Bobbu P, Gaddam SA, Tartte V (2016) Endophytic fungal isolate mediated biosynthesis of silver nanoparticles and their free radical scavenging activity and antimicrobial studies. 3 Biotech 6(2):132

Oancea S, Padureanu S, Oancea AV (2009) Growth dynamics of corn plants during anionic clays action. Luc Ştiint ific 52:212–217

Park HJ, Kim SH, Kim HJ, Choi SH (2006) A new composition of nanosized silica-silver for control of various plant diseases. Plant Pathol J 22(3):295–302

Patel N, Desa P, Pael N, Jha A, Gautam HK (2014) Agronatechlogy for plant fungal disease management: a review. Int J Cur Microbiol Appl Sci 3(10):71–84

Paula HCB, MSombra F, Cavalcante RF, Abreu FOMS, de Paula RCM (2011) Preparation and characterization of chitosan/cashew gum beads loaded with Lippia sidoides essential oil. Mater Sci Eng C 31:173–178

Paulkumar K, Gnanajobitha G, Vanaja M, Rajeshkumar S, Malarkodi C, Pandian K et al (2014) Piper nigrum leaf and stem assisted green synthesis of silver nanoparticles and evaluation of its antibacterial activity against agricultural plant pathogens. Sci World J 2014:1–9

Pereira AES, Grillo R, Mello NFS, Rosa AH, Fraceto LF (2014) Application of poly(epsilon-caprolactone) nanoparticles containing atrazine herbicide as an alternative technique to control weeds and reduce damage to the environment. J Hazard Mater 268:207–215
Peteu SF, Oancea F, Sicuia OA, Constantinescu F, Dinu S (2010) Responsive polymers for crop protection. Polym J 2:229–251
Prasad VK, Prasad KS (2014) Nanotechnology in sustainable agriculture: present concerns and future aspects. Afr J Biotechnol 13(6):706–713
Ragelle H, Vandermeulen G, Préa V (2013) Chitosan-based siRNA delivery systems. J Control Release 172:207–218
Rai M, Ingle A (2012) Role of nanotechnology in agriculture with special reference to management of insect pests. Appl Microbiol Biotechnol 94:287–293
Rai V, Acharya S, Dey N (2012) Implications of nanobiosensors in agriculture. J Biomater Nanobiotechnol 3:315–324
Rastogi A, Tripathi DK, Yadav S, Chauhan DK, Živčák M, Ghorbanpour M, El-Sheery NI, Brestic M (2019) Application of silicon nanoparticles in agriculture. 3 Biotech 9(3):90
Rauwel P, Küünal S, Ferdov S, Rauwel E (2015) A review on the green synthesis of silver nanoparticles and their morphologies studied via TEM. Adv Mater Sci Eng 2015:1–9
Agrawal S, Rathore P (2014) Nanotechnology pros and cons to agriculture. Int J Curr Microbiol App Sci 3(3):43–55
Sarkar A, Praveen G (2017) Utilization of waste biomass into useful forms of energy. In: Biofuels and bioenergy. Springer, Cham, pp 117–132
Sarlak N, Taherifar A, Salehi F (2014) Synthesis of nanopesticides by encapsulating pesticide nanoparticles using functionalized carbon nanotubes and application of new nanocomposite for plant disease treatment. J Agric Food Chem 62(21):4833–4838
Scrinis G, Lyons K, Sharmila Rahale C (2007) The emerging nano-corporate paradigm: nanotechnology and the transformation of nature, food and agri-food systems. Int J Soc Agric Food 15:22–44
Seo JW, Lee JH, Son IS, Kim YJ, Hwang Y, Chung HA, Kuswandi B, Wicaksono Y, Abdullah A, Heng LY, Ahmad M (2011) Smart packaging: sensors for monitoring of food quality and safety. Sens & Instrumen Food Qual 5:137–146
Shojaei TR et al (2009) The effect of plant growth regulators, cultivars and substrate combination on production of virus free potato minitubers. Afr J Biotechnol 8(19):4864–4871
Shrivastava S, Dash D (2012) Nanotechnology in food sector and agriculture. Proc Natl Acad Sci India Sect B Biol Sci 82(1):29–35
Shweta, Vishwakarma K, Sharma S, Narayan RP, Srivastava P, Khan AS, Dubey NK, Tripathi DK, Chauhan DK (2017) Plants and carbon nanotubes (CNTs) interface: present status and future prospects. In: Nanotechnology. Springer, Singapore, pp 317–340
Silva MS, Cocenzaa DS, Grillo R, de Meloa NFS, Tonelloa POS, de Oliveirac LC, Cassimirod DL, Rosaa AH, Fracetoa LF (2011) Paraquat-loaded alginate/chitosan nanoparticles: preparation, characterization and soil sorption studies. J Hazard Mater 190:366–374
Singh S, Vishwakarma K, Singh S, Sharma S, Dubey NK, Singh VK, Liu S, Tripathi DK, Chauhan DK (2017) Understanding the plant and nanoparticle interface at transcriptomic and proteomic level: a concentric overview. Plant Gene 11:265-272
Singh J, Vishwakarma K, Ramawat N, Rai P, Singh VK, Mishra RK, Sharma S (2019) Nanomaterials and microbes' interactions. Biotech 9(3):68
Sivamani E, DeLong RK, Qu R (2009) Protamine-mediated DNA coating remarkably improves bombardment transformation efficiency in plant cells. Plant Cell Rep 28(2):213–221
Spokas KA, Novak JM, Venterea RT (2012) Biochar's role as an alternative N-fertilizer: ammonia capture. Plant Soil 350:35–42
Steinborn A, Alder L, Spitzke M, Dork D, Anastassiades M (2017) Development of a QuEChERS-based method for the simultaneous determination of acidic pesticides, their esters, and conjugates following alkaline hydrolysis. J Agric Food Chem 65:1296–1305
Stern ST, McNeil SE (2008) Nanotechnology safety concerns revisited. Toxicol Sci 101:4–21

Takeuchi MT, Kojima M, Luetzow M (2014) State of the art on the initiatives and activities relevant to risk assessment and risk management of nanotechnologies in the food and agriculture sectors. Food Res Int 64:976–981

Tao S, Pang R, Chen C et al (2012) Synthesis, characterization and slow release properties of O-naphthylacetyl chitosan. Carbohydr Polym 88:1189–1194

Tilman D, Cassman KG, Matson PA, Naylor R, Polasky S (2002) Agricultural sustainability and intensive production practices. Nature 418:671–667

Torney F, Trewyn BG, Lin VS, Wang (2007) Mesoporous silica nanoparticles deliver DNA and chemicals into plants. Nat Nanotechnol 2(5):295–300

Tripathi DK, Singh S, Singh VP, Prasad SM, Dubey NK, Chauhan DK (2017b) Silicon nanoparticles more effectively alleviated UV-B stress than silicon in wheat (Triticum aestivum) seedlings. Plant Physiol Biochem 110:70–81

Tripathi DK, Mishra RK, Singh S, Singh S, Singh VP, Singh PK, Chauhan DK, Prasad SM, Dubey NK, Pandey AC (2017c) Nitric oxide ameliorates zinc oxide nanoparticles phytotoxicity in wheat seedlings: implication of the ascorbate-glutathione cycle. Front Plant Sci 8:1

Tripathi DK, Singh VP, Kumar D, Chauhan DK (2012) Impact of exogenous silicon addition on chromium uptake, growth, mineral elements, oxidative stress, antioxidant capacity, and leaf and root structures in rice seedlings exposed to hexavalent chromium. Acta Physiol Plant 34(1):279–289

Tripathi DK, Tripathi A, Shweta SS, Singh Y, Vishwakarma K, Yadav G, Sharma S, Singh VK, Mishra RK, Upadhyay RG, Dubey NK, Lee Y, Chauhan DK (2017a) Uptake, accumulation and toxicity of silver nanoparticle in autotrophic plants, and heterotrophic microbes: a concentric review. Front Microbiol 8(7):1–16

Vishwakarma K, Upadhyay N, Kumar N, Tripathi DK, Sharma S (2017a) Potential applications and avenues of nanotechnology in sustainable agriculture. In: Nanomaterials in plants, algae and microorganism: concepts and controversies. Elsevier, New York, pp 473–500

Vishwakarma K, Shweta, Upadhyay N, Singh J, Liu S, Singh VP, Prasad SM, Chauhan DK, Tripathi DK, Sharma S (2017b) Differential phytotoxic impact of plant mediated silver nanoparticles (AgNPs) and silver nitrate (AgNO3) on *Brassica sp*. Front Plant Sci 8:1–12

Vishwakarma K, Singh VP, Prasad SM, Chauhan DK, Tripathi DK, Sharma S (2019) Silicon and plant growth promoting rhizobacteria differentially regulate AgNPinduced toxicity in Brassica juncea: implication of nitric oxide. J Hazard Mater 121806

Vundavalli R, Vundavalli S, Nakka M, Rao DS (2015) Biodegradable nano-hydrogels in agricultural farming-alternative source for water resources. Procedia Mater Sci 10:548–554S

Wani AH, Shah MA (2012) A unique and profound effect of MgO and ZnO nanoparticles on some plant pathogenic fungi. J Appl Pharma Sci 2(3):4

Woo KS, Kim KS, Lamsal K et al (2009) An in vitro study of the antifungal effect of silver nanoparticles on oak wilt pathogen Raffaelea sp. J Microbiol Biotechnol 19:760–764

Wu L, Liu M, Liang R (2008) Preparation and properties of a double-coated slow-release NPK compound fertilizer with superabsorbent and water-retention. Bioresour Technol 99:547–554

Zhang X, Zhang J, Zhu KY (2010) Chitosan/double-stranded RNA nanoparticle-mediated RNA interference to silence chitin synthase genes through larval feeding in the African malaria mosquito (*Anopheles gambiae*). Insect Mol Biol 19:683–693

Xie L, Liu M, Ni B, Zhang X, Wang Y (2011) Slow-release nitrogen and boron fertilizer from a functional superabsorbent formulation based on wheat straw and attapulgite. Chem Eng 167:342–348

Xu G, Sun J, Shao H, Chang SX (2014) Biochar had effects on phosphorus sorption and desorption in three soils with differing acidity. Ecol Eng 62:54–60

Yang FL, Li XG, Zhu F, Lei CL (2009) Structural characterization of nanoparticle loaded with garlic essential oil and their insecticidal activity against *Tribolium castaneum* (Herbst) (Coleoptera: Tenebrionidae). J Agric Food Chem 57:10156–10162

Yang Y, Wang J, Xiu Z, Alvarez PJ (2013) Impacts of silver nanoparticles on cellular and transcriptional activity of nitrogen- cycling bacteria. Environ Toxicol Chem 32:1488–1494

Yu Z, Sun X, Song H, Wang W, Ye Z, Shi L, Ding K (2015) Glutathione-responsive carboxymethyl chitosan nanoparticles for controlled release of herbicides. Mater Sci Appl 6:591–604
Zimnitsky D, Jiang C, Xu J, Lin Z, Tsukruk VV (2007) Substrate-and time-dependent photoluminescence of quantum dots inside the ultrathin polymer LbL film. Langmuir 23(8):4509–4515
Zuo X, Zhang H, Zhu Q, Wang W, Feng J, Chen X (2016) A dual-colour fluorescent bio sensing platform based on WS2 nanosheet for detection of Hg2+ and Ag+. Biosens Bioelectron 85:464–470

17 Integrated Approach of Agri-nanotechnology: Recent Developments, Challenges and Future Trends

Prekshi Garg, Neha Srivastava, and Prachi Srivastava

Abstract

Nanotechnology is a very rousing area in the field of science and technology, which has multifaceted role in agriculture field. It is a branch of nanoscience that deals with the applications of nanomaterials and nanoparticles and provides ability to design a new material and product with improved quality, new advanced devices, smart medicine and sensors that play a promising role in agriculture and healthcare. Nanotechnology along with bioinformatics and biotechnology plays a significant part in the advancement of agriculture field. The amalgamation of these two branches helps in building tools and devices that are used in early diagnosis and treatment of plant diseases; improving the absorption of nutrients, site-specific delivery of active constituents and water treatment processes; controlling the chemical damages; and enhancing the crop quality and yield. Besides this, nanotechnology increases the production rate with low cost as well as non-toxic effect on living being and environment. The present review focusses on the various aspects of nanotechnology and nanoinformatics in the field of agriculture. This includes different approaches and applications of nanotechnology and nanoinformatics, especially in the area of agriculture.

Keywords

Nanotechnology · Nanoscience · Nanoparticles · Nanoinformatics · Agriculture

P. Garg · N. Srivastava · P. Srivastava (✉)
AMITY Institute of Biotechnology, AMITY University Uttar Pradesh, Lucknow, Uttar Pradesh, India
e-mail: psrivastava@amity.edu

M. Ghorbanpour et al. (eds.), *Biogenic Nano-Particles and their Use in Agro-ecosystems*, https://doi.org/10.1007/978-981-15-2985-6_17

17.1 Introduction

Nanoparticles usually refer to particles ranging from dimensions between 1 and 100 nm. Below 1 nm, the behaviour of these particles is explained by using known atomic, molecular and ionic interactions and forces, whereas particles above a dimension of 100 nm have properties similar to the bulk properties of the material. Particles within the range of 1–100 nm exhibit unique and variable properties due to quantum effects. Also, the particles having dimensions less than 100 nm show increased bioavailability and transportation through biological organisms, tissues, cells, organelles and membranes. The branch of biotechnology dealing with the study, analysis and use of nanoparticles is known as nanobiotechnology (Owolade and Ogunleti 2008; Guo 2009; Bhattacharyya and Bhaumik 2010; Wu and Chen 2012; Rani 2015; Manjunatha and Biradar 2016).

In contrast, nanoinformatics refers to the use of techniques of bioinformatics for analysing and processing informati0on about the structural and physiochemical properties of nanoparticles and nanomaterials, along with their interaction with the environment and their applications in different fields (González-Nilo et al. 2011). Majorly, nanoinformatics is used in the field of medicine and agriculture due to its small size, shape and ability to penetrate biological cells and tissues (Fig. 17.1). Doxil and Abraxane are two nanodrugs which are already approved and available in the market.

17.2 Integrated Approach of Nanotechnology and Nanoinformatics in Agriculture

Nanotechnology is recently emerging as a promising field to boost agricultural productivity at a large scale. With the emergence of combined approaches of nanotechnology, biotechnology and bioinformatics together can transform the current scenario

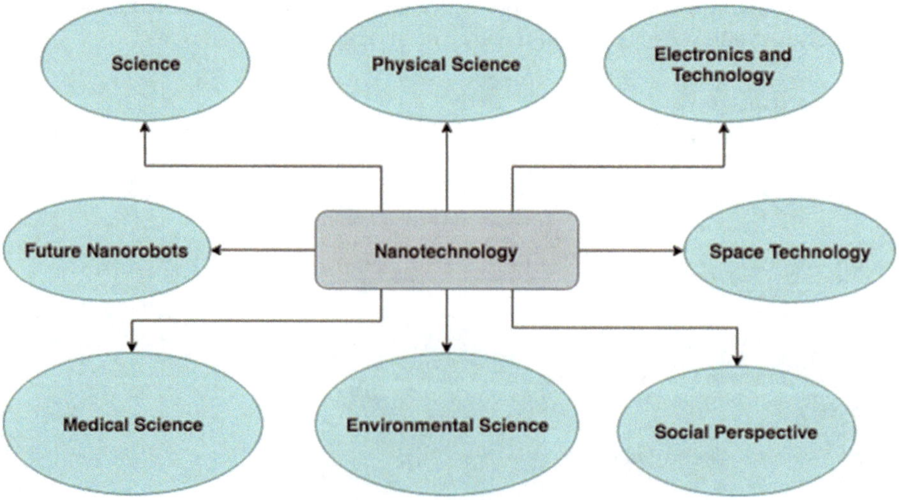

Fig. 17.1 Applications of nanotechnology in various fields

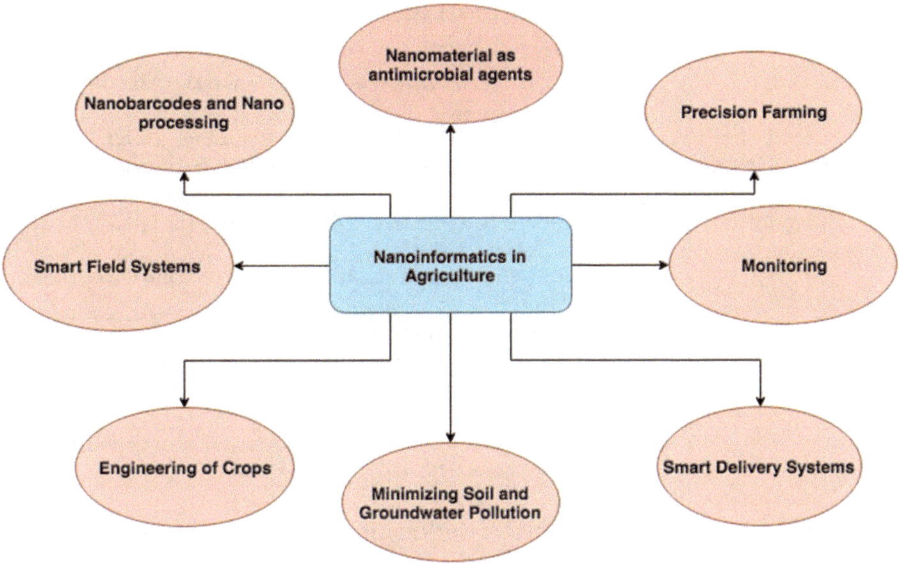

Fig. 17.2 Various integrated approaches of nanotechnology and nanoinformatics in agriculture

of agriculture. Nanosensors and nano-based smart delivery systems help ensure natural resources like water, nutrients and chemicals are used efficiently and in the appropriate amount to enhance crops in the field of agriculture. There are a number of different ways in which nanotechnology and nanoinformatics can help increase the yield of crops and also enhance the quality of products produced (Fig. 17.2).

17.3 Precision Farming

Precision farming is a technique of farming that aims to maximize the crop yield while minimizing the use of fertilizers, herbicides and pesticides. The use of less fertilizers, herbicides and pesticides will reduce the production cost of crops and will also minimize its effect on the environment (Joseph 2006). Computers, global satellite positioning systems and remote sensing devices are used to monitor localized environmental conditions and determine whether the crops are growing at maximum efficacy. Using centralized data that enables the farmers to evaluate soil conditions and the development of plants, generated from these devices, can help in limiting the quantity of fertilizers, chemicals and water that will eventually lead to high yield of plants at lower production cost (Rhodes 2014). Nanotechnology-enabled devices are increasingly used for autonomous sensors that are linked into a GPS which can provide real-time monitoring (Rathore 2014). These nanosensors are distributed throughout the field where they evaluate soil conditions and growth of crops. Wireless sensors are already being used in some parts of Australia and the USA. For example, Californian vineyards known as Pickberry, in Sonoma, have already installed wifi systems with the help of the IT company Accenture.

Table 17.1 List of nanoparticles and their antimicrobial property

Nanoparticle	Property
Polymer-based nanoparticles	Exhibit antifungal activity against various plant pathogenic fungi
Silica-silver nanoparticles (Abd-elsalam 2013)	Control plant pathogenic fungi like *Rhizoctonia solani*, *Botrytis cinerea*, *Magnaporthe grisea*, *Colletotrichum gloeosporioides* and *Pythium ultimum*
Silver nanoparticles	Have strong bactericidal activities and broad spectrum antimicrobial activities Reduce various plant diseases caused by spore-producing fungal pathogens
Zinc oxide nanoparticles (Abd-elsalam 2013)	Inhibit fungal growth of *Botrytis cinerea* by influencing cellular functions that lead to deformation in mycelial mats Inhibit the growth of conidiophores and conidia of *Penicillium expansum*
Silicon	Increase disease resistance and stress resistance by promoting the physiological activity and growth of plants

17.4 Nanomaterials as Antimicrobial Agents for Plant Pathogens

Metal nanoparticles like copper and silver nanoparticles are being studied in great detail by the researchers for their antimicrobial activities. The antimicrobial activities exhibited by various metal nanoparticles are of great application because of their capability to act against a wide range of microorganisms. There are various nanoparticles that are designed to be used as effective antimicrobial agents. Also, the products designed and prepared from nanoparticles show considerable effects against the powdery mildew disease of pumpkin within 3 days of spraying the product (Ingle 2012). The following lists nanoparticles and their antimicrobial property (Table 17.1).

17.5 Monitoring

Nanosensors are one of the most known sensors which are dispersed throughout the field to continuously monitor soil conditions and growth of crops. These integrated systems sense, monitor and generate active responses for plant production. Also, real-time monitoring is achieved by linking nanotechnology-enabled sensor devices with a GPS. Nanobiosensors are used for detection of pathogens, toxins and bacteria in food (Ghormade 2011). The following lists various nanoparticles that are used as sensors for various pesticides (Table 17.2).

Table 17.2 List of nanoparticles used as sensors

Nanoparticle used as nanosensor	Pesticide	Size of nanoparticle
Gold	Carbofuran/ triazophos	40 nm
Gold (Lisa 2009)	DDT	30 nm
Iron oxide	Dimethoate	30 nm
Zirconium oxide (Gan 2010; Wang 2009)	Organophosphate	50 nm
Silica (Ramanathan 2009)	Paraoxon	100–500 nm
Iron oxide (Kaushik 2009)	Pyrethroid	22 nm
Titanium oxide	Imidacloprid	30 nm

Table 17.3 Details of nanoparticles used for the delivery of pesticides, fertilizers, etc

Substance to be delivered	Nanoparticle used for delivery	Size of nanoparticle
Avermectin (chemical pesticide)	Porous hollow silica	15 nm
Ethiprole (chemical pesticide)	Polycaprolactone	135 nm
Gamma-cyhalothrin (chemical pesticide)	Solid lipid	300 nm
Chlorothalonil (chemical pesticide)	Polyvinylpyridine and polyvinylpyridine-co-styrene	100 nm
Plant origin biopesticide	Nanosilica	3–5 nm
Essential oil encapsulated biopesticide	Solid lipid	200–294 nm
Microorganism origin biopesticide	Silica	7–14 nm
NPK (fertilizer) (Corradini 2010)	Nano-coating of sulphur	100 nm layer
DNA (genetic material) (Elbaz and Willner 2012; Vijayakumar 2010)	Gold Starch	10–15 nm or 5–25 nm 50–100 nm
Double-stranded RNA (Zhang 2010)	Chitosan	100–200 nm

17.6 Smart Delivery Systems

Nanoscale smart devices are designed and used to identify dysfunctions in plant health before they have increased to an extent that they are visible to the farmers. Also, these devices provide counter remedies in response to dysfunctions. Therefore, these devices act as both an early warning system and as a curative device. Thus, chemical agents, such as fertilizers, pesticides and herbicides, can be delivered in a targeted and controlled manner. Nanoscale vehicles are also used for effective delivery of micronutrients and sensitive bioactives. The following lists nanoparticles that have been designed for the delivery of pesticides, fertilizers, etc. (Joseph 2006; Hussain 2017) (Table 17.3).

Table 17.4 Details of nanoparticles used to reduce soil and groundwater pollution

Nanoparticle	Function
Silver and gold nanoparticles	Destruction and mineralization of halocarbons Detection and extraction of endosulfan (a pesticide) Quantitative removal of chlorpyrifos and malathion
Silver nanoparticle-coated polyurethane foam	Antibacterial water film
Porum thin film of titanium oxide	Degradation of naphthalene and anthracene
Nanoparticles supported on alumina	Complete removal of pesticides from water

17.7 Minimizing Soil and Groundwater Pollution

Contaminated soil and groundwater can be cleaned by using iron nanoparticles as they can catalyse the oxidation of organic pollutants like PCBs, dioxins, carbon tetrachloride and trichloroethene to form less toxic carbon compounds. Iron oxide nanoparticle is also very effective in removing arsenic from groundwater. Therefore, nanoparticles remove contaminants from soil and water and provide conditions that enhance the quality and yield of crops. The following lists certain nanoparticles that are designed to reduce soil and groundwater pollution (Joshi 2010) (Table 17.4).

17.8 Engineering of Crops

Nanotechnology plays a vital role in the re-engineering of crops at genetic as well as cellular levels. This technique helps in the development of crops that have desired traits like drought resistance, salt tolerance, excess moisture tolerance, ability to withstand a wide range of temperature fluctuations and resistance to certain pests and pathogens. There is an increasing list of such traits which can be introduced into the crops to enhance the quality of crops and increase the crop yield. The following lists a few examples of nanoparticles that are designed to insert gene of interest into the plant cells in order to produce crops with desired traits (Table 17.5).

17.9 Smart Field Systems

Nanotechnology is used in the designing of smart field systems. These systems help in detecting the shortage of water in the field and locating that specific area in the field. These systems report the level of water in the soil continuously and direct the water to appropriate areas. Therefore, these systems ensure that the crops get sufficient amount of water so that the yield and quality of crops are enhanced.

Table 17.5 Details of nanoparticles used to produce crops with desired traits

Nanoparticle	Function
Carbon nanofibres	Carbon nanofibres are modified with plasmid DNA on the surface for quick, efficient and controlled delivery of genetic material into the plant cells
Chitosan nanoparticles	These are very versatile and efficient for the delivery of gene. Their efficiency can be modified by using PEGylated
Starch nanoparticles	These are fluorescently labelled and they efficiently transport gene across the plant cell wall by creating an instantaneous pore in the cell wall, cell membrane and nuclear membrane with the help of ultrasound
DNA-coated silver nanoparticle (Rad et al. 2013)	These carry plasmatic DNA into the nucleus of isolated protoplasts that have been incubated in the presence of ethylene glycol

17.10 Nanobarcodes and Nanoprocessing

The field of nanoinformatics is now widely used for the manufacturing of nanobarcodes (Tarafdar 2011). These barcodes can help monitor the quality of crops. With the help of nanobarcodes, the farmers can select the high-quality variety of crops and selectively increase the production of such crops. Therefore, nanotechnology can help in monitoring and enhancing the quality of crops and also increasing the yield of desired variety of crops.

17.11 Risk Factors and Future Considerations in the Field of Agri-nanotechnology

Future studies and researches in the field of nanotechnology should be carried out in such a manner that it would reduce the toxic effects of nanoparticles on the environment and the living beings (Sandhya Mishra 2017). The following key points should be kept under consideration:

- Future researches must be focused towards designing research work in ways that could reduce the risk factors associated with the use of nanoparticles.
- The study and synthesis of nanoparticles must not be limited to a few applications applicable in the laboratory conditions only. Nanoparticles and its applications in the field of agriculture must be expanded to bring down to the field level as well.
- The dose of nanoparticles considered safe for usage must be validated, and thereafter nanoparticles must be used within the permissible limits only. This could be achieved by conducting the concentration-dependent study in natural soil system to understand the active and non-toxic dose of nanoparticles.

- The researchers must also study the effect of nanoparticles used in agriculture on different trophic levels and on various generations. This would give a very accurate and comprehensive knowledge of nanotoxicity.
- A study of the physical and chemical properties of the soil will help in reducing the risk of nanoparticles towards plant and soil biota.
- Altering soil environment to alter the transport and bioavailability of nanoparticles to reduce the toxicity of these nanoparticles can help make nanoparticles safe and beneficial in the field of agriculture.
- Biosynthesized nanoparticles and environment-friendly approach of green synthesis of nanoparticles must be designed in a way that will lead to a lesser or no toxicity.

17.12 Conclusion

Nanotechnology is one of the most promising emerging fields due to its capability to revolutionize the existing technologies and develop new technologies that can be used in various sectors including agriculture. Nanoinformatics finds application in all fields ranging from science, physical science, environmental science, electronics and technology, space technology, social perspective and medical science to future nanorobots. Nanotechnology provides some concrete remedies for various agriculture-related problems like insect pest management, adverse effects of chemical pesticides, the development of improved crop varieties and many more. It shows some promising new developments that may lead to a better quality and high yield of crops through precision farming, monitoring of soil and environmental conditions, nanobarcodes and nanoprocessing, smart delivery systems, minimizing soil and groundwater pollution, re-engineering of crops and smart field systems. Nanoparticles are now being widely used in the formulation of new pesticides, insecticides as well as insect repellents. Nanotechnology shows some promising application with respect to nanoparticle-mediated gene transfer.

With more research in the field of nanotechnology and nanoinformatics, it is expected that the applications in these fields will help in solving some major issues like toxicity, environmental effects, climate change, decreasing quality of crops due to pests and soil and groundwater pollution.

References

Abd-elsalam K (2013) Fungal genomics & biology nanoplatforms for plant pathogenic fungi management. Fungal Genomics Biol

Bhattacharyya A, Bhaumik A (2010) Nano-particles: a recent approach to insect pest control. Afr J Biotechnol 9:3489–3493

Corradini EM (2010) A preliminary study of the incorporation of NPK fertilizer into chitosan nanoparticle express. Polym Lett

Elbaz J, Willner I (2012) DNA origami: nanorobots grab cellular control. Nat Mater 11:276–277

Gan NY (2010) Disposable organophosphorus pesticides enzyme biosensor based on magnetic composite nanoparticles modified screen printed carbon electrode. Sensors 10:625–638

Ghormade VD (2011) Perspectives for nano-biotechnology enabled protection and nutrition of plants. Biotechnol Adv 29:792–803

González-Nilo F, et al (2011) Nanoinformatics: an emerging area of information technology at the intersection of bioinformatics, computational chemistry and nanobiotechnology. Biol Res 44:43–51

Guo YR (2009) Gold immunochromatographic assay for simultaneous detection of carbofuran and triazophos in water samples. Anal Biochem 389:32–39

Hussain T (2017) Nanocides: smart delivery system in agriculture and horticultural crops. Adv Plants Agric Res 6

Ingle MR (2012) Role of nanotechnology in agriculture with special reference to management of insect pests. Appl Microbiol Biotechnol 94(2):287–293

Joseph TA (2006) Nanotechnology in agriculture and food. Nanoforum report

Joshi AA (2010) Application of nanotechnology in the remediation of contaminated groundwater: a short review. Recent Res Sci Technol 2:51–57

Kaushik AS (2009) Iron oxide-chitosan hybrid nanobiocomposite based nucleic acid sensor for pyrethroid detection. Biochem Eng J 46:132–140

Lisa MC (2009) Gold nanoparticles based dipstick immuno-assay for the rapid detection of dichlorodiphenyltrichloroethane: an organochlorine pesticide. Biosens Bioelectron 25:224–227

Manjunatha SB, Biradar DP (2016) Nanotechnology and its applications in agriculture: a review. J Farm Sci 29:1–13

Owolade OF, Ogunleti D (2008) Titanium dioxide affects disease development and yield of edible cowpea. Elec J Env Agricult Food Chem 7:2942–2947

Rad SJ, Naderi R, Alizadeh H, Yaraghi AS (2013) Silver-nanoparticle as a vector in gene delivery by incubation. IRJALS 02(21):33

Ramanathan ML (2009) Lysozyme mediated formation of protein–silica nano-composites for biosensing applications. Biointerfaces 73(1):58–64

Rani S (2015) Nanotechnology in future education. Int J E-Gov E-Bus Res:33–43

Rathore SA (2014) Nanotechnology pros and cons to agriculture: a review. Int J Curr Microbiol App Sci 3(3):43–55

Rhodes CJ (2014) Eating small: applications and implications for nanotechnology in agriculture and the food industry. Sci Prog 97:173–182

Sandhya Mishra CK (2017) Integrated approach of agri-nanotechnology: challenges and future trends. Front Plant Sci 8:471

Tarafdar KS (2011) Prospects of nanotechnology in Indian farming. Indian J Agric Sci 81:887–893

Vijayakumar PS (2010) Nanogold-loaded sharp-edged carbon bullets as plant gene carriers. Adv Funct Mater 20:2416

Wang HW (2009) EQCM immunoassay for phosphorylated acetylcholinesterase as a biomarker for organophosphate exposures based on selective zirconia adsorption and enzyme-catalytic preparation. Biosens Bioelectron 24:2377

Wu Z, Chen K (2012) Development of viral nanoparticles for efficient intracellular delivery. Nanoscale 4:3567–3576

Zhang XZ (2010) Chitosan/double-stranded RNA nanoparticle-mediated RNA interference to silence chitin synthase genes through larval feeding in the African malaria mosquito (Anopheles gambiae). Insect Mol Biol 19:683–693

Green-Nanotechnology for Precision and Sustainable Agriculture

18

Rabia Nazir, Yaseen Ayub, and Lubna Tahir

Abstract

Nanotechnology has aroused as a field that has resulted in paradigm shift in agronomic practices and given a true essence to sustainable agriculture. Nanomaterials belonging in the agriculture domain are of vast variety and had find applications in crop production, soil and water management, diagnostic measurements, controlled use of chemicals, and plant protection owing to their tailored properties, small size, and surface to volume ratio. The contribution of nanotechnology in precision farming through the development of nano-based fertilizers, pesticides, herbicides, and early pathogen diagnostic can be considered as a breakthrough. The chapter will focus on the aspects of nanotechnology that have revolutionized the agriculture field, leading to better environmental management and sustainable practices.

Keywords

Green synthesis · Nanofertilizers · Nanocides · Nano-sensors · Soil and water management

18.1 Introduction

Nanotechnology is deemed as the science for cleaner technologies having significant impact on the technological developments of the twenty-first century. Manipulating at small size and large surface area has made the miniaturization possible which resulted

R. Nazir (✉) · L. Tahir
Applied Chemistry Research Centre, Pakistan Council of Scientific and Industrial Research Laboratories Complex, Lahore, Pakistan

Y. Ayub
Department of Chemistry, Governmant Islamia College Civil Lines, Lahore, Pakistan

M. Ghorbanpour et al. (eds.), *Biogenic Nano-Particles and their Use in Agro-ecosystems*, https://doi.org/10.1007/978-981-15-2985-6_18

in materials being more efficient and less material consuming. With its emerging potential and recent pressure to look for greener aspects of this technology, scientists and engineers have diverted their efforts to design and opt ways to shift this realm to benign potential. Hence, the combination of green chemistry, green technologies, and nanotechnology has defined a new direction to nanotechnology avenue.

18.2 Green Nanotechnology

Green nanotechnology (G-NT) is a nanotechnology that is environment-friendly and can help in minimizing the environmental as well as health risks associated with the product development and consumption (Rai and Posten 2013; Safaei et al. 2014). It caters within two main aspects, i.e.:

- Production of nanomaterials without harming the environment.
- Production of nano-products that provide solutions to environmental problems, and in some cases, the third facet is also added, i.e.:
- Replacement of existing products with new environment-friendly nano-products (Figovsky and Beilin 2017).

18.3 Nano-agrotechnology

Nanotechnology has paved its way in agriculture field as well and has played an important role in transforming it to a new venture. Various subdisciplines are catered by nano-agrotechnology (N-AT) as outlined in Fig. 18.1. This engross in itself the

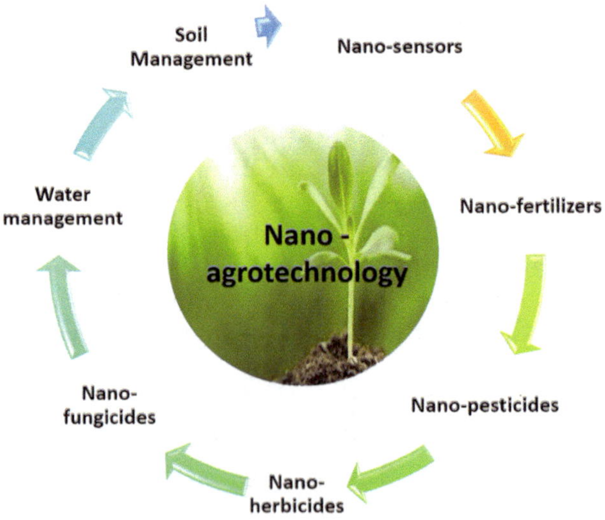

Fig. 18.1 Prospects of nanotechnology in agriculture

revolution in farming practices which deals with improving the nutrient use efficacy, less use of chemicals for control of pests and diseases, improved sensing applications, and optimized water management (Parisi et al. 2015; Prasad et al. 2017b), thereby resolving many problems associated with conventional farming along with enhancing the agriculture production in an eco-friendly way.

Nanotechnology being recognized by the European Commission as one of the "key enabling technologies" has redefined the concept of sustainable agriculture and acts as a paradigm shift in precision agriculture as well. The *precision agriculture* is defined by the Directorate-General for Internal Policies of the European Union as "A farming management concept of measuring and responding to inter and intra-field varying in crops to form a decision support system for whole farm management and to reap the maximum output from the available resources" (Duhan et al. 2017). On the other hand, the *sustainable agriculture* according to the National Research Council is a "long term maintenance of natural resources and agricultural productivity, minimal adverse impacts, adequate economic returns to farmers, optimal crop production with minimized chemical inputs, satisfaction of human needs for food and income, provision for social need of farm families and communities" (Jordan 2013).

The combination of G-NT and N-AT hence will address all the three aspects of the G-NT. Therefore, this chapter will briefly describe the green routes to the synthesis of nanomaterials and present a detailed overview about the practices that can help in agricultural precision management through these nanomaterials.

18.3.1 Biosynthesis of Nanomaterials

Nanoparticles (NPs) are generally synthesized through physical and chemical approaches, and there is a reasonable doubt about the greenness of these methods owing to the involvement of high energy and toxic chemicals. The biosynthesis of NPs involving plants, microbes, fungus, algae, etc. (Fig. 18.2) has gained attention recently because of their environment-friendly approach (Siddiqui et al. 2015).

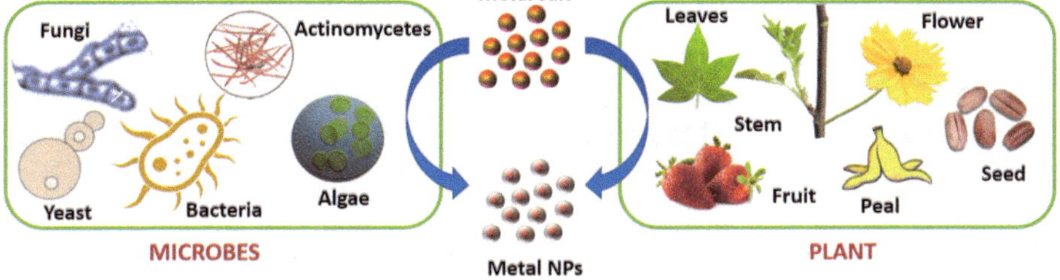

Fig. 18.2 The plant-mediated and microbial route to the green synthesis of nanoparticles

18.3.1.1 Plant Based

Plants are being rightly recognized as natural nanofactory, widely available (Jha et al. 2009) in various parts of the world, but are still waiting for an exploration of their full potential toward the green synthesis of NPs. General scheme for the synthesis of NPs using plant-based materials involves preparation of plant extract, bioreduction of metal salts to metal NPs, their confirmation via UV-vis spectrometer, and finally the separation to be used for miscellaneous applications (Singh et al. 2016).

Plant-mediated synthesis of NPs has several advantages over the microbial-based synthesis and hence is rated as better on account of the abundant availability of plants, ease of handling, enhanced stability of NPs achieved through them, fast reaction rates, and one-step synthesis making the process viable for large-scale production (Baker et al. 2013; Siddiqui et al. 2015). The presence of biomolecules in plant extracts like proteins, polysaccharides, amino acids, flavonoids, terpenoids, polyphenols, alcoholic compounds, etc. is mainly responsible for the reduction of metal salts to metal NPs. These metabolites also play an imperative role in stabilizing the NPs through capping and chelating effect as well as controlling their size and shape (Baker et al. 2013; Iravani 2011). A multitude of factors like reaction temperature, extract concentration, extract contents, and metal ion concentration have high impact on the size and shape of NPs (Iravani 2011). In this context, researchers have exploited the use of plant leaves, fruits, and even flowers from a myriad of plants in pursuit of finding the best source and conditions for application-driven synthesis of nanomaterials.

18.3.1.2 Microbes Based

The deliberate synthesis of NPs through the use of a number of possibilities among the microbes has gained popularity in the scientific community in an essence to design green routes. The microorganisms exploited in this regard are bacteria both gram positive and negative, actinomycetes, fungi, algae, and yeast (Niemeyer and Mirkin 2004), as outlined in Fig. 18.2, which generate NPs by either intracellular or extracellular activities. The former one involves the metal ions' transportation into the cell where different enzymatic activities lead to the formation of NPs, while in the latter case, metal ions are trapped on the cell surface effecting the reduction into NPs through the proteins available on the surface which also act as capping and stabilizing agents. The size and shape of NPs can be manipulated by varying the pH, temperature, metal ion concentration, and medium concentration (Durán et al. 2011; Li et al. 2011).

18.4 Applications of Nanotechnology in Precision Agriculture

The applications of nanotechnology in the field of precision agriculture are broad as outlined in Fig. 18.1. The innovations are still enjoying their infancy, but still voluminous research work had been carried out in the field of smart delivery fertilizers and sensors. This chapter will briefly describe the few important fields of precision agriculture that are greatly influenced by nanotechnology.

18.4.1 Nano-fertilizer

Conventional fertilizers, in the context of available nutrients to the plants, suffer from serious drawback, as the total crop plant uptake in terms of fertilizer use efficacy accounts to 20–50% for nitrogen, 10–25% for phosphorus, and 35–40% for potassium (Guo et al. 2018; Naderi and Danesh-Shahraki 2013). The rest of fertilizers not taken up by plants either results in atmospheric losses or leaching/runoff contributing to disturbance in soil mineral balance, soil infertility, eutrophication, greenhouse gases, and multiple other environmental problems. All these problems can be enhanced manifolds considering the fact that the world total fertilizer (NPK) demand is projected to reach 201,663 thousand tones in 2020 (FAO 2017) and hence posing a threat to irreparable damage. In this context, the nanotechnology revolution had led to the development of smart delivery nano-fertilizers that can prevent undesirable nutrient loss and henceforth put a halt to escalating consumption of fertilizers and paving a way to sustainable agriculture. The feasibility of several materials has been exploited as slow-release materials for a number of crops and proved to augment germination and increase plant growth, yield, no. of seeds, fruiting, etc. In addition to these, multiple benefits have been delineated in favor of nano-fertilizers in comparison to the conventional analogues as briefed in Fig. 18.3.

An overview of nanomaterials developed for the smart delivery of nutrients is laid out in Table 18.1. Most of the researches have opted for foliar spray applications considering the fact that NPs can be easily taken up by the plants through stomatal pores. The complexity of biological systems leads to different uptake mechanisms in roots, leaves, etc. with strong dependency on the characteristics of

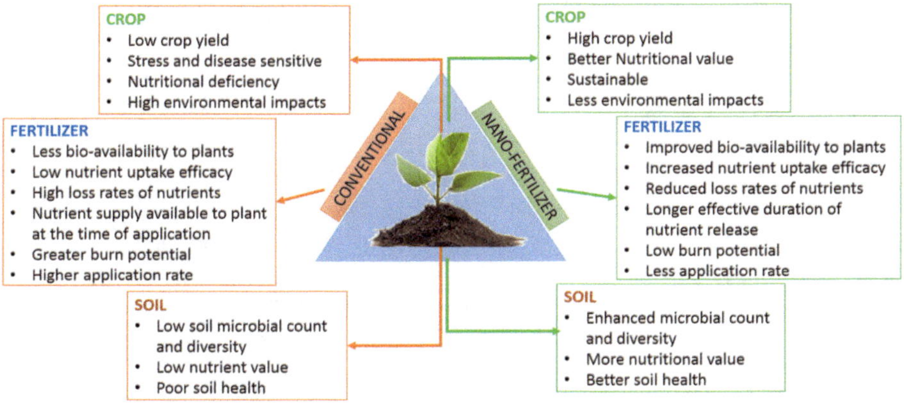

Fig. 18.3 Some advantages related to nano-fertilizer in comparison to its conventional analogues

Table 18.1 Overview of the nano-fertilizer developments and their impacts on crops studied

Nanomaterial	Crop	Application	Impacts	References
Fe	Black-eyed pea	Foliar	Increased pods/plant seeds/plant, 1000 seeds weight, chlorophyll content	Delfaniet al. (2014)
Mg				
Zn	Pearl millet	Foliar	Improved shoot length, root length, root area, chlorophyll content, total soluble leaf protein, plant dry biomass, enzyme activities, grain yield	Tarafdar et al. (2014)
B	Pomegranate	Foliar	Enhanced leaf concentrations, fruit yield, fruit quality, maturity index	Davarpanah et al. (2016)
Zn				
Zn + B	Potato	Fertigation through irrigation water	Better plant height, number of stems, number of leaves, yield, no. of days to tuberization, no. of days to flowering, no. of days to maturity	Janmohammadi et al. (2016)
Ca				
N, P, K, Fe, Zn, Ca, Mg, Mn, Cu, B, Mo				
Fe	Wheat	Foliar	Improved spike weight, 1000 grain weight, biologic yield, grain yield and grain protein content	Bakhtiari et al. (2015)

(continued)

Table 18.1 (continued)

Nanomaterial	Crop	Application	Impacts	References
Nano-fertilizer (Ca, Fe, Si, Mg. S, Zn, Fe, Al, Na, Mo, K, Cu, Ni, Co, Sr)	Black cumin	Foliar	Capsule/plant, seeds/capsule, seeds/plant seeds weight/ plant, 1000 seed weight, seed yield	
Zn/B	Coffee	Foliar	Chlorophyll content, uptake of Zn, N, P, plant height, stem dia	Wang and Nguyen (2018)
Urea-based aminopropyltrimethoxysilane	–	–	–	Hidayat et al. (2015)
Urea-modified hydroxyapatite NPs encapsulated wood				Kottegoda et al. (2011)
PVA-starch supported Cu–Zn/carbon nanofibers	Chickpea	–	Plant growth, antioxidant potential	Kumar et al. (2018)
Chitosan-based NPK	Garden pea		Reduced root elongation, upregulation of proteins (negative impacts)	Khalifa and Hasaneen (2018)
Urea-hydroxyapatite				Kottegoda et al. (2017)
Zeolite loaded with N, P, K, Ca, Mg. S, Fe, Zn, Cu				Lateef et al. (2016)
NPK				Guo et al. (2005)
Chitosan-NPK	Wheat	Foliar	Harvest index, crop index, mobilization index, plant growth	Abdel-Aziz et al. (2016)
Biochar loaded with N, P, K, Ca, Mg. S, Fe, Zn, Cu	–	–	–	Lateef et al. (2019)

(continued)

Table 18.1 (continued)

Nanomaterial	Crop	Application	Impacts	References
Phosphate-potash/neem cake/ PGPR	*Vigna radiata*	–	Germination, specific activity of enzymes, carbohydrates, protein, photosynthetic pigments, root nodule number, and microbial population	Mala et al. (2017)
Urea-kaolinite	–	–	–	Sempeho et al. (2015)
Liulitian (N, P, K, Mg, Ca, and humic acid)	Green pepper		Growth, soil nutrients, enzyme activity and microorganism	Teng et al. (2018)
Pulverized calcite-seaweed extract	Grape wine	Foliar	Enhancements in foliar development, chlorophyll concentration, berry characteristics, yield, leaf nutrient content	Sabir et al. (2014)
CNT/NPK/chitosan	French bean	Foliar	Improved water absorption, nutrient uptake, plant growth	Hasaneen et al. (2016)
Sodium alginate-g-Poly (acrylic acid-co-acrylamide)/ Clinoptilolite	–	–	–	(Rashidzadeh et al. 2014)
ZnO	Tomato	Foliar soil	Optimized NPs in response to dry biomass yield, fruit yield and nutritional quality, flower and fruit development, chlorophyll content, plant growth	Raliya et al. (2015)
TiO_2				

both plant and NPs. The translocation of NPs from the leaf to the root occurs by phloem transport mechanism (Fig. 18.4-I, II) mediated by apoplast (through the wall) as well as symplast (cell to cell) – the former being preferred for larger particles (200 nm) and the latter for smaller NPs less than 50 nm (Fig. 18.4-I). Hence, among the different particle shapes and sizes (sphere (10.9 nm), cubic (2.3 nm), rhombic dodecahedra (11.3 nm), and rod (37.4)), particles translocate and internalized differently (Raliya et al. 2016). In the case of root, uptake is initiated by strong adhesion followed by absorption and travelling to different plant tissues after chelation with microbial siderophores (Chen 2018). Another study on uptake of ceria NPs in cucumber plant through the root denotes this adherence to electrostatic interaction between positively charged NPs and negatively charged root surfaces created by the secretion of mucilage – a mixture of hydrated polysaccharides (Zhang et al. 2011). Similar findings were observed for Zn NPs translocation in *Zea mays* or corn (Zhao et al. 2012). Further, other phenomena like osmotic pressure and capillary forces play an important role in the movement of these NPs into roots and transportation to shoot with water via vascular system, transpiration, and evaporation (Fig. 18.4-II) (Raliya et al. 2015; Zhang et al. 2011) as indicated by the translocation and agglomeration in TEM images of leaf and stem (Fig. 18.4-A, B). Certain studies have also supported the direct intake of NPs through cell wall pores having limiting diameter of 3.5–5 nm or via creation of holes in the cell wall. The probability was also suggested related to the entry of NPs in roots through the apical meristematic tissue available at the tip of the root under root cap leading to diffusion of NPs into xylem (Zhu et al. 2012). Transversal cut of fluorescein isothiocyanate (FITC) stained image of the root (Fig. 18.4-C) confirms the entry of Zn NPs (dark

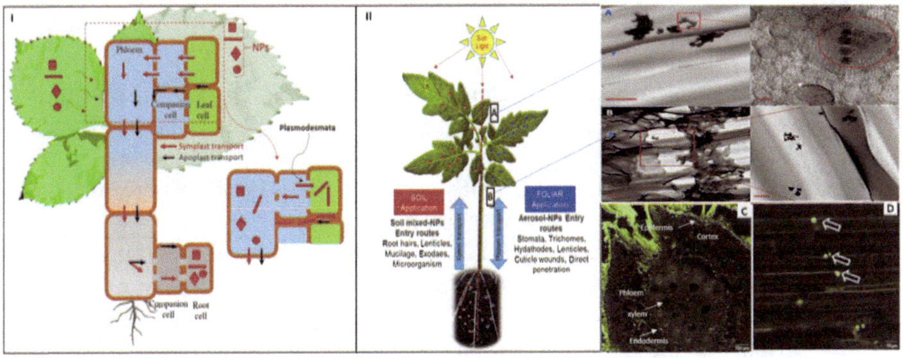

Fig. 18.4 (**I**) NPs transportation through apoplast and symplast pathways in plant cells along with the gradient or mass flow of photosynthate product. Inset represents the favorable transport of gold nanostructure (rod shaped) more through apoplast than symplast pathways. Color gradient in the phloem represents mass concentration of photosynthate with NPs (Raliya et al. 2016). (**II**) Mechanistic understanding of nanoparticle uptake, translocation, and accumulation along with TEM micrograph of leaf (*A*) and stem (*B*) for foliar application of TiO_2 (Raliya et al. 2015) and (*C*) root cross section of FITC-labeled ZnO NPs showing stained NPs aggregates in the root epidermis, cortex, endodermis, and xylem. (*D*) Confocal image of longitudinal section of corn root showing aggregation of NPs vascular cylinder attached to xylem vessel walls (Zhao et al. 2012)

green dots) into the root cortex through the epidermis, somewhat retention by the Casparian strip, and transportation to the endodermis and vascular cylinder (Fig. 18.4-D) by the apoplastic and symplastic pathways, respectively.

The studies performed on rice, ryegrass, radish, and pumpkin showed that the surface charge of NPs also play an important role in the uptake mechanism based on the binding capacity and internalization into the plant system. Positively charged Au NPs have effective binding and poor internalization, while negatively charged NPs depict poor binding to the root surfaces and high efficiency for internalization that varies from plant to plant. The same species-dependent response was observed in the transportation of NPs from the root to shoot to leaves (Zhu et al. 2012).

18.4.2 Nanocides

The development of nano-formulations of pesticides, fungicides, and herbicides, cumulatively known as nanocides, enabled better plant protection through enhanced solubility, increased bioavailability, decreased effect on nontarget organisms, improved specificity, better stability, and permeability (Fig. 18.5d). As was the case in conventional fertilizers, conventional pesticides suffer from more or less the same scenario – leaching; degradation either by photolysis, hydrolysis, or microbes; effective concentration availability; and ecotoxicity. Hence, to resolve these issues, the pesticide and related nano-formulations come up with essential traits like controlled release, effective concentration, improved targeted delivery, less number of applications, low ecotoxicity, reduced dosage, safe mode of delivery, and less human exposure (Prasad et al. 2017a; Tuteja and Gill 2012). There are four types of formulations (Fig. 18.5) available for the delivery of active ingredients, AI (pesticides, herbicides, and fungicides), i.e.:

- *Nano-emulsion* – usually oil-in-water (O/W) based in which the AI nano-droplets solubilized in oil portion are dispersed in water with surfactant molecules stabilized with emulsifiers.
- *Nano-suspensions* – also named as nano-dispersions; AI droplets are suspended in aqueous phase stabilized by surfactant molecules.
- *Nano-capsules* – having core-shell structure with AI either in the form of liquid or powder core is covered by polymeric layer shell either through encapsulation or entrapment.
- *Nanoparticles* – the nanoparticles having pesticidal, fungicidal activities, etc. can be used directly as AI. These include two types of NPs: (a) inert dusts like silica, diatomaceous earth, alumina, and clays and (b) metal NPs, e.g., Ag, TiO_2, Cu (Abd-Elsalam and Prasad 2018; Hayles et al. 2017; Kah and Hofmann 2014; Khandelwal et al. 2016; Zhao et al. 2017).

Nano-pesticides owing to their small size and large surface area are more permeable into insect as compared to conventional ones. The pest poisoning is caused by

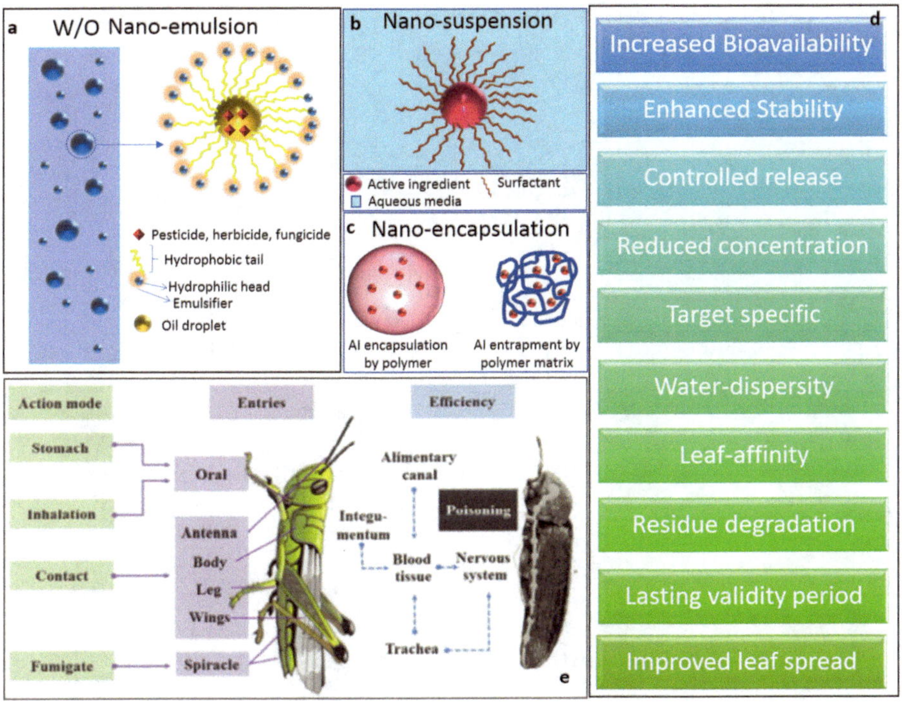

Fig. 18.5 Nanocide formulation types, i.e., (**a**) nano-emulsion, (**b**) nano-suspension, (**c**) nano-encapsulation, (**d**) along with their potential advantages over conventional counterparts (**e**) and mode of action of nano-pesticide (**e**) (Zhao et al. 2017)

four modes of entry as depicted in Fig. 18.5e, i.e., inhalation, stomach, dermal contact, and respiration system (fumigation) (Zhao et al. 2017).

The use of NPs as protectants to plant diseases has shown good nanocidal activities toward various pests, fungal strains, etc. Treatment of rice with alumina NPs dust has resulted in its protection from *Sitophilus oryzae* L. (rice weevil) and *Rhyzopertha dominica* (F.) (grain borer), as almost 100% mortality of these species was observed on the ninth day at dose rate of 500 mg Kg^{-1} (Stadler et al. 2010). Silver NPs with its antifungal activity are able to control fungal infections in the plants. Green synthesized Ag NPs dealt effectively with wheat pathogen, *Bipolaris sorokiniana* infection in plants under in vitro and in vivo conditions. A small concentration of these NPs, i.e., 2 μg/ml, was able to inhibit 100% conidial germination (Fig. 18.6a, b) (Mishra et al. 2014). Silver NPs and their core-shell structure were also found effective in inhibiting the growth of *F. oxysporum* in tomato plants (Madbouly et al. 2017) and other species, i.e., *Phytophthora capsici*, *Phytophthora nicotianae*, and *Phytophthora colocasiae* (Ho et al. 2015), respectively. In addition to that, these Ag NPs have depicted antifungal activity against various strains like *Botrytis cinerea*, *Magnaporthe grisea*, *Fusarium culmorum*, *Colletotrichum*

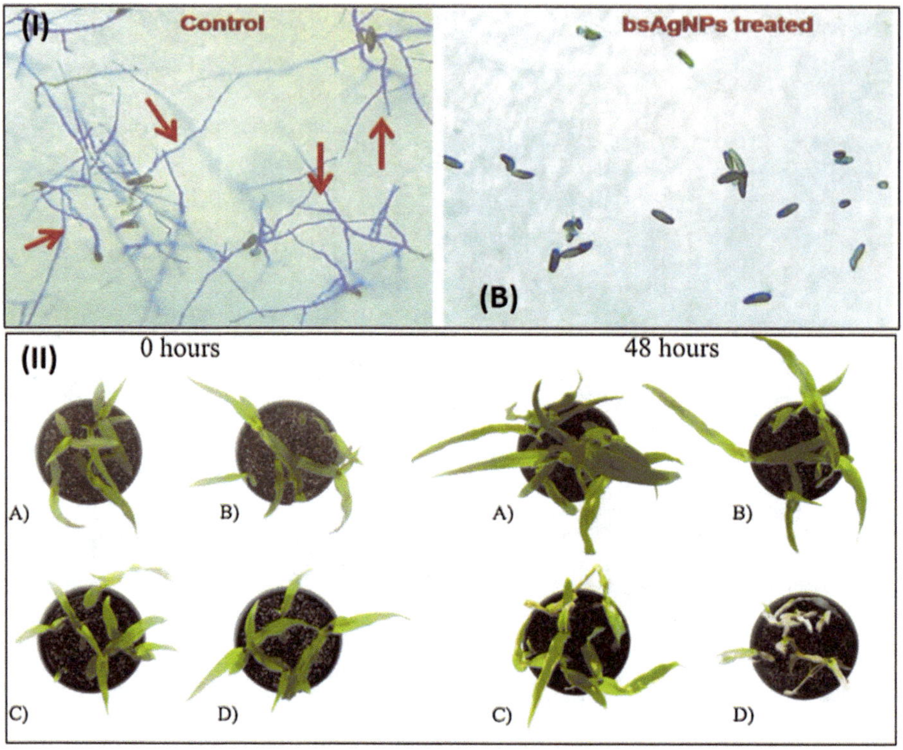

Fig. 18.6 (**I**) Antifungal effects of biosynthesized Ag NPs (bsAgNPs) on *B. sorokiniana* conidial germination in (*A*) control and (*B*) after 24 h. (**II**) Herbicidal activity of (*A*) control, (*B*) unloaded NPs, (*C*) free paraquat, (*D*) paraquat-loaded NPs in cultivations of maize after 48 h

gloeosporioides, etc. by halting growth of germinating fragments and sprout length (Abd-Elsalam 2012).

Similarly, the use of $NiFe_2O_4$ and $CoFe_2O_4$ NPs (500 ppm) has not only impeded 78.91–81.39% mycelial growth of three fungal strains, i.e., *Fusarium oxysporum*, *Colletotrichum gloeosporioides*, and *Dematophora necatrix*, but also impacted positively in controlling pathogen wilt in capsicum plant (Sharma et al. 2017). Vegetative growth as well as sporulation of *A. niger*, a common fungus, isolated from potato strain, was effectively controlled by the use of nano-sulfur. The effectiveness of functionalized NPs of ZnO, Al_2O_3, SiO_2, and TiO_2 toward controlling rice contaminated with *Sitophilus oryzae* was also checked, and it was observed that at a dose of 1 g/kg, the alumina (mean mortality 97–100%) and silica (mean mortality 86–95%) are much effective as compared to the TiO_2 (mean mortality 29–45%) and ZnO (mean mortality 36–40%). In determining the NPs activity, hydrophilic and hydrophobic characters play an important role, and the order of effectiveness is α-Al_2O_3 (hydrophobic) > γ-Al_2O_3 (hydrophilic) > SiO_2 (hydrophilic) > SiO_2 (hydrophobic) (Goswami et al. 2010). The dependence of fungicidal activity was also found dependent on the nature of composite as well as the strains. Chitosan and its NPs composites (i.e., chitosan-saponin, Cu-chitosan) were found to show in vitro mycelial inhibition activity against the three strains, i.e., *A. alternate, M. phaseolina, and R. solani*, based on the order Cu-chitosan > chitosan-saponin > chitosan, chitosan > chitosan-saponin > Cu-chitosan, Cu-chitosan > chitosan > chitosan-saponin, respectively, while Cu-chitosan was found most effective in controlling spore germination of these fungal strains (Saharan et al. 2013). Lower concentration (i.e., 1 g/L) of these Cu-chitosan was also found effective in controlling the vascular wilt disease in date palm caused by *Fusarium oxysporum* as compared to conventional fungicide Rizolex™ (dose 3 g/L) (Mohamed et al. 2018). NPs of ZrO at a concentration level of 100 μg/L have depicted significant inhibition in growth of *Rhizoctonia solani* (86.6%) relative to untreated control under laboratory conditions. Cucumber plants treated with these NPs are found to have much resistant to root rot disease under greenhouse (34–46%) as well as field conditions (52–56%) compared to nontreated control plants (Derbalah et al. 2019).

NPs have also been exploited as nano-carrier of pesticides, fungicides, herbicides, etc. The role of silica NPs as target-specific material in this aspect is well exploited by many researchers (Rastogi et al. 2019). Slow-release formulations of pesticides, e.g., avermectin and pyoluteorin, were prepared by encapsulation with porous silica NPs enabling their slow release over a period of approx. one month. The process also results in high loading of these pesticides with added advantages of pesticide resistance against UV-light degradation and better antifungal activity against *Phytophthora capsici* (Chen et al. 2010; Li et al. 2007). Some investigation have pointed out that release behavior is governed by electrostatic interactions as was the case in diquat dibromide (a herbicide) encapsulated by sulfonate-functionalized mesoporous silica NPs, while ionic strength was found to impact the bioactivity (Shan et al. 2019). Chitosan functionalized with alginate and tripolyphosphate loaded with weedicides, i.e., imazapic and imazapyc, showed a release percentage of 30 and 20%, respectively, in comparison to the unencapsulated ones, i.e., 55 and 97%. Release percentages of both herbicides vary, owing to the different modes of interaction between nanoparticles and herbicides. This slow release of herbicides has fostered the soil bacterial community, owing to the enhanced availability of nitrogen as a result of the destruction of target weed (*Bidens pilosa*) (Maruyama et al. 2016). Likewise, the loading of paraquat (herbicide) on chitosan/tripolyphosphate has shown no toxicity on *Allium cepa* (onion) seeds as against LD_{50} of 0.12 mg/mL of unloaded herbicide. Enhanced herbicidal activity of NPs loaded with herbicide was noticed in *Brassica* sp. (Fig. 18.6c) and *Zea mays* leading faster necrosis in plants in contrast to free paraquat owing to its less decomposition and better adhesion of NPs with plant leaves (Grillo et al. 2014).

In addition to NPs, other systems have also been used to encapsulate these agrochemicals like atrazine-loaded poly(ε-caprolactone) nano-capsules and metsulfuron methyl-loaded pectin NPs have same inhibitory effective against *Amaranthus viridis* (Díaz-Blancas et al. 2016), *Bidens pilosa*, and *Chenopodium album* in much lower dosages as compared to the commercial formulation applied at standard dosages (Kumar et al. 2017b; Sousa et al. 2018).

Nano-formulations are also found effective in comparison to the bulk counterparts. Nano-emulsions of pesticides (permethrin, bifenthrin) (Kumar et al. 2013; Liu et al. 2011), fungicides (tebuconazole) (Díaz-Blancas et al. 2016), herbicides (glyphosate isopropylamine) (Jiang et al. 2012), etc. were prepared and found to perform better. Nano-form of azomethine was reported to have two times higher fungicidal activity against different strains of fungi (i.e., *Rhizoctonia solani*, *Rhizoctonia bataticola*, and *Sclerotium rolfsii*) (Mondal et al. 2017).

18.4.3 Sensors

Nano-sensors, available in a variety of forms (Fig. 18.7), offer fast detection with high sensitivity, low detection limits, and response factor enabling early detection of the many of the physiochemical and biological aspects. It had vast applications in the field of soil quality, disease management, detection of contaminants and other molecules, detection of DNA and proteins, etc. (Omanović-Mikličanin and Maksimović 2016).

Nano-sensors generally consist of three main parts (Fig. 18.7). Nanomaterials usually enhance the sensitivity, while the recognition elements help in developing the specificity enabling the detection of analyte of interest based on different signal transduction techniques, i.e., electrical, mechanical, optical, or chemical detection functions (Munawar et al. 2019; Willner and Vikesland 2018).

18.4.3.1 Pesticidal Residues Determination

Conventional protocols of residual pesticides' detection are tedious, but recent advances in the field of nano-sensors enabled early and low-level on-field detection of pesticides in agriculture produce. Table 18.2 presents few of the developments done in this field.

Optical sensors owing to their colorimetric detection have gained the attention of researchers and led to various developments by the use of

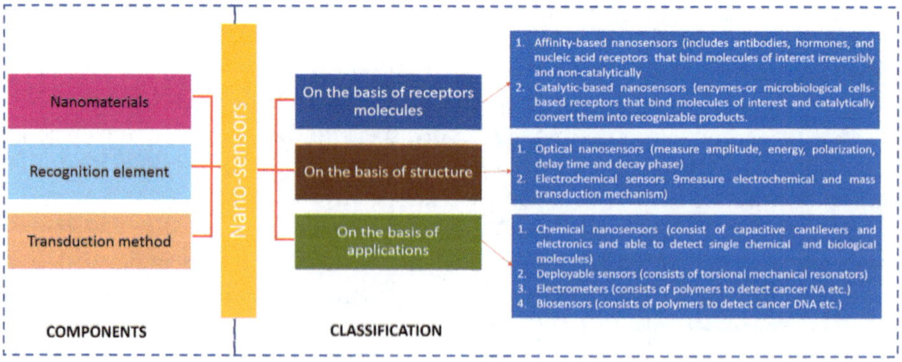

Fig. 18.7 Components and classification of nano-sensors (Saini et al. 2017; Willner and Vikesland 2018)

Table 18.2 Nano-sensors developed for pesticides/herbicides detection

Pesticide/herbicide detection	Sensing material	Detection limit	References
Carbaryl and methomyl (carbamate pesticides)	Polyaniline and multi-walled carbon nanotube core-shell modified glassy carbon electrode	1.4 μmol L^{-1} 0.95 μmol L^{-1}	Cesarino et al. (2012)
Paraoxon (organophosphorus pesticide)	Acetylcholinesterase/gold NPs (Au NPs)/graphene oxide nanosheets	0.1 pM	Wang et al. (2011)
Paraoxon (organophosphorus pesticide)	Organophosphate hydrolase-gold NPs	20 μM	Simonian et al. (2005)
Methyl parathion (organophosphorus pesticide)	ZrO_2NPs deposited on gold electrode	3 ng/mL	Liu and Lin (2005)
VX, soman, sarin, and paraoxon-ethyl (organophosphorus pesticides)	Lipoic acid-capped Au NPs	2.51 pM 15.0 pM 28.2 PM 4.52×10^4pM	Sun et al. (2011)
Sulfurazon-ethyl (herbicide)	Humic acid-capped Ag NPs	100 ppm	Dubas and Pimpan (2008)
Triazinic pesticides	Surface plasmon resonance-based polymeric material		Yılmaz et al. (2017)

acetylcholinesterase-based inhibition activity by employing gold, silver, and magnetite NPs. The low detection limits were noticed for organophosphate pesticides when using these NPs, i.e., Au (LODs: 0.1 μg/L for carbaryl and diazinon, 0.3 μg/L for malathion, and 1.0 μg/L for phorate), Ag (0.18 ng/mL for dipterex), and Fe_3O_4 (1 nM for acephate, 10 nM for methyl paraoxon, and 5 μM for sarin) (Xia et al. 2015). Florescent nano-sensors owing to their high sensitivity and rapid analysis have gained preference over the colorimetric detection approach, and the use of quantum dots, metal NPs, and surface plasmon resonance has really enabled fast detection of pesticides (Pacioni and Veglia 2007; Xia et al. 2015; Zhao et al. 2011). Fluorescent liposome-based NPs were also prepared to detect organophosphorus pesticides, e.g., dichlorvos and paraoxon (Liu et al. 2008).

In an effort to further enhance sensitivity and selectivity, the new class of nano-sensors, i.e., apta-nano-sensors, is being developed which utilized aptamers as chemical artificial receptor (recognition element). Aptamers are small nucleic acid sequences of RNA or DNA that can bind to different ligands with high binding affinity (Jokar et al. 2016; Willner and Vikesland 2018; Yan et al. 2018). Apta-nano-sensors have been extensively used for the determination of residual pesticides, i.e., acetamiprid (Jokar et al. 2016; Qi et al. 2016; Verdian 2018; Yang et al. 2015).

18.4.3.2 Diagnosis of Plant Pathogens

Diagnostic measurement of plant diseases is another milestone of the nano-sensors that enable early detection of plant pathogens and initiation of respective treatment at early stages, thereby saving the spread and damage of the crops that earlier accounted for almost 20–30% crop loss. Recent advancement in nano-diagnostic measurements and development of tailored nanomaterials that can establish covalent linkages to biological molecules (i.e., proteins, enzymes, peptides, nucleic acids) enabled rapid, sensitive, accurate on-site detections of these pathogens with added advantage of avoiding stringent sample preparation protocols (Kashyap et al. 2017). NPs are either directly used with or without modification to directly detect the pathogens, or they can detect the compounds released by plants in the disease condition (Khot et al. 2012). Such techniques hence exploit different approaches, e.g., antibody-antigen, adhesion receptor, antibiotic, and complementary DNA sequence recognitions (Khiyami et al. 2014) permit much lower detection limits as compared to the earlier used techniques like ELISA. These techniques also promoted the development of briefcase-sized kit or smart dust that can help farmers to determine the disease attack and consequently the controlled spray use and its related measures (Ali et al. 2014; Khiyami et al. 2014).

Pantoea stewartia, bacteria causing Stewart's wilt in sweet corn, was successfully diagnosed with the help of lanthanide chelate-loaded silica NPs that are able to offer very low detection limits, i.e., 103 cfu/ml (Zhang et al. 2014b). Similarly, conjugation of silica NPs with organic dye (tris-2,2′-bipyridyl dichlororuthenium (II) hexahydrate) resulted in a fluorescent biomarker helped in early detection of *Xanthomonas axonopodis* bacteria that cause spot disease in Solanaceae plant (Yao et al. 2009). Colorimetric determination of various bacterial strains (*S. typhimurium*, *S. enteritidis*, *Staph. aureus*, and *Campylobacter jejuni*) available in food was made possible by preparing the cocktail of surface-modified nanobeads that on interaction with different bacterial strains generate different colors (Alamer et al. 2017).

Among the different plant pathogenic-based diseases, viral-based diseases are the most difficult one to tackle, and once started, the use of pesticides is not much effective. Hence, their early detection before the symptomatic appearance is really advantageous for saving the crop. One such viral attack is *Cucumber mosaic virus* which, with its high transmission rate and susceptibility to a wide variety of crops, is considered a dangerous plant virus. Polypyrrole nanoribbon-based nano-sensor planted with the virus antibodies is able to detect this virus in broad range of 10 ng/ml to 100 μg/ml (Chartuprayoon et al. 2013). Another lethal vector, *Polymyxa betae*, that transmits beet necrotic yellow vein virus which causes sugar beet crop damage can be easily detected by quantum dots (QD)-based nano-sensors (Safarpour et al. 2012). A similar kind of QD-based setups was able to detect plant virus *Citrus tristeza* with much lower detection limits as compared to ELISA (Shojaei et al. 2016a, b).

Ganoderma boninense is a fungal strain causing great damage to palm oil industry by causing basal stem rot infection and having no significant treatment known since its symptomatic appearance. The said pathogen was successfully diagnosed at earlier stages of attack by the use of CdSe QD by its high interaction with DNA of *G. boninense* with reported LOD of 3.55×10^{-9} M (Bakhori et al. 2013). Ochratoxin A was detected in a concentration range of 0.3–10 mg/mL in grape juice and red wine using black phosphorene nanosheets-based sensor (Xiang et al. 2018).

Salicylic acid is a plant hormone released in stressed plants that are under attack by insects or pathogenic diseases owing to their defense mechanism. The determination of these signaling molecules accurately and in low amounts can significantly save the crop losses. One such study carried out using nano-Cu/Au electrode allowed salicylic acid detection in oilseed rape infected with the fungal pathogen *Sclerotinia sclerotiorum* (Wang et al. 2010). Another such detection system based on aptamer nano-sensor resulted in lower detection limits (0.1 μM) and early detection (30 min) with very small sample size (1 μL plant extract) in *Arabidopsis* and rice (C. Chen et al. 2019). *p-Ethylguaiacol* is a compound released by plant when it is infested by *Phytophthora cactorum*, a pathogenic fungus attacking crops belonging to the gourd family. This phytohormone can be electrochemically detected using carbon electrode screen printed by TiO_2 or SnO_2NPs in the range of 0.2 μM–0.1mM using differential pulse voltammetry with LOD of 35 nM for TiO_2 and 62 nM for SiO_2 (Fang et al. 2014).

18.4.4 Nano-barcodes

Barcodes, either overt (visible) or covert (invisible), are like fingerprint used to uniquely encode the materials in order to provide error-free identification. The NPs unique sequence is created by multiple types and multiple layers of different materials with varied properties like shape, color, and size (Fig. 18.8). In some cases, these nano-barcodes are supported on carrying materials (Shikha et al. 2017; Wang et al. 2015a). These identification codes have found their implications in agricultural field in biological and nonbiological applications. In biological applications, nano-barcodes are used as tags for carrying out multiplexed bioassay and encoding enabling better plant resistance against environmental stresses. In nonbiological applications, tagging is found helpful in developing authentication and tracking of agriculture produce (Ditta 2012). The nano-barcodes are detected under UV-light, resulting in different color combinations detectable by computer scanner (Bowles and Lu 2014).

18.4.5 Soil Management

18.4.5.1 Soil Analysis

Soil conditions like pH, moisture level, nutrient composition, salinity, etc. can be viably measured with the help of nano-sensors, thereby helping farmers with timely analysis that in turn can effectively utilize the information to control inputs and undergo site-specific treatment (Kaushal and Wani 2017).

18.4.5.2 Water Retention/Absorption Holding Capacity

The development of nanomaterials that enable water absorption and retention promotes sustainable agriculture by provision of water as and when needed. Such materials are valuable in areas of drought and also in the future considering the

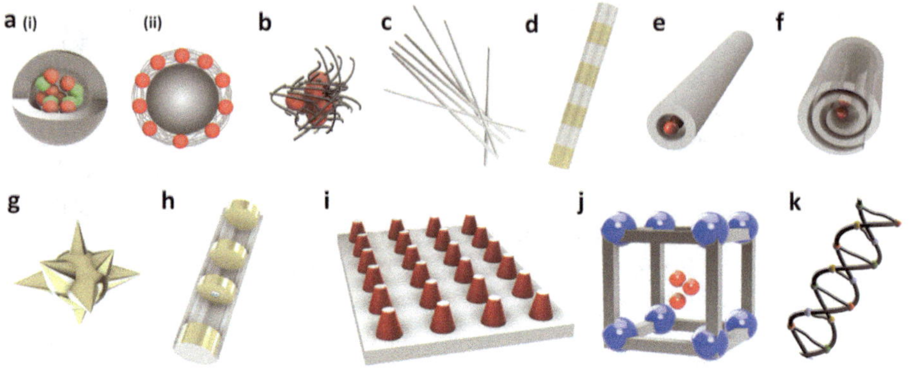

Fig. 18.8 Schemas for different barcoded nanostructures of different geometries. Encoded (**a**) nanobeads with encoding elements (*i*) encapsulated inside and (*ii*) decorated on the surface; (**b**) nanoclusters, where NPs are assembled as super-particles and stabilized by polymers; (**c**) nanowire array which has nanoscale diameter and length up to several microns; (**d**) nanorods with different metal compositions in a striped pattern; (**e**) nanotube cross section showing NPs encapsulated inside; (**f**) nanosheets with NPs loaded inside; (**g**) nanostars that are branched nanostructures with small cores prepared from metal NPs; (**h**) nanodisks with disk-shaped structures; (*i*) nanopillars with pointed top ends and tapered bases; (**j**) nanometal organic frameworks consisting of coordination bonds between metal ions and organic linkers with fluorescent NPs encapsulated inside; and (k) sequence barcodes made of oligonucleotide (or peptides) sequences as encoding elements (Shikha et al. 2017)

scarcity of water and regular downfall of water table. Superabsorbent polymers are the example of such materials that have the capacity to hold large amounts of water and can desorb it under specified conditions of temperature and pressure for longer periods of time. Silver-coated nanoclay/polymer composites act as hydrogels and are able to absorb 161 g/g of tap water in 68 h. The gel favored the soil water holding capacity that helped in retention of rain/agriculture water by 130–190 times its weight. On the other hand, the water desorption ratio of soil treated with hydrogel reached 78.6 wt% on the 30th day which is much improved as compared to the untreated (90.2 wt%) (Vundavalli et al. 2015). Another hydrogel based on poly(acrylamide-co-acrylic acid)/$AlZnFe_2O_4$ nanocomposite helped in significantly enhancing the water holding capacity of sandy loam soil, i.e., 471 g/g. In addition to that, the said treatment also improved the wheat seed germination and growth (Shahid et al. 2012). In addition to hydrogels, other materials are also exploited by researchers to enhance the water holding capacity of soil, e.g., nanocarbon (Zhou and Chen 2017), zeolite nanocomposite (Lateef et al. 2016), biochar nanocomposite (Lateef et al. 2019), agar/clay nanocomposite (Rhim 2011), etc.

18.4.5.3 Soil Decontamination

Soil remediation from toxic materials has always been a challenge especially in respect of their cost-effectiveness and ease of application. Nanotechnology has offered a better approach than the conventional techniques to remove their hazardous inorganic as well as organic contaminants from the soil.

18.4.5.3.1 Organic Contaminants' Removal

NPs have played an important role in removing the pollutants from the soil which otherwise requires complicated, expensive, and lengthy procedures. These pollutants, i.e., pesticides, organic solvents, polychlorinated biphenyls (PCBs), pharmaceutical and personal care products (PPCPs), polycyclic aromatic hydrocarbons (PAHs), etc., can be fastly reduced, photodegraded, or adsorbed from the soil after it's being injected with NPs (Fig. 18.9). Mostly used nanomaterials are zero-valent iron NPs (nZVI), TiO_2, and CNTs (Li et al. 2016). Table 18.3 details the list of pollutants removed from the soil by the use of NPs.

Agricultural soil systems are most vulnerable to pesticides contamination. The use of nZVI in soil-water system has shown some remarkable results in decontaminating the soil from pesticides. Atrazine – a pesticide that sticks in the environment for a very long time and needs reducing conditions to degrade. The presence of nZVI facilitated its degradation from soil at pH less than 7. Even better and faster degradation rates were observed by the use of bimetallic system (Fe/Pd) and addition of sulfate of Fe(II), Fe(III), and Al(III) (Satapanajaru et al. 2008). In soil, DDT on its contact with nZVI is degraded to DDD in 20 days which results in its accumulation till it is microbially degraded. Among HCHs, the degradation of β-HCH is much faster with nZVI as compared to α-HCH, δ-HCH, and γ-HCH. The reason

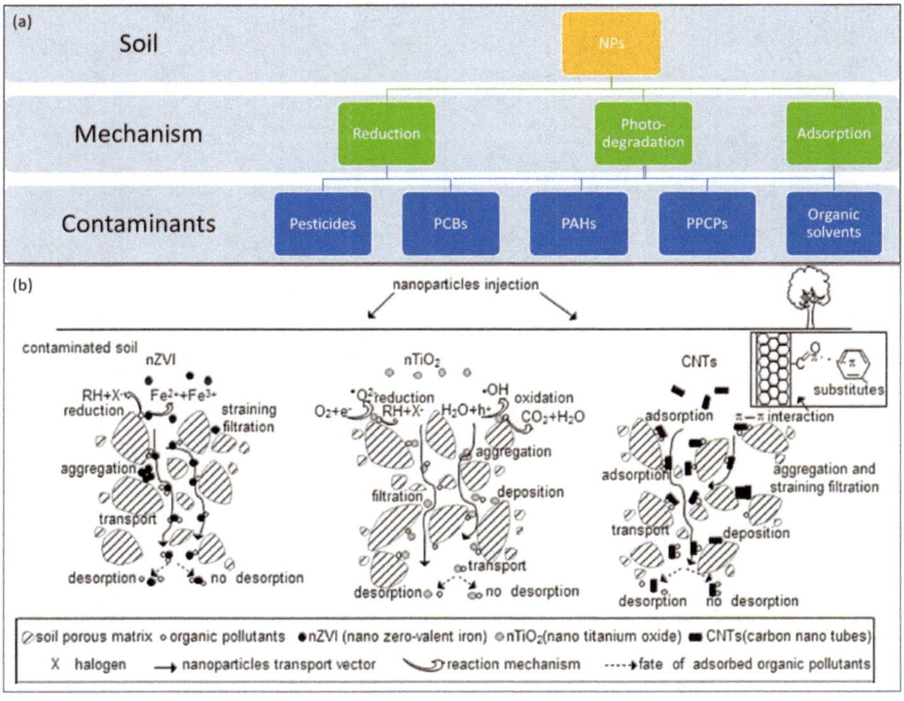

Fig. 18.9 (**a**) Soil remediation approaches of NPs from different contaminants and (**b**) its schematic diagram (Li et al. 2016)

Table 18.3 Soil remediation by nanoparticles

NPs	Contaminants	Removal	Mechanism	References
Inorganic pollutants				
nZVI	Cr(VI)	100%	Reduction to Cr(III)	Gueye et al. (2015)
		100%	Adsorption	Singh et al. (2011)
		78–91%	Adsorption	Gueye et al. (2016)
nZVI + organic acids	Pb(II)	85%	Adsorption	Wang et al. (2014)
nZVI	Zn and Pb	–	Immobilization	Mar Gil-Díaz et al. (2014)
nZVI	Chromate	76–90%	Immobilization	Xu and Zhao (2007)
Ca/CaO dispersion	As, Cd, Cr, Pb	95–90%	Adsorption	Mallampati et al. (2014)
	PCBs	97%	Dehydrochlorination	
Nano-Fe/Ca/CaO/[PO_4]	^{133}Cs, ^{134}Cs, ^{137}Cs	–	–	Mallampati et al. (2015)
Iron phosphate	Cu(II)	–	Immobilization	Liu and Zhao (2007)
Hydroxyapatite	Pb	40%	Immobilization	Jin et al. (2016)
Hydroxyapatite	Pb, Cd	0.2–3 mg/g	Immobilization	Zhang et al. (2010)
Fe_3O_4@EDTA	Ag(I), Hg(II), Mn(II), Zn(II), Pb(II), Cd(II)	96–104%	Adsorption	Ghasemi et al. (2017)
Organic pollutants				
Biochar-supported Ni/Fe	Decabromodiphenyl	87.7%	Degradation	Wu et al. (2016)
nZVI	Atrazine	–	Dechlorination and hydroxylation	Satapanajaru et al. (2008)
nZVI/Pd		–		
Re^{3+}-doped nano-TiO_2	Pesticides	54.89%	Degradation	Zeng et al. (2010)
nZVI	DDT	50%	Degradation	El-Temsah and Joner (2013)
nZVI	Chlorpyrifos	90%	Reduction	Reddy et al. (2012)

$Fe_3O_4/S_2O_8^{2-}$-system	Trichloroethylene	–	Oxidation	Yang and Yeh (2011)
Lactate-modified nZVI	PCP	61%	Reduction	Reddy et al. (2011)
nZVI		53.2%		
Pd/Fe slurry	PCP	90%	Dechlorination	Dien et al. (2013)
nZVI	Pyrene	99%	Fenton oxidation	Jorfi et al. (2013)
		62%	Reduction	Chang et al. (2007)
nZVI	PCB	95%	Reduction	Varanasi et al. (2007)
V_2O_5/TiO_2		21%		
CMC-Pd/nFe^0 and a *Sphingomonas* sp. strain	γ-HCH	100%	Reductive dechlorination and oxidative biomineralization	Singh et al. (2013)
Pd/$nFe^0$and a *Sphingomonas* sp. strain	2,3,7,8-tetrachlorodibenzo-p-dioxin			Bokare et al. (2012)
nZVI	HCHs(α-,β-,γ-, δ-)	62–80%	Reduction	Yang et al. (2010)
	DDX (DDT, DDE, DDD)			

was attributed to their different mechanisms of degradation. Anaerobic degradation of HCH leads to different degradation products, i.e., tetrachlorocytetrachlorocyclohexene, 5,6-dichloro-1,3-cyclohexadiene, monochlorobenzene, and benzene, which is prevalent in all the HCHs. The aerobic pathway, followed by all HCHs except β-HCH, results in the formation of pentachlorocyclohexene, 1,3,4,6-tetrachloro-1,4-cyclohexadiene, and 1,2,4-trichlorobenzene followed by its aromatic ring cleavage (Yang et al. 2010). The degradation efficacy is dependent on soil aging; the DDT degradation was much faster in freshly spiked soils as compared to the ones that are historically polluted (age > 50 yrs) and are rich in organic matter (El-Temsah and Joner 2013).

The high concentrations of the nZVI enable longer degradation of the DDT (El-Temsah and Joner 2013) as low concentration can be easily removed from the system owing to their high reactivity, agglomeration, and oxidation in the soil moist environment. To combat this, the stabilization of NPs was tried by various means like the use of biochar as support, CMC as stabilizer, and incorporation of other metals – all these enabled higher degradation rates. The stabilized CMC Pd/FeNPs was able to degrade 46% γ-HCH, while the nonstabilized nZVI was only 13% effective (Singh et al. 2013). Employing biochar as support of Ni/Fe bimetallic NPs (BC@Ni/Fe) reduced the agglomeration of NPs, facilitated high dispersion of NPs and high adsorption potential, and reduced bioavailability in the soil system. BC@Ni/Fe successfully decontaminated the soil from decabromodiphenyl (DBD) in just 24 h. The high degradation rate was attributed to multiple effects imparted by biochar in conjunction with NPs, i.e., (1) hydrogen bonding of BC with DBD, (2) π-π aromatic interactions between NPs and DBD, (3) aromatic structure of BC promoting electron transfer to DBD and enhancing catalytic degradation by bimetallic NPs (Wu et al. 2016). In some cases, the degradation rates were enhanced by the use of oxidizing agents like persulfate (Yang and Yeh 2011).

In spite of the positive impact, the use of nZVI has shown enhanced toxicity toward soil-residing Collembola and ostracods in terms of their mortality as well as growth inhibition (El-Temsah and Joner 2013). To overcome the negative impacts of these NPs, the concurrent treatments were experimented that make use of NPs and microbial degradation, enabling the exploitation of both biotic and abiotic phenomena. The use of such integrated systems, e.g., NPs (CMC-PD/Fe) and *Sphingomonas* sp. (bacterial strain), was observed to be highly effective and able to degrade γ-HCH almost completely in just 6 days, while at the same time span degradation achieved by independent systems is much lower, i.e., 40% with *Sphingomonas* sp. and 20% with NPs. In addition to that, a few degradation products (1,2,3,4,5-pentachlorocyclohexene, 2,4-trichlorobenzene) resulted from this combined system are anaerobically degraded by bacteria to benzene indicating the high ratio of dichlorination in this system as compared to the independent ones. Such systems are found to operate best at neutral pHs and optimal temperature of 30 °C (Singh et al. 2013).

18.4.5.3.2 Inorganic Contaminants' Removal

NPs are also found quite effective in the decontamination of soil from heavy metals. The nZVI in this case as well have remained the material of choice for many researchers. The removal mechanism depends on the standard redox potential, E°, of the contaminant metal ($E°_M$) and that of nZVI ($E°_{Fe} = -0.41$ V); based on this, the metals are removed by utilizing one of these or both mechanisms:

1. $E°_M \leq E°_{Fe}$ – Removal by adsorption
2. $E°_M > E°_{Fe}$ – Reduction and precipitation

The examples of the first case are Cd and Zn and that of second are Cr(VI), As, Cu, etc. There is a third category as well which involves both the phenomena, e.g., Pb and Ni. As a result of high involvement of E° in the removal of heavy metals, the interference effects from other metals are very phenomenal. In a study carried out to remove Cr(VI) from the soil, the presence of metals like Ni and Pb impacted the reduction of Cr(VI) owing to their higher E°. Other factors like soil organic matter have strong influence on the removal of Cr(VI) (Gueye et al. 2016). The phenomenon results in the reduction of Cr(VI) to (III), the E° of which are 1.36 V and – 0.74 V. Therefore, the Cr(VI) removal takes place through reduction and precipitation followed by removal of Cr(III) via adsorption (Singh et al. 2011).

The effect of combined systems in the case of heavy metal removal is very advantageous. In this context, the NP and organic acid system was used considering their natural source and easy biodegradability. The percentage removal of Pb(II) by the use of these systems obeyed the order: (citric acid + nZVI) – 60–80% > (malic acid + nZVI) – 50–60% > nZVI – 5–10%, while overall removal is higher in farmland soil as compared to mine soil (Wang et al. 2014). Another study has shown that different metals have different affinity with organic acids, resulting in their variable removal capacity, but generally citric acid has better affinity as compared to tartaric acid which in turn is better than oxalic acid for the three metals used in the study, Pb, Cd, and Zn (Cao et al. 2018).

Considering the difficulty of removing the metals from the soil, various researchers have used immobilization as a way forward. This will result in making the metals less mobile and hence bioavailable and hence reducing their negative impacts. Various materials are used in this aspect (Table 18.3). Nano-hydroxyapatite (nHA) has fast absorption capacity for heavy metals like Pb and Cd; hence its utilization for immobilization of these metals can give a way forward. nHA-injected soil sediments can bind these metals with high affinity, resulting in irreversible surface complexation followed by their diffusion into the nHA structure (Zizhong Zhang et al. 2010). The reduced bioavailability of heavy metals (Pb) as a result of inoculation of contaminated soil with different concentrations of nHA was depicted by the reduction in the uptake of Pb by ryegrass grown in the soil. The values of Pb concentration in aboveground portion of ryegrass reduced from 6.60–119.82 mg/kg in untreated soil to 31.85–155.09 mg/kg in nHA-treated soil (Jin et al. 2016). The implication of iron phosphate NPs suspicion on Cu(II) contaminated soil was studied by using three different kinds of soils and two NPs dosages. Results hence obtained confirmed the immobilization of the Cu(II) by the use of NPs which is directly related to their concentration and time of contact. The leachability of Cu varies with soil nature and follows the order: calcareous < neutral < acidic (Liu and Zhao 2007).

18.4.6 Water Management

The agriculture water management is also a critical aspect especially if the water used is contaminated with different pollutants and has poor quality as per requirements of the crop. For this purpose, various techniques and materials are available that can help in improving the water quality as defined in Fig. 18.10. Considering water treatment itself a very broad aspect, the chapter will just present a brief overview of the contaminants dealt using nanomaterials (nano-adsorbents and nanocatalysts) as outlined in Table 18.4. In relation to particularly irrigation system modernization and sustainability, three important approaches have really marked the revolution: (1) water aquifers or groundwater remediation using nanoparticle injection wells, (2) irrigation water with micro-nano bubbles for saving water consumption, and (3) irrigation through wireless sensor network.

In the first case, the contamination of groundwater aquifers was tackled by NPs injection into the water table through a well system (Fig. 18.11a). In this regard, nZVI had always remained a material of choice that is injected in "free form," i.e.,

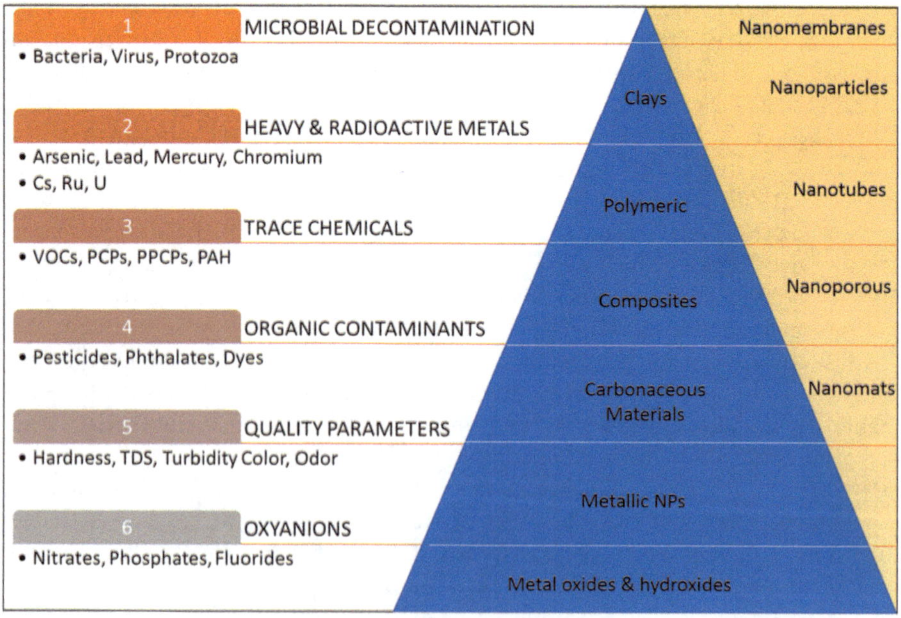

Fig. 18.10 Applications of nanotechnology in water management for the removal of various pollutants (▢) using different types (▲) and shapes (◥) of nanomaterials

Table 18.4 Overview of the water treatment using nanomaterials

Nanomaterial	Contaminant	References
Microbes		
TiO_2/ceramic	*E. coli*	He et al. (2018)
ZnO/ceramic	*E. coli*	Huang et al. (2018)
Fe_3O_4@$Ba_3(PO_4)_2$	*E. coli*	Song et al. (2018)
Heavy metals		
Graphite oxide (GO)	Ni, Zn, Pb, Cd, Cr	Sheet et al. (2014)
Silica		
Silica/GO		
Functionalized Fe_3O_4/polyacrylic acid	Cu(II), Cr(VI)	Huang and Chen (2009)
Ag (0)	Cd(II)	Al-Qahtani (2017)
nZVI	Cd(II)	Boparai et al. (2011)
nZVI/activated carbon	As(III), As(V)	Zhu et al. (2009)
nZVI/biochar	As(V)	Wang et al. (2017)
Organic ligand modified mesoporous silica	Pb(II)	Shahat et al. (2015)
nZVI	Pb(II)	Arancibia-Miranda et al. (2014)
nZVI/graphene	Pb(II)	Jabeen et al. (2013)
nHA	Pb(II)	Mousa et al. (2016)
Calcium-based nHA	Sn(II)	Ghahremani et al. (2017)
nHA	Pb(II), Cd(II), Ni(II)	Mobasherpour et al. (2012)
nHA	Cu(II)	Joshi and Manocha (2017)
Mercaptoamine-functionalized silica-coated magnetic (Fe_3O_4) nano-adsorbents	Hg(II), Pb(II)	Bao et al. (2017)
2-Hydroxyethylammonium sulfonate immobilized on γ-Fe_2O_3	Pb(II)	Khani et al. (2016)
Zirconium silicate	Cu(II), Cd(II), Pb(II)	Mahmoud et al. (2015b)
Chitosan	Cu(II), Cd(II), Hg(II), Pb(II)	Mahmoud et al. (2015a)
Chitosan-acetophenone		
Nanofiltration	Cr(VI)	Hafiane et al. (2000)
Nanopolyaniline/nanosilica	Cu(II), Pb(II), Hg(II), Cd(II)	Mahmoud et al. (2016)

(continued)

Table 18.4 (continued)

Nanomaterial	Contaminant	References
Radioactive anions and cations		
nZVI	Cs	Shubair et al. (2018b)
Fe/Cu	Cs	
Protein-based nano-traps scaffolds	U	Sun et al. (2018)
Iron oxide-urea-activated carbon	U	Mahmoud et al. (2017)
Fe/sludge carbon nanoflakes	U	Kong et al. (2016)
Cu2O/Cu	I^{-1}	Zhang et al. (2017)
Oxyanions		
Fe-Ti oxide	F^{-1}	Chen et al. (2012)
Fe-Al-Ce nano-adsorbent		Chen et al. (2009)
nHA/chitosan		Sundaram et al. (2008)
Cellulose@hydroxyapatite		Yu et al. (2013)
Fe-Ti bimetallic oxide/Fe_3O_4		Zhang et al. (2014a)
FeOOH-graphene oxide		Kuang et al. (2017)
GO/poly-amidoamines dendrimers	NO_3^{-1}	Alighardashi et al. (2018)
nZVI/sand multilayer system		Shubair et al. (2018a)
(Fe/Cu)/sand multilayer system		
nZVI/biochar		Wei et al. (2018)
Chitosan/zeolite Y/nano ZrO_2		Teimouri et al. (2016)
nZVI	PO_4^{-3}	Wen et al. (2014)
nZVI/starch		Chen et al. (2016)
Fe-Ti		Lu et al. (2015)
Hydrated Fe-Zr binary oxide		Zhou et al. (2018)
nZVI/activated carbon	NO_3^{-1}, PO_4^{-3}	Khalil et al. (2017)
Fe_3O_4/ZrO_2/chitosan		Jiang et al. (2013)
Chitosan/Al_2O_3/Fe_3O_4 nanofibers		Bozorgpour et al. (2016)
Chitosan/Al_2O_3/Fe_3O_4 beads		
Pesticides		
TiO_2	2,4-dichlorophenoxyacetic acid	Abdennouri et al. (2016)
TiO_2 pillared clay	2,4-dichlorophenoxypropionic acid	
Magnetic-silica core-shell NPs	Imidacloprid	Karimi et al. (2018)

(continued)

Table 18.4 (continued)

Nanomaterial	Contaminant	References
CNT	Fenuron	Ali et al. (2019)
MWCNT	Diazinon	Dehghani et al. (2019)
Zirconium carbide	Emodin, physcion	Zhang et al. (2019)
GO nanosheets	Ametryn	Khoshnam et al. (2019)
TiO_2/ZnO	Butachlor, diazinon	Nozhat et al. (2018)
Carbon NPs	Trifluralin, glyphosate, 2,4-D	Yousefi (2018)
Dyes		
Cobalt ferrite	Reactive Red 195	Nassar and Khatab (2016)
Clay hydrogels	Congo red, methyl violet	Bhattacharyya and Ray (2015)
Surfactant modified nano-γ-alumina	Brilliant green and crystal violet	Zolgharnein et al. (2015)
ZnO		Kataria and Garg (2017)
CuO/meso-silica	Crystal violet, methylene blue	Liang et al. (2017)
Activated carbon/γ-Fe_2O_3	Alizarin Red S	Fayazi et al. (2015)
Cuprous iodide-cupric oxide nanocomposite loaded on activated carbon	Malachite green	Nekouei et al. (2016)
Organic contaminants – others		
ZnSe-WO3	Bisphenol A	Kumar et al. (2017a)
Fe0/Fe3C@carbon spheres	Phenol	Wang et al. (2015b)
Cu/zeolite	2-Chlorophenol	Huong et al. (2016)
Zeolite	Nitrophenols	Pham et al. (2016)
MWCNT/TiO2	Tetracycline	Ahmadi et al. (2017)
WO3-TiO2 @g-C3N4	Acetylsalicylate, methyl-theobromine	Tahir et al. (2019)
Carbon nano-onions	Phenanthrene, benzopyrene	Sakulthaew et al. (2015)
NZVI/alginate polymer	Naphthalene and its related compounds	Abdel-Gawad et al. (2016)
nZVI		
Multiple nature contaminants		
Al/Fe-doped polymeric beads	F^{-1}, As(V)	Kumar et al. (2011)
Zr-MOF (metal organic framework)	PO_4^{-3}, As(V)	Shahat et al. (2018)
MgO	*E. coli*, Cd(II)	Cai et al. (2017)
Fe@MgO	Pb(II), methyl orange	Ge et al. (2018)

either as powder or slurry (Fig. 18.11c), but this suffers from the disadvantage that these make them a part of the system and can lead to hazardous impacts. Consequently, the researchers opted for the use of static NPs (Fig. 18.11b) that are bound or stabilized using some support materials or surfactants, e.g., membranes, mats, beads, porous structures, supported NPs, etc., that lead to a much improved way of decontaminating water from different targeted contaminants (Tesh and Scott 2014).

The second and interesting case of saving water consumption in irrigation system is the use of modified dip technologies that rely on the generation of small bubbles that fall in the size range of 200 nm to 50 μm and named as micro-nano bubble (MNB). The technique is now being synergistically used with the application of fertilizers for enhancing the crop production by water management. The technique helps in improving the irrigation system by enhancing the water quality through the provision of micro-nano bubble-rich water. MNB, produced by using bubble generator (Fig. 18.11d), promotes the increase in soil oxygen level to approx. 2.4–23.6% and aids the supply of oxygen to crop root zone and hence facilitates root respiration, photosynthetic activity, crop water absorption, fertilizer consumption, N accumulation, soil microbial count, and soil enzyme activity. The studies have shown the positive impact of using this MNB irrigation water on the growth of various crops like rice, tomato, and cucumber (Dahrazma et al. 2019; Liu et al. 2019; Sang et al. 2018; Zhou et al. 2019).

The third innovation in water management comes through the use of wireless sensor network (WSN) – these sensors can determine the soil moisture level or water level in the fields and are usually placed at the root level to have an idea about the moisture level of soil near the roots. The sensor gives analogue command to the control system which can then manage water content in the soil by on/off water supply mechanism (Fig. 18.12) (Ambika et al. 2019; Sadiq et al. 2019). WSN systems in conjunction with drip irrigation have resulted in positive impacts on the potato (Martinez et al. 2019), tomato (Pandey et al. 2018), and plum trees (Millán et al.

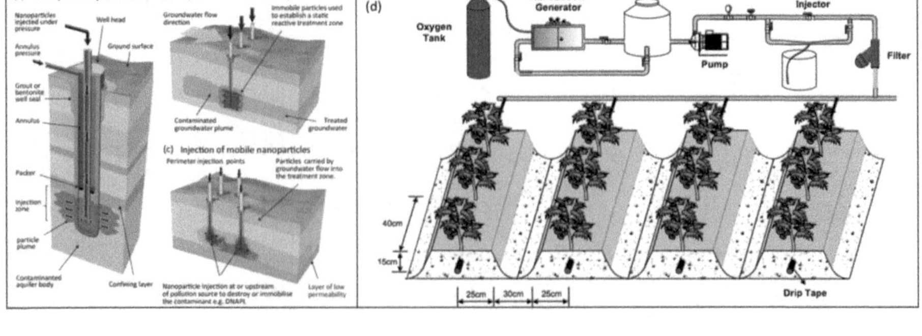

Fig. 18.11 Nanoparticle injection process for groundwater treatment representing (**a**) injection well, (**b**) injection of immobile NPs, and (**c**) injection of mobile particles (Tesh and Scott 2014). (**d**) Schematic representation of MNB technology showing water oxygenation for bubble generation and related plant cropping protocol (Zhou et al. 2019)

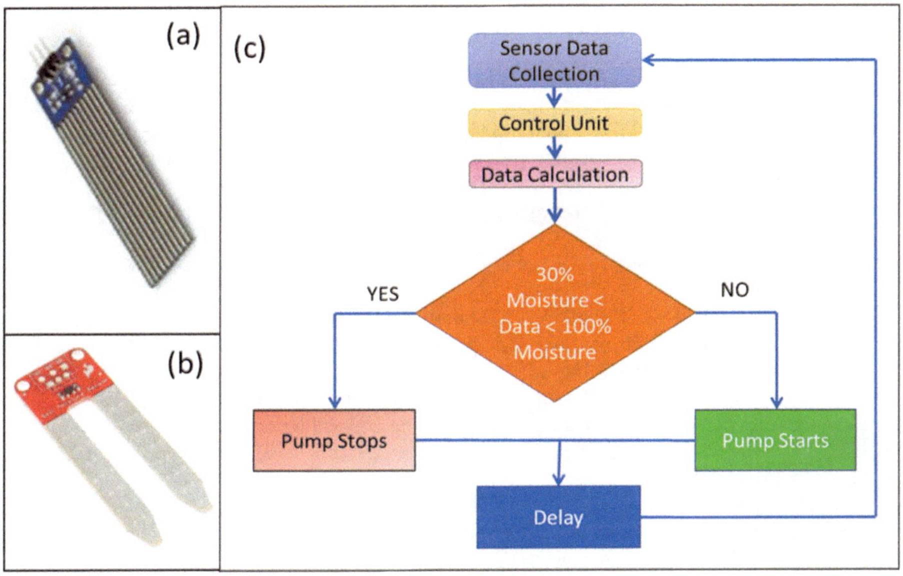

Fig. 18.12 (**a**) Water level sensor, (**b**) soil moisture sensor, (**c**) wireless sensor working principle (Sadiq et al. 2019)

2019) growth. Such systems are found effective in drought stress management (Ortiz et al. 2018).

18.5 Conclusion

Nano-agrotechnology can rightly be considered as a deliberation toward sustainability via precision agriculture approaches. It encompasses in it multitudinous disciplines that can give a new direction and essence to agronomic practices. Till today, nano-agrotechnology developments and achievements are quite promising in the fields of development of smart agrochemicals with controlled delivery options, plant disease diagnosis and pathogen detection, crop yield enhancement, crop protection, and soil and water management. In addition to that, these techniques are carrying with them the positive elements of less materials utilization, especially the toxic agrochemicals, and hence these techniques are more environment-friendly than their counterpart (conventional) techniques. In spite of great innovations and developments, the field is still enjoying its infancy because of its inability to capture the market and attract the attention of multinationals as well as consumers. The major reason of this can be attributed to the cost-effectiveness of large-scale nanomaterials' preparations and the element of toxicity attached with these small structures. Thus the need is to establish protocols and design a roadmap for enhancing the benefits of these nanotechnologies and address the issues that have hampered their way to technological success.

References

Abdel-Aziz HM, Hasaneen MN, Omer AM (2016) Nano chitosan-NPK fertilizer enhances the growth and productivity of wheat plants grown in sandy soil. Span J Agric Res 14(1):0902

Abdel-Gawad SA, Baraka AM, El-Shafei MM, Mahmoud AS (2016) Effects of nano zero valent iron and entrapped nano zero valent iron in alginate polymer on poly aromatic hydrocarbons removal. J Environ Biotechnol Res 5:18–28

Abd-Elsalam KA (2012) Nanoplatforms for plant pathogenic fungi management. Fungal Genom Biol 2:e107

Abd-Elsalam KA, Prasad R (2018) Nanobiotechnology applications in plant protection. Springer, Cham

Abdennouri M, Baâlala M, Galadi A, El Makhfouk M, Bensitel M, Nohair K et al (2016) Photocatalytic degradation of pesticides by titanium dioxide and titanium pillared purified clays. Arab J Chem 9:S313–S318

Ahmadi M, Ramezani Motlagh H, Jaafarzadeh N, Mostoufi A, Saeedi R, Barzegar G, Jorfi S (2017) Enhanced photocatalytic degradation of tetracycline and real pharmaceutical wastewater using MWCNT/TiO2 nano-composite. J Environ Manag 186:55–63. https://doi.org/10.1016/j.jenvman.2016.09.088

Alamer S, Chinnappan R, Zourob M (2017) Development of rapid immuno-based nanosensors for the detection of pathogenic bacteria in poultry processing plants. Procedia Technol 27:23–26

Ali MA, Rehman I, Iqbal A, Din S, Rao AQ, Latif A et al (2014) Nanotechnology, a new frontier in agriculture. Adv Life Sci 1(3):129–138

Ali I, Alharbi OM, ALOthman ZA, Al-Mohaimeed AM, Alwarthan A (2019) Modeling of fenuron pesticide adsorption on CNTs for mechanistic insight and removal in water. Environ Res 170:389–397

Alighardashi A, Esfahani ZK, Najafi F, Afkhami A, Hassani N (2018) Development and application of graphene oxide/poly-amidoamines dendrimers (GO/PAMAMs) nano-composite for nitrate removal from aqueous solutions. Environ Process 5(1):41–64

Al-Qahtani KM (2017) Cadmium removal from aqueous solution by green synthesis zero valent silver nanoparticles with Benjamina leaves extract. Egypt J Aquat Res 43(4):269–274

Ambika N, Meharoofa F, Menezes FE, Deepika M, Sandesha Karanth PK (2019) Agro-amigo: agro based web marketing and automated irrigation. Perspect Commun Embed Syst Signal Process PiCES 2(11):286–290

Arancibia-Miranda N, Baltazar SE, García A, Romero AH, Rubio MA, Altbir D (2014) Lead removal by nano-scale zero valent iron: surface analysis and pH effect. Mater Res Bull 59:341–348

Baker S, Rakshith D, Kavitha KS, Santosh P, Kavitha HU, Rao Y, Satish S (2013) Plants: emerging as nanofactories towards facile route in synthesis of nanoparticles. Bioimpacts 3(3):111

Bakhori N, Yusof N, Abdullah A, Hussein M (2013) Development of a fluorescence resonance energy transfer (fret)-based DNA biosensor for detection of synthetic oligonucleotide of ganoderma boninense. Biosensors 3(4):419–428

Bakhtiari M, Moaveni P, Sani B (2015) The effect of iron nanoparticles spraying time and concentration on wheat. In: Paper presented at the biological forum

Bao S, Li K, Ning P, Peng J, Jin X, Tang L (2017) Highly effective removal of mercury and lead ions from wastewater by mercaptoamine-functionalised silica-coated magnetic nano-adsorbents: behaviours and mechanisms. Appl Surf Sci 393:457–466

Bhattacharyya R, Ray SK (2015) Removal of congo red and methyl violet from water using nano clay filled composite hydrogels of poly acrylic acid and polyethylene glycol. Chem Eng J 260:269–283. https://doi.org/10.1016/j.cej.2014.08.030

Bokare V, Murugesan K, Kim J-H, Kim E-J, Chang Y-S (2012) Integrated hybrid treatment for the remediation of 2,3,7,8-tetrachlorodibenzo-p-dioxin. Sci Total Environ 435–436:563–566. https://doi.org/10.1016/j.scitotenv.2012.07.079

Boparai HK, Joseph M, O'Carroll DM (2011) Kinetics and thermodynamics of cadmium ion removal by adsorption onto nano zerovalent iron particles. J Hazard Mater 186(1):458–465
Bowles M, Lu J (2014) Removing the blinders: a literature review on the potential of nanoscale technologies for the management of supply chains. Technol Forecast Soc Chang 82:190–198
Bozorgpour F, Ramandi HF, Jafari P, Samadi S, Yazd SS, Aliabadi M (2016) Removal of nitrate and phosphate using chitosan/Al2O3/Fe3O4 composite nanofibrous adsorbent: comparison with chitosan/Al2O3/Fe3O4 beads. Int J Biol Macromol 93:557–565. https://doi.org/10.1016/j.ijbiomac.2016.09.015
Cai Y, Li C, Wu D, Wang W, Tan F, Wang X et al (2017) Highly active MgO nanoparticles for simultaneous bacterial inactivation and heavy metal removal from aqueous solution. Chem Eng J 312:158–166. https://doi.org/10.1016/j.cej.2016.11.134
Cao Y, Zhang S, Zhong Q, Wang G, Xu X, Li T et al (2018) Feasibility of nanoscale zero-valent iron to enhance the removal efficiencies of heavy metals from polluted soils by organic acids. Ecotoxicol Environ Saf 162:464–473. https://doi.org/10.1016/j.ecoenv.2018.07.036
Cesarino I, Moraes FC, Lanza MR, Machado SA (2012) Electrochemical detection of carbamate pesticides in fruit and vegetables with a biosensor based on acetylcholinesterase immobilised on a composite of polyaniline–carbon nanotubes. Food Chem 135(3):873–879
Chang M-C, Shu H-Y, Hsieh W-P, Wang M-C (2007) Remediation of soil contaminated with pyrene using ground nanoscale zero-valent iron. J Air Waste Manage Assoc 57(2):221–227
Chartuprayoon N, Rheem Y, Ng JC, Nam J, Chen W, Myung NV (2013) Polypyrrole nanoribbon based chemiresistive immunosensors for viral plant pathogen detection. Anal Methods 5(14):3497–3502
Chen H (2018) Metal based nanoparticles in agricultural system: behavior, transport, and interaction with plants. Chem Spec Bioavailab 30(1):123–134
Chen L, Wu H-X, Wang T-J, Jin Y, Zhang Y, Dou X-M (2009) Granulation of Fe–Al–Ce nano-adsorbent for fluoride removal from drinking water by spray coating on sand in a fluidized bed. Powder Technol 193(1):59–64
Chen J, Wang W, Xu Y, Zhang X (2010) Slow-release formulation of a new biological pesticide, pyoluteorin, with mesoporous silica. J Agric Food Chem 59(1):307–311
Chen L, He B-Y, He S, Wang T-J, Su C-L, Jin Y (2012) Fe—Ti oxide nano-adsorbent synthesized by co-precipitation for fluoride removal from drinking water and its adsorption mechanism. Powder Technol 227:3–8
Chen D, Gao B, Wang H, Yang K (2016) Effective removal of high concentration of phosphate by starch-stabilized nanoscale zerovalent iron (SNZVI). J Taiwan Inst Chem Eng 61:181–187. https://doi.org/10.1016/j.jtice.2015.12.007
Chen C, Feng S, Zhou M, Ji C, Que L, Wang W (2019) Development of a structure-switching aptamer-based nanosensor for salicylic acid detection. Biosens Bioelectron 140:111342
Dahrazma B, Naghedinia A, Gorji HG, Saghravani S (2019) Morphological and physiological responses of *Cucumis sativus* L. to water with micro-nanobubbles. J Agric Sci Technol 21(1):181–192
Davarpanah S, Tehranifar A, Davarynejad G, Abadía J, Khorasani R (2016) Effects of foliar applications of zinc and boron nano-fertilizers on pomegranate (*Punica granatum* cv. Ardestani) fruit yield and quality. Sci Hortic 210:57–64
Dehghani MH, Kamalian S, Shayeghi M, Yousefi M, Heidarinejad Z, Agarwal S, Gupta VK (2019) High-performance removal of diazinon pesticide from water using multi-walled carbon nanotubes. Microchem J 145:486–491
Delfani M, Baradarn Firouzabadi M, Farrokhi N, Makarian H (2014) Some physiological responses of black-eyed pea to iron and magnesium nanofertilizers. Commun Soil Sci Plant Anal 45(4):530–540
Derbalah A, Elsharkawy MM, Hamza A, El-Shaer A (2019) Resistance induction in cucumber and direct antifungal activity of zirconium oxide nanoparticles against Rhizoctonia solani. Pestic Biochem Physiol 157:230–236

Díaz-Blancas V, Medina D, Padilla-Ortega E, Bortolini-Zavala R, Olvera-Romero M, Luna-Bárcenas G (2016) Nanoemulsion formulations of fungicide tebuconazole for agricultural applications. Molecules 21(10):1271

Dien NT, De Windt W, Buekens A, Chang MB (2013) Application of bimetallic iron (BioCAT slurry) for pentachlorophenol removal from sandy soil. J Hazard Mater 252:83–90

Ditta A (2012) How helpful is nanotechnology in agriculture? Adv Nat Sci Nanosci Nanotechnol 3(3):033002

Dubas ST, Pimpan V (2008) Humic acid assisted synthesis of silver nanoparticles and its application to herbicide detection. Mater Lett 62(17):2661–2663. https://doi.org/10.1016/j.matlet.2008.01.033

Duhan JS, Kumar R, Kumar N, Kaur P, Nehra K, Duhan S (2017) Nanotechnology: the new perspective in precision agriculture. Biotechnol Rep 15:11–23

Durán N, Marcato PD, Durán M, Yadav A, Gade A, Rai M (2011) Mechanistic aspects in the biogenic synthesis of extracellular metal nanoparticles by peptides, bacteria, fungi, and plants. Appl Microbiol Biotechnol 90(5):1609–1624. https://doi.org/10.1007/s00253-011-3249-8

El-Temsah YS, Joner EJ (2013) Effects of nano-sized zero-valent iron (nZVI) on DDT degradation in soil and its toxicity to collembola and ostracods. Chemosphere 92(1):131–137

Fang Y, Umasankar Y, Ramasamy RP (2014) Electrochemical detection of p-ethylguaiacol, a fungi infected fruit volatile using metal oxide nanoparticles. Analyst 139(15):3804–3810

FAO (2017) World fertilizer trends and outlook to 2020. Retrieved from Rome

Fayazi M, Ghanei-Motlagh M, Taher MA (2015) The adsorption of basic dye (Alizarin red S) from aqueous solution onto activated carbon/γ-Fe_2O_3 nano-composite: kinetic and equilibrium studies. Mater Sci Semicond Process 40:35–43. https://doi.org/10.1016/j.mssp.2015.06.044

Figovsky O, Beilin D (2017) Green nanotechnology. Pan Stanford

Ge L, Wang W, Peng Z, Tan F, Wang X, Chen J, Qiao X (2018) Facile fabrication of Fe@MgO magnetic nanocomposites for efficient removal of heavy metal ion and dye from water. Powder Technol 326:393–401. https://doi.org/10.1016/j.powtec.2017.12.003

Ghahremani D, Mobasherpour I, Salahi E, Ebrahimi M, Manafi S, Keramatpour L (2017) Potential of nano crystalline calcium hydroxyapatite for Tin(II) removal from aqueous solutions: equilibria and kinetic processes. Arab J Chem 10:S461–S471. https://doi.org/10.1016/j.arabjc.2012.10.006

Ghasemi E, Heydari A, Sillanpää M (2017) Superparamagnetic Fe3O4@ EDTA nanoparticles as an efficient adsorbent for simultaneous removal of Ag (I), Hg (II), Mn (II), Zn (II), Pb (II) and Cd (II) from water and soil environmental samples. Microchem J 131:51–56

Goswami A, Roy I, Sengupta S, Debnath N (2010) Novel applications of solid and liquid formulations of nanoparticles against insect pests and pathogens. Thin Solid Films 519(3):1252–1257

Grillo R, Pereira AE, Nishisaka CS, de Lima R, Oehlke K, Greiner R, Fraceto LF (2014) Chitosan/tripolyphosphate nanoparticles loaded with paraquat herbicide: an environmentally safer alternative for weed control. J Hazard Mater 278:163–171

Gueye MT, Petrucci E, Di Palma L (2015) Chemical reduction of hexavalent chromium (VI) in soil slurry by nano zero valent iron. Chem Eng Trans 43:655–660

Gueye MT, Di Palma L, Allahverdiyeva G, Bavasso I, Petrucci E, Stoller M, Vilardi G (2016) The influence of heavy metals and organic matter on hexavalent chromium reduction by nano zero valent iron in soil. Chem Eng Trans 47:289–294

Guo M, Liu M, Hu Z, Zhan F, Wu L (2005) Preparation and properties of a slow release NP compound fertilizer with superabsorbent and moisture preservation. J Appl Polym Sci 96(6):2132–2138

Guo H, White JC, Wang Z, Xing B (2018) Nano-enabled fertilizers to control the release and use efficiency of nutrients. Curr Opin Environ Sci Health 6:77–83

Hafiąne A, Lemordant D, Dhahbi M (2000) Removal of hexavalent chromium by nanofiltration. Desalination 130(3):305–312. https://doi.org/10.1016/S0011-9164(00)00094-1

Hasaneen M, Abdel-aziz HMM, Omer AM (2016) Effect of foliar application of engineered nanomaterials: carbon nanotubes NPK and chitosan nanoparticles NPK fertilizer on the growth of French bean plant. Biochem Biotechnol Res 4:68–76

Hayles J, Johnson L, Worthley C, Losic D (2017) Nanopesticides: a review of current research and perspectives. In: New pesticides and soil sensors. Elsevier, London, pp 193–225

He Y, Huang G, An C, Huang J, Zhang P, Chen X, Xin X (2018) Reduction of *Escherichia coli* using ceramic disk filter decorated by nano-TiO_2: a low-cost solution for household water purification. Sci Total Environ 616–617:1628–1637. https://doi.org/10.1016/j.scitotenv.2017.10.149

Hidayat R, Fadillah G, Chasanah U, Wahyuningsih S, Ramelan AH (2015) Effectiveness of urea nanofertilizer based aminopropyltrimethoxysilane (APTMS)-zeolite as slow release fertilizer system. Afr J Agric Res 10(14):1785–1788

Ho VA, Le PT, Nguyen TP, Nguyen CK, Nguyen VT, Tran NQ (2015) Silver core-shell nanoclusters exhibiting strong growth inhibition of plant-pathogenic fungi. J Nanomater 16(1):13

Huang S-H, Chen D-H (2009) Rapid removal of heavy metal cations and anions from aqueous solutions by an amino-functionalized magnetic nano-adsorbent. J Hazard Mater 163(1):174–179. https://doi.org/10.1016/j.jhazmat.2008.06.075

Huang J, Huang G, An C, He Y, Yao Y, Zhang P, Shen J (2018) Performance of ceramic disk filter coated with nano ZnO for removing Escherichia coli from water in small rural and remote communities of developing regions. Environ Pollut 238:52–62. https://doi.org/10.1016/j.envpol.2018.03.008

Huong P-T, Lee B-K, Kim J (2016) Improved removal of 2-chlorophenol by a synthesized Cu-nano zeolite. Process Saf Environ Prot 100:272–280. https://doi.org/10.1016/j.psep.2016.02.002

Iravani S (2011) Green synthesis of metal nanoparticles using plants. Green Chem 13(10):2638–2650

Jabeen H, Kemp KC, Chandra V (2013) Synthesis of nano zerovalent iron nanoparticles–graphene composite for the treatment of lead contaminated water. J Environ Manag 130:429–435

Janmohammadi M, Pornour N, Javanmard A, Sabaghnia N (2016) Effects of bio-organic, conventional and nanofertilizers on growth, yield and quality of potato in cold steppe/Bioorganinių, tradicinių ir nanotrąšų poveikis bulvių augimui, derliui ir kokybei šaltojoje stepėje. Botanica Lithuanica 22(2):133–144

Jha AK, Prasad K, Prasad K, Kulkarni A (2009) Plant system: nature's nanofactory. Colloids Surf B Biointerfaces 73(2):219–223

Jiang LC, Basri M, Omar D, Rahman MBA, Salleh AB, Rahman RNZRA, Selamat A (2012) Green nano-emulsion intervention for water-soluble glyphosate isopropylamine (IPA) formulations in controlling *Eleusine indica* (*E. indica*). Pestic Biochem Physiol 102(1):19–29

Jiang H, Chen P, Luo S, Tu X, Cao Q, Shu M (2013) Synthesis of novel nanocomposite Fe3O4/ZrO2/chitosan and its application for removal of nitrate and phosphate. Appl Surf Sci 284:942–949. https://doi.org/10.1016/j.apsusc.2013.04.013

Jin Y, Liu W, Li X-l, Shen S-g, Liang S-x, Liu C, Shan L (2016) Nano-hydroxyapatite immobilized lead and enhanced plant growth of ryegrass in a contaminated soil. Ecol Eng 95:25–29

Jokar M, Safaralizadeh MH, Hadizadeh F, Rahmani F, Kalani MR (2016) Design and evaluation of an apta-nano-sensor to detect Acetamiprid in vitro and in silico. J Biomol Struct Dyn 34(11):2505–2517

Jordan CF (2013) An ecosystem approach to sustainable agriculture: energy use efficiency in the American South. Springer, Dordrecht

Jorfi S, Rezaee A, Moheb-ali G-a, alah Jaafarzadeh N (2013) Pyrene removal from contaminated soils by modified Fenton oxidation using iron nano particles. J Environ Health Sci Eng 11(1):17

Joshi P, Manocha S (2017) Kinetic and thermodynamic studies of the adsorption of copper ions on hydroxyapatite nanoparticles. Mater Today Proc 4(9):10455–10459. https://doi.org/10.1016/j.matpr.2017.06.399

Kah M, Hofmann T (2014) Nanopesticide research: current trends and future priorities. Environ Int 63:224–235

Karimi B, Khanaki S, Ma'mani L, Khezri SM, Karami A (2018) Efficient removal of organophosphate pesticide imidacloprid from water samples by modified magnetic-silica core-shell nanoparticles as a recoverable nano-adsorbent. J Sch Public Health Inst Public Health Res 15(4):389–400

Kashyap PL, Kumar S, Srivastava AK (2017) Nanodiagnostics for plant pathogens. Environ Chem Lett 15(1):7–13

Kataria N, Garg VK (2017) Removal of congo red and brilliant green dyes from aqueous solution using flower shaped ZnO nanoparticles. J Environ Chem Eng 5(6):5420–5428. https://doi.org/10.1016/j.jece.2017.10.035

Kaushal M, Wani SP (2017) Nanosensors: frontiers in precision agriculture. In: Nanotechnology. Springer, Singapore, pp 279–291

Khalifa NS, Hasaneen MN (2018) The effect of chitosan–PMAA–NPK nanofertilizer on *Pisum sativum* plants. 3 Biotech 8(4):193

Khalil AM, Eljamal O, Amen TW, Sugihara Y, Matsunaga N (2017) Optimized nano-scale zero-valent iron supported on treated activated carbon for enhanced nitrate and phosphate removal from water. Chem Eng J 309:349–365

Khandelwal N, Barbole RS, Banerjee SS, Chate GP, Biradar AV, Khandare JJ, Giri AP (2016) Budding trends in integrated pest management using advanced micro-and nano-materials: challenges and perspectives. J Environ Manag 184:157–169

Khani R, Sobhani S, Beyki MH (2016) Highly selective and efficient removal of lead with magnetic nano-adsorbent: multivariate optimization, isotherm and thermodynamic studies. J Colloid Interface Sci 466:198–205. https://doi.org/10.1016/j.jcis.2015.12.027

Khiyami MA, Almoammar H, Awad YM, Alghuthaymi MA, Abd-Elsalam KA (2014) Plant pathogen nanodiagnostic techniques: forthcoming changes? Biotechnol Biotechnol Equip 28(5):775–785

Khoshnam F, Zargar B, Moghadam MR (2019) Adsorption and removal of ametryn using graphene oxide nano-sheets from farm waste water and optimization using response surface methodology. J Iran Chem Soc 16(7):1383–1390

Khot LR, Sankaran S, Maja JM, Ehsani R, Schuster EW (2012) Applications of nanomaterials in agricultural production and crop protection: a review. Crop Prot 35:64–70

Kong L, Zhu Y, Wang M, Li Z, Tan Z, Xu R et al (2016) Simultaneous reduction and adsorption for immobilization of uranium from aqueous solution by nano-flake Fe-SC. J Hazard Mater 320:435–441

Kottegoda N, Munaweera I, Madusanka N, Karunaratne V (2011) A green slow-release fertilizer composition based on urea-modified hydroxyapatite nanoparticles encapsulated wood. Curr Sci 101(1):73–78

Kottegoda N, Sandaruwan C, Priyadarshana G, Siriwardhana A, Rathnayake UA, Berugoda Arachchige DM et al (2017) Urea-hydroxyapatite nanohybrids for slow release of nitrogen. ACS Nano 11(2):1214–1221. https://doi.org/10.1021/acsnano.6b07781

Kuang L, Liu Y, Fu D, Zhao Y (2017) FeOOH-graphene oxide nanocomposites for fluoride removal from water: acetate mediated nano FeOOH growth and adsorption mechanism. J Colloid Interface Sci 490:259–269

Kumar V, Talreja N, Deva D, Sankararamakrishnan N, Sharma A, Verma N (2011) Development of bi-metal doped micro-and nano multi-functional polymeric adsorbents for the removal of fluoride and arsenic (V) from wastewater. Desalination 282:27–38

Kumar RS, Shiny P, Anjali C, Jerobin J, Goshen KM, Magdassi S et al (2013) Distinctive effects of nano-sized permethrin in the environment. Environ Sci Pollut Res 20(4):2593–2602

Kumar A, Naushad M, Rana A, Inamuddin, Preeti, Sharma G et al (2017a) ZnSe-WO3 nano-hetero-assembly stacked on Gum ghatti for photo-degradative removal of Bisphenol A: symbiose of adsorption and photocatalysis. Int J Biol Macromol 104:1172–1184. https://doi.org/10.1016/j.ijbiomac.2017.06.116

Kumar S, Bhanjana G, Sharma A, Dilbaghi N, Sidhu M, Kim K-H (2017b) Development of nanoformulation approaches for the control of weeds. Sci Total Environ 586:1272–1278

Kumar R, Ashfaq M, Verma N (2018) Synthesis of novel PVA–starch formulation-supported Cu–Zn nanoparticle carrying carbon nanofibers as a nanofertilizer: controlled release of micronutrients. J Mater Sci 53(10):7150–7164. https://doi.org/10.1007/s10853-018-2107-9

Lateef A, Nazir R, Jamil N, Alam S, Shah R, Khan MN, Saleem M (2016) Synthesis and characterization of zeolite based nano–composite: an environment friendly slow release fertilizer. Microporous Mesoporous Mater 232:174–183

Lateef A, Nazir R, Jamil N, Alam S, Shah R, Khan MN, Saleem M (2019) Synthesis and characterization of environmental friendly corncob biochar based nano-composite–A potential slow release nano-fertilizer for sustainable agriculture. Environ Nanotechnol Monit Manag 11:100212

Li ZZ, Chen JF, Liu F, Liu AQ, Wang Q, Sun HY, Wen LX (2007) Study of UV-shielding properties of novel porous hollow silica nanoparticle carriers for avermectin. Pest Manag Sci Former Pestic Sci 63(3):241–246

Li X, Xu H, Chen Z-S, Chen G (2011) Biosynthesis of nanoparticles by microorganisms and their applications. J Nanomater 2011:1–16

Li Q, Chen X, Zhuang J, Chen X (2016) Decontaminating soil organic pollutants with manufactured nanoparticles. Environ Sci Pollut Res 23(12):11533–11548

Liang Z, Zhao Z, Sun T, Shi W, Cui F (2017) Enhanced adsorption of the cationic dyes in the spherical CuO/meso-silica nano composite and impact of solution chemistry. J Colloid Interface Sci 485:192–200. https://doi.org/10.1016/j.jcis.2016.09.028

Liu G, Lin Y (2005) Electrochemical sensor for organophosphate pesticides and nerve agents using zirconia nanoparticles as selective sorbents. Anal Chem 77(18):5894–5901. https://doi.org/10.1021/ac050791t

Liu R, Zhao D (2007) In situ immobilization of Cu(II) in soils using a new class of iron phosphate nanoparticles. Chemosphere 68(10):1867–1876. https://doi.org/10.1016/j.chemosphere.2007.03.010

Liu S, Yuan L, Yue X, Zheng Z, Tang Z (2008) Recent advances in nanosensors for organophosphate pesticide detection. Adv Powder Technol 19(5):419–441. https://doi.org/10.1016/S0921-8831(08)60910-3

Liu Y, Wei F, Wang Y, Zhu G (2011) Studies on the formation of bifenthrin oil-in-water nanoemulsions prepared with mixed surfactants. Colloids Surf A Physicochem Eng Asp 389(1–3):90–96

Liu Y, Zhou Y, Wang T, Pan J, Zhou B, Muhammad T et al (2019) Micro-nano bubble water oxygation: synergistically improving irrigation water use efficiency, crop yield and quality. J Clean Prod 222:835–843

Lu J, Liu D, Hao J, Zhang G, Lu B (2015) Phosphate removal from aqueous solutions by a nano-structured Fe–Ti bimetal oxide sorbent. Chem Eng Res Des 93:652–661

Madbouly AK, Abdel-Aziz MS, Abdel-Wahhab MA (2017) Biosynthesis of nanosilver using chaetomium globosum and its application to control fusarium wilt of tomato in the greenhouse. IET Nanobiotechnol 11(6):702–708

Mahmoud ME, Abou Kana MTH, Hendy AA (2015a) Synthesis and implementation of nano-chitosan and its acetophenone derivative for enhanced removal of metals. Int J Biol Macromol 81:672–680. https://doi.org/10.1016/j.ijbiomac.2015.08.063

Mahmoud ME, Nabil GM, Mahmoud SME (2015b) High performance nano-zirconium silicate adsorbent for efficient removal of copper (II), cadmium (II) and lead (II). J Environ Chem Eng 3(2):1320–1328. https://doi.org/10.1016/j.jece.2014.11.027

Mahmoud ME, Fekry NA, El-Latif MMA (2016) Nanocomposites of nanosilica-immobilized-nanopolyaniline and crosslinked nanopolyaniline for removal of heavy metals. Chem Eng J 304:679–691. https://doi.org/10.1016/j.cej.2016.06.110

Mahmoud ME, Khalifa MA, El Wakeel YM, Header MS, Abdel-Fattah TM (2017) Engineered nano-magnetic iron oxide-urea-activated carbon nanolayer sorbent for potential removal of uranium (VI) from aqueous solution. J Nucl Mater 487:13–22. https://doi.org/10.1016/j.jnucmat.2017.01.046

Mala R, Celsia Arul Selvaraj R, Barathi Sundaram V, Blessina Siva Shanmuga Rajan R, Maheswari Gurusamy U (2017) Evaluation of nano structured slow release fertilizer on the soil fertility, yield and nutritional profile of vigna radiata. Recent Pat Nanotechnol 11(1):50–62

Mallampati SR, Mitoma Y, Okuda T, Sakita S, Simion C (2014) Simultaneous decontamination of cross-polluted soils with heavy metals and PCBs using a nano-metallic Ca/CaO dispersion mixture. Environ Sci Pollut Res 21(15):9270–9277

Mallampati SR, Mitoma Y, Okuda T, Simion C, Lee BK (2015) Solvent-free synthesis and application of nano-Fe/Ca/CaO/[PO4] composite for dual separation and immobilization of stable and radioactive cesium in contaminated soils. J Hazard Mater 297:74–82. https://doi.org/10.1016/j.jhazmat.2015.04.071

Mar Gil-Díaz M, Pérez-Sanz A, Angeles Vicente M, Carmen Lobo M (2014) Immobilisation of Pb and Zn in soils using stabilised zero-valent iron nanoparticles: effects on soil properties. Clean–Soil Air Water 42(12):1776–1784

Martinez O, Arguello C, León J, Carguacundo PDC, Daga GEC (2019) Prototype of automated irrigation system improves the yield of potatoes (*Solanum tuberosum* L.) in Riobamba-Ecuador using wireless network sensors-WSN and 6LoWPAN. MASKAY 9(2):25–30

Maruyama CR, Guilger M, Pascoli M, Bileshy-José N, Abhilash P, Fraceto LF, De Lima R (2016) Nanoparticles based on chitosan as carriers for the combined herbicides imazapic and imazapyr. Sci Rep 6:19768

Millán S, Campillo C, Casadesús J, Moñino M, Vivas A, Prieto M (2019) Automated irrigation scheduling for drip-irrigated plum trees. In: Precision agriculture'19. Wageningen Academic Publishers, Wageningen, pp 59–66

Mishra S, Singh BR, Singh A, Keswani C, Naqvi AH, Singh H (2014) Biofabricated silver nanoparticles act as a strong fungicide against Bipolaris sorokiniana causing spot blotch disease in wheat. PLoS One 9(5):e97881

Mobasherpour I, Salahi E, Pazouki M (2012) Comparative of the removal of Pb^{2+}, Cd^{2+} and Ni^{2+} by nano crystallite hydroxyapatite from aqueous solutions: adsorption isotherm study. Arabian J Chem 5(4):439–446. https://doi.org/10.1016/j.arabjc.2010.12.022

Mohamed EA, Gaber MH, Elsharabasy SF (2018) Evaluating the in vivo efficacy of copper-chitosan nanocomposition for treating vascular wilt disease in date palm. Int J Environ Agric Biotechnol 3(2):447–454

Mondal P, Kumar R, Gogoi R (2017) Azomethine based nano-chemicals: development, in vitro and in vivo fungicidal evaluation against Sclerotium rolfsii, Rhizoctonia bataticola and Rhizoctonia solani. Bioorg Chem 70:153–162

Mousa S, Ammar N, Ibrahim H (2016) Removal of lead ions using hydroxyapatite nano-material prepared from phosphogypsum waste. J Saudi Chem Soc 20(3):357–365

Munawar A, Ong Y, Schirhagl R, Tahir MA, Khan WS, Bajwa SZ (2019) Nanosensors for diagnosis with optical, electric and mechanical transducers. RSC Adv 9(12):6793–6803. https://doi.org/10.1039/C8RA10144B

Naderi M, Danesh-Shahraki A (2013) Nanofertilizers and their roles in sustainable agriculture. Int J Agric Crop Sci 5(19):2229

Nassar MY, Khatab M (2016) Cobalt ferrite nanoparticles via a template-free hydrothermal route as an efficient nano-adsorbent for potential textile dye removal. RSC Adv 6(83):79688–79705

Nekouei F, Noorizadeh H, Nekouei S, Asif M, Tyagi I, Agarwal S, Gupta VK (2016) Removal of malachite green from aqueous solutions by cuprous iodide–cupric oxide nano-composite loaded on activated carbon as a new sorbent for solid phase extraction: isotherm, kinetics and thermodynamic studies. J Mol Liq 213:360–368. https://doi.org/10.1016/j.molliq.2015.07.058

Niemeyer CM, Mirkin CA (2004) Nanobiotechnology: concepts, applications and perspectives, vol 1. Wiley, New York

Nozhat, S., Fazilati, M., & Hassani, A. H. (2018). Butachlor and Diazinon elimination from aqueous solution using TiO2/ZnO nano-photocatalysts

Omanović-Mikličanin E, Maksimović M (2016) Nanosensors applications in agriculture and food industry. Bull Chem Technol Bosnia Herzegovina 47:59–70

Ortiz D, Litvin AG, Fernandez MGS (2018) A cost-effective and customizable automated irrigation system for precise high-throughput phenotyping in drought stress studies. PLoS One 13(6):e0198546

Pacioni NL, Veglia AV (2007) Determination of poorly fluorescent carbamate pesticides in water, bendiocarb and promecarb, using cyclodextrin nanocavities and related media. Anal Chim Acta 583(1):63–71. https://doi.org/10.1016/j.aca.2006.10.010

Pandey SK, Jain A, Sharda R, Sharma P, Joshi A (2018) Economic analysis of automated drip irrigation system for production of tomato crop. Indian J Econ Dev 14(3):513–520

Parisi C, Vigani M, Rodríguez-Cerezo E (2015) Agricultural nanotechnologies: what are the current possibilities? Nano Today 10(2):124–127. https://doi.org/10.1016/j.nantod.2014.09.009

Pham T-H, Lee B-K, Kim J (2016) Improved adsorption properties of a nano zeolite adsorbent toward toxic nitrophenols. Process Saf Environ Prot:104, 314–322. https://doi.org/10.1016/j.psep.2016.08.018

Prasad R, Bhattacharyya A, Nguyen QD (2017a) Nanotechnology in sustainable agriculture: recent developments, challenges, and perspectives. Front Microbiol 8:1014

Prasad R, Kumar M, Kumar V (2017b) Nanotechnology: an agricultural paradigm. Springer, Singapore

Qi Y, Xiu F-R, Zheng M, Li B (2016) A simple and rapid chemiluminescence aptasensor for acetamiprid in contaminated samples: sensitivity, selectivity and mechanism. Biosens Bioelectron 83:243–249. https://doi.org/10.1016/j.bios.2016.04.074

Rai M, Posten C (2013) Green biosynthesis of nanoparticles: mechanisms and applications. CABI, Wallingford

Raliya R, Nair R, Chavalmane S, Wang W-N, Biswas P (2015) Mechanistic evaluation of translocation and physiological impact of titanium dioxide and zinc oxide nanoparticles on the tomato (*Solanum lycopersicum* L.) plant. Metallomics 7(12):1584–1594

Raliya R, Franke C, Chavalmane S, Nair R, Reed N, Biswas P (2016) Quantitative understanding of nanoparticle uptake in watermelon plants. Front Plant Sci 7:1288

Rashidzadeh A, Olad A, Salari D, Reyhanitabar A (2014) On the preparation and swelling properties of hydrogel nanocomposite based on Sodium alginate-g-Poly (acrylic acid-co-acrylamide)/Clinoptilolite and its application as slow release fertilizer. J Polym Res 21(2):344. https://doi.org/10.1007/s10965-013-0344-9

Rastogi A, Tripathi DK, Yadav S, Chauhan DK, Živčák M, Ghorbanpour M et al (2019) Application of silicon nanoparticles in agriculture. 3 Biotech 9(3):90

Reddy KR, Khodadoust AP, Darko-Kagya K (2011) Transport and reactivity of lactate-modified nanoscale iron particles in PCP-contaminated soils. J Hazard Toxic Radioact Waste 16(1):68–74

Reddy A, Madhavi V, Reddy KG, Madhavi G (2012) Remediation of chlorpyrifos-contaminated soils by laboratory-synthesized zero-valent nano iron particles: effect of pH and aluminium salts. J Chem 2013:1–7

Rhim J-W (2011) Effect of clay contents on mechanical and water vapor barrier properties of agar-based nanocomposite films. Carbohydr Polym 86(2):691–699. https://doi.org/10.1016/j.carbpol.2011.05.010

Sabir A, Yazar K, Sabir F, Kara Z, Yazici MA, Goksu N (2014) Vine growth, yield, berry quality attributes and leaf nutrient content of grapevines as influenced by seaweed extract (Ascophyllum nodosum) and nanosize fertilizer pulverizations. Sci Hortic 175:1–8

Sadiq MT, Hossain MM, Rahman KF, Sayem AS (2019) Automated irrigation system: controlling irrigation through wireless sensor network. Int J Electr Electron Eng 7(2):33–37

Safaei Z, Azizi M, Davarynejad G, Aroiee H (2014) The effect of foliar application of humic acid and nanofertilizer (Pharmks®) on yield and yield components of black cumin (Nigella sativa L.). J Med Plants By-prod 3(2):133–140

Safarpour H, Safarnejad MR, Tabatabaei M, Mohsenifar A, Rad F, Basirat M et al (2012) Development of a quantum dots FRET-based biosensor for efficient detection of Polymyxa betae. Can J Plant Pathol 34(4):507–515

Saharan V, Mehrotra A, Khatik R, Rawal P, Sharma S, Pal A (2013) Synthesis of chitosan based nanoparticles and their in vitro evaluation against phytopathogenic fungi. Int J Biol Macromol 62:677–683

Saini RK, Bagri LP, Bajpai AK (2017) Smart nanosensors for pesticide detection. In: New pesticides and soil sensors. Elsevier, pp 519–559

Sakulthaew C, Comfort SD, Chokejaroenrat C, Li X, Harris CE (2015) Removing PAHs from urban runoff water by combining ozonation and carbon nano-onions. Chemosphere 141:265–273. https://doi.org/10.1016/j.chemosphere.2015.08.002

Sang H, Jiao X, Wang S, Guo W, Salahou MK, Liu K (2018) Effects of micro-nano bubble aerated irrigation and nitrogen fertilizer level on tillering, nitrogen uptake and utilization of early rice. Plant Soil Environ 64(7):297–302

Satapanajaru T, Anurakpongsatorn P, Pengthamkeerati P, Boparai H (2008) Remediation of atrazine-contaminated soil and water by nano zerovalent iron. Water Air Soil Pollut 192(1–4):349–359

Sempeho SI, Kim HT, Mubofu E, Pogrebnoi A, Shao G, Hilonga A (2015) Encapsulated urea-Kaolinite Nanocomposite for controlled release fertilizer formulations. J Chem 2015:17. https://doi.org/10.1155/2015/237397

Shahat A, Awual MR, Khaleque MA, Alam MZ, Naushad M, Chowdhury AS (2015) Large-pore diameter nano-adsorbent and its application for rapid lead (II) detection and removal from aqueous media. Chem Eng J 273:286–295

Shahat A, Hassan HMA, Azzazy HME, Hosni M, Awual MR (2018) Novel nano-conjugate materials for effective arsenic(V) and phosphate capturing in aqueous media. Chem Eng J 331:54–63. https://doi.org/10.1016/j.cej.2017.08.037

Shahid SA, Qidwai AA, Anwar F, Ullah I, Rashid U (2012) Improvement in the water retention characteristics of sandy loam soil using a newly synthesized poly (acrylamide-co-acrylic acid)/AlZnFe2O4 superabsorbent hydrogel nanocomposite material. Molecules 17(8):9397–9412

Shan Y, Cao L, Xu C, Zhao P, Cao C, Li F et al (2019) Sulfonate-functionalized mesoporous silica nanoparticles as carriers for controlled herbicide diquat dibromide release through electrostatic interaction. Int J Mol Sci 20(6):1330

Sharma P, Sharma A, Sharma M, Bhalla N, Estrela P, Jain A et al (2017) Nanomaterial fungicides: in vitro and in vivo antimycotic activity of cobalt and nickel nanoferrites on phytopathogenic fungi. Global Chall 1(9):1700041

Sheet I, Kabbani A, Holail H (2014) Removal of heavy metals using nanostructured graphite oxide, silica nanoparticles and silica/graphite oxide composite. Energy Procedia 50:130–138

Shikha S, Salafi T, Cheng J, Zhang Y (2017) Versatile design and synthesis of nano-barcodes. Chem Soc Rev 46(22):7054–7093

Shojaei TR, Salleh MAM, Sijam K, Rahim RA, Mohsenifar A, Safarnejad R, Tabatabaei M (2016a) Detection of Citrus tristeza virus by using fluorescence resonance energy transfer-based biosensor. Spectrochim Acta A Mol Biomol Spectrosc 169:216–222. https://doi.org/10.1016/j.saa.2016.06.052

Shojaei TR, Salleh MAM, Sijam K, Rahim RA, Mohsenifar A, Safarnejad R, Tabatabaei M (2016b) Fluorometric immunoassay for detecting the plant virus Citrus tristeza using carbon nanoparticles acting as quenchers and antibodies labeled with CdTe quantum dots. Microchim Acta 183(7):2277–2287. https://doi.org/10.1007/s00604-016-1867-7

Shubair T, Eljamal O, Khalil AM, Matsunaga N (2018a) Multilayer system of nanoscale zero valent iron and Nano-Fe/Cu particles for nitrate removal in porous media. Sep Purif Technol 193:242–254

Shubair T, Eljamal O, Khalil AME, Tahara A, Matsunaga N (2018b) Novel application of nanoscale zero valent iron and bimetallic nano-Fe/cu particles for the treatment of cesium contaminated water. J Environ Chem Eng 6(4):4253–4264. https://doi.org/10.1016/j.jece.2018.06.015

Siddiqui MH, Al-Whaibi MH, Mohammad F (2015) Nanotechnology and plant sciences: nanoparticles and their impact on plants. Springer, Cham

Simonian A, Good T, Wang S-S, Wild J (2005) Nanoparticle-based optical biosensors for the direct detection of organophosphate chemical warfare agents and pesticides. Anal Chim Acta 534(1):69–77

Singh R, Misra V, Singh RP (2011) Synthesis, characterization and role of zero-valent iron nanoparticle in removal of hexavalent chromium from chromium-spiked soil. J Nanopart Res 13(9):4063–4073

Singh R, Manickam N, Mudiam MKR, Murthy RC, Misra V (2013) An integrated (nano-bio) technique for degradation of γ-HCH contaminated soil. J Hazard Mater 258–259:35–41. https://doi.org/10.1016/j.jhazmat.2013.04.016

Singh P, Kim Y-J, Zhang D, Yang D-C (2016) Biological synthesis of nanoparticles from plants and microorganisms. Trends Biotechnol 34(7):588–599

Song J, Zhang F, Huang Y, Keller AA, Tang X, Zhang W et al (2018) Highly efficient bacterial removal and disinfection by magnetic barium phosphate nanoflakes with embedded iron oxide nanoparticles. Environ Sci Nano 5(6):1341–1349

Sousa GFM, Gomes DG, Campos EVR, Oliveira JL, Fraceto LF, Stolf-Moreira R, Oliveira HC (2018) Post-emergence herbicidal activity of nanoatrazine against susceptible weeds. Front Environ Sci 6(12). https://doi.org/10.3389/fenvs.2018.00012

Stadler T, Buteler M, Weaver DK (2010) Novel use of nanostructured alumina as an insecticide. Pest Manag Sci Former Pestic Sci 66(6):577–579

Sun J, Guo L, Bao Y, Xie J (2011) A simple, label-free AuNPs-based colorimetric ultrasensitive detection of nerve agents and highly toxic organophosphate pesticide. Biosens Bioelectron 28(1):152–157. https://doi.org/10.1016/j.bios.2011.07.012

Sun Q, Aguila B, Perman J, Ivanov AS, Bryantsev VS, Earl LD et al (2018) Bio-inspired nanotraps for uranium extraction from seawater and recovery from nuclear waste. Nat Commun 9(1):1644

Sundaram CS, Viswanathan N, Meenakshi S (2008) Uptake of fluoride by nano-hydroxyapatite/chitosan, a bioinorganic composite. Bioresour Technol 99(17):8226–8230

Tahir MB, Sagir M, Shahzad K (2019) Removal of acetylsalicylate and methyl-theobromine from aqueous environment using nano-photocatalyst WO3-TiO2 @g-C3N4 composite. J Hazard Mater 363:205–213. https://doi.org/10.1016/j.jhazmat.2018.09.055

Tarafdar J, Raliya R, Mahawar H, Rathore I (2014) Development of zinc nanofertilizer to enhance crop production in pearl millet (Pennisetum americanum). Agric Res 3(3):257–262

Teimouri A, Nasab SG, Vahdatpoor N, Habibollahi S, Salavati H, Chermahini AN (2016) Chitosan/Zeolite Y/Nano ZrO2 nanocomposite as an adsorbent for the removal of nitrate from the aqueous solution. Int J Biol Macromol 93:254–266

Teng Q, Zhang D, Niu X, Jiang C (2018) Influences of application of slow-release Nano-fertilizer on green pepper growth, soil nutrients and enzyme activity. In: Paper presented at the IOP conference series: earth and environmental science

Tesh SJ, Scott TB (2014) Nano-composites for water remediation: A review. Adv Mater 26(35):6056–6068

Tuteja N, Gill SS (2012) Crop improvement under adverse conditions. Springer

Varanasi P, Fullana A, Sidhu S (2007) Remediation of PCB contaminated soils using iron nanoparticles. Chemosphere 66(6):1031–1038

Verdian A (2018) Apta-nanosensors for detection and quantitative determination of acetamiprid – A pesticide residue in food and environment. Talanta 176:456–464. https://doi.org/10.1016/j.talanta.2017.08.070

Vundavalli R, Vundavalli S, Nakka M, Rao DS (2015) Biodegradable nano-hydrogels in agricultural farming-alternative source for water resources. Procedia Mater Sci 10:548–554

Wang S-L, Nguyen AD (2018) Effects of Zn/B nanofertilizer on biophysical characteristics and growth of coffee seedlings in a greenhouse. Res Chem Intermed 44(8):4889–4901. https://doi.org/10.1007/s11164-018-3342-z

Wang Z, Wei F, Liu S-Y, Xu Q, Huang J-Y, Dong X-Y et al (2010) Electrocatalytic oxidation of phytohormone salicylic acid at copper nanoparticles-modified gold electrode and its detection in oilseed rape infected with fungal pathogen Sclerotinia sclerotiorum. Talanta 80(3):1277–1281

Wang Y, Zhang S, Du D, Shao Y, Li Z, Wang J et al (2011) Self assembly of acetylcholinesterase on a gold nanoparticles–graphene nanosheet hybrid for organophosphate pesticide detection using polyelectrolyte as a linker. J Mater Chem 21(14):5319–5325. https://doi.org/10.1039/C0JM03441J

Wang G, Zhang S, Xu X, Li T, Li Y, Deng O, Gong G (2014) Efficiency of nanoscale zero-valent iron on the enhanced low molecular weight organic acid removal Pb from contaminated soil. Chemosphere 117:617–624. https://doi.org/10.1016/j.chemosphere.2014.09.081

Wang M, Duong B, Fenniri H, Su M (2015a) Nanomaterial-based barcodes. Nanoscale 7(26):11240–11247. https://doi.org/10.1039/C5NR01948F

Wang Y, Sun H, Duan X, Ang HM, Tadé MO, Wang S (2015b) A new magnetic nano zero-valent iron encapsulated in carbon spheres for oxidative degradation of phenol. Appl Catal B Environ 172–173:73–81. https://doi.org/10.1016/j.apcatb.2015.02.016

Wang S, Gao B, Li Y, Creamer AE, He F (2017) Adsorptive removal of arsenate from aqueous solutions by biochar supported zero-valent iron nanocomposite: batch and continuous flow tests. J Hazard Mater 322:172–181. https://doi.org/10.1016/j.jhazmat.2016.01.052

Wei A, Ma J, Chen J, Zhang Y, Song J, Yu X (2018) Enhanced nitrate removal and high selectivity towards dinitrogen for groundwater remediation using biochar-supported nano zero-valent iron. Chem Eng J 353:595–605

Wen Z, Zhang Y, Dai C (2014) Removal of phosphate from aqueous solution using nanoscale zerovalent iron (nZVI). Colloids Surf A Physicochem Eng Asp 457:433–440. https://doi.org/10.1016/j.colsurfa.2014.06.017

Willner MR, Vikesland PJ (2018) Nanomaterial enabled sensors for environmental contaminants. J Nanobiotechnol 16(1):95

Wu J, Yi Y, Li Y, Fang Z, Tsang EP (2016) Excellently reactive Ni/Fe bimetallic catalyst supported by biochar for the remediation of decabromodiphenyl contaminated soil: reactivity, mechanism, pathways and reducing secondary risks. J Hazard Mater 320:341–349

Xia N, Wang Q, Liu L (2015) Nanomaterials-based optical techniques for the detection of acetylcholinesterase and pesticides. Sensors 15(1):499–514

Xiang Y, Camarada MB, Wen Y, Wu H, Chen J, Li M, Liao X (2018) Simple voltammetric analyses of ochratoxin A in food samples using highly-stable and anti-fouling black phosphorene nanosensor. Electrochim Acta 282:490–498

Xu Y, Zhao D (2007) Reductive immobilization of chromate in water and soil using stabilized iron nanoparticles. Water Res 41(10):2101–2108. https://doi.org/10.1016/j.watres.2007.02.037

Yan X, Li H, Su X (2018) Review of optical sensors for pesticides. TrAC Trends Anal Chem 103:1–20

Yang GC, Yeh C-F (2011) Enhanced nano-Fe3O4/S2O82− oxidation of trichloroethylene in a clayey soil by electrokinetics. Sep Purif Technol 79(2):264–271

Yang S-C, Lei M, Chen T-B, Li X-Y, Liang Q, Ma C (2010) Application of zerovalent iron (Fe0) to enhance degradation of HCHs and DDX in soil from a former organochlorine pesticides manufacturing plant. Chemosphere 79(7):727–732. https://doi.org/10.1016/j.chemosphere.2010.02.046

Yang Z, Qian J, Yang X, Jiang D, Du X, Wang K et al (2015) A facile label-free colorimetric aptasensor for acetamiprid based on the peroxidase-like activity of hemin-functionalized reduced graphene oxide. Biosens Bioelectron 65:39–46. https://doi.org/10.1016/j.bios.2014.10.004

Yao KS, Li S, Tzeng K, Cheng TC, Chang CY, Chiu C et al (2009) Fluorescence silica nanoprobe as a biomarker for rapid detection of plant pathogens. In: Paper presented at the advanced materials research

Yılmaz F, Saylan Y, Akgönüllü S, Çimen D, Derazshamshir A, Bereli N, Denizli A (2017) 18- surface plasmon resonance based nanosensors for detection of triazinic pesticides in agricultural foods. In: Grumezescu AM (ed) New pesticides and soil sensors. Academic, pp 679–718

Yousefi F (2018) Prediction of the experimental data for removal of organic pesticides by carbon nanoparticle synthesized from pomegranate peel using artificial neural networks. Iran J Health Sci 6(1):43–57

Yu X, Tong S, Ge M, Zuo J (2013) Removal of fluoride from drinking water by cellulose@hydroxyapatite nanocomposites. Carbohydr Polym 92(1):269–275. https://doi.org/10.1016/j.carbpol.2012.09.045

Zeng R, Wang J, Cui J, Hu L, Mu K (2010) Photocatalytic degradation of pesticide residues with RE3+ −doped nano-TiO2. J Rare Earths 28:353–356. https://doi.org/10.1016/S1002-0721(10)60329-8

Zhang Z, Li M, Chen W, Zhu S, Liu N, Zhu L (2010) Immobilization of lead and cadmium from aqueous solution and contaminated sediment using nano-hydroxyapatite. Environ Pollut 158(2):514–519

Zhang Z, He X, Zhang H, Ma Y, Zhang P, Ding Y, Zhao Y (2011) Uptake and distribution of ceria nanoparticles in cucumber plants. Metallomics 3(8):816–822

Zhang C, Chen L, Wang T-J, Su C-L, Jin Y (2014a) Synthesis and properties of a magnetic core–shell composite nano-adsorbent for fluoride removal from drinking water. Appl Surf Sci 317:552–559

Zhang F, Zou M, Chen Y, Li J, Wang Y, Qi X, Xue Q (2014b) Lanthanide-labeled immunochromatographic strips for the rapid detection of Pantoea stewartii subsp. stewartii. Biosens Bioelectron 51:29–35

Zhang X, Gu P, Li X, Zhang G (2017) Efficient adsorption of radioactive iodide ion from simulated wastewater by nano Cu2O/Cu modified activated carbon. Chem Eng J 322:129–139. https://doi.org/10.1016/j.cej.2017.03.102

Zhang B, Ji J, Liu X, Li C, Yuan M, Yu J, Ma Y (2019) Rapid adsorption and enhanced removal of emodin and physcion by nano zirconium carbide. Sci Total Environ 647:57–65

Zhao Y, Ma Y, Li H, Wang L (2011) Composite QDs@ MIP nanospheres for specific recognition and direct fluorescent quantification of pesticides in aqueous media. Anal Chem 84(1):386–395

Zhao L, Peralta-Videa JR, Ren M, Varela-Ramirez A, Li C, Hernandez-Viezcas JA et al (2012) Transport of Zn in a sandy loam soil treated with ZnO NPs and uptake by corn plants: electron microprobe and confocal microscopy studies. Chem Eng J 184:1–8. https://doi.org/10.1016/j.cej.2012.01.041

Zhao X, Cui H, Wang Y, Sun C, Cui B, Zeng Z (2017) Development strategies and prospects of nano-based smart pesticide formulation. J Agric Food Chem 66(26):6504–6512

Zhou B, Chen X (2017) Effect of Nano-carbon on water holding capacity in a Sandy soil of the loess plateau. Earth Sci Res J 21(4):189–195

Zhou K, Wu B, Su L, Xin W, Chai X (2018) Enhanced phosphate removal using nanostructured hydrated ferric-zirconium binary oxide confined in a polymeric anion exchanger. Chem Eng J 345:640–647. https://doi.org/10.1016/j.cej.2018.01.091

Zhou Y, Zhou B, Xu F, Muhammad T, Li Y (2019) Appropriate dissolved oxygen concentration and application stage of micro-nano bubble water oxygation in greenhouse crop plantation. Agric Water Manag 223:105713. https://doi.org/10.1016/j.agwat.2019.105713

Zhu H, Jia Y, Wu X, Wang H (2009) Removal of arsenic from water by supported nano zero-valent iron on activated carbon. J Hazard Mater 172(2–3):1591–1596

Zhu Z-J, Wang H, Yan B, Zheng H, Jiang Y, Miranda OR et al (2012) Effect of surface charge on the uptake and distribution of gold nanoparticles in four plant species. Environ Sci Technol 46(22):12391–12398. https://doi.org/10.1021/es301977w

Zolgharnein J, Bagtash M, Shariatmanesh T (2015) Simultaneous removal of binary mixture of brilliant green and crystal violet using derivative spectrophotometric determination, multivariate optimization and adsorption characterization of dyes on surfactant modified nano-γ-alumina. Spectrochim Acta A Mol Biomol Spectrosc 137:1016–1028. https://doi.org/10.1016/j.saa.2014.08.115

Nanotechnology and Nutrigenomics

19

P. Janhavi, J. Natasha, R. Neelam, and P. V. Ravindra

Abstract

The rising popularity and attractive applications of nanotechnologies have impacted all areas of research, including science, agriculture, and health care. Nanoparticles are finding great potential as delivery systems to specific targets in living organisms. Recent advances in food science have revealed that food-derived bioactives significantly influence changes in the genome, epigenome, proteome, and metabolome. This concept is termed "nutrigenomics." The research in nutrigenomics is fast emerging and explored for the prevention or therapy of various lifestyle-associated disorders such as diabetes, cardiovascular diseases, cancer, and others. The major obstacle in achieving the efficacy from the bioactives is their bioavailability in the plasma and/or at the target site following consumption. The advent of various nanotechnology methods have contributed to promising tools such as nanodelivery systems, including nanocapsules, nanospheres, and biogenic nanoparticles that can enhance the bioavailability of bioactive compounds. This chapter focuses on applications of nanotechnologies in nutrigenomics with a particular focus on their applications for prevention or treatment of certain metabolic disorders.

Keywords

Nanotechnology · Nutrigenomics · Metabolic disorders · Bioactives · Nanodelivery

P. Janhavi · J. Natasha · R. Neelam · P. V. Ravindra (✉)
Department of Biochemistry, CSIR-CFTRI, Mysuru, India

M. Ghorbanpour et al. (eds.), *Biogenic Nano-Particles and their Use in Agro-ecosystems*, https://doi.org/10.1007/978-981-15-2985-6_19

19.1 Introduction

The concept of nanotechnology was first introduced by a Nobel laureate Richard P. Feynman during his famous lecture "There's Plenty of Room at the Bottom" in the year 1959 (Feynman 1960). Since then, there have been revolutionary advances in the field of science that have demonstrated Feynman's concepts of manipulating matter at the nanoscale. Nanotechnology refers to engineering matter on an atomic, molecular, as well as supramolecular scale (Mansoori 2002). It can be defined as the creation of functional materials, devices, and systems through the manipulation of matter at a length scale of ~1–100 nm, which can either be produced using synthetic or biological means (Singh 2016). The science underlying is referred to as nanoscience. Although nanotechnology is considered as one of the most exciting fields to work in recent times, the existence of functional devices and nanostructures have existed on the earth as long as life itself (Poole and Owens 2003). Thus, nanotechnology is inspired by the processes of nature, which is further being exploited to improve the quality of life, on an application level. Nanometer ranged materials have been manufactured for numerous decades. One of the best examples is the use of carbon black in the manufacture of tires. In the area of food and medicine, this technology has the capability of introducing novel functionalities through new interfacial phenomena (Chen et al. 2006). Nanotechnology comprises various techniques such as manipulation, modeling, measuring, and imaging of matter at nanoscale dimensions. This unique capacity has led to a vast range of new technologies that influence every aspect of science, environment, industry, and economy. Nanotechnology also has several applications in other technologies such as solar cells, batteries, electronics, water purification, smart materials, fabrics, paints, and chemical sensors. Nanotechnology helps in observing individual cells at molecular level by offering a variety of tools; it also helps in obtaining exact spatial information about the location of food components (nutrient or bioactive) in a tissue, cell, or even in cellular component and also increasing bioavailability of hydrophobic compounds (Srinivas et al. 2010).

Nutrigenomics is the integration of genomic science, nutrition, and the environment (Mead 2007). The study is concerned with the influence of dietary components on the genome, proteome, and metabolome. A number of questions regarding the bioavailability of dietary components (bioactives), optimal intake levels, development of food delivering matrix, product formulations, and the effect of these components on the genome are addressed in the study of nutrigenomics. Research in nutrigenomics has the potential for innumerable innovations. The field of nutrigenomics and nanotechnology has generated enormous interest in the development of functional foods as therapeutic candidates as well as delivery of drugs, but the consequences of this coalescence are not limited to the former. This chapter will focus on well-known nanotechnologies and their applications to nutrigenomics in the context of treatment of metabolic disorders.

19.2 Different Types of Nanotechnology

Nanotechnology includes a wide range of technologies such as nanobots, nanosensors, nanomaterials, nanoplasmonics, nanocoatings, nanoelectronics, and also nanoparticles (NPs) (Laurent et al. 2008). The synthesis of NPs with controlled particle size, shape, and crystalline nature is one of the major objectives in chemistry that could be used for many applications, such as biomedical, biosensor, the catalyst for bacterial biotoxin elimination, and lower cost electrode (Sharma et al. 2015). Nanoparticles can be mainly classified into three types based on dimensions, that is, one-dimension (1–100 nm thin film sized), two-dimension (carbon nanotubes), and three-dimension (dendrimers, quantum dots, fullerenes) (Bhatia 2016). They can also be categorized depending on their morphology, size, and chemical properties. Carbon-based, metal, ceramics, semiconductor, polymeric, and lipid-based NPs are some of the well-known classes of NPs (Khan et al. 2017). Various methods are used for synthesizing NPs, and these methods are broadly classified into two main classes, i.e., a bottom-up approach and a top-down approach (Wang and Xia 2004). Based on the adopted protocols, reaction condition, and operation, these are further divided into several subclasses. A top-down approach is a destructive approach, i.e., larger molecules are decomposed into smaller, and then these molecules are changed into suitable NPs. Examples of this method are mechanical milling, chemical etching, sputtering, laser ablations, and electro-explosion (Iravani 2011). In the bottom-up approach, NPs are made from comparatively simpler substances. Examples of this method are reduction and sedimentation techniques which include green synthesis, sol-gel, spinning, and biochemical synthesis (Iravani 2011). In some of the techniques, e.g., the chemical reduction method uses various hazardous chemicals such as ethylene glycol, dimethylformamide, polyol, sodium borohydride, hydrazine hydrate, sodium citrate, and N,N-dimethylformamide for the production of NPs. These chemicals later become liable for numerous health risks, also risking the environment (Gokulakrishnan et al. 2012). Thus, green and biogenic synthesis of NPs has attracted many researchers and is gaining popularity due to the feasibility and less toxic nature of methods. It is not just a choice but is essential for the current environmental scenario as these processes are environment-friendly as well as cost-effective.

19.3 Biogenic Nanoparticles - The Green Nanoparticles

In green and biogenic synthesis, biological systems such as plant parts including the leaf, bark, flower, peel, and seed are used for the synthesis of NPs. Microorganisms such as bacteria, yeast, and fungi are used for the production of NPs. Human cells are used as well. Biological sources are known to be green, precise, more effective, and convenient nanofactories. Biogenic NPs have better stability and stabilizing agents such as proteins, and other biomolecules are used for stabilization which is again from the organism itself. The additional advantage of biogenic nanoparticles includes uniform size, shape, and better stability at physiological pH, unlike that of synthesized NPs. The biomass of wheat and oat are used for the synthesis of Au NPs (Parveen et al. 2016) along with the plant extracts and microorganisms as a reducing agent (Ahmed et al. 2016). The reducing agents such as polysaccharides, proteins, enzymes, amino acids, phenolics, flavones, terpenoids, and alkaloids are significantly harmless (Salouti and Derakhshan 2019). Metals can be transformed into nanoparticles by essential biological sources such as microorganisms and plants, which are being extensively studied in recent years (Nakajima 2005). Soluble nanoscale organic compounds such as liposomes, micelles, and vesicles in plants and animals also fall into the category of nanoparticles.

19.4 Synthesis of Biogenic Nanoparticles – The Factory Within

Nanoparticles are not just synthesized from modern technologies but are also produced by natural methods such as forest fires or volcano eruptions as well as in all living organisms. Ultrafine sand grains of mineral origin are a simple example of naturally occurring nanoparticles (e.g., oxides, carbonates). The biogenic NPs may be synthesized inside the cell that will be isolated in separate intracellular compartments, and these are known as intracellular biogenic NPs. Thus, disruption of cells is essential in order to separate these NPs. On the contrary, many organisms produce NPs outside the cell or send the NPs outside post-synthesis. Such NPs are known as extracellular biogenic nanoparticles. There are several reports showing that gold and silver NPs are produced using *Aloe vera* plant (Chandran et al. 2006), *Geranium* (Shankar et al. 2004), sundried *Cinnamomum camphora*, and *Azadirachta indica* leaf extracts (Patil et al. 2012). Bacteria such as *Pseudomonas stutzeri AG259* (Slawson et al. 1994), *Marinobacter pelagius* (Sharma 2012) and *Lactobacillus* strains (Prasad et al. 2007) have been used in the synthesis of silver, gold, and titanium NPs, respectively. The silver producing *Pseudomonas stutzeri* AG259 bacteria was isolated from the silver mines as they exhibit the accumulation of silver nanoparticles. The fungus *Trichoderma viride* is used for the synthesized spherical NPs (Thakkar et al. 2010). In recent days biogenic NPs by fungi are exploited since they have higher detoxification capacity, bioaccumulation, easy synthesis method, economic, and simple downstream processing. It is reported that fungi such as *Fusarium solani* (Ingle et al. 2009), *Aspergillus niger* (Gade et al. 2008), and *Aspergillus oryzae* produce silver nanocrystals (Binupriya et al. 2010). Yeast is not an exception, where the synthesis of cadmium NPs was done using *Candida glabrata* and *Schizosaccharomyces pombe* (Dameron et al. 1989). Mourato et al. (Mourato et al. 2011) investigated the biosynthesis of silver and gold NPs using an extremophilic yeast strain from acid mine drainage. Stable lead sulfide NPs were synthesized intracellularly using marine yeast *Rhodosporidium diobovatum* (Seshadri et al. 2011). Other biological particles, such as enzymes, proteins, peptides, and viruses, have also been used for the biosynthesis of NPs (Ingale and

Chaudhari 2013). Peptides are capable of nucleating nanocrystal growth for which they have been screened, and M13 bacteriophage was used to demonstrate the same (Mao 2003). Viruses such as *Cowpea chlorotic mottle virus* and *Tobacco mosaic virus* have been used for the mineralization of inorganic materials and sulfide, respectively (Douglas et al. 2002; Shenton et al. 1999).

19.5 Nanotechnology for Delivery of Bioactives or Drugs

Nanoencapsulation

Encapsulating a substance with another secondary material in the nanoscale is nanoencapsulation. This technology is used in pharmaceutical, food, and cosmetic industries. Nanoencapsulation of bioactive compounds has versatile advantages for targeted site-specific delivery and efficient absorption through cells (Ezhilarasi et al. 2013). It also ensures the stability of bioactive compounds (polyphenols, micronutrients, enzymes, antioxidants, and nutraceuticals). Nanocarriers (NCs) protect compounds from premature degradation in the biological system, enhance the bioavailability, and prolong their presence in the blood, thus better cellular uptake (Kumari et al. 2010). There are several varieties of encapsulations for the delivery of bioactive compounds such as liposomes, micelles, carbon nanotubes, dendrimers, magnetic NCs, and nanoemulsions, some of which are described below.

Liposomes Liposomes are mainly composed of amphiphilic molecules that have a hydrophilic head and two nonpolar hydrophobic chains. The amphiphilic nature allows liposomes to encapsulate and protect sensitive bioactive compounds, both hydrophilic and hydrophobic alike. This flexibility enables targeted delivery of potentially bioactive compounds (Sharifi et al. 2019). In a study, folate-conjugated poly(L-histidine)-poly(L-lactic acid) micelles were effective in killing cancer cells (Lee et al. 2003). Feng et al. (Feng et al. 2017) demonstrate that low soluble bioactive compounds like curcumin in combination with a cancer drug encapsulated in liposomal nanoparticles could sensitize cancer cells, such as CUR and C6 ceramide in OS cell line. Liposomes mimic lipid vesicles and exist naturally in living organisms and thus have broad applications. Further, amphiphilic nature makes them economically attractive and environmentally sustainable.

Polymeric Micelles (PICM) Polymeric micelles are supramolecular structures produced as a result of self-assemblage via hydrophobic and hydrophilic effects, electrostatic interactions, hydrogen bonding, and metal complexation of amphiphilic block polymers (polymeric micelles consisting of poly(ethylene oxide)-b-poly(propylene oxide), poly(ethylene oxide)-b-poly(ester)s, and poly(ethylene oxide)-b-poly amino acids) (Aliabadi and Lavasanifar 2006). This method has gained popularity in the field of drug delivery due to their biocompatibility, low toxicity, enhanced blood circulation time of the drug, and ability to solubilize a large number of drugs in their micellar core (Mourya et al. 2011). Nishiyama and Kataoka (2006) demonstrate the ability of polymeric micelles to cause pH-sensitive release. Liu et al. (2003) described the use of poly(N-isopropyl acrylamide-co-acrylamide)-b-poly (D, L-lactide) copolymer in tumor targeting of docetaxel. They observed that hyperthermia greatly enhanced the targeting efficacy of drug-loaded micelles and also helped in the reduction of toxicity of the drug.

Carbon Nanotubes (CNTs) A Carbon nanotube is like a sheet of graphite that is rolled into a cylinder, with a distinctive hexagonal lattice work making up the sheet. Two main types of CNTs can have high structural perfection: single-walled nanotubes, consisting of a single graphite sheet seamlessly wrapped into a cylindrical tube; multi-walled nanotubes, comprising an array of nanotubes one concentrically placed inside another like rings of a tree trunk (Qian et al. 2002). Different kinds of drugs can be encapsulated into the hollow structure or inner cavities of CNTs to improve their efficacy. The nature of the functional groups attached to the surface of CNTs plays a significant role in deciding the mechanism of interaction with the cellular machinery. Thus, these principles can be studied and used to target nutraceuticals to target and to acquire effective results in gene modifications. Some of the methods used for the formation of carbon nanotubes are arc discharge (Zeng et al. 1998), laser ablation (Ma et al. 2000), carbon monoxide disproportionation, chemical vapor deposition (Benito et al. 1998), and hydrothermal method (Gogotsia and Libera 2000). Examples such as folic acid and polyethylene glycol were directly adsorbed onto the surface of CNTs to make a water-soluble and cancer cell targeting drug delivery system. The drug delivery system released the doxorubicin at a reduced pH value particular for the environment of cancerous cells and to lysosomes/endosomes. This is effectively able to induce the transcription of genes leading to cell death (Niu et al. 2013). Thus, CNTs make an efficient drug delivery system.

Dendrimers Dendrimers are the emerging polymeric architectures that are known for their defined structures, versatility in drug delivery, and high functionality whose properties resemble biomolecules (Madaanet al. 2014). These nanostructures have shown their potential abilities in entrapping and conjugating the high molecular weight hydrophilic as well as hydrophobic entities by host-guest interactions and covalent bonding (prodrug approach), respectively. Unlike traditional polymers, dendrimers have received considerable attention in biological applications due to their high-water solubility, biocompatibility, polyvalency, and precise molecular weight. Dendrimers are usually synthesized using the three techniques: (1) divergent approach (Klajnert and Bryszewska 2001), (2) convergent approach (Hawker and Frechet 1990), and (3) double-stage convergent approach (Crespo et al. 2005). Despite their extensive applications, their use in biological systems is limited due to toxicity issues associated with them (Madaan et al. 2014).

Nanoemulsions Nanoemulsions; also referred to as mini emulsions, ultrafine emulsions, and submicron emulsions are another novel drug delivery form with droplet size 10–100 nm. They are single-phase and thermodynamically stable isotropic systems that consist of emulsified oil, water, and amphiphilic molecules (Grumezescu 2017). Nanoemulsions are composed off oil droplets dispersed in an aqueous medium and stabilization by surfactants. Nanoemulsions have a higher loading capacity and bioavailability for lipophilic active ingredients and thus have added advantage over their bulkier counterparts which makes them one of the supramolecular vehicles to the effective delivery of drugs and nutraceuticals. Nanoemulsion can be formulated with a variety of techniques such as high-pressure homogenization, ultrasonication, self-emulsification, phase inversion, microfluidization, and titrimetric method. Creams, liquids, sprays, and foams are some of the examples of the dosage form in which nanoemulsion can be molded (Mohapatra et al. 2019). Encapsulated lipophilic bioactive components in pharmaceutical products, functional food, and personal care oil-in-water nanoemulsions are used as delivery systems, but only few water-in-oil emulsions are used as nanoemusion as they are very unstable (Basri et al. 2013 and Rahman et al. 2009). Better gastrointestinal tract (GI) absorption, 15–250-fold higher uptake efficiency of particles in the range of 100 nm by the GI tract, was noted, compared to that of the micrometer-sized particles (Francis et al. 2005). Nanoemulsions are used as carriers for brain delivery of risperidone which is prominently used to treat bipolar disorders, schizophrenia, and irritability associated with autism (Đorđević et al. 2015).

The small size of the droplets in nanoemulsion causes better penetration and also provides uniform distribution of nutraceuticals or drugs on the target. No creaming or sedimentation occurs on storage. The small droplets also prevent their coalescence. Since these droplets are elastic, surface fluctuations are also avoided (Matalaniset al. 2013). But like any technology, nanotechnology comes with demerits as very little is known about the absorption and excretion of nanoparticles by experimental animals or in humans. Studies are being conducted on the use of biogenic nanoparticles over synthetic ones as bioactive or drug delivery system (Ju-Nam and Lead 2008).

19.6 Advances in Nutrigenomics

Nutrigenomics, along with other "omic" sciences, aims to clarify the interaction between genes and bioactive compounds from food sources (Costa and Rosa 2011). Bioactives from food can interfere with genes in several ways and at several processes of gene modulation, for example, during transcription as transcription factors (Sales et al. 2014). There are foods that can influence the genetic changes leading to early onset and hastened progression of diseases and disorders. On the other hand, there are also food and food-derived bioactives that have potential to prevent, attenuate, or revert these changes.

Some components of food contain anti-inflammatory bioactives, such as caffeic acid, tyrosol, quercetin, and lycopene. These molecules inhibit the expression of COX2 and iNOS genes by reducing the translocation of nuclear factor kappa-B (NF-κB) from the cytoplasm to the nucleus (Dalmiel et al. 2012). α-Tocopherol in green tea is also known to decrease the chronic inflammatory process that occurs in obese individuals and thus can be used as a treatment for the same (Cozzolino and Cominetti 2013). Deficiency of micronutrients, such as folic acid; vitamins B12, B6, C, and E; selenium; niacin; and zinc, can cause alterations in the DNA similar to what is seen after radiation exposure (Cozzolino and Cominetti 2013). Molecules present in contaminated food can produce toxic metabolites that may interact with DNA, modifying their structure and inducing mutations (Moraeset al. 2009).

Furthermore, dietary compounds modulate epigenomic changes associated with age-related disorders such as diabetes, cardiovascular disease, and cancer. In the context of diabetes, several polyphenolic phytochemicals have been found to influence the expression of genes involved in processes such as glucose transport, insulin secretion, antioxidant effects, inflammation, vascular functions, and lipid metabolism. Resveratrol activates p-AKT, p-eNOS, TRX-1, HO-1, and VEGF and increases MnSOD activity in the myocardium of STZ-induced diabetic rats. Resveratrol elevates GLUT4 expression in muscle via the PI3K-AKT pathway. Quercetin regulates gene expression through NF-κB activation. Beta-carotene is used to treat patients with type 2 diabetes mellitus (T2DM). Dietary bioactives, such as genistein, curcumin, resveratrol, indole-3-carbinol, and epigallocatechin-3-gallate, regulate HDAC and histone acetyltransferase activities, suggesting that the health benefits of these compounds stem from these epigenetic mechanisms (Carlos-Reyes et al. 2019). Further, it is reported that long time consumption of trans fatty acid (TFA) and saturated fatty acid (SFA)-rich diet increases the expression of 11B-Hydroxysteriod Dehydrogenase Type-1 (11b-HSD1) gene in retroperitoneal adipose tissue of rat when compared to diet rich in polyunsaturated fatty acid (PUFA). Increased expression of 11b-HSD1 gene leads to increase conversion of inactive glucocorticoids to active form in adipose tissue increasing the possibility of developing obesity and insulin resistance (Prasad et al. 2010). Arteriosclerotic factors such as total cholesterol and low-density lipoprotein (LDL) cholesterol can be reduced and anti-atheroscleotic factor like high-density lipoprotein (HDL) cholesterol can be increased by consuming mustered oil that is rich in diacylglycerol (DAG). Thompkinson et al. (2014) have reported that omega-3 fatty acids-rich diets are helpful for cardiovascular health. Fish protein and fish oil-enriched diet decreases serum triacylglycerol and cholesterol as well as liver triacylglycerol and cholesterol along with the changes in composition of liver lipid fatty acid (Hosomi et al 2013). Diet rich in trans fatty acids is proven to have direct effect on many diseases such as breast cancer, colon cancer, diabetes, obesity, allergy risks of preeclampsia, cardiovascular diseases, shortening of pregnancy period, and disorders of nervous system (Dhaka et al. 2011). It is proven that plant-based ayurvedic medicines helps in cancer prevention (Sinha et al. 2003). Wherein, turmeric is proven to be a promising chemo-preventive agent has it possess potent antioxidant and anti-inflammatory property. Genomics and proteomics studies have also helped in better understanding of male infertility. The review by Singh and Jaiswal (2011) has revealed the use of nutrigenomics in male Infertility. Vitamin B12, folate, and zinc-rich diets were found to have a beneficial effect on spermatozoa motility as well as number (Dhillon et al. 2007). However, there is inadequate evidence to demonstrate that dietary bioactives modulate metabolic diseases-associated epigenetic alterations and prevent or delay the onset of metabolic memory-induced development of these complications. Moreover, the bioavailability of these bioactive compounds following consumption limits their efficacy in preventive or therapeutic applications. The advent of nanotechnology has contributed promising tools such as nanodelivery systems, including nanocapsules and nanospheres, which can enhance the bioavailability of bioactive compounds (Bajpai et al. 2018).

19.7 Nanotechnology and Nutrigenomics – The Combined Perspective

Though various bioactives have been found to be effective in correcting the aberrant gene expression involved in various disease conditions, their bioavailability limits their efficacy. Nanotechnology has the capacity to transport these bioactive compounds in a way that they are more bioavailable to a cell, and they can be potent chemical messengers and transcription factors that lead to alteration in gene expression. Nanotechnology also allows creating synthetic molecules or the use of biogenic particles that can alter the specific gene expression. A wide variety of encapsulation platforms, including nanoemulsions, liposomes, and biogenic production of such nanoparticles which possesses both diagnostic and therapeutic potential owing to their outstanding properties compared to their bulk counterparts, can help achieve the reverie of targeted drug delivery system. This may further solve many of the limitations faced due to hydrophobic molecules by improving their bioavailability and also help in providing tailored therapies according to the patient's genetic makeup.

Successful targeting and movement of nutraceuticals/drugs across barriers can be achieved, and this technology can bring about a revolutionary change in the field of drug targeting and delivery. It also possesses the potential to transform the food industry by changing the way food is produced, processed, packaged, transported, and consumed. Nutrigenomics is the persuasive future of health care that can be achieved through the applications of nanotechnologies.

Acknowledgments The authors are thankful to the Director, CSIR-CFTRI, Mysuru, for providing facilities and infrastructure. The corresponding author thanks the Department of Biotechnology and the Science and Engineering Research Board (SERB), Department of Science and Technology, New Delhi, for providing funds in the form of Ramalingaswami fellowship and extramural grant, respectively.

References

Ahmed J, Mulla MZ, Arfat YA (2016) Thermo-mechanical, structural characterization and antibacterial performance of solvent casted polylactide/cinnamon oil composite films. Food Control 69:196–204

Aliabadi HM, Lavasanifar A (2006) Polymeric micelles for drug delivery. Expert Opin Drug Deliv 3:139–162

Bajpai VK et al (2018) Prospects of using nanotechnology for food preservation, safety, and security. J Food Drug Anal 26(4):1201–1214

Basri M, Kassim MA, Mohamad R, Ariff AB (2013) Optimization and kinetic study on the synthesis of palm oil ester using Lipozyme TL IM. J Mol Catal B Enzym 85:214–219

Benito AM et al (1998) Carbon nanotubes production by catalytic pyrolysis of benzene. Carbon 36:681–683

Bhatia S (2016) Nanoparticles types, classification, characterization, fabrication methods and drug delivery applications. In: Natural polymer drug delivery systems. Springer, Cham, pp 33–93

Binupriya AR et al (2010) Bioreduction of trivalent aurum to nano-crystalline gold particles by active and inactive cells and cell-free extract of Aspergillus oryzae var. viridis. J Hazard Mater 177(1–3):539–545

Carlos-Reyes A et al (2019) Dietary compounds as epigenetic modulating agents in cancer. Front Genet 10:79

Chandran SP et al (2006) Synthesis of gold nanotriangles and silver nanoparticles using Aloevera plant extract. Biotechnol Prog 22(2):577–583

Chen H, Weiss J, Shahidi F (2006) Nanotechnology in nutraceutical and functional foods. Food Technol 60(3):30–36

Costa NMB, Rosa COB (2011) Functional foods: bioactive components and physiological effects. 1 Reprint. Rúbio, Rio de Janeiro

Cozzolino SMF, Cominetti C (2013) Biochemical and physiological bases of nutrition in different stages of life in health and disease, 1st edn. Monole, São Paulo

Crespo L et al (2005) Peptide, and amide bond-containing dendrimers. Chem Rev 105(5):1663–1681

Dalmiel L, Vargas T, Molina AR (2012) Nutritional genomics for the characterization of the effect of bioactive molecules in lipid metabolism and related pathways. Electrophoresis 33(15):2266–2289

Dameron CT et al (1989) Biosynthesis of cadmium sulphide quantum semiconductor crystallites. Nature 338(6216):596

Dhaka V, Gulia N, Ahlawat KS, Khatkar BS (2011) Trans fats—sources, health risks and alternative approach - a review. J Food Sci Technol 48(5):534–541

Dhillon VS, Shahid M, Husain SA (2007) Associations of MTHFR DNMT3b 4977 bp deletion in mtDNA and GSTM1 deletion, and aberrant CpG island hypermethylation of GSTM1 in nonobstructive infertility in Indian men. MHR: Basic science of reproductive medicine 13(4):213–222

Đorđević SM et al (2015) Parenteral nanoemulsions as promising carriers for brain delivery of risperidone: design, characterization and in vivo pharmacokinetic evaluation. IJ Pharm 0493(1–2):40–54

Douglas T, Strable E, Willits D, Aitouchen A, Libera M, Young M (2002) Protein engineering of a viral cage for constrained nanomaterials synthesis. Adv Mater 14(6):415–418

Ezhilarasi PN et al (2013) Nanoencapsulation techniques for food bioactive components: a review. Food Bioprocess Technol 6(3):628–647

Feng T et al (2017) Liposomal curcumin and its application in cancer. Int J Nanomed 12:6027–6044

Feynman R (1960) There's plenty of room at the bottom (reprint from speech given at annual meeting of the American Physical Society). Eng Sci 23:22–36

Francis MF et al (2005) Engineering polysaccharide-based polymeric micelles to enhance permeability of cyclosporin A across Caco-2 cells. Pharm Res 22:209–219

Gade AK et al (2008) Exploitation of Aspergillus niger for synthesis of silver nanoparticles. J Biobased Mater Bioenergy 2(3):243–247

Gogotsia Y, Libera JA (2000) Hydrothermal synthesis of multiwall carbon nanotubes. J Mater Res 15:2591–2259
Gokulakrishnan R, Ravikumar S, Raj JA (2012) In vitro antibacterial potential of metal oxide nanoparticles against antibiotic resistant bacterial pathogens. Asian Pac J Trop Dis 2(5):411–413
Grumezescu (2017) Multifunctional systems for combined delivery, biosensing and diagnostics, 1st edn Book
Hawker CJ, Frechet JMJ (1990) Preparation of polymers with controlled molecular architecture: a new convergent approach to dendritic macromolecules. J Am Chem Soc 112:7638–7647
Hosomi R, Fukunaga K, Arai H, Kanda S, Nishiyama T, Yoshida M (2013) Effect of combination of dietary fish protein and fish oil on lipid metabolism in rats. J Food Sci Technol 50(2):266–274
Ingale AG, Chaudhari AN (2013) Biogenic synthesis of nanoparticles and potential applications: an eco-friendly approach. J Nanomed Nanotechol 4(165):1–7
Ingle A, Rai M, Gade A, Bawaskar M (2009) Fusarium solani: a novel biological agent for the extracellular synthesis of silver nanoparticles. J Nanopart Res 11(8):2079
Iravani S (2011) Green synthesis of metal nanoparticles using plants. Green Chem 13(10):2638–2650
Ju-Nam Y, Lead JR (2008) Manufactured nanoparticles: an overview of their chemistry, interactions and potential environmental implications. Sci Total Environ 400(1–3):396–414
Khan I, Saeed K, Khan I (2017) Nanoparticles: properties, applications and toxicities. Arab J Chem
Klajnert B, Bryszewska M (2001) Dendrimers: properties and applications. Acta Biochim Pol 48(1):199–208
Kumari A et al (2010) Development of biodegradable nanoparticles for delivery of quercetin. Colloids Surf B Biointerfaces 80(2):184–192
Laurent S et al (2008) Magnetic iron oxide nanoparticles: synthesis, stabilization, vectorization, physicochemical characterizations, and biological applications. Chem Rev 108(6):2064–2110
Lee ES, Na K, Bae YH (2003) Polymeric micelle for tumor pH and folate-mediated targeting. J Control Release 9:103–113
Liu SQ et al (2003) Preparation and characterization of temperature-sensitive poly(N-isopropyl acrylamide)-b-poly(d,l-lactide) microspheres for protein delivery. Biomacromolecules 46:1784–1793
Ma RZ et al (2000) The morphology changes of carbon nanotubes under laser irradiation. Carbon 38:636–638
Madaan K et al (2014) Dendrimers in drug delivery and targeting: drug-dendrimer interactions and toxicity issues. J Pharm Bioallied Sci 6(3):139
Mansoori GA (2002) Advances in atomic & molecular nanotechnology, United Nations Tech Monitor; UN-APCTT Tech Monitor, pp 53–59
Mao C (2003) Viral assembly of oriented quantum dot nanowires. Proc Natl Acad Sci 100(12):6946–6951
Mead MN (2007) Nutrigenomics: the genome–food interface. Environ Health Perspect 115(12):582–589
Mourato A, Gadanho M, Lino AR, Tenreiro R (2011) Biosynthesis of crystalline silver and gold nanoparticles by extremophilic yeasts. Bioinorg Chem Appl 2011:8
Mourya VK, Inamdar NN, Choudhari YM (2011) Chitooligosaccharides: synthesis, characterization and applications. Polym Sci Ser A 53(7):583–612
Nakajima M (2005) Development of nanotechnology and materials for innovative utilization of biological functions. In: Proceedings of the 34th United States and Japan Natural Resources (UJNR) food and agriculture panel, Susono, Japan
Nishiyama N, Kataoka K (2006) Current state, achievements, and future prospects of polymeric micelles as nanocarriers for drug and gene delivery. Pharmacol Ther 112(3):630–648
Niu L et al (2013) Folate-conjugated PEG on single-walled carbon nanotubes for targeting delivery of doxorubicin to cancer cells. Macromol Biosci 13(6):735–744
Parveen K, Banse V, Ledwani L (2016) Green synthesis of nanoparticles: their advantages and disadvantages. AIP Conf Proc 1724:020048

Patil CD et al (2012) Larvicidal activity of silver nanoparticles synthesized using Plumeria rubra plant latex against Aedes aegypti and Anopheles stephensi. Parasitol Res 110(5):1815–1822
Poole CP Jr, Owens FJ (2003) Introduction to nanotechnology. Wiley, Hoboken
Prasad K, Jha AK, Kulkarni AR (2007) Lactobacillus assisted synthesis of titanium nanoparticles. Nanoscale Res Lett 2(5):248
Prasad SSSV, Kumar SSJ, Kumar PU, Qadri SS, Vajreswari A (2010) Dietary fatty acid composition alters 11β-hydroxysteroid dehydrogenase type 1 gene expression in rat retroperitoneal white adipose tissue. Lipids Health Dis 9(111):1–5
Qian D, Wagner GJ, Liu WK (2002) Mechanics of carbon nanotubes. Appl Mech Rev 55:495–433
Rahman MBA, Huan QY, Tejo BA, Basri M, Salleh AB, Rahman RNZA (2009) Self-assembly formation of palm-based esters nano-emulsion: a molecular dynamics study. Chem Phys Lett 480(4–6):220–224
Sales NM, Pelegrini PB, Goersch MC (2014) Nutrigenomics: definitions and advances of this new science. J Nutr Metab 2014:202759
Salouti M, Derakhshan FK (2019) Phytosynthesis of nanoscale materials. In: Advances in phytonanotechnology, From synthesis to application. Elsevier, pp 45–121
Seshadri S, Saranya K, Kowshik M (2011) Green synthesis of lead sulfide nanoparticles by the lead resistant marine yeast, Rhodosporidium diobovatum. Biotechnol Prog 27(5):1464–1469
Shankar SS et al (2004) Biosynthesis of silver and gold nanoparticles from extracts of different parts of the geranium plant. Appl Nano Sci 1:69–77
Sharifi F et al (2019) Generation of liposomes using a supercritical carbon dioxide eductor vacuum system: optimization of process variables. J CO2 Util 29:163–171
Sharma N (2012) Exploitation of marine bacteria for production of gold nanoparticles. Microb Cell Factories 11(1):86
Sharma T, Velmurugan N, Patel P, Chon BH, Sangwai JS (2015) Use of oil-in-water Pickering emulsion stabilized by nanoparticles in combination with polymer flood for enhanced oil recovery. Pet Sci Technol 33(17–18):1595–1604
Shenton W et al (1999) Inorganic–organic nanotube composites from template mineralization of tobacco mosaic virus. Adv Mater 11(3):253–256
Shyam Mohapatra, et al., (2019) Applications of targeted nano drugs and delivery systems, nanoscience and nanotechnology in drug delivery, Micro and Nano Technologies Book
Singh H (2016) Nanotechnology applications in functional foods; opportunities and challenges. Prev Nutr Food Sci 21(1):1–8
Singh K, Jaiswal D (2011) Human male infertility. Reprod Sci 18(5):418–425
Sinha R et al (2003) Cancer risk and diet in India. J Postgrad Med 49:222–228
Slawson RM et al (1994) Silver resistance in Pseudomonas stutzeri. Biometals 7(1):30–40
Srinivas PR et al (2010) Nanotechnology research: applications in nutritional sciences. J Nutr 140(1):119–124
Thakkar KN, Mhatre SS, Parikh RY (2010) Biological synthesis of metallic nanoparticles. Nanomedicine 6(2):257–262
Thompkinson DK, Bhavana V, Kanika P (2014) Dietary approaches for management of cardiovascular health- a review. J Food Sci Technol 51(10):2318–2330
Wang Y, Xia Y (2004) Bottom-up and top-down approaches to the synthesis of Monodispersed spherical colloids of low melting-point metals. Nano Lett 4(10):2047–2050
Zeng H et al (1998) Synthesis of various forms of carbon nanotubes by AC arc discharge. Carbon 36:259–261

Recent Developments in Nanocarrier-Based Nutraceuticals for Therapeutic Purposes

20

Ambreen Bano, Anmol Gupta, Swati Sharma, and Rolee Sharma

Abstract

Modern-day agriculture is evolving from the traditional production of raw food products to advances in novel food engineering technologies that ensure purity and functionality including health-specific products. Plants are the ultimate source of food and nutrition. Nutraceuticals are the functional foods which can provide health and medicinal benefits or can be used for prevention and treatment of various diseases along with providing basic nutrition. Nutraceuticals can be purified food nutrients, dietary supplements, herbs, cereals, milk, soups, or herbal products to genetically engineered foods enriched with vitamins and essential minerals. The components may also include phytochemicals, probiotics, vitamins, antioxidants, and essential minerals that are derived from plant and/or microbial sources. Recently, nanoparticle pharmaceutical drug delivery systems came into picture. These nanocarriers can also be used to enhance the potential of nano-formulated nutraceuticals. Presently, many nanocarrier systems have been developed such as micelles, liposomes, polymeric nanoparticles, and nanoemulsions. Some of these pharmaceutical carriers have already made their way to clinical development, while others are still under the process of preclinical development. The development of multifunctional nutraceutical nanocarriers combining several useful properties in one particle can boost up the efficacy of many therapeutic and diagnostic protocols.

Keywords

Phytochemicals · Nutraceuticals · Nanocarriers · Therapeutic

A. Bano · A. Gupta · S. Sharma · R. Sharma (✉)
Department of Biosciences, Integral University, Lucknow, Uttar Pradesh, India
e-mail: rsharma@iul.ac.in

M. Ghorbanpour et al. (eds.), *Biogenic Nano-Particles and their Use in Agro-ecosystems*, https://doi.org/10.1007/978-981-15-2985-6_20

20.1 Introduction

Let food be thy medicine and medicine be thy food – Hippocrates

The saying by Hippocrates around 2500 years ago highlights the importance of food as medicine. Agriculture products and human health go hand in hand. In the present scenario, there is a pressing need for precise and health-based, value-added food products. Ayurveda, an ancient Indian science of medicine, also lays a lot of emphasis on the role of herbal nutrition in health and treatment of various diseases. Plants contain lots of phytochemicals. Phytochemicals are the bioactive compounds, generally used as additives or nutraceuticals in food and medicine. The term "nutraceuticals" was coined by Stephen DeFelice in 1989. This term originated from words "nutrient" and "pharmaceuticals." These are also known as "functional foods" as there is huge amount of nutrients as compared to pharmaceuticals due to their expected health and medicinal benefits. The term is intended for a nutritional supplement that is sold with the intent to treat or prevent disease and does not have any regulatory definition. Thus "nutraceutical" is defined as "any substance that may be considered a food or part of a food which provides medical or health benefits, encompassing, prevention and treatment of diseases" (Dudeja and Gupta 2017).

Phytochemical nutraceuticals or functional foods are currently in the limelight as a source of alternative medicine (Chung et al. 2009). Nutraceuticals range from isolated nutrients, dietary supplements, purified food, specific diets, and herbal products such as vitamins, cereals, minerals, herbals, milk, soups, and beverages to genetically engineered foods (Zhao 2007). Recently, these have been proposed as a powerful tool in maintaining health and to act against acute and chronic diseases, thereby promoting optimal health, longevity, and quality of life (Wing Shing Ho et al. 2012).

20.2 Classification of Nutraceuticals

Nutraceuticals have been chemically classified on the basis of their nature into various classes including phytochemicals (such as flavonoids, phenols, carotenoids, tannins), probiotics, and dietary fibers. A compilation of various types of nutraceuticals, their sources, and benefits is presented in Table 20.1.

20.3 Commercial Nutraceuticals

While the terms "nutraceuticals," "dietary supplements," "pharmaceuticals," and "functional foods" are used almost interchangeably, there seems to be a thin line distinguishing these terms. "Pharmaceuticals" are used to treat diseases just like drugs, while "nutraceuticals" provide inherent protection and hence prevent disease. Pharmaceuticals are generally patent protected, while many nutrients may not be essentially protected by patent laws (Dudeja and Gupta 2017). Recent years have

Table 20.1 Different chemical classes of nutraceuticals and their sources and benefits

Class	Components	Sources	Benefits
Flavonoids	Anthocyanidins, catechins, flavanones, flavones	Fruits, tea, citrus and fruits/vegetables	Neutralizes free radicals, reduces risk of cancer
Phenols	Caffeic acid, ferulic acid	Citrus, fruits, and vegetables	Reduces heart & eye disease
Carotenoids	Carotene, lutein, lycopene, zeaxanthin	Carrots, various fruits and vegetables, green vegetables, tomato, eggs, citrus, and corn	Neutralizes free radicals, maintains healthy vision, reduces prostrate cancer
Collagen hydrolysate	Collagen hydrolysate	Gelatin	Improves osteoarthritis
Dietary fibers	Beta glucan, soluble/ insoluble fibers, whole grain	Wheat bran, oats, psyllium, and cereal grain	Reduces risk of breast and colon cancer & CVD
Fatty acids	Omega-3 fatty acid-DHA/EPA, linoleic acid	Fish and marine oils	Reduces CVD & certain cancer, improves mental & visual function
Prebiotics/ probiotics	Fructo-oligosaccharides, *Lactobacillus*	Onion powder, yogurt, and other dairy products	Improves gastrointestinal health
Glucosinolates, indoles	Isothiocyanate, sulforaphane	Cruciferous vegetables	Induces detoxification enzymes, reduces risk of cancer
Tannins	Proanthrocyanides	Cranberries, cranberry products, and cocoa	Improves urinary tract health, reduces risk of CVD
Sulfides/thiols	Diallyl sulfides, Allyll methyl trisulfide, Diathiolthiones	Onion, garlic, and cruciferous vegetables	Lower LDL cholesterol, maintain immune system
Plant sterols	Stanol ester	Corn, soy, wood oils, and wheat	Lower blood cholesterol level
Phytoestrogens	Isoflavones-daizein, genistein, lignans	Soybeans and soy-based products	Lower blood cholesterol level

Based on Chauhan et al. (2013) and Sharma et al. (2019)

witnessed a great amount of research on phytochemical nutraceuticals. Many nutraceuticals which were found to be safe for human consumption are commercially available in the market (Table 20.2).

20.3.1 Nanocarriers

The field of food nanotechnology has experienced remarkable growth over the last few years. Such growth has been fueled up by the potential of harnessing the large surface area to volume ratio of these materials, increase in bioavailability of active ingredients, introduction of controlled target release, and improvement of sensory

Table 20.2 Marketed nutraceutical products

Product	Category	Contents	Manufacturer
Calcirol D-3	Calcium supplement	Calcium and vitamins	Cadila Healthcare Limited, Ahmedabad, India
GRD	Nutritional supplement	Proteins, vitamins, minerals, and carbohydrates	Zydus Cadila Ltd., Ahmedabad, India
Proteinex®	Protein supplement	Predigested proteins, vitamins, minerals, and carbohydrates	Pfizer Ltd., Mumbai, India
Coral calcium	Calcium supplement	Calcium and trace minerals	Nature's Answer, Hauppauge, NY, USA
Chyawanprash	Immune booster	Amla, ashwagandha, pippali	Dabur India Ltd.
Omega woman	Immune supplement	Antioxidants, vitamins, and phytochemicals (e.g., lycopene and resveratrol)	Wassen, Surrey, UK
Celestial Healthtone	Immune booster	Dry fruit extract	Celestial Biolabs Limited
Amiriprash (Gold)	Good immunomodulator	Chyawannprash Avaleha, Swarnabhasma, and RasSindur	Uap Pharma Pvt. Ltd.
Kellogg's Muesli, cornflakes, oats	Immune supplement	Antioxidants, vitamins, and phytochemicals	Kellogg
PediaSure	Nutritious supplement, supports the immune system	Antioxidants (vitamins C & E and selenium), DHA, omega-3[†]	Abbott
Horlicks	Nutrition for growth and development	Calcium, vitamin, iodine, and protein	GlaxoSmithKline Consumer Healthcare
ActiPlus Dahi, CEREGROW, MILO, Baby & Me	Growth and development immune supplement	Antioxidants, vitamins, minerals, and proteins	Nestlé
Soya Industries Nutrela	Nutritious supplement	Carbohydrate, proteins, fibers, vitamins	Ruchi
Amul	Nutritious supplement	Skimmed milk, sucrose, carbohydrates, minerals	Gujarat Co-operative Milk Marketing Federation
NutriFit	Improve body composition, boost your energy	Carbohydrate, proteins, fibers, lipids, vitamins minerals, etc.	Mother Dairy Fruit & Vegetable
Yakult	Probiotics, immunity booster	*Lactobacillus casei Shirota*	Yakult

(continued)

Table 20.2 (continued)

Product	Category	Contents	Manufacturer
Quaker oats	Nutritious supplement	Salt, guar gum, calcium carbonate (thickener), vitamins, and minerals	Pepsi Co
Bournvita malt drink	Nutritious supplement	Malt extracts, cocoa, carbohydrates, emulsifiers, vitamins, minerals, and salt	Mondelez
Nutrilite – range of products	Essential supplement	All plant protein powder, calcium, co-enzymes, iron tablets, etc.	Amway
Dabur Glucose – D	Energy supplement	Dextrose, vitamin D, calcium, phosphorous, and energy	Dabur India Ltd
Calcium Sandoz	Dietary supplement	Calcium lactate-gluconate, calcium carbonate, and calcium	Novartis
Herbalife protein powder	Nutritious (protein) supplement	Soy protein, plant-based protein powder, and fructose	Herbalife International
Himalaya Pure Herbs	Healthcare Medical Equipment Personal Care Sports	Pure herbs: organic ashwagandha, amla, neem, turmeric, etc.	Himalaya Drug Company

Based on: Chauhan et al. (2013) and Sharma et al. (2019)

aspects (Chau et al. 2007; Chen et al. 2006a, b; Sanguansri and Augustin 2006). Many research groups have developed production methods for pharmaceutical drug delivery systems. Recent technological advances that make use of proteins, lipids, and polysaccharides as additives have opened the door to new applications and functionalities for nanoparticle delivery systems (Acosta 2009). To depict the next generation of nanoparticle vehicles, it is essential to optimize as well as modify the properties of these nanoparticles to enhance the bioavailability of different ingredients.

Over the past decade, both in research and clinical setting, nanoparticulate pharmaceutical carriers have been used to enhance the in vivo efficiency of many drugs. Surface modification of pharmaceutical nanocarriers is generally done to control and enhance their biological properties, resulting in prolonged half-life and biodistribution as well as precise active or passive targeting. Additionally, these can be engineered as "multifunctional pharmaceutical nanocarriers," having multiple chemical moieties assembled on the surface of nanoparticles (Rolland 1993; Gregoriadis 1988; Müller 1991; Alonso 2004) or for pH- and temperature-dependent controlled release.

Nanocarriers possess one or more of the following set of desired properties (Bernkop-Schnurch and Walker 2001; van Vlerken and Amiji 2006):

(a) Delivers drugs to specifically targeted cells or tissues
(b) Enhanced prolongevity in blood circulation
(c) Capability to accumulate specifically or nonspecifically in the required area
(d) Allows effective intracellular drug delivery
(e) Possesses surface modifications (the presence of functional groups for targeted drug delivery systems)
(f) Responsiveness toward local stimuli, such as pH and temperature changes
(g) Bears magnetic sensitivity which is helpful for oral and/or tumoral delivery

Moreover, these nanocarriers also display enhanced permeation and retention (EPR) effect through active as well as passive targeting (Blanco et al. 2015). These nanocarriers have been shown to possess the capability to administer the encapsulated agent(s) to the target specific site, in leaky vasculature of tumor endothelial cells, by binding to the suitable ligands, through "active targeting" (Saneja et al. 2014a, b; Thanki et al. 2013). Moreover, through "passive targeting," these nanocarriers can effectively penetrate targeted cells. A wide range of ligands such as folic acid, hyaluronic acid, and RGD peptides (arginylglycylaspartic acid) have been used for targeted drug delivery toward tumor site because of overexpression in cancer cells (Allen 2002; Sutradhar and Amin 2014; Zhong et al. 2014).

20.4 Classification of Nanocarriers

There are many nanosystems which can be used as carriers for nutraceuticals as indicated in Fig. 20.1. The nanocarriers can be broadly classified into three categories:

(i) Polymeric nanocarriers
(ii) Lipid-based nanocarriers
(iii) Inorganic nanocarriers

20.4.1 Polymeric Nanocarriers

Polymeric nanocarrier systems are developed from biodegradable and biocompatible polymers that are interestingly used for controlled and targeted based delivery of nutraceuticals. They are made up of colloidal particles having the diameter preferably less than <200 nm to exhibit EPR effect. Various biodegradable polymers have been used for nutraceutical delivery including poly (D,L-lactic acid) (PLA), poly(D,L-lactic-co-glycolic acid) (PLGA), poly (Ɛ-caprolactone) (PCL) as well as their copolymers with poly(ethylene glycol) (PEG) d-α-tocopheryl polyethylene glycol 1000 succinate (TPGS) (Kumari et al. 2010). Moreover, polysaccharide-based polymers as alginate, pectin, or chitosan may also be used to encapsulate nutraceuticals (Arora et al. 2016; Saneja et al. 2016). Additionally, some of the polymers such as chondroitin sulfate and hyaluronic acid have the ability to target

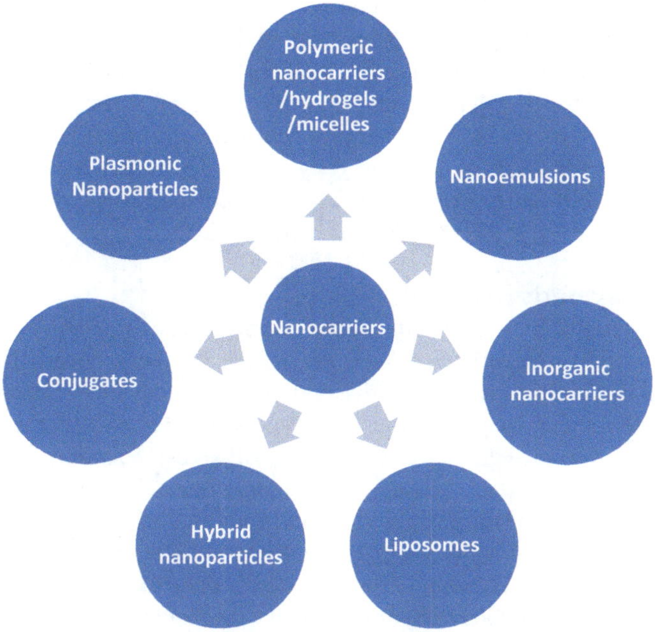

Fig. 20.1 Nanosystems used as carriers for nutraceuticals

CD44 glycoprotein due to their overexpression in the cancer cells (Platt and Szoka 2008; Lo et al. 2013). Several targeting moieties such as folic acid, peptides, aptamers, RGD, bombesin, or antibodies have been used to deliver the nutraceuticals to tumor-specific sites. For example, folate-functionalized nanoparticles of quercetin have been developed by El-Gogary et al. (2014) for its targeted delivery to folate-overexpressing cancer cells. Live fluorescence imaging in tumor-bearing mice exhibited altered intra-tumoral distribution with increased folate-functionalized nanoparticles uptake (El-Gogary et al. 2014). A recent study showed cytotoxicity enhancement due to the efficient cellular uptake of the nanoparticles via caveolar endocytosis (Jiang et al. 2016).

20.4.1.1 Polymeric Micelles

Encapsulation methods as micelles, layer-by-layer self-assembled capsules, nanogels, liposomes, and polymersomes have been used for drug delivery. To ensure further effective targeted drug delivery, these capsules are coated with specific antigen. One of the most promising encapsulation methods is electrostatic layer-by-layer (LbL) self-assembled procedure, in which the opposite-charged components are adsorbed on the surface to form nanoshells. These charged components may be enzymes, linear antibodies, polyelectrolytes, or inorganic nanoparticles. For targeted release, the core of these charged particles should be soluble in specific conditions such as low pH which prevail in target cells only (Ai et al. 2003; Sutton et al. 2007).

An alternative encapsulation approach in drug delivery is drug loading into micelle nanocarriers (Deng et al. 2012; Owen et al. 2012). Micelles are amphiphilic macromolecules having two distinctive block domains, i.e., hydrophilic and hydrophobic. While encapsulating the nutraceuticals in these micelles, hydrophilic blocks hold at the interior hydrophobic domain which was composed of the hydrophobic segments, forming the micellar corona. They are formed by the copolymeric amphiphiles and their self-aggregation above a certain concentration, which is known as critical micellar concentration (CMC) (Talelli et al. 2015).

20.4.1.2 Polymeric Hydrogels

These are cross-linked hydrophilic polymer networks which provide sustained, local delivery of nutraceuticals. These polymeric networks have a large affinity toward water and are prevented from dissolving by covalent cross-linkages, by non-covalent attractions, or by physical entanglements (Hoare and Kohane 2008; Ladet et al. 2008). Polymeric hydrogels can also be modified using active targeting ligands and made thermo-reversible, i.e., they can change their phase as per the environmental temperature (Saneja et al. 2016), for example, folate-functionalized PEG cross-linked acrylic polymer (FA-CLAP) hydrogel for site-specific delivery of curcumin (Pillai et al. 2014), which demonstrated higher uptake in HeLa cell lines compared to non-functionalized hydrogels. In an another study, curcumin hydrogels have being developed using gelatin, chitosan, and hyaluronan which showed increased in vitro cytotoxicity against A549 lung adenocarcinoma cells than native curcumin (Teong et al. 2015).

Beta-glucans are the vital component of cell wall in various organisms including mushrooms, yeast, oats, barley, and *Candida* which are polymeric carbohydrates with a glucose units having beta-(1, 3) linkages and an occasional beta-(1, 6) branches (Buckeridge et al. 2004; Bohn and Miller 1995). Beta-glucans boost up the innate immune system and pharmacologically classified as a biological response modifiers (Bohn and Miller 1995). These have also been granted the "generally recognized as safe (GRAS)" status by USFDA. β-glucans act like a pathogen-associated molecular pattern (PAMP), and they are mainly recognized by dectin-1 and complement receptors 3 (CR3) that are present on immune cells such as dendritic cells, macrophages, and neutrophils (Goodridge et al. 2011; Huang et al. 2012).

The particulate beta-1, 3 and 1, 6 glucan preparations are porous, 2–4 μm particles derived by alkaline and acidic extraction from the cell walls of *Saccharomyces cerevisiae* (Baker's yeast). The porous nature of the glucan enables encapsulation and targeted delivery of several types of cargo molecules including proteins (Yu et al. 2015), siRNA (Aouadi et al. 2009), DNA (Soto and Ostroff 2008), and drugs (Soto et al. 2010, 2012). Small molecules such as rifabutin and rifampicin have been encapsulated in GPs by using hydrogels to seal pores of glucan (Soto et al. 2010; Upadhyay et al. 2017) where the encapsulation has shown to enhance the antimycobacterial effectiveness of these antibiotics (Upadhyay et al. 2019).

Glucan particles have shown better efficacy by enhancing the immunotherapy in carcinoma models (mouse mammary gland), by inducing a protective Th1 cellular response (IFN-γ and IL-12) (Baran et al. 2007). β-glucans also have been shown to possess anti-inflammatory effects (Du et al. 2015).

The diffusion properties and dispersion stability of glucan particles with hydrophilic molecules of different nature as well as molecular weight, including vitamin B12, bovine serum albumin, (BSA) and caffeine, have been studied (Saloň et al. 2016). In vitro encapsulation of curcumin in yeast-derived glucan particles has been shown to promote its anti-inflammatory potential (Plavcová et al. 2019), while in vivo systemic suppression of inflammation was achieved by siRNA encapsulated GPs delivery (Aouadi et al. 2009).

20.4.1.3 Polymeric Conjugates

Polymeric conjugates have been designed for nutraceuticals delivery. These are not encapsulated but attached covalently to a macromolecular polymeric carrier via hydrolyzable bond which is generally stable in the systemic circulation and breaks at the target-specific site due to specific stimuli, such as change in pH, temperature, enzymes, or light (Luo et al. 2014; Pang et al. 2014). Many different types of spacers including adipic dihydrazide, succinic anhydride, and disulfide have been used to connect the nutraceuticals with the help of polymeric conjugates functionalized with targeting ligands as hyaluronic acid (Chang et al. 2016).

Glucan particles that are derived from yeast have also been used as an *antigen-targeted delivery systems* by their conjugation with the monoclonal antibodies for the treatment of diseases including enteric infections (Baert et al. 2015) and cancer (Yan et al. 2005).

20.4.2 Lipid-Based Nanocarriers

Many solid lipid nanoparticles, liposomes, self-emulsifying systems, and nanoemulsions have been used as lipid-based drug delivery systems for enhancing the efficacy and bioavailability of nutraceuticals.

20.4.2.1 Liposomes

Liposomes consist of a lipid bilayer membrane surrounding an aqueous interior compartment (Perche and Torchilin 2013). Their precursors are usually from naturally occurring phospholipids and cholesterol, which make them biodegradable. These liposomes can be surface modified with specific ligands such as aptamers, sialic acid, folic acid, etc. (Sercombe et al. 2015) and made to persist for long in circulation by incorporating polyethylene glycol (PEG) on their surface. For example, a recent study demonstrated ursolic acid (modified PEG liposome) which enhanced in vitro cytotoxicity in EC-304 cells as compared to native ursolic acid (Zhao et al. 2015b).

20.4.2.2 Solid Lipid Nanoparticles

These are lipids consisting colloidal carriers (e.g., cholesterol, glyceryl monostearate, stearic acid), which are dispersed in an aqueous surfactant solution or water.

These nanoparticles have the ability to bypass P-glycoprotein (P-gp)-mediated efflux and inhibit P-gp (Saneja et al. 2014a; Weber et al. 2014). Recently, Chen et al. demonstrated the aloe-emodin encapsulation into solid lipid nanoparticles which enhanced its in vitro cytotoxicity against human breast cancer MCF-7 cells and human hepatoma HepG2 cells and increased the cellular uptake of solid lipid nanoparticles as compared to native aloe-emodin solution (Chen et al. 2015).

20.4.2.3 Nanoemulsions

Nanoemulsions are made up of two or more immiscible heterogeneous mixtures in which one phase is dispersed while the second phase is partially immiscible/miscible (Blanco et al. 2015), widely used for delivering nutraceuticals with poor solubility. These nanoemulsions have numerous advantages including small globule size, thermodynamic stability, improved solubilizing capacity, and the use of GRAS (generally recognized as safe) excipients. They enhance the bioavailability of these systems through their protective effect in the gastrointestinal tract (GIT) and improve solubilization (Porter et al. 2008; Porter et al. 2007). Further, due to their small size of these droplets, they increase the aqueous medium of the gut and the interfacial area between the lipophilic droplet facilitating the homogenous distribution of the administered nutraceutical in the GIT. Recently, there has been substantial interest toward an isotropic mixture of a surfactant(s), oil, and cosurfactant (modified form of nanoemulsions) known as self-emulsifying drug delivery systems (Gursoy and Benita 2004) which enhances the bioavailability of nutraceuticals. 2-iminothiolane-modified chitosan coated with curcumin nanoemulsions exhibited 33-fold improvement in the bioavailability as compared to native curcumin (Vecchione et al. 2016).

20.4.2.4 Self-Assembled and Polymeric Nanocarriers

Since the early work of Sessa and Weissmann in 1970, the self-assembled lipid-based nanocarriers such as liposomes have been the tremendous amount of object work. Polymeric micelles, an important class of self-assembled drug nanocarriers, are also currently undergoing innovative clinical trials. As reviewed by Torchilin, many biocompatible amphiphilic polymers have also been explored (Bromberg 2008; Torchilin 2007; Matsumura 2007; Matsumura 2008). During the self-assembly process, the poorly water-soluble drugs are physically entrapped within the hydrophobic core of micellar nanocarriers. Polymeric nanoparticles have been widely used toward the design of novel drugs nanocarriers which are composed of either biopolymers including chitosan, dextran, gelatin, hyaluronan, or synthetic ones such as poly(D,L-lactic-co-glycolic acid) (PLGA), albumin, N-(2-hydroxypropyl)methacrylamide (HPMA), and polyglutamate. An albumin-based paclitaxel nanocarrier, i.e., Abraxane, has received FDA approval and has been currently used for metastatic breast cancer treatment (Gradishar 2006).

20.4.3 Inorganic Nanocarriers

Inorganic nanocarriers include magnetic nanomaterials, carbon nanotubes, nanosilica, quantum dots, and gold nanoparticles (Anselmo and Mitragotri 2015; Santos et al. 2015). These find the application in nutraceutical delivery due to their unique physiochemical properties, such as the ability for surface fictionalization, size, shape, higher surface to volume ratios, and chemical composition. Gold nanoparticles (AuNPs) have been widely studied for nutraceuticals delivery due to well-defined surface chemistry, excellent biocompatibility, and their ease of synthesis (Zhao et al. 2015a).

20.4.3.1 Carbon Nanotubes

Carbon atoms which are composed of tubular hydrophobic networks having length and diameter of approximately 1–4 nm and 1–100 nm, respectively, explored for their nutraceutical delivery (Bianco et al. 2011; Nagai et al. 2011). While nanotubes are practically insoluble in all solvents and are associated with toxicity, they can be chemically modified to make them water-soluble carriers, decreasing their toxicity and enhancing their biocompatibility (Pérez-Herrero and Fernández-Medarde, 2015). Many other nanoparticles (magnetic nanoparticles) have also been demonstrated in nutraceuticals delivery due to their magnetic properties and small size. Recently, magnetic nanoparticles of curcumin (MNP-CUR) were developed which increase the serum bioavailability by 2.5-fold as compared to native curcumin in order to improve its bioavailability and efficacy (Yallapu et al. 2013). Furthermore, the nanoparticle formulation suppressed pancreatic tumor growth, delaying tumor growth, and enhanced the survival of mice in an HPAF-II xenograft mouse model (Yallapu et al. 2013).

20.4.3.2 Other Inorganic Nanocarriers

Other inorganic nanoparticles such as calcium phosphate silicon and silica with or without internal porosity, present an interesting substitute to the more conventional organic ones, such as liposomes and micellar systems, for their targeted based delivery of therapeutic agents which are biocompatible with little or no toxic products toward immune response (Slowing et al. 2008; Kester et al. 2008; Morgan et al. 2008) where the drugs can be physiosorbed either loaded within the inorganic matrix of the nanocarriers. For example, a therapeutic molecule such as hexanoyl-ceramide and a range of small organic diagnostic can retain their activities by incorporating it within calcium phosphate nanoparticles (Morgan et al. 2008).

The striking surface area increase associated with the use of colloidal nanosized materials has also been put upon/used to prepare therapeutically active nanoparticles. The large surface area enables the conjugation of a high payload of molecules. The excellent biocompatibility of gold colloids and existence of well-established synthetic routes have been exploited to design nontoxic drugs and gene carriers.

20.4.4 Functional Therapeutic Nanoparticles (Hybrid Nanocarriers)

20.4.4.1 Plasmonic Nanoparticles

Plasmonic colloidal nanostructures, including nanoparticles nanoholes, nanoshells, and nanorods, are arguably the most promising nanomaterials in the medicine field because of their ability to support localized surface plasmons as they have coherent oscillations of electrons conduction on a metal surface that are excited by electromagnetic radiation (Atwater 2007; Ozbay 2006; Maier and Atwater 2005). In oncology, plasmonic nanoparticles-based sensing approaches are being widely investigated (Qian et al. 2008). These nanoparticles can dramatically increase the fluorescence of organic fluorophores,which are located within an inorganic or organic shell surrounding the metal core, which can be used in designing novel fluorescence-based biosensing approaches (Aslan et al. 2007). Alternatively, the gold nanoparticles have intrinsic properties that can be used to design various novel in vivo and in vitro biosensing/imaging strategies. In recent years, therapeutic applications involving gold colloids have come up. Thus, the optical properties of the plasmonic nanoparticles are indeed strongly investigated for designing of novel light-activated therapeutic applications (Yu et al. 2007; Durr et al. 2007; Popovtzer et al. 2008).

20.4.4.2 Chemotherapeutic Drugs Combination with Nutraceuticals Using Nanocarriers

In recent years, nanomedicines have been used for a variety of cancer therapies including hyperthermia, tumor-targeted drug delivery, and photodynamic therapy (Rezvantalab et al. 2018). Co-encapsulation of nutraceutical with chemotherapeutic agent is combination strategy, where a single nanocarrier has achieved wide recognition because of its synergistic effect, reduced toxicity, and enhanced bioavailability. The combination of both the agents is useful in preclinical studies to address solubility problem associated with nutraceutical and chemotherapeutic agent. However, targeted drug delivery can be achieved by these nanocarriers through surface modification with targeting ligands which results in lower adverse effects (Parhi et al. 2012; Saneja et al. 2014a). Presently, combinatorial nanoparticles of epigallocatechin gallate (EGCG) and paclitaxel (Ptx) in a targeted core/shell poly(lactic-co-glycolic acid) (PLGA) nanoparticle were developed to address the adversities by Ptx chemotherapy (Narayanan et al. 2015).

The co-delivery of curcumin and doxorubicin using lipid-coated PLGA nanoparticles was investigated against human osteosarcoma (cell lines) (Wang et al. 2016). The co-encapsulation of aspirin and curcumin into methoxy poly(ethylene glycol)-poly (lactide-co-glycolide) (mPEG-PLGA) nanoparticles has demonstrated synergistic anticancerous effects in human ovarian carcinoma cells as well as in activation of the mitochondrial apoptosis pathway (Zhou et al. 2015). In order to maximize the therapeutic efficacy of daunorubicin (DNR) and to reverse multidrug resistance (MDR), Xu and his coworkers developed the co-encapsulation of polyethylene glycol–polylactic-co-glycolic acid–poly-L-lysine (PEG-PLGA-PLL) delivery system for the co-delivery of DNR and gambogic acid (GA) (Xu et al. 2015).

20.5 Mechanism of Action

20.5.1 Typical Nanocarriers for Encapsulation

For drug delivery, many encapsulation methods have been used; some of the typical types are micelles, layer-by-layer self-assembled capsule, nanogels liposomes, and polymersomes that are covered as the nanocarriers. To ensure further effective targeted drug delivery, these capsules are coated with specific antigen.

One of the most promising encapsulation methods is electrostatic layer-by-layer (LbL) self-assembled procedure, where the opposite-charged components are adsorbed on the surface to form nanoshells. These charged components may be enzymes, linear antibodies, polyelectrolytes, or inorganic nanoparticles (Ai et al. 2003), but the core should be soluble in conditions where controlled release is needed as well as insoluble in certain conditions such as low pH.

Another alternative encapsulation approach in drug delivery is drug loading into nanocarrier micelles. They are amphiphilic macromolecules having two distinctive block domains that are hydrophilic and hydrophobic as well as copolymers.

20.5.2 Factors Controlling Release from Nanocarriers

Pores size in the nanocarriers is one of the most important mechanisms that controls the release of drug molecules. The duration of drug release depends upon the pore size as well as the effective volume. The change in effective volume of nanocarrier through external stimuli enables control over the opening and closing mechanism as well as change in internal permeability and effective volume of the nanocarriers that allow controlled drug delivery (Gao et al. 2011; Kommareddy and Amiji 2007). The nonprotein nanocarriers like polymer nanoparticles, polymeric micelles, and stabilized liposomes may be designed to contain pH responsive block copolymers, which undergo conformation changes due to pH changes. In another aspect, nanocarriers like polymer vesicles and nanoemulsion droplet use ultrasound to control the drug release mechanism, whereas the release rate is related to the ultrasound's ability toward physically stimulate of the nanocarrier (Rapoport et al. 2009). Similarly, liposomes that are responsive toward temperature can be used as nanocarriers for the thermally stimulated drug release mechanism, as they allow different types of molecules such as thermal responsive ones to be concurrently encapsulated (Chen et al. 2012).

20.5.3 Basic Property of Pharmaceutical Nanocarriers: Longevity in the Blood

The "basic" property for any multifunctional nanocarrier is its longevity, as well as long-circulating pharmaceuticals and pharmaceutical carriers (Torchilin 1998; Lasic and Martin 1995; Cohen and Bernstein 1996; Moghimi and Szebeni 2003). Since for body defense system, pharmaceutical nanocarriers usually represent foreign particles, as they become easily opsonized and removed from the circulation before the completion of their function.

The important reason for producing such long-circulating drugs and drug carriers is to maintain a required level of pharmaceutical agent in the blood for a longer time intervals, as long-circulating drugs contain microparticulates or large macromolecules that can slowly accumulate in pathological sites with leaky and affected vasculature including inflammations, tumors, and infarcted areas and facilitate drug delivery in those areas (Maeda et al. 2000; Maeda 2001; Gabizon 1995). Thus, the prolonged circulation can be helpful in achieving a desirable targeting effect for specific ligand-modified drugs and the drug carriers for allowing more time for interaction with the target (Torchilin 1996).

In addition, there are other polymers that have been suggested as alternative steric protectors for nanoparticle drug carriers (Torchilin and Trubetskoy 1995) and are expected to be biocompatible, hydrophilic, soluble, and possess a highly flexible main chain. The single-end lipid-modified poly(acryl amide) and poly(vinyl pyrrolidone) (Chonn et al. 1992; Lasic et al. 1991; Torchilin et al. 2001), together with other amphiphilic polymers which possess fairly soluble and flexible hydrophilic moiety, such as phospholipid(PE)-modified poly(2-methyl-2-oxazoline) or poly(2-ethyl-2-oxazoline) (Woodle et al. 1994), poly(acryloyl morpholine) (PAcM), phosphatidyl polyglycerols, and polyvinyl alcohol, have been successfully used as a liposome steric protectors. The surface modification of hydrophobic polymeric nanoparticles may be performed by chemical grafting of polymer chains onto a particle or physical adsorption of a protecting polymer on a particle surface.

Recently, a lot of research has been performed for surface modification of superparamagnetic nanoparticles, which are now contemplated as promising agents for drug delivery into targeted site and for diagnostic imaging purposes (Torchilin 2006).

20.5.4 Nanoparticle Vehicles in Nutrient and Nutraceutical Delivery

Nanoparticle systems can be generated by

1. "Top-down" approach: Different mechanical size reduction processes are used to make small particles.

2. "Bottom-up" approach: A chemical processes where the nanoparticle is produced by the self-assembly of the smaller molecules such as proteins and lipids (Shimomur and Sawadaishi 2001; Whitesides and Grzybowski 2002; Inoue et al. 2007).

However, there is a leading trend to combine bottom-up and top-down approaches to develop nanoparticle systems (Horn and Rieger 2001).

20.6 Bioavailability Enhancement with Nanoparticles

The term bioavailability indicates the dose segment that is available at the site where drug acts. On the other hand, the uptake (or intestinal absorption) refers to fraction of the dose which is absorbed through the intestinal walls. Although the entire dose is related to both definitions, it is necessary to understand and design the effective nutraceuticals, active ingredients, and nanoparticle delivery systems for nutrients in the biological processes that regulate the nutrient uptake and bioavailability.

Thus, drug delivery system based on nanocarriers has emerged as an effective vehicle due to favorable physicochemical characteristics, high surface-to-volume ratio, and nanoscale size. Nanoscale drug delivery enhances the shelf life, protects the food components against moisture, enables controlled release and aqueous solubility, and influences texture and flavor. Further, nanocarriers have the ability to modulate pharmacodynamic as well as pharmacokinetic profiles of nutraceuticals (Díaz and Vivas-Mejia 2013). Many nanocarriers such as micelles, liposomes, polymeric nanoparticles, etc. have been used to enhance the efficacy of nutraceuticals and the bioavailability in recent years. So, it is beneficial to develop phytochemical-based nanocarriers which help in nutraceutical delivery for various purposes such as cancer chemotherapy and thus enhance their pharmacokinetics as well as pharmacodynamics outcomes.

20.7 Future Prospective and Challenges

The applications of nanocarriers in the therapeutic delivery systems make it feasible for proper delivery and release of nutraceuticals inside the targeted cells. Thus, to achieve effective controlled nutraceutical/drug delivery, specific molecules and ligands can play an important role in assisting the process of cell targeting and controlled drug release. For a therapeutic delivery system, inappropriate drug release leads to failure of the drug in disease mitigation. Furthermore, if the therapeutic system requires different stages to be synthesized, then the large-scale production of the medication may be difficult.

20.8 Conclusion

In recent years, the explosion of research and development in the field of functional nanoparticles and nanocarriers has drastically increased the knowledge required for their advancement in the clinical practice. There are many therapeutic nutraceutical nanomaterials that have been already proposed, and it is not easy to predict which of these will be most successful, but the future looks bright for several of these applications, including that of nanocarriers for drugs or molecules.

Acknowledgments We acknowledge the support from the DST-FIST, Department of Science and Technology as well as Uttar Pradesh Council of Science and Technology and Integral University, Lucknow.

References

Acosta E (2009) Bioavailability of nanoparticles in nutrient and nutraceutical delivery. Curr Opin Colloid Interface Sci 14(1):3–15
Ai H, Jones SA, De Villiers MM et al (2003) Nano-encapsulation of furosemide microcrystals for controlled drug release. J Control Release 86(1):59–68
Allen TM (2002) Ligand-targeted therapeutics in anticancer therapy. Nat Rev Cancer 2(10):750
Alonso MJ (2004) Nanomedicines for overcoming biological barriers. Biomed Pharmacother 58(3):168–172
Anselmo AC, Mitragotri S (2015) A review of clinical translation of inorganic nanoparticles. AAPS J 17(5):1041–1054
Aouadi M, Tesz JG, Nicoloro SM et al (2009) Orally delivered siRNA targeting macrophage Map 4k4 suppresses systemic inflammation. Nature 458:1180–1184
Arora D, Sharma N, Sharma V et al (2016) An update on polysaccharide-based nanomaterials for antimicrobial applications. Appl Microbiol Biotechnol 100(6):2603–2615
Aslan K, Wu M, Lakowicz JR et al (2007) Fluorescent core– shell Ag@ SiO2 nanocomposites for metal-enhanced fluorescence and single nanoparticle sensing platforms. J Am Chem Soc 129(6):1524–1525
Atwater HA (2007) The promise of plasmonics. Sci Am 296(4):56–62
Baert K, de Geest BG, de Rycke R (2015) β-glucan microparticles targeted to epithelial APN as oral antigen delivery system. J Control Release 220:149–159
Baran J, Allendorf DJ, Hong F et al (2007) Oral Beta-glucan adjuvant therapy converts nonprotective Th2 response to protective Th1 cell-mediated immune response in mammary tumor-bearing mice. Folia Histochem Cytobiol 45(2):107–114
Bernkop Schnurch A, Walker G (2001) Multifunctional matrices for oral peptide delivery. Crit Rev Ther Drug Carrier Syst 18(5):459–501
Bianco A, Kostarelos K, Prato M (2011) Making carbon nanotubes biocompatible and biodegradable. Chem Commun 47(37):10182–10188
Blanco E, Shen H, Ferrari M (2015) Principles of nanoparticle design for overcoming biological barriers to drug delivery. Nat Biotechnol 33(9):941
Bohn JA, Miller JN (1995) (1-3)-beta-D-glucans as biological response modifiers: a review of structure-functional activity relationships. Carbohydr Polym 28:3–141
Bromberg L (2008) Polymeric micelles in oral chemotherapy. J Control Release 128(2):99–112
Buckeridge MS, Rayon C, Urbanowicz B et al (2004) Mixed linkage (1→3), (1→4)-Beta-D-glucans of grasses. Cereal Chem 81:115
Chang M, Zhang F, Wei T et al (2016) Smart linkers in polymer–drug conjugates for tumor-targeted delivery. J Drug Target 24(6):475–491

Chau CF, Wu SH, Yen GC (2007) The development of regulations for food nanotechnology. Trends Food Sci Technol 18(5):269–280
Chauhan B, Kumar G, Kalam N (2013) Current concepts and prospects of herbal nutraceutical: a review. J Adv Pharm Technol Res 4(1):4
Chen H, Weiss J, Shahidi F (2006a) Nanotechnology in nutraceuticals and functional foods. Food Technol 60(7):73–75
Chen L, Remondetto GE, Subirade M (2006b) Food protein-based materials as nutraceutical delivery systems. Trends Food Sci Technol 17(5):272–283
Chen KJ, Liang HF, Chen HL et al (2012) A thermoresponsive bubble-generating liposomal system for triggering localized extracellular drug delivery. ACS Nano 7(1):438–446
Chen R, Wang S, Zhang J (2015) Aloe-emodin loaded solid lipid nanoparticles: formulation design and in vitro anti-cancer study. Drug Deliv 22(5):666–674
Chonn A, Semple SC, Cullis PR (1992) Association of blood proteins with large unilamellar liposomes in vivo. Relation to circulation lifetimes. J Biol Chem 267(26):18759–18765
Chung HY, Cesari M, al AS (2009) Molecular inflammation: underpinnings of aging and age-related diseases. Ageing Res Rev 8(1):18–30
Cohen S, Bernstein H (1996) Microparticulate systems for the delivery of proteins and vaccines, vol 77. CRC Press, Brand
Deng C, Jiang Y, Cheng R et al (2012) Biodegradable polymeric micelles for targeted and controlled anticancer drug delivery: promises, progress and prospects. Nano Today 7(5):467–480
Díaz M, Vivas-Mejia P (2013) Nanoparticles as drug delivery systems in cancer medicine: emphasis on RNAi-containing nanoliposomes. Pharmaceuticals 6(11):1361–1380
Du B, Lin C, Bian Z et al (2015) An insight into anti-inflammatory effects of fungal beta-glucans. Trends Food Sci Technol 41(1):49–59
Dudeja P, Gupta RK (2017) Nutraceuticals. In: Food safety in the 21st century. Academic Press, San Diego, pp 491–496
Durr NJ, Larson T, Smith DK (2007) Two-photon luminescence imaging of cancer cells using molecularly targeted gold nanorods. Nano Lett 7(4):941–945
El-Gogary RI, Rubio N, Wang JT et al (2014) Polyethylene glycol conjugated polymeric nanocapsules for targeted delivery of quercetin to folate-expressing cancer cells in vitro and in vivo. ACS Nano 8:1384–1401
Gabizon AA (1995) Liposome circulation time and tumor targeting: implications for cancer chemotherapy. Adv Drug Deliv Rev 16(2–3):285–294
Gao Y, Chen Y, Ji X et al (2011) Controlled intracellular release of doxorubicin in multidrug-resistant cancer cells by tuning the shell-pore sizes of mesoporous silica nanoparticles. ACS Nano 5(12):9788–9798
Goodridge HS, Reyes CN, Becker CA et al (2011) Activation of the innate immune receptor dectin-1 upon formation of a 'phagocytic synapse'. Nature 472(7344):471–U541
Gradishar WJ (2006) Albumin-bound paclitaxel: a next-generation taxane. Expert Opin Pharmacother 7(8):1041–1053
Gregoriadis G (1988) Liposomes as a drug delivery system: optimization studies. In: Biotechnological applications of lipid microstructures. Springer, New York, pp 151–159
Gursoy RN, Benita S (2004) Self-emulsifying drug delivery systems (SEDDS) for improved oral delivery of lipophilic drugs. Biomed Pharmacother 58(3):173–182
Hoare TR, Kohane DS (2008) Hydrogels in drug delivery: progress and challenges. Polymer 49(8):1993–2007
Horn D, Rieger J (2001) Organic nanoparticles in the aqueous phase—theory, experiment, and use. Angew Chem Int Ed 40(23):4330–4361
Huang H, Ostroff GR, Lee CK et al (2012) Relative contributions of dectin-1 and complement to immune responses to particulate beta-glucans. J Immunol 189(1):312–317
Inoue Y, Yoshimura S, Tozuka Y et al (2007) Application of ascorbic acid 2-glucoside as a solubilizing agent for clarithromycin: solubilization and nanoparticle formation. Int J Pharm 331(1):38–45

Jiang Y, Lu H, Dag A et al (2016) Albumin–polymer conjugate nanoparticles and their interactions with prostate cancer cells in 2D and 3D culture: comparison between PMMA and PCL. J Mater Chem B 4(11):2017–2027
Kester M, Heakal Y, Fox T et al (2008) Calcium phosphate nanocomposite particles for in vitro imaging and encapsulated chemotherapeutic drug delivery to cancer cells. Nano Lett 8(12):4116–4121
Kommareddy S, Amiji M (2007) Poly (ethylene glycol)–modified thiolated gelatin nanoparticles for glutathione-responsive intracellular DNA delivery. Nanomedicine 3(1):32–42
Kumari A, Yadav SK, Yadav SC (2010) Biodegradable polymeric nanoparticles based drug delivery systems. Colloids Surf B: Biointerfaces 75(1):1–18
Ladet S, David L, Domard A (2008) Multi-membrane hydrogels. Nature 452(7183):76
Lasic DD, Martin FJ (1995) Stealth liposomes, vol 20. CRC Press, Boca Raton
Lasic DD, Martin FJ, Gabizon A et al (1991) Sterically stabilized liposomes: a hypothesis on the molecular origin of the extended circulation times. Biochim Biophys Acta (BBA)-Biomembranes 1070(1):187–192
Lo YL, Sung KH, Chiu CC et al (2013) Chemically conjugating polyethylenimine with chondroitin sulfate to promote CD44-mediated endocytosis for gene delivery. Mol Pharm 10(2):664–676
Luo C, Sun J, Sun B et al (2014) Prodrug-based nanoparticulate drug delivery strategies for cancer therapy. Trends Pharmacol Sci 35(11):556–566
Maeda H (2001) The enhanced permeability and retention (EPR) effect in tumor vasculature: the key role of tumor-selective macromolecular drug targeting. Adv Enzym Regul 41:189
Maeda H, Wu J, Sawa T et al (2000) Tumor vascular permeability and the EPR effect in macromolecular therapeutics: a review. J Control Release 65(1–2):271–284
Maier SA, Atwater HA (2005) Plasmonics: localization and guiding of electromagnetic energy in metal/dielectric structures. J Appl Phys 98(1):10
Matsumura Y (2007) Preclinical and clinical studies of anticancer drug-incorporated polymeric micelles. J Drug Target 15(7–8):507–517
Matsumura Y (2008) Poly (amino acid) micelle nanocarriers in preclinical and clinical studies. Adv Drug Deliv Rev 60(8):899–914
Moghimi SM, Szebeni J (2003) Stealth liposomes and long circulating nanoparticles: critical issues in pharmacokinetics, opsonization and protein-binding properties. Prog Lipid Res 42(6):463–478
Morgan TT, Muddana HS, Altinoglu EI et al (2008) Encapsulation of organic molecules in calcium phosphate nanocomposite particles for intracellular imaging and drug delivery. Nano Lett 8(12):4108–4115
Müller RH (1991) Colloidal carriers for controlled drug delivery and targeting: modification, characterization and in vivo distribution. Taylor & Francis, Boca Raton
Nagai H, Okazaki Y, Chew SH et al (2011) Diameter and rigidity of multiwalled carbon nanotubes are critical factors in mesothelial injury and carcinogenesis. Proc Natl Acad Sci 108:E1330–E1338
Narayanan S, Mony U, Vijaykumar DK et al (2015) Sequential release of epigallocatechin gallate and paclitaxel from PLGA-casein core/shell nanoparticles sensitizes drug-resistant breast cancer cells. Nanomedicine 11(6):1399–1406
Owen SC, Chan DP, Shoichet MS (2012) Polymeric micelle stability. Nano Today 7(1):53–65
Ozbay E (2006) Plasmonics: merging photonics and electronics at nanoscale dimensions. Science 311(5758):189–193
Pang X, Yang X, Zhai G (2014) Polymer-drug conjugates: recent progress on administration routes. Expert Opin Drug Deliv 11(7):1075–1086
Parhi P, Mohanty C, Sahoo SK (2012) Nanotechnology-based combinational drug delivery: an emerging approach for cancer therapy. Drug Discov Today 17:1044–1052
Perche F, Torchilin VP (2013) Recent trends in multifunctional liposomal nanocarriers for enhanced tumor targeting. J Drug Deliv 2013:1

Pérez-Herrero E, Fernández-Medarde A (2015) Advanced targeted therapies in cancer: drug nanocarriers, the future of chemotherapy. Eur J Pharm Biopharm 93:52–79

Pillai JJ, Thulasidasan AKT, Anto RJ et al (2014) Folic acid conjugated cross-linked acrylic polymer (FA-CLAP) hydrogel for site specific delivery of hydrophobic drugs to cancer cells. J Nanobiotechnol 12(1):25

Platt VM, Szoka FC Jr (2008) Anticancer therapeutics: targeting macromolecules and nanocarriers to hyaluronan or CD44, a hyaluronan receptor. Mol Pharm 5(4):474–486

Plavcová Z, Šalamúnová P, Saloň I et al (2019) Curcumin encapsulation in yeast glucan particles promotes its anti-inflammatory potential in vitro. Int J Pharm 568:118532

Popovtzer R, Agrawal A, Kotov NA et al (2008) Targeted gold nanoparticles enable molecular CT imaging of cancer. Nano Lett 8(12):4593–4596

Porter CJ, Trevaskis NL, Charman WN (2007) Lipids and lipid-based formulations: optimizing the oral delivery of lipophilic drugs. Nat Rev Drug Discov 6(3):231

Porter CJ, Pouton CW, Cuine JF et al (2008) Enhancing intestinal drug solubilisation using lipid-based delivery systems. Adv Drug Deliv Rev 60(6):673–691

Qian X, Peng XH, Ansari DO et al (2008) In vivo tumor targeting and spectroscopic detection with surface-enhanced Raman nanoparticle tags. Nat Biotechnol 26(1):83

Rapoport NY, Kennedy AM, Shea JE et al (2009) Controlled and targeted tumor chemotherapy by ultrasound-activated nanoemulsions/microbubbles. J Control Release 138(3):268–276

Rezvantalab S, Drude NI, Moraveji MK et al (2018) PLGA-based nanoparticles in cancer treatment. Front Pharmacol 9:1260

Rolland A (ed) (1993) Pharmaceutical particulate carriers. Therapeutic applications, vol 61. Marcel Dekker, New York

Saloň I, Hanuš J, Ulbrich P et al (2016) Suspension stability and diffusion properties of yeast glucan microparticles. Food Bioprod Process 99(Supplement C):128–135

Saneja A, Dhar Dubey R, Alam N et al (2014a) Co-formulation of P-glycoprotein substrate and inhibitor in nanocarriers: an emerging strategy for cancer chemotherapy. Curr Cancer Drug Targets 14(5):419–433

Saneja A, Khare V, Alam N et al (2014b) Advances in P-glycoprotein-based approaches for delivering anticancer drugs: pharmacokinetic perspective and clinical relevance. Expert Opin Drug Deliv 11(1):121–138

Saneja A, Nehate C, Alam N (2016) Recent advances in chitosan-based nanomedicines for cancer chemotherapy. In: Chitin and chitosan for regenerative medicine. Springer, New Delhi, pp 229–259

Sanguansri P, Augustin MA (2006) Nanoscale materials development–a food industry perspective. Trends Food Sci Technol 17(10):547–556

Santos HA, Bimbo LM, Peltonen L et al (2015) Inorganic nanoparticles in targeted drug delivery and imaging. In: Targeted drug delivery: concepts and design. Springer, Cham, pp 571–613

Sercombe L, Veerati T, Moheimani F et al (2015) Advances and challenges of liposome assisted drug delivery. Front Pharmacol 6:286

Sharma S, Bano A, Pathak N (2019) Pre- and probiotics: using functional foods in the fight against microbial resistance to antibiotics. Antibacterial drug discovery to combat MDR, 978-981-13-9870-4, 459963_1_En, (18)

Shimomur M, Sawadaishi T (2001) Bottom-up strategy of materials fabrication: a new trend in nanotechnology of soft materials. Curr Opin Colloid Interface Sci 6(1):11–16

Slowing II, Vivero-Escoto JL, Wu CW et al (2008) Mesoporous silica nanoparticles as controlled release drug delivery and gene transfection carriers. Adv Drug Deliv Rev 60(11):1278–1288

Soto ER, Ostroff GR (2008) Characterization of multilayered nanoparticles encapsulated in yeast cell wall particles for DNA delivery. Bioconjug Chem 19(4):840–848

Soto E, Kim YS, Lee J et al (2010) Glucan particle encapsulated rifampicin for targeted delivery to macrophages. Polymers 2:681–689

Soto ER, Caras AC, Kut LC et al (2012) Glucan particles for macrophage targeted delivery of nanoparticles. J Drug Deliv 13:14352

Sutradhar KB, Amin ML (2014) Nanotechnology in cancer drug delivery and selective targeting. ISRN Nanotechnol 2014:1

Sutton D, Nasongkla N, Blanco E et al (2007) Functionalized micellar systems for cancer targeted drug delivery. Pharm Res 24(6):1029–1046

Talelli M, Barz M, Rijcken CJ et al (2015) Core-crosslinked polymeric micelles: principles, preparation, biomedical applications and clinical translation. Nano Today 10(1):93–117

Teong B, Lin CY, Chang SJ et al (2015) Enhanced anti-cancer activity by curcumin-loaded hydrogel nanoparticle derived aggregates on A549 lung adenocarcinoma cells. J Mater Sci Mater Med 26(1):49

Thanki K, Gangwal RP, Sangamwar AT et al (2013) Oral delivery of anticancer drugs: challenges and opportunities. J Control Release 170(1):15–40

Torchilin VP (1996) How do polymers prolong circulation time of liposomes? J Liposome Res 6(1):99–116

Torchilin VP (1998) Polymer-coated long-circulating microparticulate pharmaceuticals. J Microencapsul 15(1):1–19

Torchilin VP (2006) Multifunctional nanocarriers. Adv Drug Deliv Rev 58(14):1532–1555

Torchilin VP (2007) Micellar nanocarriers: pharmaceutical perspectives. Pharm Res 24(1):1

Torchilin VP, Trubetskoy VS (1995) Which polymers can make nanoparticulate drug carriers long-circulating? Adv Drug Deliv Rev 16(2–3):141–155

Torchilin VP, Levchenko TS, Whiteman KR et al (2001) Amphiphilic poly-N-vinylpyrrolidones: synthesis, properties and liposome surface modification. Biomaterials 22(22):3035–3044

Upadhyay TK, Fatima N, Sharma D et al (2017) Preparation and characterization of beta-glucan particles containing a payload of nanoembedded rifabutin for enhanced targeted delivery to macrophages. EXCLI 16:210–228

Upadhyay TK, Fatima N, Sharma A et al (2019) Nano-rifabutin entrapment within glucan microparticles enhances protection against intracellular Mycobacterium tuberculosis. Artif Cells Nanomed Biotechnol 47(1):427–435

Van Vlerken LE, Amiji MM (2006) Multi-functional polymeric nanoparticles for tumour-targeted drug delivery. Expert Opin Drug Deliv 3(2):205–216

Vecchione R, Quagliariello V, Calabria D et al (2016) Curcumin bioavailability from oil in water nano-emulsions: in vitro and in vivo study on the dimensional, compositional and interactional dependence. J Control Release 233:88–100

Wang L, Wang W, Rui Z et al (2016) The effective combination therapy against human osteosarcoma: doxorubicin plus curcumin co-encapsulated lipid-coated polymeric nanoparticulate drug delivery system. Drug Deliv 23(9):3200–3208

Weber S, Zimmer A, Pardeike J (2014) Solid lipid nanoparticles (SLN) and nanostructured lipid carriers (NLC) for pulmonary application: a review of the state of the art. Eur J Pharm Biopharm 86(1):7–22

Whitesides GM, Grzybowski B (2002) Self-assembly at all scales. Science 295(5564):2418–2421

Wing Shing Ho J, Wan Man Cheung M, Wai Lam Yu V (2012) Active phytochemicals from Chinese herbs as therapeutic agents for the heart. Cardiovasc Hematol Agents Med Chem 10(3):251–255

Woodle MC, Engbers CM, Zalipsky S (1994) New amphipatic polymer-lipid conjugates forming long-circulating reticuloendothelial system-evading liposomes. Bioconjug Chem 5(6):493–496

Xu P, Wang R, Li J et al (2015) PEG–PLGA–PLL nanoparticles in combination with gambogic acid for reversing multidrug resistance of K56 New amphipatic polymer-lipid conjugat2/A02 cells to daunorubicin. RSC Adv 5:61051–61059

Yallapu MM, Ebeling MC, Khan S (2013) Novel curcumin-loaded magnetic nanoparticles for pancreatic cancer treatment. Mol Cancer Ther 12(8):1471–1480

Yan J, Allendorf DJ, Brandley B (2005) Yeast whole glucan particle (wgp) beta-glucan in conjunction with antitumour monoclonal antibodies to treat cancer. Expert Opin Biol Ther 5(5):691–702

Yu C, Nakshatri H, Irudayaraj J (2007) Identity profiling of cell surface markers by multiplex gold nanorod probes. Nano Lett 7(8):2300–2306

Yu M, Chen Z, Guo W et al (2015) Specifically targeted delivery of protein to phagocytic macrophages. Int J Nanomedicine 10:1743–1757
Zhao J (2007) Nutraceuticals, nutritional therapy, phytonutrients, and phytotherapy for improvement of human health: a perspective on plant biotechnology application. Recent Pat Biotechnol 1(1):75–97
Zhao J, Lee P, Wallace JM et al (2015a) Gold nanoparticles in cancer therapy: efficacy, biodistribution, and toxicity. Curr Pharm Des 21:4240–4251
Zhao T, Liu Y, Gao Z et al (2015b) Self-assembly and cytotoxicity study of PEG-modified ursolic acid liposomes. Mater Sci Eng 53:196–203
Zhong Y, Meng F, Deng C (2014) Ligand-directed active tumor-targeting polymeric nanoparticles for cancer chemotherapy. Biomacromolecules 15(6):1955–1969
Zhou L, Duan X, Zeng S et al (2015) Codelivery of SH-aspirin and curcumin by mPEG-PLGA nanoparticles enhanced antitumor activity by inducing mitochondrial apoptosis. Int J Nanomedicine 10:5205

Current Status of Biologically Produced Nanoparticles in Agriculture

21

S. K. Singh, R. C. Kasana, R. S. Yadav, and R. Pathak

Abstract

The potential applications of nanomaterial and nanotechnologies in modern agriculture can provide state-of-the-art solutions to improve the quality of human life in the near future. The impending applications of nanotechnology in the fields of food and agriculture, viz., nano-fertilizer, plant nutrition, plant protection, and agrifood, have been reviewed. Nevertheless the indiscriminate use of nanoparticles due to their changed physicochemical properties may be toxic and risky to biological systems and the environment. The benefits of advances in nanotechnology and probable risks have been reviewed and future approaches highlighted.

Keywords

Nanoparticles · Nanofertilizer · Plant protection · Nutrition · Agrifood · Environment · Biological system

Besides utilizing various technological interventions for improvement of crop production, researchers are now exploring nanotechnology for increasing agricultural production (Parisi et al. 2015). Nanotechnology is the science and technology which deals with very minute material in nanometer scale. A nanoparticle (NP) is a minute particle with at least one dimension that ranges between 1 and 100 nm. They are different from bulk particles in terms of their physical, chemical, and biological properties and have been used in improving agriculture (Khandelwal and Joshi 2018). The NPs have tremendous scope of novel applications in all walks of life due

S. K. Singh (✉) · R. C. Kasana · R. S. Yadav · R. Pathak
ICAR-Central Arid Zone Research Institute, Jodhpur, Rajasthan, India

M. Ghorbanpour et al. (eds.), *Biogenic Nano-Particles and their Use in Agro-ecosystems*, https://doi.org/10.1007/978-981-15-2985-6_21

to unique functional physicochemical properties acquired with increased large surface to volume ratio, surface area, and higher catalytic activity and reactivity (Yang and Watts 2005; Adhikari et al. 2010; Fakruddin et al. 2012; Siddiqui et al. 2015; Raliya et al. 2017). Due to acquired novel chemical, physical, or biological properties different from those of their bulk counterparts, NPs have tremendous opportunities on the one hand but may create safety issues on the other (Agrawal and Rathore 2014; Mukhopadhyay 2014; Prasad et al. 2017).

Various metal NPs have been synthesized by numerous workers using two basic approaches, i.e., "top-down and bottom-up" using different physical, chemical, and biological methods. The physical and chemical synthesis of NPs is being widely used due to their specificity and formation of monodispersed NPs (Iravani et al. 2014) using various techniques like reducing agent, sol-gel, solvo-thermal, microwave assisted, ion sputtering, laser ablation (Papp et al. 2007; Wani et al. 2011; Ramesh 2013; Dar et al. 2014), etc. However, these processes of NPs synthesis are less preferred and having disadvantages for high cost of production and the use of hazardous chemicals and radiations. Therefore biological synthesis of NPs is increasingly gaining popularity as a rapid, eco-friendly, and easily scaled-up technology (Singh et al. 2016).

Biological synthesis of NPs is highly preferred over physical and chemical synthesis due to multiple advantages like simple, rapid, stable, cost-effective, biocompatibility, and eco-friendly production methodologies, and there is no need for external stabilizing agent and safe applicability without any hazardous effects on plants and animals (BegümKarakoçak et al. 2016). The capping agents play an important role in functionalization and stabilization of NPs which are not externally required in biologically synthesized NPs. The biomolecules and polysaccharides produced in the biological systems would work as capping and stabilizing agents. Though, despite all these advantages for biologically synthesized NPs, the polydispersity of the NPs has to be controlled by improving factors affecting NPs formation like reaction conditions, pH, temperatures, mixing ratio, etc. (Gurunathan et al. 2014). Microorganisms and plants have been considered as an effective and potential source for biological synthesis of NPs. The biological synthesis of NPs involves the use of biological system which includes various agents like bacteria (Raliya et al. 2014a; Das et al. 2014; Wang et al. 2016), fungi (Jain et al. 2013; Ottoni et al. 2017), actinomycetes (Karthik et al. 2014; Waghmare et al. 2014), algae (Azizi et al. 2013; Rajeshkumar et al. 2014), yeast (Moghaddam et al. 2015), organic waste material (de Barros et al. 2018), plant extract (Ramesh et al. 2014; Zahir et al. 2015; Naseem and Farrukh 2015; Suresh et al. 2015), etc. Various plant parts like roots, stems, fruits, leaves, and their extracts are being used for synthesis of NPs.

Pestovsky and Antonio (2017) reviewed the potential applications of silica, selenium, silver, copper, gold, palladium, manganese, zinc oxide, ferric oxide, titanium dioxide, hydroxyapatite, sulfur, nickel, chitosan, calcium alginate, polyethylene glycol, and zeolite NPs as nanofertilizers, growth stimulants, nanopesticides, pesticide carriers, and antimicrobial agents and nanoformulations in agriculture. Prasad et al. (2017) reviewed the recent developments of nanotechnology in sustainable agriculture; future challenges and perspectives with reference to applications of

NPs as nanobiosensors, nanofertilizers, nanopesticides, and nanotechnologies in food industry; and their ecotoxicological implications.

The European Commission has recognized nanotechnology as "Key Enabling Technologies" for sustainable competitiveness and growth in several industrial sectors (Parisi et al. 2015). Several experimental studies reveal that nanotechnology will have potential long-term impact on agriculture and food production. The pros and cons of nanotechnology in the field of agriculture need to be taken into account before releasing any such technology, and its impacts have been comprehensively reviewed (Agrawal and Rathore 2014). Nanomaterials can have remarkable usages in precision and sustainable agricultural production (Abobatta 2018; Ali et al. 2018; Rawat et al. 2018). NPs having enhanced activity different from corresponding bulk materials have positive morphophysiological effects in crop plants, plant disease resistance, plant growth, etc. (Sharon et al. 2010; Misra et al. 2013; Abd-elsalam 2013) but at the same time may have deleterious effects on biological systems and the environment due to the toxicity of free radicals which may result into lipid peroxidation and DNA damage (Moore 2006; Lin and Xing 2007; Xingmao et al. 2010; Rathore et al. 2012).

Nanotech materials have been developed for slow release and efficient dosages of fertilizers for the plant (Chinnamuthu and Boopathi 2009; Singh 2012; Suppan 2013). The nanoencapsulated fertilizers have been widely used to minimize fertilizer consumption and environmental pollution (Chinnamuthu and Boopathi 2009; DeRosa et al. 2010). The germination and growth of *Glycine max* enhanced using nanometer materials (Lu et al. 2002). The chlorophyll contents in cluster bean leaves increased by spraying biologically produced magnesium oxide (MgO) NPs (Raliya et al. 2014b). Zinc nanofertilizer improved crop production in pearl millet (Tarafdar et al. 2014). Application of zinc oxide (ZnO) NPs of 25 nm at 1000 ppm concentration promoted seed germination and seedling vigor and showed early flowering and higher leaf chlorophyll content. Pod yield per plant was 34% higher compared to bulk zinc sulfate ($ZnSO_4$), but at higher concentration of 2000 ppm, ZnO NPs showed inhibitory effects (Prasad et al. 2012). The studies carried out on influence of ZnO NPs on onion showed that plants treated with NPs at the concentration of 20 and 30 $\mu g\ m\ L^{-1}$ showed better growth and flowered 12–14 days earlier than the control. Treated plants also showed significantly higher values for seeded fruit per umbel, seed weight per umbel, and 1000 seed weight over control plants (Laware and Raskar 2014). Biosynthesized NPs based Zn NPs were developed to enhance crop production in pearl millet (*Pennisetum americanum*) (Tarafdar et al. 2014).

The NPs of ZnO exhibited greater reactivity and were found to be more soluble and available source of Zn to plant in Zn fertilizers as compared to the bulk particles (Milani et al. 2015). Studies on the effect of iron oxide (FeO) and ZnO NPs on growth and yield of carrot showed that among various concentrations of ZnO NPs at 100 ppm + FeO NPs at 50 ppm was the best combination resulting in maximum vegetative growth and yield (Elizabath et al. 2017).

The TiO_2 NPs have been shown to have enhanced the germination and growth of *Brassica napus* (Mahmoodzadeh et al. 2013). Studies on the effect of titanium dioxide (TiO_2) NPs on the growth of spinach seeds showed that plants produced by seed treatment of TiO_2 NPs had 73% more dry weight, increase in chlorophyll-a formation, and three folds higher photosynthetic rate compared to the control (Zheng et al. 2005). Treatment of Canola seeds with different concentrations of TiO_2 NPs of ~20 nm showed that 2000 mg L^{-1} concentration promoted both seed germination and seedling vigor when compared with control (Mahmoodzadeh et al. 2013). Different concentrations from 10 to 50 μg m L^{-1} of TiO_2NPs used for the treatment in onion seeds indicated that lower concentrations 10 μg m L^{-1} to 40 μg m L^{-1} enhance seed germination, promptness index, and seedling growth, while concentration of 50 μg m L^{-1} and above can be inhibitory for seed germination and seedling growth (Raskar and Laware 2013). Application of TiO_2NPs at 60 mg L^{-1} promoted sage (*Salvia officinalis)* seed germination percentage, increased vigor index, and lowest mean germination time, but higher concentrations did not improve mean germination time (Feizi et al. 2013). Mung bean production enhanced when TiO_2NPs were used as plant fertilizer (Raliya et al. 2015). Foliar application of nanomicronutrient fertilizers and TiO_2 NPs significantly enhanced both growth and yield in barley (Janmohammadi et al. 2016).

Studies on the influence of metal NPs using 3-aminopropyl functionalized silica (Si) NPs, palladium (Pd) NPs entrapped in an aluminum hydroxide matrix, dodecanethiol functionalized gold (Au) NPs, and copper (Cu) nanosize activated powder on germination of lettuce seeds showed that NPs (Pd, Au at low concentrations; Si, Cu at higher concentrations; and combination of Au and Cu) had a positive influence on seed germination (Shah and Belozerova 2009). The effect of Cu NPs at different concentrations on growth of *Phaseolus radiates* and *Triticum aestivum* showed that seedling lengths of test species were negatively related to the exposure concentration of Cu NPs (Lee et al. 2008). The effect of Cu NPs on germination and growth of seeds of *Glycine max* L. and *Cicer arietinum* showed that germination occurred up to 2000 ppm of CuO NPs, but the root growth was prevented above 500 ppm Cu (Adhikari et al. 2012). Treatment with Cu NPs under controlled laboratory conditions showed plant growth inhibition for *Raphanus sativus*, perennial *Lolium perenne*, and annual *Lolium rigidum* (Atha et al. 2012). The effect of Cu NPs on *Arabidopsis thaliana* showed a significant reduction in plant biomass and total chlorophyll content on exposure to 2, 5, 10, 20, 50, and 100 mg L^{-1}Cu NPs, but there was an increase in anthocyanin content at 10, 20, 50, and 100 mg L^{-1}; lipid peroxidation at 5, 10, and 20 mg L^{-1}; and amino acid proline content at 10 and 20 mg L^{-1} concentrations of Cu NPs (Nair and Chung 2014).

From the review of the literature on impact of various NPs on plants, it is clear that both positive and negative effects of NPs on germination and growth of plants were observed in living plants (Aslani et al. 2014). Metal NPs, under low concentrations, play a key role at the limit of plant tolerance in the development of plants. If plants absorb an excess of metals, toxic effects can obviously occur resulting in decrease of growth and irregularities in cell division. But in some cases, the excess metal NPs can act as cofactor for enzymes, which are involved in the formation of intermediate metabolites and hence help in plant growth promotion. However, the response of plants to metal NPs varies with the nature of the metal, the type of plant species, and the stage of growth.

The application of nanofertilizers can help in discharging nutrients in soil in a controlled way and avoid water pollution (Naderi and Abedi 2012). The use of nanofertilizers results in increased elemental efficiency and reduced soil toxicity (Naderi and Shahraki 2013). The use of nanosensors in developing nanofertilizers is a step forward in developing smart agriculture (Rameshaiah et al. 2015). Combined application of biochars and chemical fertilizers improved wheat productivity and soil quality (Sadaf et al. 2017).

Plant diseases cause about 40% losses in crop production (Flood 2010). A large number of bacterial and fungal pathogens attack various agricultural and horticultural crops resulting in huge yield losses. The most common bacterial pathogens belong to genera *Erwinia*, *Pseudomonas*, *Corynebacterium*, *Xanthomonas*, *Ralstonia*, *Pectobacterium*, *Agrobacterium*, and *Xylella*, while fungal pathogens causing spoilage of crops are the species belonging to genera *Alternaria*, *Aspergillus*, *Cladosporium*, *Colletotrichum*, *Phomopsis*, *Fusarium*, *Penicillium*, *Phoma*, *Phytophthora*, *Pythium*, *Rhizopus*, *Botrytis*, *Ceratocystis*, *Rhizoctonia*, *Sclerotinia*, etc. (Mansfield et al. 2012; Chowdappa and Gowda 2013).

Various nanoparticles exhibit antibacterial and antifungal activities (Aziz et al. 2015; Patra and Baek 2017; Singh et al. 2019). The use of TiO_2 has been found to have certain yield attributing effects in crops such as promotion of plant growth, enhancement of photosynthetic rate, and reduction in disease severity. TiO_2 as a growth promoter was found to play the function of antibiotics in *Vigna unguiculata* production (Owolade and Adenekan 2008). The TiO_2 NPs reduced *Curvularia* leaf spot and bacterial leaf blight disease incidence and severity in maize (Owolade et al. 2008). The silicate and water-soluble polymer upon exposure to radioactive rays have been demonstrated with antifungal activity against phytopathogenic fungi (Park et al. 2006). Silicon is absorbed into plants and increases disease resistance by promoting growth of plants but has no effect on pathogenic microorganisms. Nevertheless nanosized silica-silver particles are quite effective against certain plant diseases (Sharon et al. 2010). Gold nanoparticles (Au NPs) have been shown to act as delivery systems of DNA and pesticides in plant cells of rice and tobacco plants (Ghormade et al. 2011). The nano-ZnO synthesized from zinc nitrate checked conidiophores and conidial development and showed cell wall deformity of a fungal pathogen *Aspergillus fumigatus* due to de novo synthesis of hydroxyl and superoxide radicals and eventually killed the fungal hyphae (Patra and Goswami 2012). Studies on antifungal activities of ZnO NPs with size of 70 ± 15 nm against two postharvest pathogenic fungi (*Botrytis cinerea* and *Penicillium expansum*) showed that ZnO NPs at concentrations greater than 3 mmol L^{-1} can significantly inhibit the growth of *B. cinerea* and *P. expansum* and *P. expansum* was more sensitive to the treatment with ZnO NPs than *B. cinerea* (He et al. 2011).

Studies on using TiO_2NPs formulation containing zinc for management of bacterial leaf spot on red rose showed that field applications of TiO_2/Zn at 500–800 ppm significantly reduced disease severity compared with the untreated control and the activity was better or at par with other standards used for management of rose diseases (Paret 2013). ZnO NPs inhibited the fungal growth of *Botrytis cinerea* and conidia of *Penicillium expansum* (Abd-elsalam 2013). Controlled release matrices of chitosan NPs exhibited antimicrobial potential which is one of the least exploited areas and has allowed new opportunities for microbial control in agriculture and food areas for the sustainable management of viruses, bacteria, and fungi (Cota-Arriola et al. 2013). The Cu NPs demonstrated a significant inhibitory activity against plant pathogenic fungi, *Fusarium culmorum*, *Fusarium oxysporum*, and *Fusarium graminearum* (Shende et al. 2015).

Green synthesis of Ag NPs and their application in management of fungal pathogens have been summarized (Nair et al. 2010; Krishnaraj et al. 2012; Rafique et al. 2017). The antibiotic activities of a nanosized silica-silver exhibited cent per cent growth inhibition of several pathogens, namely, *Pythium ultimum*, *Magnaporthe grisea*, *Colletotrichum gloeosporioides*, *Botrytes cinerea*, and *Rhizoctonia solani* at 3 ppm concentration causing various plant diseases (Park et al. 2006). In agriculture Ag NPs have been applied as antimicrobial coatings in seeds and woods. For instance, Ag NPs exhibited antifungal effect against *Raffaelea* sp. causing wilt in oak trees (Kim et al. 2009). The effect of Ag NPs on the growth of sclerotium-forming species *Rhizoctonia solani*, *Sclerotinia sclerotiorum*, and *S. minor* revealed that Ag NPs effectively inhibit the hyphal growth and the value of hyphal growth rate for *R. solani*, *S. sclerotiorum*, and *S. minor* was 12%, 36%, and 41% at 7 ppm of Ag NPs supplemented medium, respectively, compared to control (Min et al. 2009). Studies on the effect of Ag NPs against six *Colletotrichum* species associated with pepper anthracnose under different culture conditions showed that the application of 100 ppm concentration of Ag NPs inhibited the growth of fungal hyphae as well as conidial germination in vitro when compared to the control (Lamsal et al. 2011a). Studies on the effect of Ag NPs against powdery mildew at various concentrations showed that the application of 100 ppm Ag NPs resulted in the highest inhibition rate for both before and after the outbreak of disease on cucumbers and pumpkins (Lamsal et al. 2011b). Silver-doped TiO_2 NPs as fertilizer possess bactericidal properties and can inactivate viruses (Liga et al. 2011). The silver nanoparticles exhibited a dose-dependent fungistatic activity on *Colletotrichum gloeosporioides*, the causal agent of anthracnose disease in many fruit plants with inhibition of the fungus reaching almost 90% with a low Ag NPs concentration of 56 μg silver m L^{-1} in potato dextrose agar (Aguilar-Méndez et al. 2011). The antifungal activity of silver and Cu NPs nanoparticles against two plant pathogenic fungi *Alternaria alternata* and *Botrytis cinerea* was checked, and it was found that the application of 15 mg L^{-1} concentration of Ag NPs produced maximum inhibition of the growth of fungal hyphae (Ouda 2014). The foliar spray of Ag NPs resulted in complete suppression of sun hemp rosette virus in bean and bean yellow mosaic virus in faba bean (Jain and Kothari 2014; Elbeshehy et al. 2015). During the ex vitro assays carried out to assess antifungal activity of the IAA- and IBA-stabilized AgNPs against pathogenic fungal strains, the IBA-AgNPs show better activity with various growth inhibition rates against *Curvularia lunata*, *Rhizoctonia solani*, and *Colletotrichum gloeosporioides* compared to the controls (Thangavelu et al. 2018).

The potential of NPs as modern approaches in insect pest management has been reviewed by Rai and Ingle (2012). The nano-silica as NPs to control a range of agricultural insect pests (Ulrichs et al. 2005; Barik et al. 2008), polyethylene glycol-coated NPs to control *Tribolium castaneum* insect (Yang et al. 2009), applications of Ag, aluminum oxide (Al_2O_3), ZnO and TiO_2 in the control of rice weevil and silkworm (*Bombyx mori*) and baculovirus are some of the examples of insect pest management (Goswami et al. 2010). Pesticides are often used to reduce yield losses due to pest attack. The potential of NPs in insects and their use in pest control have been reported (Bhattacharyya et al. 2010). The efficiency of pesticides has been improved by the use of Ag ions, Au NPs, and iron oxide (Fe_2O_3) NPs to manage plant pests (Jo et al. 2009; Al-Samarrai 2012; Vinutha et al. 2013; Nuruzzaman et al. 2016).

The nanoencapsulated formulation of pesticide is released slowly and is more soluble, specific, and stable (Bhattacharyya et al. 2016). The nanosized particles of active ingredients of pesticides with protective coating are environmentally safe and require reduced dosage (Nuruzzaman et al. 2016). Elmer and White (2018) reviewed the future of nanotechnology in plant pathology highlighting the use of engineered NPs as bactericides, fungicides, nanofertilizers, and biosensors for plant disease diagnostics. The NPs as protectants, carriers for pesticides, and RNA interference-mediated protection have been reviewed (Balaure et al. 2017; Sinha et al. 2017; Worrall et al. 2018). Nanoencapsulation enhanced the postemergence herbicidal activity of atrazine against mustard plants. The mechanisms of interaction of the effect of atrazine NPs system on mustard, nontarget organisms, maize, and risk assessment were studied to address the safety issues (Oliveira et al. 2015a, b).

Nanotechnology-based hybridized microarrays, nanopore sensors, and metal oxide semiconductor chip for rapid detection of biological entities have been reviewed (Bhattacharya et al. (2007). The carbon nanotubes have shown to regulate seed germination and plant growth. The application of multiwalled carbon nanotubes in tomato seeds resulted in an increase in the rate of seed germination to 90% compared to control which showed 71% seed germination (Khodakovskaya et al. 2009). Multiwalled carbon nanotubes enhanced the growth of tobacco cell culture as compared to control at 5–500 μg/ml concentrations (Khodakovskaya et al. 2012). The applications of nanotechnology in frontier areas such as biocompatible nanotubes (BNTs) (Sadeghi et al. 2013), entrapped bioactive compounds in nanodelivery systems (Sadeghi et al. 2014), nanocrystals (Tzoumaki et al. 2015), and bioconjugation of quantum dots to antibodies for detection of proteins, DNA, oligopeptides, and nucleotides (Sozer and Kokini 2014; Bonilla et al. 2016) are of great significance in food and agriculture sectors. The new perspective of nanotechnology as precision agricultural techniques might promote increased crop yields and reduced leaching and emissions thereby reducing loss of nutrients (Duhan et al. 2017).

Biosynthesized ZnO NPs enhanced exopolysaccharide production by *Bacillus subtilis* strain JCT1 for arid soil applications. The amelioration of exopolysaccharide was stable and resulted in enhanced soil aggregation, moisture, and soil organic carbon (Raliya et al. 2014a). Synthesized NPs of Au improved water purification by improving the efficiency of membrane filters (Bharathi et al. 2016).

The positive impact of nanotechnology in agrifood sector by way of nanomaterials for controlled release of nutrients, pesticides, and fertilizers and application of pesticide and nanosensors have been reviewed (Dasgupta et al. 2015; Parisi et al. 2015; Fraceto et al. 2016; Rawat et al. 2018; Singh et al. 2019). Yata et al. (2018) reviewed the potential applications of nanomaterial in agrifood sector, industrialization, and patented technologies. The European Food Safety Authority presented an inventory of current and potential future applications of nanotechnology in the agrifood sector to review the regulation of nanomaterials in the EU as well as in non-EU countries (RIKILT and JRC 2014).

Substantial production and utilization of nanomaterial are likely to cause contamination of air, water, and/or soil (Das et al. 2009; Bernhardt et al. 2010; Meetoo 2011; Gottschalk and Nowack 2011). The indiscriminate use of NPs due to its changed physicochemical properties is risky to biological systems and the environment (Warheit et al. 2008; Mukhopadhyay 2014). Nanomaterial is almost similar in size to cellular structures, and it can easily enter the cell and may alter vital cellular functions or enter the bloodstream and may be transported to organs, and accumulate and may result in toxicity (Shvedova et al. 2010; Bertrand and Leroux 2012). Toxicity of NPs applied in the field to the ecosystem and human is a major concern. Studies conducted with Ag NPs have reported that the citrate-coated colloidal Ag NPs were not genotoxic (genetic), cytotoxic (cell), and phototoxic (toxicity through photodegradation) to humans; however, citrate-coated Ag NPs in powder form were toxic (Khot et al. 2012). Another interesting observation was that the phototoxicity of the powdered Ag NPs was repressed by coating them with biocompatible polyvinylpyrrole. Such biocompatible coatings should be explored to reverse the toxicity of nanomaterials to increase the chances of applying nanomaterials in plant germination and growth of plants (Khot et al. 2012). Studies are also required to be conducted to investigate the adverse effect of such coatings on the desired seeds and plants properties and the effectiveness of nanomaterial. The benefits and risks of advances in nanotechnology relevant to food and agriculture have been reviewed (Agrawal and Rathore 2014; Sadeghi et al. 2017).

21.1 Conclusions

The indiscriminate and extensive application of fertilizers and pesticides for boosting agricultural production to feed ever-increasing human population has greatly contributed in polluting air, water, food, and soil. The use of NPs as pesticides and fertilizers with sustained slow release is anticipated to reduce the dosage of agrochemicals that has given expectation as the future technology. Continuous innovations are required to meet the challenges of increasing global food security and to mitigate climate change. However the changed physicochemical properties of nanomaterial provide easy excess to biological systems and environment which raises safety issues. The assessment of the probable hazards by the interaction of NPs with biological systems and the environment needs to be considered before adopting particular nanomaterial for the application in the field. The potential applications of nanomaterial in the fields of food and agriculture with probable risks have been reviewed as future perspectives of nanotechnology.

References

Abd-elsalam KA (2013) Fungal genomics and biology nanoplatforms for plant pathogenic fungi management. Fungal Genomics Biol 2:e107

Abobatta WF (2018) Nanotechnology application in agriculture. Acta Sci Agric 2(6):99–102

Adhikari T, Biswas AK, Kundu S (2010) Nanofertilizer- a new dimension in agriculture. Indian J Fertil 6(8):22–24

Adhikari T, Kundu S, Biswas AK et al (2012) Effect of copper oxide nano particle on seed germination of selected crops. J Agri. Sci Technol 2:815–823

Agrawal S, Rathore P (2014) Nanotechnology pros and cons to agriculture: a review. Int J Curr Microbiol Appl Sci 3(3):43–55

Aguilar-Méndez MA, Martín-Martínez ES, Ortega-Arroyo L et al (2011) Synthesis and characterization of silver nanoparticles: effect on phytopathogen *Colletotrichum gloeosporioides*. J Nanopart Res 13:2525–2532

Ali S, Shafique O, Mahmood T et al (2018) A review about perspectives of nanotechnology in agriculture. Pak J Agric Res 31(2):116–121

Al-Samarrai AM (2012) Nanoparticles as alternative to pesticides in management plant diseases-a review. Int J Sci Res Publ 2(4):1–4

Aslani F, Bagheri S, MuhdJulkapli N et al (2014) Effects of engineered nanomaterials on plants growth: an overview. Sci World J 2014:641759

Atha DH, Wang H, Petersen EJ et al (2012) Copper oxide nanoparticle mediated DNA damage in terrestrial plant models. Environ Sci Technol 46(3):1819–1827

Aziz N, Faraz M, Pandey R et al (2015) Facile algae-derived route to biogenic silver nanoparticles: synthesis, antibacterial and photocatalytic properties. Langmuir 31:11605–11612

Azizi S, Namvar F, Mahdavi M et al (2013) Biosynthesis of silver nanoparticles using brown marine macroalga, *Sargassum muticum* aqueous extract. Materials (Basel) 6:5942–5950

Balaure PC, Gudovan D, Gudovan I (2017) Nanopesticides: a new paradigm in crop protection. New Pestic Soil Sens 12:129–192

Barik T, Sahu B, Swain V (2008) Nanosilica- from medicine to pest control. Parasitol Res 103(2):253–258

BegümKarakoçak B, Raliya R, Davis JT et al (2016) Biocompatibility of gold nanoparticles in retinal pigment epithelial cell line. Toxicol Vitro 37:61–69

Bernhardt ES, Colman BP, Hochella MF Jr et al (2010) An ecological perspective on nanomaterial impacts in the environment. J Environ Qual 39(6):1954–1965

Bertrand N, Leroux JC (2012) The journey of a drug-carrier in the body: an anatomo-physiological perspective. J Control Release 161(2):152–163

Bharathi P, Balasubramanian N, Anitha S et al (2016) Improvement of membrane system for water treatment by synthesized gold nanoparticles. J Environ Biol 37:1407–1414

Bhattacharya S, Jang J, Yang L et al (2007) Biomems and nanotechnology-based approaches for rapid detection of biological entities. J Rapid Methods Automat Microbiol 15:1–32

Bhattacharyya A, Bhaumik A, Rani PU et al (2010) Nanoparticles – a recent approach to insect pest control. Afr J Biotechnol 9(24):3489–3493

Bhattacharyya A, Duraisamy P, Govindarajan M et al (2016) Nano-biofungicides: emerging trend in insect pest control. In: Prasad R (ed) Advances and applications through fungal nanobiotechnology. Springer, Cham, pp 307–319. https://doi.org/10.1007/978-3-319-42990-8_15

Bonilla JC, Bozkurt F, Ansari S et al (2016) Applications of quantum dots in food science and biology. Trends Food Sci Technol 53:75–89

Chinnamuthu CR, Boopathi PM (2009) Nanotechnology and agroecosystem. Madras Agric J96:17–31
Chowdappa P, Gowda S (2013) Nanotechnology in crop protection: status and scope. Pest Manag Horticult Ecosyst 19(2):131–151
Cota-Arriola O, Cortez-Rocha MO, Burgos-Hernández A et al (2013) Controlled release matrices and micro/nanoparticles of chitosan with antimicrobial potential: development of new strategies for microbial control in agriculture. J Sci Food Agric 93:1525–1536
Dar MI, Chandiran AK, Gratzel M et al (2014) Controlled synthesis of TiO2 nanoparticles and nanospheres using a microwave assisted approach for their application in dyesensitized solar cells. J Mater Chem 2:1662–1667
Das M, Saxena N, Dwivedi PD (2009) Emerging trends of nanoparticles application in food technology: safety paradigms. Nanotoxicology 3(1):10–18
Das VL, Thomas R, Varghese RT et al (2014) Extracellular synthesis of silver nanoparticles by the Bacillus strain CS 11 isolated from industrialized area. 3 Biotech 4:121–126
Dasgupta N, Ranjan S, Mundekkad D et al (2015) Nanotechnology in agro-food: from field to plate. Food Res Int 69:381–400
de Barros CHN, Cruz FGC, Mayrink W et al (2018) Bio-based synthesis of silver nanoparticles from orange waste: effects of distinct biomolecule coatings on size, morphology, and antimicrobial activity. Nanotechnol Sci Appl 11:1–14
DeRosa MC, Monreal C, Schnitzer M et al (2010) Nanotechnology in fertilizers. Nat Nanotechnol 5:91–94
Duhan JS, Kumar R, Kumar N et al (2017) Nanotechnology: the new perspective in precision agriculture. Biotechnol Rep 15:11–23
Elbeshehy EKF, Elazzazy AM, Aggelis G (2015) Silver nanoparticles synthesis mediated by new isolates of Bacillus spp., nanoparticle characterization and their activity against Bean Yellow Mosaic Virus and human pathogens. Front Microbiol 6:453
Elizabath A, Bahadur V, Misra P et al (2017) Effect of different concentrations of iron oxide and zinc oxide nanoparticles on growth and yield of carrot (*Daucus carota* L.). J Pharma Phytochem 6(4):1266–1269
Elmer W, White JC (2018) The future of nanotechnology in plant pathology. Annu Rev Phytopathol 56:111–133
Fakruddin M, Hossain Z, Afroz H (2012) Prospects and applications of nano biotechnology: a medical perspective. J Nanobiotechnol 10(1):31
Feizi H, Shahram A, Farzin A et al (2013) Comparative effects of nanosized and bulk titanium dioxide concentrations on medicinal plant *Salvia officinalis* L. Ann Rev Res Biol 3(4):814–824
Flood J (2010) The importance of plant health to food security. Food Secur 2:215–231
Fraceto LF, Grillo R, de Medeiros GA et al (2016) Nanotechnology in agriculture: which innovation potential does it have? Front Environ Sci 4:20
Ghormade V, Deshpande MV, Paknikar KM (2011) Perspectives for nano-biotechnology enabled protection and nutrition of plants. Biotechnol Adv 29(6):792–803
Goswami A, Roy I, Sengupta S et al (2010) Novel applications of solid and liquid formulations of nanoparticles against insect pests and pathogens. Thin Solid Films 519:1252–1257
Gottschalk F, Nowack B (2011) The release of engineered nanomaterials to the environment. J Environ Monit 13(5):1145–1155
Gurunathan S, Han JW, Park JH et al (2014) A green chemistry approach for synthesizing biocompatible gold nanoparticles. Nanoscale Res Lett 9:248
He L, Liu Y, Mustapha A et al (2011) Antifungal activity of zinc oxide nanoparticles against *Botrytis cinerea* and *Penicillium expansum*. Microbiol Res 166:207–215
Iravani S, Korbekandi H, Mirmohammadi SV et al (2014) Synthesis of silver nanoparticles: chemical, physical and biological methods. Res Pharma Sci 9(6):385–406
Jain D, Kothari S (2014) Green synthesis of silver nanoparticles and their application in plant virus inhibition. J Mycol Plant Pathol 44(1):21–24
Jain N, Bhargava A, Tarafdar JC et al (2013) A biomimetic approach towards synthesis of zinc oxide nanoparticles. Appl Microbiol Biotechnol 97:859–869

Janmohammadi M, Amanzadeh T, Sabaghnia N et al (2016) Impact of foliar application of nano-micronutrient fertilizers and titanium dioxide nanoparticles on the growth and yield components of barley under supplemental irrigation. Acta Agric Slov 107(2):265–276

Jo YK, Kim BH, Jung G (2009) Antifungal activity of silver ions and nanoparticles on phytopathogenic fungi. Plant Dis 93(10):1037–1043

Karthik L, Kumar G, Vishnu Kirthi A et al (2014) *Streptomyces* sp. LK3 mediated synthesis of silver nanoparticles and its biomedical application. Bioprocess Biosyst Eng 37:261–267

Khandelwal A, Joshi R (2018) Synthesis of nanoparticles and their application in agriculture. Acta Sci Agric 2(3):10–13

Khodakovskaya M, Dervishi E, Mahmood M et al (2009) Carbon nanotubes are able to penetrate plant seed coat and dramatically affect seed germination and plant growth. ACS Nano 3:3221–3227

Khodakovskaya MV, de Silva K, Biris AS et al (2012) Carbon nanotubes induce growth enhancement of tobacco cells. ACS Nano 6:2128–2135

Khot LR, Sankaran S, Maja JM et al (2012) Applications of nanomaterials in agricultural production and crop protection: a review. Crop Prot 35:64–70

Kim SW, Kim KS, Lamsal K et al (2009) An in vitro study of the antifungal effect of silver nanoparticles on oak wilt pathogen *Raffaelea* sp. J Microbiol Biotechnol 19(8):760–764

Krishnaraj C, Ramachandran R, Mohan K et al (2012) Optimization for rapid synthesis of silver nanoparticles and its effect on phytopathogenic fungi. Spectrochim Acta Part A Mol Biomol Spectrosc 93:95–99

Lamsal K, Kim SW, Jung JH et al (2011a) Application of silver nanoparticles for the control of *Colletotrichum* species *in vitro* and pepper anthracnose disease in field. Mycobiology 39:194–199

Lamsal K, Kim SW, Jung JH et al (2011b) Inhibition effects of silver nanoparticles against powdery mildews on cucumber and pumpkin. Mycobiology 39:26–32

Laware SL, Raskar S (2014) Influence of zinc oxide nanoparticles on growth, flowering and seed productivity in onion. Int J Curr Microbiol Appl Sci 3(7):874–881

Lee WM, An YJ, Yoon H et al (2008) Toxicity and bioavailability of copper nanoparticles to the terrestrial plants mung bean (*Phaseolus radiatus*) and wheat (*Triticum aestivum*): plant agar test for water-insoluble nanoparticles. Environ Toxicol Chem 27:1915–1921

Liga MV, Bryant EL, Colvin VL et al (2011) Virus inactivation by silver doped titanium dioxide nanoparticles for drinking water treatment. Water Res 45(2):535–544

Lin D, Xing B (2007) Phytotoxicity of nanoparticles: inhibition of seed growth. Environ Pollut 150:243–250

Lu CM, Zhang CY, Wen JQ et al (2002) Research on the effect of nanometer materials on germination and growth enhancement of *Glycine max* and its mechanism. Soybean Sci 21(3):168–171

Mahmoodzadeh H, Nabavi M, Kashefi H (2013) Effect of nanoscale titanium dioxide particles on the germination and growth of canola (*Brassica napus*). J Ornamental Hort Plants 3(1):25–32

Mansfield J, Genin S, Magori S et al (2012) Top 10 plant pathogenic bacteria in molecular plant pathology. Mol Plant Pathol 13(6):614–629

Meetoo D (2011) Nanotechnology and the food sector: from the farm to the table. Emirates J Food Agric 23(5):387–403

Milani N, Hettiarachchi GM, Kirby JK et al (2015) Fate of zinc oxide nanoparticles coated onto macronutrient fertilizers in an alkaline calcareous soil. PLoS One 10(5):e0126275

Min JS, Kim KS, Kim SW et al (2009) Effects of colloidal silver nanoparticles on sclerotium-forming phytopathogenic fungi. J Plant Pathol 25:376–380

Misra AN, Misra M, Singh R (2013) Nanotechnology in agriculture and food industry. Int J Pure Appl Sci Technol 16:1–9

Moghaddam AB, Namvar F, Moniri M et al (2015) Nanoparticles biosynthesized by fungi and yeast: a review of their preparation, properties, and medical applications. Molecules 20:16540–16565

Moore MN (2006) Do nanoparticles present ecotoxicological risks for the health of the aquatic environment ? Environ Int 32:967–976

Mukhopadhyay SS (2014) Nanotechnology in agriculture: prospects and constraints. Nanotechnol Sci Appl 7:63–71

Naderi MR, Abedi A (2012) Application of nanotechnology in agriculture and refinement of environmental pollutants. J Nanotechnol 11(1):18–26

Naderi MR, Danesh-Shahraki A (2013) Nanofertilizers and their roles in sustainable agriculture. Int J Agric Crop Sci 5(19):2229–2232

Nair PMG, Chung IM (2014) Impact of copper oxide nanoparticles exposure on *Arabidopsis thaliana* growth, root system development, root lignification and molecular level changes. Environ Sci Pollut Res 21:12709–12722

Nair R, Varghese SH, Nair BG et al (2010) Nanoparticulate material delivery to plants. Plant Sci 179:154 163

Naseem T, Farrukh MA (2015) Antibacterial activity of green synthesis of iron nanoparticles using *Lawsonia inermis* and *Gardenia jasminoides* leaves extract. J Chem 2015:912342

Nuruzzaman M, Rahman MM, Liu Y et al (2016) Nanoencapsulation, nano-guard for pesticides: a new window for safe application. J Agric Food Chem 64:1447–1483

Oliveira HC, Stolf-Moreira R, Martinez CBR et al (2015a) Nanoencapsulation enhances the post-emergence herbicidal activity of atrazine against mustard plants. PLoS One 10:e0132971

Oliveira HC, Stolf-Moreira R, Martinez CBR et al (2015b) Evaluation of the side effects of poly (epsilon- caprolactone) nanocapsules containing atrazine to ward maize plants. Front Chem 3:61

Ottoni CA, Simoes MF, Fernandes S et al (2017) Screening of filamentous fungi for antimicrobial silver nanoparticles synthesis. AMB Express 7:31

Ouda SM (2014) Antifungal activity of silver and copper nanoparticles on two plant pathogens, *Alternaria alternata* and *Botrytis cinerea*. Res J Microbiol 9(1):34–42

Owolade OF, Adenekan MO (2008) Effects of titanium dioxide on the diseases, development and yield of edible cowpea. J Plant Prot Res 48(3):329–335

Owolade OF, Ogunleti DO, Adenekan MO (2008) Titanium dioxide affects diseases, development andyield of edible cowpea. J Plant Protect Res 7(5):2942–2947

Papp S, Patakfalvi R, Dékánya I (2007) Formation and stabilization of noble metal nanoparticles. Croat Chem Acta 80(3–4):493–502

Paret ML (2013) Evaluation of a light-activated nanoparticle formulation of titanium dioxide with zinc for management of bacterial leaf spot on rosa 'Noare'. Hortic Sci 48(2):189–192

Parisi C, Vigani M, Rodríguez-Cerezo E (2015) Agricultural nanotechnologies: what are the current possibilities? Nano Today 10:124–127

Park HJ, Kim SH, Kim HJ et al (2006) A new composition of nanosized silica-silver for control of various plant diseases. Plant Pathol J 22(3):295–302

Patra JK, Baek KH (2017) Antibacterial activity and synergistic antibacterial potential of biosynthesized silver nanoparticles against foodborne pathogenic bacteria along with its anticandidal and antioxidant effects. Front Microbiol 8:167

Patra P, Goswami A (2012) Zinc nitrate derived nano ZnO: fungicide for disease management of horticultural crops. Int J Innovat Hortic 1:79–84

Pestovsky YS, Antonio AM (2017) The use of nanoparticles and nanoformulations in agriculture. J Nanosci Nanotechnol 17:8699–8730

Prasad TNVKV, Sudhakar P, Sreenivasulu Y et al (2012) Effect of nanoscale zinc oxide particles on the germination, growth and yield of peanut. J Plant Nutr 35:905–927

Prasad R, Bhattacharyya A, Nguyen QD (2017) Nanotechnology in sustainable agriculture: recent developments, challenges, and perspectives. Front Microbiol 8:1014

Rafique M, Sadaf I, Rafique MS et al (2017) A review on green synthesis of silver nanoparticles and their applications. Artif Cells Nanomed Biotechnol 45:1272–1291

Rai M, Ingle A (2012) Role of nanotechnology in agriculture with special reference to management of insect pests. Appl Microbiol Biotechnol 94:287–293

Rajeshkumar S, Malarkodi C, Paulkuma K et al (2014) Algae mediated green fabrication of silver nanoparticles and examination of its antifungal activity against clinical pathogens. Int J Met 2014:692643

Raliya R, Tarafdar JC, Mahawar H et al (2014a) ZnO nanoparticles induced exopolysaccharide production by *B. subtilis* strain JCT1 for arid soil applications. Int J Biol Macromol 65:362–368
Raliya R, Tarafdar JC, Singh SK et al (2014b) MgO nanoparticles biosynthesis and its effect on chlorophyll contents in the leaves of clusterbean (*Cyamopsis tetragonoloba* L.). Adv Sci Eng Med 6:1–8
Raliya R, Biswas P, Tarafdar JC (2015) TiO2 nanoparticle biosynthesis and its physiological effect on mung bean (*Vigna radiata* L.). Biotechnol Rep 5:22–26
Raliya R, Saharan V, Dimkpa C, Biswas P (2017) Nanofertilizer for precision and sustainable agriculture: current state and future perspectives. J Agric Food Chem 66(26):6487. https://doi.org/10.1021/acs.jafc.7b02178
Ramesh S (2013) Sol–gel synthesis and characterization of nanoparticles. J Nanosci 2013:929321
Ramesh P, Rajendran A, Meenakshisundaram M (2014) Green synthesis of zinc oxide nanoparticles using flower extract Cassia auriculata. J Nanosci Nanotechnol 1(1):41–45
Rameshaiah GN, Pallavi J, Shabnam S (2015) Nano fertilizers and an attempt for developing smart agriculture. Int J Eng Res Gen Sci 3(1):314–320
Raskar S, Laware SL (2013) Effect of titanium dioxide nano particles on seed germination and germination indices in onion. Plant Sci Feed 3(9):103–107
Rathore P, Phanse N, Patel B (2012) Nanomaterials: a future concern. Int J Res Chem Environ 2(2):1–7
Rawat A, Kumar R, Bhatt B et al (2018) Nanotechnology in agriculture- a review. Int J Curr Microbiol Appl Sci 7(8):969–978
RIKILT, JRC (2014) Inventory of nanotechnology applications in the agricultural, feed and food sector. EFSA supporting publication 2014: EN-621, 125 pp
Sadaf J, Shah GA, Shahzad K et al (2017) Improvements in wheat productivity and soil quality can accomplish by co-application of biochars and chemical fertilizers. Sci Tot Environ 607–608:715–724
Sadeghi R, Kalbasi A, Emam-jomeh Z et al (2013) Biocompatible nanotubes as potential carrier for curcumin as a model bioactive compound. J Nanopart Res 15(11):1931
Sadeghi R, Moosavi-Movahedi AA, Emam-jomeh Z et al (2014) The effect of different desolvating agents on BSA nanoparticle properties and encapsulation of curcumin. J Nanopart Res 16(9):2565
Sadeghi R, Rodriguez RJ, Yao Y et al (2017) Advances in nanotechnology as they pertain to food and agriculture: benefits and risks. Annu Rev Food Sci Technol 8:467–492
Shah V, Belozerova I (2009) Influence of metal nanoparticles on the soil microbial community and germination of lettuce seeds. Water Air Soil Pollut 197:143–148
Sharon M, Choudhary AK, Kumar R (2010) Nanotechnology in agricultural diseases and food safety. J Phytology 2(4):83–92
Shende S, Ingle AP, Gade A et al (2015) Green synthesis of copper nanoparticles by *Citrus medica* Linn. (Idilimbu) juice and its antimicrobial activity. World J Microbiol Biotechnol 31(6):865–873
Shvedova AA, Kagan VE, Fadeel B (2010) Close encounters of the small kind: adverse effects of man-made materials interfacing with the nano-cosmos of biological systems. Annu Rev Pharmacol Toxicol 50:63–88
Siddiqui MH, Al-Whaibi MH, Firoz M, Al-Khaishany MY (2015) Role of nanoparticles in plants. Nanotechnol Plant Sci:19–35. https://doi.org/10.1007/978-3-319-14502-0_2
Singh S (2012) Achieving second green revolution through nanotechnology in India. Agric Situations India 3:545–572
Singh P, Kim YJ, Zhang D et al (2016) Biological synthesis of nanoparticles from plants and microorganisms: Review. Trends Biotechnol 34(7):588–599
Singh SK, Pathak R, Kalia RK (2019) Perspectives of nanomaterial in sustainable agriculture. In: Raliya R (ed) Nanoscale engineering in agricultural management. CRC Press/Taylor & Francis Group, Boca Raton, pp 64–79

Sinha K, Ghosh J, Sil PC (2017) New pesticides: a cutting-edge view of contributions from nanotechnology for the development of sustainable agricultural pest control. In: Grumezescu AM (ed) New pesticides and soil sensors. Academic Press, Cambridge, MA, pp 47–49

Sozer N, Kokini JL (2014) Use of quantum nanodot crystals as imaging probes for cereal proteins. Food Res Int 57:142–151

Suppan S (2013) Nanomaterials in soil: our future food chain? Institute of Agriculture and Trade Policy, Minneapolis, pp 1–16

Suresh D, Shobharani RM, Nethravathi PC et al (2015) *Artocarpus gomezianus* aided green synthesis of ZnO nanoparticles: luminescence, photocatalytic and antioxidant properties. Spectrochim Acta A Mol Biomol Spectrosc 141:128–134

Tarafdar JC, Raliya R, Mahawar H (2014) Development of zinc nanofertilizer to enhance crop production in pearl millet (*Pennisetum americanum*). Agric Res 3(3):257–262

Thangavelu RM, Gunasekaran D, Jesse MI et al (2018) Nanobiotechnology approach using plant rooting hormone synthesized silver nanoparticle as "nanobullets" for the dynamic applications in horticulture – An in vitro and ex vitro study. Arab J Chem 11:48–61

Tzoumaki MV, Karefyllakis D, Moschakis T et al (2015) Aqueous foams stabilized by chitin nanocrystals. Soft Matter 11(31):6245–6253

Ulrichs C, Mewis I, Goswami A (2005) Crop diversification aiming nutritional security in West Bengal- biotechnology of stinging capsules in nature's water-blooms. Annu Tech Issue State Agric Technol' Service Assoc 10:1–18

Vinutha JS, Bhagat D, Bakthavatsalam N (2013) Nanotechnology in the management of polyphagous pest *Helicoverpa armigera*. J Acad Indus Res 1(10):606–608

Waghmare SS, Deshmukh AM, Sadowski Z (2014) Biosynthesis, optimization, purification and characterization of gold nanoparticles. Afr J Microbiol Res 8:138–146

Wang C, Kim YJ, Singh P et al (2016) Green synthesis of silver nanoparticles by Bacillus methylotrophicus, and their antimicrobial activity. Artif Cells Nanomed Biotechnol 44(4):1127–1132

Wani IA, Gangulya A, al AJ (2011) Silver nanoparticles: ultrasonic wave assisted synthesis, optical characterization and surface area studies. Mater Lett 65:520–522

Warheit DB, Sayes CM, Reed KL et al (2008) Health effects related to nanoparticle exposures: environmental, health and safety considerations for assessing hazards and risks. Pharmacol Ther 120(1):35–42

Worrall EA, Hamid A, Mody KT et al (2018) Nanotechnology for plant disease management. Agronomy 8:285

Xingmao M, Geiser-lee J, Deng Y et al (2010) Interactions between engineered nanoparticles ENPs and plants: Phytotoxicity, uptake and accumulation. Sci Total Environ 408:3053–3061

Yang L, Watts DJ (2005) Particle surface characteristics may play an important role in phytotoxicity of alumina nanoparticles. Toxicol Lett 158(2):122–132

Yang FL, Li XG, Zhu F et al (2009) Structural characterization of nanoparticles loaded with garlic essential oil and their insecticidal activity against *Tribolium castaneum* (Herbst) (Coleoptera: Tenebrionidae). J Agric Food Chem 57:10156–10162

Yata VK, Tiwari BC, Ahmad I (2018) Nanoscience in food and agriculture: research, industries and patents. Environ Chem Lett 16(1):79–84

Zahir AA, Chauhan IS, Bagvan A et al (2015) Green synthesis of silver and titanium dioxide nanoparticles using Euphorbia prostrata extract shows shift from apoptosis to G0/G1 arrest followed by necrotic cell death in *Leishmania donovani*. Antimicrob Agents Chemother 59:4782–4799

Zheng L, Hong F, Lu S et al (2005) Effect of nano-TiO_2 on strength of naturally aged seeds and growth of spinach. Biol Trace Element Res 104:83–91

A Missing Dilemma on Nanoparticle Producer Microorganisms

22

R. Soner Silme and Ömür Baysal

Abstract

The development of eco-friendly biological methods in material synthesis has been reported with chemically well-defined variety of inorganic nanoparticles (NP) that are produced by using various microorganisms. In last decades, lots of research articles have suggested required conditions to control and particle stability on biosynthesized nanoparticles, besides their applications in a wide spectrum of potential fields including target oriented drug delivery, cancer therapy, gene therapy and DNA based diagnosis, using of antimicrobial agents, biosensors, enhancing enzymatic reaction capacity with advanced medical visualization technology. Even the present limitations and future prospects for the production of inorganic nanoparticles by microorganisms are dramatically studied, their disadvantages in practice concerning their negative effects on micro- and macroorganisms are attracting the attention of researchers. As another concept, the behaviours of microorganism change depending on available concentration of nanomolecules containing inorganic chemical structures in environment affecting their antibacterial compounds secretion. The review highlights particularly ignored or missed cases on the usage of nanoparticle producer microorganisms. We briefly discuss here, as an another concept; enhancing anti-phytopathogen potential capacity of soil can negatively be affected by NP synthesizing microorganisms that may drastically impair microflora balance and its own biocontrol capacity besides in contrary to their expected positive advantages in purpose of their antimicrobial property.

R. S. Silme
Center for Research and Practice in Biotechnology and Genetic Engineering, Istanbul University, Istanbul, Turkey
e-mail: rsoner@istanbul.edu.tr

Ö. Baysal (✉)
Faculty of Science, Department of Molecular Biology and Genetics/Molecular Microbiology Unit, Muğla Sıtkı Koçman University, Muğla, Turkey
e-mail: omurbaysal@mu.edu.tr

M. Ghorbanpour et al. (eds.), *Biogenic Nano-Particles and their Use in Agro-ecosystems*, https://doi.org/10.1007/978-981-15-2985-6_22

Keywords
Biological nanoparticles · Microorganisms · Biocontrol · Ecological approach

22.1 Introduction

Nanoparticles—particles having one or more dimensions and a ranging width of 100 nm or less—have attracted attention due to their useful properties and advantages for easy application (Kato 2011). Different types of nanoparticles are synthesized via a large number of physical, chemical and biological methods (Luechinger et al. 2010; Liu et al. 2011). Even though physical and chemical methods are more popular in the synthesis of nanoparticles, the use of toxic chemicals restricts their applications. Therefore, development of safe, non-poisonous, and eco-friendly methods for the synthesis of nanoparticles is of importance to expand their application areas. One of the ways in purpose of this goal is to synthesize nanoparticles by microorganisms to diminish their possible negative impact on the environment in the long term.

22.2 Biological Synthesis of Nanoparticles by Microorganisms

In earlier years, union of nanoparticles utilizing microorganisms has expanded thoroughly because of its enormous practice. *Bacillus* species has portrayed to use of metal nanoparticles; analysts demonstrated the capacity of microorganisms to diminish silver and create extracellularly, reliably circled nanoparticles, with 10–20 nm in size (Sunkar and Nachiyar 2012). A new *Bacillus subtilis* (EU07) strain (Baysal et al. 2008, 2013; Baysal and Silme 2018) shows also similar property on nanoparticle production (unpublished data). The silver delivering microscopic organisms detached from the silver mine show the silver nanoparticles collected in the periplasmic side of *Pseudomonas stutzeri* AG259 (Slawson et al. 1994). Microbes are additionally familiar with blending gold nanoparticles. Sharma et al. (2012) revealed that a novel strain of *Marinobacter pelagius* has been acknowledged as appropriate for durable, monodisperse gold nanoparticle arrangement. Prasad et al. (2007) have been accounted for utilization of *Lactobacillus* strains to blend the titanium nanoparticles.

22.3 Soil Property and Condition Affecting Nanoparticle Production

In light of the previously mentioned reports and our previous studies, we accept that a comprehension of the natural effect of nanoparticles discharged in agrobiological system that must be incorporated into the examination of essential hazard appraisal factors during the tripartite associations of nanoparticles with plant, soil, and soil microbial network (Table 22.1). We assume that decomposing inorganic elements in soil have active function on microbial balance, particularly on anti-phytopathogen

Table 22.1 The possible interactions and effect of nanomaterials on soil microbes and plant under varying soil physico-chemical properties (Mishra et al. 2017)

Soil parameters	Nanomaterials	Major findings	References
Soil types			
Silty clay	TiO_2	Significantly lowered carbon mineralization	Simonin et al. (2015)
Sandy loam	TiO_2	Adverse impact on soil microbial community	Simonin et al. (2015)
	CuO, Fe_3O_4	Negative effect on soil microbial community	Frenk et al. (2013)
	ZnO	No toxicity on *Cucumis sativus* with soil pH 5.5 and at concentration of 2000 mg/kg	Kim et al. (2011)
	CuO, ZnO	Toxic effect on *Triticum aestivum*	Dimkpa et al. (2013)
	AgNPs	Reduced microbial biomass	Hänsch and Emmerling (2010)
		Reduced soil enzymatic activities and substrate induced	Colman et al. (2013)
	CeO_2, Fe_3O_4, SnO_2	No effect on microbial biomass C and N	Vittori Antisari et al. (2013)
	TiO_2	Reduced bacterial diversity	Ge et al. (2013)
	TiO_2 and ZnO	Reduced microbial biomass and substrate-induced respiration	Ge et al. (2011)
	TiO_2; ZnO	Altered soil bacterial community with reduced taxa	Ge et al. (2012)
Loamy clay	ZnO	Toxic effect on *Triticum aestivum* in soil pH 7.36 and at concentration of 45.45 mg/kg	Du et al. (2011)
pH			
Acidic	AgNPs, ZnO	Enhanced toxicity towards *Eisenia fetida* adverse effect on ammonification, respiration, and dehydrogenase activity of soil microbes	Shoults-Wilson et al. (2011) and Shen et al. (2015)
Alkaline	TiO_2	Significant reduction in soil microbial community	Simonin et al. (2015)
	AgNPs	Declined toxicity towards soil microbial activity	Schlich and Hund-Rinke (2015)
Organic matter			
High	AgNPs	Reduced toxicity towards biofilm-forming communities	Sheng and Liu (2011) and Wirth et al. (2012)
	TiO_2	Toxic effect on microbial activity, i.e. carbon mineralization	Simonin et al. (2015)
	ZnO	Positive impact on *Zea mays* when alginate added at concentration of 400–800 mg/kg	Zhao et al. (2013)

(continued)

Table 22.1 (continued)

Soil parameters	Nanomaterials	Major findings	References
Low	CuO, Fe_3O_4	Enhanced toxicity towards microbial community	Frenk et al. (2013)
Cation exchange capacity			
High	AgNPs	Reduced toxic impact on soil bacterium *Pseudomonas chlororaphis* O6	Calder et al. (2012)
	ZnO	Non-toxic effect on *Lepidium sativum*	Jośko and Oleszczuk (2013)
Low	AgNPs	Enhanced toxicity towards soil microbes	Calder et al. (2012)

potential changes and limit the growth of pathogenic invasion. Without a doubt, the main soil factors are soil type, pH, natural balance, and contamination quantity of discharged nanoparticles towards plants and microorganisms.

22.4 Efficiency of Biocontrol Agents Through Nanoparticle Production for Crop Protection

22.4.1 Nano-pesticides

Pesticides are chemical compounds that generally used to dispose and control the harmful organisms causing huge financial losses in crop production. The plant irritations influencing the refined items lead to smart misfortunes by restricting the item yield. Auxiliary metabolites (alkaloids, phenolics, terpenoids, etc.) discharged by plant as self-conservation instrument of nature give safeguard and defensive capacity against plant pathogens. In control strategies, use of too much pesticide causes noteworthy financial loses (Sharon et al. 2010). In spite of the fact that there are a few advantages of pesticide using, that can also cause serious issues on abiotic and biotic conditions, nature and human wellbeing because of potential contamination efficiency (environmental and water resources pollution, toxicological remnants etc.).

In this way, researchers are taking efforts to comprehend and investigate the degree of these significant cases to increase utility of learning about untargetted contamination effect and the likely effect of discharged nanoparticles on soil condition and crop management. The plant-soil communication is the primary main thrust for agricultural activities, which is affected by any adjustment in physicochemical properties of soil framework. Strikingly, soil is really the fundamental flora of discharged nanoparticles, and thus, their resulting communication with diverse soil segments could have a significant effect on the targeting, transporting, and conducting of nanoparticles. For example, previous reports about antimicrobial property of the most prominent and most concentrated silver nanoparticles (AgNPs) have plainly shown the significance of soil pH, natural issue substance, and cation trade limit in controlling their destiny, lethality, and bioavailability (Shoults-Wilson et al. 2011).

As noted above, nanoparticles present on earth eventually give rise to contamination; once the convenient conditions have not been appropriate for decomposing, these transported soil particles accumulate on plants. Moreover, the vast majority of the current studies have additionally focused on the immediate effect of discharged nanoparticles on soil microbial network structure (Simonin and Richaume 2015). In such manner, at first, Ge et al. (2012) imagined the possible effect of TiO_2 and ZnO-NPs on soil bacterial network in a portion subordinate way. Utilizing from DNA-based fingerprinting examination, they watched bacterial assorted variety decreasing taxa of *Rhizobiales*, *Bradyrhizobiaceae*, and *Bradyrhizobium* (identified with nitrogen obsession) in light of these nanoparticle treatments. In any case, some positive effects on the growth of *Sphingomonadaceae* and *Streptomycetaceae* was noted. It is intriguing to note here that TiO_2 and ZnO-NPs have together changed the bacterial network structure with unmistakable effect on ecological procedures. For example, the declining population are intently connected with nitrogen obsession process though expanding of community that is probably going to affect the deterioration procedure of natural poisons and biopolymers. Further, Shahrokh et al. (2014) additionally uncovered portion subordinate effect of AgNPs on nitrate reductase movement of *Rhizobium* and *Azotobacter*, where low portion of AgNPs (0.2 ppm) encouraged to decrease of nitrate action in *Azotobacter*. In view of such examinations, it has been foreseen that the denitrifying bacterial network is thought to be very powerless to nanoparticles harmfulness (VandeVoort and Arai 2012). Regardless of the obviously known effect of nanoparticles on soil microbial network, there exists a lack of research findings giving clear association between soil factors and dangerous conduct of nanoparticles towards soil biota (Chunjaturas et al. 2014; Shah et al. 2014; Mishra and Singh 2015). In this specific situation, Frenk et al. (2013) prove the effect of copper oxide (CuO) and magnetite (Fe_3O_4) nanoparticles on soil bacterial network in two different soil types (sandy topsoil and sandy dirt soil). Curiously, increasingly antagonistic effects of both nanoparticles in sandy soil with CuO show generally solid impact on network synthesis and bacterial action. Hence *Rhizobiales* and *Sphingobacteriaceae*, being the most focused genus in microbial community, were adversely affected by CuO in sandy soils. Still constrained effects were also seen in sandy soil where 1% CuO diminished oxidative potential and Fe_3O_4 nanoparticles did not change the bacterial community structure (Schlich and Hund-Rinke 2015). Shen et al. (2015) exhibited ecotoxicological effects of ZnO-NPs on soil organisms based on parameters including ammonification, O_2 consumption, dehydrogenase movement, and fluorescein diacetate hydrolase action. The negative effects of ZnO-NPs on soil microorganisms were observed to be progressive in acidic and non-decomposed soil than that of ZnO, which had generally low harm in soluble soil. Likewise, danger of TiO_2 NPs was observed to be principally affected by soil pH and its natural impact has been reported by Simonin et al. (2015). The researcher observed remarkable decrease on carbon mineralization (parameter to contemplate microbial community) in soil with high pH and natural issue content. Above all, increasing our knowledge on the correct information on nanomaterial contamination level would help us to understand of nanoparticles contamination in the soil supporting its harmfulness.

Smart methodology on powerful control techniques and solutions in the yield by possibility of these conventions are vague and petulant because of natural dangers of contamination and dispersing of dangerous synthetic compounds (Singh et al. 2016a). Thus, picking eco-accommodating, non-poisonous, and maintainable techniques for manufacturing a horde of nanoparticles is the present zone of worldwide intrigue. Along these lines, a few organic operators, for example, microbes, parasites, actinomycetes, plants, and green growth, have been misused for the biosynthesis of nanoparticles (Mishra et al. 2014).

The prevalence of natural strategy for nanoparticle combination could be evaluated by how fast and stable is the entire procedure of combining, which requires a wide scope of non-dangerous biomolecules of minimal effort and above all it gives rise to increase of stable nanoparticles (Singh et al. 2016a; Hussain et al. 2016). In addition, shape and size of the nanoparticles can likewise be managed by altering the pH and temperature of the response blend (Gericke and Pinches 2006). Nanoparticles present an incredibly dazzling stage for a different scope of natural applications, as it draws more attention on the single-step procedure to biosynthesize nanoparticles for future improvements in the territory of electrochemical sensor, biosensors, medication, and agrobiotechnology.

In this survey, we combine beneficial side of nanoparticles as natural techniques. These techniques are condition well-disposed and financially monetary. Relationship of different combination strategies, to be specific physical, complex and biological techniques seems valid here. Further advances are alluring to spin the impression of nanoparticle innovation into a judicious reasonable methodology. The misuse of nanoparticles in beautifying agents and drug covering is generally expanded step by step. The metal oxide content in nanoparticle, for example, zinc oxide and titanium dioxide, is presently recorded on different items, as varied as cosmetics, sunscreens, toothpaste, and prescription drugs (Yu and Li 2011).

22.4.2 Nanoparticles in Horticulture

Nanotech conveyance frameworks for bugs, supplements, and plant hormones: In the capable utilization of agrarian regular resources like water, supplements, and synthetic compounds during accuracy cultivating, nanosensors and nano-based savvy conveyance frameworks are easy to use. Through utilizing nanomaterials and worldwide situating frameworks with satellite imaging of fields, farmers may now indirectly recognize pests or reasons for stress, especially during dry season. Nanosensors dispersed in the field can detect the presence of plant infections and the dimension of soil supplements. To set aside manure utilization and to limit ecological contamination, nanoencapsulated moderate discharge composts have additionally turned into a style (DeRosa et al. 2010). To check the nature of horticultural production, nanobarcodes and nanoprocessing could be utilized. Li et al. (2005) utilized the possibility of basic food item standardized identifications for conservative, capable, fast, and easy unravelling and acknowledgment of maladies. They made infinitesimal tests or nanobarcodes that may maybe label various pathogens in a ranch, which may essentially be identified utilizing any fluorescent-based apparatuses (Li et al. 2005). Right through nanotechnology, researchers are skilled to concentrate plant's guideline of hormones, for example, auxin, which is responsible for root development and seedling association. Nanosensors have been invented to follow the respond to auxin. This is a stage forward for auxin, as it enables researchers to realize how plant roots adjust to their condition, especially to negligible soils (McLamore et al. 2010).

Nanotechnology for yield biotechnology: Nanocapsules can encourage effective attack of herbicides, permitting moderate and ordinary release of the dynamic effective substances. This can go about as "enchantment shots", containing herbicides, synthetics, or qualities which target demanding plant parts to free their substance (Pérez-de-Luque and Rubiales 2009). Torney et al. (2007) have abused a 3 nm mesoporous silica nanoparticle in conveying DNA and synthetic substances into detached plant cells.

The biomaterial is a substance or combination of substances other than medication mixes which might be engineered or common in inception that can be utilized for any time allotment, which increments or replaces for the most part or completely any tissue, organ, or capacity, to keep up or improve the personal satisfaction. The word "biomaterial" limits any substance thought to cooperate with the biotic framework on living organism. The origin of biomaterials started in 1950, under the idea that the biomaterials ought to be idle or diminish negative response (side effects) of host tissue when embedded (Zavaglia and Prado da Silva 2016). From that point forward, there has been significant innovative work in the field of biomaterials prompting the creation of second- and third-age biomaterials. Moreover, the convergence of various research fields, for example, natural science, immunology, materials science, physical science, and building has opened a totally new measurement and innovation which is called "nanotechnology" (Prasad 2016).

Fast progression in the field of nanotechnology has reformed each part of sciences, including the biomaterials prompting the discovery of bionanomaterials. Improved physical systems like helium particle and electron infinitesimal strategies and nanofabrication have permitted nanosized gadget generation and investigation. They can be portrayed as atomic materials made out of natural mixes (e.g. antibodies, proteins, lipids, DNA, RNA, infections, and cell segments). The produced bionanomaterials may impact as novel strands and sensors. These sorts of structures may take into consideration of manufacturers as complex tools independent from known and ones previously used under delicate exploratory conditions (Honek 2013).

The expanding desire for sustainable condition is getting attention by researchers on biodegradable and biocompatible materials with the idea of "green" materials; in this specific circumstance, nanobiomaterials from horticultural waste might be considered as an alluring option having biodegradable, sustainable, or biocompatible properties other than being precisely solid, hardened, and profoundly crystalline with extraordinary warm dependability. Biodegradability, basic openness, and remarkable mechanical property of nanocellulose have pulled in a great deal of interest as a novel wellspring of nanometre-sized materials. Planned studies of nanocellulose for biodegradable materials and wide utilizations of nanocellulose in especially ecological field make nanocellulose as a key player in contamination. Nanocellulose extraction from agrarian squanders, similar to citrus and orange, is by all account promising substitutes for waste treatment. Moreover, a successful method for extraction and a couple of all the more wide utilizations of nanocellulose in organic science are greatly expected sooner rather than before. The high surface region, the rich wealth of practical gatherings, and other amazing properties make graphene oxide as one of the first class carbon mixes. The potential utilization of graphene oxide (GO) in different fields of science has been investigated. Composites of GO from sugarcane and rice remains has made eminent improvement in large-scale manufacturing of GO. Hence, accessible conventions for the combination of GO from agrowastes are in the earliest reference point of production; further, the cutting edge techniques for composites of eco-accommodating, financially savvy, agrowaste cause GO mixes that should be grown adequately. Creation of formless silicon nanoparticles has been discovered utilizing diverse agrowastes, for example, corn centre point, rice husk, and other plant sources. The upgrade in the harvest yield by diminishing the utilization of manure and pesticides (SiO_2NP) is one of the first class components for the improvement of crops.

Carbon being a crucial segment has been investigated for its potential use in the field of restorative and mechanical purposes. The wrong use of manufactured crude materials for the generation of nanostructured materials like carbon nanotubes and carbon nanofibers are of essential significance, and the use of characteristic substances that are of auxiliary or lesser significance is largely supported. At recent studies, one such methodology is by the usage of agricultural waste for the creation of carbon nanomaterials, which has been clarified previously. The discourse likewise underlines the significance of sorts of carbon and their uses in the field of therapeutics and nanoscience. The dialog likewise manages the segments of rural waste that could be used as promising instruments for the development of carbon nanoparticles, along these lines getting the most ideal of yield out of waste from farming area. Disregarding these potential focal points, combination of bionanomaterials from agrarian squanders is still equivalently overlooked and has not yet made it to the market segment to any degree in the examination with other present-day modern divisions. The deluge of research disclosures is by all account mostly declared by the little undertakings or the scholarly area; thus, there should be some positive contemplation to follow later on for the commercialization of these tip-top bionanomaterials in enormous scale.

22.5 Outlook

Noteworthy, the literature sources show some microorganisms causing plant disease and economical yield losses on crops. They are also capable of synthesizing NP and form new alloy components. (Please check ones showing with asterisk in Tables 22.2, 22.3, 22.4, and 22.5.) In these circumstances, we should consider the capacity of pathogenic microorganisms in view of NP production that may lower the beneficial microorganism community directing anti-phytopathogen potential of the soil in farming areas. As known there is no detailed information on network existing within NP producers-pathogenic microorganisms and beneficial (conferring the control of pathogenic species) microorganism profiles. In our previous review, a requirement of a detailed metagenomics data is emphasized to understand the complexity of soil microbial interactions occurring between NP concentration and microbial viability that is required for sustainable soil property (Baysal and Silme 2018). Additionally, NP concentration changing in soil does not depend on remnants source of NP application in purpose of plant protective implementations based on fungicides, insecticides and herbicides, residue of NP are processed by metallophilic microorganisms that converts to another form that is able to uptake by bacterial cells. As noted, sulphide and miscellaneous and oxide nanoparticles synthesized by microorganisms have been acknowledged. Within these microorganisms, major soilborne pathogens, *Fusarium oxysporum*, *Aspergillus* spp. and *Verticillium* spp. are producing NP by using heavy metals (Tables 22.2, 22.3, 22.4, 22.5). These properties have shown moderate approaches and microorganisms consist of numerous metal resistance gene clusters, which are able to cell detoxification via a number of mechanisms at high concentrations of mobile heavy metal ions. However, heavy metal ions, Hg^{2+}, Cd^{2+}, Ag^{+}, Co^{2+}, CrO_4^{2+}, Cu^{2+}, Ni^{2+}, Pb^{2+} and Zn^{2+} cause toxic effects, which is detrimental to the survival of beneficial microorganisms in soil that help us protecting our crops from phytopathogenic microorganisms as well. Treatments of NPs on soil lead to these negative effects that can be taken into account for diminishing of soil microflora capacity which should not be missed/ ignored playing major role in maintainable ecological balance. While NPs are suggested as an effective control of plant pathogens, in a long period introduction of NPs on soil microflora might prepare convenient deployment for emerging aggressive / resistant pathogenic strains against to known and registered chemical compounds commonly used in plant protection. Non-target application of NPs causing constraints on microflora in favour of pathogen might be the reason of uncontrolled dissemination of nanoparticles in the environment that create adversely possibility for the pathogens attacking to host plants. Inversely, the pathogenic growth occurs very rapidly that suppresses beneficial community of soil microflora since they have much easier ability to cell detoxification of metal ions compared to non-pathogenic individuals of soil microflora. The potential side effect of NP residues in soil varies according to soil physicochemicals depending on microflora profile. In case of emerging this convenient microflora, the pathogenic behaviour will take over the control making favour condition for its own growth independently. Therefore, in Table 22.1, the soil types and organic content affecting soil microflora are summarized.

Table 22.2 Metal nanoparticles synthesized by microorganisms (Li et al. 2011). Plant pathogens are marked with an asteriks (*)

Microorganisms	Products	Culturing temperature (°C)	Size (nm)	Shape	Location	References
Sargassum wightii	Au	Not available	8–12	Planar	Extracellular	Singaravelu et al. (2007)
Rhodococcus sp.	Au	37	5–15	Spherical	Intracellular	Ahmad et al. (2003a, b)
Shewanella oneidensis	Au	30	12 ± 5	Spherical	Extracellular	Suresh et al. (2011)
Plectonema boryanum	Au	25–100	<10–25	Cubic	Intracellular	Lengke et al. (2006a)
Plectonema boryanum UTEX 485	Au	25	10 nm–6 μm	Octahedral	Extracellular	Lengke et al. (2006b)
Candida utilis	Au	37	Not available	Not available	Intracellular	Gericke and Pinches (2006)
V. luteoalbum	Au	37	Not available	Not available	Intracellular	Gericke and Pinches (2006)
Escherichia coli	Au	37	20–30	Triangles, hexagons	Extracellular	Du et al. (2007)
Yarrowia lipolytica	Au	30	15	Triangles	Extracellular	Agnihotri et al. (2009)
Pseudomonas aeruginosa	Au	37	15–30	Not available	Extracellular	Husseiny et al. (2007)
Rhodopseudomonas capsulate	Au	30	10–20	Spherical	Intracellular	He et al. (2007)
Shewanella algae	Au	25	10–20	Not available	Intracellular	Konishi et al. (2007a)
Brevibacterium casei	Au, Ag	37	10–50	Spherical	Extracellular	Kalishwaralal et al. (2010)
Trichoderma viride	Ag	27	5–40	Spherical	Extracellular	Fayaz et al. (2010)
Phanerochaete chrysosporium	Ag	37	50–200	Pyramidal	Extracellular	Vigneshwaran et al. (2006)
Bacillus licheniformis	Ag	37	50	Not available	Extracellular	Kalimuthu et al. (2008)
Escherichia coli	Ag	37	50	Not available	Extracellular	Gurunathan et al. (2009)

Corynebacterium glutamicum	Ag	30	5–50	Irregular	Extracellular	Sneha et al. (2010)
Trichoderma viride	Ag	10–40	2–4	Not available	Extracellular	Mohammed Fayaz et al. (2009)
Ureibacillus thermosphaericus	Au	60–80	50–70	Not available	Extracellular	Juibari et al. (2011)
Bacillus cereus	Ag	37	4–5	Spherical	Intracellular	Babu and Gunasekaran (2009)
Aspergillus flavus∗	Ag	25	8.92 ± 1.61	Spherical	Extracellular	Vigneshwaran et al. (2007)
Aspergillus fumigates∗	Ag	25	5–25	Spherical	Extracellular	Bhainsa and D'Souza (2006)
Verticillium sp.∗	Ag	25	25 ± 8	Spherical	Extracellular	Senapati et al. (2004)
Fusarium oxysporum∗	Ag	25	5–50	Spherical	Extracellular	Senapati et al. (2004)
Neurospora crassa	Au, Au/Ag	28	32, 20–50	Spherical	Intracellular, extracellular	Castro-Longoria et al. (2011)
Shewanella algae	Pt	25	5	Not available	Intracellular	Konishi et al. (2007b)
Enterobacter sp.	Hg	30	2–5	Spherical	Intracellular	Sinha and Khare (2011)
Shewanella sp.	Se	30	181 ± 40	Spherical	Extracellular	Lee et al. (2007)
Escherichia coli	CdTe	37	2.0–3.2	Spherical	Extracellular	Bao et al. (2010)
Yeast	Au/Ag	30	9–25	Irregular polygonal	Extracellular	Zheng et al. (2010)
Fusarium oxysporum∗	Au-Ag alloy	25	8–14	Spherical	Extracellular	Senapati et al. (2005)
Pyrobaculum islandicum	U(VI), Tc(VII), Cr(VI), Co(III), Mn(IV)	100	N/A	Spherical	Extracellular	Kashefi and Lovley (2000)
Desulfovibrio desulfuricans	Pd	25	50	Spherical	Extracellular	Lloyd et al. (1998)

Table 22.3 Oxide nanoparticles synthesized by microorganisms (Li et al. 2011). Plant pathogens are marked with an asteriks (*)

Microorganisms	Products	Culturing temperature (°C)	Size (nm)	Shape	Location	References
Shewanella oneidensis	Fe_3O_4	28	40–50	Rectangular, rhombic, hexagonal	Extracellular	Perez-Gonzalez et al. (2010)
QH-2	Fe_3O_4	22–26	81 ± 23 × 58 ± 20	Rectangular	Intracellular	Zhu et al. (2010)
Recombinant AMB-1	Fe_3O_4	28	20	Cuboctahedral	Intracellular	Amemiya et al. (2007)
Yeast cells	Fe_3O_4	36	Not available	Wormhole-like	Extracellular	Zhou et al. (2009a)
Yeast cells	$FePO_4$	36	Not available	Nanopowders	Extracellular	Zhou et al. (2009b)
WM-1	$Fe_3 O_4$	28	54 ± 12.3 × 43 ± 10.9	Cuboidal	Intracellular	Li et al. (2007)
Shewanella oneidensis MR-1	Fe_2O_3	25	30–43	Pseudohexagonal/irregular or rhombohedral	Intracellular	Bose et al. (2009)
HSMV-1	$Fe_3 O_4$	63	113 ± 34 × 40 ± 5	Bullet-shaped	Intracellular	Lefèvre et al. (2010a)
Saccharomyces cerevisiae	Sb_2O_3	25–60	2–10	Spherical	Intracellular	Jha et al. (2009a)
Lactobacillus sp.	$BaTiO_3$	25	20–80	Tetragonal	Extracellular	Jha and Prasad (2010)
Lactobacillus sp.	TiO_2	25	8–35	Spherical	Extracellular	Jha et al. (2009b)
*Fusarium oxysporum**	TiO_2	300	6–13	Spherical	Extracellular	Bansal et al. (2005)
*Fusarium oxysporum**	$BaTiO_3$	25	4–5	Spherical	Extracellular	Bansal et al. (2006)
*Fusarium oxysporum**	ZrO_2	25	3–11	Spherical	Extracellular	Bansal et al. (2004)

Table 22.4 Sulphide nanoparticles synthesized by microorganisms (Li et al. 2011). Plant pathogens are marked with an asteriks (*)

Microorganisms	Products	Culturing temperature (°C)	Size (nm)	Shape	Location	References
Multicellular prokaryotes	Fe_3S_4	25	Not available	Not available	Intracellular	Lefèvre et al. (2010b)
Uncultured magnetotactic bacterium	Probably polyphosphate	Not available	Not available	Rectangular	Extracellular	Arakaki et al. (2010)
Rhodopseudomonas palustris	CdS	30	8	Cubic	Intracellular	Bai et al. (2009)
Coriolus versicolor	CdS	25	100–200	Spherical	Extracellular	Sanghi and Verma (2009)
Lactobacillus	CdS	25–60	4.9 ± 0.2	Spherical	Intracellular	Prasad and Jha (2010)
Yeast	CdS	25–60	3.6 ± 0.2	Spherical	Intracellular	Prasad and Jha (2010)
E. coli	CdS	25	2–5	Wurtzite crystal	Intracellular	Sweeney et al. (2004)
Schizosaccharomyces pombe	CdS	Not available	1–1.5	Hexagonal lattice	Intracellular	Kowshik et al. (2002)
Schizosaccharomyces pombe and *Candida glabrata*	CdS	Not available	2	Hexagonal lattice	Intracellular	Dameron et al. (1989)
Rhodobacter sphaeroides	CdS	Not available	8	Hexagonal lattice	Intracellular	Bai et al. (2006)
Desulfobacteraceae	CdS	Not available	2–5	Hexagonal lattice	Intracellular	Labrenz et al. (2000)
Rhodobacter sphaeroides	ZnS	Not available	10.5 ± 0.15	Spherical	Extracellular	Bai and Zhang (2009)
*Fusarium oxysporum**	CdS	Not available	5–20	Spherical	Extracellular	Ahmad et al. (2002)
Sulphate-reducing bacteria	FeS	Not available	2	Spherical	Extracellular	Watson et al. (1999)

Table 22.5 Other miscellaneous nanoparticles synthesized by microorganisms (Li et al. 2011). Plant pathogens are marked with an asteriks (*)

Microorganisms	Products	Culturing temperature (°C)	Size (nm)	Shape	Location	References
*Fusarium oxysporum**	$PbCO_3$, $CdCO_3$	27	120–200	Spherical	Extracellular	Sanyal et al. (2005)
*Fusarium oxysporum**	$SrCO_3$	27	10–50	Needle-like	Extracellular	Rautaray et al. (2004)
Brevibacterium casei	PHB	37	100–125	Not available	Intracellular	Pandian et al. (2009)
Yeasts	$Zn_3(PO_4)_2$	25	10–80 × 80–200	Rectangular	Extracellular	Yan et al. (2009)
*Fusarium oxysporum**	CdSe	10	9–15	Spherical	Extracellular	Kumar et al. (2007)

22.5.1 A Possible Induction of Bacterial and Fungal Resistance by NPs

In recent studies, the positive effect of metallic nanoparticles (silver, copper, titanium, zinc, and iron) that are used against multiple drug resistance (MDR) of microorganisms has been underlined due to their antimicrobial nature (Huh and Kwon 2011; Fernandez-Moure et al. 2017). The mechanisms behind the antimicrobial effect of these nanoparticles are well-known (Fig. 22.1; Singh et al. 2018). There are several green metallic nanoparticles obtained from microorganisms, which have been suggested for antimicrobial treatments against many pathogenic microorganisms (Singh et al. 2015a, b, 2016b; Sathiyavimal et al. 2018). These prominent nanoparticles showed increase of antimicrobial efficacy by conventional antibiotics such as lincomycin, oleandomycin, vancomycin, novobiocin, penicillin G, and rifampicin. The findings on zinc oxide displayed very effective antibacterial activity against *S. aureus*, *E. coli*, and *P. aeruginosa* (Pasquet et al. 2014). Even comparative studies between biological and chemical nanoparticles demonstrated higher antimicrobial effect of synthesized nanoparticles; they do not seem an alternative indispensable method for struggling with bacterial pathogens alone. Suggesting silver NPs and nanocomposites is too early whether they might be recommended to enhance the effectiveness of antibiotics or fully replace them to control local and systemic infections. Recent data show potential development of bacterial resistance to silver NPs that should be considered (Panáček et al. 2018). On the other hand, the ABC is the biggest protein transporter superfamily existing in all organisms.

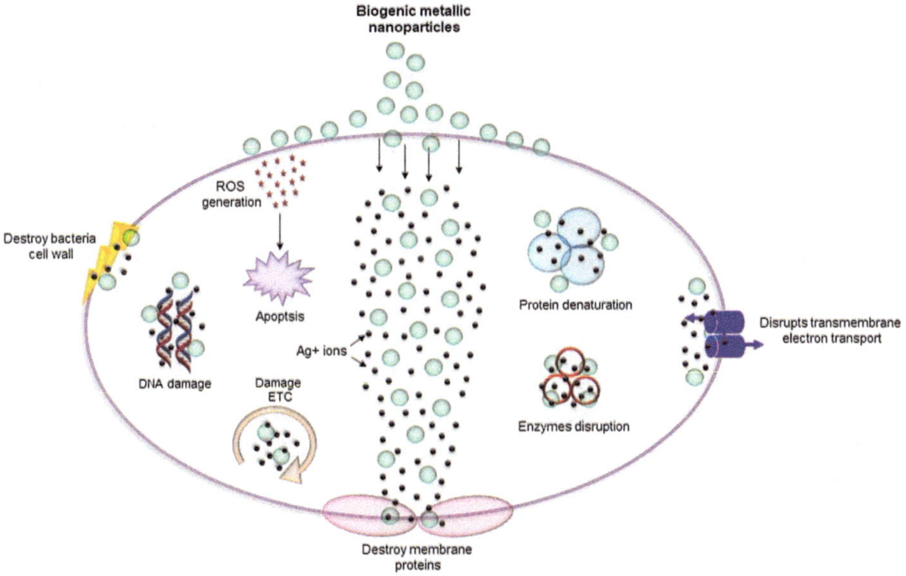

Fig. 22.1 Various mechanisms of antimicrobial activity of biogenic metallic nanoparticles. (*ROS* reactive oxygen (Singh et al. 2018))

Different proteins related to transduction cases of micromolecules and signals are encoded for translocation of various substrates with primary and secondary metabolites in cell (sugars, amino acids, ions, peptides, proteins, etc.) (Benadiba and Maor 2016). They are also valid in both prokaryotes and eukaryotes. In prokaryotes, they carry nutrients into cells, as well as efflux proteins, exhausting toxins and drugs out of the cell. In eukaryotes, they play a role in expressing efflux transporter proteins protecting the cell from toxins (Videira et al. 2014).

Recent data has shown unfavourable results related to growth of antibiotic resistance strains in n-ZnO and n-TiO2 particle producer *Bacillus subtilis*. This case strongly correlated with physiological adaptation of bacterial strains to antibiotics besides expected positive inhibitory effect of nanoparticles such as death or oxidative stress. Interestingly, nanoparticles may alter bacterial physiology and lead to dissemination of antibiotic resistance in bacteria (Eymard-Vernain et al. 2018).

In fungi, ABC transporters have been reported as target point of many pathogenic fungi-causing diseases on human and agricultural products. But it seems to be sceptic, because there are some fungi causing plant diseases (such as *F. oxysporum*, *Verticillium*) that are able to synthesize of NPs (Tables 22.2, 22.3, 22.4, and 22.5). On the contrary to findings of Yang et al. (2016), we should also be aware of raising of resistant-pathogen individuals to systemic fungicides consisting NPs and development of new drug delivery systems in pathogenic cells that may bypass the ABC family of transporters system on the cell wall such as the way of remnant non-target nanoparticles.

22.6 Conclusion

Nature comprises amazing mysterious waiting for us to explore and microorganisms that have tremendously genomic capacity ready for easily stress regulation. Therefore, it will be not wrong to say that any microorganism can overcome all negative issues and convert ongoing cases into favour of its own benefit. In this review, we have tried to imply that even we have new synthesized nanoparticles that will be used for different purposes, it does not mean there will be no any negative effect on environment. That is also possible accidentally introducing of nanoparticles onto non-target individuals and to be the reason for unexpected negative effect on balanced biological cycles of micro-/macro organisms. Because if we ignore the survival effort and quickly adaptation capacity of microorganism in inconvenient conditions, we may encounter with unexpected cases such as high antimicrobial resistance and dynamic population of harmful pest out of our control and we may create favourable conditions for these pathogens. In brief, we should say "let's not break one side while doing other side".

References

Agnihotri M, Joshi S, Kumar AR, Zinjarde S, Kulkarni S (2009) Biosynthesis of gold nanoparticles by the tropical marine yeast *Yarrowia lipolytica* NCIM 3589. Mater Lett 63(15):1231–1234. https://doi.org/10.1016/j.matlet.2009.02.042

Ahmad A, Mukherjee P, Mandal D, Senapati S, Khan MI, Kumar R, Sastry M (2002) Enzyme mediated extracellular synthesis of CdS nanoparticles by the fungus, *Fusarium oxysporum*. J Am Chem Soc 124(41):12108–12109. https://doi.org/10.1021/ja027296o

Ahmad A, Mukherjee P, Senapati S, Mandal D, Khan MI, Kumar R, Sastry M (2003a) Extracellular biosynthesis of silver nanoparticles using the fungus *Fusarium oxysporum*. Colloids Surf B Biointerfaces 28:313–318. https://doi.org/10.1016/S0927-7765(02)00174-1

Ahmad A, Senapati S, Khan MI, Kumar R, Ramani R, Srinivas V, Sastry M (2003b) Intracellular synthesis of gold nanoparticles by a novel alkalotolerant actinomycete, *Rhodococcus* species. Nanotechnology 14(7):824–828. https://doi.org/10.1088/0957-4484/14/7/323

Amemiya Y, Arakaki A, Staniland SS, Tanaka T, Matsunaga T (2007) Controlled formation of magnetite crystal by partial oxidation of ferrous hydroxide in the presence of recombinant magnetotactic bacterial protein Mms6. Biomaterials 28(35):5381–5389. https://doi.org/10.1016/j.biomaterials.2007.07.051

Arakaki A, Shibusawa M, Hosokawa M, Matsunaga T (2010) Preparation of genomic DNA from a single species of uncultured magnetotactic bacterium by multiple-displacement amplification. Appl Environ Microbiol 76(5):1480–1485. https://doi.org/10.1128/AEM.02124-09

Babu MMG, Gunasekaran P (2009) Production and structural characterization of crystalline silver nanoparticles from *Bacillus cereus* isolate. Colloids Surf B Biointerfaces 74(1):191–195. https://doi.org/10.1016/j.colsurfb.2009.07.016

Bai HJ, Zhang ZM (2009) Microbial synthesis of semiconductor lead sulfide nanoparticles using immobilized *Rhodobacter sphaeroides*. Mater Lett 63(9–10):764–766. https://doi.org/10.1016/j.matlet.2008.12.050

Bai HJ, Zhang ZM, Gong J (2006) Biological synthesis of semiconductor zinc sulfide nanoparticles by immobilized *Rhodobacter sphaeroides*. Biotechnol Lett 28(14):1135–1139. https://doi.org/10.1007/s10529-006-9063-1

Bai HJ, Zhang ZM, Guo Y, Yang GE (2009) Biosynthesis of cadmium sulfide nanoparticles by photosynthetic bacteria *Rhodopseudomonas palustris*. Colloids Surf B Biointerfaces 70(1):142–146. https://doi.org/10.1016/j.colsurfb.2008.12.025

Bansal V, Rautaray D, Ahmad A, Sastry M (2004) Biosynthesis of zirconia nanoparticles using the fungus *Fusarium oxysporum*. J Mater Chem 14(22):3303–3305. https://doi.org/10.1039/B407904C

Bansal V, Rautaray D, Bharde A, Ahire K, Sanyal A, Ahmad A, Sastry M (2005) Fungus-mediated biosynthesis of silica and titania particles. J Mater Chem 15(26):2583–2589. https://doi.org/10.1039/B503008K

Bansal V, Poddar P, Ahmad A, Sastry M (2006) Room temperature biosynthesis of ferroelectric barium titanate nanoparticles. J Am Chem Soc 128(36):11958–11963. https://doi.org/10.1021/ja063011m

Bao H, Lu Z, Cui X, Qiao Y, Guo J, Anderson JM, Li CM (2010) Extracellular microbial synthesis of biocompatible CdTe quantum dots. Acta Biomater 6(9):3534–3541. https://doi.org/10.1016/j.actbio.2010.03.030

Baysal Ö, Silme RS (2018) The ecological role of biodiversity for crop protection. In: Dunea D (ed) Plant competition in cropping systems. Intech, New York. https://doi.org/10.5772/intechopen.78228

Baysal Ö, Calışkan M, Yeşilova Ö (2008) An inhibitory effect of a new *Bacillus subtilis* strain (EU07) against *Fusarium oxysporum* f. sp. *radicis lycopersici*. Physiol Mol Plant Pathol 73:25–32. https://doi.org/10.1016/j.pmpp.2008.11.002

Baysal Ö, Lai D, Xu HH, Siragusa M, Calışkan M, Carimi F, Teixeira da Silva JA, Tor M (2013) A proteomic approach provides new insights into the control of soil-borne plant pathogens by *Bacillus* species. PLoS One 8:e53182. https://doi.org/10.1371/journal.pone.0053182

Benadiba M, Maor Y (2016) Importance of ABC transporters in drug development. Curr Pharm Des 22:5817–5829. https://doi.org/10.2174/1381612822666160810120359

Bhainsa KC, D'Souza SF (2006) Extracellular biosynthesis of silver nanoparticles using the fungus *Aspergillus fumigatus*. Colloids Surf B Biointerfaces 47(2):160–164. https://doi.org/10.1016/j.colsurfb.2005.11.026

Bose S, Hochella MF, Gorby YA, Kennedy DW, McCready DE, Madden AS, Lower BH (2009) Bioreduction of hematite nanoparticles by the dissimilatory iron reducing bacterium *Shewanella oneidensis* MR-1. Geochim Cosmochim Acta 73(4):962–976. https://doi.org/10.1016/j.gca.2008.11.031

Calder AJ, Dimkpa CO, McLean JE, Britt DW, Johnsonc W, Anderson AJ (2012) Soil components mitigate the antimicrobial effects of silver nanoparticles towards a beneficial soil bacterium, *Pseudomonas chlororaphis* O6. Sci Total Environ 429:215–222. https://doi.org/10.1016/j.scitotenv.2012.04.049

Castro-Longoria E, Vilchis-Nestor AR, Avalos-Borja M (2011) Biosynthesis of silver, gold and bimetallic nanoparticles using the filamentous fungus *Neurospora crassa*. Colloids Surf B Biointerfaces 83(1):42–48. https://doi.org/10.1016/j.colsurfb.2010.10.035

Chunjaturas W, Ferguson JA, Rattanapichai W, Sadowsky MJ, Sajjaphan K (2014) Shift of bacterial community structure in two Thai soil series affected by silver nanoparticles using ARISA. World J Microbiol Biotechnol 30:2119–2124. https://doi.org/10.1007/s11274-014-1633-0

Colman BP, Arnaout CL, Anciaux S, Gunsch CK, Hochella MF Jr, Kim B et al (2013) Low concentrations of silver nanoparticles in biosolids cause adverse ecosystem responses under realistic field scenario. PLoS One 8:e57189. https://doi.org/10.1371/journal.pone.0057189

Dameron CT, Reese RN, Mehra RK, Kortan AR, Carroll PJ, Steigerwald ML, Brus LE, Winge DR (1989) Biosynthesis of cadmium sulphide quantum semiconductor crystallites. Nature 338(6216):596–597

DeRosa MC, Monreal C, Schnitzer M, Walsh R, Sultan Y (2010) Nanotechnology in fertilizers. Nat Nanotechnol 5:91

Dimkpa CO, Latta DE, McLean JE, Britt DW, Boyanov MI, Anderson AJ (2013) Fate of CuO and ZnO nano-and microparticles in the plant environment. Environ Sci Technol 47:4734–4742. https://doi.org/10.1021/es304736y

Du L, Jiang H, Liu X, Wang E (2007) Biosynthesis of gold nanoparticles assisted by *Escherichia coli* DH5α and its application on direct electrochemistry of haemoglobin. Electrochem Commun 9(5):1165–1170. https://doi.org/10.1016/j.elecom.2007.01.007

Du W, Sun Y, Ji R, Zhu J, Wu J, Guo H (2011) TiO2 and ZnO nanoparticles negatively affect wheat growth and soil enzyme activities in agricultural soil. J Environ Monit 13:822–828. https://doi.org/10.1039/c0em00611d

Eymard-Vernain E, Luche S, Rabilloud T, Lelong C (2018) Impact of nanoparticles on the *Bacillus subtilis* (3610) competence. Sci Rep 8:2978. https://doi.org/10.1038/s41598-018-21402-0

Fayaz AM, Balaji K, Girilal M, Yadav R, Kalaichelvan PT, Venketesan R (2010) Biogenic synthesis of silver nanoparticles and their synergistic effect with antibiotics: a study against gram-positive and gram-negative bacteria. Nanomed-Nanotechnol 6(1):e103–e109. https://doi.org/10.1016/j.nano.2009.04.006

Fernandez-Moure JS, Evangelopoulos M, Colvill K, Van Eps JL, Tasciotti E (2017) Nanoantibiotics: a new paradigm for the treatment of surgical infection. Nanomed (Lond) 12:1319–1334. https://doi.org/10.2217/nnm-2017-0401

Frenk S, Ben-Moshe T, Dror I, Berkowitz B, Minz D (2013) Effect of metal oxide nanoparticles on microbial community structure and function in two different soil types. PLoS One 8:e84441. https://doi.org/10.1371/journal.pone.0084441

Ge Y, Schimel JP, Holden PA (2011) Evidence for negative effects of TiO_2 and ZnO nanoparticles on soil bacterial communities. Environ Sci Technol 45:1659–1664. https://doi.org/10.3390/s100100625

Ge Y, Schimel JP, Holden PA (2012) Identification of soil bacteria susceptible to TiO_2 and ZnO nanoparticles. Appl Environ Microbiol 78:6749–6758. https://doi.org/10.1128/AEM.00941-12
Ge Y, Priester JH, Van De Werfhorst LC, Schimel JP, Holden PA (2013) Potential mechanisms and environmental controls of TiO2 nanoparticle effects on soil bacterial communities. Environ Sci Technol 47:14411–14417. https://doi.org/10.1021/es403385c
Gericke M, Pinches A (2006) Biological synthesis of metal nanoparticles. Hydrometallurgy 83(1–4):132–140. https://doi.org/10.1016/j.hydromet.2006.03.019
Gurunathan S, Kalishwaralal K, Vaidyanathan R, Venkataraman D, Pandian SRK, Muniyandi J, Hariharan N, Eom SH (2009) Biosynthesis, purification and characterization of silver nanoparticles using *Escherichia coli*. Colloids Surf B Biointerfaces 74(1):328–335. https://doi.org/10.1016/j.colsurfb.2009.07.048
Hänsch M, Emmerling C (2010) Effects of silver nanoparticles on the microbiota and enzyme activity in soil. J Plant Nutr Soil Sci 173:554–558. https://doi.org/10.1002/jpln.200900358
He S, Guo Z, Zhang Y, Zhang S, Wang J, Gu N (2007) Biosynthesis of gold nanoparticles using the bacteria *Rhodopseudomonas capsulate*. Mater Lett 61(18):3984–3987. https://doi.org/10.1016/j.matlet.2007.01.018
Honek JF (2013) Bionanotechnology and bionanomaterials: John Honek explains the good things that can come in very small packages. BMC Biochem 14:29. https://doi.org/10.1186/1471-2091-14-29
Huh AJ, Kwon YJ (2011) "Nanoantibiotics": a new paradigm for treating infectious diseases using nanomaterials in the antibiotics resistant era. J Control Release: Off J Control Release Soc 156:128–145. https://doi.org/10.1016/j.jconrel.2011.07.002
Hussain I, Singh NB, Singh A, Singh H, Singh SC (2016) Green synthesis of nanoparticles and its potential application. Biotechnol Lett 38:545–560. https://doi.org/10.1007/s10529-015-2026-7
Husseiny MI, El-Aziz MA, Badr Y, Mahmoud MA (2007) Biosynthesis of gold nanoparticles using *Pseudomonas aeruginosa*. Spectrochim Acta A 67(3–4):1003–1006. https://doi.org/10.1016/j.saa.2006.09.028
Jha AK, Prasad K (2010) Ferroelectric $BaTiO_3$ nanoparticles: biosynthesis and characterization. Colloids Surf B Biointerfaces 75(1):330–334. https://doi.org/10.1016/j.colsurfb.2009.09.005
Jha AK, Prasad K, Prasad K (2009a) A green low-cost biosynthesis of Sb_2O_3 nanoparticles. Biochem Eng J 43(3):303–306. https://doi.org/10.1016/j.bej.2008.10.016
Jha AK, Prasad K, Kulkarni AR (2009b) Synthesis of TiO_2 nanoparticles using microorganisms. Colloids Surf B Biointerfaces 71(2):226–229. https://doi.org/10.1016/j.colsurfb.2009.02.007
Jośko I, Oleszczuk P (2013) Influence of soil type and environmental conditions on ZnO, TiO_2 and Ni nanoparticles phytotoxicity. Chemosphere 92:91–99. https://doi.org/10.1016/j.chemosphere.2013.02.048
Juibari MM, Abbasalizadeh S, Jouzani GS, Noruzi M (2011) Intensified biosynthesis of silver nanoparticles using a native extremophilic *Ureibacillus thermosphaericus* strain. Mater Lett 65(6):1014–1017. https://doi.org/10.1016/j.matlet.2010.12.056
Kalimuthu K, Suresh Babu R, Venkataraman D, Bilal M, Gurunathan S (2008) Biosynthesis of silver nanocrystals by *Bacillus licheniformis*. Colloids Surf B Biointerfaces 65(1):150–153. https://doi.org/10.1016/j.colsurfb.2008.02.018
Kalishwaralal K, Deepak V, Ram Kumar Pandian S, Kottaisamy M, BarathManiKanth S, Kartikeyan B, Gurunathana S (2010) Biosynthesis of silver and gold nanoparticles using *Brevibacterium casei*. Colloids Surf B Biointerfaces 77(2):257–262. https://doi.org/10.1016/j.colsurfb.2010.02.007
Kashefi K, Lovley DR (2000) Reduction of Fe(III), Mn(IV), and toxic metals at 100°C by *Pyrobaculum islandicum*. Appl Environ Microbiol 66(3):1050–1056
Kato H (2011) In vitro assays: tracking nanoparticles inside cells. Nat Nanotechnol 6(6):139–140
Kim S, Kim J, Lee I (2011) Effects of Zn and ZnO nanoparticles and Zn2C on soil enzyme activity and bioaccumulation of Zn in *Cucumis sativus*. Chem Ecol 27:49–55. https://doi.org/10.1080/02757540.2010.529074

Konishi Y, Tsukiyama T, Tachimi T, Saitoh N, Nomura T, Nagamine S (2007a) Microbial deposition of gold nanoparticles by the metal-reducing bacterium *Shewanella algae*. Electrochim Acta 53(1):186–192. https://doi.org/10.1016/j.electacta.2007.02.073

Konishi Y, Ohno K, Saitoh N, Nomura T, Nagamine S, Hishida H, Takahashi Y, Uruga T (2007b) Bioreductive deposition of platinum nanoparticles on the bacterium *Shewanella algae*. J Biotechnol 128(3):648–653. https://doi.org/10.1016/j.jbiotec.2006.11.014

Kowshik M, Deshmukh N, Vogel W, Urban J, Kulkarni SK, Paknikar KM (2002) Microbial synthesis of semiconductor CdS nanoparticles, their characterization, and their use in the fabrication of an ideal diode. Biotechnol Bioeng 78(5):583–588. https://doi.org/10.1002/bit.10233

Kumar SA, Ansary AA, Abroad A, Khan MI (2007) Extracellular biosynthesis of CdSe quantum dots by the fungus, *Fusarium oxysporum*. J Biomed Nanotechnol 3(2):190–194. https://doi.org/10.1166/jbn.2007.027

Labrenz M, Druschel GK, Thomsen-Ebert T, Gilbert B, Welch SA, Kemner KM, Logan GA, Summons RE, De Stasio G, Bond PL, Lai B, Kelly SD, Banfield JF (2000) Formation of sphalerite (ZnS) deposits in natural biofilms of sulfate-reducing bacteria. Science 290(5497):1744–1747. https://doi.org/10.1126/science.290.5497.1744

Lee JH, Han J, Choi H, Hur HG (2007) Effects of temperature and dissolved oxygen on Se(IV) removal and Se(0) precipitation by *Shewanella* sp. HN-41. Chemosphere 68(10):1898–1905. https://doi.org/10.1016/j.chemosphere.2007.02.062

Lefèvre CT, Abreu F, Schmidt ML, Lins U, Frankel RB, Hedlund BP, Bazylinski DA (2010a) Moderately thermophilic magnetotactic bacteria from hot springs in Nevada. Appl Environ Microbiol 76(11):3740–3743. https://doi.org/10.1128/AEM.03018-09

Lefèvre CT, Abreu F, Lins U, Bazylinski DA (2010b) Nonmagnetotactic multicellular prokaryotes from low-saline, nonmarine aquatic environments and their unusual negative phototactic behaviour. Appl Environ Microbiol 76(10):3220–3227. https://doi.org/10.1128/AEM.00408-10

Lengke MF, Fleet ME, Southam G (2006a) Morphology of gold nanoparticles synthesized by filamentous cyanobacteria from gold(I)-thiosulfate and gold(III)-chloride complexes. Langmuir 22(6):2780–2787. https://doi.org/10.1021/la052652c

Lengke MF, Ravel B, Fleet ME, Wanger G, Gordon RA, Southam G (2006b) Mechanisms of gold bioaccumulation by filamentous cyanobacteria from gold(III)-chloride complex. Environ Sci Technol 40(20):6304–6309. https://doi.org/10.1021/es061040r

Li Y, Cu YT, Luo D (2005) Multiplexed detection of pathogen DNA with DNA-based fluorescence nanobarcodes. Nat Biotechnol 23:885–889. https://doi.org/10.1038/nbt1106

Li W, Yu L, Zhou P, Zhu M (2007) A *Magnetospirillum* strain WM-1 from a freshwater sediment with intracellular magnetosomes. World J Microbiol Biotechnol 23(10):1489–1492

Li X, Xu H, Chen ZS, Chen G (2011) Biosynthesis of nanoparticles by microorganisms and their applications. J Nanomater 2011:270974. https://doi.org/10.1155/2011/270974

Liu J, Qiao SZ, Hu QH, Lu GQ (2011) Magnetic nanocomposites with mesoporous structures: synthesis and applications. Small 7(4):425–443. https://doi.org/10.1002/smll.201001402

Lloyd JR, Yong P, Macaskie LE (1998) Enzymatic recovery of elemental palladium by using sulfate-reducing bacteria. Appl Environ Microbiol 64(11):4607–4609

Luechinger NA, Grass RN, Athanassiou EK, Stark WJ (2010) Bottom-up fabrication of metal/metal nanocomposites from nanoparticles of immiscible metals. Chem Mater 22(1):155–160. https://doi.org/10.1021/cm902527n

McLamore ES, Diggs A, Calvo Marzal P, Shi J, Blakeslee JJ, Peer WA, Murphy AS, Porterfield DM (2010) Non-invasive quantification of endogenous root auxin transport using an integrated flux microsensor technique. Plant J 63:1004–1016. https://doi.org/10.1111/j.1365-313X.2010.04300.x

Mishra S, Singh HB (2015) Biosynthesized silver nanoparticles as a nanoweapon against phytopathogens: exploring their scope and potential in agriculture. Appl Microbiol Biotechnol 99:1097–1107. https://doi.org/10.1007/s00253-014-6296-0

Mishra S, Singh BR, Singh A, Keswani C, Naqvi AH, Singh HB (2014) Biofabricated silver nanoparticles act as a strong fungicide against *Bipolaris sorokiniana* causing spot blotch disease in wheat. PLoS One 9:e97881. https://doi.org/10.1371/journal.pone.0097881

Mishra S, Keswani C, Abhilash PC, Fraceto LF, Singh HB (2017) Integrated approach of agri-nanotechnology: challenges and future trends. Front Plant Sci 8:471. https://doi.org/10.3389/fpls.2017.00471
Mohammed Fayaz A, Balaji K, Kalaichelvan PT, Venkatesan R (2009) Fungal based synthesis of silver nanoparticles—an effect of temperature on the size of particles. Colloids Surf B Biointerfaces 74(1):123–126. https://doi.org/10.1016/j.colsurfb.2009.07.002
Panáček A, Kvítek L, Smékalová M, Večeřová R, Kolář M, Röderová M, Dyčka F, Šebela M, Prucek R, Tomanec O, Zbořil R (2018) Bacterial resistance to silver nanoparticles and how to overcome it. Nat Nanotechnol 13:65–71. https://doi.org/10.1038/s41565-017-0013-y
Pandian SRK, Deepak V, Kalishwaralal K, Muniyandi J, Rameshkumar N, Gurunathan S (2009) Synthesis of PHB nanoparticles from optimized medium utilizing dairy industrial waste using *Brevibacterium casei* SRKP2: a green chemistry approach. Colloids Surf B Biointerfaces 74(1):266–273. https://doi.org/10.1016/j.colsurfb.2009.07.029
Pasquet J, Chevalier Y, Pelletier J, Couval E, Bouvier D, Bolzinger MA (2014) The contribution of zinc ions to the antimicrobial activity of zinc oxide. Colloids Surf A Physicochem Eng Asp 457:263–274. https://doi.org/10.1016/j.colsurfa.2014.05.057
Pérez-de-Luque A, Rubiales D (2009) Nanotechnology for parasitic plant control. Pest Manag Sci 65:540–545. https://doi.org/10.1002/ps.1732
Perez-Gonzalez T, Jimenez-Lopez C, Neal AL, Rull-Perez F, Rodriguez-Navarro A, Fernandez-Vivas A, Iañez-Pareja E (2010) Magnetite biomineralization induced by *Shewanella oneidensis*. Geochim Cosmochim Acta 74(3):967–979. https://doi.org/10.1016/j.gca.2009.10.035
Prasad R (2016) Advances and applications through fungal nanobiotechnology. Springer, Cham. ISBN: 978-3-319-42989-2
Prasad K, Jha AK (2010) Biosynthesis of CdS nanoparticles: an improved green and rapid procedure. J Colloid Interface Sci 342(1):68–72. https://doi.org/10.1016/j.jcis.2009.10.003
Prasad K, Jha AK, Kulkarni AR (2007) *Lactobacillus* assisted synthesis of titanium nanoparticles. Nanoscale Res Lett 2:248–250. https://doi.org/10.1007/s11671-007-9060-x
Rautaray D, Sanyal A, Adyanthaya SD, Ahmad A, Sastry M (2004) Biological synthesis of strontium carbonate crystals using the fungus *Fusarium oxysporum*. Langmuir 20(16):6827–6833. https://doi.org/10.1021/la049244d
Sanghi R, Verma P (2009) A facile green extracellular biosynthesis of CdS nanoparticles by immobilized fungus. Chem Eng J 155(3):886–891. https://doi.org/10.1016/j.cej.2009.08.006
Sanyal A, Rautaray D, Bansal V, Ahmad A, Sastry M (2005) Heavy-metal remediation by a fungus as a means of production of lead and cadmium carbonate crystals. Langmuir 21(16):7220–7224. https://doi.org/10.1021/la047132g
Sathiyavimal S, Vasantharaj S, Bharathi D, Saravanan M, Manikandan E, Kumar SS, Pugazhendhi A (2018) Biogenesis of copper oxide nanoparticles (CuONPs) using *Sida acuta* and their incorporation over cotton fabrics to prevent the pathogenicity of Gram negative and Gram positive bacteria. J Photochem Photobiol B Biol 188:126–134. https://doi.org/10.1016/j.jphotobiol.2018.09.014
Schlich K, Hund-Rinke K (2015) Influence of soil properties on the effect of silver nanomaterials on microbial activity in five soils. Environ Pollut 196:321–330. https://doi.org/10.1016/j.envpol.2014.10.021
Senapati S, Mandal D, Ahmad A, Khan MI, Sastry M, Kumar R (2004) Fungus mediated synthesis of silver nanoparticles: a novel biological approach. Indian J Phys 78A(1):101–105
Senapati S, Ahmad A, Khan MI, Sastry M, Kumar R (2005) Extracellular biosynthesis of bimetallic Au-Ag alloy nanoparticles. Small 1(5):517–520. https://doi.org/10.1002/smll.200400053
Shah V, Collins D, Walker VK, Shah S (2014) The impact of engineered cobalt, iron, nickel and silver nanoparticles on soil bacterial diversity under field conditions. Environ Res Lett 9:024001. https://doi.org/10.1088/1748-9326/9/2/024001
Shahrokh S, Hosseinkhani B, Emtiazi G (2014) The impact of nano-silver on bacterial aerobic nitrate reductase. J Bioprocess Biotechnol 4:162. https://doi.org/10.4172/2155-9821.1000162

Sharma N, Pinnaka AK, Raje M, Fnu A, Bhattacharyya MS, Choudhury AR (2012) Exploitation of marine bacteria for production of gold nanoparticles. Microb Cell Factories 11:86. https://doi.org/10.1186/1475-2859-11-86

Sharon M, Choudhary AK, Kuma R (2010) Nanotechnology in agricultural diseases and food safety. J Phytol 2(4):83–92

Shen Z, Chen Z, Hou Z, Li T, Lu X (2015) Ecotoxicological effect of zinc oxide nanoparticles on soil microorganisms. Front Environ Sci Eng 9:912–918. https://doi.org/10.1007/s11783-015-0789-7

Sheng Z, Liu Y (2011) Effects of silver nanoparticles on wastewater biofilms. Water Res 45:6039–6050. https://doi.org/10.1016/j.watres.2011.08.065

Shoults-Wilson WA, Reinsch BC, Tsyusko OV, Bertsch PM, Lowry GV, Unrine JM (2011) Role of particle size and soil type in toxicity of silver nanoparticles to earthworms. Soils Sci Soc Am J 75:365–377. https://doi.org/10.1016/j.envpol.2016.08.016

Simonin M, Richaume A (2015) Impact of engineered nanoparticles on the activity, abundance, and diversity of soil microbial communities: a review. Environ Sci Pollut Res 22:13710–13723. https://doi.org/10.1007/s11356-015-4171-x

Simonin M, Guyonnet JP, Martins JM, Ginot M, Richaume A (2015) Influence of soil properties on the toxicity of TiO_2 nanoparticles on carbon mineralization and bacterial abundance. J Hazard Mater 283:529–535. https://doi.org/10.1016/j.jhazmat.2014.10.004

Singaravelu G, Arockiamary JS, Ganesh Kumar V, Govindraju K (2007) A novel extracellular synthesis of monodisperse gold nanoparticles using marine alga, *Sargassum wightii* Greville. Colloid Surf B Biointerface 57:97–101. https://doi.org/10.1016/j.colsurfb.2007.01.010

Singh P, Kim YJ, Singh H, Mathiyalagan R, Wang C, Yang DC (2015a) Biosynthesis of anisotropic silver nanoparticles by *Bhargavaea indica* and their synergistic effect with antibiotics against pathogenic microorganisms. J Nanomater 2015:10. https://doi.org/10.1155/2015/234741

Singh P, Kim YJ, Singh H, Wang C, Hwang KH, Farh Mel A, Yang DC (2015b) Biosynthesis, characterization, and antimicrobial applications of silver nanoparticles. Int J Nanomedicine 10:2567–2577. https://doi.org/10.2147/IJN.S72313

Singh P, Kim YJ, Zhang D, Yang DC (2016a) Biological synthesis of nanoparticles from plants and microorganisms. Trends Biotechnol 34:588–599. https://doi.org/10.1016/j.tibtech.2016.02.006

Singh P, Singh H, Kim YJ, Mathiyalagan R, Wang C, Yang DC (2016b) Extracellular synthesis of silver and gold nanoparticles by *Sporosarcina koreensis* DC4 and their biological applications. Enzym Microb Technol 86:75–83. https://doi.org/10.1016/j.enzmictec.2016.02.005

Singh P, Garg A, Pandit S, Mokkapati V, Mijakovic I (2018) Antimicrobial effects of biogenic nanoparticles. Nanomaterials (Basel, Switzerland) 8(12):1009. https://doi.org/10.3390/nano8121009

Sinha A, Khare SK (2011) Mercury bioaccumulation and simultaneous nanoparticle synthesis by *Enterobacter* sp. cells. Bioresour Technol 102:4281–4284. https://doi.org/10.1016/j.biortech.2010.12.040

Slawson RM, Lohmeier-Vogel EM, Lee H, Trevors JT (1994) Silver resistance in *Pseudomonas stutzeri*. Biometals 7:30–40

Sneha K, Sathishkumar M, Mao J, Kwak IS, Yun YS (2010) *Corynebacterium glutamicum*-mediated crystallization of silver ions through sorption and reduction processes. Chem Eng J 162(3):989–996. https://doi.org/10.1016/j.cej.2010.07.006

Sunkar S, Nachiyar CV (2012) Microbial synthesis and characterization of silver nanoparticles using the endophytic bacterium *Bacillus cereus*: a novel source in the benign synthesis. Glob J Med Res 12:43–49

Suresh AK, Pelletier DA, Wang W, Broich ML, Moon JW, Gu B, Allison DP, Joy DC, Phelps TJ, Doktycz MJ (2011) Biofabrication of discrete spherical gold nanoparticles using the metal reducing bacterium *Shewanella oneidensis*. Acta Biomater 7(5):2148–2152. https://doi.org/10.1016/j.actbio.2011.01.023

Sweeney RY, Mao C, Gao X, Burt JL, Belcher AM, Georgiou G, Iverson BL (2004) Bacterial biosynthesis of cadmium sulfide nanocrystals. Chem Biol 11(11):1553–1559. https://doi.org/10.1016/j.chembiol.2004.08.022

Torney F, Trewyn BG, Lin VS, Wang K (2007) Mesoporous silica nanoparticles deliver DNA and chemicals into plants. Nat Nanotechnol 2:295–300. https://doi.org/10.1038/nnano.2007.108

VandeVoort AR, Arai Y (2012) Effect of silver nanoparticles on soil denitrification kinetics. Ind Biotechnol 8:358–364. https://doi.org/10.1089/ind.2012.0026

Videira M, Reis RL, Brito MA (2014) Deconstructing breast cancer cell biology and the mechanisms of multidrug resistance. Biochim Biophys Acta 1846:312–325. https://doi.org/10.1016/j.bbcan.2014.07.011

Vigneshwaran N, Kathe AA, Varadarajan PV, Nachane RP, Balasubramanya RH (2006) Biomimetics of silver nanoparticles by white rot fungus, *Phanerochaete chrysosporium*. Colloids Surf B Biointerfaces 53(1):55–59. https://doi.org/10.1016/j.colsurfb.2006.07.014

Vigneshwaran N, Ashtaputre NM, Varadarajan PV, Nachane RP, Paralikar KM, Balasubramanya RH (2007) Biological synthesis of silver nanoparticles using the fungus *Aspergillus flavus*. Mater Lett 61(6):1413–1418. https://doi.org/10.1016/j.matlet.2006.07.042

Vittori Antisari L, Carbone S, Gatti A, Vianello G, Nannipieri P (2013) Toxicity of metal oxide (CeO_2, Fe_3O_4, SnO_2) engineered nanoparticles on soil microbial biomass and their distribution in soil. Soil Biol Biochem 60:87–94. https://doi.org/10.1016/j.soilbio.2013.01.016

Watson JHP, Ellwood DC, Soper AK, Charnock J (1999) Nanosized strongly-magnetic bacterially-produced iron sulphide materials. J Magn Magn Mater 203(1–3):69–72. https://doi.org/10.1016/S0304-8853(99)00191-2

Wirth SM, Lowry GV, Tilton RD (2012) Natural organic matter alters biofilm tolerance to silver nanoparticles and dissolved silver. Environ Sci Technol 46:12687–12696. https://doi.org/10.1021/es301521p

Yan S, He W, Sun C, Zhang X, Zhao H, Li Z, Zhou W, Tian X, Sun X, Han X (2009) The biomimetic synthesis of zinc phosphate nanoparticles. Dyes Pigments 80(2):254–258. https://doi.org/10.1016/j.dyepig.2008.06.010

Yang L, Li W, Kirberger M, Liao W, Ren J (2016) Design of nanomaterial based systems for novel vaccine development. Biomater Sci 4:785–802. https://doi.org/10.1039/c5bm00507h

Yu JX, Li TH (2011) Distinct biological effects of different nanoparticles commonly used in cosmetics and medicine coatings. Cell Biosci 1:19

Zavaglia CA, Prado da Silva MH (2016) Biomaterials. Ref Module Mater Sci Mater Eng. https://doi.org/10.1016/B978-0-12-803581-8.04109-6

Zhao L, Hernandez-Viezcas JA, Peralta-Videa JR, Bandyopadhyay S, Peng B, Munoz B, Keller AA, Gardea-Torresdey JL (2013) ZnO nanoparticle fate in soil and zinc bioaccumulation in corn plants (*Zea mays*) influenced by alginate. Environ Sci Processes Impacts 15:260–266. https://doi.org/10.1039/C2EM30610G

Zheng D, Hu C, Gan T, Dang X, Hu S (2010) Preparation and application of a novel vanillin sensor based on biosynthesis of au-ag alloy nanoparticles. Sensors Actuators B Chem 148(1):247–252. https://doi.org/10.1016/j.snb.2010.04.031

Zhou W, He W, Zhong S, Wang Y, Zhao H, Li Z, Yan S (2009a) Biosynthesis and magnetic properties of mesoporous Fe_3O_4 composites. J Magn Magn Mater 321(8):1025–1028. https://doi.org/10.1016/j.jmmm.2008.10.007

Zhou W, He W, Zhang X, Yan S, Sun X, Tian X, Han X (2009b) Biosynthesis of iron phosphate nanopowders. Powder Technol 194(1–2):106–108. https://doi.org/10.1016/j.powtec.2009.03.034

Zhu K, Pan H, Li J, Yu-Zhang K, Zhang SD, Zhang WY, Zhou K, Yue H, Pan Y, Xiao T, Wu LF (2010) Isolation and characterization of a marine magnetotactic spirillum axenic culture QH-2 from an intertidal zone of the China Sea. Res Microbiol 161(4):276–283. https://doi.org/10.1016/j.resmic.2010.02.003

Detection and Degradation of Pesticides Using Nanomaterials

23

Jayesh Bhatt, Monika Jangid, Neha Kapoor, Rakshit Ameta, and Suresh C. Ameta

Abstract

Pesticides are those chemical substances which are meant to kill pests. The use of pesticides has become so common these days that this term is often treated as synonymous with plant protection product. Pesticides are commonly used to eliminate or control different agricultural pests that can damage crops and livestock and reduce the productivity of farm. The assessment of the risks of pesticides cannot be easily monitored as adverse effects of a pesticide may take a long time after exposure or after repeated or prolonged exposure. Such effects are chronic in nature and may cause cancer, impaired immune defense, lower reproductive ability, and reduced growth. Such factors can affect the size of population and entire ecosystem in the long term. Pesticides are meant to kill a particular pest, but quite a large percentage of pesticides reach some other destination than their basic target because they enter the air, water, soil, or sediments and ultimately in our food. Pesticides are known to have some human health hazards, short-term impacts like headaches and nausea to chronic impacts like cancer, loss of reproductive ability, etc. The uses of pesticides also adversely affect the general biodiversity in the soil. If there are no harmful chemicals in the soil, then that soil is of higher quality. Hence, the chapter focuses on the use of nanoparticles in the detection and degradation of pesticides.

J. Bhatt · N. Kapoor · S. C. Ameta
Department of Chemistry, PAHER University, Udaipur, Rajasthan, India

M. Jangid
Department of Chemistry, RNT College, Kapasan, Chittorgarh, Rajasthan, India

R. Ameta (✉)
Department of Chemistry, JRN Rajasthan Vidyapeeth (Deemed to be University), Udaipur, Rajasthan, India

M. Ghorbanpour et al. (eds.), *Biogenic Nano-Particles and their Use in Agro-ecosystems*, https://doi.org/10.1007/978-981-15-2985-6_23

Keywords
Nanomaterial · Pesticide · Insecticide · Herbicide · Rodenticide · Bactericide · Fungicide · Larvicide

23.1 Introduction

Pesticides are not a newer development. Many ancient civilizations were using pesticides to protect their crops from insects and pests. Commonly used pesticides at that time include sulfur, oil, ash, mercury, arsenic, lead, etc. on common crops. In the nineteenth century, more focus was there on compounds made with the roots of some tropical vegetables and chrysanthemums. Dichlorodiphenyltrichloroethane (DDT) was discovered in 1939, which proved to be an extremely effective pesticide, and it was frequently used as the insecticide all over the globe. However, DDT was banned in most of the developed and developing countries later due to biological effects and human safety.

These are grouped according to the types of pests, which they usually kill. These are:

- Insecticides – insects
- Herbicides – plants
- Rodenticides – rodents (rats and mice)
- Bactericides – bacteria
- Fungicides – fungi
- Larvicides – larvae

23.1.1 Biodegradable Pesticides

The biodegradable pesticides are those which can be broken down by microbes and other living beings into harmless or almost harmless compounds, while there are some persistent pesticides also, which may take months or sometimes even years to break down.

The major advantage of pesticides is that they can help farmers in protecting their crops from insects and other pests. Some major benefits are:

- Controlling pests and plant disease vectors
- Controlling human/livestock disease vectors and nuisance organisms
- Controlling organisms that harm other human activities and structures

Pesticides may have a different mechanism of action. Foliar pesticides are taken up by leaves or green plant parts, while soil-acting pesticides normally act via the soil on growing seeds, shoots, or roots. Contact pesticides are used to affect those parts of a weed or pests that directly come in contact with spray liquid. Systemic pesticides usually affect the entire target organism (weed or insect) through some active ingredient, which is transported in the plant's vascular system following uptake by the roots or other parts of the plant treated with the compound. It ensures that the pesticide has reached all parts of the plant.

The studies on the effects of pesticides are based on laboratory experiments carried out under controlled conditions (e.g., a temperature of 25 °C), often with one experimental organism and one pesticide at a time. It is difficult to decide the consequences of the results available on the laboratory scale to possible consequences in natural environments, because a more complex system exists in the actual environment.

Several pesticides may occur together in agricultural streams as evident from environmental monitoring. Sometimes two or more chemicals can interact synergistically, and the ultimate effect is more than double. Pesticides may also interact antagonistically. That is the best scenario, when they counteract each other in environments.

23.2 Insecticides

Insecticides are formulated to kill, harm, repel, or mitigate one or more species of insects. Some insecticides disrupt the nervous system, while others may damage their exoskeletons, repel them, or control them by some other means.

23.2.1 Atrazine

Atrazine belongs to the triazine class. It is used to prevent pre- and post-emergence broadleaf weeds in crops like maize (corn) and sugarcane and on turfs, such as golf courses and residential lawns. It is the most commonly detected pesticide contaminating drinking water. It is also an endocrine disruptor. Death of plants results from starvation and oxidative damage caused by a breakdown in the electron transport process. Atrazine remains in the soil for months (although in some soils, it can persist for years also). It can migrate from the soil to groundwater. It degrades in the soil basically by the action of microbes.

Nsibande and Forbes (2019) used quantum dots (QDs) as sensitive fluorescent probes, because QDs have attractive and unique optical properties. The coupling of QDs to molecularly imprinted polymers (MIPs) was used to develop QD@MIPs-based fluorescence sensors, which provide monitoring atrazine in water. Here, highly fluorescent CdSeTe/ZnS QDs were fabricated using the conventional organometallic synthesis approach, and these were then encapsulated with MIPs. The sensor showed a good response time (5 min) upon interaction with atrazine, and the fluorescence intensity was quenched linearly within the range of 2–20 mol L^{-1}. This sensor was applied in real water samples, which showed satisfactory recoveries (92–118%).

Zhao et al. (2019) reported a novel dual-chemosensor coupling, a separation (molecularly imprinted polymers), an instrumental-free detection (gold nanoparticle-based colorimetric assay), and an instrument-based quantification (surface-enhanced Raman spectroscopy, SERS) method for sensitive determination of atrazine in apple juice. MIPs were used to effectively extract atrazine from apple juice with high recoveries (~93%). It was reported that large AuNPs provided the highest sensitivity in colorimetric analysis (< 0.01 mg L^{-1}) as compared to medium and small AuNPs. On the other hand, medium AuNPs achieved the lowest limit of detection (0.0012 mg L^{-1}) and quantification (0.0040 mg L^{-1}) in SERS analysis.

A novel and efficient photocatalyst (Ag@$Mg_4Ta_2O_9$ nanoparticles) was prepared by Alkayal and Hussein (2019) via hydrothermal technique. The percentage of Ag in the Ag@$Mg_4Ta_2O_9$ nanoparticles has been investigated on the properties of $Mg_4Ta_2O_9$ (physical and chemical). It was indicated that pore size distribution around 24 nm for 2.0 wt% Ag@$Mg_4Ta_2O_9$ nanocomposite was there. These nanoparticles were then applied for atrazine degradation photocatalytically. The 2.0 wt% of Ag@$Mg_4Ta_2O_9$ displayed the best photocatalytic efficiency for the degradation of atrazine, which may be due to the high BET surface area and low band gap. It was also reported that this nanocomposite can be reused for atrazine degradation many times without any significant loss of efficiency.

Truca et al. (2019) incorporated Cu into ZnO lattice so as to reduce its band gap and also to extend its response to visible radiation. As-obtained Cu-ZnO was continuously integrated with g-C_3N_4 to prepare Cu-ZnO/g-C_3N_4 Z-direct scheme photocatalyst. Cu-ZnO utilized the h^+ only for atrazine degradation (direct and indirect through formation of hydroxyl radicals), while dissolved g-C_3N_4 utilized only generated e^- (indirectly via reaction with dissolved O_2 to form superoxide anion, which continuously reacts with H_2O to form $^{\bullet}OH$). It was also reported that photocatalytic degradation of atrazine by synthesized Cu-ZnO material was greater than that by synthesized g-C_3N_4 material. Cu-ZnO/g-C_3N_4 utilized both: generated e^- and h^+ for degradation of atrazine. Recycling experiments also indicated greater stability of as-synthesized Cu-ZnO/g-C_3N_4 during long-term atrazine degradation.

23.2.2 Dichlorvos

Dichlorvos is an organophosphate, which is widely used as an insecticide to control household pests and also in protecting stored products from insects. Dichlorvos is effective against mushroom flies, aphids, spider mites, caterpillars, thrips, and whiteflies in greenhouses and also in outdoor crops. It has become controversial because of its prevalence in urban waterways and due to the fact that its toxicity extends well beyond insects. It acts against insects as both a contact and a stomach poison.

Hou et al. (2016) reported an ultrasensitive fluorescence resonance energy transfer (FRET) biosensor of dichlorvos (organophosphorus insecticide). This sensor is based on carbon dots and the traditionally colorimetric Ellman's test. 5-Thio-2-nitrobenzoic acid anion (TNB^-) was used as energy acceptors to construct a FRET sensing system of acetylcholinesterase (AChE) and its inhibitor. Here, quaternized carbon dots (Q-CDs) were chosen as energy donors to form an effective FRET system. It was revealed that TNB^- can quench the fluorescence of Q-CDs to a larger extent. Dichlorvos, which is also an inhibitor of AChE, can be detected using this sensing system. Linear range of dichlorvos was reported to be 5.0×10^{-11} to 1.0×10^{-7} M under optimal conditions. As this sensing system has high sensitivity and green fluorescent material with high quantum yield, it can be used for rapid determination of dichlorvos.

Dong et al. (2016) constructed a simple and sensitive fluorescent sensor for dichlorvos, which was based on carbon dots-Cu(II) system. They obtained these carbon dots by simple hydrothermal reaction of feather. It was reported that fluorescence of these carbon dots can be selectively quenched by Cu^{2+} ion, which was restored on introducing acetylcholinesterase and acetylthiocholine in the system, because there thiocholine is produced, which reacts with Cu^{2+} ion. As dichlorvos is one kind of acetylcholinesterase inhibitor, it can be detected with a linear range of 6.0×10^{-9} to 6.0×10^{-8} M. This type of sensor can be successfully used for the analysis of cabbage and fruit juice samples.

Rao et al. (2012) synthesized undoped and magnesium doped samples of TiO_2 by sol-gel method with magnesium weight percentages in the range of 0.75–1.5 wt%. It was revealed that there is a red shift for doped TiO_2. The doped catalyst had smaller particle size and higher surface area than its undoped counterpart. The photocatalytic efficiency of as-synthesized catalysts was investigated by the photocatalytic degradation of aqueous dichlorvos in the presence of visible light irradiation. Better catalytic activity was found with Mg^{2+}-doped catalysts than undoped TiO_2, which may be attributed to more efficient electron-hole creation in Mg^{2+}-TiO_2 in visible light.

Senthilnathan and Philip (2011) prepared N-doped TiO_2. Different organic compounds such as triethylamine, urea, ethylamine, and ammonium hydroxide were used as source of nitrogen. It was revealed that nitrogen doped TiO_2 from triethylamine precursor had better photocatalytic activity for the degradation of dichlorvos under visible and solar radiation. N-doped TiO_2 showed better photocatalytic activity under solar radiation as compared to UV and visible light. It was reported that commercial grade dichlorvos produced intermediates such as 2,2-dichlorovinyl-O-methyl phosphate and O,O,O-trimethyl phosphonic ester during the photocatalytic degradation. However, no intermediates remained present at the end of the reaction.

23.2.3 Endosulfan

Endosulfan is an off-patent organochlorine insecticide and acaricide. It has two isomers, endo and exo, which are known popularly as I and II or α and β. Endosulfan became a highly controversial agrichemical. It has been used in agriculture to control insect pests such as whiteflies, aphids, leafhoppers, Colorado potato beetles, and cabbage worms.

A glassy carbon substrate was covalently modified by Liu et al. (2012) with a mixed layer of 4-aminophenyl and phenyl in situ electrografting of their aryldiazonium salts in acidic media. Single-walled carbon nanotubes (SWCNTs) were covalently and vertically anchored on the surface of electrode via the formation of amide bonds. These bonds are formed by the reaction between the amines present on the modified substrate and the carboxylic groups at the ends of the nanotubes. Then ferrocenedimethylamine (FDMA) was attached to the ends of SWCNTs through amide bonding followed by the attachment of an epitope, i.e., endosulfan hapten to which an antibody would bind. Antibody-complexed electrodes were exposed to samples containing spiked endosulfan (unbound target analyte) in water and analyzed it using the square-wave voltammetry (SWV) technique. As-fabricated electrochemical immunosensor was successfully used for the detection of endosulfan in the range of 0.01–20 ppb. The lowest detection limit of this immunosensor was found to be 0.01 ppb endosulfan in 50 mM phosphate buffer at pH 7.0.

Li et al. (2014) reported a modified rapid, easy, low-cost, effective, rugged, and safe sample preparation method with magnetic nanoparticles (Fe_3O_4; MNPs). They established a new method for the determination of multiple pesticides present in vegetables and fruits. Recoveries were evaluated in four representative matrices (tomato, cucumber, orange, and apple) under optimum conditions, with three spiked concentrations of 10, 50, and 200 μg kg^{-1} for all. It was revealed that the recovery of pesticides ranged between 71.5% and 111.7%.

Thomas et al. (2011) prepared silver nanoparticles doped anatase TiO_2 nanocrystals through a low-temperature hydrothermal route. As-synthesized nanocrystals were found to be highly efficient solar photocatalysts, and they had higher photocatalytic activity as compared to pure nano-TiO_2 and commercial photocatalyst Degussa P25, in the presence of sunlight. Silver nanoparticles were prepared by a single-step chemical reduction and stabilization employing L-Dopa. It was revealed that nanosilver doping on TiO_2 induces a red shift of absorption edge due to the narrowing of band gap. As-synthesized photocatalysts were used for the degradation of the organochlorine pesticide, endosulfan, and they could achieve almost complete degradation with highly active calcined nanosilver doped TiO_2 catalyst.

The removal of multi-pesticides was studied by Shabeer et al. (2015) via a combined treatment process with coagulation-adsorption on nano-clay. Different nano-clays (nano-bentonite, nano-halloysite, and organically modified nano-montmorillonite) were used as the adsorbent, while alum and polyaluminium chloride (PAC) were used as coagulants. It was observed that only the coagulation method was not sufficient to purify water, whereas a combination of coagulation with adsorption provided better purification. Out of the nano-clays used, organically modified nano-montmorillonite gave the best result for the removal of pesticides from water. Freundlich isotherm indicated that adsorption of pesticides on different nano-clays depends on its type, presence and absence of coagulants, as well as the properties of pesticides. They observed that alum-PAC coagulation aided by nano-clay as an adsorbent was the superior process for the removal of many pesticides simultaneously (atrazine, aldrin, metribuzin, β-endosulfan, α-endosulfan, endosulfan sulfate, lindane, chlorpyrifos, pendimethalin, DDT, and cypermethrin) from water.

23.2.4 Parathion

Parathion or methyl parathion is an organophosphate insecticide and acaricide. It is highly toxic to nontarget organisms, including humans. Methyl parathion is somewhat less toxic than parathion. It is often applied to cotton, rice, and fruit trees. Parathion is a cholinesterase inhibitor. It generally disrupts the nervous system by inhibiting acetylcholinesterase. It is absorbed via the skin and mucous membranes and also orally. Parathion (after metabolic activity gets converted into paraoxon) exposure can result in headaches, convulsions, poor vision, vomiting, abdominal pain, severe diarrhea, unconsciousness, tremor, dyspnea, and finally lung edema as well as respiratory arrest.

Tang and Xiang (2016) prepared surface molecularly imprinted CdTe nanoparticles via a reverse microemulsion polymerization. Parathion was selected as template molecule, 3-aminopropyltriethoxysilane and tetramethoxysilane as the polymerization precursors, and cross-linkers. As-synthesized materials show distinguished selectivity and high binding affinity to parathion as compared to chlorpyrifos, diazinon, and pyrimithate. It was observed that the relative fluorescence intensity of polymers decreased with the increase in concentration of parathion in the range 0.05–1000 μmol L^{-1}. The proposed method was used for the detection of parathion in water samples, where recoveries were in the range of 97.72–100.59%.

A selective and sensitive novel electrochemical sensor was developed by Prasad et al. (2015) for the detection of methyl parathion. They used carbon dots (C-dots)/ ZrO_2 nanocomposite. This was fabricated using electrochemical deposition onto a glassy carbon electrode. The C-dots/ZrO_2 modified glassy carbon electrode could rapidly and selectively determine methyl parathion in rice samples using adsorptive stripping voltammetry. It was reported that stripping response was highly linear for concentrations of methyl parathion in the range of 0.2–48 ng mL^{-1} having a detection limit of 0.056 ng mL^{-1}.

Hou et al. (2015) developed a simple and sensitive fluorescent sensor for methyl parathion, which is based on L-tyrosine methyl ester functionalized carbon dots (Tyr-CDs) and tyrosinase system. These carbon dots were obtained via a simple hydrothermal route using citric acid as a carbon resource and L-tyrosine methyl ester as modification reagent. It was revealed that tyrosinase can catalyze the oxidation of tyrosine methyl ester on the surface of carbon dots to corresponding quinone products, which are responsible for the quenching of fluorescence of carbon dots. When methyl parathion was introduced in the system, they decrease the enzyme activity, resulting in the decrease of fluorescence quenching rate. It was reported that the enzyme inhibition rate was proportional to the logarithm of the methyl parathion concentration in the range 1.0×10^{-10} to 1.0×10^{-4} M with detection limit of 4.8×10^{-11} M (S/N = 3). The present method had a low detection limit, wide linear range, good selectivity, and high reproducibility. This sensing system has been successfully applied for the analysis of samples of cabbage, milk, and fruit juice.

The degradation of parathion (PTH) was investigated by Liu et al. (2019) using ferrate (VI). It was reported that the removal of PTH (5 mg L^{-1}) could reach 99% only in 300 s under the optimum conditions as [Fe(VI)]:[PTH] = 15:1; T = 25 °C and pH = 7.0. They also used real water samples to evaluate the feasibility of this Fe(VI) oxidation method. Fe(VI) may attack PS double bond and the PO single bond connecting the nitrophenol or the ethyl group in PTH molecule. Six products were identified; the major ones are paraoxon, thiophosphates, and phosphates.

Ag-TiO_2 nanoparticulate film was synthesized by Ramacharyulu et al. (2015) via dip coating. It was indicated TiO_2 (anatase) particles were in size ranging from 5 to 15 nm, while Ag nanoparticles were in size 10–20 nm. They carried out photocatalytic degradation of methyl parathion in aqueous solution using this Ag-TiO_2 nanoparticulate film, and results were compared with TiO_2 nanoparticulate film. Methyl parathion was first degraded initially to paraoxon, which on further degradation gives *p*-nitrophenol, trimethyl ester of phosphoric acid, and finally phosphate ion; however, minute amounts of carbon dioxide and acetaldehyde were also detected.

Nano-sized titanium dioxide (rutile) powders were used by Wang et al. (2007) as the catalysts for sonocatalytic degradation of methyl parathion. It was observed that the degradation of methyl parathion was more rapid sonocatalytically in the presence of TiO_2 particles than without TiO_2 catalyst. They also evaluated the effect of reaction parameters, such as species of TiO_2 particles, concentration of methyl parathion, different amounts of TiO_2, pH, intensity, and frequency of ultrasonics and temperature. It was revealed that methyl parathion in aqueous solution was completely mineralized to ions, like NO_2^-, NO_3^-, PO_4^{3-}, SO_4^{2-}, etc. The kinetics of the degradation was found to follow the first-order reaction. It was reported that more than 95% degradation of methyl parathion could be achieved within 80 min under optimal conditions.

23.2.5 Chlorpyrifos

Chlorpyrifos is an insecticide, which is used to control many kinds of pests, including termites, mosquitoes, and roundworms. It is also used on golf courses and crops such as cotton, corn, almonds, and fruit trees (oranges, bananas, and apples). Its exposure may lead to acute toxicity, with symptoms such as runny nose, tears, and increased saliva or drooling. More serious exposures can cause vomiting, abdominal muscle cramps, muscle twitching, tremors, weakness, difficulty in breathing, and paralysis.

Capoferri et al. (2018) reported that electrochromic behavior of iridium oxide nanoparticles (IrOx NPs) as physicochemical transducer together with a molecularly imprinted polymer (MIP) as recognition layer resulted in a fast and efficient translation of the detection event. They fabricated a sensor using screen-printing technology with indium tin oxide as a transparent working electrode. IrOx NPs were electrodeposited onto this electrode followed by thermal polymerization of polypyrrole in the presence of chlorpyrifos. Direct visual detection and smartphone imaging were used to detect and quantify this pesticide. It was observed that the application of different oxidation potentials for 10 s resulted in color changes, which is directly related to the concentration of the chlorpyrifos. Its concentration was found to be dependent on the color intensity of the electrode for smartphone imaging, at fixed potential. The proposed electrochromic sensor detects chlorpyrifos, a highly toxic compound with a 100 fM and 1 mM dynamic range.

Hunde et al. (2017) developed sensitive, cost-effective, and selective electrochemical sensor iron oxide nanoparticles as modifier in the paste of graphite powder. Cyclic voltammetry (CV) and differential pulse voltammetry (DPV) were used for the determination of chlorpyrifos from its aqueous solutions. Both these techniques were used to know the electrochemical interactions between chlorpyrifos and modified carbon paste sensor. It was revealed that the peak current varies directly to the pesticide concentration in lower concentration range of chlorpyrifos (1.0–100 μM) with detection limit of 2.8×10^{-6} mol L^{-1}. The relative stability of the modified sensor was found to be fine, and the reproducibility of the results was up to 98%, even after a period of 2 months. This proposed method is quite successful, when applied for the quantification of chlorpyrifos from its aqueous samples.

A highly sensitive piezoelectric biosensor was developed by Halamek et al. (2005) for the detection of cholinesterase inhibitors. It was reported that the regeneration of the sensor surface could be achieved with 1 mol L^{-1} formic acid, which enabled 40 measurements with one sensor. The total measurement time (binding + regeneration) was only 25 min, and the detection limit for chlorpyrifos was 10^{-10} mol L^{-1} (0.02 μg L^{-1}). The present sensor was used for the determination of its levels in river water samples.

23.2.6 Pirimicarb

Pirimicarb is a carbamate insecticide and it is used on a broad range of crops, which includes vegetable, cereal, and orchard crops. This acts by inhibition of acetylcholinesterase activity. It causes reproductive and developmental toxicity, neurotoxicity, and acute toxicity. It is also known to cause severe respiratory irritation, pulmonary edema, and slower heartbeat.

An electrochemical study of the pirimicarb (PMC) was performed by Selva et al. (2017) using a boron-doped diamond working electrode. Cyclic, differential pulse, and square-wave voltammetry experiments were carried out over a wide range of pH (2.0–8.0), which showed three irreversible oxidation processes. It was reported that the two first processes were pH-dependent, while the third was independent of pH. These three oxidation processes were found to be independent of each other, involving the transfer of one electron each. They proposed an analytical method for PMC quantification in water samples using differential pulse (DP) voltammetry. An analytical curve was obtained from 2.0 to 219 μmol L^{-1} with a detection limit of 1.24 μmol L^{-1}. The accuracy of this method was evaluated with recoveries ranging between 88.6% and 96.3%. Some advantages of this method are its simplicity, reliability, and portability.

23.2.7 Cyhalothrin

Cyhalothrin is a pyrethroid class of insecticides. It is used to control insects in cotton crops, and pyrethroids have been developed also for the control of household and agricultural insects and human lice. Some major toxic effects are paresthesia, nausea, headache, vomiting, dizziness, fatigue, CNS depression, fever, blurred vision, coma, respiratory failure, etc.

Wang et al. (2017) prepared a novel core-shell fluorescence probe for pyrethroids by precipitation polymerization. Here, ZnO quantum dots and cyhalothrin were used as the substrate and template, respectively. Such polymer microspheres were used sensitively and selectively to determine cyhalothrin. It was observed that a linear relationship was there between the concentration of cyhalothrin (0–80 μmol L^{-1}) and the fluorescence intensity. This method was used for the determination of cyhalothrin in milk.

23.2.8 Carbaryl

Carbaryl belongs to the carbamate family of insecticides. It is normally used to control aphids, ticks, fire ants, spiders, fleas, and many other pests. Carbaryl is used as an insecticide on cotton, corn, fruit, soybean, nut, vegetable, etc. crops and also in home yards and gardens. Carbaryl is a cholinesterase inhibitor. It causes headache, cramps, muscle weakness, memory loss, and anorexia.

Wang et al. (2014) developed a novel nonenzymatic sensor based on cobalt (II) oxide (CoO)-decorated reduced graphene oxide (rGO). It was then used for the detection of carbofuran (CBF) and carbaryl (CBR). It was reported that two separate differential pulse voltammetric peaks for these pesticides were obtained in a mixed solution, thus making their simultaneous detection possible. This nonenzymatic sensor shows a linear relationship over a wide concentration range of 0.2–70 μM for CBF and 0.5–200 μM for CBR. It was revealed that the lower detection limit of this sensor was 4.2 and 7.5 μg L^{-1} for CBF and CBR, respectively. This sensor was used to detect these pesticides in fruit and vegetable samples with satisfactory results.

An original and versatile dual-readout (colorimetric and fluorometric) protocol has been proposed by Zhao et al. (2016) using silver nanoparticles (Ag NPs) and fluorescent carbon dots (CDs), which can be used for the assay of acetylcholinesterase (AChE) activity and its inhibitors. The first dual-mode proposal has been made for assessing AChE activity using a CDs-based IFE sensing strategy, where the detection limit was found to be as low as 0.021 and 0.016 mU mL^{-1} by colorimetric and fluorometric measurements, respectively. The proposed assay is useful to screen AChE inhibitors such as tacrine and carbaryl. It has advantages such as simplicity, rapidity, flexibility, and diversity. This method was used successfully for the quantitative detection of spiked carbaryl in apple juice samples.

23.3 Herbicides

Herbicide is an agent, which is commonly used for killing or inhibiting the growth of unwanted plants, like residential or agricultural weeds and invasive species.

23.3.1 2,4-Dichlorophenoxyacetic Acid

2,4-Dichlorophenoxyacetic acid is a systemic herbicide, which selectively kills most broadleaf weeds by causing uncontrolled growth in them. On the other hand, cereals, lawn turf, and grassland are relatively unaffected. It is also called 2,4-D. It is one of the oldest and most widely available herbicides and defoliants. It is widely used as a weedkiller on cereal crops, pastures, and orchards.

A fluorescent optosensor containing a molecularly imprinted polymer (MIP) in combination with electrospinning was fabricated by Limaee et al. (2019) and then used for selective recognition of 2,4-dichlorophenoxyacetic acid. The first free radical polymerization was carried out on the surface of polyethersulfone (PES) nanofibers using a monomer, initiator, template, and a polymerizable 1,8-naphthalimide derivative as the fluorogenic monomer. PES nanofiber@fluorescent molecularly imprinted polymer (PES nanofiber@FMIP) was then produced by UV curing. It was reported that as-developed sensor was able to selectively determine 2,4-D in a linear range of 1×10^{-7} to 1×10^{-3} M with limit of detection (LOD) 1.01×10^{-8} M. It was indicated that PES nanofiber@FMIP could satisfactorily determine even trace concentrations of 2,4-D.

The effect of pH (3–9), contact time (3–90 min), amount of adsorbent (0.1–0.4 g), and herbicide initial concentration (0.5–3 ppm) was evaluated by Dehghani et al. (2014) on 2,4-D removal efficiency in the presence of granular activated carbon. It was reported that pH 3 and contact time of 60 min were found to be optimal for 2,4-D removal. The removal rate of 2,4-D increased rapidly on adding adsorbent, but it decreased by herbicide initial concentration (63%). It was also significantly enhanced by decreasing pH and increasing the contact time. It was revealed that adsorption of 2,4-D onto the granular activated carbon is best fitted to type II Langmuir model, and this reaction follows the second-order kinetics.

Cai et al. (2020) observed the degradation of 2,4-dichlorophenoxyacetic acid using anodic oxidation (AO) with a boron-doped diamond (BDD) anode, and it was found to be greatly enhanced by the combination with electro-Fenton (EF) using carbon black modified graphite felt (CB-GF) as cathode. It was reported that a high current density, low initial pH, and high concentration of Na_2SO_4 favored the degradation of 2,4-D by AO. In EF reaction, the degradation rate of 2,4-D was increased eightfold after using CB-GF cathode with a higher mineralization current efficiency (53%) and a lower energy consumption (71.8 kWh kg^{-1} TOC) due to the higher generation of H_2O_2. It was indicated that the major oxidizing radicals responsible for 2,4-D degradation by AO were $^{\bullet}OH$ (57.5%) radicals, while its contribution improved to 92% in the case of EF. EF was more efficient as compared with AO for 2,4-D degradation. It was suggested that EF using BDD anode and high-performance CB-GF cathode was efficient for the degradation of 2,4-D, and it could be a promising electrochemical advanced oxidation process for organic wastewater treatment.

Highly efficient hetero-assemblies were prepared with acidified g-C_3N_4 (ACN), polyaniline (PANI), reduced graphene oxide (rGO), and biochar. Kumar et al. (2019) used these organic semiconductors to synthesize g-C_3N_4/ACN/RGO@

Biochar (GARB), g-C_3N_4/PANI/RGO@Biochar (GPRB), and ACN/PANI/RGO@Biochar (APRB) nano-assemblies. As-synthesized catalysts were used for the visible light assisted degradation of 2,4-dichlorophenoxyacetic acid. It was reported APRB was the best to degrade 99.7% of 2,4-D on the exposure to Xe lamp within 50 min. It also retained high activity in natural sunlight. Here, rGO acts as an electron mediator and it protects higher positioned bands of PANI and ACN in APRB for its significant photocatalytic activity. It was observed that 42% of total organic carbon was removed in 2 h for 2,4-D.

Cymbopogon nardus (C.N) was first extracted via simultaneous ultrasonic-hydrodistillation (UAE-HD) extraction method, and then it was used to synthesize silver nanoparticles (Kamarudin et al. 2019). It was confirmed that spherical shape Ag nanoparticles were formed with size ranging between 10 and 50 nm. It was proven that a large number of phenolic compounds were greatly involved in the nanoparticle synthesis process. They also observed the catalytic activity of as-synthesized Ag nanoparticles toward the degradation of 2,4-dichlorophenoxyacetic acid with higher degradation performance (98%).

A photocatalyst was prepared by Mehrabadi and Faghihian (2019) via doping of titanium dioxide on clinoptilolite nanoparticles, and it was used for simultaneous degradation of 2,4-dichlorophenoxyacetic acid and 2-methyl-4-chlorophenoxyacetic acid (MCPA) mixture in the presence of ultraviolet and sunlight irradiations. It was indicated that the efficiency of photocatalyst was significantly increased by the immobilization of titanium dioxide on the clinoptilolite surface. It was also found that the band gap of photocatalyst shifted to the visible region, and the recombination of electron hole was also significantly reduced. They estimated the degradation efficiency by the determination of total organic carbon and the degradation products.

Hydrothermal synthesis of TiO_2 was carried out by Sandeep et al. (2018) using the sol-gel method, which resulted in material with good crystallinity and particle size. It was reported that hydrothermal catalyst (H-TiO_2) showed 96% degradation under UV light and 83% under sunlight.

Jia et al. (2017) constructed a novel molecular imprinting fluorescence sensor by anchoring mesoporous structured imprinting microspheres on the surfaces of quantum dots (QDs) surface. It was used for selective and sensitive detection of 2,4-dichlorophenoxyacetic acid (2,4-D). This sensor exhibited a satisfactory linearity within 0.66–80 μM, with a low detection limit of 2.1 nM within 20 min, under optimized conditions. It was successfully applied for the detection of 2,4-D in bean sprout samples. High recoveries at three spiking levels of 2,4-D ranging from 95.0% to 110.1% with precisions below 4.9% were attained. It exhibited high sensitivity and good selectivity for the separation, enrichment, and detection of 2,4-D in real food samples and ensuring food safety.

23.3.2 Clopyralid

Clopyralid is a selective herbicide used for the control of broadleaf weeds, especially thistles and clovers. Clopyralid belongs to the picolinic acid family of herbicides. It controls creeping thistle, *Cirsium arvense*, a noxious, perennial weed. It is particularly damaging to peas, tomatoes, and sunflowers and can render potatoes, lettuce, and spinach inedible; however, it does not affect grasses of the family Poaceae.

A solid-phase extraction (SPE) based on a dummy molecularly imprinted polymer (MIP) and liquid chromatography-tandem quadrupole mass spectrometry (LC-MS/MS) was developed by Tan et al. (2019) for the selective determination of four pyridine carboxylic acid herbicides (aminopyralid, picloram, fluroxypyr, and clopyralid) in samples of milk. They used adsorption isotherms and kinetics to determine the adsorption performance and specific recognition mechanism of both: MIPs and non-molecularly imprinted polymers (NIPs). It was reported that recovery at three spiking levels of 10, 20, and 50 $\mu g \cdot L^{-1}$ ranged between 75.3% and 89.8% with relative standard deviations (RSDs) <14.3%. The limit of detection (LOD) was also estimated to be 0.124 $\mu g \cdot L^{-1}$. The feasibility of this method was successfully applied to quantify clopyralid in milk.

Celis et al. (2019) prepared montmorillonite (SWy-2)-chitosan bionanocomposites (SW-CH) following different methodologies and used as an adsorbent for the removal of clopyralid in aqueous solution and soil/water suspensions. They assessed the potential of these materials to prevent and remediate soil and water contamination by anionic pesticides. The SW-CH bionanocomposites were found to be good adsorbents for the herbicide at pH level, particularly where anionic form of the herbicide ($pK_a = 2.3$) and the cationic form of CH ($pK_a = 6.3$) predominated. The performance of this bionanocomposites as adsorbents of clopyralid depends on the amount and arrangement of chitosan in the samples. It was observed that clopyralid adsorption was rapid and linear up to herbicide concentrations of 0.5 mM. High salt concentrations (0.1 M NaCl) promoted desorption of the adsorbed clopyralid from SW-CH, which strongly suggests that its adsorption occurred mainly through an ion exchange mechanism on positively charged CH sites at the surface of montmorillonite. An acidic soil (pH = 4.5) with SW-CH at rates of 5% and 10% led to a significant increase in clopyralid adsorption, but this effect was found negligible in the case of alkaline soil (pH = 8.0). It reflects the absence of positively charged sites in SW-CH at higher pH values.

23.3.3 Glyphosate

Glyphosate is a broad-spectrum systemic herbicide. It acts by inhibiting the plant enzyme 5-enolpyruvylshikimate-3-phosphate synthase, and it is normally used to kill weeds, particularly annual broadleaf weeds and grasses that compete with crops.

Mirmohseni et al. (2019) used a mixture of polydimethylsiloxane and polyacrylic resins as adsorbent for the detection of glyphosate (GLY). They developed a simple and cost-effective analysis method, which was based on quartz crystal nanobalance technique (QCN) to detect the presence of GLY. A thin layer of mixture (polydimethylsiloxane/acrylic resins) was coated on the surface of gold-coated quartz crystal electrodes. The frequency shifts were found to be linear for the concentration of 0.2–8.6 mgL^{-1} with sensitivity factor of 70.73 Hz mg^{-1} L^{-1}. It was reported that no major interference was observed in the performance of the sensor with some possible interferences tested (NaCl, $CaSO_4$, $MgSO_4$, and $NaHCO_3$).

A photoelectrochemical (PEC) sensor based on hierarchically porous Cu-BTC/g-C_3N_4 nanosheet (Cu-BTC/CN-NS, BTC = Benzene-1,3,5-tricarboxylic acid) material was constructed by Cao et al. (2019a). Hierarchically porous Cu-BTC is a composite material, which can help to efficiently capture suitable pesticide molecules and accelerate signal transmission. The CN-NS had a good optical performance. The Cu metal center can coordinate with added glyphosate to form Cu-glyphosate complexes on exposure to visible light, which leads to increased steric hindrance of electron transfer, and as a result, a decrease in photocurrent was observed. This sensor can realize the detection of glyphosate from non-electroactive to electroactive. It was indicated that this photoelectrochemical sensor has a lower detection limit of 1.3×10^{-13} mol L^{-1} and a wide range of detection (1.0×10^{-12} to 1.0×10^{-8} mol L^{-1} and 1.0×10^{-8} to 1.0×10^{-3} mol L^{-1}). Apart from this, Cu-BTC/CN-NS-based sensor has the characteristics of short detection time and easy operation and, therefore, can find potential applications in photoelectrochemical analysis.

Magnetic BiOBr/Fe_3O_4 nanocomposite photocatalysts were successfully prepared by Cao et al. (2019b) via a facile solvothermal process. As-prepared catalysts exhibited excellent photocatalytic activity toward glyphosate degradation in water in the presence of visible light. It was observed that the rate of glyphosate degradation reached up to 97% within 60 min, which was higher than that of pure BiOBr (85%). Photogenerated holes (h^+) were confirmed to be the major reactive oxidizing species in the photodegradation of glyphosate. As-prepared BiOBr/Fe_3O_4 photocatalysts have magnetic properties and also good recyclability. It was reported that after five repeated trials, the percentage of degradation of glyphosate was still more than 90%, which indicates that nanocomposites have excellent reusability as well as great potential in the treatment of industrial wastewater.

Eliana et al. (2019) prepared SBA type nanostructured catalysts modified with iron and cobalt. These catalysts were used for the degradation of glyphosate through catalytic wet air oxidation at atmospheric pressure and room temperature. The reaction products were analyzed as acetate, nitrate, nitrite, and phosphate ions. The stability of catalyst and possibility of its recycling were also studied. It was found that both Fe and Co ions can form complexes with glyphosate, but only Fe-SBA(20) catalyst could lead to the oxidative degradation of the herbicide. It was reported that Co-SBA(20) material acts as an adsorbent of herbicide, but it was not found to be active in its degradation. It was revealed that this complex was capable of activating O_2 from an airflow and generates oxoiron intermediates, which promoted the degradation of the glyphosate into short chain ions, and less toxic and more biodegradable products at ambient conditions.

23.3.4 Paraquat

Paraquat is one of the most widely used herbicides. It is quick-acting and nonselective and kills green plant tissue on contact. It is also toxic to human beings and animals due to its redox activity. It kills a wide range of annual grasses, broadleaf weeds, and the tips of established perennial weeds. Paraquat is highly toxic to mammals on ingestion including humans and leads to acute respiratory distress syndrome (ARDS).

Photocatalytic degradation of paraquat was investigated by Sorolla et al. (2012) using mesoporous-assembled Cu-TiO_2/SBA15 in the presence of UV and visible light. This catalyst was prepared by impregnation of Cu-TiO_2 colloids onto SBA-15. Colloids of Cu-TiO_2 were obtained via sol-gel method, but its mesoporous support was prepared via hydrothermal route. It was revealed that Cu-doped TiO_2 had a small crystalline size and it was well dispersed on SBA-15. The TiO_2/SBA-15 had the highest degradation of paraquat for all pH under UV illumination. It was also observed that 2 wt% Cu-TiO_2/SBA-15 show the highest activity under visible light.

The N,S/TiO_2 thin films were prepared by Zahedi et al. (2015) via sol-gel dip coating method. The 84.39% of decomposition was achieved in 5 h at optimum pH 5.8. It was revealed that nanostructure of N,S/TiO_2 thin films can be recycled, and this is stable, with a good potential for practical application.

The TiO_2 nanoparticles were synthesized by Phuinthiang and Kajitvichyanukul (2018) from the extract of *Coffea arabica* L. (TiO_2/C). These were used for the removal of paraquat from contaminated water photocatalytically. As the value of pH_{pzc} of TiO_2/C was 2.9, the surface of catalyst was highly acidic. Paraquat is easily removed in alkaline medium because of the adsorption ability of paraquat on its surface. This degradation followed the pseudo-first-order model, and the order of the rate in the presence of ultraviolet (UV) and H_2O_2 was:

$$TiO_2 / C > TiO_2 > \text{without } TiO_2$$

It was reported that using combined TiO_2/C, a maximum of 66.3% degradation was achieved within 90 min at pH 10.

23.3.5 Mesotrione

Mesotrione is a herbicide which is a member of the class of 4-hydroxyphenylpyruvate dioxygenase (HPPD) inhibitors. It prevents carotenoid from being made, so chlorophyll degrades and the plant dies. It is a herbicide used on field corn, seed corn, sweet corn, yellow popcorn, and grain sorghum. The major adverse effects are ocular lesions and liver and kidney effects.

A new, simple, sensitive, and fast analytical method was developed by Deroco et al. (2017) for the estimation of the mesotrione (MST). They used square-wave voltammetry (SWV) and a modified glassy carbon electrode with carbon black (CB/GCE) in the presence of the cationic surfactant cetyltrimethylammonium bromide (CTAB) for this purpose. It was reported that MST exhibited three well-defined irreversible reduction peaks, at −242, −710, and −1105 mV. An electrocatalytic effect was observed for the first peak, and the apparent heterogeneous electron-transfer rate constant obtained using electrochemical impedance spectroscopy for a 5.0 mmol L^{-1} $K_3[Fe(CN)_6]$/5.0 mmol L^{-1} $K_4[Fe(CN)_6]$ (1:1) solution was 9.77×10^{-3} cm s^{-1}, which was almost two orders of magnitude greater than the rate constant obtained for the GCE electrode (2.15×10^{-4} cm s^{-1}). It was reported that analytical curve for MST was linear in the MST concentration range from 0.040 to 7.2 μmol L^{-1} with a limit of detection of 0.026 μmol L^{-1}. The proposed method was used to determine the concentration of MST in natural lake and tap water samples and also in sugarcane juice samples, satisfactorily.

23.4 Rodenticides

Rodenticides are heterogeneous group of compounds exhibiting different toxicities to humans and rodents. They are among the most toxic substances found in homes. Earlier, only heavy metals (arsenic, thallium) were the often-used rodenticides.

23.4.1 Warfarin

Coumarins are used as rodenticides for controlling rats and mice in residential, industrial, and agricultural areas. Warfarin is both odorless and tasteless, and it is effective, when mixed with food bait, because the rodents will return to the bait and continue to feed over a period of days until a lethal dose is accumulated (considered to be 1 mg kg^{-1} day^{-1} over about 6 days). The use of warfarin as a rat poison is now declining, because many rat populations have developed resistance to it. People can be exposed to warfarin in the workplace by breathing it in, swallowing it, skin absorption, and eye contact.

Poly(MImEO8BS)-Ni nanocomposite was synthesized by Molaakbari et al. (2017). They applied it to modify a glassy carbon electrode along with conductive polymeric ionic liquids. It was observed that oxidation of warfarin at the surface of modified electrode occurs at a potential of about 230 mV, which is relatively less positive than that of an unmodified glassy carbon electrode. Square-wave voltammetry (SWV) showed a linear dynamic range from 1.0×10^{-6} to 1.0×10^{-4} M with a detection limit of 1.5×10^{-7} M for warfarin. This modified electrode was successfully used for the simultaneous determination of warfarin and tramadol in pharmaceutical compounds.

A simple and sensitive electrochemical sensor based on magnetic Fe_3O_4 nanoparticles modified carbon paste electrode (CPE) was developed by Gholivand et al. (2015) to detect warfarin at low level of 0.21 μM. The curve obtained was linear for the concentration of warfarin ranging between 0.5 and 1000 μM with a limit of detection of 0.21 μM. Such sensor showed excellent stability, and it can be used for the determination of warfarin in tablet, human serum, and urine with satisfactory results.

Gholivand and Solgi (2017) also developed a new sensor by electrodepositing cobalt oxide nanoparticles on multi-walled carbon nanotubes modified glassy carbon electrode (MWCNTs/GCE). The presence of cobalt oxide nanoparticles on the electrode surface was found to enhance the accumulation of warfarin experimentally and resulted in improved electrochemical response. The effects of different parameters such as pH, scan rate, accumulation potential, time, and pulse amplitude on the sensor response were evaluated. Under optimal conditions, differential pulse adsorptive anodic stripping voltammetric (DPASV) response of such modified electrode was linear in a wide range of 8 nM–50 μM with a limit of detection of 3.3 nM. This sensor was successfully applied to determine warfarin in urine and plasma samples.

A chemically modified electrode was prepared by Taei et al. (2016). They incorporated $MnFe_2O_4$ into multi-walled carbon nanotubes paste matrix (MWCNTs/$MnFe_2O_4$/CPE). The electrochemical behavior of warfarin was studied at chemically modified electrode. The MWCNTs/$MnFe_2O_4$/CPE had higher electrocatalytic activity for oxidation as it produces a sharp oxidation peak current at about +0.91 vs. Ag/AgCl reference electrode at pH 4.0. It was found that peak current depends linearly on the concentration of warfarin over the range of 0.10–447.0 μmol L^{-1} with the detection limit of 0.08 μmol L^{-1}. This method can be used as a rapid, highly selective, simple, and precise method for the determination of warfarin in biological fluids.

It was reported by Vadaei and Faghihian (2018) that the ability of SnTe photocatalyst was enhanced in the degradation of warfarin, while difficulties in the separation of used photocatalyst from solutions were also overcome by the immobilization of the photocatalyst on a suitable porous support. They prepared a novel nano-sized photocatalyst by coupling of SnTe on the surface of SBA-15 support. It was indicated that the activity of SnTe photocatalyst was improved significantly after the immobilization on the support. On the other hand, only a lower catalyst dose was needed. The visible light irradiation was found to be more effective than UV irradiation. The equilibrium was also established within 10 min. The separation of as-synthesized photocatalyst from the solution was much easier than the bulk SnTe, and the regenerated photocatalyst was retaining more than 90% of its initial efficiency.

Murgolo et al. (2017) deposited new supported catalyst composed of a nanostructured TiO_2 film on a stainless steel mesh (nanoTiO_2-SS) via metal organic chemical vapor deposition (MOCVD) technique. It was reported that the rate of removal of warfarin with this new catalyst was found almost twofold than that of TiO_2 Degussa P25.

23.4.2 Bromadiolone

A highly sensitive and specific fluorescence immunoassay was developed by Li et al. (2019) for the rapid detection of bromadiolone (BRD). It is a widely used second-generation rodenticide, and poisoning due to this in humans is relatively common. Two novel BRD haptens were first synthesized, and a sensitive and specific mAb for BRD (15C1) was produced. Then mAb 15C1 was incorporated as a detection reagent and gold nanoclusters (AuNCs) as a fluorescent probe into an alkaline phosphatase-based competitive fluorescence immunoassay. The fluorescence immunoassay exhibited high specificity and achieved a limit of detection of

0.047 ng mL^{-1} for BRD, which was almost more than tenfold better than earlier methods. It was reported that recoveries in spiked human serum were found to be in the range between 77.9% and 85.6%. They claimed that this fluorescence immunoassay could provide a good clinical tool for the rapid diagnosis of BRD poisoning.

23.5 Bactericides

A bactericide or bacteriocide (biocidal) is a substance that kills bacteria. These are disinfectants, antiseptics, or antibiotics. These can be used to control sulfate-reducing bacteria, slime-forming bacteria, iron-oxidizing bacteria, etc.

23.5.1 Bismerthiazol

Bismerthiazol is a fungicide for the control of a plant's bacterial diseases. It is used to control bacterial leaf blight, cercospora spot on rice, and canker on orange. Cheng et al. (2018) assayed bismerthiazol by a rapid and portable method employing protein-capped gold nanoclusters as probes. It was reported that the luminescent intensity of the nanoclusters showed a correlative response toward bismerthiazol from 5 to 4000 μg mL^{-1} with a linear relation in the range of 5–100 μg mL^{-1}. Even 5 μg mL^{-1} of bismerthiazol could be detected using this method. A high affinity of bismerthiazol to interact with the soybean protein-capped gold nanoclusters is responsible for the excellent selectivity of this method. They could quantify bismerthiazol in several cabbage samples, which indicated its good performance in practical applications. This method has several advantages such as simple operation, fast response, visual readout, and good selectivity.

23.6 Fungicides

Fungicides are pesticides which are used to kill or prevent the growth of fungi and their spores. They can control fungi, damaging plants, including rusts, mildews, and blights, or control mold and mildew.

23.6.1 Captan

Captan belongs to the phthalimide class of fungicides. It is a nonsystemic fungicide, which is used to control diseases of many fruit, ornamental, and vegetable crops. Captan is used as a preservative for awnings, leather, and draperies and as a root dip and seed treatment. It is also incorporated into paints, wallpaper pastes, plastic, and leather goods. Captan residues in apples cause eye and skin irritation, dermatitis, conjunctivitis, and vomiting in humans. Phthalimide fungicides include captan and captafol, which are skin sensitizers and can cause cancer and liver, reproductive, and developmental toxicity to human health.

Composite nanoparticles were synthesized by Khan et al. (2014) using an eco-friendly hydrothermal process. It was reported that well crystalline optically active composite nanoparticles were there with an average diameter of 30 nm. As-synthesized nanoparticles were utilized for the development of chemical sensor, which was fabricated by coating the nanoparticles on silver electrode. The proposed sensor exhibited high sensitivity (1.7361 $mA.mM^{-1}.cm^{-2}$), lower detection limit (8.0 mM), and a long range of detection (77.0 mM–0.38 M). It was observed that the resistance of composite nanoparticles based sensor was found to be 2.7 MV, which changed from 2.7 to 1.7 on changing fungicide concentration. As-designed sensor has advantages such as simple technique, low cost, lower detection limit, high sensitivity, and long range of detection. It can detect phthalimide (captan and captafol) even at the trace level and sense over a wide range of concentrations.

Nesakumar et al. (2015) developed an electrochemical biosensor based on acetylcholinesterase (AChE) immobilized on ZnO nanorod interface. The Pt/ZnO/AChE bioelectrode exhibited a high sensitivity of 0.538 $\mu A\ cm^{-2}\ \mu M^{-1}$ in the linear range between 0.05 and 25.0 μM with a limit of detection of 107 nM. The recovery results were appreciable (98.4% and 102.4%) in apple. The present work has a great potential for the detection of captan in apple samples.

The removal of captan from aqueous solution was observed by Tiwari and Bind (2014) using super-paramagnetic nano-iron oxide loaded poly(styrene-co-acrylic acid) hydrogel (as adsorbent) by batch as well as column method. Adsorption equilibrium was investigated as a function of initial pH, adsorbent dose, time, concentration, bed height, and flow rate. It was reported that equilibrium data fitted well to Langmuir equilibrium model. As-synthesized copolymer was magnetized in situ, and the size, structure, and coating of magnetic nanoparticles were characterized.

23.6.2 Carbendazim

Carbendazim is a member of the class of benzimidazoles. It is a broad-spectrum systemic fungicide with protective and curative action. It controls ascomycetes, fungi imperfecti, and basidiomycetes on a wide variety of crops. The main crops are cereals, fruits (banana, pome, citrus, strawberries, stone, pineapples, mangoes, avocados, etc.), vegetables, vines, mushrooms, cotton, hops, ornamentals, pasture, and turf. It produces reproductive and developmental toxic effects at high oral doses, which include decreased sperm count and histopathological changes.

Gao et al. (2019) selected nanoporous gold (NPG) with unique structural and functional properties as a recognition element to fabricate an electrochemical sensor. The simultaneous electrochemical detection of methyl parathion (MP) and carbendazim (CBM) could be achieved using NPG having a large peak potential separation of 0.70 V. It was reported that good linear responses were observed for the detection of MP and CBM, in large concentration range 3.0–120 μM, high sensitivity of 484.51 $\mu A\ mM^{-1}\ cm^{-2}$, and low detection limit of 0.24 μM for CBM. This

NPG/GCE electrode exhibited strong specificity, selectivity, and anti-interference capability in the simultaneous detection of MP and CBM. Satisfactory results were obtained for the analysis of environmental water sample.

Cyclodextrin-graphene hybrid nanosheets (CD-GNs) have been used by Guo et al. (2011) for the ultrasensitive detection of carbendazim by electrochemical method. It was reported that peak currents of carbendazim on the GNs modified glassy carbon electrode (GNs/GCE) and the CD-GNs/GCE were increased by 11.7 and 82.0 times as compared to the bare GCE, respectively. Nanocomposite film not only exhibited the excellent electrical properties of GNs, but it also showed the high supramolecular recognition capability of CDs. It was observed that the peak currents increase linearly with the concentration of carbendazim in the range of 5 nM–0.45 μM at the CD-GNs/GCE. It was found that the limit of detection for carbendazim was 2 nM (S/N = 3) and recoveries were ranging from 98.9% to 104.5%. As-developed electrochemical sensor exhibited good stability and reproducibility for the detection of carbendazim. The proposed CD-GNs-based electrochemical sensor was also successfully used for the detection of carbendazim in water samples.

Santana et al. (2019) synthesized a new nanocomposite based on ZnCdTe semiconductor nanocrystals (NCs) in situ on reduced graphene oxide (rGO). It was indicated that nanocrystals grow onto the rGO matrix. Carbon paste electrodes modified with the nanocomposite (QD-rGO/CPE) were prepared to evaluate the electrochemical performance, which showed high sensitivity in the electroanalytical detection of the carbendazim. A linear curve was obtained in the range of 9.98×10^{-8} to 1.18×10^{-5} mol L^{-1} under optimal conditions. The limit of detection (LOD) and the limit of quantification (LOQ) for carbendazim were reported to be 9.16×10^{-8} and 2.78×10^{-7} mol L^{-1}, respectively. The present electrode was successfully used for the determination of carbendazim in orange juice samples.

Zeng et al. (2010) studied the photocatalytic degradation of carbofuran solution (0.2, 0.4, 0.8 g L^{-1}) Re^{3+}-doped nano-TiO_2. They could achieve a highest degradation rate of 54.89% after 4 h degradation keeping the concentration of nano-TiO_2 0.4 g L^{-1}. Carbendazim residues on tomato leaves were determined. The photocatalytic degradation of carbendazim in tomato leaves and soil was studied with different concentrations of catalyst (0, 0.2, 0.4, 0.6, and 0.8 g L^{-1}). It was observed that the degradation rate of pesticide residues could be increased by 20–30% on the tomato leaves and 15–20% in soil, where best concentration of photocatalytic degradation required was 0.2–0.4 g L^{-1}.

Kaur et al. (2014) used a batch-type photoreactor for the degradation of carbendazim. The effects of catalyst loading, initial concentration, area/volume ratio, pH, and light conditions on the rate of degradation were studied. It was reported Aeroxide P25 exhibited better degradation efficiency than LR grade TiO_2 in the presence of both UV and sunlight. The degradation was negligible with TiO_2 under UV without catalyst. The addition of H_2O_2 did not show an appreciable increase in the rate of degradation for both the catalysts. The optimal parameters for the degradation of carbendazim are catalyst loading 1 g L^{-1}, area/volume ratio 0.919 cm^2

m L^{-1}, intensity 30 W m^{-2}, and pH 6.5, where around 85% mineralization of carbendazim could be achieved.

Iron doped TiO_2 nanoparticles were synthesized by Kaur et al. (2016a) via surface impregnation method. It was revealed that the size of crystallite was in the range of 25–34 nm. A narrowing of band gap from 3.2 to 2.8 eV was observed. The photocatalytic efficiency of as-prepared catalyst was evaluated for the degradation of carbendazim. Fe improved the photocatalytic activity of TiO_2 synergistically in the presence of sunlight. The optimum Fe loading in doped TiO_2 was found to be 2 wt%, which gives 98.5% degradation of carbendazim.

Iron and silica doped TiO_2 nanoparticles were also synthesized by Kaur et al. (2016b) via surface impregnation method. They varied the concentration of dopants Fe (1%, 2%, 3%, and 4%) and Si (3%, 5%, 7%, and 9%). It was reported that the particle size was in the range of 25–35 nm for these doped TiO_2. The 2% Fe or 5% Si doped TiO_2 showed about 98% degradation of carbendazim under sunlight. It was indicated that the absorption band edge of Fe and Si dopants were red shifted toward visible wavelength. It was also revealed that Fe and Si doped TiO_2 degrades the fungicide much faster than the undoped TiO_2 at optimum catalyst loading of 1 g L^{-1}.

23.6.3 Mancozeb

Mancozeb is a dithiocarbamate fungicide, which shows multisite protective action on contact. It is a combination of two other dithiocarbamates. This mixture controls many fungal diseases in a wide range of field crops, fruits, nuts, vegetables, and ornamentals. Mancozeb reacts with the sulfhydryl groups of amino acids and enzymes, resulting in the disruption of lipid metabolism, respiration, and production of adenosine triphosphate. It has a potential to cause goiter, a condition in which the thyroid gland is enlarged and can produce birth defects. It has also been classified as a probable human carcinogen.

The use of dopamine dithiocarbamate functionalized silver nanoparticles (DDTC-Ag NPs) has been reported by Rohit et al. (2014). They used it as a colorimetric probe for the detection of mancozeb in environmental water as well as fruit juice samples. It was observed that mancozeb induces the aggregation of DDTC-Ag NPs via Michael addition and enamine formation between DDTC-Ag NPs and mancozeb, which resulted in a color change from brownish orange to bluish. It was revealed that the response is linearly dependent on the concentration of mancozeb in the range from 5×10^{-5} to 3×10^{-4} M, and its limit of detection was 21.1×10^{-6} M. The proposed method is a very fast and efficient procedure to detect mancozeb in various samples of water (tap, canal, and river) and fruit juice (apple, grape, and tomato).

A novel and economical electrochemical sensor was developed by Kumar et al. (2016) to detect and remove mancozeb from soil and vegetable samples. They use superparamagnetic iron oxide nanoparticles (SPIONs) and molecularly imprinted star polymers. The SPIONs were synthesized via a hydrothermal method, and then these were coated with vinyl silane, using it as a platform for the synthesis of mancozeb-imprinted polymers (MISPs). The resulting MISPs exhibited a specific recognition ability and high adsorption capacity, and it can be easily extracted from complex matrices using an external magnetic field. The proposed sensor presented good electrochemical response for mancozeb in the range from 5.96 to 257.0 μg L^{-1} under optimal experimental conditions. Its detection limit was calculated as 0.96 μg L^{-1} (S/N = 3). As-fabricated MISPs have a great potential to become a selective and sensitive approach for the detection of mancozeb even at the trace level as well as its removal.

Nanomaterials are widely used in the detection of different pesticides even at lower concentrations using various types of sensors. These nanostructured materials are also able to degrade a variety of pesticides to almost harmless or less harmful compounds. These will provide a pathway of the detection and degradation of pesticides to keep our earth clean and green for years to come.

References

Alkayal NS, Hussein MA (2019) Photocatalytic degradation of atrazine under visible light using novel Ag@$Mg_4Ta_2O_9$ nanocomposites. Sci Rep 9:7470. https://doi.org/10.1038/s41598-019-43915y

Cai J, Zhou M, Lu YPX (2020) Degradation of 2,4-dichlorophenoxyacetic acid by anodic oxidation and electro-Fenton using BDD anode: influencing factors and mechanism. Sep Purif Technol 230:115867. https://doi.org/10.1016/j.seppur.2019.115867

Cao Y, Wang L, Wang C, Hu X, Liu Y, Wang G (2019a) Sensitive detection of glyphosate based on a Cu-BTC MOF/g-C_3N_4 nanosheet photoelectrochemical sensor. Electrochim Acta 317:341–347

Cao L, Ma D, Zhou Z, Xu C, Cao C, Zhao P et al (2019b) Efficient photocatalytic degradation of herbicide glyphosate in water by magnetically separable and recyclable BiOBr/Fe_3O_4 nanocomposites under visible light irradiation. Chem Eng J 368:212–222

Capoferri D, Diduk RA, Carlo MD, Compagnone D, Merkoçi A (2018) Electrochromic molecular imprinting sensor for visual and smartphone-based detections. Anal Chem 90(9):5850–5856

Celis R, Adelino MA, Hermosín MC, Cornejo J (2019) Montmorillonite–chitosan bionanocomposites as adsorbents of the herbicide clopyralid in aqueous solution and soil/water suspensions. J Hazard Mater 209–210:67–76

Cheng Y, Zhang Y, Pei R, Xie Y, Yao W, Guo Y et al (2018) Fast detection of bismerthiazol in cabbage based on fluorescence quenching of protein-capping gold nanoclusters. Anal Sci 34:414–419

Dehghani M, Nasseri S, Karamimanesh M (2014) Removal of 2,4-dichlorophenolyxacetic acid (2,4-D) herbicide in the aqueous phase using modified granular activated carbon. J Environ Health Sci Eng 12(1):28. https://doi.org/10.1186/2052-336X-12-28

Deroco PB, Lourencao BC, Filho OF (2017) The use of modified electrode with carbon black as sensor to the electrochemical studies and voltammetric determination of pesticide mesotrione. Microchem J 133:188–194

Dong G, Hou J, Tian Z, Lu J, Wang Q, Ai S et al (2016) A sensitive fluorescent sensor for selective determination of dichlorvos based on the recovered fluorescence of carbon dots-Cu(II) system. Food Chem 202:81–87

Eliana G, Vaschetto EG, Sicardi MI, Elías VR, Ferrero GO, Carraro PM et al (2019) Metal modified silica for catalytic wet air oxidation (CWAO) of glyphosate under atmospheric conditions. Adsorption 25:1–8. https://doi.org/10.1007/s10450-019-00090-w

Gao X, Gao Y, Bian C, Ma H, Liu H (2019) Electroactive nanoporous gold driven electrochemical sensor for the simultaneous detection of carbendazim and methyl parathion. Electrochim Acta 310:78. https://doi.org/10.1016/j.electacta.2019.04.120

Gholivand MB, Solgi M (2017) Sensitive warfarin sensor based on cobalt oxide nanoparticles electrodeposited at multi-walled carbon nanotubes modified glassy carbon electrode (Co_xO_yNPs/MWCNTs/GCE). Electrochim Acta 246:689–698

Gholivand MB, Torkashvand M, Yavari E (2015) Electrooxidation behaviour of warfarin in Fe_3O_4 nanoparticles modified carbon paste electrode and its determination in real samples. Mater Sci Eng 48:235–242

Guo Y, Guo S, Li J, Wang E, Dong S (2011) Cyclodextrin–graphene hybrid nanosheets as enhanced sensing platform for ultrasensitive determination of carbendazim. Talanta 84(1):60–64

Halamek J, Pribyl J, Makower A, Skladal P, Scheller FW (2005) Sensitive detection of organophosphates in river water by means of a piezoelectric biosensor. Anal Bioanal Chem 382(8):1904–1911

Hou J, Dong J, Zhu H, Teng X, Ai S, Mang M (2015) A simple and sensitive fluorescent sensor for methyl parathion based on L-tyrosine methyl ester functionalized carbon dots. Biosens Bioelectron 68:20–26

Hou J, Tian Z, Xie H, Tian Q, Ai S (2016) A fluorescence resonance energy transfer sensor based on quaternized carbon dots and Ellman's test for ultrasensitive detection of dichlorvos. Sensors Actuators B Chem 232:477–483

Hunde T, Berhe M, Tadese A, Tirfu M, Woldu A, Menasbo B et al (2017) Nano Fe_3O_4–graphite paste modified electrochemical sensor for phosphatic pesticide–chlorpyrifos. Momona Ethiop J Sci 9(1):76–89

Jia M, Zhang Z, Li J, Shao H, Chen L, Yang X (2017) A molecular imprinting fluorescence sensor based on quantum dots and a mesoporous structure for selective and sensitive detection of 2,4-dichlorophenoxyacetic acid. Sensors Actuators B Chem 252:934–943

Kamarudin NS, Jusoh R, Setiabudi HD, Jusoh NWC, Jaafar NF, Sukor NF (2019) Cymbopogon nardus mediated synthesis of Ag nanoparticles for the photocatalytic degradation of 2,4-dicholorophenoxyacetic acid. Bull Chem React Eng Catal 14(1):173–181

Kaur T, Toor AP, Wanchoo RK (2014) Parametric study on degradation of fungicide carbendazim in dilute aqueous solutions using nano TiO_2. Desalin Water Treat 54(1):122–131

Kaur T, Sraw A, Toor AP, Wanchoo RK (2016a) Utilization of solar energy for the degradation of carbendazim and propiconazole by Fe doped TiO_2. Sol Energy 125:65–76

Kaur T, Sraw A, Wanchoo RK, Toor AP (2016b) Visible–light induced photocatalytic degradation of fungicide with Fe and Si doped TiO_2 nanoparticles. Mater Today Proc 3(2):354–361

Khan SB, Rahman MM, Akhtar K, Asiri AM (2014) Detection and monitoring of toxic chemical at ultra trace level by utilizing doped nanomaterial. PLoS One 9(10):e109423. https://doi.org/10.1371/journal.pone.0109423

Kumar S, Karfa P, Patra S, Madhuri R, Sharma PK (2016) Molecularly imprinted star polymer-modified superparamagnetic iron oxide nanoparticle for trace level sensing and separation of mancozeb. RSC Adv 6(43):36751–36760

Kumar A, Sharma G, Naushad M, Muhtaseb AH, Kumar A, Hira I et al (2019) Visible photodegradation of ibuprofen and 2,4-D in simulated waste water using sustainable metal free-hybrids based on carbon nitride and biochar. J Environ Manag 231:1164–1175

Li Y-F, Qiao L-Q, Li F-W, Ding Y, Yang Z-J, Wang M-L (2014) Determination of multiple pesticides in fruits and vegetables using a modified quick, easy, cheap, effective, rugged and safe method with magnetic nanoparticles and gas chromatography tandem mass spectrometry. J Chromatogr A 1361:77–87

Li H, Wen K, Dong B, Zhang J, Bai Y, Liu M et al (2019) Novel inner filter effect-based fluorescence immunoassay with gold nanoclusters for bromadiolone detection in human serum. Sensors Actuators B 297:126787. https://doi.org/10.1016/j.snb.2019.126787

Limaee NY, Rouhani S, Olya ME, Najafi F (2019) Selective 2,4-dichlorophenoxyacetic acid opto-sensor employing a polyethersulfone nanofiber-coated fluorescent molecularly imprinted polymer. Polymer 177:73–83
Liu G, Wang S, Liu J, Song D (2012) An electrochemical immunosensor based on chemical assembly of vertically aligned carbon nanotubes on carbon substrates for direct detection of the pesticide endosulfan in environmental water. Anal Chem 84(9):3921–3928
Liu H, Pan X, Chen J, Qi Y, Qu R, Wang Z (2019) Kinetics and mechanism of the oxidative degradation of parathion by ferrate(VI). Chem Eng J 365:142–152
Mehrabadi Z, Faghihian H (2019) Clinoptilolite modified with TiO_2 for simultaneous elimination of two herbicides; 2,4-D and MCPA by UV and sunlight-assisted photocatalytic degradation. Mater Res Bull 119:110569. https://doi.org/10.1016/j.materresbull.2019.110569
Mirmohseni A, Farhadi K, Jahangiri S (2019) Application of polydimethylsiloxane/acrylic resins coated quartz crystal nano balance sensor for detection of glyphosate pesticide. Int J Environ Anal Chem 2:1–13. https://doi.org/10.1080/03067319.2019.1639687
Molaakbari E, Mostafavi A, Beitollahi H, Tohidiyan Z (2017) Synthesis of conductive polymeric ionic liquid/Ni nanocomposite and its application to construct a nanostructure based electrochemical sensor for determination of warfarin in the presence of tramadol. Talanta 171:25–31
Murgolo S, Petronella F, Ciannarella R, Comparelli R, Agostiano A, Curri ML et al (2017) A new supported TiO_2 film deposited on stainless steel for the photocatalytic degradation of contaminants of emerging concern. Chem Eng J 318:103–111
Nesakumar N, Sethuraman S, Krishnan UM, Rayappan JBB (2015) Cyclic voltammetric acetylcholinesterase biosensor for the detection of captan in apple samples with the aid of chemometrics. Anal Bioanal Chem 407(16):4863–4868
Nsibande SA, Forbes PBC (2019) Development of a quantum dot molecularly imprinted polymer sensor for fluorescence detection of atrazine. Luminescence 34:480–488
Phuinthiang P, Kajitvichyanukul P (2018) Degradation of paraquat from contaminated water using green TiO_2 nanoparticles synthesized from *Coffea arabica* L. in photocatalytic process. Water Sci Technol 79(5):905–910
Prasad PR, Naidoo EB, Sreedhar NY (2015) Electrochemical preparation of a novel type of C-dots/ZrO_2 nanocomposite onto glassy carbon electrode for detection of organophosphorus pesticide. Arab J Chem. https://doi.org/10.1016/j.arabjc.2015.02.012
Ramacharyulu PVRK, Praveen Kumar J, Prasad GK, Srivastava AR (2015) Synthesis, characterization and photocatalytic activity of Ag–TiO_2 nanoparticulate film. RSC Adv 5(2):1309–1314
Rao TS, Segne TA, Susmitha T, Kiran AB, Subrahmanyam C (2012) Photocatalytic degradation of dichlorvos in visible light by Mg^{2+}-TiO_2 nanocatalyst. Adv Mater Sci Eng 2012:1. https://doi.org/10.1155/2012/168780
Rohit JV, Solanki JN, Kailasa SK (2014) Surface modification of silver nanoparticles with dopamine dithiocarbamate for selective colorimetric sensing of mancozeb in environmental samples. Sensors Actuators B Chem 200:219–226
Sandeep S, Nagashree KL, Maiyalagan T, Keerthiga G (2018) Photocatalytic degradation of 2,4-dichlorophenoxyacetic acid – a comparative study in hydrothermal TiO_2 and commercial TiO_2. Appl Surf Sci 449:371–379
Santana P, Lima J, Santana T, Santos L, Matos C, da Costa L et al (2019) Semiconductor nanocrystals-reduced graphene composites for the electrochemical detection of carbendazim. J Braz Chem Soc 30(6):1317–1325. https://doi.org/10.21577/0103-5053.20190026
Selva TMG, Araujo WR, Bacil RP, Paixao TRLC (2017) Study of electrochemical oxidation and quantification of the pesticide pirimicarb using a boron-doped diamond electrode. Electrochim Acta 246:588–596
Senthilnathan J, Philip L (2011) Photodegradation of methyl parathion and dichlorvos from drinking water with N-doped TiO_2 under solar radiation. Chem Eng J 172(2–3):678–688
Shabeer TPA, Saha A, Gajbhiye VT, Gupta S, Manjaiah KM, Varghese E (2015) Exploitation of nano-bentonite, nano-halloysite and organically modified nano-montmorillonite as an adsorbent and coagulation aid for the removal of multi-pesticides from water: a sorption modelling approach. Water Air Soil Pollut 226(3):41. https://doi.org/10.1007/s11270-015-2331-8

Sorolla MG, Dalida ML, Khemthong P, Grisdanurak N (2012) Photocatalytic degradation of paraquat using nano-sized Cu-TiO_2/SBA-15 under UV and visible light. J Environ Sci 24(6):1125–1132
Taei M, Hasanpour F, Basiri F, Tavakkoli N, Rasouli N (2016) Highly selective differential pulse voltammetric determination of warfarin in pharmaceutical and biological samples using $MnFe_2O_4$/MWCNT modified carbon paste electrode. Micro Chem J 129:166–172
Tan S, Yu H, He Y, Wang M, Liu G, Hong S et al (2019) A dummy molecularly imprinted solid-phase extraction coupled with liquid chromatography-tandem mass spectrometry for selective determination of four pyridine carboxylic acid herbicides in milk. J Chromatogr 1108:65–72
Tang J, Xiang L (2016) Development of a probe based on quantum dots embedded with molecularly imprinted polymers to detect parathion. Pol J Environ Stud 25(2):787–793
Thomas T, Kumar J, Praveen K, Chitra KR (2011) Synthesis of Ag doped nano TiO_2 as efficient solar photocatalyst for the degradation of endosulfan. Adv Sci Lett 4(1):108–114
Tiwari A, Bind A (2014) Adsorption of pesticide (captan) onto super paramagnetic poly (styrene-co-acrylic acid) hydrogel from aqueous solution using batch and column studies. Anal Chem Lett 4(4):267–278
Truca NTT, Duc SD, Thauan DV, Tahtamouni TA, Pham T-D, Hanh NT et al (2019) The advanced photocatalytic degradation of atrazine by direct Z-scheme Cu doped ZnO/g-C_3N_4. Appl Surf Sci 489:875–822
Vadaei S, Faghihian H (2018) Enhanced visible light photodegradation of pharmaceutical pollutant, warfarin by nano-sized SnTe, effect of supporting, catalyst dose, and scavengers. Environ Toxicol Pharmacol 58:45–53
Wang J, Sun W, Zhang Z, Zhang X, Li R, Ma T, Zhang P, Li Y (2007) Sonocatalytic degradation of methyl parathion in the presence of micron-sized and nano-sized rutile titanium dioxide catalysts and comparison of their sonocatalytic abilities. J Mol Catal A Chem 272(1–2):84–90
Wang MY, Huang JR, Wang M, Zhang DE, Chen J (2014) Electrochemical nonenzymatic sensor based on CoO decorated reduced graphene oxide for the simultaneous determination of carbofuran and carbaryl in fruits and vegetables. Food Chem 151:191–197
Wang Q, Jiang J, Sui W, Lin X, Liu B (2017) Sensitive molecularly imprinted fluorescence determination of pyrethroids using green zinc oxide quantum dots. Anal Lett 50(7):1139–1149
Zahedi F, Behpour M, Ghoreishi SM, Khalilian H (2015) Photocatalytic degradation of paraquat herbicide in the presence TiO_2 nanostructure thin films under visible and sun light irradiation using continuous flow photoreactor. Sol Energy 120:287–295
Zeng R, Wang J, Cui J, Hu L, Mu K (2010) Photocatalytic degradation of pesticide residues with Re^{3+}- doped nano-TiO_2. J Rare Earths 28:353–356
Zhao D, Chen C, Sun J, Yang X (2016) Carbon dots-assisted colorimetric and fluorometric dual-mode protocol for acetylcholinesterase activity and inhibitors screening based on the inner filter effect of silver nanoparticles. Analyst 141(11):3280–3288
Zhao B, Feng S, Hu Y, Wang S, Lu X (2019) Rapid determination of atrazine in apple juice using molecularly imprinted polymers coupled with gold nanoparticles-colorimetric/SERS dual chemosensor. Food Chem 276:366–375

Use of Nanomaterials in Food Science 24

Satish Kumar Ameta, Avinash Kumar Rai, Divya Hiran, Rakshit Ameta, and Suresh C. Ameta

Abstract

The current global population is nearly 6 billion; due to this rapid population growth, there is a need to produce food in a more efficient, safe, and sustainable way, and it should be safe from the adverse effects of pathogenic organisms. A large proportion of population living in developing countries face daily food shortages as a result of environmental impacts or some other reasons like political instability, etc., while in the developed countries, food is surplus. For developing countries, the objective is to develop drought- and pest-resistant crops, with maximized yield. In developed countries, the food industry depends on consumer's demand for fresher and healthier foodstuffs. The present chapter describes the use of nanoparticles in food science.

Keywords

Nanomaterials · Food science · Food safety · Packaging

S. K. Ameta
Department of Environmental Science, Mewar University, Chittorgarh, Rajasthan, India

A. K. Rai · S. C. Ameta
Department of Chemistry, PAHER University, Udaipur, Rajasthan, India

D. Hiran
Department of Home Science, Government Meera Girls College, Udaipur, Rajasthan, India

R. Ameta (✉)
Department of Chemistry, J. R. N. Rajasthan Vidyapeeth (Deemed to be University), Udaipur, Rajasthan, India

M. Ghorbanpour et al. (eds.), *Biogenic Nano-Particles and their Use in Agro-ecosystems*, https://doi.org/10.1007/978-981-15-2985-6_24

24.1 Introduction

The food market demands new technologies, which are essential to keep market leadership in the food processing industry to produce fresh, authentic, convenient, and flavorful food products, prolonging the product's shelf life and freshness with improved quality food (Alfadul and Elneshwy 2010). The new materials, products, and applications are anticipated to bring lots of advancements and improvements to the food and relevant sectors, impacting agriculture and food production, food processing, distribution, storage, nanoadditives, cleaning, and sensors for the detection of contaminants and developments of innovative products. Nanotechnology is an area of rising attention and unwraps new possibilities for the food industry. Nanotechnology integrates several disciplines, including physics, chemistry, biotechnology, and engineering.

There is a public opinion in general about nanotechnology applications, and it ranged from neutral to slightly positive one. There are suggestions that consumers should remain cautious about using nanofoods. The application of nanotechnology to the agricultural and food industries was first addressed by the US Department of Agriculture road map in September 2003 (Rashidi and Khosravi-Darani 2011). It has been predicted that nanotechnology will transform the complete food industry, changing all the way food is produced, processed, packaged, transported, and consumed. Food undergoes a variety of modifications in postharvest and processing that are likely to affect its biological and biochemical features, and here, nanotechnology developments could eventually also influence the food industry.

According to a definition in a report by the European Nanotechnology Gateway, a food is called nanofood when nanoparticles, or nanotechnology techniques, are used during cultivation, production, processing, or packaging of the food (Nanowerk 2019). It should be made clear here that it does not mean that it is atomically modified food or produced by nanomachines.

Applications of nano include smart packaging, on-demand preservatives, and interactive foods. Building on the concept of on-demand food, the idea of interactive food was to allow consumers to modify food depending on their own nutritional needs or tastes. There are many positive impacts of nanotechnology in the food industry, and these are expected to grow in the near future. Applications of nanomaterials will continue to affect the food industry commercially because of their unique and novel properties. New nanomaterials are developed that will make not only a difference in the taste of food but also safety and the health benefits that food is bound to deliver. This rapidly developing technology is concerned about every aspect of the food system starting from growth, packaging, processing, transportation, shelf life, and availability of nutrients. Therefore, nanotechnology may play a potential role in promoting the food industry.

24.2 Applications of Food Nanotechnology

Nanotechnology is the science of very small materials, and it can have a big impact in the food industry. Nanotechnology possesses a great potential in ensuring modification of color and flavor and nutritional values, increasing the shelf life of food, and monitoring the integrity of food via barcodes such as cold chain, i.e., whenever there is a slight change in food storage conditions because of its submicroscopic nature (Aigbogun et al. 2017). It is an emerging area of science with potentials to generate radical new products and processes in the food sector. It is commonly distinguished between two forms of nanofood applications:

- Food additives (nano inside)
- Food packaging (nano outside)

Concepts in nanotechnology provide a sound framework for better understanding of the interactions and assembly behavior of food components into microstructures, which is likely to affect food structure, rheology, and functional properties at the submicroscopic scale (Sanguansri and Augustin 2006). It can also modify permeation of materials by the incorporation of synthesized nanoparticles (zinc, silver, gold, etc.) for improved packaging system (Fig. 24.1).

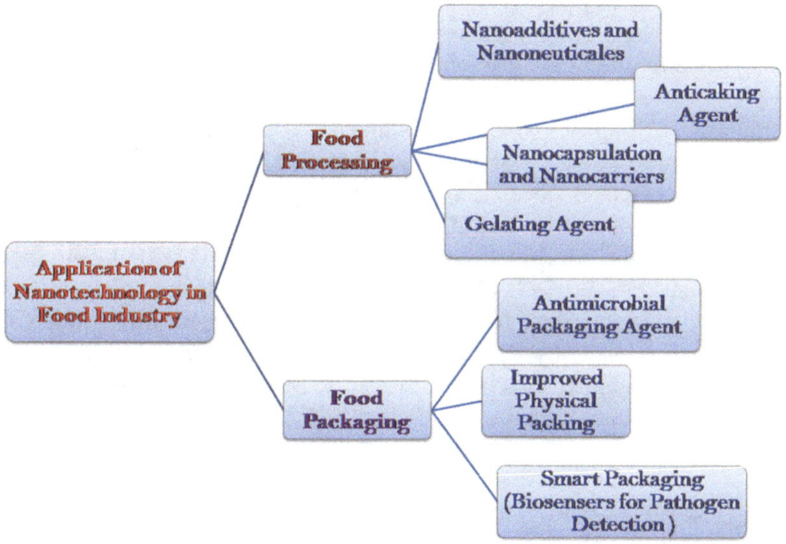

Fig. 24.1 Application of food nanotechnology

24.2.1 Food Processing

Food processing deals with transformation of agricultural/animal products into food. It also includes many forms of processing foods ranging from grinding grain to make raw flour for home cooking to complex industrial methods used to make convenience foods. Food processing can be improved much in the aspects of smart delivery of nutrients, nanoencapsulation of nutraceuticals, bioseparation of proteins, rapid sampling of biological and chemical contaminants, solubilization, delivery, and color in food systems. These are some of the emerging uses of nanotechnology in food science (Ravichandran 2010). Nanotechnology can assist in the development of functional or interactive foods, which respond to requirements of the human body and can deliver these nutrients more efficiently. Various research groups are also developing new on-demand foods, which will remain dormant in the body and deliver nutrients to cells as and when needed. The concept is that thousands of nanocapsules containing flavor or color enhancers, or added nutritional elements (such as vitamins), would remain dormant in the food and these will be released only, when triggered by the consumer (Amin et al. 2015). A key element in this sector is the development of nanocapsules that can be incorporated into food to deliver nutrients timely. Other developments in food processing include the addition of nanoparticles to existing foods to enable increased absorption of nutrients. One of the bakeries in Western Australia is quite successful in incorporating tuna fish oil (a source of omega-3 fatty acids) in nanocapsules in their top-selling product "Tip-Top" Up bread (Bund 2008). The microcapsules are designed to break open only when they have reached the stomach, thus avoiding the unpleasant taste of the fish oil.

In this aspect, food and cosmetic companies are already working together to develop newer mechanisms to deliver vitamins directly to the skin. Nestlé, having 49% stake in L'Oréal, is developing transparent sun creams so that vitamin E is delivered directly to the skin. The major objective is to manufacture a cream which is first absorbed by the skin and then releases vitamin E slowly, apart from providing protection against UV. While Estée Lauder is manufacturing antiaging formulations making use of nanoparticles (Joseph and Morrison 2006), Unilever is developing ice creams with low fat by decreasing the size of emulsion particles, which is responsible for its texture. It is hoped that up to 90% less of the emulsion will be used and that will decrease fat content from 16% to about 1% (Verma and Gupta 2017).

24.2.1.1 Nanosensors

Nanosensors are devices that can measure physical quantities and convert those quantities to respective signals that can be detected and analyzed. These nanosensors are diagnostic devices to monitor the quality of food processes and also safety of food. Such nanosensors are used for the detection of very minute amounts of any chemical pollutant, virus, or bacteria in food systems. There is a possibility of combining biology with nanoscale technology in fabrication of sensors with a great potential of higher sensitivity and reduced response time. Nanoparticles may selectively attach themselves to food pathogens, and even traces of harmful pathogens could be detected with sensors using either infrared light or magnetic materials. It is advantageous that numerous nanoparticles can be placed on a single nanosensor so as to detect the presence of different bacteria and pathogens rapidly and accurately also. Another advantage is that nanosensors can gain access into the tiny crevices because of its small size, where the pathogens often used to hide, and nanotechnology will reduce the time taken for this detection from days to a few hours, minutes, or even seconds (Choudhury and Goswami 2012). Nanotechnology utilizes biological molecules like sugars or proteins as target-recognition groups for nanostructures as biosensors (Charych et al. 1996). These biosensors could serve as detectors of food pathogens and other contaminants and also to track food products.

Nanosensors may play a significant role in detection of any traces of pesticide available in various food products, providing a potent solution to food safety. These are of significant importance as such devices are capable for finding out and detect very minute quantity of organic compounds, very low concentration of pathogens, and other harmful chemicals. These devices show extraordinary sensitivity, quick response, and rescue (Otles and Yalcin 2010; Yalcin and Otles 2010). Mostly, nanosensors are used in recognition of pesticide as organophosphate in plants, fruits, and aquatics. It is known that pesticides are highly permeable and soluble and these are harmful. These are extensively used in agronomy. There is significant importance in analysis of residue of highly sensitive pollutants (Mclaren et al. 2009). Nanosensors devices have benefits over other techniques like gas/liquid chromatography and mass spectroscopy, because they contain high surface to volume ratios, primarily loading of more antibody/enzymes (great sensitivity interface), less recognition limits, exceptional selectivity with small size, and quick response. It has been known that the use of these nanosized materials is likely to enhance the sensitive transducer indication or signals. Carbon nanotubes semiconductor and metal nanostructures have exceptional electrical or optical characteristics (Hwang et al. 2011). Intelligent packing is another aspect, where covering a nano-biosensor makes these to fluoresce in various colors on interaction with different pathogens of food materials. Various types of devices have been developed to identify contaminants, chemicals, and pathogens in food materials. Such device is highly sensitive in identifying *E. coli* and *Salmonella* (Chen et al. 2008).

Traditional methods to screen food to find disease-causing microbes can take as long a day, which are normally very slow so that they may or may not be able to efficiently catch tainted products. However, some faster methods are known, but they have their own limitations. Magnetic resonance can detect extremely low levels of bacteria, but it is almost ineffective at higher bacteria concentrations, while fluorescence is opposite. A hybrid nanosensor incorporating magnetic resonance and fluorescence has been developed to rapidly detect the presence of *E. coli* (Banerjee et al. 2016). It has been reported that detector could sense different concentrations of a pathogenic strain of *E. coli* known as O157: H7 during test of milk in less than an hour. Such sensors will find the use in detecting environmental contamination control in the food sectors.

24.2.1.2 Interactive "Smart" Food Using Encapsulation

Nanoencapsulation is a technology to pack substances in miniature using techniques like nanocomposite, nanoemulsification, and nanostructuration providing final product functionality that includes controlled release of the core (Sekhon 2010). Nanocapsules can be incorporated into food to deliver any nutrient. Addition of nanoparticles to existing food can also enable increased absorption of these nutrients (Jampilek et al. 2019). Nanoparticle additives could easily be absorbed by the body, and these could increase shelf life of the product. Nanosized dispersions, emulsions, and filled micelles have an advantage that they are not subjected to sedimentation resulting in better life span and storage of the product. As size of these nanoadditives is much smaller as compared to wavelength of light, they can be easily incorporated even in clear and transparent foods without causing problems of colors. Substances, which are difficult to dissolve by the body, can more easily be absorbed in nanoscale size, because of their larger surface area. If any active substance is to be protected during storage or its passage through the intestines, the nanotechnology can provide perfect protective layers. It is also possible to tailor these protective layers so as to release active substances in an intelligent way (caused by a change of pH value).

It may also be useful in protection against environmental factors and used in the design of food ingredients flavors and antioxidants (Imafidon and Spanier 1994). The main aim is to improve the functionality of such ingredients and keeping their concentration minimum. As the requirement of novel ingredients into foods is gaining popularity, better delivery and controlled release systems for nutraceuticals will be needed (Haruyama 2003; Lawrence and Rees 2000).

Bioactives, like coenzyme Q10 (CoQ10), vitamins, iron, calcium, curcumin, etc., have been widely tested in nanodelivery systems (He and Hwang 2016). Different nanodelivery vehicles have been developed such as association colloids, lipid-based nanoencapsulators/nanocarriers, nanoemulsions, biopolymeric nanoparticles, nanolaminates, nanofibers, etc. These nanodelivery systems can increase the bioavailability of bioactives by different pathways. Nanoencapsulation can enhance bioavailability of bioactive compounds after oral administration through targeted delivery systems. Such nanoencapsulation enables to control the release of flavors at the desired time and also to protect the degradation of these flavors during processing and storage (Yu et al. 2018).

Nowadays people are requiring more nutritional supplements because of the fact that many nutrients in food are being destroyed in the digestive tract. Each part presents a completely different environment, from oral cavity to the colon. In other words, there are a number of factors which decide the absorption of food in the body for infants, children, adults, old people, and those who are suffering from any type of gastrointestinal diseases. A nutrition delivery system is a system or nanocarrier that delivers nutrition to specific places. Chitosan is such a nanocarrier (Maestrelli et al. 2006). Although a delivery system has numerous functions, one of them is to transport a functional ingredient to its desired site. Just like taste, texture, and shelf life, major functions of a delivery system for a food product are that it should protect an ingredient from chemical or biological degradation, such as oxidation, and controlling the rate of release of functional ingredient under specific environmental conditions. Nanodispersions and nanocapsules are ideal mechanisms for delivery of functional ingredients because they can effectively perform all these tasks.

One of important part of the food industry is extracting nutrition from raw materials. Conventional methods for food processing are being replaced by newer techniques like nanotechnology, which will play a major role here. These techniques may improve food processing yields and decrease waste or spoilage of nutrition. Nutrition delivery systems must be prepared with biodegradable materials to prevent adverse effects on health of consumers. Some of the nanodelivery vehicles are as follows.

24.2.1.2.1 Association Colloids

A colloid system of a substance contains small particles dispersed throughout. An association colloid is the colloid, when particles are made up of even smaller molecules. It is already used to deliver polar, nonpolar, and amphiphilic functional ingredients (Golding and Sein 2004). The size of association colloids ranges between 5 and 100 nm, and there are usually transparent solutions. Vesicles, bilayers, micelles, reverse micelles, and liquid crystals are some of the examples of association colloids. The major disadvantages of such colloids are that they may compromise with the flavor of the ingredients and these can spontaneously dissociate on dilution.

24.2.1.2.2 Nanoemulsions

An emulsion is a mixture of two or more liquids (such as oil and water) that are immiscible and, as such, do not easily combine. A nanoemulsion is an emulsion, when the diameters of the dispersed droplets is about 500 nm or even less than that. Nanoemulsions can encapsulate functional ingredients within their droplets facilitating a reduction in their chemical degradation (McClements and Decker 2000). Different types of nanoemulsions having more complex properties such as nanostructured multiple emulsions or nanostructured multilayer emulsions have multiple encapsulating abilities from a single delivery system that can carry several functional components. Here, a functional component is encased within one component of a particular multiple emulsion system, which could be released in response to a specific environmental trigger.

24.2.1.2.3 Biopolymeric Nanoparticles

Food-grade biopolymers like polysaccharides or proteins can be used to produce nanosized particles (Tak et al. 2015). A single biopolymer separates into smaller nanoparticles with aggregative (net attraction) or segregative (net repulsion) interactions. Nanoparticles can then be used to encapsulate functional ingredients and release them in response to particular environmental triggers. Most common components used biodegradable biopolymeric nanoparticles such as polylactic acid (PLA). PLA is quite commonly used to encapsulate and deliver drugs, vaccines, and proteins, but it has certain limitations:

- It is quickly removed from the bloodstream.
- It remains isolated in the liver and kidneys.

Therefore, PLA needs an associative compound such as polyethylene glycol to be successful in delivering active components to other areas of the body (Taylor et al. 2005).

24.2.1.2.4 Nanolaminates

Apart from nanodispersions and nanocapsules, nanolaminate is also a nanoscale technique, which is commercially used in the food industry. It consists of two or more layers of material with nano-dimensions, and it is an extremely thin food-grade film (1–100 nm per layer), which has physically bonded or chemically bonded dimensions (Ravichandran 2010). A nanolaminate has a number of important applications in the food industry due to these advantages in the preparation of edible films. Such edible films are present on wide range of foods such as chocolate, candies, fruits, vegetables, meats, baked goods, French fries, etc. (Pavlath and Orts 2009). These films protect foods from gases, humidity, and lipids. They can also improve the textural properties of foods and serve as carriers of colors, flavors, nutrients, antioxidants, and antimicrobials.

Presently, these edible nanolaminates are prepared from polysaccharides, proteins, and lipids. Polysaccharide and protein-based films provide good protection against oxygen and carbon dioxide, but these are poor to protect against moisture. While lipid-based nanolaminates are significantly good to protect food from moisture, they have limited resistance to gases and show poor mechanical strength. Neither polysaccharides and proteins nor lipids provide all the properties in an edible coating and identify additives. Newer additives are searched that can improve these properties, e.g., polyols. Foods can be coated with nanolaminates either by dipping them into a series of solutions containing substances or by spraying substances onto the food surface (Shit and Shah 2014). The degree of adsorption of any substance depends on the nature of surface of food as well as on the nature of the adsorbing substance. Various adsorbing substances can form different layers of a nanolaminate; some of these are polyelectrolytes (proteins and polysaccharides), charged lipids, and colloidal particles. Different nanolaminates include varying functional agents such as flavors, colors, antimicrobials, anti-browning agents, antioxidants, enzymes, etc.

24.2.1.2.5 Nanofibers and Nanotubes

Nanotechnology has two more materials, which are likely to have an impact on the food industry. These are nanofibers and nanotubes. Nanofibers are usually not composed of food-grade substances; therefore, nanofibers have only fewer applications in the food industry. Nanofibers have small diameters in the size from 10 to 1000 nm, which makes them ideal to serve as a platform for bacterial cultures (Guo 2015). These could also serve as the structural matrix for artificial foods and food packaging material, which are eco-friendly. As the efforts are growing in the area of producing nanofibers from food-grade materials, their use is likely to increase in the future. Similar to nanofibers, the use of nanotubes is mainly for nonfood applications. Carbon nanotubes are commonly used as low-resistance conductors and catalytic reaction vessels. Certain globular milk proteins can self-assemble into similarly structured nanotubes in some appropriate environmental conditions (Graveland-Bikker and de Kruif 2006).

24.2.1.3 Antimicrobial Properties

Microbial contamination is there due to pathogenic infections and poor nutrition associated with major types of food, but in particular weaning foods. Bacterial deterioration is one of the most discussed subjects as far as the production, processing, transport, and storing of food are concerned. Newer nano-antimicrobials have shown potential in safeguarding deterioration of food, and as a result, the shelf life of food is extended (Mitura and Zarzycki 2018). A number of metal and metal oxide nanomaterials have been reported to be effective as antimicrobials. Their physicochemical properties are considered responsible for excessive formation of reactive oxygen species (ROS), which leads to oxidative stress and subsequent cell damage (Fu et al. 2014; Wu et al. 2014).

Release of metal ions at all the places can affect cellular structure or function, whether it is outside the cell, at the cell surface, or within the cell. Metal or metal oxide-based nanocomposites are used in food packaging and coating or sometimes even as ingredients of food. Silver nanoparticles and its nanocomposites are most commonly used nanomaterials as antimicrobials in the food industry (He and Hwang 2016). The use of a number of silver-containing zeolites or similar substances as food contact materials has been approved by the USFDA for the purpose of disinfection (Duncan 2011). These nanoparticles are used as a source of Ag^+ ions, which binds to membrane proteins, forming pits and/or causing other morphological changes (Morones et al. 2005). These also catalyze the generation of ROS in bacterial cells, which, in turn, leads to cell death through oxidative stress (Kim et al. 2007). It has been suggested that silver nanocomposites are quite safe for packaging of food, as no detectable or negligible levels of silver nanoparticles are released, which migrate from containers to actual food samples and food stimulants (Ntim et al. 2015). Nanocomposites have an advantage of enhanced stability, which is very much required for sustaining antimicrobial activity and reducing the possibility of migration of metal ions into stored foods. Some polymers are designed to form nanocomposites with metal/metal oxide nanomaterials for various applications in food science. Polymers most widely used in nanocomposites include gelatin, polylactic acid, isotactic polypropylene, and low-density polyethylene (LDPE). Nanocomposites of LDPE with metals (Ag/LDPE) and metal oxide (CuO/LDPE, TiO_2/LDPE, and ZnO/LDPE) are used in some food applications. Apart from these ZnO/gelatin, ZnO/polylactic acid, and ZnO/graphene oxide/polylactic acid, ZnO/polycarbonate, ZnO/isotactic polypropylene, Ag/OMteLDPE, and

Ag/poly(3-hydroxybutyrate-co-18 mol%-3-hydroxyvalerate) are also used specifically for food packaging applications (He and Hwang 2016). Polystyrene, polyvinylpyrrolidone, and poly(vinyl chloride) are also reported along with chitosan, as nanocomposite films, which binds to Cu or ZnO nanomaterials to control the growth of food pathogens or inactivate these (Li et al. 2009).

24.2.1.4 Protection Against Chemical Ingredients

24.2.1.4.1 Antioxidants

As some metal/metal oxide nanomaterials cause oxidative stress through the formation of ROS (Manke et al. 2013), therefore, efforts were made to develop less reactive nanomaterials so that they can act as antioxidant carriers. Polymeric nanoparticles are considered suitable for the encapsulation of bioactive compounds such as flavonoids and vitamins, which are released in the stomach having acidic environments (Pool et al. 2012). SiO_2-gallic acid nanoparticles were also developed, and these were tested as antioxidant with its scavenging capacity of DPPH (2,2-diphenyl-1-picrylhydrazyl) radicals (Deligiannakis et al. 2012). Browning of fresh-cut fruits is another problem, which can be controlled by the application of antioxidant treatments in association with edible coating, because browning of fresh-cut fruits is an undesirable effect due to conversion of phenolic compounds into some dark-colored pigments in the presence of oxygen (air), during storage and marketing (Garcia and Barrett 2019). But only some applications of nanomaterials directly as anti-browning agents have been known. The shelf life of Fuji apples (as a fresh-cut product) was enhanced using nano-ZnO-coated active packaging (Li et al. 2011).

24.2.1.5 Enhancement of Physical Properties

24.2.1.5.1 Color Additives

A wide range of nanoscale color additives has been prepared and studied. These additives must be approved by the Office of Cosmetics and Colors in the Center for Food Safety and Applied Nutrition and the USFDA and used only for approved purpose, specifications, and of course restrictions (He and Hwang 2016). Certain such nanomaterial products have been duly approved to be used as food color additives. TiO_2 is approved as a food color additive with the limit that it should not exceed 1% w/w (Shi et al. 2013). It has also been permitted to use color additive mixtures made with TiO_2 containing SiO_2 and/or Al_2O_3 as dispersing aids, but their limit should not be more than 2% of the total.

24.2.1.5.2 Flavors

Flavors provide sensory perception of taste along with smell to stimulate appetite and eating experience, and therefore, it is one of the most important parts of the food system. Nanoencapsulation technique has been quite commonly used to improve release and retention of flavor and to deliver dietary balance (Nakagawa 2014). SiO_2 nanomaterials can also act as carriers of these fragrances or flavors in food as well as nonfood products (Dekkers et al. 2011).

24.2.1.5.3 Anticaking Agents

SiO_2 is also used for thickening of pastes (as an anticaking agent) so as to maintain flow properties in powdered products along with as a carrier of fragrances or flavors in food and nonfood products. As a result, it has been used in food products and is registered within the EU as a food additive E551. However, still there is a debate regarding the health and safety issues related to the use of such engineered nanoparticles in consumer products (Athinarayanan et al. 2014).

24.2.2 Packaging and Food Safety

Food is a perishable item. It can be contaminated and/or degraded at any stage of the food chain. The process may be chemical, physical, or biological. The introduction/presence of any pathogen in food can result in its poisoning, which can be deadly. Therefore, it is of utmost importance that the food must be protected at all levels. A good-quality packaging material is thus required, which is safe, nontoxic, and cost-effective. Packaging using nanomaterial controls pH, temperature, moisture, and freshness of the material kept inside the packet. It also contains information for consumers, along with controlling the environment to enhance the shelf life of the food material. It provides smart packaging and extends the shelf life of a product so that food material can be transported to a long distance also. Nanosensors are developed for smart packaging so as to detect spoilage of food and release nano-antimicrobials as and when required, to extend shelf life. It helps markets and in keeping the food fresh and that too for a longer period. There is a great demand for nano-enabled packaging for food, beverage, and pharmaceutical industries due to the regularly changing consumption patterns. Such nanopackaging systems can repair small losses like holes/tears because of environmental conditions (temperature and moisture) and make the customer alert if the food is being contaminated due to some or the other reasons. Nanoscience can provide solutions for such problems, change in permeation behavior of foils, enhancing barrier properties (like thermal, chemical, mechanical, and microbial), improving upon mechanical and heat resistance properties, introducing active antimicrobial and antifungal surfaces, and sensing/signaling any microbiological and biochemical changes that have occurred (Alfadul and Elneshwy 2010). Not only this, the cost of food additive ingredients can be lowered and the shelf life of food products can be increased using this technology. It is potential frontier of material science in packaging using nanomaterials. It has been estimated that advancement in nanotechnology supported by increased global investments has thrusted the nano-enabled packaging market all over the world in the past few years (Nano-enabled Packaging Market 2019).

About 400 companies and more than 400,000 scientists in the world are developing different nanotechnology applications in food and food packaging (Neethirajan and Jayas 2011). It has been estimated that nanotechnologies are projected to make use of nanomaterials of about US dollar 3 trillion by the year 2020 (Wesley et al. 2014). Here, smart packaging industry is growing relatively faster than predicted and is reaching maturity. The demand of consumer today is much more from packaging, particularly from protecting the quality, freshness, and safety of foods as well as convenience.

Several organizations have developed smart packaging systems. Researchers of Rutgers University have developed an electronic tongue for inclusion in packaging for Kraft Foods, which consists of an array of nanosensors extremely sensitive to gases released by spoiling food. They make sensor strip to change its color giving a clear signal, whether the food is fresh or not (World Bank 2017). A packaging film was developed by the Durethan KU2–2601, which is relatively lighter, stronger, and more heat resistant than those available currently in the market (Rani et al. 2017; Hamad et al. 2018). The main purpose of these food packaging films is to protect contents from drying out, moisture and oxygen. Such film is known as a hybrid system that contains number of silicate nanoparticles, thus reducing the entrance of oxygen and other gases and the exit of moisture, preventing food from spoiling.

Plastic bottles are normally used by breweries in shipping of beer, as these bottles are lighter than glass and low cost than metal cans. But, alcohol in beer reacts with the plastic of the bottles, which shortens the shelf life of alcohol to a great extent. Voridan has developed a nanocomposite in association with Nanocor, containing clay nanoparticles and named as Imperm (Srinivas 2016). These bottles are having both the qualities; these are lighter and stronger than glass and also less likely to shatter. It has been suggested that such nanocomposites structure minimizes the loss of carbon dioxide from the beer and the ingress of oxygen to the bottle, so as to keep the beer fresh up to a 6-month shelf life (Lua and Bowles 2013). Whatever may be the impacts of nanotechnology on the food industry and products entering the market, as the safety of food is the main concern, there is an urgent need to find new sensors, which will not only ensure food safety and security, but they will make customers and shopkeepers alert that a food is going to reach the end of its shelf life. Antimicrobial coatings and dirt repellent plastic bags are used ensuring the safety and security of packaged food. This will solve food shortage crises by ensuring that food reaches to masses in time and with good qualities (Hamad et al. 2018).

Nanosensors used in different food packaging industries include time-temperature integrator and gas detector (Pradhan et al. 2015). Nanoparticle in solution, nanoparticle-based sensors, array biosensors, electronic noses, nano-test strips, and nanocantilevers are among the different types of nanosensors used (Tang et al. 2009). Electronic noses are the type of sensor, which uses several chemical sensors attached to a data processing system (Vidhyalakshmi et al. 2009). Such sensor behaves just like our human nose, and therefore, this sensor is termed as electronic nose (e-nose). There are also reports of electronic tongue (e-tongue) sensors that are based on the principle of an electronic nose. The color is changed in contact with any sign of spoilage in the food material indicating that the food does not remain fit for consumption (Yuan et al. 2008). Packaging with such nanosensors can identify conditions of food and containers (internal and external) throughout the supply chain. Nanosensors can also detect gases in food, spoiled in plastic packaging, where a change in color of packaging alerts the consumer.

A number of types of packaging materials are used in food sectors, which include active packaging, smart/intelligent packaging, edible coating, and biobased (biodegradable) polymeric films (Rai et al. 2018). Nanotechnology-driven food packaging has been categorized as follows.

24.2.2.1 Active Packaging

Active packaging means the use of active nanomaterials like antimicrobials and oxygen scavenging materials. The use of such nanomaterials is beneficial to interact directly with food to provide better protection to the product. Some nanomaterials can provide antimicrobial properties to food packaging. Some of them are nanosilver, nano-titanium dioxide, nano-magnesium oxide, nano-copper oxide, carbon nanotubes, etc. Active packaging utilizes the packaging materials, which interact with the environment, and food also and plays an active role in increasing the shelf life of products. It allows packages to play a dynamic role in food preservation. Advances made in active packaging leads to delayed oxidation, controlled respiration rate, microbial growth, and moisture migration (Brody 2006).

These packaging technologies include absorbers of carbon dioxide, odor, and ethylene and emitters of CO_2 and aroma. But purging moisture control and oxygen removal remain prominent in active packaging, and out of these, purge control has been most successful on commercial scale. Best example is the use of drip-absorbing pad in the poultry industry (Suppakul et al. 2003). Active packaging technology also involves change in permselectivity (selective permeation) of package materials to different gases. Some nanocomposite materials have been used in active packaging so as to prevent oxygen, carbon dioxide, and moisture from reaching the food (Brody et al. 2008). An active system involving moisture scavenging has been quite commonly used particularly for packaging of dried and moisture-sensitive foods, while oxygen scavengers are normally inserted into the package in the form of small sachets just to reduce the oxygen level within the package, because an environment free from oxygen prevents oxidation of food as well as the growth of aerobic bacteria and mold. Ethylene-absorbing material is incorporated in packaging materials with the object of slowing the process of ripening and senescence of fruits and vegetables triggered by ethylene (Brody 2009).

Different antimicrobial agents are also incorporated in packaging so as to prevent the growth of spoilage and pathogenic microorganisms, which can directly influence the shelf life of products. It is more advantageous compared to direct addition of some antimicrobial agents onto foods either by sprays or drips. This packaging permits slow but a continuous release of antimicrobial agent from this packaging material to food surfaces, so that a high concentration of these agents is not there over a long period (Quintavalla and Vicini 2002). Different enzymes, bacteriocins, essential oils, anhydrides, and weak organic acids have also been investigated in terms of antimicrobial activity of food packaging systems (Corrales et al. 2014).

Improved packaging by nanomaterials involves mixing them into the polymer matrix, so as to improve the gas barrier properties and also resistance of the packaging against temperature and humidity. The US Food and Drug Administration has also approved the use of these nanocomposites in contact with food.

24.2.2.2 Smart/Intelligent Packaging

Smart packaging is designed to sense any microbial or biochemical changes in the food products as it can detect the growth of any pathogens in the food. Some such smart packaging has been developed to be used as a tracking device for food safety. Presently, British Airways, MonoPrix supermarket, and Nestlé are using chemical sensors, which can quickly detect any color change (Pehanich 2006).

Nanotechnology has been used to manufacture a "smart" packaging, which can extend the shelf life of food dramatically, which permits it to be transported to a longer distance. Intelligent or smart packaging has been designed to monitor and communicate information about quality of food. It includes time-temperature indicators (TTIs), ripeness indicators, biosensors, and radio frequency identification. Such smart devices may be either incorporated in package materials itself or these are attached inside or outside of a package (Kerry et al. 2006; Yam et al. 2005; Kerry and Butler 2008). Smart packaging either responds to change in environmental conditions, repairs, or alerts the consumer about such contamination and/or the presence of some harmful pathogens. It is capable of detecting spoilage of food and release nanoantimicrobes to extend shelf life of food, so that supermarkets can keep their food materials even for a longer time periods before it is sold. Nanosensors are used as tiny chips invisible to the human eye, embedded into food products that act as electronic barcodes (Sekhon 2010).

Intelligent packaging is also developed, with specific preservative, which starts releasing preservatives as soon as food starts spoiling. Such "release on command" preservative packaging is based on a bio-switch. These are "smart" food packaging, which will warn, when oxygen has got inside or if food is being spoiled. These packaging are already in use in brewing and dairy industries. These consist of nanofilters, which can filter microorganisms and sometime even viruses. In some experiments, color was successfully removed from beetroot juice, but the flavor is not affected. Similarly red wine was turned into colorless or white in color. Lactose can also be filtered from milk, and it is replaced with some other sugar so that milk can be used by the lactose intolerant also. Nanoceramic particles are used for clustering of dirt molecules, so as to keep cooking oil fresh (Llobet et al. 2007; Wooster 2010). Nanotechnology has proved it worth in food safety by developing highly sensitive and low-cost nanosensors, which can respond to changes in environmental conditions during storage, degradation products, or contamination by microbes. Such nanosensors can be effectively used in packaging materials (Bouwmeester et al. 2014; Liao et al. 2005).

Time-temperature indicators (TTIs nanosensors) have been designed to monitor, record, and translate the safety of food. These TTIs allowed consumers to know about the quality of purchased materials. They also allow manufacturers to locate their foods along the supply line. A system based on gold nanoparticles for chilled foods has been developed by Timestrip. The system looks red above freezing temperature, but red color is lost because of agglomeration of the gold nanoparticles, when accidental freezing occurs (Robinson and Morrison 2010).

Gas sensors are also used for identification and quantification of various microorganisms, because they emit gas. Metal oxide gas nanosensor is most commonly used due to their high sensitivity and stability (Setkus 2002). Conducting polymers based nanosensors are also used because of their capability for identification and quantification of such microorganisms based on their gas emissions (Ahuja et al. 2007). Nontoxic and irreversible oxygen sensors were developed to assure the absence of oxygen in oxygen-free food packaging systems. Here, a UV-activated oxygen indicator was used, which is almost colorless on UV exposure, but blue color is restored in the presence of oxygen (Lee et al. 2002). Rapid, sensitive, and relatively low-cost diagnostic methods for detection of pathogens are being developed making use of unique magnetic, electrical, luminescent, and catalytic properties of nanomaterials (Merkoci 2010; Ayala-Zavala et al. 2014).

An electronic tongue or nose device has been fabricated, which consists of an array of nanosensors. Electronic nose (e-nose) has been developed for detecting freshness of fish (Oconnell et al. 2001), dairy products (off-flavor and rancidity of milk) (Ampuero and Bosset 2003; Marsili 1999, 2000; Capone et al. 2001), spoilage of red wine (Berna et al. 2008), red meat (El Barbri et al. 2008; Längkvist et al. 2013; Musatov et al. 2010), wine aging (Lozano et al. 2008), classification of different brands of coffee (Pardo et al. 2000), etc. These are extremely sensitive to gases released by spoiling microorganisms and produce change in color, indicating that the food is deteriorated. Such nanosensors could be placed directly into the packaging material (Liu et al. 2007; Lange et al. 2002). An electronic tongue developed by Kraft Foods (Smart Packaging systems, Glenview, IL, USA) may be incorporated in packaging. It also consists of an array of nanosensors, which are sensitive to gases released by spoiling food, and in that case, the sensor changes its color, giving a visible signal about freshness or adulteration of food (Momin et al. 2013). Electronic tongue (e-tongue) has been fabricated for determination of concentration of nitrite, nitrate, and chloride in minced meat (Campos et al. 2010), ripening of grapes (Campos et al. 2013), quality of tea (Kumar et al. 2016), polyphenols in wine (Andrei et al. 2016; Cetó et al. 2012; Magro et al. 2016), etc.

Silicate nanoparticles-packed films can control flow of oxygen into the pack as well as leakage of moisture out of the packings. It protects the package from being spoiled. The amount of packaging waste associated with processed food can also be reduced by using nanotechnology assisting in the preservation of fresh food.

24.2.2.3 Carbon Nanotubes

Food packaging materials are available ranging from films, carbon nanotubes, to waxy nano-coatings. Carbon nanotubes (CNTs) are available in both forms, single-walled nanotube (SWCNT) and multiwalled nanotube (MWCNT). SWCNT is generally one atom thick, while MWCNT comprises of several concentric tubes with very high aspect ratios and elastic modulus. Asgari et al. (2014) reported that CNTs infused with polyethylene films can be used for the packaging of Mazafati dates, which can prevent fungal invasion up to 90 days. Carbon nanotubes have also been developed in packaging to pump out carbon dioxide or absorb undesirable flavors (Sinha et al. 2006). It has been known that CNTs also have antibacterial properties, which may be due to their direct penetration through microbial cells (Sharma et al. 2017). It has been reported that antimicrobial activity of silver nanoparticles against

E. coli and *B. cereus* spores is significantly enhanced in combination with titanium dioxide and carbon nanotubes, respectively (Krishna et al. 2005). Ionic nanocomposites of carbon ceramic electrode with multiwalled carbon nanotubes may be used for the electrochemical determination of the adulterants in food and beverages such as food dyes, like sunset yellow, tartrazine, etc. In addition, carbon nanotubes have many other properties that may be exploited later on to develop the next generation of nanosensors. An important role is played by CNTs in food packaging and processing, but there may be some pitfalls, like these may migrate into food and contaminate it and can lead to toxic effects in human beings. The toxicity levels of CNTs are considerably high, and therefore, their use is limited.

24.2.2.4 Biobased Packaging

Nanotechnology can also be used for improving plastic substance barrier, incorporation of bioactive, sensing and signaling of important information about the food, for the change of the pervasion action of foils, growing different barrier characters (microbial, chemical, thermal, and mechanical), enhancing heat resistance and also mechanical characters (Berekaa 2015). It may also decrease the environmental contamination by making use of decomposable packaging based on biodegradable plastics.

These biodegradable plastics are basically polymeric materials, where at least one step in the degradation process is through metabolic process in the presence of some naturally occurring organisms. Disintegration or fragmentation of the plastics leads to biodegradation under desired conditions of moisture, temperature, and oxygen availability, without producing toxic or environmentally harmful materials (Chandra and Rustgi 1998). Such biodegradable polymers can be classified depending upon their source:

- Polymers, which are directly extracted or removed from biomass (i.e., polysaccharides, polypeptides, proteins, polynucleotides, etc.)
- Polymers, which are produced by classical chemical synthesis using some renewable biobased monomers or mixed sources of biomass and petroleum (polylactic acid or biopolyester)
- Polymers, which are produced by microorganism or genetically modified bacteria (polyhydroxybutyrate, bacterial cellulose, xanthan, curdian, pullan)

Various kinds of biodegradable polymer nanocomposites have been prepared with desired properties for a wide range of applications (Ray and Bousmina 2005). At present, the most commonly used biodegradable nanocomposites include starch and derivatives, polylactic acid (PLA), polybutylene succinate (PBS), polyhydroxybutyrate (PHB), and aliphatic polyester polycaprolactone, which are suitable for packaging purpose.

24.2.2.4.1 Starch and Their Derivatives

Starch is a potential raw material, because it is available from many plants. Its large production may fulfill current requirements, and it is low cost (Gonera and Cornillon

2002), but it cannot form films with appropriate mechanical strength for a packaging material. So, it is first plasticized or chemically modified. It is converted to a thermoplastic material on treating in an extruder by application of both thermal and mechanical energy. Plasticizers play an important role in efficiently reducing intramolecular hydrogen bonds and also provide stability to properties of product, when thermoplastic starches are produced. Hence, there are ample opportunities that thermoplastic starches may be used as packaging material (Kim and Pometto 1994).

24.2.2.4.2 Polylactic Acid (PLA)

A wide range of biopolyesters can be prepared by conventional chemical synthesis. Presently, polylactic acid is the polymer, which has the highest potential as renewable packaging material, and it is commercially produced on major scale because lactic acid, a monomer of PLA, is easily produced by fermentation of carbohydrate feedstock (biomass). This feedstock can be obtained from a variety of agricultural products like wheat, maize, molasses, and whey. Biodegradable polylactic acid (PLA) polymer was evaluated for its use as a material for antimicrobial food packaging (Jin and Zhang 2008). They incorporated nisin in PLA films for controlling foodborne pathogens. Antimicrobial activity of PLA/nisin films was evaluated in liquid foods (orange juice and liquid egg white) against *Listeria monocytogenes*, *Escherichia coli* O157:H7, and *Salmonella* Enteritidis. Ramos et al. (2014) prepared nano-biocomposite films based on polylactic acid by incorporating thymol (as the active additive) and modified montmorillonite (D43B) at two different concentrations. It was observed that thermal stability was not significantly affected by the addition of thymol, but the incorporation of D43B improved its mechanical properties and reduced the oxygen transmission rate by the formation of intercalated structures. It was suggested that the formulated nano-biocomposites could be considered a potential antioxidant active packaging material.

24.2.2.4.3 Polyhydroxybutyrate (PHB)

Polylactic acid and polyhydroxybutyrate both offer a lot of opportunities in food packaging applications, because they are compatible with many foods, including dairy products, beverage, ready meals, and fresh meat products. It is accumulated by a large number of bacteria in the form of energy and carbon reserves. This biopolyester may also find industrial applications easily because of its biodegradability and biocompatibility (Van der Walle et al. 2001). Poly(lactic acid) and poly(hydroxybutyrate) were blended and plasticized by Arrieta et al. (2014) with a natural terpene D-limonene (LIM) so as to increase PLA crystallinity and to also obtain flexible films for food packaging applications. As-prepared materials were melt-blended and processed in transparent films. Disintegrability under composting conditions was also worked out, and it was reported that PHB delays the PLA disintegrability, but on the contrary, D-limonene speeds it up. Ma et al. (2018) developed poly(lactic acid)-poly(hydroxybutyrate) (PLA-PHB) based films containing bioactive elements and prepared seven formulations containing different contents of plasticizers (mono-caprylin glycerate (GMC) or glycerol monolaurate (GML)). Two formulations (PLA-PHB-based films with 0.5% GMC or GML) were selected, and 5% cinnamaldehyde was added into each of these. It was revealed that PLA-PHB-based films possessed better mechanical properties and better active properties on application to high lipid food simulant. This study showed that it is possible to use biodegradable active packing as an alternative to replace nonbiodegradable packaging for chilled salmon.

24.2.2.4.4 Polycaprolactone (PCL)

It is a biodegradable polyester having a low melting point around 60 °C only. It has some interesting applications in the fields of medical and agricultural areas (Nakayama et al. 1997). It has high elongation at break and low modulus. Apart from it, its physical properties and availability on commercial scale made it very attractive material for commodity applications. An antimicrobial nanopackaging was developed by Ahmed et al. (2019) for food application by incorporating zinc oxide nanoparticles and clove essential oil (CEO) into polylactide/polyethylene glycol polycaprolactone (PLA/PEG/PCL) blend. Here, CEO acts as an efficient plasticizer, which facilitates the chain mobility in the blend, as evident from tensile and thermal properties. The efficacy of these composite films was confirmed by using *Staphylococcus aureus* and *Escherichia coli* inoculated in scrambled egg. It was indicated that the PLA/PEG/PCL/ZnO/CEO film exhibited the highest antibacterial activity during 21 days storage at 4 °C. Cesur et al. (2018) prepared antimicrobial and biodegradable food packaging films with polycaprolactone (PCL). The 0.4 wt% of organo nanoclay (C) and 25, 50, 75 wt% chitosan (K) and glycerol monooleate (GMO) or oleic acid (OA) as a plastifier (5, 10, 20, and 30 wt%) were added, and 12 polymeric composite films were prepared. The samples were coded as PCL (P), organo nanoclay (C), oleic acid (O), and glycerol monooleate (G). The antimicrobial properties of these films were evaluated against *Escherichia coli, Pseudomonas aeruginosa, Bacillus cereus*, and *Candida albicans*. Polycaprolactone (PCL)/starch/pomegranate rind (PR) hybrids were developed by Khalid et al. (2018) for antimicrobial packaging applications. PR was used as an antimicrobial compound, and it was incorporated directly in PCL matrix, without the extraction of any active compound from the fruit rind. It was revealed that PCL/PR films show reasonably good antimicrobial activity at higher concentrations. Addition of starch was found to enhance the antimicrobial activity of PR. As all the materials used here are biodegradable and food contactable, it has been suggested that the as-developed material could be used as food-grade antimicrobial packaging material.

Newer and newer nanomaterials have also been regularly developed, so that their physical and mechanical properties of packaging are improved, particularly in terms of tensile strength, water resistance, gas permeability, rigidity, flame resistance, etc. Due to such interesting properties, polymer nanocomposites are emerging as promising materials with a large capability for their applications in the active food packaging industry (Youssef 2013).

24.2.3 Types of Nanomaterials in Food

Nanoparticles present in foods can be easily categorized based on their composition, (organic or inorganic), as this factor has a major impact on their gastrointestinal fate and potential toxicity (McClements and Xiao 2017).

24.2.3.1 Inorganic Nanoparticles

A number of nanoparticles used in food materials are mainly composed of some or the other inorganic materials, such as silver, silicon dioxide, iron oxide, titanium dioxide, or zinc oxide (Pietroiusti et al. 2016). These nanoparticles are either crystalline in nature or amorphous solids at ambient temperature. These may be spherical or nonspherical with different surface characteristics and sizes depending on precursor materials and conditions of preparation while these were fabricated. Such inorganic nanoparticles have different tendencies to dissolve under specific solution conditions (pH and ionic strength) and also chemical reactivities, which have a major impact on their gastrointestinal fate and toxicity.

24.2.3.1.1 Silver Nanoparticles

Silver nanoparticles (AgNPs) are commonly used as antimicrobial agents in food packaging, chopping boards, storage containers, refrigerators, and health supplements. Silver nanoparticles are used for their antimicrobial effects in certain types of food containers by some manufacturer in the United States such as Kinetic Go Green basic nanosilver food storage container, Oso fresh food storage container, and FresherLonger™ Plastic Storage bags. It is also possible that some of these silver nanoparticles may migrate into foods from these containers and they could be ingested by humans (Echegoyen and Nerin 2013). Emamifar et al. (2011) prepared nanocomposite LDPE films containing Ag and ZnO nanoparticles via melt mixing in a twin screw extruder. Orange juice was sterilized and then inoculated with 8.5 log cfu/mL of *Lactobacillus plantarum*. They filled packages prepared from nanocomposite films with orange juice and then stored at 4 °C. Microbial stability of the juice was evaluated after 7, 28, 56, 84, and 112 days of storage. It was reported that microbial growth rate significantly reduced on using this nanocomposite packaging material. A potential role of quantum sensors (QS) in food spoilage and food safety has been indicated (Naik and Kowshik 2014). Anti-QS materials like ATNPs were proposed as efficient models for controlling spoilage of food. Incorporation of ATNPs in food packaging materials could play an important role in preservation of food and ensure its safety by prolonging their shelf life. They proposed ATNPs as QS inhibitors with their potential use as an antipathogenic but nontoxic bioactive material. Ag/TiO_2-SiO_2-coated food packaging film was developed also (Peter et al. 2015). Its ability to inactivate *Botrytis cinerea* was evaluated during the storage of fresh lettuce. Packaging film was prepared by coating the Ag/TiO_2-SiO_2 ethanol suspension on polyethylene film. As-prepared packaging film was used for storage of green lettuce in a vegetation room. It was revealed that the shelf life of the lettuce stored in single- and double-layer film modified with TiO_2 was extended by 4 and 2 days, respectively. It was revealed that the spoilage of the lettuce in double-layer film modified with Ag/TiO_2-SiO_2 was lower after 5 days of storage than that of the lettuce stored in film modified with ethanol and unmodified film. Microbiological and chemical characteristics of white bread during storage in paper packages modified with Ag/TiO_2–SiO_2, Ag/N–TiO_2, or Au/TiO_2 have been investigated (Peter et al. 2016). The whiteness and the water retention of the modified packages were found to be slightly superior. The water retention was also observed to be very good, especially for the Ag/TiO_2–SiO_2 paper. These improvements can be associated with the high specific surface area and with the low agglomeration tendency of Ag nanoparticles in comparison with the Au ones. Their use extends the shelf life of bread by 2 days except Au/TiO_2 as compared with the unmodified paper package.

24.2.3.1.2 Zinc and Zinc Oxide Nanoparticles

Zinc and zinc oxide nanoparticles may be used as an additive in supplements and functional foods for nutrition, because this is an essential trace element required to maintain human health and well-being. ZnO nanoparticles can also be used in food packaging as antimicrobial agents so that contamination of foods with harmful bacteria can be prevented (Sirelkhatim et al. 2015). These are also used as ultraviolet (UV) light absorbers to protect foods from UV light exposure, if food is sensitive toward it. The preparation of ZnO nanoparticles loaded starch-coated polyethylene film was reported (Tankhiwale and Bajpai 2012). This ZnO-loaded film was tested for its biocidal action against *E. coli.* As-developed material has a great potential to be used as food packaging material to prevent foodstuff from bacterial contamination. TiO_2 and ZnO are biocompatible nanomaterials, and their biocompatibility was established through toxicity studies on cell lines (Venkatasubbu et al. 2016). Titanium dioxide and zinc oxide nanoparticle were synthesized by wet chemical process. The antibacterial activities of these materials were evaluated as food preservatives against *Salmonella typhi*, *Klebsiella pneumoniae*, and *Shigella flexneri*, and it was indicated that TiO_2 and ZnO nanoparticles inhibited the growth of *Salmonella*, *Klebsiella*, and *Shigella.* It was revealed that the mode of their action is through generation of ROS in the case of *Salmonella and Klebsiella*, but it is still unclear in the case of *Shigella.*

24.2.3.1.3 Titanium Dioxide Nanoparticles

TiO_2 nanoparticles are used as ingredients in foods to provide its characteristic optical properties, so that lightness and brightness are enhanced (Weir et al. 2012). It is widely used as food additive and antimicrobial agent for food packaging and storage containers. TiO_2 ingredients utilized in the food industry as lightening agents should have particle sizes in the range of 100–300 nanometers, so that their light-scattering properties are increased. Six different coating suspensions were prepared, through mixing TiO_2 (Aeroxide®P-25) nanoparticles (NPs) with three different types of binders [Shellac (A), polyuretahne (B), and polycrylic (C)] at a 1:4 to 1:16 NP to binder weight ratio (Yemmireddy and Hung 2015). They evaluated bactericidal activity of these TiO_2 coatings against *Escherichia coli* O157:H7 at three different UV-A light intensities. TiO_2 coatings with binder polyuretahne showed highest adhesion strength and scratch hardness as compared to coatings with other binders. It was found that TiO_2 coatings with binder polycrylic were found to be physically more stable and able to retain their original bactericidal property on repeated use experiments (1, 3, 5, and 10 times).

24.2.3.1.4 Silicon Dioxide Nanoparticles

Silicon dioxide nanoparticles are used in certain powdered foods as anticaking agents. Silicon dioxide and carbon having particle size in the range of a few hundred nm are used as food additives and for food packaging.

24.2.3.2 Organic Nanoparticles

These nanoparticles are basically composed of organic substances, like carbohydrates, proteins, or lipids. These substances are liquids, semisolids, or solids (crystalline or amorphous) at ambient temperatures, which depends on their composition and processing conditions.

24.2.3.2.1 Lipid Nanoparticles

Lipid nanoparticles are present in a wide range within many commercial food products. These lipids and lipid nanoparticles are mostly used as oral delivery systems for different drugs and other active ingredients. Lipids usually increase drug absorption in the gastrointestinal tract (GIT). These molecules in the form of nanoparticles improve mucosal adhesion because of their small particle size and increase their residence time in GIT. Lipid nanoparticles will also protect the loaded drugs from different degradations (chemical as well as enzymatic) and release drug molecules from the lipid matrix gradually into blood, thus resulting in enhanced therapeutic profiles as compared to free drug (Severino et al. 2012).

24.2.3.2.2 Protein Nanoparticles

Protein nanoparticles and other protein assemblies have shown a great potential recently in the field of catalysis, materials synthesis, drug and gene delivery, and bio-imaging (Rong et al. 2011). Protein nanoparticles are also found in foods in the form of casein micelles, which are available in bovine milk and other dairy products. These are nothing but small clusters of casein molecules and calcium phosphate ions.

24.2.3.2.3 Carbohydrate Nanoparticles

Carbohydrate nanoparticles are either digestible or indigestible polysaccharides, like starch, cellulose, xanthan, carrageenan, alginate, and pectin.

24.2.3.2.4 Complex Nanoparticles

Nanoparticles utilized in foods are many a time fabricated using combinations of these three ingredients, such as lipids, proteins, and carbohydrates. Coacervates are formed by electrostatic complexation of oppositely charged proteins and polysaccharides.

24.2.4 Role in Tracking, Tracing, Nanolithography, and Brand Protection

Nanotechnology is also helping food industries in providing authentication and track and trace features of a food product so that adulteration and diversion of products can be prevented (Nam et al. 2003). It is simply done by generating some complex invisible nanobarcodes with desired information, and it can be encrypted onto the food products and packaging. Such a nanobarcode detection system was created by Li et al. (2005) that produced fluorescence on exposure to ultraviolet light in a combination of color, which can be read by a computer scanner. This system has been tested on the food and some biological samples containing various pathogens like *E. coli*, anthrax, tularemia bacteria, and Ebola SARS viruses. These are also clearly indicated simultaneously by different color codes. Different codes can be created in the technology by altering the stripe orders, where every food item is assigned brand so that food batches can be traced.

24.2.5 Implication and Safety Concerns

Although there had been a rapid development in food nanotechnology using nanoparticles, little is known about the toxicity due to nanoparticles. Nanomaterials have some unique properties like high surface area, which makes them more active chemically than their bulk counterparts, and therefore, they could participate in most of the biological reaction having harmful effects on human health and/or environment. It is very much desired that nanostructures in food or related industries should not damage them directly or indirectly. Food and related industries have seen major changes due to unique properties of nanomaterials. But these unique properties may occasionally lead to dangerous side effects to ecosystems and even in people. There are two main safety concerns on using nanoparticles, and these are allergies and heavy metal release. At present, they are being used into food products at a relatively faster rate without desired knowledge and regulations, which can affect health and environment (Ranjan et al. 2014). It is necessary to take extra care while using nanomaterials as they may have potential toxic effects and their use in food science is increasing day by day. A report by the British Royal Society notes that we may face a nanotoxicity crisis in the future (Amini et al. 2014). Only with a proper detailed understanding of the properties of nanomaterials like size, solubility, surface chemistry, composition, etc., we will be in a position to find useful and safe food products. Of course, some of these unique properties of nanomaterials are making them wonderful materials, but sidewise, their use is also questionable at some or the other side.

24.2.5.1 Size

The size of nanoparticles is a very important feature for its unique properties as the surface area of nanoparticles depends upon its size. The effect of surface area on the respiration has been known as some nanoparticles are reported to cause pulmonary inflammation (Qiao et al. 2015). The toxicity of these nanoparticles not only depends on its chemical component, but it also depends on the quantity as well as position of the deposition. The size of particles is an important factor, which can be deciding for observing dermal-cell cytotoxicity in vitro. Such absorbed nanoparticles in different absorption routes could trigger an immune system response. The smaller size of these nanoparticles permits them to pass through different biological barriers. They can then settle in tissues like the central nervous system (CNS). Thus, the size of the nanoparticles is very important for safety purpose, whether such nanomaterials can be used in food and food-related industries or not.

24.2.5.2 Chemical Composition

Reagents used in the production of nanoparticles may be toxic. Some may remain in the final product and result in exposure to toxins that are unrelated to the nanomaterials themselves. For instance, some observed toxic effects of carbon nanotubes and semiconductor nanoparticles are related to residual reagents during synthesis. The remaining reagents and impurities may hinder our understanding of possible side effects of carbon nanotubes. Iron ions and impurities can accelerate the oxidative stress in cells. Crystallinity is another important aspect of chemical composition. Titanium oxide has three different levels of crystallinity that each has different cytotoxic effects (Suker and Albadran 2013).

24.2.5.3 Surface Structure

Cytotoxicity may also be affected from surfaces of nanostructures. Roughness, charge, hydrophobicity, and surface chemistry are the major factors that could affect the toxicological nature of absorbed nanoparticles in the human body (Kirchner et al. 2005). The toxic effects of nanoparticles can be controlled to some extent by coating nanoparticles with hydrophilic polymer like polyethylene glycol (PEG). It was indicated that positively charged nanoparticles were found to be more toxic as compared to negative or neutral nanoparticles.

24.2.5.4 Solubility

Solubility is also important in the toxicity of nanoparticles. Soluble titanium oxide nanoparticles (hydrophilic) are more toxic as compared to insoluble titanium oxide nanoparticles (Oberdörster 2001). Solubility of the toxicity of oxide nanoparticles has also been reported (Brunner et al. 2006).

24.2.5.5 Routes of Nanoparticle Exposure

There are different entry routes on exposure to nanoparticles such as dermal, respiratory, and digestive routes. These nanoparticles may enter the bloodstream after absorption and settle in different tissues like the brain or trigger some immune responses. Some genetic alteration has been also reported due to nanoparticles in food or nanoengineering of food (Bowman and Fitzharris 2007). Although there had been a long debate and it will go on, nanotechnology has entered into food packaging and food processing, so some safety measures are also required from governments and food producers.

Some nanoparticles may be dispersed in the air during the production of nanoparticles used in food and related industries. This should be taken into consideration, and workers' health must be protected from respiratory tract uptake of nanoparticles. The digestive path is another major route of uptake of nanoparticles. Some nanoparticles may also enter in the respiratory tract and then the digestive system through mucociliary clearance. The skin is an alternate main route of contact between human and nanomaterials.

There are several diseases that are associated with exposure to nanoparticles due to their accumulation or contact with cells and its internal parts like the mitochondrion, nucleus, cytoplasm, membrane, and lipid vesicle (Buzea et al. 2007). Various diseases may be caused by nanoparticles, and these are

- Through inhalation
  - Parkinson's disease
  - Alzheimer's disease
  - Asthma
  - Bronchitis
  - Cancer
  - Arteriosclerosis
  - Vasoconstriction
  - Thrombus
  - High blood pressure
  - Heart disease
  - Disease of unknown etiology in the kidneys and liver
  - Podoconiosis
  - Kaposi's sarcoma
- Through ingestion
  - Crohn's disease
  - Colon cancer
- Through skin contact
  - Autoimmune diseases
  - Dermatitis
  - Urticaria
  - Vasculitis

24.3 Emerging Challenges and Potential Solutions

With the developments in nanotechnology, its applicability to the food industry is likely to increase in the future. Of course, the success will depend on consumer acceptance. Various nanostructured materials (NSMs) ranging from inorganic metal, metal oxides, and their nanocomposites to nano-organic materials with bioactive agents have been applied in a wide range of food materials (Bajpai et al. 2018). Human exposure to these nanomaterials is regularly increasing, and its impact on health of the human and environment has become a point of public concern and interest. Although huge benefits are being offered by nanotechnology, the accumulation of nanostructured materials in human bodies and also in the environment has aroused several health and safety issues. This will require a uniform international regulatory framework for nanotechnology.

Different preparation technology could produce nanoparticles with different physical properties for their application in food. But, public perception regarding this new technology is still uncertain. Multiple guidelines of potential risks posed by nanomaterials have been released by different regulatory bodies like the US Environmental Protection Agency (USEPA), International Organization for Standardization and the Organization for Economic Cooperation and Development (IOSOECD), National Institute for Occupational Safety and Health (NIOSH), Health and Consumer Protection Directorate of the European Commission (HCPDEC), and Food and Drug Administration (FDA). It has been reported that these nanomaterials can improve food safety by increasing the efficacy of food packaging, shelf life, and nutritional value of food using additives without affecting the taste and physical characteristics of food products.

Nanotechnology is regularly gaining momentum; thus, it has become a very important tool for the food and bioprocessing industry to meet demands of increasing population growth all over the world. It has almost revolutionized conventional food science and food industry (He and Hwang 2016). It is also important to ascertain the toxicity of nanoparticles and the possible environmental and health hazards it may cause. Improvement in inorganic nanosubstance and microfluid manufacturing has permitted the preparation of effective and competent sensors to quickly detect/identify pathogens, microbes, or pesticides. The nanosensor or nanobiosensor should also be used in environmental contamination control in the food sectors. Functionalized food and nanosubstance should improve food value and protection as flavor and nutrient transporter.

It is still challenging to develop a healthy and sustainable food industry making use of nanoparticles in some or other forms. A large strength of the public is having a fear in using food engineered and genetically modified materials. Of course, one should be cautious in using nano-based materials in food science and the food industry but not afraid of its application. Although the fate and potential toxicity of nanomaterials are not fully known at this stage and such concerns require the education of public, ultimate success of such products will depend on acceptance by consumer. Time is not far off when nanofood technology will become a new frontier of this century, provided its harmful effects are fully controlled so that health and the environment are not adversely affected.

References

Ahmed J, Mulla M, Jacob H, Luciano G, Almusallam A (2019) Polylactide/poly(ε-caprolactone)/zinc oxide/clove essential oil composite antimicrobial films for scrambled egg packaging. Food Packag Shelf Life 21:100355. https://doi.org/10.1016/j.fpsl.2019.100355

Ahuja T, Mir IA, Kumar D, Rajesh (2007) Biomolecular immobilization on conducting polymers for biosensing applications. Biomaterials 28:791–805

Aigbogun IE, Mohammed SSD, Orukotan AA, Tanko JD (2017) The role of nanotechnology in food industries- a review. J Adv Microbiol 7:1–9

Alfadul SM, Elneshwy AA (2010) USE of nanotechnology in food processing, packaging and safety – review. Afr J Food Agric Nutr Dev 10(6):2719–2739

Amin AA, Salama MF, Seliem EI, Mahmoud KF (2015) Encapsulation of nano carotenoids; evaluation of stability and safety. Int J Curr Microbiol App Sci 4(2):1017–1029

Amini SM, Gilaki M, Karchani M (2014) Safety of nanotechnology in food industries. Electron Physician 6(4):962–968

Ampuero S, Bosset JO (2003) The electronic nose applied to dairy products: a review. Sensors Actuators B Chem 94:1–12

Andrei V, Sharpe E, Vasilescu A, Andreescu S (2016) A single use electrochemical sensor based on biomimetic nanoceria for the detection of wine antioxidants. Talanta 156–157:112–118

Arrieta MP, López J, Hernández A, Rayón E (2014) Ternary PLA–PHB–limonene blends intended for biodegradable food packaging applications. Eur Polym J 50:255–270

Asgari P, Moradi O, Tajeddin B (2014) The effect of nanocomposite packaging carbon nanotube base on organoleptic and fungal growth of Mazafati brand dates. Int Nano Lett 4:1–5

Athinarayanan J, Periasamy VS, Alsaif MA, Al-Warthan AA, Alshatwi AA (2014) Presence of nanosilica (E551) in commercial food products: TNF-mediated oxidative stress and altered cell cycle progression in human lung fibroblast cells. Cell Biol Toxicol 30(2):89–100

Ayala-Zavala JF, González-Aguilar GA, Ansorena MR, Alvarez-Párrilla E, de la Rosa L (2014) Nanotechnology tools to achieve food safety. Chapter 17. In: Bhat R, Gomez-Lopez V (eds) Practical food safety: contemporary issues and future directions. Wiley-Blackwell, Oxford, pp 341–353

Bajpai VK, Kamle M, Shukla S, Mahato DK, Chandra P, Hwang SK, Kumar P (2018) Prospects of using nanotechnology for food preservation, safety, and security. J Food Drug Anal 26:1201–1214

Banerjee T, Sulthana S, Shelby T, Heckert B, Jewell J, Woody K, Karimnia V et al (2016) Multiparametric magneto-fluorescent nanosensors for the ultrasensitive detection of Escherichia coli O157:H7. ACS Infect Dis 2:667–673. https://doi.org/10.1021/acsinfecdis.6b00108

Berekaa MM (2015) Nanotechnology in food industry; advances in food processing, packaging and food safety. Int J Curr Microbiol App Sci 4(5):345–357

Berna AZ, Trowell S, Cynkar W, Cozzolino D (2008) Comparison of metal oxide based electronic nose and mass spectrometry based electronic nose for the prediction of red wine spoilage. J Agric Food Chem 56:3238–3244

Bouwmeester H, Brandhoff P, Marvin HJP, Weigel S, Peters RJB (2014) State of the safety assessment and current use of nanomaterials in food and food production. Trends Food Sci Technol 40(2):200–210

Bowman DM, Fitzharris M (2007) Too small for concern? Public health and nanotechnology. Aust N Z J Public Health 31(4):382–384

Brody A (2006) Nano and food packaging technologies converge. Food Technol 60(3):92–94

Brody AL (2009) Innovations in fresh prepared meal delivery systems. Food Technol 63:84–86

Brody AL, Bugusu B, Han JH, Sand CK, Mchugh TH (2008) Scientific status summary—innovative food packaging solutions. J Food Sci 73(8):R107–R116

Brunner TJ, Wick P, Manser P, Spohn P, Grass RN, Limbach LK, Bruinink A et al (2006) In vitro cytotoxicity of oxide nanoparticles: comparison to asbestos, silica, and the effect of particle solubility. Environ Sci Technol 40(14):4374–4381

Bund (2008) Ausdem Labor auf den Teller: Die Nutzung der Nanotechnologie im Lebensmittelsektor. https://www.bund.net/fileadmin/user_upload_bund/publikationen/chemie/nanotechnologie_aus_dem_labor_auf_den_teller.pdf. Accessed on 25 May 2019

Buzea C, Pacheco II, Robbie K (2007) Nanomaterials and nanoparticles: sources and toxicity. Biointerphases 2(4):17–71

Campos I, Masot R, Alcañiz M, Gil L, Soto J, Vivancos JL, Martínez-Mañez R et al (2010) Accurate concentration determination of anions nitrate, nitrite and chloride in minced meat using a voltammetric electronic tongue. Sensors Actuators B Chem 149(1):71–78

Campos I, Bataller R, Armero R, Gandia JM, Soto J, Martínez-Máñez R, Gil-Sánchez L (2013) Monitoring grape ripeness using a voltammetric electronic tongue. Food Res Int 54(2):1369–1375
Capone S, Epifani M, Quaranta F, Siciliano P, Taurino A, Vasanelli L (2001) Monitoring of rancidity of milk by means of an electronic nose and a dynamic PCA analysis. Sensors Actuators B Chem 78:174–179
Cesur S, Köroğlu C, Yalçın HT (2018) Antimicrobial and biodegradable food packaging applications of polycaprolactone/organo nanoclay/chitosan polymeric composite films. J Vinyl Addit Technol 24:376–387
Cetó X, Céspedes F, Pividori MI, Gutiérrez JM, del Valle M (2012) Resolution of phenolic antioxidant mixtures employing a voltammetric bio-electronic tongue. Analyst 137:349–356
Chandra R, Rustgi R (1998) Biodegradable polymers. Prog Polym Sci 23:1273–1335
Charych D, Cheng Q, Reichert A, Kuziemko G, Stroh N, Nagy J, Spevak W et al (1996) A 'litmus test' for molecular recognition using artificial membranes. Chem Biol 3:113–120
Chen Y, Cheng JJ, Creamer KS (2008) Inhibition of anaerobic digestion process: a review. Bioresour Technol 99(10):4044–4064
Choudhury SR, Goswami A (2012) Supramolecular reactive sulphur nanoparticles: a novel and efficient antimicrobial agent. J Appl Microbiol 114:1–10
Corrales M, Fernandez A, Han JH (2014) Antimicrobial packaging systems. In: Han J (ed) Innovations in food packaging. Elsevier, Amsterdam, pp 133–169
Dekkers S, Krystek P, Peters RJ, Lankveld DX, Bokkers BG, van Hoeven-Arentzen PH, Bouwmeester H et al (2011) Presence and risks of nanosilica in food products. Nanotoxicology 5:393–405
Deligiannakis Y, Sotiriou GA, Pratsinis SE (2012) Antioxidant and antiradical SiO2 nanoparticles covalently functionalized with gallic acid. ACS Appl Mater Interfaces 4:6609–6617
Duncan TV (2011) Applications of nanotechnology in food packaging and food safety: barrier materials, antimicrobials and sensors. J Colloid Interface Sci 363:1–24
Echegoyen Y, Nerin C (2013) Nanoparticle release from nano-silver antimicrobial food containers. Food Chem Toxicol 62:16–22
El Barbri N, Llobet E, El Bari N, Correig X, Bouchikhi B (2008) Electronic nose based on metal oxide semiconductor sensors as an alternative technique for the spoilage classification of red meat. Sensors 8:142–156
Emamifar A, Kadivar M, Shahedi M, Soleimanian-Zad S (2011) Effect of nanocomposite packaging containing Ag and ZnO on inactivation of Lactobacillus plantarum in orange juice. Food Control 22:408–413
Fu PP, Xia Q, Hwang HN, Ray PC, Yu H (2014) Mechanisms of nanotoxicity: generation of reactive oxygen species. J Food Drug Anal 22:64–75
Garcia E, Barrett DM (2019) Preservative treatments for fresh cut fruits and vegetables. http://www.fruitandvegetable.ucdavis.edu/files/217061.pdf. Accessed on 23 May 2019
Golding M, Sein A (2004) Surface rheology of aqueous casein-monoglyceride dispersions. Food Hydrocoll 18:451–461
Gonera A, Cornillon P (2002) Gelatinization of starch/gum/sugar systems studied by using DSC, NMR, and CSLM. Starch-Starke 54(11):508–516
Graveland-Bikker JF, Kruif de CG (2006) Unique milk protein based nanotubes: food and nanotechnology meet. Trends Food Sci Technol 17(5):196–203
Guo KW (2015) Current relevant nanotechnologies for the food industry. In: Bagchi D, Swaroop A, Bagchi M (eds) Genomics, proteomics and metabolomics in nutraceuticals and functional foods. Wiley, Chichester, p 629
Hamad AF, Han JH, Kim BC, Rather IA (2018) The intertwine of nanotechnology with the food industry. Saudi J Biol Sci 25:27–30
Haruyama T (2003) Micro- and nanobiotechnology for biosensing cellular responses. Adv Drug Deliv Rev 55:393–401
He X, Hwang HM (2016) Nanotechnology in food science: functionality, applicability, and safety assessment. J Food Drug Anal 24:671–681

Hwang SH, Park YB, Yoon KH, Bang DS (2011) Smart materials and structures based on carbon nanotube composites. In: Yellampalli S (ed) Carbon nanotubes – synthesis, characterization, applications. InTech, Rijeka. ISBN: 978-953-307-497-9, Available from: file:///C:/Users/hp/Downloads/16812.pdf. Accessed 20 May 19

Imafidon GI, Spanier AM (1994) Unraveling the secret of meat flavor. Trends Food Sci Technol 5:315–321

Jampilek J, Kos J, Kralova K (2019) Potential of nanomaterial applications in dietary supplements and foods for special medical purposes. Nano 9:pii: E296. https://doi.org/10.3390/nano9020296

Jin T, Zhang H (2008) Biodegradable polylactic acid polymer with nisin for use in antimicrobial food packaging. J Food Sci 73(3):M127–M134. https://doi.org/10.1111/j.1750-3841.2008.00681.x

Joseph T, Morrison M (2006) Nanoforum report: nanotechnology in agriculture and food. https://nanotech.law.asu.edu/Documents/2009/09/nanotechnology_in_agriculture_and_food_234_2644.pdf. Accessed on 25 May 2019

Kerry J, Butler P (2008) Smart packaging technologies for fast moving consumer goods. Wiley, Chichester

Kerry JP, O'Grady MN, Hogan SA (2006) Past, current and potential utilization of active and intelligent packaging systems for meat and muscle-based products: a review. Meat Sci 74:113–130

Khalid S, Yu L, Feng M, Meng L, Bai Y, Ali A, Chen L et al (2018) Development and characterization of biodegradable antimicrobial packaging films based on polycaprolactone, starch and pomegranate rind hybrids. Food Packag Shelf Life 18:71–79. https://doi.org/10.1016/j.fpsl.2018.08.008

Kim M, Pometto AL (1994) Food packaging potential of some novel degradable starch-polyethylene plastics. J Food Prot 57(11):1007–1012

Kim JS, Kuk E, Yu KN, Kim JH, Park SJ, Lee HJ, Kim SH et al (2007) Antimicrobial effects of silver nanoparticles. Nanomed Nanotechnol Biol Med 3:95–101

Kirchner C, Liedl T, Kudera S, Pellegrino T, Muñoz JA, Gaub HE, Stölzle S et al (2005) Cytotoxicity of colloidal CdSe and CdSe/ZnS nanoparticles. Nano Lett 5(2):331–338

Krishna V, Pumprueg S, Lee SH, Zhao J, Sigmund W, Koopman B, Moudgil BM (2005) Photocatalytic disinfection with titanium dioxide coated multi- wall carbon nanotubes. Process Saf Environ Prot 83:393–397

Kumar AS, Shanmugam R, Nellaiappan S (2016) Chemical tea quality assessment by analyzing key polyphenolic functional groups using flow injection analysis coupled with a dual electrochemical detector. Sensors Actuators B Chem 227:352–361

Lange D, Hagleitner C, Hierlemann A, Brand O, Baltes H (2002) Complementary metal oxide semiconductor cantilever arrays on a single chip: mass-sensitive detection of volatile organic compounds. Anal Chem 74(13):3084–3095

Längkvist M, Coradeschi S, Loutfi A, Rayappan JBB (2013) Fast classification of meat spoilage markers using nanostructured ZnO thin films and unsupervised feature learning. Sensors (Basel) 13:1578–1592

Lawrence MJ, Rees GD (2000) Microemulsion-based media as novel drug delivery systems. Adv Drug Deliv Rev 45:89–121

Lee SW, Mao C, Flynn CE, Belcher AM (2002) Ordering of quantum dots using genetically engineered viruses. Science 296:892–895

Li Y, Cu Y, Luo D (2005) Multiplexed detection of pathogen DNA with DNA based fluorescence nanobarcodes. Nat Biotechnol 23:885–889

Li X, Xing Y, Jiang Y, Ding Y, Li W (2009) Antimicrobial activities of ZnO powder-coated PVC film to inactivate food pathogens. Int J Food Sci Technol 44:2161–2168

Li X, Li W, Jiang Y, Ding Y, Yun J, Tang Y, Zhang P (2011) Effect of nano-ZnO-coated active packaging on quality of fresh-cut 'Fuji' apple. Int J Food Sci Technol 46:1947–1955

Liao F, Chen C, Subramanian V (2005) Organic TFTs as gas sensors for electronic nose applications. Sensors Actuators 107:849–855

Liu Y, Chakrabartty S, Alocilja E (2007) Fundamental building blocks for molecular biowire based forward error-correcting biosensors. Nanotechnology 18:1–6

Llobet E, Gualdrón O, Vinaixa M (2007) Efficient feature selection for mass spectrometry based electronic nose applications. Chemom Intell Lab Syst 85(2):253–261

Lozano J, Arroyo T, Santos JP, Cabellos JM, Horrillo MC (2008) Electronic nose for wine ageing detection. Sensors Actuators B Chem 133:180–186

Lua J, Bowles M (2013) How will nanotechnology affect agricultural supply chains? Int Food Agribusiness Manag Rev 16(2):21–42

Ma Y, Li L, Wang Y (2018) Development of PLA-PHB-based biodegradable active packaging and its application to salmon. Packag Technol Sci 31(11):739–746

Maestrelli F, Garcia-Fuentes M, Mura P, Alonso MJ (2006) A new drug nanocarrier consisting of chitosan and hydoxypropylcyclodextrin. Eur J Pharm Biopharm 63(2):79–86

Magro M, Bonaiuto E, Baratella D, de Almeida Roger J, Jakubec P et al (2016) Electrocatalytic nanostructured ferric tannates: characterization and application of a polyphenol nanosensor. ChemPhysChem 17:3196–3203

Manke A, Wang L, Rojanasakul Y (2013) Mechanisms of nanoparticle-induced oxidative stress and toxicity. Biomed Res Int 2013:942916. https://doi.org/10.1155/2013/942916

Marsili RT (1999) SPME-MS-MVA as an electronic nose for the study of off-flavors in milk. J Agric Food Chem 47:648–654

Marsili RT (2000) Shelf-life prediction of processed milk by solid-phase microextraction, mass spectrometry, and multivariate analysis. J Agric Food Chem 48:3470–3475

McClements DJ, Decker EA (2000) Lipid oxidation in oil-in-water emulsions: impact of molecular environment on chemical reactions in heterogeneous food systems. J Food Sci 65:1270–1282

McClements DJ, Xiao H (2017) Is nano safe in foods? Establishing the factors impacting the gastrointestinal fate and toxicity of organic and inorganic food-grade nanoparticles. NPJ Sci Food 1:6. https://doi.org/10.1038/s41538-017-0005-1

Mclaren A, Valdes-Solis T, Li G, Tsang SC (2009) Shape and size effects of ZnO nanocrystals on photocatalytic activity. J Am Chem Soc 131(35):12540–12541

Merkoci A (2010) Nanoparticles-based strategies for DNA, protein and cell sensors. Biosens Bioelectron 26(4):1164–1177

Mitura KA, Zarzycki PK (2018) Biocompatibility and toxicity of allotropic forms of carbon in food packaging. In: Grumezescu AM, Holban AM (eds) Role of materials science in food bioengineering. Academic, London, p 93

Momin JK, Jayakumar C, Prajapati JB (2013) Potential of nanotechnology in functional foods. Emirates J Food Agric 25(1):10–19

Morones JR, Elechiguerra JL, Camacho A, Holt K, Kouri JB, Ramírez JT, Yacaman MJ (2005) The bactericidal effect of silver nanoparticles. Nanotechnology 16:2346–2353

Musatov VY, Sysoev VV, Sommer M, Kiselev I (2010) Assessment of meat freshness with metal oxide sensor microarray electronic nose: a practical approach. Sensors Actuators B Chem 144:99–103

Naik K, Kowshik M (2014) Anti-quorum sensing activity of AgCl-TiO_2 nanoparticles with potential use as active food packaging material. J Appl Microbiol 117:972–983

Nakagawa K (2014) Nano- and microencapsulation of flavor in food systems. In: Kwak HS (ed) Nano- and microencapsulation for foods. Wiley, Oxford, pp 249–272

Nakayama A, Kawasaki N, Maeda Y, Arvanitoyannis I, Ariba S, Yamamoto N (1997) Study of biodegradability of poly (3- valerolactone-co-L-lactide). J Appl Polym Sci 66:741–748

Nam JM, Thaxton CS, Mirkin CA (2003) Nanoparticle-based bio- bar codes for the ultrasensitive detection of proteins. Science 301(5641):1884–1886

Nano-enabled Packaging Market by Type (2019) (Active, Intelligent, and Others) and Industry Vertical (Food & Beverages, Pharmaceuticals, Cosmetics, Consumer Electronics, and Others) – Global Opportunity Analysis and Industry Forecast, 2018–2025. https://www.alliedmarketresearch.com/nano-enabled-packaging-market. Accessed on 21 May 2019)

Nanowerk Nanotechnology in Food (2019). https://www.nanowerk.com/nanotechnology-in-food.php. Accessed 13 May 2019

Neethirajan S, Jayas DS (2011) Nanotechnology for the food and bioprocessing industries. Food Bioprocess Technol 4:39–47

Ntim SA, Thomas TA, Begley TH, Noonan GO (2015) Characterisation and potential migration of silver nanoparticles from commercially available polymeric food contact materials. Food Addit Contam 32:1003–1011

Oberdörster G (2001) Pulmonary effects of inhaled ultrafine particles. Int Arch Occup Einviron Health 74:1–8

Oconnell M, Valdora G, Peltzer G, Negri RM (2001) A practical approach for fish freshness determinations using a portable electronic nose. Sensors Actuators B Chem 80(2):149–154

Otles S, Yalcin B (2010) Nano-biosensors as new tool for detection of food quality and safety. LogForum 6(4):67–70

Pardo M, Niederjaufner G, Benussi G, Comini E, Faglia G, Sberveglieri G, Holmberg M, Lundstrom I (2000) Data preprocessing enhances the classification of different brands of Espresso coffee with an electronic nose. Sensors Actuators B Chem 69:397–403

Pavlath AE, Orts W (2009) Edible films and coatings: why, what, and how? In: Embuscado ME, Huber KC (eds) Edible films and coatings for food applications. Springer, Berlin. https://doi.org/10.1007/978-0-387-92824-11

Pehanich M (2006) Small gains in processing and packaging. Food Process 11:46–48

Peter A, Tegla D, Giurgiulescu L, Cozmuta AM, Nicula C, Cozmuta LM, Vagelas I (2015) Development of Ag/TiO_2-SiO_2-coated food packaging film and its role in preservation of green lettuce during storage. Carpathian J Food Sci Technol 7:88–96

Peter A, Mihaly-Cozmuta L, Mihaly-Cozmuta A, Nicula C, Ziemkowska W, Basiak D et al (2016) Changes in the microbiological and chemical characteristics of white bread during storage in paper packages modified with Ag/TiO_2–SiO_2, Ag/N–TiO_2 or Au/TiO_2. Food Chem 197:790–798

Pietroiusti A, Magrini A, Campagnolo L (2016) New frontiers in nanotoxicology: Gutmicrobiota/microbiome-mediated effects of engineered nanomaterials. Toxicol Appl Pharmacol 299:90–95

Pool H, Quintanar D, Figueroa JDD, Mano CM, Bechara JEH, Godínez LA, Mendoza S (2012) Antioxidant effects of quercetin and catechin encapsulated into PLGA nanoparticles. J Nanomater 2012:1–12. https://doi.org/10.1155/2012/145380

Pradhan N, Singh S, Ojha N, Shrivastava A, Barla A, Rai V, Bose S (2015) Facets of nanotechnology as seen in food processing, packaging, and preservation industry. Biomed Res Int 2015:1–17. https://doi.org/10.1155/2015/365672

Qiao H, Liu W, Gu H, Wang D, Wang Y (2015) The transport and deposition of nanoparticles in respiratory system by inhalation. J Nanomater 2015:1–8. https://doi.org/10.1155/2015/394507

Quintavalla S, Vicini L (2002) Antimicrobial food packaging in meat industry. Meat Sci 62:373–380

Rai M, Ingle AP, Gupta I, Pandit R, Paralikar P, Gade A, Chaud MV et al (2018) Smart nano-packaging for the enhancement of food shelf life. Environ Chem Lett 17:277–290. https://doi.org/10.1007/s10311-018-0794-8

Ramos M, Jiménez A, Peltzer M, Garrigós MC (2014) Development of novel nano-biocomposite antioxidant films based on poly (lactic acid) and thymol for active packaging. Food Chem 162:149–155. https://doi.org/10.1016/j.foodchem.2014.04.026

Rani A, Singh R, Shukla G (2017) Nanotechnology in agriculture and food sector. In: Applications of nanotechnology: an introduction. Horizon Books, New Delhi, p 29

Ranjan S, Dasgupta N, Chakraborty AR, Melvin SS, Shanker R, Ramalingam C, Kumar A (2014) Nanoscience and nanotechnologies in food industries: opportunities and research trends. J Nanopart Res 16(6):2464. https://doi.org/10.1007/s11051-014-2464-5

Rashidi L, Khosravi-Darani K (2011) The applications of nanotechnology in food industry. Crit Rev Food Sci Nutr 51:723–730

Ravichandran R (2010) Nanotechnology applications in food and food processing: innovative green approaches, opportunities and uncertainties for global market. Int J Green Nanotechnol Phys Chem 1:72–96

Ray SS, Bousmina M (2005) Biodegradable polymers and their layered silicate nanocomposites: in greening the 21st century materials world. Prog Mater Sci 50(8):962–1079

Robinson DKR, Morrison MJ (2010) Nanotechnologies for food packaging: reporting the science and technology research trends. Report for the observatory NANO. Available from: http://www.observatorynano.eu/project/filesystem/files/Food%20Packaging%20Report%202010%20DKR%20Robinson.pdf

Rong J, Niu Z, Lee LA, Wang Q (2011) Chemistry and materials development of protein-based nanoparticles. Compr Nanosci Technol 2:153–174

Sanguansri P, Augustin MA (2006) Nanoscale materials development – a food industry perspective. Trends Food Sci Technol 17(10):547–556

Sekhon BS (2010) Food nanotechnology – an overview. Nanotechnol Sci Appl 3:1–15

Setkus A (2002) Heterogeneous reaction rate based description of the response kinetics in metal oxide gas sensors. Sensors Actuators 87:346–357

Severino P, Andreani T, Macedo AS, Fangueiro JF, Santana MHA, Silva AM, Souto EB (2012) Current state-of-art and new trends on lipid nanoparticles (SLN and NLC) for oral drug delivery. J Drug Deliv 2012:1–10. https://doi.org/10.1155/2012/750891

Sharma C, Dhiman R, Rokana N, Panwar H (2017) Nanotechnology: an untapped resource for food packaging. Front Microbiol 8:1735. https://doi.org/10.3389/fmicb.2017.01735

Shi H, Magaye R, Castranova V, Zhao J (2013) Titanium dioxide nanoparticles: a review of current toxicological data. Part Fibre Toxicol 10:1–33. https://doi.org/10.1186/1743-8977-10-15

Shit SC, Shah PM (2014) Edible polymers: challenges and opportunities. J Polym 2014:1–13. https://doi.org/10.1155/2014/427259

Sinha N, Ma J, Yeow JTW (2006) Carbon nanotube-based sensors. Nanosci Nanotechnol 1(6):573–590

Sirelkhatim A, Mahmud S, Seeni A, Noor HMK, Ling CA, Bakhori SKM, Hasan H et al (2015) Review on zinc oxide nanoparticles: antibacterial activity and toxicity mechanism. Nano Micro Lett 7:219–242

Srinivas K (2016) The role of nanomaterials in food industry with advantages and risks on human health. Res J Pharm, Biol Chem Sci 7(5):2382–2399

Suker DK, Albadran RM (2013) Cytotoxic effects of titanium dioxide nanoparticles on rat embryo fibroblast REF-3 cell line *in vitro*. Eur J Exp Biol 3(1):354–363

Suppakul P, Miltz J, Sonneveld K, Bigger SW (2003) Active packaging technologies with an emphasis on antimicrobial packaging and its application. J Food Sci 68(2):408–420

Tak L, Singh S, Bais B, Pandey A, Soni A, Singh PK (2015) Nanotechnology: an advanced way to meat processing and packaging. Glacier J Sci Res:1–16

Tang D, Sauceda JC, Lin Z, Ott S, Basova E, Goryacheva I, Biselli S et al (2009) Magnetic nanogold microspheres-based lateral-flow immunodipstick for rapid detection of aflatoxin B_2 in food. Biosens Bioelectron 25(2):514–518

Tankhiwale R, Bajpai SK (2012) Preparation, characterization and antibacterial applications of ZnO-nanoparticles coated polyethylene films for food packaging. Colloids Surf B: Biointerfaces 90:16–20

Taylor TM, Davidson PM, Bruce BD, Weiss J (2005) Liposomal nanocapsules in food science and agriculture. Crit Rev Food Sci Nutr 45:587–605

Van der Walle GA, de Koning GJ, Weusthuis RA, Eggink G (2001) Properties, modifications and applications of biopolyesters. Adv Biochem Eng Biotechnol 71:263–291

Venkatasubbu GD, Baskar R, Anusuya T, Seshan CA, Chelliah R (2016) Toxicity mechanism of titanium dioxide and zinc oxide nanoparticles against food pathogens. Colloids Surf B: Biointerfaces 148:600–606

Verma S, Gupta S (2017) Nanotechnology and food science: tomorrow design the food. Int J Curr Microbiol App Sci 6(9):3553–3561

Vidhyalakshmi R, Bhakyaraj R, Subhasree RS (2009) Encapsulation the future of probiotics- a review. Adv Biol Res 3(3–4):96–103

Weir A, Westerhoff P, Fabricius L, Hristovski K, Goetz VN (2012) Titanium dioxide nanoparticles in food and personal care products. Environ Sci Technol 46:2242–2250

Wesley SJ, Raja P, Sunder Raj AA, Tiroutchelvamae D (2014) Review on- nanotechnology applications in food packaging and safety. Int J Eng Res 3:645–651

Wooster TJ (2010) Nanoemulsions for Beverage Applications. Nano and Microtechnologies in the Food and Health Food Industries, 25th and 26th October, Amsterdam, The Netherlands. Available from: www.nano.org.uk/conferences/food_health/abstracts.html. Accessed 26 Feb 2010

World Bank (2017) ICT in agriculture (updated edition): connecting smallholders to knowledge, networks and institutions. World Bank Publications, Washington, DC, p 303

Wu HH, Yin JJ, Wamer WG, Zeng MY, Lo YM (2014) Reactive oxygen species related activities of nano-iron metal and nano-iron oxides. J Food Drug Anal 22:86–89

Yalcin B, Otles S (2010) Nanobiosensor and food pathogen interaction mechanisms. Elec J Env Agricult Food Chem 9(7):1257–1273

Yam KL, Takhistov PT, Miltz J (2005) Intelligent packaging: concepts and applications. J Food Sci 70:R1–R10

Yemmireddy VK, Hung Y-C (2015) Effect of binder on the physical stability and bactericidal property of titanium dioxide (TiO_2) nanocoatings on food contact surfaces. Food Control 57:82–88

Youssef AM (2013) Polymer nanocomposites as a new trend for packaging applications. Polym-Plast Technol Eng 52:635–660

Yu H, Park JY, Kwon CW, Hong SC, Park KM, Chang PS (2018) An overview of nanotechnology in food science: preparative methods, practical applications, and safety. J Chem 2018:1–10. https://doi.org/10.1155/2018/5427978

Yuan Y, Gao Y, Zhao J, Mao L (2008) Characterization and stability evaluation of β-carotene nanoemulsions prepared by high pressure homogenization under various emulsifying conditions. Food Res Int 41(1):61–68

Biogenic Nanomaterials and Their Applications in Agriculture

25

Yasmin M. Heikal and Heba M. M. Abdel-Aziz

Abstract

Nanotechnology leads as a pioneer technology in all science areas. It opens up a broad range of possibilities in different areas such as medicine, pharmaceuticals, electronics, and farming. Nanotechnology's potential for revolutionizing healthcare, textiles, equipment, data and communication technology, and energy industries has been well publicized. Attention is also being paid to the implementation of nanotechnology to agriculture and food industries nowadays. A convergence has been taking place in latest years between biological techniques, green chemistry, and nanotechnology. The aim of this convergence is to produce new nanomaterials and procedures of production that decrease or prevent the use of toxic materials. Recently, extensive study has been carried out to synthesize metal nanoparticles using microorganisms and crops and has been acknowledged as a green and effective manner to further exploit microorganisms as useful nanofactories. Here, we discuss a thorough summary of the prospective uses of distinct biological sources for the synthesis of nanoparticles, types, physiochemical properties, and nanoparticle characterization which exist in multiple forms in agriculture. In addition, we highlight latest milestones accomplished by monitoring critical parameters for the biogenic synthesis of nanoparticles, including biological source selection, incubation period, pH, and temperature. We also define opportunities for future growth of biological nanoparticles and their applications in agriculture.

Keywords

Metal nanoparticles · Nanotechnology · Biogenic synthesis · Plants · Microorganisms · Agriculture · Characterization · Stabilization

---

Y. M. Heikal (✉) · H. M. M. Abdel-Aziz
Botany Department, Faculty of Science, Mansoura University, Mansoura, Egypt
e-mail: Yasminheikal@mans.edu.eg

M. Ghorbanpour et al. (eds.), *Biogenic Nano-Particles and their Use in Agro-ecosystems*, https://doi.org/10.1007/978-981-15-2985-6_25

25.1 Introduction

Since the last century, nanotechnology has been a recognized field of many studies. Nanotechnology is a mixture of biological, physical, and chemical principles that create nanosized particles with specific functions (Kumar et al. 2014). At nanoscale level, nanotechnology generated materials of different kinds, and the term nano is modified from the Greek word meaning "dwarf." A nanometer (nm) is about one billionth of a meter, or about three atoms in length side by side (Thakkar et al. 2010). Nanoparticles (NPs) are a broad class of materials that include particulate matter having at least one dimension below 100 nm (Laurent et al. 2010).

Nanomaterials can be widely categorized into two kinds in terms of structure, namely, organic and inorganic. Organic nanomaterials are based on carbon, while noble metal (e.g., gold, platinum, silver, titanium, zinc, cerium, iron, and thallium), magnetic materials, and semiconductors such as titanium dioxide and zinc oxide are inorganic nanomaterials. The physiochemical characteristics of large-scale materials are mainly known, and latest findings have concentrated on materials between the atomic scale and the much bigger bulk scale that exist in the region. The greatest contributing factor that affects the physiochemical characteristics is the bigger surface area to volume ratio observed in nanometer-scale products (Mansoori et al. 2007). Thus, important modifications in surface chemistry as well as chemical reactivity occur at the nanometer scale (Jefferson 2000). Reports on the characteristics of nanoparticles and other nanostructured materials are sometimes founded on insufficient characterization (Grainger and Castner 2008).

There are two basic methods to synthesize metal nanoparticles: "bottom-up" and "top-down." Both of the abovementioned fundamental approaches to nanoparticle synthesis were accomplished through various physical, chemical, and biological techniques (Golinska et al. 2014). Nanoparticles have been physically and chemically generated for a long time, but latest advances demonstrate that microorganisms and biological systems play a critical role in nanoparticles in manufacturing (Ankamwar et al. 2005). Because of their increasing achievement and easy formation of nanoparticles, the use of bacteria in this region is evolving quickly. The organisms used in the synthesis of nanoparticles range from simple bacterial prokaryotic cells to higher eukaryotes (Korbekandi et al. 2009). In fact, organisms' capacity to produce metal nanoparticles has opened up a fresh interesting strategy to the growth of these natural nanofactories. However, different physiochemical approaches to metal nanoparticle synthesis are restricted by the pollution induced by heavy metals. In addition, metal nanoparticle biosynthesis is an eco-friendly technique (green chemistry) without harsh, poisonous, and costly chemicals being used (Huang et al. 2007). Thus, biologically synthesizing nanoparticles with the benefits of nontoxicity, reproducibility in manufacturing, simple scaling-up, and well-defined morphology has become a fresh trend in the manufacturing of nanoparticles (Baker et al. 2013; Makarov et al. 2014; Iv et al. 2015).

Accordingly, in latest years, study has concentrated on producing nanomaterials through nanotechnology-based procedures that encourage green chemistry values and decrease or eliminate the use of dangerous chemicals altogether. Thus, eco-friendly green nanotechnology-based nanoparticles manufacturing procedures have drawn significant global interest (Kulkarni and Muddapur 2014; Shah et al. 2015). Recent studies have concentrated on using biological entities to synthesize a broad range of nanoparticles to emphasize this alternative method. Biosynthesis via biological unicellular and multicellular entities such as actinomycetes (Ahmad et al. 2003), bacteria (Lengke and Southam 2006), fungi (Ahmad et al. 2005), marine algae (Rajathi et al. 2012), plants (Philip 2010), viruses (Lee et al. 2002), and yeast (Kowshik et al. 2003) gives alternative, environmentally friendly routes to nanoparticle production. All of these biological organisms can act as biological factories to produce specific nanoparticles to varying degrees. Each biological entity has active substances which act as reduction agents and stabilizers to produce nanoparticles of various shapes, sizes, physicochemical properties, and compositions (Mohanpuria et al. 2008).

In the process of generating extremely stable and well-characterized nanoparticles, the significant aspects that could be regarded are:

(i) Selection of the best organisms: Researchers focused on the significant inherent characteristics of organisms such as enzyme operations and biochemical pathways to select the best applicants for the manufacturing of metal nanoparticles.
(ii) Optimum conditions for cell growth and enzyme activity: It is very essential to optimize development circumstances. It is necessary to optimize nutrients, inoculum size, light, temperature, pH, mixing velocity, and buffer power.
(iii) Optimal conditions for response: Concentration of the substrate, concentration of the biocatalyst, donor and concentration of the electron, pH, exposure time, temperature, buffer strength, mixing velocity, and light must be regulated and optimized. We must therefore optimize the circumstances for bioreduction in the mixture reaction (Ahmed et al. 2003; Korbekandi et al. 2009).

Nanotechnology is now becoming an allied science that has been used most frequently for many years in other areas of science such as physics, electronics, and engineering. This multidisciplinary science also includes a number of applications in other fields, including molecular biology, biophysics, and biotechnology (Bhatia 2016). Nanoparticles have been promoted for a broad spectrum of applications due to the distinctive and novel size-dependent characteristics. Some of these applications are in agriculture (Prasad et al. 2014), as catalysts (Akhavan and Ghaderi 2010), biosensors (Miller et al. 2002), labeling for immunoassays (Liu et al. 2008), environmental remediation (Njagi et al. 2011; Mahdavi et al. 2013), and vector delivery of cancer therapeutic drugs (Wen et al. 2011).

The following phase of hereditarily altered yield development, contributions from creature generation, synthetic pesticides, and accuracy cultivating methods is probably going to be advanced and confined by nanotech innovative work in the agrarian segment. The utilization of nanotechnology has been for the most part hypothetical in agribusiness; however, it has started and will keep on significantly affecting the development of new practical materials, item improvement, and procedure plan and instrumentation for sanitation and biosecurity in the principle nourishment industry (Joseph and Morrison 2006). The consequences for society all in all will be emotional. Ongoing advances in materials science have delivered dominance in the innovation of nanoparticles, with wide implications in farming. One territory specifically is the cotton area where current strategies of turning cotton are very inefficient. Over 25% of cotton fiber is lost to scrap or waste from the development of cotton to the finish of its texture (Kumar 2009).

Present-day farming has been impacted as a significant factor by adjustments in rural innovation. Nanotechnology plays a conspicuous position among the most recent line of mechanical developments in the transformation of farming and sustenance generation. The development of nano-gadgets and nanomaterials could open up new applications in plant biotechnology and agribusiness. Nanotechnology studies are concentrating principally on applications in the zones of hardware, vitality, drug, and life sciences (Scrinis and Lyons 2007), since horticulture is not viewed as a solid part. While nano-compound pesticides are as of now being used, different applications are still in their beginning periods, and showcasing or contacting the normal individual may take years. These applications are principally expected to address a portion of the confinements and difficulties looked by huge-scale, substance, and capital-serious cultivating frameworks. This incorporates better soil executives including: i) focused utilization of information sources, ii) effective control of poisons, iii) new yield and creature attributes, iv) broadening and separation of cultivating strategies and items with regards to huge scale (Prasad et al. 2014).

This review summarizes the various kinds of nanomaterials using plants and microorganisms and the characterization and biogenic synthesis of metal and metal oxide nanoparticles. It also describes the factors affecting the process of synthesis and prospective applications for nanoparticles in agriculture.

25.2 Classification and Types of NPs

Depending on their size, morphology, and chemical characteristics, NPs are widely split into different classifications. A review of the most notable NPs is provided as follows based on physical and chemical features.

25.2.1 Carbon-Based NPs

Carbon nanotubes (CNTs) and fullerenes are two principle classes of carbon-based NPs. Fullerenes comprise nanomaterials delivered from empty globular enclosures, for example, allotropic sorts of carbon. They have additionally made noteworthy organization enthusiasm for nanocomposites for some organization applications, for example, fillers (Saeed and Khan 2016), proficient gas adsorbents for ecological remediation (Ngoy et al. 2014), and as a help vehicle for numerous inorganic and natural impetuses (Mabena et al. 2011).

25.2.2 Metal NPs

Metal NPs are produced by the precursors of metals. These NPs have distinctive optoelectrical characters because of notable localized surface plasmon resonance (LSPR) features. Alkali and noble metal NPs, that is to say, Au, Ag, and Cu, have an absorption band in the electromagnetic energy spectrum's visible area overseas. Metal NPs regulated size, facet, and shape synthesis is essential in state-of-the-art products today (Dreaden et al. 2012). Metal NPs have multiple uses in various study fields due to their sophisticated optical characteristics.

25.2.3 Ceramic NPs

Earthenware production NPs are nonmetallic inorganic solids that are combined by warmth and cooling. In shapeless, polycrystalline, thick, permeable, or empty structures, they can be found (Sigmund et al. 2006). These NPs are along these lines getting amazing consideration from researchers because of their utilization in, for example, photocatalysis, catalysis, imaging, and shading corruption (Thomas et al. 2015).

25.2.4 Semiconductor NPs

Because of this property, semiconductor materials have characteristics between metals and nonmetals; they have discovered different applications in the literature (Ali et al. 2017). Semiconductor NPs have broad bandgaps and thus showed important changes with bandgap tuning in their characteristics. In photocatalysis, picture optics, and electronic devices, they are therefore very significant materials (Sun 2000).

25.2.5 Polymeric NPs

They are organic NPs and a collective of special polymer nanoparticles (PNPs) used for them in the literature. Most of them are nanospheres or nanocapsules (Mansha et al. 2017). Nanospheres are matrix particles which general mass is usually strong, and at the outer limit of the spherical surface, the other molecules are adsorbed. In nanocapsular, the solid mass is completely encapsulated in the particle (Rao and Geckeler 2011).

25.2.6 Lipid-Based NPs

They comprise lipid molecules which are used in various biomedical applications efficiently. A lipid NP with a diameter varying from 10 to 1000 nm is typically spherical. Like polymeric NPs, lipid NPs have a strong lipid nucleus and a matrix which includes lipophilic molecules that are soluble. The internal core of these NPs has been stabilized by surfactants or emulsifiers (Rawat et al. 2011). A unique field that centers on lipid NP synthesis and design for different uses such as drug carriers is lipid nanotechnology (Puri et al. 2009).

25.3 Various Approaches in Nanoparticle Synthesis

Several techniques such as chemical, physical, biological, and enzymatic are applied for the production of nanoparticles (NPs). Physical methods include heat evaporation, spray pyrolysis, plasma arcing, ball milling, ultrathin films, layer-by-layer development, lithographic techniques, sputter deposition, pulsed laser desorption, epistaxis of molecular beams, and synthesis of nanoparticles with diffusion flame (Joerger et al. 2000). So also, concoction techniques are utilized to integrate NPs by electro-testimony, sol-gel process, synthetic arrangement affidavit, compound vapor statement (Oliveira et al. 2005), delicate synthetic strategy, Langmuir-Blodgett technique, reactant pathway, hydrolysis (Pileni 1997), coprecipitation technique, and wet concoction technique (Gan et al. 2012). In physical and synthetic techniques, high radiation and profoundly focused diminishing specialists and balancing out operators that are unsafe to the air and human well-being have been utilized. In this way, natural nanoparticle blend is a solitary advance bioreduction technique, and less vitality is utilized to integrate earth inviting NPs (Sathishkumar et al. 2009).

25.3.1 Toxicology of Nanoparticles and Advantage of Biological Nanoparticles

The toxicological effects of nanoparticles have developed anxiety and concerns in the wider society for human health and the environment. For instance, parameters such as morphology, particle size, structure, chemical reactivity, concentration, aggregation, and dispersion can all directly affect nanoparticles' conduct and relations with particular settings. It was shown that particles of around 10 nm can cause higher mortality rates of human cells compared to bigger particles varying from 50 to 100 nm (Carlson et al. 2008; Gorth et al. 2011). Hazardous chemicals such as surfactants tend to be used in conventional chemical and physical manufacturing procedures. These surfactants function as agents of sculpture that guide the development of particles, while usually capping agents are used to stabilize and stop aggregation of nanoparticles. Toxicity problems occur because the removal of all chemical toxic materials from the nanoparticle surface is highly hard (Gautam and Van Veggel 2013). Thus toxicity appears to result not only from the morphology, size, structure, and surface reactivity of nanoparticles but also from the presence of toxic materials on the surface. This added problem requires understanding the relations between different chemicals during the production of NPs and eventually the relations that occur in the setting to know future toxicity problems (Nel et al. 2009).

Nanoparticles acquired from biogenic courses are free of dangerous side effects appended to nanoparticles during physiochemical synthesis contrasted with physicochemically inferred nanoparticles, which thus limits the biomedical uses of the resulting nanoparticles (Baker et al. 2013). Nanoparticles' biological synthesis has a few advantages, including quick and eco-friendly assembling approaches and manufactured nanoparticles' cost-effective and biocompatible nature. Furthermore, extra stabilizing agents are not required since microorganism and plant components themselves carry on as topping and balancing out operators (Makarov et al. 2014). Moreover, the surfaces of biogenic nanoparticles continuously and specifically adsorb biomolecules when reaching complex natural liquids, shaping a crown that connects with organic frameworks. These crown layers give extra adequacy over bare biological nanoparticles (Monopoli et al. 2012). In this way, biological nanoparticles are progressively viable on the grounds that they are joined to the outside of combined nanoparticles by naturally dynamic parts from natural sources, for example, plants and microorganisms. There are bottomless pharmacologically dynamic metabolites, especially in restorative plants, which are conjectured to be associated with combined nanoparticles, giving extra advantage by improving the adequacy of nanoparticles (Mukherjee et al. 2012; Makarov et al. 2014). The extra advantage of nanoparticles' biological blend is that it can lessen the quantity of steps required, including connecting certain useful gatherings to the nanoparticles' surface to make them naturally dynamic (Baker et al. 2013). When thinking of biological synthesis, there is a long list of microorganisms and plants to do the job.

25.3.2 Synthesis Mechanism of Nanoparticles

Different techniques may be used to synthesize NPs, but these techniques are widely split into two primary groups, i.e., (1) bottom-up approach and (2) top-down approach (Wang and Xia 2004), as shown in Fig. 25.1. These methods further split according to the procedure, response condition, and adopted protocols into different subclasses.

25.3.2.1 Top-Down Syntheses

The top-down approach is to mill or attribute big macroscopic particles, and it is also called destructive approach. Starting from a bigger molecule, which is reduced to nanoscale level and then the produced units are changed to specific NPs. This procedure includes the synthesis by self-assembly of nanoparticles from nuclear parts that are already miniaturized. This involves physical and chemical formation (Imtiaz et al. 2017). It is a strategy that is relatively inexpensive. The imperfection of the surface structure is a significant drawback of the top-down strategy. Such surface structural failures can have an important effect on metallic nanoparticles' physical characteristics and surface chemistry owing to the high aspect ratio (Thakkar et al. 2010). Examples of this technique include grinding/milling, chemical vapor deposition (CVD), deposition of physical vapor (PVD), and other methods for decomposition (Iravani 2011). To synthesize coconut shell (CS) NPs, this

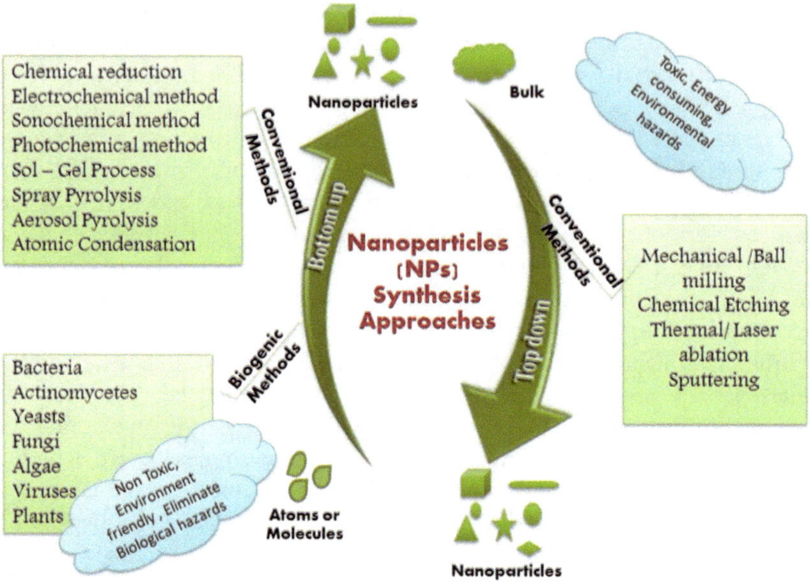

Fig. 25.1 Scheme of different synthetic approaches for nanoparticles: (**a**) top-down and (**b**) bottom-up approaches

method is used. The milling technique was used for this purpose, and the raw CS powders were finely milled with the assistance of ceramic balls and a well-known planetary mill for distinct intervals (Bello et al. 2015).

25.3.2.2 Bottom-Up Syntheses

The nanostructured construction blocks (nanoparticles) are first formed by the bottom-up strategy of nanomaterial synthesis and then assembled into the final product; hence this method is also called the build-up approach. Sedimentation and decrease methods are examples of this situation. It involves sol-gel, green synthesis, spinning, and synthesis of biochemical (Iravani 2011). A separate benefit of the bottom-up strategy is the increased ability to obtain metallic nanoparticles with relatively smaller defects and more homogeneous chemical composition(s).

25.3.3 Biogenic or Green Synthesis Using Plants and Microorganisms

25.3.3.1 Plants

Phytonanotechnology has created new avenues for nanoparticle synthesis and is an environmentally friendly, easy, fast, cost-effective, and stable technique. Phytonanotechnology has benefits, including scalability, biocompatibility, and the medical applicability of synthesizing nanoparticles using as a reduction medium the universal solvent, water (Noruzi 2015). Thus, plant-derived nanoparticles generated from easily accessible plant products and plant nontoxicity are appropriate to meet the increased requirements for NPs in the environmental and medical fields. What's more, different plant segments were utilized for the combination of metal nanoparticles, including stems, leaves, roots, and products of the soil extricated. The exact system and parts in charge of plant-intervened engineered nanoparticles stay to be explained. Proteins, amino acids, natural acids, nutrients, and optional metabolites, for example, polyphenols, flavonoids, terpenoids, alkaloids, heterocyclic mixes, and polysaccharides, have been proposed to have a critical influence in decreasing metal salt and, likewise, to go about as topping and balancing out specialists for manufactured nanoparticles (Duan et al. 2015). Philip et al. (2011) demonstrated the amalgamation and steadiness of biomolecule connection of silver and gold nanoparticles in *Murraya koenigii* leaf extract. Reports likewise demonstrate that particular components for incorporating nanoparticles exist in independent plant species (Baker et al. 2013). For instance, explicit components, for example, emodin, a laxative gum with quinone mixes found in xerophyte crops (plants adjusted to get by in deserts or low-water situations), are in charge of the union of silver nanoparticles: cyperoquinone, dietchequinone, and remirin in mesophyte crops (earthly plants not adjusted to especially dry or wet conditions) (Makarov et al. 2014).

25.3.3.2 Nanoparticle Synthesis Using Microorganisms

Microorganisms have been demonstrated to be significant nanofactories with tremendous potential as eco-agreeable and financially saving instruments, maintaining a strategic distance from harmful, brutal synthetic compounds and the intense interest for physiochemical amalgamation. The component and exploratory strategies utilized in microorganisms to blend nanoparticles are portrayed in Fig. 25.2. In the course of recent years, microorganisms, including microbes, actinomycetes, parasites, and yeasts, have been extra and intracellularly ready for blend of metal nanoparticles. A scope of organic conventions utilizing bacterial biomass (supernatant) has been uncovered. Extracellular union has gotten a great deal of consideration among the different systems as it disposes of the downstream preparing advances required to recuperate nanoparticles in intracellular techniques, including sonication to separate the phone divider, a few centrifugations and washing steps required to clean nanoparticles, and others. Moreover, metal-safe qualities, proteins, peptides, chemicals, diminishing of cofactors, and natural materials assume critical jobs by going about as declining specialists. These components help to give regular topping to nanoparticle amalgamation, hence preventing nanoparticles from conglomerating and helping them balance out for a long minute (Singh et al. 2016).

25.3.3.2.1 Bacteria

The production of nanoparticles has been demonstrated in latest studies using several *Bacillus* and other species, including *Bacillus amyloliquefaciens*, *Bacillus licheniformis*, *Rhodobacter sphaeroides* (Singh et al. 2011; Elbeshehy et al. 2015), *Bacillus subtilis*, *Streptomyces anulatus*, and *Listeria monocytogenes* (Elbeshehy et al. 2015; Soni et al. 2015). Different genera of microorganisms were recorded for the synthesis of metal nanoparticles, including *Bacillus*, *Streptomyces*, *Klebsiella*, *Weissella*, *Enterobacter*, *Rhodopseudomonas*, *Corynebacterium*, *Lactobacillus*,

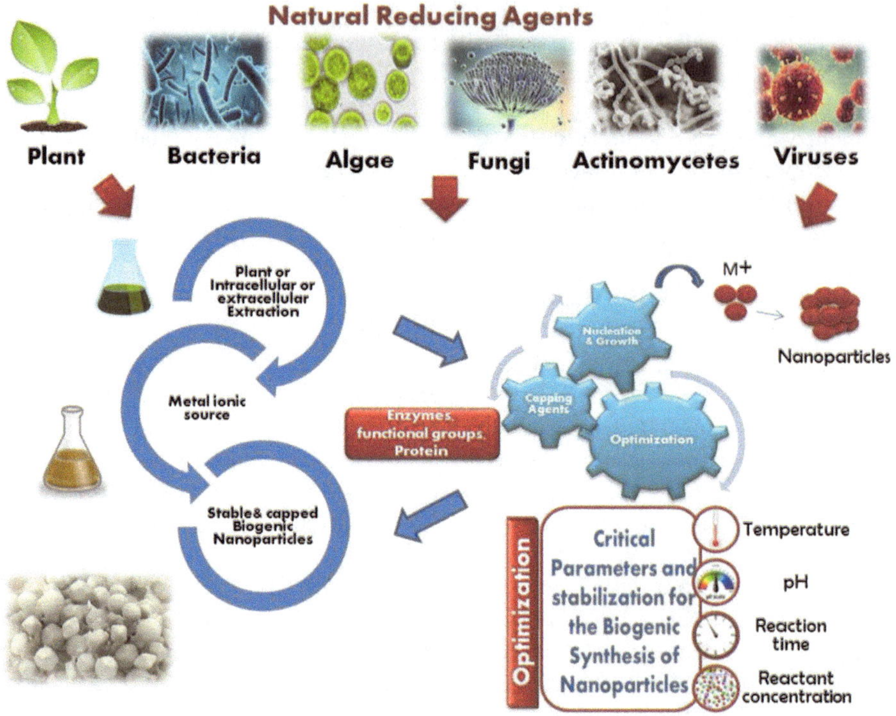

Fig. 25.2 Schematic presentation of the mechanism of biogenic nanoparticle synthesis from various sources (plants and microorganisms) and the parameters affecting the process

Pseudomonas, *Escherichia*, *Pyrobaculum*, *Aeromonas*, *Brevibacterium*, *Shewanella*, *Trichoderma*, *Sargassum*, *Desulfovibrio*, *Plectonema boryanum*, *Rhodococcus*, *Rhodobacter*, and others (Li et al. 2011a, b). These studies indicate that the primary mechanism for synthesizing nanoparticles using bacteria depends on enzymes (Zhang et al. 2011); for example, it has been discovered that the nitrate reductase enzyme is accountable for the synthesis of silver nanoparticles in *B. licheniformis*.

25.3.3.2.2 Fungi

Most organisms with significant metabolic substances with higher accumulation capacity and straightforward downstream preparing are anything but difficult to create for proficient, minimal effort generation of nanoparticles (Alghuthaymi et al. 2015). Besides, parasites have higher resistances and capacity to ingest metals in respect to microscopic organisms, especially as to the high divider restricting limit of contagious biomass metal salts for the generation of high return nanoparticles (Castro-Longoria et al. 2011; Alghuthaymi et al. 2015). Three potential systems were suggested to clarify the production of metal NPs from fungi: nitrate reductase activity, electron transport quinones, or both (Alghuthaymi et al. 2015). Contagious compounds, for example, *Penicillium* and *Fusarium oxysporum* reductase catalysts, nitrate reductase, and NADPH-subordinate reductases, have been found to assume a huge job in nanoparticle blend (Anil Kumar et al. 2007), tantamount to the system found in organisms.

25.3.3.2.3 Actinomycetes

Nanoparticles dependent on actinomycetes have not been all around looked into, in spite of the fact that actinomycetes-interceded nanoparticles have extraordinary mono-dispersibility and solidness, just as noteworthy biocidal movement against different pathogens (Golinska et al. 2014). Numerous examinations for the combination of zinc, copper, and silver nanoparticles utilizing *Streptomyces* sp. have been revealed. Numerous examinations for combination of zinc, copper and silver nanoparticles utilizing *Streptomyces* sp. have been revealed by reductase compound which assumes a key job in the decrease of metal salt (Karthik et al. 2014). Like different microorganisms, yeasts have likewise been broadly examined with simple downstream preparing for the extracellular blend of nanoparticles on a huge scale (Apte et al. 2013; Waghmare et al. 2015).

25.3.3.2.4 Algae

Algae are amphibian eukaryotic oxygenic photoautotrophs and can aggregate different substantial metals in some of them. In any case, there are not very many reports of respectable metal nanoparticles utilizing organic blend to utilize green growth. The dried alga *Chlorella vulgaris*, a solitary cell green alga, was shown to be a solid restricting capacity to frame algal-bound gold to tetrachloroaurate particles, which was in this manner diminished to Au(0). Roughly 88% of algal-bound gold was metallic and gold precious stones were amassed in the inward and external pieces of the cell surfaces with icosahedral, decahedral, and tetrahedral structures (Luangpipat et al. 2011). The dried green growth of *Spirulina platensis*; palatable blue-green algae; was used for extracellular amalgamation of silver and gold nanoparticles (Govindaraju et al. 2009). Singaravelu et al. (2007) and Rajasulochana et al. (2010) utilized *Kappaphycus alvarezii* and *Sargassum wightii* to record the combination of extracellular metal bionanoparticles. Senapati et al. (2012) additionally announced the intracellular creation of gold nanoparticles utilizing *Tetraselmis kochinensis*.

25.3.3.2.5 Viruses

Viruses can be utilized to combine nanowires with utilitarian parts for various applications, for example, battery terminals, photovoltaic gadgets, and super capacitors. The amalgamation and recuperation of nanoparticles based on viruses were delayed with low profitability. What's more, the issues related with microorganism-based nanoparticle amalgamation likewise incorporate entangled estimates, for example, microbial examining, segregation, development, and support (Nam et al. 2006).

25.4 Critical Parameters and Stabilization for the Biological Synthesis of Nanoparticles

In spite of several merits for nanoparticles from an approach of biological production, the polydispersity of the formed NPs remains an encounter. Consequently, many researchers attempted to create a relatively stable scheme for generating homogeneous size and morphology nanoparticles. The control of metal nanoparticles' form and size has been demonstrated either by restricting their environmental development or by modifying functional molecules (Kathiresan et al. 2009; Singh et al. 2015). For example, *Ganoderma* spp. were used to synthesize monodispersed (20 nm) and biocompatible gold nanoparticles. Improved reaction conditions include temperature, pH, incubation time, aeration, salt concentration, redox conditions, irradiation and mixing ratio (Gurunathan et al. 2014).

25.4.1 Temperature

For the union of nanoparticles utilizing microorganisms, developing microorganisms at the most astounding conceivable temperature is proposed as the catalyst in charge of nanoparticle combination is progressively dynamic at raised temperatures (Gurunathan et al. 2009).

25.4.2 pH

pH is one of the critical factors which affects NPs production. Changes in pH change the charge of characteristic phytochemicals for harvests and further influence their coupling capacity and diminish metal particles during nanoparticle amalgamation. This thusly can influence the yield of nanoparticles and morphology. For instance, the *Avena sativa* concentrate shaped various little estimated gold nanoparticles at pH 3.0 and 4.0, while the nanoparticle collection was seen at pH 2.0. Thus, it was recommended that conglomeration of nanoparticles commands the decrease strategy at corrosive pH esteems. This impact might be identified with the way that increasingly utilitarian gatherings are open at pH 3.0 and 4.0 contrasted with pH 2.0 official and nucleating metal particles. On the other hand, in the blend of *Curcuma longa* tuber powder silver nanoparticles at basic pHs, concentrates may include all the more contrarily stacked utilitarian gatherings that are able to do adequately official and diminishing silver particles, bringing about the union of more nanoparticles (Sathishkumar et al. 2010). Soluble pH (for *Isaria fumosorosea*), (Banu et al. 2014) and acidic pH (for *Fusarium acuminatum*) have been demonstrated to be reasonable for nanoparticle combination. Additionally, different conditions are subject to species and concentrates, for example, terms, salt focuses, and places of blend of nanoparticles (Pereira et al. 2015).

25.5 Characterization of Nanoparticles

For the assessment of different physicochemical characteristics of NPs, different characterization methods have been performed. These include methods such as X-ray photoelectron spectroscopy (XPS), X-ray diffraction (XRD), assessment of particle size, and scanning electron microscope (SEM), transmission electron microscope (TEM), and Brunauer-Emmett-Teller (BET) (Table 25.1, Fig. 25.3).

25.5.1 Morphological Characterizations

The characteristics of morphology of NPs are of excellent concern as morphology always affects most of the NPs characteristics. Different characterization methods are the most significant methods for morphological research such as TEM and SEM. The SEM method is based on the concept of electron scanning and offers all accessible nanoscale data on the NPs. There is a wide literature available, where individuals used this method not only to study the morphology of their nanomaterials but also the dispersion of NPs in the bulk or matrix. The morphological characteristics of ZnO altered metal-organic frameworks (MOFs) have been explored using SEM method, indicating the dispersion of ZnO NPs and the morphologies of MOFs under distinct circumstances of reaction (Mirzadeh and Akhbari 2016). Likewise, TEM is based on the concept of electron transmission, so it can provide data from very small to greater magnification on the bulk material. Bar et al. (2009) used aqueous *Jatropha curcas* seed extract to synthesize silver nanoparticles

Table 25.1 Nanoparticle parameter and the consistent characterization techniques

Parameter characterized	Characterization techniques
Size (structural properties)	TEM, XRD, DLS, NTA, SAXS, HRTEM, SEM, AFM, EXAFS, FMR, DCS, ICP-MS, UV-vis, MALDI, NMR, TRPS, EPLS, magnetic susceptibility
Shape	TEM, HRTEM, AFM, EPLS, FMR
Crystal structure	XRD, EXAFS, HRTEM, electron diffraction, STEM
Size distribution	DCS, DLS, SAXS, NTA, ICP-MS, FMR, DTA, TRPS, SEM
Elemental-chemical composition	XRD, XPS, ICP-MS, ICP-OES, SEM-EDX, NMR, MFM, LEIS
Magnetic properties	SQUID, VSM, MFM, FMR, XMCD, magnetic susceptibility
Optical properties	UV-vis-NIR, PL, EELS-STEM
Density	DCS, RMM-MEMS
Agglomeration state	Zeta potential, DLS, DCS, UV-vis, SEM, Cryo-TEM, TEM
Single-particle properties	Sp-ICP-MS, MFM, HRTEM, liquid TEM
Concentration	ICP-MS, UV-vis, RMM-MEMS, PTA, DCS, TRPS
Growth kinetics	SAXS, NMR, TEM, cryo-TEM, liquid-TEM
Surface charge	Zeta potential, EPM
Surface area	BET, liquid NMR

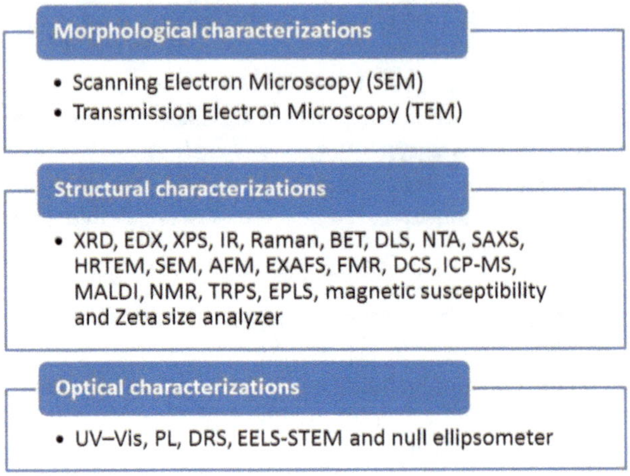

Fig. 25.3 Various methods for nanoparticle characterization: (**a**) morphological, (**b**) structural, (**c**) optical

without any toxic chemicals used to stabilize the produced NPs. For the reddish-yellow colored silver nanoparticles synthesized from 10^{-3} M $AgNO_3$, the surface plasmon absorption bands are characteristic at 425 nm. The particles in HRTEM are predominantly spherical in form varying from 15 to 25 nm in diameter. It also produces larger and irregular particles with a diameter of 30–50 nm.

25.5.2 Structural Characterizations

In studying the structure and nature of bonding materials, the structural features are of main significance. It offers various data about the subject material's bulk characteristics. The popular methods used to study structural characteristics of NPs are energy-dispersive X-ray (EDX), XRD, IR, XPS, BET, Raman, and Zeta size analyzer. XRD is one of the most significant methods for characterizing the structural characteristics of NPs. It provides sufficient data on the crystallinity and stage of NPs (Ullah et al. 2017).

Generally fixed with field outflow examining electron microscopy (FE-SEM) or TEM gadget, EDX is often utilized with a harsh thought of percent to appreciate the fundamental structure. The electron shaft concentrated by SEM or TEM on a solitary NP through the program accelerates to procure the knowledge data being contemplated from the NP. NP involves parts, every one of which radiates attributes of vitality X-beams by illuminating electron bars. The particular force of X-beams is straightforwardly relative to the grouping of the particular segment in the molecule. This strategy is much of the time utilized by specialists to help SEM and different methods check their parts in arranged items (Iqbal et al. 2016). Also, the essential affirmation and graphene impregnation of In2O3/graphene heterostructure NPs were completed utilizing comparable techniques that demonstrated C, In, and O as contributing components (Mansha et al. 2016).

XPS is the most sensitive method and is used to know the exact elementary ratio of the elements in NP materials and the exact bonding nature. It is surface-sensitive method and can be used to determine the general composition structure and depth variation of the structure in profiling research. XPS is based on the fundamental values of spectroscopy, and the typical XPS spectrum consists of the Y-axis plot amount of electrons versus the X-axis electron binding energy (eV). Each component has its own fingerprint of binding energy value and therefore provides specific XPS peaks (Lykhach et al. 2015). The characterization of nanoparticle vibration is usually researched through spectroscopies of FT-IR and Raman. These techniques are the most advanced and viable compared to other elementary analytical methods. The fingerprint region is the most significant range for NPs, which provides data on the material signature (Dablemont et al. 2008).

Due to its signal-enhancing ability via SPR phenomenon, Raman's recently enhanced surface spectroscopy (SERS) is evolving as a vibration-conforming tool (Muehlethaler et al. 2016). One study disclosed SERS' technique of studying nanostructured and quantum dots vibrational features with phonon modes in TiO_2, ZnO, and PbS NPs. They discovered that plasmonic resonances in semiconductor devices could be attributed to the enhanced range (Ma et al. 2011).

25.5.3 Particle Size and Surface Area Characterization

It is possible to use various methods to assess the size of the NPs. TEM, SEM, AFM, XRD, and dynamic light scattering (DLS) are included. AFM, TEM, SEM, and XRD may offer a better concept of particle size (Kestens et al. 2016), but it is possible to use the zeta prospective size analyzer/DLS to discover the NP size at exceptionally low levels. In addition, the analysis of nanoparticle tracking (NTA) is a relatively new and special procedure that can be useful in DNA and proteins (biological systems). In the NTA technique, we can see and evaluate the NPs in liquid media related to the particle size of the Brownian motion rate. This method enables us to discover in a liquid medium the size distribution profile of NPs with a diameter between 10 and 1000 nm. This method generated some excellent outcomes relative to DLS and was discovered to be very accurate for sizing both monodisperse and polydisperse samples with significantly better maximum resolution (Filipe et al. 2010).

25.5.4 Optical Characterizations

In photocatalytic applications, optical properties are of great concern and photochemists have therefore gained excellent understanding of this method to show the mechanism of their photochemical processes. These characteristics are based on the well-known Beer-Lambert law and principles of fundamental light (Swinehart

1962). These methods provide data on the characteristics of NPs in terms of absorption, reflection, luminescence, and phosphorescence. NPs, particularly metallic and semiconductor NPs, are best suited for apps related to photography due to their distinct colors. Thus, to understand the mechanism for each use, it is important to know the reflection and absorption of these materials. The well-known optical tools are photoluminescence (PL), ultraviolet-visible (UV-vis), and null ellipsometer, which are used to study the optical characteristics of NPs. The UV/vis diffuse reflectance spectrometer (DRS) is a fully equipped tool for measuring optical absorption, transmittance, and reflectance. PL also believes in useful method for studying photoactive NPs (optical properties) and other nanomaterials in relation to UV. This method provides extra data on the materials' absorption or emission ability and their impact on the photoexciton's general excitation moment. It therefore offers important data on the load recombination and half-life of the excited materials in their conductance band, which are helpful for all apps linked to photography and imaging. Depending on the nature of the study, the PL spectrum can be recorded as emission or absorption (Yu et al. 2013).

25.6 Applications of Nanoparticles in Agriculture

In recent years, nanotechnology had entered every aspect of life. In recent agricultural research works, nanotechnology proved to transform conventional farming to precision farming methods. The following lines will address recent applications of nanotechnology in agriculture.

25.6.1 Nanomaterials as Nanofertilizers

There is a significant increase in the use of nanotechnology in agriculture and crop science agriculture. As a consequence of advances in nanotechnology, new techniques are being suggested for producing large-scale nanoparticles of physiologically significant metals. These techniques are used to alter fertilizer formulations in order to improve plant cell uptake to minimize nutrient loss (Abdel-Aziz et al. 2019). Some distinctive characteristics of nanoparticles are the ability of controlled release kinetics to specific locations. These make these materials considered as "smart delivery system." Mechanisms such as slow or controlled release and targeted delivery can upturn the efficacy of nutrient use by using nanostructured fertilizers. It has been noted that in reaction to environmental triggers and biological requirements, their active ingredients are exactly released. According to the laboratory-scale studies, the improvement of crop productivity by enhancing seed germination rate, seedling development, photosynthetic activity, nitrogen metabolism, and carbohydrate and protein synthesis is regarded as a consequence of nanofertilization (Rai et al. 2016; Abdel-Aziz et al. 2019).

25.6.1.1 Different Types of Nanofertilizers

25.6.1.1.1 Nitrogen Nanofertilizers

There are many commercial plant fertilizers that can supply the three components (nitrogen, phosphorus, and potassium – NPK) that are needed. NPK numbers show each element's quantity. Many sources of nitrogen, including ammonia (NH_3), are used in fertilizers such as ammonium nitrate (NH_4NO_3), phosphate of diammonium ($(NH_4)_2HPO_4$), calcium cyanamide ($CaCN_2$), ammonium sulfate ($(NH_4)_2SO_4$), sodium nitrate ($NaNO_3$), calcium nitrate ($Ca(NO_3)_2$) and urea (N_2H_4CO). Phosphorus is usually provided as a phosphate such as diammonium phosphate ($(NH_4)_2HPO_4$) or calcium dihydrogen phosphate ($Ca(H_2PO_4)_2$). Potassium comes from potassium chloride (KCl) or potassium sulfate (K_2SO_4) (Shakhashiri 2010).

25.6.1.1.2 Potash Nanofertilizers

The crops can take potash fertilizer as a natural substance as K^+, and it can assist in the process of photosynthesis, the control of storage of water, and the stomatal opening in leaves. Pellet-based polyacrylamide coating was used slowly to release potash fertilizer. For an hour, a mixture of potash and clay was dried and toothpaste was used to coat it, so that the polymer was properly bonded. This polymer has been plunged into polyacrylamide (Rameshaiah et al. 2015).

25.6.1.1.3 Zinc Nanofertilizers

Micronutrient zinc inadequacy has been accounted for as a difficult issue in the globe. Day by day sustenance can supply substantially less zinc; accordingly, there is less shot of backhanded supply to people utilizing zinc-containing manure on the grounds that the equivalent nanoparticles can be utilized to coat zinc for dispersion and solvency of zinc (Milani et al. 2010; Rameshaiah et al. 2015). As indicated by studies, zinc dissolvability diminishes by expanding the pH (Bickel and Killorn 2001). Similar proportions between their surface zones ought to be considered for the structure of nanoparticles; if not, it can negatively affect the all-out solvency of zinc.

25.6.1.1.4 Nanoporous Zeolites

Zeolites and nano clays are used to improve the effectiveness of fertilizer use. They are a group of naturally minerals with a honeycomb-like crystal layered framework (Chinnamuthu and Boopathi 2009a, b). It can be used to fill its network with nitrogen, potassium, phosphorus, calcium, and a full collection of minor and trace nutrients. They are like slowly released nutrients "according to nutrient demand." The primary use of zeolites in agriculture is the storage, capture, and release of nitrogen. There are many contamination factors in groundwater, including the use of soluble N fertilizers. The ionic form's nitrogen release dynamics is much quicker than (in zeolites) the absorbed form (Abdel-Aziz et al. 2019).

25.6.2 Nanoherbicides

First, nanotechnology has only been used in the areas of medicine and pharmacology, and its implementation has only evolved in crop protection afterward. Technologies linked to the release of herbicides and pesticides have essentially altered in a controlled and encapsulated way. "Smart Seed" refers to crops that are immersed in nanoencapsulations of specific bacterial strain. Nanoparticles play a significant role in targeting and uploading substances across plants in specific areas as intelligent delivery systems (González-Melendi et al. 2008; Corredor et al. 2009). Pérez-de-Luque and Rubiales (2009) revealed the control of nanocapsulated herbicide parasitic weeds lowering phytotoxicity. Different formulations of herbicides are discerned, with special reference to slow release herbicides, systemic application, and microencapsulation, in order to enhance their various methods of action, including in combination with parasitic weed nanoparticle carrier (Dhillon and Mukhopadhyay 2015).

25.6.3 Nanopesticides

Nanosized pesticides may contain either small-scale structures with useful pesticide properties or very small particles of pesticide active ingredients (Bergeson 2010a). Nanopesticides can increase the diffusion of agricultural formulations and the unwanted motion of pesticides (Bergeson 2010b). Nanomaterials and biocomposites have revealed many exclusive characteristics. Permeability of rigidity, thermal stability, crystallinity, biodegradability, and solubility (Bouwmeester et al. 2009; Bordes et al. 2009) required for the formulation of nanopesticides are regarded as some of these significant benefits. Another characteristic of nanopesticides is the high specific surface, and thus the affinity to the goal can be improved with the use of these materials (Jianhui et al. 2005).

25.7 Environmental Toxicity of Nanoparticles

Different examinations on the natural poisonous quality of nanoparticles have been performed. Nanotypes of realized inorganic metals practically identical to nanotitanium dioxide (TiO_2) and different nanometals were made as the primary segments for the assembling of ecological uneasiness. A few government studies of nanotoxicology are accessible (Ma et al. 2010; Li et al. 2011a, b). In spite of the fact that nanomaterial transmission systems are not comprehended crosswise over cell dividers and layers, nanomaterials may change the layers and other cell structures inside cells (Chomoucká et al. 2010). For the viability of their accumulation, the surface portrayal of designed nanomaterial is of significant essentialness, prompting watery development and transport. Ecotoxicology study on designed nanomaterials uncovered poisonous impacts on spineless creatures and fish (Handy et al. 2008). Little data on common plant species are available (Panda et al. 2011). Nanotechnology advancements and continuous development of nanoparticles were seen in the previous decade. Another examination by Klaine et al. (2008) found that there was a need for a database of the creation, attitude, and lethality of engineered nanomaterials in plants. Lin and Xing (2007) detailed an investigation of five sorts of nanoparticles in six diverse plant species. The phytotoxicity effects of multi-walled carbon nanotubes (MWNTs); which resemble zinc, zinc oxide, aluminum and alumina nanoparticles; have been tested on seed germination and root development of corn, lettuce, radish, ryegrass and cucumber. Most nanoparticles demonstrated low results in these species, except for Zn NPs and ZnO NPs (Lin and Xing 2007). There was no self-important draw of seed, except for 2000 mg/L Zn NPs sway on ryegrass and ZnO nanoparticles on corn. Around 20 mg/L were unsurprising inhibitory focuses (IC50) of Zn NPs and ZnO nanoparticles for ryegrass and 50 mg/L for radish. These outcomes recommended that significant natural operators could result from the unseemly use and expulsion of such built nanoparticles. Contrasted with their mass material, an ongoing overview played out a broad examination of a few nanoparticles (Lin and Xing 2007). So as to assess some data from five conventional nanoparticles, this examination utilized three unmistakable trial types of gear to assess harmfulness. Both the nanoparticles and the subsequent mass materials (Stampoulis et al. 2009) assessed the impacts of the contribution or absence of association of test blends on the germination of zucchini seeds. A raised variety was seen when the nanomaterials were segregated through a surfactant. The sodium dodecyl sulfate surfactant seemed, by all accounts, to be more productive in water-insoluble example materials than the edifices tried and troubled. The subsequent test contained impacts of test materials on root advancement, with little effect on any material being perceived. The last trial utilized biomass in general to assess the impacts of plant development. As a result of the issue of characterizing levels of nanoparticles with respect to metal particles, these preliminaries were somewhat quantitatively troublesome. A fascinating outcome – the impact of carbon nanotubes on the general biomass of zucchini seedlings – was evaluated. Carbon nanotubes defer the development of seedling as indicated by data (De La Torre-Roche et al. 2013).

25.8 Conclusion

The capacity to utilize nanoparticles in particular fields builds the need to deliver them on a mechanical scale and in stable plans with naturally well-disposed procedures. A lot of exertion is in this way being made to use characteristic assets and execute organic amalgamation techniques with demonstrated favorable circumstances, for example, being eco-accommodating, simple to scale up, and financially saving; along these lines, the potential for green nanoparticles utilizing natural assets is exceptional. The natural way of incorporating nanoparticles has numerous advantages, for example, the steady assembling of nanoparticles with directed measurements and structures, the nonattendance of resulting confused synthetic combination, the nonappearance of dangerous contaminants, and the ability to orchestrate quickly utilizing different therapeutic plants and microorganisms. Significantly, the yield of orchestrated nanoparticles comparing to the grouping of metal salt and accessible organic assets stays to be clarified, and the parameters that can defeat the issues of polydispersity of natural nanoparticles in various natural frameworks still require enhancement. Moreover, the absence of learning of the suitable synthetic segments and the fundamental components for the amalgamation, activity, and soundness of natural nanoparticles stay open issues for the union of nanoparticles in the misuse of plants and smaller-scale creatures. In this way, the bounty of microorganisms and plants proficiently utilized for metal nanoparticles' organic union prompts higher investigation of natural nanofactories to address the issue for nanoproducts in particular fields. The biogenic nanoparticles produced in agroecosystems have various applications, including nanofertilizers, nanoherbicides, and nanopesticides.

References

Abdel-Aziz HMM, Hasaneen MNA, Omer AM (2019) Impact of engineered nanomaterials either alone or loaded with NPK on growth and productivity of French bean plants: seed priming vs foliar application. S Afr J Bot 125:102–108. https://doi.org/10.1016/j.sajb.2019.07.005

Ahmad A, Senapati S, Khan MI, Kumar R, Sastry M (2003) Extracellular biosynthesis of monodisperse gold nanoparticles by a novel extremophilic actinomycete, *Thermomonospora* sp. Langmuir 19(8):3550–3553

Ahmad A, Senapati S, Khan MI, Kumar R, Sastry M (2005) Extra–/intracellular biosynthesis of gold nanoparticles by an alkalotolerant fungus, *Trichothecium* sp. J Biomed Nanotechnol 1(1):47–53

Akhavan O, Ghaderi E (2010) Cu and CuO nanoparticles immobilized by silica thin films as antibacterial materials and photocatalysts. Surf Coat Technol 205(1):219–223

Alghuthaymi MA et al (2015) Myconanoparticles: synthesis and their role in phytopathogens management. Biotechnol Biotechnol Equip 29:221–236

Ali S, Khan I, Khan SA, Sohail M, Ahmed R, Rehman A, Ur Ansari MS, Morsy MA (2017) Electrocatalytic performance of Ni@Pt core–shell nanoparticles supported on carbon nanotubes for methanol oxidation reaction. J Electroanal Chem 795:17–25. https://doi.org/10.1016/j.jelechem.2017.04.040

Anil Kumar S et al (2007) Nitrate reductase-mediated synthesis of silver nanoparticles from AgNO3. Biotechnol Lett 29:439–445

Ankamwar B, Damle C, Ahmad A, Sastry M (2005) J Nanosci Nanotechnol 5:1665–1671

Apte M et al (2013) Psychrotrophic yeast Yarrowia lipolytica NCYC 789 mediates the synthesis of antimicrobial silver nano-particles via cell-associated melanin. AMB Express 3:32

Baker S et al (2013) Plants: emerging as nanofactories towards facile route in synthesis of nanoparticles. Bioimpacts 3:111–117

Banu AN et al (2014) Optimization and synthesis of silver nanoparticles using Isaria fumosorosea against human vector mosquitoes. Parasitol Res 113:3843–3851

Bar H, Bhui DK, Sahoo GP, Sarkar P, De SP, Misra A (2009) Green synthesis of silver nanoparticles using latex of *Jatropha curcas*. Colloids Surf A Physicochem Eng Asp 339(1–3):134–139

Bello SA, Agunsoye JO, Hassan SB (2015) Synthesis of coconut shell nanoparticles via a top down approach: assessment of milling duration on the particle sizes and morphologies of coconut shell nanoparticles. Mater Lett. https://doi.org/10.1016/j.matlet.2015.07.063

Bergeson LL (2010a) Nanosilver pesticide products: what does the future hold. Environ Qual Manag 19(4):73e82

Bergeson LL (2010b) Nanosilver: US EPA's pesticide office considers how best to proceed. Environ Qual Manag 19(3):79e85
Bhatia S (2016) Nanoparticles types, classification, characterization, fabrication methods and drug delivery applications. In: Natural polymer drug delivery systems. Springer, Cham. https://doi.org/10.1007/978-3-319-41129-3_2
Bickel A, Killorn R (2001) Spatial response of corn to banded zinc sulfate fertilizer in Iowa. In: 31st north central extension industry soil fertility conference, Potash and Phosphate Institute West Lafayette, 605/692–6280, 17, pp 14–15
Bordes P, Pollet E, Avérous L (2009) Nano-biocomposites: biodegradable polyester/nanoclay systems. Prog Polym Sci 34:125e155
Bouwmeester H, Dekkers S, Noordam MY, Hagens WI, Bulder AS, de Heer C, ten Voorde SECGS, Wijnhoven WP, Marvin HJP, Sips AJAM (2009) Review of health safety aspects of nanotechnologies in food production. Regul Toxicol Pharmacol 53:52e62
Carlson C, Hussein SM, Schrand AM et al (2008) Unique cellular interaction of silver nanoparticles: size-dependent generation of reactive oxygen species. J Phys Chem B 112(43):13608–13619
Castro-Longoria E et al (2011) Biosynthesis of silver, gold and bimetallic nanoparticles using the filamentous fungus Neurospora crassa. Colloids Surf B: Biointerfaces 83:42–48
Chinnamuthu CR, Boopathi PM (2009a) Nanotechnology and agroecosystem. Madras Agric J 96:17–31
Chinnamuthu CR, Boopathi PM (2009b) Review nanotechnology and agroecosystem. Madras Agric J 96(1–6):17–31
Chomoucká J, Drbohlavová J, Hubálek J, Babula P, Adam V, Kizek R (2010) Toxicity of nanoparticles for plants. List Cukrova Repa 126:400
Corredor E, Testillano PS, Coronado MJ, González-Melendi P, Fernández-Pacheco R, Marquina C, Ibarra R, de-la-Fuente JM, Rubiales D, Perez-de-Luque A, Risueno MC (2009) Nanoparticle penetration and transport living pumpkin plants: in situ subcellular identification. BMC Plant Biol 9:45
Dablemont C, Lang P, Mangeney C, Piquemal J-Y, Petkov V, Herbst F, Viau G (2008) FTIR and XPS study of Pt nanoparticle functionalization and interaction with alumina. Langmuir 24:5832–5841. https://doi.org/10.1021/la7028643
De La Torre-Roche R, Hawthorne J, Deng Y, Xing B, Cai W, Newman LA, Wang Q, Ma X, Hamdi H, White JC (2013) Multiwalled carbon nanotubes and C60 fullerenes differentially impact the accumulation of weathered pesticides in four agricultural plants. Environ Sci Technol 47:12539–12547. https://doi.org/10.1021/es4034809
Dhillon NK, Mukhopadhyay SS (2015) Nanotechnology and allelopathy: synergism in action. J Crop Weed 11(2):187–191
Dreaden EC, Alkilany AM, Huang X, Murphy CJ, El-Sayed MA (2012) The golden age: gold nanoparticles for biomedicine. Chem Soc Rev 41:2740–2779. https://doi.org/10.1039/C1CS15237H
Duan H et al (2015) Green chemistry for nanoparticle synthesis. Chem Soc Rev 44:5778–5792
Elbeshehy EK et al (2015) Silver nanoparticles synthesis mediated by new isolates of Bacillus spp., nanoparticle characterization and their activity against bean yellow mosaic virus and human pathogens. Front Microbiol 6:453
Filipe V, Hawe A, Jiskoot W (2010) Critical evaluation of nanoparticle tracking analysis (NTA) by nanosight for the measurement of nanoparticles and protein aggregates. Pharm Res 27:796–810. https://doi.org/10.1007/s11095-010-0073-2
Gan PP, Ng SH, Huang Y, Li SFY (2012) Green synthesis of gold nanoparticles using palm oil mill effluent (POME): a low-cost and eco-friendly viable approach. Bioresour Technol 113:132–135
Gautam A, Van Veggel FCJM (2013) Synthesis of nanoparticles, their biocompatibility, and toxicity behavior for biomedical applications. J Mater Chem B 1(39):5186–5200
Golinska P et al (2014) Biogenic synthesis of metal nanoparticles from actinomycetes: biomedical applications and cytotoxicity. Appl Microbiol Biotechnol 98:8083–8097

González-Melendi P, Fernández-Pacheco R, Coronado MJ, Corredor E, Testillano PS et al (2008) Nanoparticles as smart treatment-delivery systems in plants: assessment of different techniques of microscopy for their visualization in plant tissues. Ann Bot 101:187–195

Gorth DJ, Rand DM, Webster TJ (2011) Silver nanoparticle toxicity in Drosophila: size does matter. Int J Nanomedicine 6:343–350

Govindaraju K, Kiruthiga V, Kumar VG, Singaravelu G (2009) Extracellular synthesis of silver nanoparticles by a marine alga, *Sargassum wightii Grevilli* and their antibacterial effects. J Nanosci Nanotechnol 9(9):5497–5501

Grainger DW, Castner DG (2008) Nanobiomaterials and nanoanalysis: opportunities for improving the science to benefit biomedical technologies. Adv Mater 20:867–877

Gurunathan S et al (2009) Biosynthesis, purification and characterization of silver nanoparticles using Escherichia coli. Colloids Surf B: Biointerfaces 74:328–335

Gurunathan S et al (2014) A green chemistry approach for synthesizing biocompatible gold nanoparticles. Nanoscale Res Lett 9:248

Handy RD, Owen R, Valsami-Jones E (2008) The ecotoxicology of nanoparticles and nanomaterials: current status, knowledge gaps, challenges, and future needs. Ecotoxicology 17:315–325. https://doi.org/10.1007/s10646-008-0206-0

Huang J, Li Q, Sun D, Lu Y, Su Y, Yang X, Wang H, Wang Y, Shao W, Ning H, Hong J, Chen C (2007) Nanotechnology 18(105):104

Imtiaz M, Shahzad T, Shahid M, Ismail IMI, Mustafa G, Almeelbi T (2017) Zinc oxide nanoparticles affect carbon and nitrogen mineralization of phoenix dactylifera leaf litter in a sandy soil, J. Hazard Mater 324:298–305. https://doi.org/10.1016/j.jhazmat.2016.10.063

Iqbal N, Khan I, Yamani ZH, Qurashi A (2016) Sonochemical assisted solvothermal synthesis of gallium oxynitride nanosheets and their solar-driven photo-electrochemical water-splitting applications. Sci Rep 6:32319. https://doi.org/10.1038/srep32319

Iravani S (2011) Green synthesis of metal nanoparticles using plants. Green Chem 13:2638. https://doi.org/10.1039/c1gc15386b

Iv M et al (2015) Clinical applications of iron oxide nanoparticles for magnetic resonance imaging of brain tumors. Nanomedicine (London) 10:993–1018

Jefferson DA (2000) The surface activity of ultrafine particles. Philos Trans R Soc A Math Phys Eng Sci 358(1775):2683–2692

Jianhui Y, Kelong H, Yuelong W, Suqin L (2005) Study on anti-pollution nanopreparation of dimethomorph and its performance. Chin Sci Bull 50(2):108e112

Joerger R, Klaus T, Granqvist CG (2000) Biologically produced silver–carbon composite materials for optically functional thin-film coatings. Adv Mater 12:407–409

Joseph T, Morrison M (2006) Nanotechnology in agriculture and food. Nanoforum Report, Institute of Nanotechnology, April 2006, pp 1–13.

Karthik L et al (2014) Streptomyces sp. LK3 mediated synthesis of silver nanoparticles and its biomedical application. Bioprocess Biosyst Eng 37:261–267

Kathiresan K et al (2009) Studies on silver nanoparticles synthesised by a marine fungus, Penicillium fellutanum isolated from coastal mangrove sediment. Colloids Surf B: Biointerfaces 71:133–137

Kestens V, Roebben G, Herrmann J, Jämting A, Coleman V, Minelli C, Clifford C, De Temmerman P-J, Mast J, Junjie L, Babick F, Cölfen H, Emons H (2016) Challenges in the size analysis of a silica nanoparticle mixture as candidate certified reference material. J Nanopart Res 18:171. https://doi.org/10.1007/s11051-016-3474-2

Klaine SJ, Alvarez PJ, Batley GE, Fernandes TF, Handy RD, Lyon DY, Mahendra S, McLaughlin MJ, Lead JR (2008) Nanomaterials in the environment: behavior, fate, bioavailability, and effects. Environ Toxicol Chem 27:1825–1851. https://doi.org/10.1897/08-090.1

Korbekandi H, Iravani S, Abbasi S (2009) Crit Rev Biotechnol 29:279–306

Kowshik M, Ashtaputre S, Kharrazi S et al (2003) Extracellular synthesis of silver nanoparticles by a silver-tolerant yeast strain MKY3. Nanotechnology 14(1):95–100

Kulkarni N, Muddapur U (2014) Biosynthesis of metal nanoparticles: a review. J Nanotechnol 2014, Article ID 510246, 8 pages

Kumar R (2009) Nanotechnology in agriculture. G.B. Pant University of Agriculture and Technology, Pantnagar
Kumar DA, Palanichamy V, Roopan SM (2014) Green synthesis of silver nanoparticles using *Alternanthera dentata* leaf extract at room temperature and their antimicrobial activity. Spectrochim Acta, Part A 127:168–171
Laurent S, Forge D, Port M, Roch A, Robic C, Vander Elst L, Muller RN (2010) Magnetic iron oxide nanoparticles: synthesis, stabilization, vectorization, physicochemical characterizations, and biological applications. Chem Rev 110:2574–2574. https://doi.org/10.1021/cr900197g
Lee S-W, Mao C, Flynn CE, Belcher AM (2002) Ordering of quantum dots using genetically engineered viruses. Science 296(5569):892–895
Lengke M, Southam G (2006) Bioaccumulation of gold by sulfate-reducing bacteria cultured in the presence of gold(I)- thiosulfate complex. Geochim Cosmochim Acta 70(14):3646–3661
Li M, Zhu L, Lin D (2011a) Toxicity of ZnO nanoparticles to Escherichia coli: mechanism and the influence of medium components. Environ Sci Technol 45:1977–1983. https://doi.org/10.1021/es102624t
Li X et al (2011b) Biosynthesis of nanoparticles by microorganisms and their applications. J Nanomater 2011:16
Lin D, Xing B (2007) Phytotoxicity of nanoparticles: inhibition of seed germination and root growth. Environ Pollut 150:243–250. https://doi.org/10.1016/j.envpol.2007.01.016
Liu X, Dai Q, Austin L et al (2008) A one-step homogeneous immunoassay for cancer biomarker detection using gold nanoparticle probes coupled with dynamic light scattering. J Am Chem Soc 130(9):2780–2782
Luangpipat T, Beattie IR, Chisti Y, Haverkamp RG (2011) Gold nanoparticles produced in a microalga. J Nanopart Res 13:6439–6445
Lykhach Y, Kozlov SM, Skála T, Tovt A, Stetsovych V, Tsud N, Dvořák F, Johánek V, Neitzel A, Mysliveček J, Fabris S, Matolín V, Neyman KM, Libuda J (2015) Counting electrons on supported nanoparticles. Nat Mater. https://doi.org/10.1038/nmat4500
Ma X, Geiser-Lee J, Deng Y, Kolmakov A (2010) Interactions between engineered nanoparticles (ENPs) and plants: phytotoxicity, uptake and accumulation. Sci Total Environ 408:3053–3061. https://doi.org/10.1016/j.scitotenv.2010.03.031
Ma S, Livingstone R, Zhao B, Lombardi JR (2011) Enhanced Raman spectroscopy of nanostructured semiconductor phonon modes. J Phys Chem Lett 2:671–674. https://doi.org/10.1021/jz2001562
Mabena LF, Sinha Ray S, Mhlanga SD, Coville NJ (2011) Nitrogen-doped carbon nanotubes as a metal catalyst support. Appl Nanosci 1:67–77. https://doi.org/10.1007/s13204-011-0013-4
Mahdavi M, Ahmad MB, Haron MJ, Gharayebi Y, Shameli K, Nadi B (2013) Fabrication and characterization of SiO2/(3-Aminopropyl)triethoxysilane-coated magnetite nanoparticles for lead(II) removal from aqueous solution. J Inorg Organomet Polym Mater 23(3):599–607
Makarov VV et al (2014) "Green" nanotechnologies: synthesis of metal nanoparticles using plants. Acta Nat 6:35–44
Mansha M, Qurashi A, Ullah N, Bakare FO, Khan I, Yamani ZH (2016) Synthesis of In2O3/graphene heterostructure and their hydrogen gas sensing properties. Ceram Int 42:11490–11495. https://doi.org/10.1016/j.ceramint.2016.04.035
Mansha M, Khan I, Ullah N, Qurashi A (2017) Synthesis, characterization and visible-light-driven photoelectrochemical hydrogen evolution reaction of carbazole-containing conjugated polymers. Int J Hydrog Energy. https://doi.org/10.1016/j.ijhydene.2017.02.053
Mansoori GATFG, Assoufid L, Zhang G (2007) Molecular building blocks for nanotechnology, vol. 109 of topics in applied physics. Springer, New York
Milani N, McLaughlin M, Hettiaratchchi GM, Beak DK, Kirby JK, Stacey S (2010) Fate of nanoparticulate zinc oxide fertilisers in soil: solubility, diffusion and solid phase speciation. Soil Solut Chang World:1–6
Miller MM, Prinz GA, Cheng S-F, Bounnak S (2002) Detection of a micron-sized magnetic sphere using a ring shaped anisotropic magneto resistance-based sensor: a model for a magnetoresistance-based biosensor. Appl Phys Lett 81(12):2211–2213

Mirzadeh E, Akhbari K (2016) Synthesis of nanomaterials with desirable morphologies from metal–organic frameworks for various applications. Cryst Eng Commun 18:7410–7424. https://doi.org/10.1039/C6CE01076H

Mohanpuria P, Rana NK, Yadav SK (2008) Biosynthesis of nanoparticles: technological concepts and future applications. J Nanopart Res 10(3):507–517

Monopoli MP et al (2012) Biomolecular coronas provide the biological identity of nanosized materials. Nat Nanotechnol 7:779–786

Muehlethaler C, Leona M, Lombardi JR (2016) Review of surface enhanced Raman scattering applications in forensic science. Anal Chem 88:152–169. https://doi.org/10.1021/acs.analchem.5b04131

Mukherjee S et al (2012) Green chemistry approach for the synthesis and stabilization of biocompatible gold nanoparticles and their potential applications in cancer therapy. Nanotechnology 23:455103

Nam KT et al (2006) Virus-enabled synthesis and assembly of nanowires for lithium ion battery electrodes. Science 312:885–888

Nel AE, Mädler L, Velegol D et al (2009) Understanding biophysicochemical interactions at the nano-bio interface. Nat Mater 8(7):543–557

Ngoy JM, Wagner N, Riboldi L, Bolland O (2014) A CO_2 capture technology using multi-walled carbon nanotubes with polyaspartamide surfactant. Energy Procedia 63:2230–2248. https://doi.org/10.1016/j.egypro.2014.11.242

Njagi EC, Huang H, Stafford L et al (2011) Biosynthesis of iron and silver nanoparticles at room temperature using aqueous sorghum bran extracts. Langmuir 27(1):264–271

Noruzi M (2015) Biosynthesis of gold nanoparticles using plant extracts. Bioprocess Biosyst Eng 38:1–14

Oliveira MM, Ugarte D, Zanchet D, Zarbin AJG (2005) Influence of synthetic parameters on the size, structure, and stability of dodecanethiol-stabilized silver nanoparticles. J Colloid Interface Sci 292:429–435

Panda KK, Achary VM, Krishnaveni R, Padhi BK, Sarangi SN, Sahu SN, Panda BB (2011) In vitro biosynthesis and genotoxicity bioassay of silver nanoparticles using plants. Toxicol in Vitro 25:1097–1105. https://doi.org/10.1016/j.tiv.2011.03.008

Pereira L et al (2015) Metallic nanoparticles: microbial synthesis and unique properties for biotechnological applications, bioavailability and biotransformation. Crit Rev Biotechnol 35:114–128

Pérez de Luque A, Rubiales D (2009) Nanotechnology for parasitic plant control. Pest Manag Sci 65:540–545

Philip D (2010) Green synthesis of gold and silver nanoparticles using *Hibiscus rosa sinensis*. Phys E 42(5):1417–1424

Philip D et al (2011) *Murraya Koenigii* leaf-assisted rapid green synthesis of silver and gold nanoparticles. Spectrochim Acta A Mol Biomol Spectrosc 78:899–904

Pileni MP (1997) Nanosized particles made in colloidal assemblies. Langmuir 13:3266–3276

Prasad R, Kumar V, Prasad KS (2014) Nanotechnology in sustainable agriculture: present concerns and future aspects. Afr J Biotechnol 13(60):705–713

Puri A, Loomis K, Smith B, Lee J-H, Yavlovich A, Heldman E, Blumenthal R (2009) Lipid-based nanoparticles as pharmaceutical drug carriers: from concepts to clinic. Crit Rev Ther Drug Carrier Syst 26:523–580

Rai M, Ribeiro C, Mattaso L, Duran N (2016) Nanotechnology in food and agriculture. Springer, Cham. https://doi.org/10.1007/978-3-319-14024-7_4

Rajasulochana P, Dhamotharan R, Murugakoothan P, Murugesan S, Krishnamoorthy P (2010) Biosynthesis and characterization of gold nanoparticles using the alga *Kappaphycus alvarezii*. Int J Nanosci 9:511–516

Rajathi Arockiya Aarthi F, Parthiban C, Ganesh Kumar V, Anantharaman P (2012) Biosynthesis of antibacterial gold nanoparticles using brown alga, Stoechospermum marginatum (kützing). Spectrochim Acta A Mol Biomol Spectrosc 99:166–173

Rameshaiah GN, Pallavi J, Shabnam S (2015) Nano fertilizer and nano sensor- an attempt for developing smart agriculture. J Eng Res Gen Sci 3:314–320

Rao JP, Geckeler KE (2011) Polymer nanoparticles: preparation techniques and size-control parameters. Prog Polym Sci 36:887–913. https://doi.org/10.1016/j.progpolymsci.2011.01.001
Rawat MK, Jain A, Singh S, Mehnert W, Thunemann AF, Souto EB, Mehta A, Vyas SP (2011) Studies on binary lipid matrix based solid lipid nanoparticles of repaglinide: in vitro and in vivo evaluation. J Pharm Sci 100:2366–2378. https://doi.org/10.1002/jps.22435
Saeed K, Khan I (2016) Preparation and characterization of single walled carbon nanotube/nylon 6,6 nanocomposites. Instrum Sci Technol 44:435–444. https://doi.org/10.1080/10739149.2015.1127256
Sathishkumar M, Sneha K, Won SW, Cho CW, Kim S, Yun YS (2009) Cinnamon zeylanicum bark extract and powder mediated green synthesis of nano-crystalline silver particles and its bactericidal activity. Colloids Surf B Biointerfaces 73:332–338
Sathishkumar M et al (2010) Immobilization of silver nanoparticles synthesised using Curcuma longa tuber powder and extract on cotton cloth for bactericidal activity. Bioresour Technol 101:7958–7965
Scrinis G, Lyons K (2007) The emerging nano-corporate paradigm: nanotechnology and the transformation of nature, food and agri-food systems. Int J Sociol Agric Food 15:22–44
Senapati S, Syed A, Moeez S, Kumar A, Ahmad A (2012) Intracellular synthesis of gold nanoparticles using alga *Tetraselmis kochinensis*. Mater Lett 79:116–118
Shah M, Fawcett D, Sharma S, Tripathy SK, Poinern GEJ (2015) Green synthesis of metallic nanoparticles via biological entities. Materials 8(11):7278–7308
Shakhashiri (2010) Apply a fertilizer to the soil to keep cultivated plants healthy. 103–1. www.scifun.org
Sigmund W, Yuh J, Park H, Maneeratana V, Pyrgiotakis G, Daga A, Taylor J, Nino JC (2006) Processing and structure relationships in electrospinning of ceramic fiber systems. J Am Ceram Soc 89:395–407. https://doi.org/10.1111/j.1551-2916.2005.00807.x
Singaravelu G, Arockiamary JS, Kumar VG, Govindaraju K (2007) A novel extracellular synthesis of monodisperse gold nanoparticles using marine alga, *Sargassum wightii Greville*. Colloids Surf B: Biointerfaces 57(1):97–101
Singh BR et al (2011) Synthesis of stable cadmium sulfide nanoparticles using surfactin produced by Bacillus amyloliquifaciens strain KSU-109. Colloids Surf B: Biointerfaces 85:207–213
Singh P et al (2015) Microbial synthesis of flower-shaped gold nanoparticles. Artif Cells Nanomed Biotechnol. Published online May 6, 2015. https://doi.org/10.3109/21691401.2015.1041640
Singh P, Kim YJ, Zhang D, Yang DC (2016) Biological synthesis of nanoparticles from plants and microorganisms. Trends Biotechnol 34(7):588–599. https://doi.org/10.1016/j.tibtech.2016.02.006
Soni N et al (2015) Antimicrobial and mosquitocidal activity of microbial synthesised silver nanoparticles. Parasitol Res 114:1023–1030
Stampoulis D, Sinha SK, White JC (2009) Assay-dependent phytotoxicity of nanoparticles to plants. Environ Sci Technol 43:9473–9479. https://doi.org/10.1021/es901695c
Sun S (2000) Monodisperse FePt nanoparticles and ferromagnetic FePt nanocrystal superlattices. Science 80(287):1989–1992. https://doi.org/10.1126/science.287.5460.1989
Swinehart DF (1962) The Beer-Lambert law. J Chem Educ 39:333. https://doi.org/10.1021/ed039p333
Thakkar KN, Mhatre SS, Parikh RY (2010) Biological synthesis of metallic nanoparticles. Nanomedicine 6(2):257–262
Thomas S, Harshita BSP, Mishra P, Talegaonkar S (2015) Ceramic nanoparticles: fabrication methods and applications in drug delivery. Curr Pharm Des 21:6165–6188. https://doi.org/10.2174/1381612821666151027153246
Ullah H, Khan I, Yamani ZH, Qurashi A (2017) Sonochemical-driven ultrafast facile synthesis of SnO2 nanoparticles: growth mechanism structural electrical and hydrogen gas sensing properties. Ultrason Sonochem 34:484–490. https://doi.org/10.1016/j.ultsonch.2016.06.025
Waghmare SR et al (2015) Ecofriendly production of silver nanoparticles using Candida utilis and its mechanistic action against pathogenic microorganisms. 3 Biotech 5:33–38

Wang Y, Xia Y (2004) Bottom-up and top-down approaches to the synthesis of monodispersed spherical colloids of low melting-point metals. Nano Lett 4:2047–2050. https://doi.org/10.1021/nl048689j

Wen Z-Q, Li G, Ren D (2011) Detection of trace melamine in raw materials used for protein pharmaceutical manufacturing using surface-enhanced Raman spectroscopy (SERS) with gold nanoparticles. Appl Spectrosc 65(5):514–521

Yu ZB, Xie YP, Liu G, Lu GQ, Ma XL, Cheng H-M (2013) Self-assembled CdS/Au/ZnO heterostructure induced by surface polar charges for efficient photocatalytic hydrogen evolution. J Mater Chem A 1:2773. https://doi.org/10.1039/c3ta01476b

Zhang X et al (2011) Synthesis of nanoparticles by microorganisms and their application in enhancing microbiological reaction rates. Chemosphere 82:489–494

Biosensors and Nanobiosensors in Environmental Applications

26

Mojtaba Salouti and Fateme Khadivi Derakhshan

Abstract

There is a big demand for fast, reliable, and low-cost systems for the detection, monitoring, and diagnosis of pollutant in the environment and agriculture. Quantitative analysis of environmental samples is usually carried out using traditional analytical methods such as chromatographic and spectroscopic techniques to identify various environmental contaminants. These methods, although accurate and sensitive, require sophisticated and expensive instrumentation, expert personnel for their operation, and multistep and complicated sample preparation. These techniques are also labor-intensive and time-consuming, and it is hard to monitor contaminants on site, in real time, and at high frequency. To overcome the issues associated with current diagnostic techniques, a wide range of new biosensors (an analytical device for the quantitative detection of analyte with a biologically active element) are being developed. Several of these biosensors rely on nanotechnological platforms. Hence, in this chapter, an emphasis has been given on the deployment of nanobiosensor in detection of pollutant in the environment.

Keywords

Environmental pollution · Detection · Biosensors · Nanobiosensors

M. Salouti (✉)
Biology Research Center, Zanjan Branch, Islamic Azad University, Zanjan, Iran

F. Khadivi Derakhshan
Department of Biology, Urmia Branch, Islamic Azad University, Urmia, Iran

M. Ghorbanpour et al. (eds.), *Biogenic Nano-Particles and their Use in Agro-ecosystems*, https://doi.org/10.1007/978-981-15-2985-6_26

26.1 Introduction

The detection of pollutants, heavy metals, and toxic intermediates from waste streams and the monitoring of soil, water conditions, and many other vital features are highly detailed and comprehensive tasks (Malik et al. 2013; Bellan et al. 2011; Long et al. 2013a; Shahbazi et al. 2018). However, nanobiosensors are revolutionizing this field with prospective solutions by minimizing the load of conventional laboratory techniques and protocols and by enhancing sensitivity, robustness, and point-of-use portability (Amini et al. 2017; Kaushal and Wani 2017; Khan and Fatima 2014; Tripathi et al. 2017). A nanobiosensor is able to detect any biophysical and biochemical signal associated with a particular analyte (molecule) (Kabariya and Ramani 2017). The long-term use of nanobiosensors is to assess the existence and concentration in soil, water, and wastewater of toxic chemicals and pollutant (Srivastava et al. 2018). The applications of nanobiosensors were also extended for environmental monitoring of pollutants, toxicants, microorganisms and detection and prevention of bioterrorism in military, the net soil contaminant such as pesticides, herbicides, and heavy metals (Amini et al. 2018; Girigoswami and Akhtar 2019). Because of their submicron size, nanosensors have revolutionized in the fields of chemical and biological analysis, to enable the rapid analysis of multiple substances in vivo and environment samples (Steffens et al. 2017). This chapter emphasizes various applications of biosensors and nanobiosensors in the environmental and agriculture area. The major implication in the area of agriculture is physical monitoring of soil quality and fertility; indicator for seed viability; precision agriculture; detection of residual pesticides and herbicide, fertilizers, and toxins; and detection of microbiological pathogens in plants (Álvarez et al. 2016; Antonacci et al. 2018; Bagde and Borkar 2013; Choudhary et al. 2015; DeRosa et al. 2010; Duhan et al. 2017; Kaushal and Wani 2017). Biosensors and nanobiosensors can provide quick and specific information on contaminated locations for environmental control (detection of heavy metals, pesticides, organic compounds, biochemical oxygen demand (BOD), toxicity, and tracking (Álvarez et al. 2016; Antonacci et al. 2018; Bagde and Borkar 2013; Choudhary et al. 2015; DeRosa et al. 2010; Duhan et al. 2017; Kaushal and Wani 2017; Long et al. 2013a; Rajkumar et al. 2017; Salgado et al. 2011; Suresh and Periasamy 2014; Touhami 2014; Turdean 2011; Wan Jusoh and Ling Shing 2014; Yildirim 2016)). In this regard, some of the recent biosensor and nanobiosensor configurations are presented here.

26.2 Biosensors

Biosensors are analytical devices that convert a biological sensing element in close proximity or integrated with a signal transducer in order to quantify a compound condition. Biosensor is a probe that integrates an abiological one with a physicochemical transducer component to yield a measurable signal (Suresh and Periasamy 2014).

Researchers from various fields such as physics, chemistry, biology, engineering, and medicine are interested in developing, constructing, and manufacturing new sensing devices to get more efficient and reliable information. Biosensors are being developed for different applications, including quality control of food, environmental and bioprocess control, agriculture, medical applications, and military (Rodríguez-Mozaz et al. 2004).

For environmental control and monitoring, biosensors can provide fast and specific data of contaminated sites (biological/ecological quality or for the chemical monitoring of inorganic/organic priority pollutants). They give other benefits over present analytical methods, such as portability and on-site operation, and the capacity to measure pollutants in complicated matrices with minimal sample preparation (Salgado et al. 2011).

Traditional analytical methods employed for the environmental monitoring of pollutants include various chromatographic techniques, enzyme-linked immunosorbent assay, capillary electrophoresis, and surface-enhanced Raman scattering spectroscopy; although these methods have low limits of detection and good selectivity, they require expensive reagents, time-consuming sample pretreatment, expensive equipment, and professional technicians (Lang et al. 2016). Traditional methods are not effective for in situ measurements as in the case of accidental release of pesticides or acute poisoning (and so on) and limit their application in on-site analysis, where rapid, miniaturized, and portable equipment is needed such as environmental monitoring biosensors (Guo et al. 2017). For example, harmful algal blooms (HABs) are global phenomena throughout the world's oceans that have led to increased concerns in terms of human health, environmental preservation, and economic challenges. Six major marine biotoxin groups are associated with HABs. Biosensors have been developed to detect such biotoxins produced from algae and other microorganisms in the harvested fish/shellfish from oceans (McPartlin et al. 2016).

26.2.1 Components of a Biosensor

The working of any biosensor can be explained with the following example. In a biosensor, a particular "bio" element can recognize a particular analyte followed by the "sensor" element which can transduce the change of the biomolecule into a measurable signal in the form of electrical, optical, piezoelectric, calorimetric signals, and so on (Girigoswami and Akhtar 2019). A typical biosensor construct has three main features: a cognition element (enzyme, antibody, DNA, microorganisms, tissues, or even synthetic molecules), a signal-transducing structure (electrical, optical, piezoelectric, or thermal), and an amplification/processing element, the same models including also a permselective membrane which controls transport of analyte to the bioreceptor (Turdean 2011). The work of any biosensor can be explained as follows:

1. Bioreceptors that bind the specific form to the sample: The bioreceptor must be highly specific for the purpose of the analysis, stable under normal storage conditions, and show a low variation between assays.
2. An electrochemical interface where specific biological processes occur giving rise to a signal: Most commonly, in a biosensor, a biorecognition phase (e.g., enzyme, antibody, receptor, and single-stranded DNA) interacts with the analyte

to produce a signal, which may be due to (i) a change in proton concentration, (ii) a release or uptake of gases such as ammonia or oxygen, (iii) a release or uptake of electrons, (iv) a light emission, absorption, or reflectance, (v) a heat emission, or (vi) a mass change and so forth.
3. A transducer that converts the specific biochemical reaction in an electrical signal: The function of the transducer is to convert the signal into an appropriate measurable response (e.g., current, potential, or temperature change).
4. A signal processor for converting the electronic signal into a meaningful physical parameter.
5. A proper interface to display the results to the operator (Girigoswami and Akhtar 2019; Pandit et al. 2016). The complete biosensor should be cheap, small, portable, and simple enough to be used by semiskilled operators (Di Lorenzo 2016).

26.2.2 Types of Biosensors

Biosensors were developed in the 1960s by the pioneers Clark and Lyons. Biosensors are categorized based on how the signals are transmitted from samples to different groups such as electrochemical, optical, thermal, physiometric, immunochemical, magnetic, enzyme, and DNA base (Mehrotra 2016).

For classifications, several approaches can be utilized:

(a) Depending upon the used transduction principle, biosensors could be distributed into groups of electrochemical, mass-dependent, optical, radiation-sensitive, and so on (Thevenot et al. 2001).
(b) Enzyme, nucleic acid, proteins, saccharides, oligonucleotides, ligands, etc., are the various sets of biosensors which could be acquired if bioelement is considered as the basis of categorization (Mohanty and Kougianos 2006).
(c) Biosensors could be achieved based on DNA, toxins, glucose, mycotoxins, enzymes, or drugs according to the type of analyte to be detected (Turner 2000).

Biosensors are generally classified based on the bioreceptor element involved in the biological recognition process (enzymatic, nucleic acid-based, antibody-based, whole cell–based) or y the physicochemical transducer used (electrochemical, optical, acoustic, piezoelectric, thermal, colorimetric). This classification is described below.

26.2.2.1 Enzyme-Based Biosensors

Enzyme-based biosensors have emerged as a valuable technique for qualitative and quantitative analysis of a variety of target analytes in the biomedicine, environmental control, food quality control, and agricultural and pharmaceutical industries (Ispas et al. 2012). Enzymatic biosensors are used to measure food ingredients (sugars, acids, amino acids, inorganic ions, alcohols, and carbohydrates), contaminants (pesticide residues and heavy metals), food additives (sorbitol, benzoic acid, sulfites), and food freshness indicators (such as biogenic amines). Biosensors are also used in medical fields to detect glucose, urea, cholesterol, etc. (Economou et al. 2017; Girigoswami and Akhtar 2019). The problem with biosensors based on enzymatic inhibition is that only a few enzymes are sensitive to heavy metals (Turdean 2011). The first enzyme-based sensor was reported by Updike and Hicks in 1967 (Mehrotra 2016). The majority of existing enzymatic biosensors are based on either electrochemical or optical transduction, while other types of transducers are less frequently used. The two most important classes of electrochemical transducers for biosensing are the amperometric and potentiometric/ion-selective field effect transistor devices (Economou et al. 2017; Ispas et al. 2012).The commonly used enzymes for this purpose are peroxidases, oxidoreductases, aminooxidases, polyphenol oxidases, L-lactate dehydrogenase, tyrosinase, and nitrate reductase (Economou et al. 2017).

Adsorption, covalent binding to solid surfaces and supported films, entrapment in polymer hydrogels, and microencapsulation have been used for a long period to immobilize enzymes (Turdean 2011). The choice of enzyme/analyte system is based on the fact that these toxic analytes inhibit normal enzyme function. In general, the development of these biosensing systems relies on a quantitative measurement of the enzyme activity before and after exposure to a target analyte. Typically, the percentage of inhibited enzyme (I%) that results after exposure to the inhibitor is quantitatively related to the inhibitor (i.e., analyte) concentration and the incubation time. Consequently, the residual enzyme activity is inversely related to the inhibitor concentration. Given that pollutant compounds selectively inhibit the activity of certain enzymes, their activity and the resulting product concentration are affected (Amine et al. 2006).

26.2.2.2 DNA-Based Biosensors

DNA biosensors, based on nucleic acid recognition methods, are being developed toward the assay of rapid, simple, and economical testing of genetic and infectious diseases. Moreover, the detection of specific DNA sequence is of significance in numerous areas including clinical, environmental, and food analysis (Kavita 2017). These sensors offer advantages as better compatibility with micro-fabrication technology and can be constructed with a broad spectrum of conducting/semiconducting matrices such as gold, platinum, conducting polymers, etc. (Uniyal and Sharma 2018). Major types of DNA biosensors include electrochemical, optical, acoustic and piezoelectric ones (Zhai et al. 1997).

Nucleic acid biosensors are either based on the highly specific hybridization of complementary strands of DNA/RNA molecules or play the role of a highly specific receptor of biochemical/chemical species (DNA hybridization, SPR-DNA, electrochemical DNA, label-based or indirect detection, label-free or direct detection). The interaction is due to the formation of stable hydrogen bonds between the two nucleic acid strands. Nucleic acid biosensors are of major interest owing to their great promise for obtaining sequence-specific information in a faster, simpler, and cheaper manner compared to the traditional ones (Kavita 2017; Mehrotra 2016).

26.2.2.3 Immunosensors

Immunosensors are affinity ligand-based biosensor solid-state devices in which the immunochemical reaction is coupled to a transducer. The fundamental basis of all immunosensors is the specificity of the molecular recognition of antigens by antibodies to form a stable complex. Immunosensors can be categorized based on the detection principle applied. The main developments are electrochemical, optical, and microgravimetric immunosensors (Shofiul Azam et al. 2014). Immunoassay technique performs a fast, simple, reliable, and sensitive analysis of different compounds being applied in several areas of interest such as clinical analysis for medical diagnosis, as well as in environmental analysis, and food quality control (Balahura et al. 2019).

The most widely used immunosensors apply three different kinds of signal and transduction methods, following the formation of the complex antigen–antibody: optical, in which changes of the optical properties of the surrounding media are produced (e.g., color, luminescence, changes in refractive index); electrochemical, based on electrical signals (current, voltage differences, resistance); or piezoelectric, which relies on the changes in mass detected by piezoelectric devices (Lara and Perez-Potti 2018; Ricci et al. 2012). Generally, the basic principles of immunosensors with regard to the different transducer systems included electrochemical, mass detecting, heat detecting, and optical immunosensors (Morgan et al. 1996). However, the low chemical/physical stability limits the application of antibodies in harsh environments such as acids, organic solvents, and high temperature (Uniyal and Sharma 2018).

26.2.2.4 Whole-Cell-Based Biosensors

A whole-cell-based biosensor is an analytical device which integrates whole cells or organelles, which are responsible for its selectivity, with a physical transducer to generate a measurable signal proportional to the concentration of analytes (Turdean 2011). The cells are cheaper, have longer active lifetime, and are less sensitive to inhibition, pH, and temperature variations than enzymes (Bagde and Borkar 2013). While entire cell-based biosensors are not as sensitive to environmental changes as molecular-based ones, these biosensors can be modified using simple genetic engineering methods to detect a series of complex responses within a living cell (Gui et al. 2017). According to the classical definition, the whole cell must be integrated with a transducer (photometer, luminometer, charge-coupled device, and liquid scintillation counter) to function as a true biosensor (Turdean 2011). The main mechanism of a typical whole cell biosensor is the detection and amplification of a particular type of analyte into an electrical and optical signal via a processor. This process can be detected by immobilization and the use of living cells or bacteria as a unit providing molecular recognition elements. In contrast to a standard biosensor, whole-cell-based biosensors can detect a wider variety of substances, making them more susceptible to changes in a sample's electrochemical state (Gui et al. 2017). Whole-cell biosensors often are less costly than the corresponding enzymatic biosensors because some microorganisms can be cultivated and isolated rather easily, which is not the case for many enzymes. On the other side, they often lack the specificity for the respective analytes. Whole-cell biosensors mostly are self-contained, do not require the addition of cofactors, and are the biorecognition elements of choice when the total amount of hazardous substances or pollutants is to be determined (Damborský et al. 2016).

Microbial biosensors are generally defined as analytical devices composed of a microorganism that detects a target substrate and converts the detected signal to a quantifiable response in a physiological, electrical, or biochemical manner. Microbial biosensor sensing and recognition mechanisms include different kinds of standard optical, electrochemical, and sensory-regulated systems. Cell behavior regulation or metabolic pathways can be detectable because microorganisms not only detect environmental variables including nutrients, temperature, and pH but also sense their own metabolic status (Lim et al. 2015a). Whole-cell bacterial biosensors consist of genetically engineered bacteria containing a contaminant-sensing gene capable of detecting the presence of an analyte, coupled with a reporter gene capable of producing a detectable response (Strosnider 2003). For example, *Pseudomonas diminuta* or *Flavobacterium* sp. is usually used for organophosphorus hydrolase (direct organophosphate determination) isolation only in specialized laboratories.

The pigments present in the chlorophylls convert the light absorbed into energy (photons), and a small amount of energy is emitted as fluorescence. Currently, there are a few studies that utilize cyanobacteria for the development of whole-cell biosensor. Algal biosensors are commonly used for the detection of heavy metals, biocides, and BOD (Aisyah et al. 2014). An algal biosensor for toxicity assessment of estuarine waters is proposed by Campanella et al. in 2001. The sensor was obtained by coupling a suited algal bioreceptor (Cyanobacterium *Spirulina subsalsa*) to an amperometric gas diffusion electrode. The analytical device allows the monitoring of the evolution of photosynthetic O_2 and the detection of alterations due to toxic effects caused by environmental pollutants (heavy metals, triazine herbicides, carbamate insecticides) present in the medium (Campanella et al. 2001).

26.2.2.5 Electrochemical Biosensors

Electrochemical biosensors combine the sensitivity of electroanalytical methods with the inherent bioselectivity of the biological component. The biological component in the sensor recognizes its analyte resulting in a catalytic or binding event that ultimately produces an electrical signal monitored by a transducer that is proportional to analyte concentration. The two classes of electrochemical biosensors are biocatalytic devices and affinity sensors (Ronkainen et al. 2010). The sensor substrate usually contains three electrodes: a reference electrode, an active electrode, and a sink electrode. An auxiliary electrode may also be present as an ion source. The target analyte is involved in the reaction that takes place on the active electrode surface, and the ions produced create a potential which is subtracted from that of the reference electrode to give a signal (Bagde and Borkar 2013). The basic principle for this class of biosensors is that chemical reactions between immobilized biomolecule and target analyte produce or consume ions or electrons, which affects measurable electrical properties of the solution, such as electric current or potential. The electrochemical signal produced is then used to relate quantitatively to the amount of analyte present in a sample solution. Potentiometry (which measures the potential of the biosensor electrode with respect to a reference electrode), amperometry (which measures the current produced during oxidation or reduction of electroactive product or reactant), conductometry (which measures the change in conductance arising due to the biochemical reaction), and, more recently, electrochemical impedance spectroscopic measurements are among the electrochemical detection techniques often used in conjunction with immunoassay systems and immunosensors, leading to their respective categories according to the type of signal measured (Sawant 2017; Touhami 2014). The types of electrochemical biosensors are as follows:

(a) Amperometric biosensors are self-contained integrated devices based on the measurement of the current resulting from the oxidation or reduction of an electroactive biological element providing specific quantitative analytical information. Generally speaking, amperometric biosensors work by producing a current when a potential is applied between two electrodes and the analyte undergoes or is engaged in a redox reaction which can be followed by evaluating the current in an electrochemical cell. The analyte being measured or the biomolecule involved with it changes its oxidation state at the electrode. The electron transfer signal is then measured which is proportional to the amount of the redox-active species at the electrode (Sadeghi 2013). In wastewater oxygen monitoring, organophosphate detection, and phenol, cyanide, and heavy metal ion detection, amperometric biosensors have been commonly used (Tsopela et al. 2014). The disadvantages of these biosensors are they require a reference electrode, multiple membranes, or enzymes (Giannoudi et al. 2006).
(b) A potentiometric biosensor can be described as a device that incorporates a biological sensing element linked to a potential electrochemical transducer. Potentiometric biosensors usually rely on a biochemical reaction leading to a simpler chemical specie and its subsequent electrochemical detection (NH_4OH, CO_2, pH, H_2O_2, etc.). The analytical signal generated by a potentiometric biosensor is an electrical potential (Yunus et al. 2013). Potentiometry is widely used for pH monitoring, and glass membrane electrodes are commonly used. However, glass fragility and size limitation impede in situ measurement in complex configurations and mediums (Shitanda et al. 2009; Tsopela et al. 2014). The main types of potentiometric sensors are membrane-based ion-selective electrodes, screen-printed electrodes (SPE), ion-selective field effect transistors, solid-state devices, and chemically modified electrodes (e.g., used as sensitive layers of metal oxides or electrodeposited polymers) (Yunus et al. 2013). Although pH electrodes have been widely used for detection of pesticides in agriculture, other ion-selective electrodes, such as chloride ion-selective electrode, have also generated major interest in wastewater treatment domain (Tsopela et al. 2014). The disadvantages of potentiometric biosensor are that they require reference electrode, have limited linear range, and are often pH-sensitive (Giannoudi et al. 2006).

(c) The conductometric biosensors are based on the fact that almost all enzymatic reactions involve either consumption or production of charged species and, therefore, lead to a global change in the ionic composition of the tested sample. Biosensors based on the conductometric principle present a number of advantages: transducers are not light-sensitive, thin-film electrodes are suitable for miniaturization and large-scale production using inexpensive technology, they do not require any reference electrode, large spectrum of compounds of different nature can be determined on the basis of various reactions and mechanisms, and the driving voltage can be sufficiently low to decrease significantly the power consumption (Jaffrezic-Renault and Dzyadevych 2008). Conductometric biosensors were used to detect heavy metal ions and pesticides for multiple applications in environmental monitoring (Tsopela et al. 2014).

(d) Voltammetry is the most versatile technique in electrochemical analysis. Voltammetry belongs to a category of electroanalytical methods, through which information about an analyte is obtained by varying a potential and then measuring the resulting current. It is, therefore, an amperometric technique (Touhami 2014). Differential pulse voltammetry was applied to screen the inhibition effect of different concentrations of fisetin in the micromolar range. Voltammetry can detect multiple compounds, which have different peak potentials, in a single electrochemical experiment (or scan), thus offering the simultaneous detection of multiple analytes (Grieshaber et al. 2008).

26.2.2.6 Optical Biosensors

Optical biosensors represent the most common type of biosensor. Optical biosensors consist of a light source, as well as numerous optical components, to generate a light beam with specific characteristics and to beeline this light to a modulating agent, a modified sensing head along with a photodetector (Mehrotra 2016). Optical biosensors widely use the principle of surface plasmon resonance (SPR), bioluminescent optical fiber, evanescent wave fluorescence, interferometry, ellipsometry, reflectrometric interference spectroscopy, and surface-enhanced Raman scattering (Girigoswami and Akhtar 2019). Optical detection is performed by taking advantage of the interaction between the optical field and the biorecognition element. Optical biosensing can be widely divided into two general modes: label-free and label-based. In short, the detected signal is generated directly in a label-free mode by the interaction between the analyzed material and the transducer. In contrast, label-based sensing involves the use of a label, and the optical signal is then generated by a colorimetric, luminescent, or fluorescent method. The optical biosensor can use various biological materials, including nucleic acids, enzymes, antibodies, whole cells, and tissues, as biorecognition elements (Damborský et al. 2016). Optical-based transduction systems encompass a wide variety of sensors based on the change in a property of light upon interaction of the recognition molecule and the target analyte. These properties include amplitude (adsorption), wavelength (fluorescence, surface plasmon), polarization, and time dependence (time-resolved fluorescence). Modern optical biosensors have distinct properties and advantages such as high sensitivity, easy adaptation to multiplexed systems, and portable operation capability (Biran and Walt 2002).

The main components of fiber-optic biosensors (FOBs), which influence sensitivity and detection limit, are light source, optical transmission medium (fiber, wave guide, etc.), immobilized biological recognition element (nucleic acid, enzymes, antibodies, or microbes), optical probes (such as fluorescent markers) for transduction, and an optical detection system. One of the main benefits of using optical biosensors in combination with optical fibers is that it allows long-distance sample assessment, and this has significant consequences for field monitoring (Narsaiah et al. 2012).

The fluorescent sensing technique is based on the measurement of fluorescence intensity which is proportional to the concentration of the target analyte. Fluorescent biosensors have been widely applied in analytical chemistry due to their easy construction using standard molecular biology techniques. Fluorescent biosensors can probe ions, metabolites, and protein biomarkers with great sensitivity and can also report the presence, activity, or status of the target (serum, cell extracts) in complex solution (Su et al. 2011). There are three types of fluorescence biosensing. The first is direct sensing when a specific molecule is detected before and after a change or reaction takes place. The second form is indirect biosensing when a dye is added that will optically transduce the presence of a specific target molecule such as green fluorescent protein (GFP). A third type of fluorescence biosensing, called fluorescence energy transfer (FRET), can be used, and it generates a unique fluorescence signal (Touhami 2014). This type of biosensor is user-friendly – easy to engineer, manipulate, and transfer into cells. Single-chain FRET biosensor is another example. They consist of a pair of AFPs, which are able to transfer fluorescence resonance energy between them when brought close together. Different methods may be used to regulate changes in FRET signals based on intensity, ratio, or lifetime of AFPs (Mehrotra 2016).

Surface plasmon resonance (SPR) biosensors can directly sense biomolecular interactions without labeling, allowing real-time measurements of analyte concentration and kinetics as well as thermodynamic binding parameters. SPR has been used in interaction studies and the screening of a variety of moieties, including proteins, carbohydrates, cells, and nucleic acids, leading to applications in areas such as military defense, clinical diagnostics, and pharmaceutics (Abdulhalim et al. 2008). A practical SPR instrument combines an optical detector part, usually measuring intensity shift, a sensor chip with a gold surface and a layer enabling ligand immobilization, which is integrated with a fluidics system enabling a flow-through operation (Damborský et al. 2016). Depending on the size of the target sample, four distinct assay formats were illustrated with SPR: direct assay, competitive assay, sandwich assay, and inhibition assay formats (Abdulhalim et al. 2008).

FOBs are based on the transmission of light along silica glass fiber or plastic optical fiber to the site of analysis. FOBs can be used in combination with different types of spectroscopic technique, e.g., absorption, fluorescence, phosphorescence, SPR, etc. (Touhami 2014). FOBs have the potential to detect bacteria in aerosolized form with greater speed and specificity than current microbiological methods. In 1999, Ferreira et al. demonstrated one of the first applications of FOBs to the detection of aerosolized bacteria, a common cause of nosocomial infections (Hayman 2008).

Bioluminescent biosensors use enzyme-catalyzed exothermic biochemical reactions to generate excited-state emitters based on the principle of light emission by viable bacteria in response to any chemical, biological, or physical changes in the analyte (Axelrod et al. 2016).

The simplest form of biosensor is the colorimetric test strips. Colorimetric techniques have demonstrated excellent potential for cheap daily-life apps due to their apparent benefits of simplicity, low cost, and no need for any costly tools. Besides, it measures the change in color or optical density of the test sample upon a chemical reaction, and the color change of substrates can be observed easily by naked eyes (Aldewachi et al. 2018). Colorimetric biosensor for detection of toxicant in water can be used as an early warning sign to monitor water toxicity in the environment (Wasito et al. 2019).

26.2.2.7 Mass Spectrometry Biosensors (Piezoelectric Biosensors)

Mass spectrometry (MS) is an analytical technique that can be used by ionizing chemical compounds to assess the mass-to-charge ratio of charged particles, the particle mass, the elemental structure of a sample, and the chemical structures of molecules. MS is generally composed of three fundamental parts: the ionization source (a small sample is ionized, usually to cations, by loss of an electron), the mass analyzer (the ions are sorted and separated according to their mass and charge), and the detector (which registers the number of ions at each m/z value). The MS can be used to identify and, progressively, accurately quantify thousands of proteins from complicated samples, which is thought to have a wide effect on biology and medicine. But, in particular, the size of these systems is big, making them unfeasible for field applications requiring portable devices, particularly for biosensors (Sang et al. 2013).

26.2.2.8 Acoustic Biosensors

Acoustic wave sensors operate by monitoring the change in the physical properties of an acoustic wave in response to the measure. In acoustic detectors, piezoelectric materials are frequently used to produce acoustic waves in strong metals using suitably tailored electrical areas and to detect the acoustic waves by the load produced by the mechanical deformation caused (Fogel et al. 2016). Acoustic biosensors can be made with piezoelectric crystals such as quartz, lithium niobate, or lithium tantalate since they are robust and environmentally stable. In addition, such sensors are versatile and can detect, in principle, various biomolecules (Durmuş et al. 2015). The gravimetric nature of the signal will ensure that this will remain a routine method of characterizing layer–layer interactions in a label-free fashion, which lends itself very well to investigation of biorecognition immobilization, particularly for the design of other biosensors (Fogel et al. 2016). Acoustic wave sensors, such as bulk or surface acoustic waves, can be classified according to the waves they produce. Bulk acoustic wave biosensors use either longitudinal or shear waves, although the latter are often preferred in the medium of interest to decrease acoustic radiation. Surface acoustic wave sensors have been used for years in measuring temperature, viscosity, pressure, concentration, acceleration, and chemical/biological entities. They are also used for signal processing activities and are environmentally delicate (Durmuş et al. 2015).

26.2.2.9 Thermal Biosensors

Thermal biosensors are based on measurement of the thermal changes occurring on biochemical recognition. Most of the biochemical reactions involve a change in enthalpy, and the heat changes can be measured by sensitive thermistors. Thermal biosensors were developed in the early 1970s and, by the end of the decade, were used for continuous measurements and enzyme reactor control. Another enhanced version of the thermal biosensor, the enzyme thermistor, has been created for analytical assays, and this instrument has been used to analyze roughly distinct analytes (Ramanathan et al. 1999). There has been less consideration for thermal biosensors. In addition, negative remarks such as complex thermostating, very weak sensitivity, or nonspecific heating impacts led to bad reputation. However, the enzyme thermistor has enhanced our knowledge of immobilized multi-enzyme systems for signal amplification and the use of immobilized coenzymes and various immobilization techniques (Lammers and Scheper 1999). Enzyme thermistors are biosensors that use thermal resistors to measure the heat change caused by an enzymatic reaction. They combine the selectivity of enzymes with the sensitivity of biosensors and allow continuous analysis in a flow-injection mode. They can be used to monitor fermentation systems, biocatalysis, enzyme-catalyzed synthesis, and clinical and food technology (Sawant 2017). There are several advantages and disadvantages associated with the enzyme thermistors as an online biosensor (Ramanathan et al. 1999). Thermal biosensors have numerous advantages: (a) thermistors have very excellent long-term stability due to no chemical interaction between transducer and sample; (b) thermistors are goods of inexpensive bulk; (c) different optical or ionic sample features do not disturb measurements; (d) thermal biosensors operate in some instances without complex and interference-prone multienzyme systems, e.g., analysis of disaccharides; and (e) multiple applications were discovered for thermal biosensors (Lammers and Scheper 1999; Ramanathan et al. 1999).

26.2.3 Biosensors Environmental Applications

Environmental safety is the key requirement of our well-being. Rapid urbanization and industrialization have greatly contributed several pollutants such as heavy metals, inorganic and organic compounds, pesticides, toxins, endocrine-disrupting hormones, etc. to the environment. Such pollutants are becoming a significant global challenge for the safety of the environment and human health (Gui et al. 2017). The elevated cost and slow turnaround times typically associated with controlled pollutant measurement obviously indicate a need for quick, portable, and cost-effective environmental screening and surveillance techniques (Andrea Medeiros Salgado et al. 2011).

In latest years, the development of biosensors for pollutant identification has gained significant attention. Such sensors offer a huge benefit in detecting the minimal contaminant amount in complicated matrices, such as river, wastewater, and solid (Ejeian et al. 2018). Biosensors can be used as environmental quality-monitoring tools in the assessment of biological/ecological quality or for the chemical monitoring of both inorganic and organic priority pollutants (Andrea Medeiros Salgado et al. 2011). Several biosensor developments for marine measurements of eutrophication, pesticides, anti-biofouling agents, polycyclic aromatic hydrocarbons (PAHs), endocrine disruptors, trace metals, organism detection, and algal toxins have been reported in literature (Moro et al. 2018). For example, the use of photosynthetic material and in particular the photosystem II (PSII) sub-membrane fraction as the biological receptor in a biosensor provides an excellent tool for the detection of toxic metal cations. Copper (Cu) has become a widespread pollutant due to its use as algaecide and fungicide in agriculture, the sensitivity of PSII to this metal could be exploited for the development of sensors and assays. Different photosynthetic biosensors are used to detect heavy metals (Rouillon et al. 2006). Photoelectrochemical cell using thylakoid membranes (immobilized in a cross-linked albumin-glutaraldehyde matrix) isolated from spinach leaves was used to elaborate a phytotoxicity biosensor to detect PbC12 and CdCli in solution. It is shown that the photocurrent generated in the cell is linearly correlated with the electron transport activity of the photosynthetic membranes. Measurements with tin biosensor are performed rapidly (< 5 min) and require only small volumes (80 μl) (Carpentier et al. 1991).

26.2.3.1 Biosensors in Agriculture

Agriculture, food, and natural resources are a part of those challenges like sustainability, susceptibility, human health, and healthy life. Agricultural production also constrained by a number of abiotic and biotic factors like insect pests, diseases, and weeds causes substantial damage to agricultural production. These biosensors may have a huge impact on the precision farming methods (Sekhon 2014).

Due to the fast development of the population, the use of pesticides in agriculture has increased to satisfy the increasing demand for food (Martinazzo et al. 2018). Pesticides have become essential in modern agriculture, but pesticides can contaminate foods, soil, air, and water. Pesticides enter into the aquatic environment in several ways: through the use in agricultural, industrial, and municipal sewage when applied to the control of aquatic herbs and insects, by accident in warehouses, during transport, or even by the improper disposal of packaging (Steffens et al. 2017). Pesticides may be classified according to the following criteria:

(a) Target pests: herbicides (simazine, alachlor, cybutryne, atrazine, etc.), fungicides (pentachlorobenzene, hexachlorobenzene, etc.), insecticides (chlorpyrifos, endosulfan, dichlorvos, etc.), and acaricides (dicofol)
(b) Chemical nature of the active ingredients: pyrethroids, organochlorines, organophosphates, etc.
(c) Action spectra: selective and nonselective
(d) Action mode: systemic and nonsystemic or contact

(e) Toxicity: extremely hazardous, highly hazardous, moderately hazardous, slightly hazardous, Unlikely to present acute hazard
(f) Timing of application: before plantation, preemergence, and postemergence (Martinazzo et al. 2018)

In agriculture, the use of insecticides is strictly regulated by complicated and contradictory laws for each compound (Bucur et al. 2018). Biosensors can be used to measure the levels of pesticides and herbicide in the soil, food, air, and groundwater.

Parathion and chlorpyrifos (CP) are broad-spectrum organophosphate (OP) pesticides having a wide range of applications against numerous insect species on several crops. These contaminants are widely spread in different environmental matrices and are highly toxic causing human fatalities if exposed by any routes such as ingestion, skin adsorption, and inhalation (Salgado et al. 2011). The toxicity mechanism of the neurotoxic insecticides is based on the inhibition of acetylcholinesterase (AChE; EC 3.1.1.7), and the reproduction of this inhibition in vitro can be used for multianalyte selective monitoring. Numerous biosensors are reported to be built for the identification of neurotoxic OP and CP based on cholinesterase inhibition that have been examined with regard to their overall characteristics, the parameters affecting the enzymatic inhibition. Strategies for the building of biosensors using different techniques of immobilization and the roles of different matrices were used. In the case of biosensors based on AChE, the use of chemometric techniques is based on two facts: (a) the inhibition of AChE produced by insecticide mixtures is higher than the individual inhibition percentages produced by each individual inhibitor and (b) the enzymes extracted from various organisms or genetically engineered have variable sensitivities for different insecticides (Bucur et al. 2018).

Double-stranded calf thymus deoxyribonucleic acid (DNA biosensor) entrapped polypyrrole-polyvinyl sulfonate films fabricated onto indium-tin-oxide-coated glass plates have been used to detect CP. CP detection strategy was based on the change in voltammetric patterns of the modified electrode due to oxidation of guanine. These biosensing electrodes have a response time of 30 s, are stable for about 5 months when stored in desiccated conditions at 25 °C, and can be used to amperometrically detect chlorpyrifos and malathion (Prabhakar et al. 2007).

Various enzyme (AChE, organophosphorus hydrolase (OPH) enzymes, methyl parathion hydrolase (MPH), and organophosphorus acid anhydrolase (OPAA))-based biosensors have been reported with slight modifications in immobilization method to ensure sensitive CP detection such as self-assembled monolayer-based, polymer-based, graphene-based, carbon nanotube-based, gel-based, and metal- and metal oxide-based biosensors (Uniyal and Sharma 2018Pogačnik and Franko (1999) used a photothermal sensor (based on thermal lens spectrometry) to determine OP and carbamate pesticides in spiked drinking water and fruit juices. Pesticide detection in a single sample by the biosensor consisting of the AChE (placed in a flow injection analysis (FIA)) was performed within 15 min (Pogačnik and Franko 1999). A solid-phase fluoroimmunoassay (labeled antiparaquat antibodies) combined with an optical transducer chemically modified with an analyte derivative coupled to a FIA system was reported by Mallat et al. in 2001. The fluorescence signal was indirectly related to the paraquat concentration in the sample. Although the immunosensor achieved a lower paraquat detection limit of 0.06 mg l^{-1} in river water (in 15 min), the procedure was quite elaborate, requiring labeling of antiparaquat antibodies (Mallat et al. 2001).

Crew et al. described a biosensor array based on six AChE for use in an automated instrument incorporating a neural network program (electrochemical analysis using chronoamperometry) to detect the OPs (dichlorvos, malaoxon, chlorpyrifos-oxon, chlorpyrifos-methyl-oxon, chlorfenvinphos, and pirimiphos-methyl-oxon) in water and food samples. The biosensor arrays and automated instrument were evaluated in situ in field experiments where the instrument was successfully applied to the analysis of a range of environmental samples (Crew et al. 2011). Istamboulie et al. is a bi-enzymatic amperometric sensor designed by immobilizing both AChE (*Drosophila melanogaster* AChE) and phosphotriesterase in a polyvinylalcohol matrix. The biosensor was shown to be able to discriminate between chlorpyrifos and chlorfenvinphos inhibitions in real water samples (Istamboulie et al. 2009).

On the other side, different biosensors based on amperometric, potentiometric and fiber-optic AChE were recorded. Potentiometric AChE biosensors detect the pH change in the presence of OP pesticides arising from a reduction in the acid produced during the enzyme-catalyzed hydrolysis of the cholinesters. The amperometric AChE biosensors are based on measuring the shift in concentration of the thiocholine electroactive product generated as a consequence of acetylthiocholine hydrolysis (Songa and Okonkwo 2016). The fiber-optic biosensor monitors the pH change using a fluorescein label attached to AChE or dextran, or chemiluminescence (Mulchandani et al. 2001). Although these biosensors are highly sensitive, they often lack selectivity because they respond with any inhibitor of cholinesterase, and scientists investigate enzymes that are capable of direct and selective recognition and hydrolysis of ops such as OPH, MPH, and OPAA (Songa and Okonkwo 2016). The enzyme was used to detect OPs in microbial structures as well as a purified enzyme using methods such as optical, acoustic, potentiometric, and amperometric (Songa and Okonkwo 2016). A pH electrode modified with an immobilized purified OPH layer formed by cross-linking OPH with bovine serum albumin and glutaraldehyde with pH meter was the basic element of the potentiometric OPH biosensor, a measuring cell placed on a mixing magnetic stirrer and a chart recorder. Two distinct biosensors based on OPH were built. The first is based on measuring the decline in fluorescence intensity of isothiocyanate fluorescein, which is covalently immobilized into the enzyme. Second, the assessment was based on the connection between the hydrolyzed quantity of OP and the quantity of chromophoric product created by the catalyzed hydrolysis enzyme. Screen-printed thick-film carbon electrodes have been modified by amperometric OPH-enzyme which has been deposited on the electrode in Nafion film (Mulchandani et al. 2001).

A significant drawback is also the complicated, long-lasting, and costly operation conducted in dedicated microbiological laboratories for the extraction and purification of OPH (to note that these enzymes are not accessible commercially). Another disadvantage which limits the application of the developed biosensors is the long response time (2–3 min) (Margarita et al. 2016).

The MPH-specific substrate, methyl parathion, is still widely used as a potent insecticide and acaricide in agriculture. The MPH catalytic hydrolysis of methyl parathion produces dimethyl phosphate and 4-nitrophenol with a high turnover rate, and the latter is an electroactive compound that makes the construction of an amperometric biosensor possible (Songa and Okonkwo 2016).

Among the most widespread pesticides is photosynthesis-inhibiting herbicides, such as atrazine, metribuzin, diuron, bromacil, ioxynil, and dinoseb. Chromatographic methods (high-performance liquid chromatography, gas chromatography, gas chromatography–mass spectrometry, and capillary electrophoresis) are the most commonly reliable and sensitive methods used to monitor the presence of herbicides. Nevertheless, the requirement for expensive equipment, organic solvents, and laborious sample preparation prevents the use of chromatography for rapid screening and prescreening of a large number of samples (Scognamiglio et al. 2009).

They all belong to distinct families but have a prevalent mode of action: binding specifically to the chloroplast D1 protein with subsequent electron and proton flow disruption through photosystem II (Piletska et al. 2006). Several detection systems were tested in combination with D1 protein: electrochemical (amperometry and cyclic voltammetry) and optical (surface plasmon resonance and ellipsometry). The main mechanisms of Dl action are either on the ability of herbicides to replace the plastoquinone molecule in Dl protein and in this way change the electrochemical and optical properties of the system (Piletska et al. 2006). PSII is the supramolecular pigment–protein complex in the chloroplast (located in the thylakoid membrane of algae, cyanobacteria, and higher plants), which catalyzes the light-induced transfer of electrons from water to plastoquinone (PQ) in a process that evolves oxygen. The PSII complex is also known to bind some groups of (photosynthetic) herbicides, heavy metals, and other chemical substances that affect its activity (Giardi et al. 2001). Several groups of pesticides and herbicides are also known to bind the PSII complex. PSII drives electron transmission under illumination that is inhibited by herbicides. Approximately 30% of herbicides target the PSII plant. Thus, the immobilized and stabilized photosynthetic membrane will serve as the biomediator for the biosensor. Amperometric or optical systems can then translate and monitor the impact of compounds that change or inhibit photosynthetic activity, measured as oxygen evolution, electron transport, or fluorescence. One of the benefits of using PSII-based biosensors is the simplicity of biological transduction, which can be monitored directly without requiring additional markers or molecules for transducer. Another benefit of herbicide binding is its extreme susceptibility and selectivity (Giardi and Pace 2006). Rizzuto et al. in 2000 tested a biosensor consisting of PSII particles of the cyanobacterium *Synechococcus elongatus* for detection of herbicides in samples for three rivers. By washing the sensor, they were able to reuse it for several assays after removal of the toxic agent (Rizzuto et al. 2000). Naessens et al. (2000) detected a response to atrazine, simazine, and diuron using a *Chlorella vulgaris* biosensor. A fluorescence biosensor based on mutants resistant to various herbicide subclasses was developed; it makes possible to distinguish between subclasses of herbicides (Rizzuto et al. 2000).

Biosensors based on photosynthetic enzymes have low selectivity; while some PSII-based biosensors can detect phenylurea and phenolic herbicides at ppb concentrations, other biosensors are much less susceptible in accordance with present European regulations on peak concentrations of pesticide residues in water. In addition, to this day, the amount of practical apps for surface and groundwater assessment continues to be very restricted. Their performance recommends these biosensors as a "complete toxicity" screening tool and indicator. Photosynthetic biosensors were combined with other enzyme biosensors to broaden the purpose of the assessment and address the detection of various analytes (Bucur et al. 2018). Scognamiglio et al. (2013) reported a biosensor was based on microelectrode array technology and employed an array of biological recognition elements intimately integrated to an electrochemical–optical transduction system. The modification of gold microelectrode transducer surface allowed the simultaneous determination of selected target analytes by means of amperometric and fluorescence signal variation of immobilized enzymes and microorganisms able to recognize various food components, like sugars and phenols, and contaminants, like pesticides (Scognamiglio et al. 2013).

As illustrated by the literature reports compiled in Table 26.1, the development of biosensors for pesticides is the subject of considerable interest, particularly in the areas of food and environment monitoring.

Plant diseases are responsible for major economic losses in the agricultural industry worldwide. Monitoring plant health and detecting pathogen early are essential to reduce disease spread and facilitate effective management practices. Although DNA-based and serological methods have revolutionized plant disease detection, they are not very reliable at asymptomatic stage, especially in case of pathogen with systemic diffusion. They need at least 1–2 days for sample harvest, processing, and analysis. New technologies offer opportunity to assess disease with greater objectivity. The practical application of each traditional or innovative method depends on the cost and availability of instruments, specialization level (plant, field, farm, or agricultural district), rapidity of analysis, and the stage of disease at which detection is possible (Martinelli et al. 2015). The first report of biosensor-based plant pathogen detection deals with epitope mapping of mAbs on tobacco mosaic virus (TMV) and cowpea mosaic virus (CPMV). Individual virus-specific mAbs were immobilized via an Fc-specific rabbit anti-mouse pAb on a CM5 surface after which viral solutions were injected and captured using a Biacore system. Boltoverts and coworkers used a plasmon SPR sensor and a pAb for the detection of TMV. The authors tested different ways of immobilizing pAb and found protein A-based immobilization to be superior (Skottrup et al. 2008). A quartz crystal microbalance (QCM) immunosensor was developed for the detection of both *Cymbidium mosaic virus* and *Odontoglossum ringspot ovirus* by pre-coating the QCMs with virus-specific antibodies. The QCM was able to detect as low as 1 ng each of the two orchid viruses (Eun et al. 2002). Skuttrop et al. describes a biosensor for detection of fungal spores using SPR and a mouse monoclonal

Table 26.1 Examples of biosensors developed for the detection of pesticides and herbicides

Bioreceptor	Components	Pesticide/herbicide	References
AChEdm[a] (incorporate six histidine tails.)	Nickel-modified thick-film electrodes	Paraoxon, dichlorvos, chlorpyrifos ethyl-oxon	Andreescu et al. (2001)
AChE and six mutants (B03; B03-23; B05; B06-23; B07, B08-23.29.44)	Amperometric-tetracyanoquinodimethane graphite layer and Ag/AgCl	Methamidophos	Nunes et al. (2001)
dmAChE (E69Y Y71D), eeAChE	Amperometry-FIA, carbon pellet	Dichlorvos	Sotiropoulou et al. (2005)
AChE (Dm, E69W mutant)	PVA-SbQ polymer- FIA	Neurotoxic insecticides	Bucur et al. (2005)
AChE (B349 mutant)	Magnetic microbeads Ni–His, polyvinyl alcohol-based	Chlorpyriphos-oxon, chlorfenvinphos	Istamboulie et al. (2007)
AChE	Electrochemical, amperometric, gold screen printed electrode and cysteamine SAM, ferricyanide	Paraoxon	Arduini et al. (2013)
AChE and choline oxidase (ChO)	Optical, colorimetric, iodine starch	Paraoxon	Guo et al. (2017)
AChE	Electrochemical, amperometric, graphite, and macroalgae *Cladophoropsis membranous*	Methyl parathion	Nunes et al. (2014)
AChE	Electrochemical, impedimetric carbon paste electrode, and reticulated sphere structures of $NiCo_2S_4$	Methyl parathion	Peng et al. (2017)
AChE	Electrochemical, impedimetric, interdigitated array microelectrodes with chitosan	Carbaryl	Gong et al. (2014)
AChE	Electrochemical, amperometric, porous glassy carbon electrode with graphene oxide network	Carbaryl	Li et al. (2017a)
OPH	Amperometric probe, carbon paste	Paraoxon, methyl parathion	Wang et al. (1999)
OPH	Amperometric-FIA-thin-film gold	Paraoxon, methyl parathion	Wang et al. (2003)
OPH	Amperometric/ potentiometric- FIA-thin-film gold	Paraoxon, parathion, dichlorvos, diazinon	Schöning et al. (2003)
Multienzyme OPH/ AChE	Amperometric- FIA, layer-by-layer multi walled CNT_DNA	Paraoxon, non-OP Pesticide carbaryl	Zhang et al. (2015)

(continued)

Table 26.1 (continued)

Bioreceptor	Components	Pesticide/herbicide	References
OPAA	pH-FET[b] - silica gel- Ag/ AgCl	Diisopropyl fluorophosphate, paraoxon, demeton-S	Simonian et al. (2001)
PSII *Synechococcus bigranulatus*	Amperometry, gold screen-printed electrodes (AuSPE)	Diuron	Maly et al. (2005)
PSII *Synechococcus elongatus*	Amperometry, dialysis membrane, Teflon membrane of the Clark oxygen electrode	Diuron, atrazine, simazine, ioxynil, bromoxynil, dinoseb	Koblizek et al. (1998)
PSII isolated from *Synechococcus elongatus*	Amperometry, SPE-graphite-working electrode, Ag/AgCl	Diuron, atrazine, simazine ioxynil, bromoxynil, dinoseb	Koblížek et al. (2002)
Thylakoid from *Spinacia oleracea* L., *Senecio vulgaris* mutant resistant to atrazine	Amperometry, SPE-graphite graphite-working electrode, Ag/AgCl, bovine serum albumin/glutaraldehyde	Diuron, atrazine, simazine terbuthyl-azine, diethylterbuthylazine	Touloupakis et al. (2005)
PSII-enriched thylakoid fractions from spinach	Colorimetry, quartz substrates, layer by layer, poly ethylenimine	Terbutryn	Ventrella et al. (2011)
Mutant strains of *Chlamydomonas reinhardtii* with engineered D1 protein	Fluorescence, silicon septum	Atrazine prometryne terbuthyl-azine diuron linuron	Giardi et al. (2009)
"BBY"-crude PSII preparation from spinach leaves	Amperometry, AuSPE	Diuron	Bhalla et al. (2011)
Thylakoids from spinach	Biosolar cell, air-breathing platinum cathode	Atrazine bromacil diuron	Rasmussen and Minteer (2013)
Thylakoids from spinach	Amperometry, carbon-based SPE, one shot system	Diuron atrazine ioxynil	Bettazzi et al. (2007)
C. reinhardtii mutants	Fluorescence	Atrazine, prometryn, diuron	Scognamiglio et al. (2009)
Pure PS II cores and BBY particles from spinach	Amperometry, SPE, mercaptopropionic acid	Atrazine, picric acid	Bhalla and Zazubovich (2011)
PSII complex from *Synechococcus elongatus f. thermalis*	Amperometry	Atrazine, isoproturon, diuron	Masojídek et al. (2011)
Thylakoids from spinach	Amperometry, SPE, laser printing	Diuron, linuron	Touloupakis et al. (2012)
AChE/BChE whole cells of *Chlamydomonas reinhardtii*	Optical- amperometric, multi-array, SPE	Atrazine, diuron, paraoxon, chlorpyrifos	Scognamiglio et al. (2012)

(continued)

Table 26.1 (continued)

Bioreceptor	Components	Pesticide/herbicide	References
Whole cells of *C. reinhardtii*	Fluorescence-amperometric, micro-electrodes array,	Atrazine, prometryn, diuron	Scognamiglio et al. (2013)
Whole-cell *E. coli ATCC 25922, Bacillus subtilis, S. cerevisiae*	Electrochemical, entrapment in calcium alginate on a polyethylene terephthalate sheet	Ametryn, acephate	Gao et al. (2016)
Synechocystis sp. PCC6803 cyanobacteria	Photo-amperometry, platinum electrodes, poly(vinylalcohol) bearing styrylpyridinium groups,	Diuron	Avramescu et al. (1999)
Thylakoids from mutant spinach plants	Fluorescence, silicio septum inside a series of flow cells	Urea, diamine, triazine, phenols	Giardi et al. (2005)
Chlorella mirabilis algae	Fluorescence	Diuron, simazine irgarol	Moro et al. (2018)
Thylakoids from spinach	Biosolar cell, carbon paper electrodes	Diuron	Rasmussen et al. (2014)
Anabaena variabilis	Amperometric, carbon felt electrode, p-benzoquinone, alginate	Atrazine, diuron	Tucci et al. (2019)
C. reinhardtii (green algae)	Electrochemical, silicon chip, platinum, platinum black, tungsten/tungsten oxide, iridium oxide	Diuron	Tsopela et al. (2014)
C. reinhardtii	Optical, fluorescence, glass based microfluidic chip	Diuron, simazine, atrazine	Tahirbegi et al. (2017)
Dictyosphaerium chlorelloides strain Dc1M	Fiber-optics luminescent O_2 transducer, Adsorption on ImmobaSil™ porous silicone disks	Simazine	Haigh-Flórez et al. (2014)

[a]*dm Drosophila melanogaster, ee Electric eel, be Bovine erythrocytes, he* Human erythrocytes, *FIA* flow injection analysis
[b]Field effect transistor

antibody (Pst mAb8) for label-free detection of urediniospores from the model organism *Puccinia striiformis* f. sp. *tritici*. Assay conditions were optimized, and a detection limit of 3.1 × 105 urediniospores/ml was achieved (Skottrup et al. 2007). The antibody-based biosensors provide several advantages such as fast detection, improved sensitivity, real-time analysis, and potential for quantification. Antibody-based biosensors hold great value for agricultural plant pathogen detection (Fang and Ramasamy 2015). Zezza et al. used a Biacore X SPR sensor for detection of *Fusarium culmorum* DNA and found a detection limit of 0.06 pg DNA. The drawback of this system is that nucleotide extraction is needed prior to analysis, thereby making it less attractive for point-of-care analysis (Fang and Ramasamy 2015). But et al. designed four molecular beacons specific to the RNA-dependent RNA polymerase and coat protein genes of two orchid viruses, namely, *Cymbidium mosaic virus* (CymMV) and *Odontoglossum ringspot virus* (ORSV). CymMV and ORSV have been detected (as low as 0.5 ng of viral RNA in 100 mg orchid leaves) with specific oligonucleotide probes with a fluorescent moiety attached to one end of the DNA, while a quenching moiety is attached to the opposite end (Eun and Wong 2000). The constraints of DNA-based biosensors incorporate the necessity for the blend of explicit DNA test, intensification of DNA, surprising expense (DNA-based subatomic reference points), and unsuitability for real-time detection (DNA-based piezoelectric biosensor) (Fang and Ramasamy 2015). Enzymatic biosensors could be used for plant pathogen identification if the target VOC could be obtained as a liquid sample. Past examinations have shown that several phytohormones are catabolized by redox enzymes and offer opportunities to use these enzymes to develop extremely selective enzyme-based biosensors for the detection of plant chemicals (Thomas et al. 1999). Many of the volatile organic compounds (VOCs) generated by infected plants are alcohols and aldehydes that can be catalyzed by alcohol dehydrogenase enzymes, such as cis-3-hexen-1-ol and trans-2-hexanal. These enzymes can therefore be used to develop biosensors for the identification of VOCs based on alcohol or aldehyde that are specific to the infection. Oxidases could also deactivate prevalent phytohormones such as auxin, cytokinins, and gibberellins that are indicative of plant health in relation to these particular volatile organic compounds. Gibberellin is deactivated by GA-2-oxidases providing the potential for plant disease prediction for gibberellin detection. Although enzyme-based biosensors usually provide high sensitivity and specificity for the detection, stability of enzymes is of major concern. In addition, the enzyme catalysis varies with factors such as temperature and pH which compromise the accuracy of the biosensor (Fang and Ramasamy 2015; Kulagina et al. 1999; Thomas et al. 1999). Biophotonic-based sensors have three strengths: (a) they can effectively detect early stages of infection, (b) results are rapid and presented directly in the orchard, and (c) they could be highly integrated with other systems, providing good spatialization. The major weakness is low availability (Martinelli et al. 2015).

26.2.3.2 Environmental Monitoring

In the past, nutrients, heavy metals, bacteria, and priority pollutants (compounds such as industrial chemicals, pesticides, steroids, hormones, and known health impacts of petroleum hydrocarbons) were the focus of water and soil pollution. These compounds are discovered in small levels (microgram per liter and nanogram per liter) but with potential damaging effects on human health and the surrounding environment. Because they are not biodegradable and can accumulate in organisms, heavy metals (lead (Pb), cadmium (Cd), mercury (Hg), and nickel (Ni)) are toxic and carcinogenic, even at low levels. Hormones (natural and synthetic) are derived from national effluents, which can trigger some impacts in humans, such as sperm decrease, enhanced breast, testicular and prostate cancer, and endometriosis (Steffens et al. 2017). Requirements for applying most traditional analytical methods to analyze environmental pollutants are often a major impediment to their frequent implementation. The need for disposable devices or instruments for environmental applications has motivated the creation of fresh techniques and more appropriate methodologies, particularly for environmental monitoring. In this context, biosensors appear as a suitable alternative or as a complementary analytical tool (Salgado et al. 2011). In this section we provide an overview of biosensor systems for environmental applications, and in the following sections, we describe the various biosensors that have been developed for environmental monitoring, considering the pollutants and analysis that are usually mentioned in the literature.

The most frequently observed heavy metal contaminants in the setting are Pb, chromium, Zn, Hg, Cd, and copper (Cu). Biosensors are being created and used in environmental samples to monitor heavy metal levels. In addition, their biological foundation makes them suitable for measuring heavy metal toxicologically, while standard methods can only measure levels (Salgado et al. 2011). For the inhibitive determination of trace Hg in electrochemical biosensor based on enzyme, a large number of enzymes have been used: horseradish peroxidase (HRP), urease, glucose oxidase, alcohol oxidase, glycerol 3-phosphate oxidase, and invertase (Amine et al. 2006).

Urease has been entrapped in both PVC and cellulose triacetate layers on the surface of pH-sensitive iridium oxide electrodes and used for determination of Hg (Amine et al. 2006). Other enzymes such as invertase and HRP are used for detection of phenyl Hg and methyl Hg, respectively (Han et al. 2001; Mohammadi et al. 2005). Moreover, the whole-cell biosensor detection Hg makes use of lux genes from *Vibrio fischeri*. This lux gene is fused with a Hg-inducible mergene and is introduced in *Escherichia coli* (CM2624) (Gayathri and Braganca 2009).

Cd ion could be tracked by enzyme sensors since several enzymes such as urease and BChE were discovered to be inhibited (Amine et al. 2006). The binding of Cd to urease and the consequent changes of the enzyme structure are the basis of the SPR biosensing system reported by Lee and Russel in 2003. To facilitate the formation of a SAM of the urease on gold-coated glass SPR sensor disks, the enzyme has been modified with N-succinimidyl 3-(2-pyridyldithiol) propionate and which is detected Cd ion concentration in the range of 0–10 mg l^{-1}(Lee and Russell 2003). Mourzina et al. (2004) immobilized urease and BChE on the insulator surface of a laser-scanned semiconductor transducer (LSST). This biosensor was then tested and compared for the development of an enzyme-based light-addressed potentiometric sensor (LAPS). The enzyme LAP sensors with photocurable membranes demonstrate a degree of sensitivity close to the theoretical value and working ranges of $6.3 \cdot 10^{-5}$–$1.1 \cdot 10^{-2}$ and $1 \cdot 10^{-4}$–$1 \cdot 10^{-1}$ mol L^{-1} urea for acrylamide- and acrylate-based membrane matrices, respectively, and $2.5 \cdot 10^{-4}$–$2 \cdot 10^{-1}$ mol L^{-1} BChE for an acrylamide membrane matrix (Mourzina et al. 2004). A screen-printed three-electrode amperometric biosensor based on urease and the glutamic dehydrogenase system was developed by Rodriguez et al. and applied to the screening of heavy metals (Cu, Hg, Cd, and Pb) in water and soil samples. Immobilization of urease on the surface of SPE was performed by entrapment in alginate gel and adsorption on the electrode in a nafion film. Analysis of water and soil samples with the developed nafion-based sensor produced inhibition on urease activity according to their metal contents (Rodriguez et al. 2004). Recently, an array-based urease optical biosensor based on the sol–gel (urease and AChE) immobilization has been used for heavy metal determination in tap and river water. The analytical ranges of Cd(II), Cu(II), and Hg(II) were between 10 nM and 100 Mm (Tsai and Doong 2005).

Whole-cell biosensors showed promising results in detecting various pollutants in the areas that were affected. These reactions may be due to interactions between living cells in the biosensors with organic substances such as xenobiotics, heavy metals, changes in pH, or radiation in water and soil. For the identification of Cu, Zn, Ag, Hg, Cd, Pb, etc., with the resistance characteristics to these metals, some of the bacterial structures have been assessed as feasible. Some biosensors were built by fusing the resistance mechanism genes with the genes responsible for expressing bioluminescent proteins such as luciferin to detect metals in samples (Wan Jusoh and Ling Shing 2014). Due to their quick reaction, high growth rate, and low price, bacteria are preferable as sensing components. Bacterial biosensors are dependent on promoter–reporter expression systems consisting of a transcription regulator and promoter or operator and an open reading frame for measurable activity proteins (Arias-Barreiro et al. 2010; Tecon and Van der Meer 2008). Filamentous fungi and yeast in the setting occupy a broad variety of niches, and it would be sensible to expect them to react jointly to a broad range of substrates. The use of yeast can provide distinct benefits over other eukaryotic cells: (a) ease of cultivation and manipulation, (b) open to distinct methods of transducer (Farré and Barceló 2009). For instance, naturally occurring filamentous fungi such as *Mycena citricolor* and *Armillaria mellea* have demonstrated their potential by producing luminescent light in evaluating environmental toxicity. They demonstrated a strong reaction to Cu, Zn, and pesticides (Hollis et al. 2000). Table 26.2 summarizes the characteristics of various biosensors for heavy metal ion and organic waste sensing.

BOD or BOD5 is a parameter widely used to indicate the amount of biodegradable organic material in water (Andrea Medeiros Salgado et al. 2011). The BOD assessment is an empirical test in which the oxygen demand in analyzed water samples is determined by a standardized laboratory procedure. The BOD test is also widely used for assessing the biodegradation efficiency in wastewater purification processes at wastewater treatment facilities. The traditional BOD test has some advantages; it is a universal way of testing most wastewater samples and water bodies; it also requires no expensive equipment. However, in terms of analytical time, the test has serious limitations. This traditional technique is largely devalued by its low responsiveness. Operational analysis is made possible by developing BOD assessment methods based on the use of biosensor analyzers (Reshetilov et al. 2013). Most BOD sensors rely on the measurement of the bacterial respiration rate in close proximity to a transducer, commonly of the Clark type (an amperometric sensor developed by Clark in 1956 for measuring dissolved oxygen) (Andrea Medeiros Salgado et al. 2011). The receptor element of the biosensor may contain one or several cultures. Therefore, the oxidation of organic compounds by the culture occurring in the receptor element will always be lower than by activated sludge cultures (Reshetilov et al. 2013).

Most BOD sensors described are film-type microbial sensors based on whole cells and some BOD sensor based on bioreactor-type sensor systems. The principle of film-type microbial system operation is based on the measurement of oxygen

Table 26.2 Examples of biosensors developed for the detection of heavy metals

Bioreceptor	Transduction method	Analyte	References
Urease	Optical, SAM on the AuSPR	Cd	Lee and Russell (2003)
Urease	Optical fiber, immobilization in Ultra Bond membrane	Hg, Ag, Cu, Ni(II), Zn, Co, Pb	Kuswandi (2003)
Urease	Optical, alginate beads	Hg	Prakash et al. (2008)
Urease	Electrochemical, amperometric, three-electrode screen-printed	Hg, As, Cd	Pal et al. (2009)
Urease	Conductometric, sol gel, cellulose swab	Cr	Nepomuscene et al. (2007)
Alcohol oxidase (AlOx) and horseradish peroxidase (HRP)	Optical, chemiluminescence	Hg	Deshpande et al. (2010)
Alkaline phosphatase (ALP)	Conductometry; Au/ ALP-glutaraldehyde-bovine serum albumin-glycerol	Cd, Zn, Co, Ni, Pb	Berezhetskyy et al. (2008)
ALP	Amperometry; SPE/ALP, sol-gel/chitosan film; Ag/ AgCl	Hg, Cd, Ag, Zn, Cu	Shyuan et al. (2008)
Luciferase	Optical, luminescent, silicon photomultiplier	Cu	Lukyanenko et al. (2019)
Sol–gel-immobilized-urease	Electrochemical, conductometric, thick-film interdigitated electrode	Cd, Cu, Pb	Ilangovan et al. (2006)
Single-stranded DNA	Electrochemical, voltammetric, Au electrode	Cd	Wong et al. (2007)
DNA (direct metal binding DNA sequence)	Optical, fluorescent	Cd	Zhu et al. (2017)
DNA/exonuclease III (triple-helix molecular switch)	Optical, fluorescent, label-free	As	Pan et al. (2018)
Aptamers	Optical, fluorescence, micro-spin column	Pb	Chen et al. (2018)
Aptamer (T−Hg2+−T)	Optical, fluorescence, magnetic beads	Hg	Sun et al. (2018)
Aptamer (double-stranded DNA)	Optical, fluorescence, SYBR green I	Cd	Zhou et al. (2019)
DNA hybrid system	Optical, fluorescence, metal organic framework (UiO-66-NH2)	Hg	Wu et al. (2016)
Guanine-rich sequence and N-methyl mesoporphyrin	Optical, fluorescence	Hg, Pb	Zhu et al. (2018)

(continued)

Table 26.2 (continued)

Bioreceptor	Transduction method	Analyte	References
Nucleic acids (NH2-(CH2)6-GTACAAACAA-3)	Optical, evanescent-wave optical fiber, optical fiber platform	Hg	Long et al. (2013b)
DNA probes with eight T-T mismatches	Optical, luminescence, Ru(phen)(2)(dppz)](2+)	Hg	Zhang et al. (2010)
DNAzymes	Optical, fluorescence, carboxylated magnetic beads	Pb	Ravikumar et al. (2017)
DNAzyme	Optical, fluorescence, graphene oxide,	Pb	Zhao et al. (2011)
Cell-based *Shewanella oneidensis* MR-1	Electrochemical, solid-state thin-film Ag/AgCl	3,5-dichlorophenol (DCP)	Yang et al. (2016)
Whole cell, *E.coli*	Colorimetric, calcium alginate beads, Prussian blue	Hg, Pb, Cd, As,	Wasito et al. (2019)
Whole-cell *E.coli* RBE23-17, Cd-responsive promoter	Electrochemical	Cd	Biran et al. (2000)
Whole-cell *E. coli* ATCC 25922, *Bacillus subtilis*, *S. cerevisiae*	Electrochemical, entrapment in calcium alginate on a polyethylene terephthalate sheet	Cd, Cu, Phenol	Gao et al. (2016)
Pseudomonas sp. B4251, *Bacillus cereus* B4368, and *E. coli* 1257	Electrochemical, electrolyte-dielectric-semiconductor, and an ion-sensitive layer of silicon nitride	Zn, Co, Cu	Gruzina et al. (2007)
Shewanella oneidensis (transformed with plasmid-encoded mtrB integrated with arsenic-inducible promoter)	Bioelectrochemical, biofilm formation	As	Webster et al. (2014)
Whole-cell *E. coli* TV1061 (luxCDABE integrated with heat-shock *grpE* promoter)	Optical, bioluminescence, calcium alginate matrix	Hg, formaldehyde and ammonium hydroxide	Axelrod et al. (2016)
Whole-cell *E. coli* DH5α (plasmids pHK194 and pHK200 expressing GFP)	Optical, fluorescence, chemostat-like microfluidic platform	Cd, Pb	Kim et al. (2015)
Whole-cell *E. coli* (CadC-controlled T7 RNA transcription systems)	Optical, fluorescence, microbial culture in a microfluidic device immobilized	As, Cd, Pb, Zn	Kim et al. (2016)

(continued)

Table 26.2 (continued)

Bioreceptor	Transduction method	Analyte	References
Whole-cell *S. cerevisiae* (purine synthesis)	Colorimetric, entrapment in alginate beads modification of AMP pathway	Cu	Vopálenská et al. (2015)
Whole-cell *E. coli* (lacZ with arsenic-responsive promoter)	Colorimetric, change in pH	As	de Mora et al. (2011)
Whole-cell *Deinococcus radiodurans* (lacZ, Cd-inducible genesDR_0659, crtI)	Colorimetric, red pigment	Cd	Joe et al. (2012)
Whole-cell *E. coli* (zraP and cusC promoters were fused to reporter proteins, GFP and RFP)	Optical, fluorescence, monitoring bacterial system	Zn, Cu	Ravikumar et al. (2012)
*Pseudomonas fluorescens*10586s pUCD607 with *c* the lux insertion on a plasmid	Optical, bioluminescence (luminometer)	Zn, Cu, Cd, Ni	McGrath et al. (1999)
*Alcaligenes eutrophus (*AE1239), pMOL 90 + Tn4431/luxCDABE	Optical, bioluminescence	Cu	Leth et al. (2002)
Tetraselmis chui	Electrochemical, voltammetric non-living biomass in carbon paste electrodes	Cu	Alpat et al. (2007)
Rhodotorula mucilaginosa	Electrochemical, voltammetric carbon paste electrode, differential pulse stripping voltammetry	Cu	Yüce et al. (2010b)
Rhizopus arrhizus	Electrochemical, voltammetric carbon paste electrode, differential pulse stripping voltammetry	Pb	Yüce et al. (2010a)
Aptamers	Optical, evanescent-wave optical fiber, optical fiber surface	Bisphenol A	Yildirim et al. (2014)
Whole-cell *E. coli* (plasmid pColD-C23, cda promoter, luxAB)	Optical, bioluminescence	Benzene, toluene, xylene	
Whole-cell (toluene-benzene utilization (tbu), promoter PtbuA1, GFP)	Optical, fluorescence	Benzene, toluene, ethylbenzene	Stiner and Halverson (2002)
Whole-cell *E. coli* DH5α (xylR binds a subset of toluene-like compounds from the TOL plasmid, reporter geneLuc)	Optical, bioluminescence,	Benzene, toluene, xylene	Willardson et al. (1998)

consumption by microorganisms immobilized on the surface of the transducer, and bioreactor-type sensor systems have found wide use at wastewater treatment facilities for continuous control of the extent of purification. A feature of film-type microbial biosensors is that there is a layer of microbial film forming the biological recognition element between the porous (most commonly, cellulose) membrane and the gas-permeable membrane of the oxygen electrode. The oxidation of organic compounds contained in the sample consumes part of the oxygen that occurs in the layer of immobilized microorganisms. Oxygen remaining penetrates the gas-permeable Teflon membrane and is reduced in the oxygen electrode cathode. The current strength in the system is directly proportional to the electrode's reduced oxygen magnitude. The equilibrium (background) current is recorded after a balance is established between the diffusion of oxygen to the layer of immobilized microorganisms and the endogenous respiration rate of immobilized microorganisms. Other BOD biosensors include optical (Pang et al. 2007; Sakaguchi et al. 2007), biofuel cell (Chang et al. 2004; Moon et al. 2004), and calorimetric BOD biosensor (Reshetilov et al. 2013). In 1977, Karube developed the first BOD sensor which consisted of a dissolved oxygen electrode and a membrane impregnated with omnivorous yeast, *Trichosporon cutaneum*. By using luminescent bacteria, a photoluminescence BOD biosensor has been developed and produced by Ishikawa Seisakusho and has been developed by Yang et al. for on-site monitoring. However, responses of these sensors were influenced by the limited amount of dissolved oxygen in the sample solution. To overcome this problem a new type of BOD sensor using *Pseudomonas fluorescens* Biovar V, with ferricyanide as mediator was utilised. To determine low-BOD samples (below 5 mg L^{-1}) such as river water, or secondary effluents from industry or sewerage plants, highly sensitive BOD sensors have been developed using *Pseudomonas putida*. This type of sensor has also been used with an optical fiber system, using an optode (Nakamura and Karube 2003). The cyanobacteria *Anabaena torulosa* entrapped on a cellulose membrane and fixed into a cylindrical well connected to a fluorescence spectrometer with OF indicated the presence of heavy metals (Cu, Pb, and Cd), 2,4-dichlorophenoxyacetate, and chlorpyrifos. When the organisms are exposed to toxicants, major photosynthetic transport pathways are inhibited. Thus, the fluorescence emission will increase as a way to diffuse the energy, which has been absorbed. The presence of the toxicants was indicated by the change of fluorescence emission, before and after the exposure (Pospíšilová et al. 2015). Table 26.3 summarizes the characteristics of various biosensors for BOD sensing. In conclusion, biosensor BOD analyzers are robust, simple, and inexpensive analytical tools that can be used with traditional BOD determination methods to successfully control aqueous ecosystems.

The important areas of H_2O_2 application include industrial (pharmaceutical, food, clinical) and environmental analyses. Its use as an antibacterial agent added to milk, and the environmental need to avoid halogenated substances for disinfection purposes makes H_2O_2 an important substance in the food and beverage industry. It plays an important role in natural oxidation processes as it is found in air, solids, and water (Chen et al. 2012; Rouillon et al. 2006). Various materials such as Prussian blue (PB), heme proteins, carbon nanotubes (CNTs), and transition metals have

Table 26.3 Examples of biosensors developed for BOD detection

Bioreceptor	Transduction method	Sample	References
Microbial	Immobilized on a nylon membrane	Beverage wastewater	Dhall et al. (2008)
Yeast	Microbial membrane on an oxygen electrode; inorganic Al2O3 sol–gel matrix	River waters	Chen et al. (2002)
Activated sludge	Electrochemical, glucose glutamic acid, MB, chitosan-bovine serum albumin Cryogel	Wastewater	Niyomdecha et al. (2017)
Activated sludge and *Bacillus subtilis*	Optical, luminescence, oxygen sensing film tris(4,7-diphenyl-1,10-phenanthroline) ruthenium(II) dye (Ru(dpp), glass sample	Artificial wastewater	Kwok et al. (2005)
Mixed-culture 16S-rRNA gene (including *Geobacter*)	Electrochemical, biofilm formation	Wastewater, non-aerating conditions	Yamashita et al. (2016)
Chromobacterium violaceum R1	Electrochemical, amperometric, entrapment in calcium alginate (enclosure with polyamide membrane), ferricyanide	Wastewater, real water	Khor et al. (2015)
S. cerevisiae – (sequential injection analysis)	Optical, chemiluminescence, ferricyanide, quinone	Ionic liquids	Costa et al. (2018)
S. cerevisiae	Optical, spectrophotometry, 2,6-dichlorophenolindophenol, ferricyanide	Sea and river water	Nakamura et al. (2007)
Klebsiella pneumoniae cells,	Amperometric, ferricyanide mediator	Wastewater	Bonetto et al. (2011)

been used for the construction of H_2O_2 sensors. PB also possesses good catalytic specificity to H_2O_2 due to the polycrystal structure of PB that can allow only small molecules to penetrate its lattice. Screen printing technology and enzyme-based electrochemical biosensor were also used to manufacture PB-based biosensors. Other metal hexacyanoferrate (Cu, Ni, Co, chromium, vanadium, ruthenium, and manganese)-based sensors have similar or lower electrocatalytic reduction capabilities for H_2O_2 but have more electrochemical stability over a wide range of pH compared to PB-based electrodes (Chen et al. 2012; Giannoudi et al. 2006).

The most efficient practice of producing electrochemical H_2O_2 biosensors based on redox enzyme–protein (heme proteins) is to establish direct transmission of electrons between the protein and the electrode. The enzymes that are used intensively in development of hydrogen peroxide sensors are horseradish peroxidase (HRP), catalase (CAT), hemoglobin (Hb), microperoxidase (MP), myoglobin (Mb), and cytochrome c (Cyt c). Various strategies such as silica sol–gel, conducting polymer ionic liquid, self-assembly monolayer, and layer-by-layer assembly have been successfully proved to be effective in building this hydrogen peroxide sensor. (Chen et al. 2012; Giannoudi et al. 2006). A variety of compounds have been used as mediators for the detection of hydrogen peroxide. These compounds are phenothiazine molecules such as methylene blue, methylene green, Meldola's Blue (MDB), thionine, toluidine blue, ferrocene carboxylic acid, aminoethyl ferrocene, and phenothiazines. Mediators are able to shuttle electron between electrodes and enzymes in several configurations such as soluble matrices, associated in monolayer or multilayer, or incorporated in porous matrices (Chen et al. 2012; Rouillon et al. 2006).

26.2.3.3 Microbial Biosensors for Environmental Detection

Microbial and cellular biosensors play an important role among the analytical methods used to monitor the environment. In ecotoxicity testing, microbial biosensors have numerous advantages. Microorganisms are generally cheaper for cultivation than higher organisms and can be produced in large lots, subjected to strict quality control procedures, and dried for storage freezing. They react quickly to toxic compounds and indicate the bioavailability of compounds in a manner that is impossible for chemical analysis. (1) The use of engineered variants of microorganisms, like algae and bacteria, and macromolecules, like enzymes and proteins, represents a crucial approach in multi-response biosensing system design (Scognamiglio et al. 2013).

Many of the bacterial biosensors developed in environmental samples for the analysis of heavy metals make use of specific genes, such as biological receptors, responsible for bacterial resistance to these elements. Bacterial strains resistant to a number of metals such as Zn, Cu, tin, Ag, Hg, and Co have been isolated as possible biological receptors. These genes' metal resistance is caused only when the element hits the cytoplasm of the bacteria. The specificity of this resistance mechanism adds to the building of metal detection cell biosensors from the fusion of these resistance genes with genes, such as luciferin, which encode bioluminescent proteins (Salgado et al. 2011). On the other hand, most BOD sensors are described as whole-cell microbial sensor and luminescent bacterial optical sensor. Whole cells of bacteria (*Pseudomonas putida*, *Bacillus polymyxa*, *B. subtilis*) or yeasts (*S. cerevisiae*, *Hansenula anomala*, *Candida*, *Trichosporon*, *Arxula adeninivorans*, *Serratia marcescens*) are known to be used as biocatalysts in BOD sensors (Reshetilov et al. 2013).

Immobilization is a technique that allows the microorganism (special bacteria) to be used in the long term without reducing its ability as a bioreceptor. In most cases, microbial cells on the physicochemical transducer surface are retained by simple adsorption, i.e., cells are mostly placed on a porous membrane by hydrogel suction or water retention, a polyvinyl alcohol aqueous solution, or polycarbonyl sulfonate (Reshetilov et al. 2013). Three methods can be used to immobilize microorganisms in a biosensor: (a) covalent binding on a support, (b) physical adsorption on a membrane-like cellulose (hydrogen bonds, van der Waals links, etc.), and (c) encapsulation or inclusion of the microorganisms in a reticulated neutral and biocompatible aqueous matrix (Charrier et al. 2011b). Moreover, trapping method is one method which can be used for cell immobilization. In this method, the bacteria will be in a trap in the form of a matrix polymer, for example, in the form of a bead.

Polymers commonly used are agarose, acrylamide, chitosan, and alginate (Wasito et al. 2019). Immobilization of whole-cell microbial in sol–gel matrices is a promising modern trend of making biorecognition elements (Chen et al. 2002). For analyzed samples, these components are extremely permeable and have excellent strength and stability as well as low toxicity for immobilized microorganisms (Reshetilov et al. 2013).

In recent times, microbe-based sensors have been used in the diagnosis of heavy metals and toxicity in soils and water samples (Kanjana 2017). In chromate whole-cell biosensor and Hg whole-cell biosensors, a number of promoter regions have been identified as sensing elements, sequenced as luxCDABE reporter system, and cloned as promoter cassettes of the *Alcaligenes eutrophus* (Gayathri and Braganca 2009).

As far as biological identification components are concerned, algae are often used for trace identification of environmental toxic compounds in biochemistry because they are susceptible to modifications in their surroundings caused by the presence of toxicants and can therefore provide data on the toxic effects of pollutants on living organisms. In the biosensor industry, microalgae and photosynthetic cyanobacteria are used as they provide data on pollution concentrations based on changes in photosynthetic or metabolic activity. Monitoring of species such as reactive oxygen species, H_2O_2 (the concentration of which is related to potential stress exerted on algae), and pH-related ions that participate in algae's metabolism and photosynthesis can indicate the presence of herbicides, heavy metals, pathogens, and pollutants (Tsopela et al. 2014).

Two types of amperometric environmental sensors incorporating whole-cell eukaryotic algae (*Chlorella*, *Scenedesmus*, *Selenastrum*) were investigated for use in monitoring pollution of aquatic systems by Pandard et al. in 1993. Both sensors permitted photosynthetic events to be monitored, one by evaluating the decrease rate of a redox mediator by the illuminated biocatalyst and the other by tracking its output of photosynthetic oxygen using a semi-protected oxygen electrode. The oxygen electrode-based biosensor provided good sensitivity with long operating life and proved to be a better approach than the mediator system for monitoring algal biocatalysts for the detection of Hg (nitrate salt) or Cu (sulfate salt) showing toxicity to photosynthetic organisms, but this method took a long time (Giardi et al. 2001; Koblizek et al. 1998). Durrieu and Tran-Minh developed an optical biosensor which is constructed to detect Pb and Cd from inhibition of AP present on the external membrane of *C. vulgaris* microalgae. The microalgal cells are immobilized on removable membranes placed in front of the tip of an optical fiber bundle inside a homemade microcell. *C. vulgaris* was cultivated in the laboratory, and its AP activity is strongly inhibited in the presence of heavy metals (Durrieu and Tran-Minh 2002). On the other hand, a biosorption-based biosensor using *Tetraselmis chui* microalgae was developed for the voltammetric measurement of Cu^{2+} in real sample. Cu^{2+} was accumulated, at open circuit, on the algal biosensor, and the voltammetric measurements were carried out by differential pulse cathodic stripping voltammetry (Alpat et al. 2007).

The relatively poor sensitivity and selectivity of microbial biosensors are still critical issues, and this can be attributed to the nature of biological sensing mechanisms. Another intrinsic limitation of microbial biosensors is the slow response caused by decelerated diffusion of substrates and products through the cell wall (Lim et al. 2015b). Gravimetric, nuclear-based, electromagnetic, tensiometer-based, and hygrometric methods have been developed for measuring soil moisture content in the field (Dwevedi et al. 2017). In the next subparagraphs, the newly developed biosensors based on these recognition elements will be described.

26.2.3.3.1 In-site and Online Monitoring

Efficient tools for online and in situ monitoring of environmental pollutants are required to provide early warning systems. Currently, a large spectrum of microbial biosensors have been developed that enable the monitoring of pollutants by measuring light, fluorescence, color, or electric current (Paitan et al. 2003). Biosensors for determining pollutants have been shown to be delicate, low cost, and readily adjusted for internet surveillance. Furthermore, biosensors using whole cells are of concern in ecotoxicity because many toxics target these bioreceptors. Online biosensors that use bioluminescent bacteria for measuring have also been developed with an objective of detecting pollutants, and it has been possible to observe a technological approach (Charrier et al. 2011a).

The promoter-based whole-cell biosensor can identify nanomolar levels of cadmium in water, seawater, and soil samples within minutes and can be used for online and in situ tracking continuously. This electrochemical biosensor has a cadmium-responsive promoter from *Escherichia coli* fused to a promoter-less *lacZ* gene, which was monitored using an electrochemical assay of β-galactosidase activity (Biran et al. 2000). In addition to Cd, this biosensor can detect the presence of a variety of heavy metals such as Hg, Zn, and Cu (Paitan et al. 2003).

Some biosensing systems for herbicide detection use isolated chloroplasts or intact cells of algae to measure changes in chlorophyll fluorescence. Unicellular microalgal strains, S*elenastrum*, *Desmodesmus*, and *Chlorella*, with a cell size of 3–10 mm, have frequently been used in bioassays (estimating the toxicity of liquid sample surface waters, soil extracts, etc.) due to their fast growth (a doubling time of several hours) in liquid growth media, where tested samples can be easily added. Biosensors, based on PSII complex, have been used for the detection of photosynthetic herbicides within the last two decades (Masojídek et al. 2011). These reagentless biosensors are compatible for in-field use or online monitoring. An optical whole-cell biosensor using *Chlorella vulgaris* microalgae entrapped on a quartz microfiber filter and placed in a five-membrane homemade flow cell was designed by Védrine et al. (2003) for determination of herbicides as aquatic contaminants. *C. vulgaris* cell biosensor designed for monitoring herbicides in drinking water was developed by Rodriguez et al. (2002). It has also been shown that the reaction centers of photosynthetic microorganisms bind herbicide and could possibly be used as part of sensing equipment based on Langmuir–Blodgett monolayer movies or liposome-forming artificial membrane. Even isolated DI protein has been embedded on a working electrode for the potentiometric monitoring of the specific interaction between a protein and an herbicide. A fluorescence biosensor based on mutants resistant to various herbicide subclasses was developed; it makes possible to distinguish between subclasses of herbicides (Giardi et al. 2001; Giardi and Pace 2006; Piletska et al. 2006).

Multicell devices incorporating various photosynthetic biorecognition elements are increasingly tested with real samples and in in-field conditions (Bucur et al. 2018). For example, BOD sensors based on the original principle, using *T. cutaneum*, have been developed by Yang et al. for on-site monitoring (Nakamura and Karube 2003). A lab-on-chip device for water toxicity analysis consists in a portable system for onsite detection composed of three-electrode electrochemical microcells, integrated on a fluidic platform constructed on a glass substrate by Tsopela et al. in 2016. The basic detection principle consisted in electrochemically monitoring disturbances in metabolic photosynthetic activities of algae induced by the presence of Diuron herbicide. Superior sensitivity results (limit of detection of around 0.1 μM) were obtained with an organic light-emitting diode, having an emission spectrum adapted to algal absorption spectrum and assembled on the final system (Tsopela et al. 2016).

26.2.3.3.2 Reporter Genes

A reporter gene is fused to another gene or a promoter so that the expression of that gene or promoter may be assayed. The product of the reporter gene is typically more stable and easier to detect than the gene to which it is fused (Charrier et al. 2011b). A reporter gene encodes a mechanism that generates a cellular reaction that can be detected. It determines the biosensor's sensitivity and detection limits. For the reporter gene to be used in a biosensor, specific features are required. The gene needs an expression or activity that can be measured using a straightforward assay, reflecting the quantity of chemical or physical change. The biosensor must also be free from any gene expression or activity comparable to the gene expression or activity being evaluated. Ensuring that the biosensor is free from any comparable gene expression or activity protects the reaction from being misinterpreted and ensures that the measurement represents the required chemical or physical change directly (Strosnider 2003). Some bacterial biosensors (such as whole cell) are designed by combining a reporter gene that produces a signal with a contaminant-sensing element that reacts to chemical or physical changes, such as exposure to a particular analyte. The sensing element stimulates the reporter gene through a biochemical pathway in the cell when the biosensor is subjected to such a shift. The reporter gene then generates a measurable reaction, such as emitting visible light, indicating the degree of chemical or physical change (Strosnider 2003). The most commonly used reporter gene is the luc operon from the firefly *Photinus pyralis*. There are some commonly used genes of reporters that have been shown to be effectively incorporated into whole biosensors based on cells. These include *lux* (bacterial luciferase), *luc* (firefly luciferase), *lacZ* (β-galactosidase), and GFP. GFP is a GFP reporter gene coding and autofluorescence, so it does not need to emit a substratum or an ATP. But with this reporter, the intrinsic fluorescence of certain host cells generally increases the background fluorescence, and this can cause signal interference. Another reporter, β-galactosidase (*lacZ*), a well-characterized bacterial enzyme, was also frequently used in molecular biology as it is an excellent monitor of transfection efficiency. *LacZ* has some distinctive detection benefits, using either colorimetric or fluorescent techniques, as it is easy and quick to use with a sample. Fujimoto and coworkers have created a novel reporter gene, *crtA*, which is accountable for carotenoid synthesis, in another reporter scheme. The *crtA*-based entire cell-based biosensors, when applied to a sample, change the color of the culture media from yellow to red without adding a supporting substratum and are therefore considered a good choice for rapid detection in emergency situations (Gui et al. 2017). For example, the bacterial biosensors for the detection of arsenic are engineered by pairing a luc operon reporter gene from firefly *P. pyralis* encoding the enzyme luciferase, which produces a detectable cellular response and a contaminant-sensing gene that detects the contaminant and in turn triggers the reporter gene (Gayathri and Braganca 2009).

Biosensors based on fusion of responsive promoters to reporter genes are sensitive and specific. Several such biosensors were constructed that can detect heavy metals and hydrocarbons. The use of reporters whose activity can be monitored electrochemically offers several advantages: monitoring is rapid and can be performed online and in situ with high sensitivity and reproducibility, monitoring in crude or turbid solutions, simultaneous measurements of several samples and compatibility with other types of reporter genes using color, and light of fluorescence; moreover, the sensitivity of electrochemical measurements can also be used for amperometric determination of target bacteria in water (Paitan et al. 2003).

26.2.3.3.3 Luminescent Bacteria and Fluorescence Microbial Biosensors

Currently, the most commonly used reporter proteins for optical detection in microbial systems are GFP for fluorescence and bacterial luciferase for luminescence. Bioluminescence is the light produced by some organisms mediated by the enzyme luciferase (luciferase genes luxCDABE) or a photoprotein mediated by an oxidation reaction (Charrier et al. 2011b).

Like bioluminescent reporter lux gene, gfp gene coding for the GFP has also been widely applied as reporters and fused to the host gene that allows reporter activity to be examined in individual cells. Because GFP is very stable and not known to be generated by microorganisms that are native to terrestrial habits, it offers excellent benefit and flexibility in assessing reporter activity. GFP as a reporter protein's main disadvantage is the delay between protein manufacturing and protein fluorescence (Su et al. 2011). Fluorescent proteins may continue to accumulate for many hours and enable detection even after cell death due to their elevated stability. Furthermore, GFP does not involve a substrate or ATP, thus reducing the cell burden (Woutersen et al. 2011).

Most lux strains have sensitivities varying from milligrams per liter to micrograms per liter with detection limits, generally with greater sensitivities in compound-specific strains. Although multiple molecular manipulations may enhance the sensitivity of lux strains, most reported detection thresholds are still too large to identify concentrations of individual contaminants as they happen in

European drinking water today. However, lux strains sensing specific toxic effects have the advantage of being able to respond to mixtures of contaminants inducing the same effect and thus could be used as a sensor for the sum effect, including the effect of compounds that are as yet not identified by chemical analysis (Woutersen et al. 2011). Biosensors using microorganisms such as luminescent and florescent bacteria have short life cycles and can provide an excellent response to pollution or toxins that can be used within the scope of health research related to ecosystem or environment (Wasito et al. 2019). Assays using bioluminescent bacteria can be divided into two groups: constituent expression systems and inducible expression systems. Normally, bacteria with constitutive expression have an elevated luminescence expression that reduces under toxic circumstances. They are generally natural bioluminescent bacteria, such as *Aliivibrio fischeri*, and are often used to detect acute cytotoxicity because the reaction is not specific to compounds. By comparison, inducible devices have a small baseline luminescence that improves after exposure to particular compounds ("lights on"). Both promoter and reporter genes can be inserted from other bacteria in these systems to provide an ideal reaction to interesting compounds (Woutersen et al. 2011).

On the other hand, luminescent bacteria are used as optical BOD sensor to measure BOD for water and wastewater. Optical BOD detectors can be designed using two methods: by using luminescent bacteria in the sensor's biorecognition component and by using a luminescent biomaterial assistance. In the former situation, the principle of measurement is based on the relationship between the luminescence intensity generated by bacteria and the cell assimilation of organic compounds from samples of wastewater. The optical biosensors of the luminescent bacteria have a high sensitivity and therefore allow the determination of low BOD values. An important advantage of such systems is that they enable microprinted circuit boards, microsensors, and on-chip biosensors (Reshetilov et al. 2013).

The potential for detection of heavy metals with fluorescent and luminescent bacterial biosensors based on promoter fusion to a reporter gene has been demonstrated (Paitan et al. 2003). Maderova et al. measured bioavailability and toxicity of Zn in soil (laboratory soil amended and field samples) using the luminescence marked constitutively expressed *Escherichia coli* HB101 (pUCD607) and the Zn-specific *E. coli* MG1655 (pZNT-lux) sensors. Zn is a metal ubiquitous in the environment and essential to biological systems. Elevated concentrations of Zn in soil, however, can pose a threat to biota (Maderova and Paton 2013). A Zn-specific fluorescent biosensor, *Pseudomonas putida* X4 (pczcR3GFP), was constructed by fusing a promoterless enhanced GFP gene with the czcR3 promoter in the chromosome of *P. putida* X4 by Liu et al. in 2012. In water extracts of four different soils amended with Zn, the reporter strain detected about 90% of the Zn content of the samples. The authors concluded that the biosensor constitutes an alternative system for the convenient evaluation of Zn toxicity in the environment (Liu et al. 2012).

Biosensor strategies for pesticide detection in marine ecosystems are mainly based on the use of enzymes, antibodies, or microorganisms, such as bacteria (Moro et al. 2018). A whole-cell luminescent cyanobacterial (*Synechocystis* sp. strain

PCC6803) biosensor which responds to a range of compounds, including different herbicide types and nonherbicide toxicants, has been developed by Shao et al. in 2002. The full-cell luminescent cyanobacterial biosensor has been shown to be easier, quicker, more precise, and more economical than other techniques for identifying herbicide toxicity, such as photosystem-based whole-cell and tissue biosensors (Shao et al. 2002). Frense et al. (1998) reported the use of an optical biosensor that incorporated the green alga *Scenedesmus subspicatus* living cell (immobilized on filter paper and covered with alginate) for detection of herbicides in wastewater. The measuring principle was the determination of chlorophyll fluorescence by fiber-optic electronic tools depending on the load of water samples with toxic compounds (Frense et al. 1998).

26.2.4 Advantages of Biosensors

The greatest advantage is the ability of biosensors to detect the bioavailable fraction of the contaminant, as opposed to the total concentration. Knowing the bioavailable fraction allows a more accurate assessment of the site and the potential risks involved. Biosensors create a clearer picture by providing physiologically relevant data in response to a contaminant. This response, usually luminescence, is quick and easy to measure, resulting in real-time data. Biosensors are also quick, cheaper, and less labor-intensive than other traditional techniques such as atomic absorption spectrometry, inductively coupled nuclear electron spectrometry, and sequential extraction. The results obtained from biosensors are compatible with and comparable to chemical analysis, while being free of chemical extractions and analytical procedures. Biosensors can also be more sensitive than chemical methods (Kaushal and Wani 2017; Strosnider 2003).

26.3 Nanobiosensor

Nanotechnology refers to a nanoscale technology, which has promising applications in day-to-day life. This technology emphasizes the implications of individual atoms or molecules or submicron dimensions in terms of their applications to physical, chemical, and biological systems and eventually their integration into larger complex systems (Dehnad et al. 2015; Khadivi Derakhshan et al. 2012; Patel et al. 2014). According to the different components, engineered nanomaterials can be generally grouped into four types: (a) carbon-based materials such as CNTs and fullerenes; (b) metal-based materials such as Au, Ag, metal oxides, and quantum dots (QD) like TiO_2, ZnO, and Fe_2O_3; (c) dendrimers or nanosized polymers; and (d) composites combining nanoparticles with other nanoparticles and/or larger bulk-type materials (He and Feng 2017). When biosensors are investigated using nanoscale, such devices are then call nanobiosensors. With the progression in sciences, nanobiosensors with superbly dedicated miniature sensors with high miniaturization were designed and developed in the twenty-first century based on the ideas of nanotechnology (Rai et al. 2012). The use of nanotechnology and nanoscience in biosensors has led to the production and introduction of mechanisms of new signal converter that greatly increases the sensitivity of their identification. With the increasing advances in nanoscience and the ability to produce very small electrodes, the production of new types of nanobiosensors is possible (Mohammadi-Aloucheh et al. 2018). By miniaturizing such devices and/or using nanomaterials as sensing layers, the sensitivity and performance of nanobiosensors can be increased. Nanobiosensors work on the size of the nanoscale and can trace analytes with fast and accurate biological identification. (Steffens et al. 2017). Ideally, nanobiosensors should possess the following characteristics: (a) high stability, (b) specificity for particular analytes of interest in the intended environment of use, (c) fast dynamics, and (d) accuracy and reproducibility over the useful analytical range. Recent advances in nanotechnology have led to the development of nanoscale sensors that have exquisite sensitivity and versatility (Kwak et al. 2017). By adding biomaterials and converters that work with nanomaterials, we can identify new, high-quality biosensors that can be used to identify biomolecules, and these biosensors can detect environmental contamination at high speeds and high levels of pollution (Mohammadi-Aloucheh et al. 2018).

Engineered nanomaterials are materials between 1 and 100 nm and exist as metalloids, metallic oxides, nonmetals, and carbon nanomaterials and as functionalized dendrimers, liposomes, and QD. Their small size, large surface area, and high reactivity have enabled their use as bactericides/fungicides and nanofertilizers. Nanoparticles can be designed as biosensors for plant disease diagnostics and as delivery vehicles for genetic material, probes, and agrichemicals (Wade Elmer and White 2018).

26.3.1 Constituents of Nanobiosensors

In terms of the conceptual and fundamental mode of operation, these components are, namely, bioreceptor, transducer, and detector. Bioreceptor (antibodies, proteins, enzymes, immunological molecules, and so on) is that component of a biosensor which serves as a template for the material to be detected. The second component is the transducer system. The main function of this device is to convert the interaction of bioanalyte and its corresponding bioreceptor into an electrical form. The third component is the detector system. This receives the electrical signal from the transducer component and amplifies it suitably so that the corresponding response can be read and studied properly (Malik et al. 2013).

Nanobiosensors are classified based on thefollowing: (a) method used to identify required interactions, (b) type of interaction between the sensor component and the analyte element, (c) transduction system, and (d) nature of compound recognition (biological or nonbiological) (Steffens et al. 2017).

26.3.2 Types of Nanobiosensors

Nanobiosensor is a compact analytical device/unit and called as modified version of a biosensor in which the immobilized layer of biological material like proteins, DNA/RNA, viruses, cellular lipid bilayers, microbial cells, and others are in contact with the sensor that analyzes the biological signal and converts it into electrical signal (Kanjana 2017).

26.3.2.1 Mechanical Nanobiosensors

The biomolecular interaction is measured using mechanical nanoscale biosensors. Based on the modifications in surface stress produced by the relationships between the probe and target molecules on their surfaces, chemical vapors at very small levels can be identified. The magnitude of the shift in surface stress relies on the sort of interaction that occurs, including the forces of hydrogen bonding, electrostatic, van der Waals, etc. There are three mechanisms for transforming the recognition of the interest analyte into the cantilever's micromechanical bending: (a) bending in response to a surface stress, (b) bending in response to a mass loading, and (c) bending as a result of a temperature change (Choudhary et al. 2015). The advantage of nano-mechanical devices is that they are highly mass-sensitive. The more the size decreases, the more the mass reduces, and hence the addition of bound analyte molecules results in an increased relative change to the main mass (Rai et al. 2012). Because these nanobiosensors are instruments that allow particular molecules to be immobilized, they also provide a concentrated and specific recognition to avoid cross-connections that are not specific. They also have the advantage of detecting multiple analytes and can be reused depending on the design of the nanobiosensor and mimic organic environments (Steffens et al. 2017).

26.3.2.2 Optical Nanobiosensors

Optical nanobiosensors are commonly used for the detection of pathogen based on fluorescence and surface plasmon resonance (SPR). Generally speaking, this method is based on tracking the shift in the optical signal between the functionalized pathogen and nanomaterial. The biggest benefit of this method is that with negligible cell disruption, the sensor can integrate into the deeper portion of the cell (Kabariya and Ramani 2017).

Optical transducers are mostly appealing to develop a powerful, user-friendly, portable, and cost-effective device (Kabariya and Ramani 2017).

26.3.2.3 Nanowire Nanobiosensors

One of the nanobiosensor classes is nanowire biosensors. Nanowires are cylindrical arrangements having lengths in the order of few micrometers to centimeters and diameters within the nanorange. Nanowires are the one-dimensional nanostructures with very good electron transport properties (Malik et al. 2013). The nanowire consists of sensing elements and is coated by biomolecules like DNA, proteins, polypeptides, filamentous bacteriophages, and fibrin (Kabariya and Ramani 2017). This is a hybrid of two molecules that are highly susceptible to external signals: single-stranded DNA (which serves as the "sensor") and a nanotube of carbon which serves as the transmitter. Using chemical or biological molecular ligands, the surface properties of nanowires can be easily modified, making them independent analysts. This translates the chemical binding event on their surface into an intense sensitivity, real-time, and quantitative fashion shift in the conductance of the nanowire (Choudhary et al. 2015).

26.3.2.4 Electronic Nanobiosensors

Electronic nanobiosensors work by electronically detecting a target DNA binding which actually forms a bridge on a microchip between two electrically separated wires. Each chip contains multiple sensors that can be addressed independently with capture samples from the same or different organisms for different target DNA molecules (Choudhary et al. 2015).

26.3.2.5 Nanoshell Biosensors

Due to the specific compound coated into the core, i.e., shell or outer coating layer, the shell shape is spherical or round cores, and the shell thickness is a few nanometers (Kabariya and Ramani 2017).

26.3.3 Environmental Application of Nanobiosensors

For environmental applications, the main advantages offered by nanobiosensors over conventional analytical techniques are the possibility of portability, miniaturization, and work. Nanobiosensors can be used as environmental quality monitoring tools in the assessment of biological/ecological quality or for the chemical monitoring of both inorganic and organic priority pollutants (Andrea Medeiros Salgado et al. 2011).

The main advantages offered by nanobiosensors are the possibility of continuous monitoring, work on-site, and the ability to measure pollutants in complex matrices with minimal sample. Nanobiosensors can be effectively used for sensing a wide variety of fertilizers, herbicide, pesticide, insecticide, pathogens, moisture, soil, pH, and their controlled use can support sustainable agriculture for enhancing crop productivity (Sekhon 2014). Furthermore, crop productivity is daily endangered by pests, weeds, and pathogens that influence the relative farm economy; therefore, plants need to be protected by proper action. In this sense, nanostructured biosensors can contribute to intelligent farming not only by tracking soil conditions and plant development across large fields but also by identifying infectious diseases in crops before noticeable symptoms happen (Antonacci et al. 2018).The nanotechnology-based biosensors are at the early stage of development.

26.3.3.1 Role of Nanobiosensors in Agriculture

Using nanobiosensors, farmers can monitor environmental conditions closely for plant growth and protection. Nanobiosensors can be effectively used in agriculture for sensing a wide variety of fertilizers, herbicides, pesticides, pathogen, moisture, soil pH, and others for enhancing the crop productivity (Kanjana 2017). In parallel with typical chromatography, nanobiosensors were employed as an alternative for pesticide measurement by virtue of its high selectivity, sensitivity, excellent detection, and reliability as well as its rapidity. Nanobiosensors showed great capabilities on the recognition of biomolecular interactions through changes in their surface stress or mass (Álvarez et al. 2016; Wade Elmer and White 2018). Nanosensors can be connected to a GPS and spread across the field to monitor disease, plant health, soil conditions, and potential issues such as depletion of soil nutrients and water deficit in real time. Such data and signals include the optimum times for planting and harvesting plants and the time and amount of water, fertilizers, pesticides, herbicides, and other treatments required considering the particular physiology, pathology, and environmental circumstances of the plant (Kanjana 2017).

At present, several methods for plant pathogen detection are available, but some of them are time-consuming like traditional culture-based methods; others are specific of reactions, such as the high cost of nucleic-acid-based PCR methods (time-consuming and unable to detect early infections), the ELISA (time-consuming and low potential for spatialization), DNA fingerprinting, and amplification of the internal transcribed spacer region from rRNA gene increase specificity of identification. Intelligent nanobiosensors can assist, identify, and treat nutrient deficiencies in soils and plants by providing macro- and micronutrients in accordance with the growing season's temporal and spatial nutrient demands (Duhan et al. 2017; Kanjana 2017). The detection technology based in nanobiosensor is a novel microbial detection that starts to revolutionize agriculture (Álvarez et al. 2016).

26.3.3.1.1 As an Agent to Promote Sustainable Agriculture

New nutrient delivery systems that exploit the nanoscale porous domains on plant surfaces can be developed. A nanofertilizer relates to a product which in one of three respects supplies nutrients to plants. The nutrient can be encapsulated in nanomaterials such as nanotubes or nanoporous materials, covered with a thin protective polymer film or delivered as nanoscale particles or emulsions. Ideally, nanotechnology could provide devices and mechanisms to synchronize the release of nutrient (from fertilizers) with its uptake by crops; the nanofertilizers should release the nutrients on demand while preventing them from prematurely converting into chemical/gaseous forms that cannot be absorbed by plants (DeRosa et al. 2010). To achieve this, biosensor could be attached to this nanofertilizer that allows selective nitrogen release linked to time, environmental, and soil nutrient condition (Rai et al. 2012). Smart nanofertilizer delivery platforms have been developed for micronutrients such as Zn and Fe, where the primary nutrient release mechanism is based on the recognition and binding of a particular plant signal by a nanobiosensor occurs in a polymer film that coats nanoparticles or salts of micronutrients (Zn, Fe, Cu, and Mn) (Duhan et al. 2017; Kanjana 2017).

Nanobiosensors can promote more sustainable practices through precise tracking of fertilizers in water and soil, thereby encouraging farmers to acquire data on spatial and temporal differences in field fertilizer levels. In fact, in recent years, sensors have demonstrated their strong potential for assessing soil organic matter or complete carbon content, soil salinity, potassium content, residual nitrate, phosphate, and urea (Antonacci et al. 2018). Mura et al. developed a rapid colorimetric assay using cysteamine-modified AuNPs for the direct detection of nitrates in water samples. AuNPs stabilized with citrate have been modified with cysteamine, which has an excellent affinity to nitrates, and its ability to capture nitrates has been evaluated and quantified by naked eye color variations at a concentration of 35 ppm (Mura et al. 2015). Ali and coworkers reported nitrate detection in soil with microfluidic impedimetric sensor using graphene oxide nanosheets and poly(3,4 ethylenedioxythiophene) nanofibers. The oxygenated functional groups available at GO allows an increased charge transfer resistance of the electrochemical electrode. The sensor provides a sensitivity of 61.15 Ω/(mg/L)/cm^2 within a wide concentration range of 0.44–442 mg/L for nitrate ions in agricultural soils (Ali et al. 2017).

Urea is also mainly utilized in agriculture as a nitrogen fertilizer, but being quickly hydrolyzed to ammonium carbonate creates many dangerous impacts such as harm to germinating seedlings and young crops or toxicity to nitrite. For these reasons, the availability of satisfactory methods to quantify urea in soils becomes essential (Antonacci et al. 2018). AuNP-catalyzed 3,3′,5,5′-tetramethylbenzidine-H_2O_2 reporting system is used as an ultrasensitive pH indicator as reported by Deng and coworkers. This nanobiosensor with HRP enzyme sensing platform is used to detect urea, urease, and urease inhibitor. The limit of detection for urea and urease was 5 μM and 1.8 U/L, respectively (Deng et al. 2016a).

Aptamer-based nanobiosensors appear as a helpful instrument for studying the origin and responses of metabolites in plant rhizospheres generated by living cells and for investigating the controlled released materials used in agriculture. The inclusion of nanodevices such as aptamers in layer-by-layer polymer movies may have potential for targeted apps (Kanjana 2017).

GreenSeeker is an excellent indicator of biomass. This tiny device makes plants speak up for their nitrogen needs. GreenSeeker calculates the normalized difference in vegetative index using red and near-infrared light. It is based on the simple principle that plant chlorophyll absorbs red light as an energy source during photosynthesis (Duhan et al. 2017).

26.3.3.1.2 Nanobiosensors for Seed Storage

Stored grain bulks are ecological systems where communities of insects, mites, and microflora interact with abiotic variables to cause spoilage. The survival and reproduction of biological agents in grain are dependent on the temperature and moisture levels. Increased levels of CO_2 indicate that insects, mold, or excessive respiration is present. During spoilage, volatile carbon dioxide and odor from biological pests rise and can be used as a reliable indicator of incipient spoilage of grain. Analysis of volatile compounds in the headspace above the bulk of grain is a promising and quick method for identifying fungal spoilage (Neethirajan and Jayas 2007).

Seeds emit several volatile aldehydes during storage that determine the aging degree. Even other seeds are detrimental to these gases. Such volatile aldehydes can be identified, and seeds with indications of decay can be separated and reinforced before use (Choudhary et al. 2015).

26.3.3.1.3 Nanobiosensors for Fungal Plant Pathogen Detection

Plant disease prediction is a technique of management used to predict plant disease opportunities or severity and to assist farmers make cost-effective disease control choices. Research is currently being conducted using nanosensors in crop systems to enhance pathogen detection techniques. (Khiyami et al. 2014). The full use of nanobiosensor in plant disease diagnostics research has only begun to be realized.

Fungal diseases (majority *Ascomycetes* and *Basidiomycetes*) cause significant economic agricultural losses around the world. Fungal plant diseases are generally managed with the applications of chemical fungicides. Chemical control has been found very effective for some fungal diseases, but it leaves several nonspecific effects that destroy beneficial organisms along with pathogens. Such ecological disturbances open the route to undesirable health, safety, and several environmental risks (Patel et al. 2014). For biosensing application in plant diseases management, the limit of detection and the overall performance of a biosensor can be greatly improved by using nanomaterials for their construction (Fang and Ramasamy 2015). Nanobiosensor as a novel tool could improve actual delivering techniques to manage common plant diseases (Álvarez et al. 2016). NPs can provide early disease detection diagnostic instruments and can be used in the part of the plant that has been assaulted by disease or pest. Different methods have been found for particular identification between target phytopathogenic cells and biofunctionalized nanomaterials, such as antibody–antigen, adhesion–receptor, antibiotic, and complementary sequence DNA. These techniques of detection include polymer conductive nanowires, nanoporous silicon, CNTs, and AuNP (Khiyami et al. 2014).

AuNPs are excellent markers for use in biosensors as multiple optical or electrochemical procedures can be modified to identify pathogens. A number of studies based on NPs have been conducted to create biomolecular detection with AuNPs functionalized with DNA or protein, which are used as target-specific samples (Khiyami et al. 2014). AuNPs-based optical immunosensors (using anti-teliospore antibodies) have been developed for onsite testing and detection of Karnal bunt (KB) disease (*Tilletia indica*) in wheat using SPR (Singh et al. 2010). CuONP and nanolayers were synthesized by sol–gel and spray pyrolysis methods, respectively. Both CuONP and nanostructural layer biosensors were used for detecting the *A. niger* fungi (Etefagh et al. 2013). NPs of TiO_2 or SnO_2 on screen-printed carbon (SPC) electrodes have been developed for evaluating their potential application in the electrochemical sensing of p-ethylguaiacol, a fingerprint compound present in the volatile signature of fruits and plants infected with a pathogenic fungus *Phytophthora cactorum*. The electroanalytical data obtained using cyclic voltammetry and differential pulse voltammetry showed that both SnO_2 and TiO_2 exhibited high sensitivity (174–188 $\mu A\ cm^{-2}\ mM^{-1}$) and low detection limits (35–62 nM) for p-ethylguaiacol detection. These results demonstrate that metal oxides are a reasonable alternative to expensive electrode materials such as gold or platinum for amperometric sensor applications (Fang et al. 2014). The genus *Trichoderma* is a soil-borne fungi which in numerous reports has been successfully used as a biological control agent against various plant pathogens. The identification of *Trichoderma* species worldwide is currently deduced from micromorphological descriptions which are tedious and prone to error. Siddiquee et al. successfully developed an electrochemical DNA biosensor based on ionic liquid, ZnO nanoparticles, and a chitosan (CHIT) nanocomposite membrane on a modified gold electrode (AuE). A single-stranded DNA probe was immobilized on this electrode. Methylene blue (MB) was used as the hybridization indicator to monitor the hybridization reaction of the *Trichoderma harzianum* target DNA. This nanobiosensing system was capable of detecting the target analyte at concentration ranges of 1.0×10^{-18}-1.82×10^{-4} mol L^{-1}, with a LOD of 1.0×10^{-19} mol L^{-1} (Siddiquee et al. 2014). Hashimoto et al. constructed a system for the rapid diagnosis of soil-borne diseases (*Ralostonia* and *Fusarium*), consisting of two biosensors. The system was constructed using equal quantities of two different microbes, each individually immobilized on an electrode. When microbial respiration increased with the assimilation of organic compounds in the sample, the decrease of the dissolved oxygen concentration was measured with an oxygen electrode (Hashimoto et al. 2008).

26.3.3.1.4 Nanobiosensors for Viral and Bacterial Plant Pathogen Detection

Plant pathogenic bacteria, phytoplasmas, viruses, and viroids are difficult to control, and preventive measures are essential to minimize the losses they cause each year in different crops (Lopez et al. 2009). In addition to single probe sensors, nano-chips made of microarrays which contain fluorescent oligo probes were also reported for detecting single nucleotide change in the bacteria and viruses with high sensitivity and specificity based on DNA hybridization (Fang and Ramasamy 2015). The bioconjugated NPs provide an extremely high fluorescent signal for bioanalysis and can be easily incorporated into an antibody specific to a surface antigen of the microbe of interest; the method is sensible for the detection of bacterial plant pathogens. For a bacterium, there are many surface antigens available for specific recognition by using antibody-conjugated nanoparticles (Zhao et al. 2004). Several reports have established the capability of antibody-based biosensors for detection of plant pathogens. A fiber-optic particle plasmon resonance immunosensor is developed for label-free detection of orchid viruses that use Au nanorods as the sensing material. The Au nanorods are employed to create a near-infrared sensing window to solve the color interference problem of sample matrix for direct sensing of target analyte. The Au nanorods are immobilized on the unclad fiber core surface and functionalized by antibodies which can specifically recognize the corresponding *Cymbidium mosaic virus* or *Odontoglossum ringspot virus* for rapid (in 10 min) viral infection diagnosis (Lin et al. 2014). In addition, quantum dot (QD) has been reported to have capabilities to overcome the limitations of organic dyes. QD-immunofluorescence resonance energy transfer-based sensors have been developed to detect witches' broom disease of lime caused by *Phytoplasma aurantifolia*. The immunosensor developed showed a high sensitivity, specificity of 100%, and a detection limit of 5 ca. *P. aurantifolia* per Ml (Rad et al. 2012). QD biosensors were useful in detecting *rhizomania* in *Polymyxa betae*. *Rhizomania* is the most destructive disease in sugar beet caused by beet necrotic yellow vein virus. *Polymyxa betae* (Keskin), the only known vector of beet necrotic yellow vein virus, for transmission of the virus to the plants was successfully detected by QD-immunofluorescence resonance energy transfer-based sensor (Safarpour et al. 2012). Oil palm has become an important source for vegetable oil in Asia, especially for Malaysia and Indonesia. This valuable resource is facing a serious problem on infection called basal stem rot caused by a species known as *Ganoderma boninense*. If not detected at early stage, the plants will end up dying. Bakhori et al. developed an optical DNA biosensor based on fluorescence resonance energy transfer (FRET) utilizing synthesized QD (5–8 nm) for the detection of specific sequence of DNA for *G. boninense*, an oil palm pathogen. The FRET signal that can be observed after hybridization with the target DNA can be a marker for the existence of *G. boninense*. The developed nanobiosensor has shown high sensitivity with detection limit of 3.55×10^{-9} M (Bakhori et al. 2013).

Silica-based NPs (60 nm) were filled with a fluorescent dye and conjugated to an antibody specific to a surface antigen of the microbe of interest. Detection of a single bacterial cell was possible using this technique by Zhao et al. in 2004. This method has potential for sensitive detection of plant pathogens (Jatav and Nirmal 2013). Moreover, the fluorescent silica NPs were conjugated with the secondary antibody of goat anti-rabbit immunoglobulin G (IgG) and successfully detected plant pathogen such as *Xanthomonas axonopodis* pv. vesicatoria that causes bacterial spot disease in tomatoes and peppers. These results demonstrated that the fluorescent silica nanoprobe biomarker will have been potential for rapid diagnosis applications on plant diseases (Yao et al. 2009).

Seo et al. in 2008 reported the fabrication process of a SEPTIC (SEnsing of Phage-Triggered Ion Cascades) chip, consisting of two Ti contact pads and a 150-nm-wide Ti nanowell device on $LiNbO_3$ substrate. The use of this chip as nanoscale electric field probe was demonstrated by detecting the transitory ion efflux from bacteria being infected by phage. When the bacteria were resistant to the phages (uninfected bacteria), small voltage fluctuations were observed in the nanowell displaying a power spectral density. This technology could prove invaluable veterinary and agriculture practice, as well as in applications to microbiological threat detection and reduction in biodefense applications (Seo et al. 2008). In addition to single biosensors directed at one pathogen, research has been ongoing to develop nano-chip microarrays that contain multiple fluorescent oligo probes to detect small nucleotide changes in plant pathogenic bacteria and viruses (Elmer and White 2018).

Pantoea stewartii-NCPPB 449 is urgently required for international shipments due to tremendous agricultural economic losses. Zhao and colleagues presented an electrochemical enzyme-linked immunoassay using AuNPs tags with antibodies of HRP to detect *P. stewartii* subsp. *stewartii* plant bacterial pathogen, reaching a detection limit of 7.8×10^3 CFU/ml (Zhao et al. 2014). AuNPs have been also largely employed for sensor functionalization in sensing systems for pathogen detection thanks to their high surface-to-volume ratios, offering lower detection limits and higher specificity in comparison with conventional strategies (Antonacci et al. 2018). Firrao et al. reported the use of a diagnostic probe made of a specific DNA bearing a fluorescein at its 5′ end and a 2 nm AuNP at its 3′ end, which acts as a quencher. The nanobiotransducer performs as a molecular beacon and emits a stronger fluorescence signal when hybridized to target DNA and in the diagnosis of the *phytoplasma* associated with the flavescence dorée of grapevine (Firraro et al. 2005).

26.3.3.1.5 Nanobiosensors for Herbicide Detection

Herbicides are widely used in agriculture for weed control based on different modes of action. Herbicides belonging to phenylurea (e.g., diuron and linuron), triazine (atrazine, simazine cyanazine, etc.), diazine (bromacil, lenacil, etc.), and phenol chemical groups (dinoseb, ioxynil, bromoxynil, etc.) inhibit photosynthesis in plants, cyanobacteria, algae, and diatoms (Bucur et al. 2018). Nanobiosensors for herbicide residue detection have several benefits, such as compact models, sensitivity, low range of detection, super selectivity, and quick reactions. Nanotubes, nanowires, nanoparticles, or nanocrystals are mostly used to determine the transduction of the signal as they have distinctive heat, electrical, and optical characteristics and are useful in increasing sensitivity, decreasing response time, and changing the detection variety (Kaushal and Wani 2017).

Some nanobiosensors have created the very mode of action of these herbicides to detect photosynthesis-inhibiting herbicides. PSII activity is inhibited not only by some herbicides but also by disruptive endocrine compounds, heavy metals, explosives such as TNT, or ionizing radiation. Herbicide sensitivity depends not only on the particular interaction between a particular herbicide and photosynthetic enzymes but also on the photosynthetic component type and preparation. Isolated PSII structures have been shown to be more susceptible than thylakoid membranes or whole cells, as cells possess protective intracellular processes, in relation to the cell membrane acting as a diffusion barrier, that assist to prevent the impact of herbicides to some extent (Bucur et al. 2018). Table 26.4 summarizes some of the most representative examples of herbicide detection nanobiosensors.

Inkjet printing of feasible photosynthetic cyanobacteria has been proved, maintaining their photosynthetic activity on CNT-modified paper, while intensive study is being undertaken in the field of biosolar cells and self-powered biosensors, concentrating in particular on effective cabling of photosynthetic enzymes to various supports. This stands as proof of the interest in photosynthetic enzymes and the huge potential of combining all these new ideas toward herbicide detection (Bucur et al. 2018).

Monitoring of H_2O_2 in the metabolism and photosynthesis of algae, the concentration of which is associated with potential stress on algae, may be a helpful measure of herbicide presence (Tsopela et al. 2014). Shitanda et al. in 2009 described screen-printed/CNTs-sodium alginate algal (*Chlorella vulgaris* cells) biosensor which was fabricated for evaluation of 6-chloro-N-ethyl-N-isopropyl-1,3,5-atrazine and 3-(3,4-dichlorophenyl)-1,1-diethylurea (DCMU). The concentrations that gave 50% inhibition of the oxygen reduction current (IC_{50}) for atrazine and DCMU were 12 and 1 μmol dm^{-3}, respectively. In comparison with the conventional algal biosensors, in which the algal cells were entrapped in an alginate gel and immobilized on the surface of a transparent indium tin oxide electrode, the present sensor is much smaller and less expensive, with the shorter assay time (Shitanda et al. 2009).

Fluoroimmunoassay based on the fluorescent property of cadmium telluride (CdTe) QD was used along with immunoassay to detect

Table 26.4 A summary of different types of nanobiosensors for the detection of herbicides

Transduction method	Bioreceptor	Herbicide	References
Electrochemical, amperometry, alginate gel, screen-printed electrode CNTs	*C. reinhardtii* cells	Linuron, imazine	Husu et al. (2013)
Electrochemical, amperometry, magnetic NP, SPE	*Chlorella pyrenoidosa* microalgae whole-cell	Atrazine, propazine	Zamaleeva et al. (2011)
Electrochemical, impedimetric, microwires platinum NP	Aptamers	Atrazine	Madianos et al. (2018)
Electrochemical, amperometric, complex magnetic beads functionalized with protein G	M13 phage/antibody anti-atrazine monoclonal	Atrazine	González-Techera et al. (2015)
Electrochemical immunosensor, biotinylated-fab fragment	K47 antibody	Atrazine	Hleli et al. (2006)
Electrochemical, voltammetric, AuNP	Antibodies anti-atrazine monoclonal	Atrazine	Liu et al. (2014)
Electrochemical, field effect transistor, single-walled CNT	Antibodies anti-atrazine monoclonal	Atrazine	Belkhamssa et al. (2016b)
Piezoelectric immunosensor, self-assembled layer of cysteamine	Monoclonal antibody D6F3	Atrazine	Přibyl et al. (2003)
Conductimetric immunosensor, antibodies labeled with AuNP	Antiprostate-specific antigen	Atrazine	Valera et al. (2008)
Cantilever sensor, antibody–antigen	Hapten-specific anti-atrazine antibodies	Atrazine	Suri et al. (2008)

herbicide2,4-dichlorophenoxyacetic acid (2,4-D). The detection of 2,4-D was carried out by fluoroimmunoassay-based biosensor using competitive binding between conjugated 2,4-D–alkaline phosphatase–CdTe and free 2,4-D with immobilized anti 2,4-D antibodies in an immunoreactor column. It was possible to detect 2,4-D up to 250 pg mL^{-1}(Vinayaka et al. 2009). Nanobiosensor based on an atomic force microscopy tip functionalized with the acetolactate synthase enzyme was successfully detected for the herbicide metsulfuron-methyl (an acetolactate synthase inhibitor) through the acquisition of force curves.

26.3.3.1.6 Nanobiosensors for Pesticide Detection

In view of the alarming levels of pesticides being used in agriculture practices, there is a need for their rapid, sensitive, and specific detection in food and environmental samples, as pesticides are harmful to living beings even in trace levels (Vinayaka et al. 2009). A big amount of food and environmental samples contaminated with minute amounts of distinct insecticides/metabolites in complicated matrices need to be screened quickly and sensitively (Bucur et al. 2018). Currently unimolecular and array type of nanomaterial-based biosensors are being developed for the detection of pesticides. However, the format of biosensors varies, from free biomolecules to those conjugated to a substrate such as NPs, nanowires, nanotubes, and thin films. Interaction of the target with the biosensor can be measured either directly or indirectly by recording the changes in color, fluorescence, or electrical potential (Jatav and Nirmal 2013). Organophosphorus pesticides such as dichlorvos and paraoxon, at very low levels could be monitored by liposome, nanoparticles, CNTs, and iMono-based nanobiosensors.

Enzyme nanobiosensors are created to verify pesticide presence on the enzyme inhibition computation included in the enzyme response. Some biosensors are developed on the basis of the AChE inhibition and chemometric result calculations utilizing artificial neural networks (Kaushal and Wani 2017). An electrochemical biosensor for the determination of methyl parathion and chlorpyrifos is described by Viswanathan et al. Used to prepare nanosized polyaniline matrix for AChE enzyme immobilization, the SAMs of single-walled CNT wrapped in thiol terminated single-strand oligonucleotide (ssDNA) on Au were used. The main phase of this biosensor was the enzymatic reaction of AChE-acetylcholine that causes small changes in local pH in the vicinity of an electrode surface and which is to determine carbamate pesticides in chicken, broccoli, and apple specimens without any spiking operation (Viswanathan et al. 2009).

Nanobiosensors use QD as recognition elements for fluorescence detection of pesticides in different media. Luan et al. demonstrated that the presence of OPs can inhibit the AChE activity and thus change the fluorescent intensity of layer-by-layer microarrays of QDs/AChE microscopic dot arrays. Therefore, the QDs/AChE microscopic dot arrays were used for the sensitive visual detection of OPs. Linear calibration for parathion and paraoxon was obtained in the range of 5–100 $\mu g\ L^{-1}$ under the optimized conditions with the limit of detection (LOD) of 10 $\mu g\ L^{-1}$. The arrays have been successfully used for detection of OPs in fruits and real water samples (Luan et al. 2016). Based on AChE inhibition, a biosensing electrode involving reduced graphene oxide (RGO) which supported zirconium oxide (ZrO_2/RGO) nanoparticles is fabricated for chlorpyrifos detection. Mogha et al. indicated that AChE/ZrO_2/RGO is an efficient and biocompatible electrode that can be used for chlorpyrifos detection in ultralow concentrations (Mogha et al. 2016). DNA aptamers that selectively bind to acetamiprid have been recently used for detection of this insecticide with QDs as signal reporters. However, screening and selecting the appropriate aptamer for the target pesticide can be challenging which may limit their application as recognition elements (Nsibande and Forbes 2016). Nucleic acid sensor has been fabricated via immobilization of single-standard calf thymus deoxyribose nucleic acid (ssCT-DNA) onto chitosan (CH)-iron oxide (Fe_3O_4) NPs-based hybrid nanobiocomposite film deposited onto indium-tin-oxide (ITO)-coated glass for pyrethroids [cypermethrin (CM) and permethrin (PM)] detection by Kaushik et al in 2009. This disposable ssCT-DNA/CH- Fe_3O_4 nanobiocomposite/ITO bioelectrode is stable for about 2 months under refrigerated conditions and can detect CM (0.0025–2 ppm) within 25 s and PM (1–300 ppm) within 40 s using DPV technique (Kaushik et al. 2009).

NP-based biosensors are particularly attractive because they can be easily synthesized in bulk using standard chemical techniques and do not require advanced fabrication approaches (Suresh and Periasamy 2014). AuNPs (30 nm)-based dipstick competitive immunoassay (anti-DDT antibodies (IgY)) was developed to detect organochlorine pesticide such as DDT at nanogram level (ppb) by Lisa et al. in 2009. The lowest detection limit of DDT was determined to be 27 ng mL^{-1} with the optimized conditions. AuNPs have the property of agglomeration associated with color production, and AuNPs nanobiosensor can be suitable for the detection of organochlorine pesticide in food and environmental samples and applied for rapid onsite testing of pesticides (Lisa et al. 2009). Fluorine-doped tin-oxide (FTO)-based electrochemical nanosensor was developed for chlorpyrifos detection with AuNPs and anti-chlorpyrifos antibodies (immunosensors). The FTO-AuNPs sensor was successfully employed for the detection of chlorpyrifos in standard as well as in real samples up to 10 nM for apple and cabbage and 50 nM for pomegranate. The proposed FTO-AuNPs nanosensor can be used as a quantitative tool for rapid, onsite detection of chlorpyrifos traces in real samples when miniaturized due to its excellent stability, sensitivity, and simplicity (Talan et al. 2018). Chen et al. (Charrier et al. 2011a, b) reported the fabrication of a nanocomposite biosensor for the sensitive and specific detection of methyl parathion. The nanocomposite sensing film was prepared via the formation of AuNPs on silica particles, mixing with multi-walled CNTs and subsequent covalent immobilization of MPH. The composite of the individual materials was finely tuned to offer the sensing film with high specific surface area and high conductivity. A significant synergistic effect of nanocomposites on the biosensor performance was observed in biosensing methyl parathion. The square wave voltammetric responses displayed well-defined peaks, linearly proportional to the concentrations of methyl parathion in the range from 0.001 μg mL^{-1} to 5.0 μg mL^{-1} with a detection limit of 0.3 ng mL^{-1}. The application of this biosensor in the analysis of spiked garlic samples was also evaluated. The proposed protocol can be used as a platform for the simple and fast construction of biosensors with good performance for the determination of enzyme-specific electroactive species (Chen et al. 2011b).

Carbon materials have received great attention in the last two decades with the emergence of nanoscience. These include the modification of electrodes with different nanocarbons, such as carbon powder, CNTs, graphene sheets, and carbon capsules (Suresh and Periasamy 2014). The nanocomposite consisting of CuO nanoflowers (NFs) and carboxyl-functionalized single-walled CNTs was prepared to improve the sensing performance for chlorpyrifos detection. Changes in the differential pulse voltammetric patterns of the fabricated biosensor were used to detect CP (7×10^{-5} μg mL^{-1}). This aptasensor also exhibited good selectivity and outstanding repeatability and was successfully applied to the determination of chlorpyrifos in spiked apple and celery cabbage with satisfactory recoveries (Uniyal and Sharma 2018; Xu et al. 2018). Despite the high sensitivity for CP detection, aptamer-based biosensors still require more exploration of immobilization strategies to increase the ease of biosensor fabrication (Uniyal and Sharma 2018).

The contemporary approach used to improve the analytical performance of the amperometric sensors based on OPH is based on the functionalization of the interface of the biosensors using nanomaterials. Data reported in the studies show that electrode-reducing capacity is accomplished owing to the electrocatalytic characteristics of the nanostructures and electrode fouling is prevented, while big surface-to-volume ratio, structural robustness, and nanomaterial biocompatibility favor the sensitivity and stability improvement of biosensors. The CNTs are appropriate for alteration of transducers because they combine chemical inertia and mechanical strength and rigidity with elevated transfer rate characteristics (Margarita et al. 2016). Table 26.5 summarizes some of the most representative examples of pesticide detection nanobiosensors.

26.3.3.2 Environmental Monitoring

Many physical–chemical properties of nanoscale materials are used in the field of biosensor development. The nanomaterials such as metal NPs (gold, silver, cobalt, etc.), CNT, magnetic NPs, and QDs have been actively investigated for their applications in biosensors which have become a new interdisciplinary frontier between biological detection and material science (Prasad et al. 2017). It is an extremely thorough and comprehensive task to determine pollutants, toxic intermediate, and heavy metals from waste streams and to monitor weather conditions such as humidity assessment and many other essential characteristics (Sekhon 2014).

26.3.3.2.1 Nanobiosensors to Detect Contaminants in Soil

Protection of the soil health and the environment requires the rapid, sensitive detection of pollutants and pathogens with molecular precision (Choudhary et al. 2015). Diagnosis of soil with this biosensor was based on the concept of calculating the relative activity of favorable and unfavorable soil microbes chosen during respiration on the basis of differential use of oxygen. Biosensors incorporating NPs are defined as nanobiosensors/nanosensors. The presence of NPs boosts the overall efficiency of biosensors probably due to the increased surface for reaction (Kaushal and Wani 2017). Metal and metal oxide NPs are widely used nanomaterials due to their high electroactivity and electronic conductivity for electron transfer (Fang and Ramasamy 2015). Electrochemical sensing of methyl salicylate, a key plant volatile (released by plants during infections), has been achieved by Umasankar et al., using a AuNP-modified screen-printed carbon electrode (SPCE) (Umasankar and Ramasamy 2013).

Ralstonia solanacearum, the devastating causal agent of potato bacterial wilt, is a soil-borne bacterium that can survive in the soil for a long time. The development of sensitive on-field detection methods for this pathogen is highly desirable due to its widespread host range and distribution. A nanobiosensor used AuNPs functionalized with single-stranded oligonucleotides to detect as little as 15 ng of *R. solanacearum* genomic DNA in farm soil. The advantages of this strategy include rapidity, facile usage, and being a visual colorimetric method (Khaledian et al. 2017).

During the past decade, several publications have demonstrated the capability of antibody-based biosensors for detection of plant pathogens (Elmer and White 2018). A fiber-optic particle plasmon resonance immunosensor based on Au nanorods is developed for label-free detection of orchid viruses. The AuNRs are immobilized on the unclad fiber core surface and functionalized by antibodies which can specifically recognize the corresponding *Cymbidium mosaic virus* (48 pg/mL) or *Odontoglossum ringspot virus* (42 pg/mL) for rapid viral infection diagnosis (in 10 min) (Lin et al. 2014).

Table 26.5 The recently reported nanobiosensors for pesticide detection

Transduction method	Pesticide	Bioreceptor	References
Cantilever sensor, polyethyleneimine	Paraoxon, diisopropyl fluorophosphates, parathion	OPH	Karnati et al. (2007)
Fluorescence, liposomes	Paraoxon, dichlorvos	AChE	Vamvakaki and Chaniotakis (2007)
Fluorescence, n(CdSe)ZnS core–shell QD	Paraoxon	OPH	Ji et al. (2005)
Optical, fluorescence, QD and acetylcholine	Dichlorvos	AChE, choline oxidase	Meng et al. (2013)
Fiber-optic, AuNPs covalently coupled with	Paraoxon	AChE	Lin et al. (2006)
Optical, cells microplate with silica NPs and PEi hybrid	Methyl parathion	*Sphingomonas* sp.	Mishra et al. (2017)
Optical, colorimetric, AuNP	Acetamiprid	Artificial antibody acetamiprid-binding aptamer	Shi et al. (2013)
Electrochemical, voltammetric, carbon paste electrode with multiwalled CNT	Pirimicarb	Laccase	
Electrochemical, voltammetric, screen-printed electron with carbon black nanoparticles	Paraoxon	BChE	Arduini et al. (2015)
Electrochemical, impedimetric, SPCE, IrOx NP	Chlorpyrifos	Tyrosinase	Mayorga-Martinez et al. (2014)
Electrochemical, impedimetric, ionic liquids-AuNP porous carbon composite	Dichlorvos	AChE	Wei and Wang (2015)
Electrochemical, impedimetric, Au electrode nanostructured monolayer of fourth-generation poly amidoamine with a cystamine core	Carbaryl	AChE	Santos et al. (2015)
Electrochemical, impedimetric, AuNP, multiwalled CNT, reduced graphene oxide nanoribbons	Acetamiprid	Aptamers	Fei et al. (2015)
Electrochemical, impedimetric, AgNP and nitrogen-doped graphene oxide	Acetamiprid	Aptamers	Jiang et al. (2015)

(continued)

Table 26.5 (continued)

Transduction method	Pesticide	Bioreceptor	References
Electrochemical, impedimetric, microwires platinum NP	Acetamiprid and atrazine	Aptamers	Madianos et al. (2018)
Electrochemical, impedimetric, screen-printed electron with Fe_3O_4 and AuNP	Methyl parathion	Hydrolase	Zhao et al. (2013)
Electrochemical, carbon paste electrode with chitosan, AuNP, nafion	Methyl parathion	AChE	Deng et al. (2016b)
Electrochemical, voltammetric, boron-doped diamond electrode with AuNP, carbon spheres	Chlorpyrifos	AChE	Wei et al. (2014)
Electrochemical, voltammetric, platinum electrode with ZnO nanospheres	Dichlorvos	AChE	Sundarmurugasan et al. (2016)
Electrochemical, voltammetric, carbon black and graphene oxide/Fe_3O_4 nanocomposite, chitosan	Chlorpyrifos	Aptamers	Jiao et al. (2017)
Electrochemical, voltammetric, IrOx-chitosan nanocomposite with glassy carbon electrode	Carbofuran	AChE	Jeyapragasam and Saraswathi (2014)
Electrochemical, amperometric, glassy carbon electrode with graphene oxide and multiwalled CNT	Carbofuran	AChE	Li et al. (2017b)
Electrochemical, amperometric, glassy carbon electrode with NiONP-carboxylic graphene-Nafion composite	Carbofuran	AChE	Yang et al. (2013)
Electrochemical, amperometric- single-walled CNT	Paraoxon, methyl parathion, nitrophenol	OPH	Deo et al. (2005)
Electrochemical, amperometric, PB multiwalled CNT, SPE	Pirimicarb	AChE	Chai et al. (2013)
Electrochemical, amperometric, glassy carbon electrode, Au nanorods	Paraoxon	AChE	Lang et al. (2016)
Electrochemical, amperometric, glassy carbon electrode with NiO NP-carboxylic graphene-nafion	Chlorpyrifos	AChE	Yang et al. (2013)

(continued)

Table 26.5 (continued)

Transduction method	Pesticide	Bioreceptor	References
Electrochemical, amperometric, single and multiwalled CNT, FIA	Paraoxon	OPH	Pedrosa et al. (2010)
Electrochemical, amperometric, Fe_3O_4/Au/CNTs/ZrO_2/Prussian blue/Nafion membrane screen-printed carbon electrodes (SPCE)	Dimethoate	AChE	Gan et al. (2010)
Electrochemical, amperometric, multiwalled CNT and graphene oxide nanoribbons structure	Carbaryl	AChE	Liu et al. (2015)
Electrochemical, amperometric, multiwalled CNT, cellulose acetate composite on a SPCE	Carbaryl	AChE	Cai and Du (2008)
Electrochemical, amperometric, Au–Pt bimetallic NPs, glassy carbon electrode, glutaraldehyde	Paraoxon ethyl, aldicarb, sarin	AChE-choline oxidase	Upadhyay et al. (2009)
Electrochemical, chronoamperometry, polyaniline, multiwalled CNT, glassy carbon electrode	Carbaryl, methomyl	AChE	Cesarino et al. (2012)
Electrochemical, graphene QD, glassy carbon electrode	Paraoxon	Pralidoxime	Dong et al. (2015)

KB of wheat incited by *Tilletia indica* is an economically important quarantined fungal pathogen. Similarity in teliospore configuration makes it difficult to differentiate KB teliospores from the teliospores of other bunt fungi using conventional approaches. In order to determine the correct identity of KB teliospore as an infectious entity, it is essential to develop specific diagnostic probes and high quality of immunological reagents against infectious entities. For onsite testing of KB, Singh et al. developed SPR electrochemical immunosensor (using anti-teliospore antibodies) based on a modified AuNP for rapid detection of Karnal bunt in the field (Singh et al. 2010, 2014).

26.3.3.2.2 Nanobiosensors to Detect Heavy Metals in Soil and Water

The pollution of natural water and soil environment by heavy metals and respective ions can cause severe hazards to human health, and portable, low-cost, and fast heavy metal analyses are a priority issue worldwide (Campanella et al. 2001).

Integrating the electrochemical biosensors with nanomaterials such as nanoparticles could significantly improve the device performance for detection of heave metals (Ejeian et al. 2018). Domínguez-Renedo et al. developed urease amperometric biosensor-modified SPC electrodes and AuNPs for the measurement of Hg^{+2}.

The amperometric response of urea was affected by the presence of Hg (II) ions which caused a decrease in the current intensity (Domínguez-Renedo et al. 2009). Recent studies have established the fact that CNT can enhance the electrochemical reactivity of important biomolecules and can promote the electron transfer reactions of proteins (Suresh and Periasamy 2014). An electrochemical single-walled CNT biosensor was developed for Hg^{2+}detection in water and serum samples (limit of detection 0.84 pM). An efficient Hg2+ recycling strategy was designed using exonuclease III (Shi et al. 2017). Taking into account the fine progress made in the fields of DNA-based devices and nanotechnology, the development of high-precision nanobiosensor based on the FRET is an effective method for multiplexed heavy metal detection (Ejeian et al. 2018). Wu and coworkers designed a new innovative FRET sensor based on QDs and DNAzyme. The QD−DNAzyme nanobiosensors were constructed by conjugating quencher-labeled DNAzymes onto the surface of carboxyl-silanized QDs. In the presence of Pb and Cu, the emission is restored due to the cleavage of DNAzymes. The detection could be completed within 25 min with a single laser excitation source (Wu et al. 2010). Against various metal ions, the QD reflected different fluorescence signals and exhibited differentials between various metal ions (Ejeian et al. 2018). An ultrasensitive method for surface-enhanced Raman scattering (SERS) detection of Hg2+ was developed based on the AuNPs chain induced by single-stranded DNA to form double helical DNA by T–Hg2+–T base pairs. This method could achieve a low limit of detection of 0.45 pg mL^{-1} in the range of 0.001–0.5 ng mL^{-1}. The practicability of the developed method was favorable in the analysis of real samples (Xu et al. 2015). The selectivity of the biosensor was tested for other metal cations such as Zn2+, Cu2+, Ni2+, Pb2+, and Cr2+ at concentrations up to 20 μM, and no significant response (lower than 15% in comparison to the response to Hg2+) was observed. Recently, a SERS biosensor was proposed for simple and sensitive detection of Hg2+ between 1 pM and 100 nM using magnetic substrate ($CoFe_3O_4$@Ag) conjugated with single-stranded DNA and single-walled CNT. The SERS-based biosensor exhibited good performance of the detection of Hg2+ (Yang et al. 2017).

Pb2+ detection technologies are quite important in environment monitoring and human health protection. The fabrication and evaluation of a glassy carbon electrode (GCE) modified with self-doped polyaniline nanofiber mesoporous carbon nitride and bismuth for simultaneous determination of trace Cd2+ and Pb2+ by square wave anodic stripping voltammetry was reported by Zhang et al. (2016). The fabricated electrode exhibited linear calibration curves ranging from 5 to 80 nM for Cd2+ and Pb2+. The limits of detection (LOD) were 0.7 nM for Cd2+ and 0.2 nM for Pb2+ (Zhang et al. 2016). Recently, optical florescent biosensors with DNAzyme based on graphene QDs and AuNPs for Pb^{2+} detection have been reported by Niu et al. 2018. The nanobiosensor possesses an extremely broad detection range of Pb2+ from 50 nM to 4 μM, with a detection limit of 16.7 nM (Niu et al. 2018). Table 26.6 summarizes the recently reported nanobiosensors in the literature for the detection of heavy metals.

Table 26.6 Summary of recent designed nanobiosensors for heavy metals detection

Bioreceptor	Transduction method	Analyte	Sample	References
Whole-cell *Saccharomyces cerevisiae* S288C	Electrochemical, boron-doped nanodiamond-chitosan hydrogel polymer on glassy carbon electrode, menadione and ferricyanide	Cu, Cd, Ni, Pb, 3,5-dichlorophenol, 4-chlorophenol, phenol	Wastewater	Gao et al. (2017)
Whole-cell *E. coli* ATCC 25922	Electrochemical, thionine, chitosan-entrapped carbon nanodots film, glassy carbon	Cd, Cu, Pb, Zn	Wastewater	Fang et al. (2016)
Whole-cell *E. coli* DH5α (plasmid pT7cadO1945 that expresses GFP on exposure to Pb2+)	Fluorescence, microchemostat platform, nanoscale hydrodynamic film	Pb	On-site	Bae et al. (2018)
Nucleic acids	Electrochemical, voltammetric, Au substrate with vertically aligned single-walled CNT	Hg	Water and serum samples	Shi et al. (2017)
Nucleic acids	Optical, SERS, single-walled CNT and Co Fe_3O_4@Ag Substrate	Hg	Environmental sample	Yang et al. (2017)
DNA	Piezoelectric, quartz crystal microbalance with dissipation monitoring, AuNPs	Hg	Lake-water	Chen et al. (2011a)
DNA	Optical, colorimetric, AuNPs	Pb, Hg	Water	Knecht and Sethi (2009)
DNA	Optical, fluorescent, 3-mercaptopropionic acid, Ag nanoclusters	Cu	Soil and pond water	Su et al. (2010)
DNA- G4 (direct metal binding DNA sequence)	Optical, photoluminescent graphene oxide, QD	Pb	Surface water	Park et al. (2015)
DNA (direct metal binding DNA sequence)	Optical, fluorescent, carbon-based QD	Hg, Ag	Real water	Song et al. (2015)
DNA G4 (direct metal binding DNA sequence)	Electrochemiluminescent, glassy carbon electrode, CdSe-QD, AuNPs	Pb	Surface water	Lu et al. (2016)
DNA (direct metal binding DNA sequence)	Optical, colorimetry- fluorescence AuNPs,	Ag, Hg, Cr, Sn, Cd, Pb, Zn, Mn	Surface water	Tan et al. (2016)
DNA (T−Hg2+−T) (direct metal binding DNA sequence)	Colorimetry, biotin-labeled and thiolated DNA-modified AuNPs	Hg	Surface water	Guo et al. (2012)

(continued)

Table 26.6 (continued)

Bioreceptor	Transduction method	Analyte	Sample	References
DNA (T–Hg2+–T)	Colorimetry, AuNPs	Hg	Surface water	Duan and Guo (2012)
DNA (T–Hg2+–T)(Direct metal binding DNA sequence)	Colorimetry, AuNPs	Hg	Surface water	Chen et al. (2014)
(T)-rich Hairpin DNA (T–Hg2+–T); Antibody	Colorimetry, AuNPs, lateral flow strip biocomponent	Hg	Surface water	He et al. (2011)
ssDNA (T–Hg2+–T)	Optical, SERS, Au nanostar	Hg	Tap water	Ma et al. (2013)
dsDNA	Optical, SERS, single Au nanowire on film	Hg	Liquid	Kang et al. (2011)
ssDNA (T–Hg2+–T)	Optical, fluorescence, AuNPs	Hg	Tap water	Wang et al. (2015)
Aptamer (direct metal binding DNA sequence)	Optical, SERS, Au@Ag shell–core NPs	As	Surface water	Song et al. (2016)
Aptamer (T–Hg^{2+}–T)	Optical, chemiluminescence, AuNPs	Hg	Tap water, surface water	Qi et al. (2017)
Aptamer (guanine-rich)	Electrochemical, impedimetric, multiwalled CNTs, AuNPs	Pb	Standard solutions	Zhu et al. (2014)
Aptamer (cytosine-rich)	Optical, fluorescent nanoprobe, graphene oxide	Ag	Water	Wen et al. (2010)
Aptamer	Optical, fluorescent, CdSe/ZnS- QD, graphene oxide	Pb	Liquid	Li et al. (2013)
Aptamer	Colorimetric resonance scattering, AuNPs	As	Real water	Wu et al. (2012)
DNAzyme	Optical, fluorescence, graphene QD, and AuNP	Pb		Niu et al. (2018)

DNAzymes	Optical, fluorescent, CdSe-ZnS, QDs	Cu, Pb	Liquid	Wu et al. (2010)
DNAzymes	Electrochemiluminescence, Au nanodendrites, silver/zinc oxide	Pb	Surface water	Li et al. (2015)
Peptide (cysteine monomers in the presence of diphenylphosphoryl azide)	Electrochemical, potentiometric, multiwalled CNT, Au electrode	Cd, Hg	Wastewater	Abdul Rahman et al. (2012)
Glutathione/4-mercaptopyridine	Optical, surface-enhanced Raman scattering, AgNPs	As	Real water	Li et al. (2011)
Phosphatase and esterase/	Conductometric, immobilizing *Arthrospira platensis* cells, bovine serum albumin/AuNPs	Cd, Hg	Mixture and in pure solution	Tekaya et al. (2014)
Laccase and HRP	Optical, fast Fourier transform, nanostructured porous Si and SiO_2, Fabry–Pérot thin films	Ag, Pb, Cu	Water samples	Shtenberg et al. (2015)
Phytochelatins	Optical, AuNPs-modified optical fiber, SPR	Cd	Standard solutions	Lin and Chung (2009)
Mercaptopropionic acid – Cys	Optical, bio-inspired colorimetric, AuNP s	Hg	Aqueous solution	Darbha et al. (2008)

26.3.3.2.3 Nanobiosensors for the Detection of Toxin, Pollution, and Pathogens in Water

The presence of pathogens in environmental matrices, and mainly in water compartments, could constitute a serious danger for human health, and some bio- and nanobiosensors were recently proposed for their environmental monitoring. The nanobiosensor allows various procedures to be integrated into a single device to design various components in real time tracking, which is very helpful for tracking water and wastewater structure (Ejeian et al. 2018). For example, rapid and specific optical biosensors based on SPR were proposed for the detection of metabolically active *Legionella pneumophila* in complex environmental water samples (Justino et al. 2017). Nanobiosensor effectiveness relies on the selectivity interaction of tiny biomolecules as components of biorecognition and the hypersensitivity of nanostructures when subjected to pollutants (Ejeian et al. 2018). An amperometric magnetoimmunoassay, based on the use of core–shell magnetic (Fe_3O_4) NPs and screen-printed carbon electrodes, was developed for the selective determination of *L. pneumophila* SG1 by Martín et al. (2015). A specific capture antibody was linked to the poly (dopamine)-modified magnetic NPs and incubated with bacteria, and bacteria were sandwiched using the antibody labeled with HRP. The possibility of detecting *L. pneumophila* at 10 CFU mL^{-1} level in less than 3 h, after performing a membrane-based preconcentration step, was also demonstrated (Martín et al. 2015). Park et al. in 2010 reported that multiwalled CNT electrode functionalized with oxygen plasma treatment was prepared and characterized, and its DNA-sensing ability for *L. pneumophila* detection was examined using electrochemical measurement (Park et al. 2010). This DNA nanobiosensor has limitation in the environmental samples. An ultrasensitive electrochemical immunosensor based on a ZnO nanorod matrix electrode was developed for detecting *L. pneumophila.* The peptidoglycan-associated lipoprotein of *L. pneumophila*, as a component of *Legionella* antigen, was bound to the primary antibody, and secondary antibody conjugated to HRP was then bound to the antigen. This nanobiosensor detection limit is ~1 pg/mL with excellent selectivity (Park et al. 2014).

Escherichia coli (*E. coli*) is a typical inhabitant in the intestinal tracts of humans and warm-blooded animals, which is often preferred as an indicator organism because it is specific for water pollution and reliably reflects fecal contamination. Some strains of E. coli such as *E. coli* 0157:H7 can cause diarrhea, urinary tract infections, inflammations, and peritonitis in immunosuppressed patients such as children and elderly people. Most biosensors for *E. coli* detection used antibodies as recognition elements. Electrochemiluminescence biosensor was developed for *E. coli* O157:H7 quantitative detection based on a polydopamine surface imprinted polymer (SIP) and nitrogen-doped graphene QDsin water sample. The linear relationships between the electrochemiluminescence intensity and *E. coli* O157:H7 concentration were obtained from 10^1 colony-forming units (CFU) mL^{-1} to 10^7 CFU mL^{-1} with a limit of detection of 8 CFU mL^{-1} (Chen et al. 2017). Micro- and nanoscale sensors are suitable for detecting waterborne pathogens, and common nanoscale materials such as CNTs and QD are now extensively applied for quantitative detection of microorganisms including bacteria and protozoa (Samendra et al. 2014). Recently, nucleic acid biosensor-based researches have been increasingly focused. Nucleic acid biosensors offer desirable sensitivity for detecting particularly waterborne pathogens even at low levels (Koedrith et al. 2015).

Harmful toxins such as brevetoxins and microcystins are produced from the algal blooms of cyanobacteria provided by the eutrophication of aquatic systems, and reliable and cost-effective systems are thus needed for the early detection of such toxins (Justino et al. 2017). Molybdenum disulfide and Au nanorod nanocomposites were employed to provide a large surface area, and excellent biocompatibility in electrochemical immunosensor (HRP-labeled anti-microcystin-LR antibody) was developed for the detection of microcystin-LR in water. The immunosensor exhibited a linear response to MC-LR ranging from 0.01 to 20 μg L^{-1} with a detection limit of 5 ng L^{-1} (Zhang et al. 2017). Okadaic acid (OA) is a representative diarrhetic shellfish poisoning toxin which is highly toxic and carcinogenic to human and rich in polluted shellfish. Sensitive fluorescence immunosensor (anti-OA monoclonal antibody) based on magnetic beads and QDs had been developed for OA detection. The limit of detection was 0.05 μg/L with a linear range of 0.2–20 μg/L for OA detection, which was far lower than traditional id-ELISA technique, and OA detection for the real sample could be completed within 1 h (Pan et al. 2017). Domoic acid associated with amnesic shellfish poisoning was also detected in seawater samples with an effect of transistor immunobiosensors based on CNTs. Good reproducibility (0.52–1.43%) and low limit of detection (10 ng mL^{-1}) were found in a working range between 10 and 500 ng mL^{-1} (Marques et al. 2017).

Bisphenol is a key monomer in the production of polycarbonate plastic and epoxy resins, which has been widely used in a variety of common consumer goods, e.g., water bottles, food cans, and medical devices. However, bisphenol A is an endocrine-disrupting compound especially found in water that can mimic the functions of estradiol and may have negative effects on human health (He et al. 2017). Aptamer (fluorescein amidite)-based "turn-off" fluorescent biosensor with AuNP for ultra-sensitive detection of small molecules such as bisphenol A in water samples was developed by Ragavan et al. 2013. Another aptasensor (anti-bisphenol A) based on fluorescence with molybdenum carbide nanotube (label-free) for detection of bisphenol A in real water sample was developed by He et al. in 2017. This method shows a linear range of 2–20 nM with a detection limit of 2 nM for detecting bisphenol A (He et al. 2017). Another endocrine-disrupting chemical, the 4-nonylphenol, was recently analyzed in seawater samples by electrochemical lable-free immunosensors based on field effect transistors (FET), with single-walled CNTs. The immunosensor show an excellent analytical performance with reproducibility of 0.56±0.08%, repeatability of 0.5±0.2%, limit of detection for 4-nonylphenol as low as 5 μg L^{-1}, and average recovery between 97.8% and 104.6% (Belkhamssa et al. 2016a).

Estrogen is considered to be carcinogenic and has a tumor promotion effect. Its level related to the risk of breast cancer is evident. It is well documented that estrogen pollution causes death and deformation in birds, fishes, animals, as well as human beings. It should be noted that 17 β-estradiol is considered a key estrogen pollution (Dai and Liu 2017). 17-Estradiol was detected in lake water by photoelectrochemical sensing platform based on anti-E2 aptamer as the biorecognition element was developed onto CdSe nanoparticle-modified TiO_2 nanotube arrays. The designed PEC aptasensor exhibits excellent performances in determining E2 with a wide linear range of 0.05–15 pM (Fan et al. 2014). Table 26.7 provides a summary of recent nanobiosensors for environmental monitoring, whose applications and analytical performances are discussed and compared in this section.

Table 26.7 Summary of recent designed biosensors for detect toxin, pollution, and pathogen in water

Analyte detected	Transduction method	Recognition element	References
L. pneumophila	Optical, SPR, Au substrate with streptavidin-conjugated QD	Nucleic acids (16s rRNA)	Foudeh et al. (2015)
L. pneumophila	Electrochemical, voltammetric, immunosensor based on a ZnO nanorod with Au working electrode	Antibody	Park et al. (2014)
L. pneumophila	Electrochemical, amperometric, screen-printed carbon electrodes with Fe_3O_4@ polydopamine complex	Antibody (polyclonal)	Martín et al. (2015)
L. pneumophila	Electrochemical, well-patterned multiwalled CNT, oxygen plasma treatment prior	Nucleic acids (DNA)	Park et al. (2010)
E. coli	Optical, electrochemiluminescence, glassy carbon electrode with polydopamine imprinted polymer and nitrogen-doped QD	Antibody (polyclonal)	Chen et al. (2017)
Bacillus subtilis	Electrochemical, amperometric, Au electrode with single-walled CNT	Antibodies (polyclonal)	Yoo et al. (2017)
Bisphenol A	Optical, fluorescence, AuNP	Aptamers	Ragavan et al. (2013)
Bisphenol A	Optical, fluorescence, molybdenum carbide nanotubes	Aptamers	He et al. (2017)
Phenol, m-cresol, catechol	Electrochemical, amperometric, glassy carbon (GC) electrode, multiwalled CNT	DNA	Zheng et al. (2009)
Phenol	Electrochemical, core shell magnetic MgFe2O4 NPs, carbon paste electrode	Tyrosinase	Liu et al. (2005)
Nonylphenol	Electrochemical, field effect transistor, single-walled CNT	Antibodies (monoclonal)	Belkhamssa et al. (2016a)
17 -estradiol	Photo-electrochemical, CdSe NPs and TiO_2 Nanotubes	Aptamers	Fan et al. (2014)
Nitrate	Electrochemical, amperometric, polypyrrole/CNT film,	Nitrate reductase	Can et al. (2012)
Nitrate, nitrite	Electrochemical, Cyclic voltammetry, Cu, Zn superoxide dismutase, CNT–polypyrrole nanocomposite modified platinum electrode	Nitrate reductase	Madasamy et al. (2014)

26.4 Future Perspectives

Biosensors have versatile applications in the field of medicine, engineering, biomedical engineering, toxicology, ecotoxicology, etc., and their miniaturization becomes useful for a cost-effective, rapid, and sensitive detection. Nanobiosensors have now captured the field of biosensors as the different nanostructures at ultralow size exhibit novel properties which are not shown by their bulk counterpart. Although environmental biosensors can be constructed using the improved characteristics of nanomaterials and novel nanocomposites, an increased attention has been focused on the in situ and real-time monitoring of pollutants by other technologies. Biosensors cover well-established bioanalytical techniques, while nanobiosensors with nanotechnology inclusion are revolutionizing this field with prospective solutions by minimizing the load of conventional laboratory methods and protocols, combined with rapid response time, enhanced sensitivity, robustness, and point-of-use portability. Nanobiosensor can measure more variables with greater sensitivity and less sample material is required. The use of biosensors is simple and the technique is rapid and cost-effective. Because nanobiosensors work on an atomic scale, they have the highest multianalyte and high-throughput efficiency, and so they can be readily used to detect pathogens, pollutants, and environmental toxicity. The main limitation of recent environmental biosensors is due to the lack of application in real environmental samples since the majority of identified "environmental biosensors" have been applied to tap water samples or synthetic samples. There is a high demand for quick, reliable, and low-cost technologies to detect, monitor, and diagnose pollutants and toxins in the setting. However, several limitations still hamper the wide use of this technology in the real field, including, for example, the low storage capacity and working stability of the biocomponents.

New exciting approach of using nano-enabled biosensors can be coupled with robotics and GPS systems to create smart delivery systems that detect, map, and treat specific areas in a field prior to or during the onset of symptoms. This technology could reduce agrochemical inputs and increase yield and profits. Growers and scouts could perform diagnostics in situ once portable devices with biosensors are developed. An extremely valuable use for fast and sensitive biosensors is at ports of entry, where quarantined pathogens could be intercepted with greater efficiency. The value of rapid analysis in detecting food pathogens and mycotoxins is obvious. One of the important trends in nanobiosensors is to develop biosensors for applications in extreme conditions, such as highly acidic, alkaline, saline, extreme temperature, and organic solvent environment because more and more detections will involve such unfriendly conditions, particularly important for the growing nanobiosensor industry because of the great need for low-cost, sensitive, selective, and fast response biosensor in the market. The numerous advances in nanotechnology-based biosensor technology have generated tremendous technology push, as evident from the exponentially increased number of publications, patent applications, projects, and focused nanotechnology initiatives/themes.

References

Abdul Rahman N, Yusof N, Amirah Mohd Maamor N, Mariam Mohd Noor S (2012) Development of electrochemical sensor for simultaneous determination of Cd(II) and Hg(II) Ion by exploiting newly synthesized cyclic dipeptide. Int J Electrochem Sci 7:186–196

Abdulhalim I, Zourob M, Lakhtakia A (2008) Surface plasmon resonance for biosensing: a mini-review. Electromagnetics 28:214–242

Aisyah WN, Jusoh W, Wong LS (2014) Exploring the potential of whole cell biosensor: a review in environmental applications. Int Sci Acad Eng Technol 21:52–56

Aldewachi H, Chalati T, Woodroofe MN, Bricklebank N, Sharrack B, Gardiner P (2018) Gold nanoparticle-based colorimetric biosensors. Nanoscale 10:18–33

Ali MA, Jiang H, Mahal NK, Weber RJ, Kumar R, Castellano MJ, Dong L (2017) Microfluidic impedimetric sensor for soil nitrate detection using graphene oxide and conductive nanofibers enabled sensing interface. Sensors Actuators B Chem 239:1289–1299

Alpat SK, Alpat Ş, Kutlu B, Özbayrak Ö, Büyükışık HB (2007) Development of biosorption-based algal biosensor for Cu(II) using *Tetraselmis chui*. Sensors Actuators B Chem 128:273–278

Álvarez SP, López NEL, Lozano JM, Negrete EAR, Cervantes MES (2016) Plant fungal disease management using nanobiotechnology as a tool. In: Prasad R (ed) Advances and applications through fungal nanobiotechnology. Springer, Cham, pp 169–192

Amine A, Mohammadi H, Bourais I, Palleschi G (2006) Enzyme inhibition-based biosensors for food safety and environmental monitoring. Biosens Bioelectron 21:1405–1423

Amini B, Kamali M, Salouti M, Yaghmaei P (2017) Fluorescence bio-barcode DNA assay based on gold and magnetic nanoparticles for detection of Exotoxin A gene sequence. Biosens Bioelectron 92:679–686

Amini B, Kamali M, Salouti M, Yaghmaei P (2018) Spectrophotometric, colorimetric and visually detection of Pseudomonas aeruginosa ETA gene based gold nanoparticles DNA probe and endonuclease enzyme. Spectrochim Acta A Mol Biomol Spectrosc 199:421–429

Andrea Medeiros Salgado, Silva LM, Melo AF (2011) Biosensor for environmental applications. In: Somerset V (ed) Environmental biosensors. IntechOpen, Rijeka, pp 3–16

Andreescu S, Magearu V, Lougarre A, Fournier D, Marty J-L (2001) Immobilization of enzymes on screen-printed sensors via an histidine tail. application to the detection of pesticides using modified cholinesterase. Anal Lett 34:529–540

Antonacci A, Arduini F, Moscone D, Palleschi G, Scognamiglio V (2018) Nanostructured (Bio) sensors for smart agriculture. TrAC Trends Anal Chem 98:95–103

Arduini F, Guidone S, Amine A, Palleschi G, Moscone D (2013) Acetylcholinesterase biosensor based on self-assembled monolayer-modified gold-screen printed electrodes for organophosphorus insecticide detection. Sensors Actuators B Chem 179:201–208

Arduini F, Forchielli M, Amine A, Neagu D, Cacciotti I, Nanni F, Moscone D, Palleschi G (2015) Screen-printed biosensor modified with carbon black nanoparticles for the determination of paraoxon based on the inhibition of butyrylcholinesterase. Microchim Acta 182:643–651

Arias-Barreiro CR, Okazaki K, Koutsaftis A, Inayat-Hussain SH, Tani A, Katsuhara M, Kimbara K, Mori IC (2010) A bacterial biosensor for oxidative stress using the constitutively expressed redox-sensitive protein roGFP2. Sensors (Basel, Switzerland) 10:6290–6306

Avramescu A, Rouillon R, Carpentier R (1999) Potential for use of a cyanobacterium Synechocystis sp. immobilized in poly(vinylalcohol): application to the detection of pollutants. Biotechnol Tech 13:559–562

Axelrod T, Eltzov E, Marks RS (2016) Bioluminescent bioreporter pad biosensor for monitoring water toxicity. Talanta 149:290–297

Bae J, Lim J-W, Kim T (2018) Reusable and storable whole-cell microbial biosensors with a microchemostat platform for in situ on-demand heavy metal detection. Sensors Actuators B Chem 264:372–381

Bagde VL, Borkar DB (2013) Biosensor: use in agriculture. Int J Sci Res 2:1–3

Bakhori NM, Yusof NA, Abdullah AH, Hussein MZ (2013) Development of a fluorescence resonance energy transfer (FRET)-based DNA biosensor for detection of synthetic oligonucleotide of *Ganoderma boninense*. Biosensors 3:419–428

Balahura LR, Stefan-Van Staden RL, Van Staden JF, Aboul-Enein HA (2019) Advances in immunosensors for clinical applications. J Immunoass Immunochem 40:40–51

Belkhamssa N, da Costa JP, Justino CIL, Santos PSM, Cardoso S, Duarte AC, Rocha-Santos T, Ksibi M (2016a) Development of an electrochemical biosensor for alkylphenol detection. Talanta 158:30–34

Belkhamssa N, Justino CIL, Santos PSM, Cardoso S, Lopes I, Duarte AC, Rocha-Santos T, Ksibi M (2016b) Label-free disposable immunosensor for detection of atrazine. Talanta 146:430–434

Bellan LM, Wu D, Langer RS (2011) Current trends in nanobiosensor technology. Wiley Interdiscip Rev Nanomed Nanobiotechnol 3:229–246

Berezhetskyy AL, Sosovska OF, Durrieu C, Chovelon JM, Dzyadevych SV, Tran-Minh C (2008) Alkaline phosphatase conductometric biosensor for heavy-metal ions determination. IRBM 29:136–140

Bettazzi F, Laschi S, Mascini M (2007) One-shot screen-printed thylakoid membrane-based biosensor for the detection of photosynthetic inhibitors in discrete samples. Anal Chim Acta 589:14–21

Bhalla V, Zazubovich V (2011) Self-assembly and sensor response of photosynthetic reaction centers on screen-printed electrodes. Anal Chim Acta 707:184–190

Bhalla V, Zhao X, Zazubovich V (2011) Detection of explosive compounds using Photosystem II-based biosensor. J Electroanal Chem 657:84–90

Biran I, Walt DR (2002) Chapter 1 – optrode-based fiber optic biosensors (bio-optrode). In: Ligler FS, Rowe Taitt CA (eds) Optical biosensors. Elsevier Science, Amsterdam, pp 5–56

Biran I, Babai R, Levcov K, Rishpon J, Ron EZ (2000) Online and in situ monitoring of environmental pollutants: electrochemical biosensing of cadmium. Environ Microbiol 2:285–290

Bonetto MC, Sacco NJ, Ohlsson AH, Cortón E (2011) Assessing the effect of oxygen and microbial inhibitors to optimize ferricyanide-mediated BOD assay. Talanta 85:455–462

Bucur B, Dondoi M, Danet A, Marty J-L (2005) Insecticide identification using a flow injection analysis system with biosensors based on various cholinesterases. Anal Chim Acta 539:195–201

Bucur B, Munteanu FD, Marty JL, Vasilescu A (2018) Advances in enzyme-based biosensors for pesticide detection. Biosensors 8:27

Cai J, Du D (2008) A disposable sensor based on immobilization of acetylcholinesterase to multiwall carbon nanotube modified screen-printed electrode for determination of carbaryl. J Appl Electrochem 38:1217–1222

Campanella L, Cubadda F, Sammartino MP, Saoncella A (2001) An algal biosensor for the monitoring of water toxicity in estuarine environments. Water Res 35:69–76

Can F, Korkut Ozoner S, Ergenekon P, Erhan E (2012) Amperometric nitrate biosensor based on Carbon nanotube/Polypyrrole/Nitrate reductase biofilm electrode. Mater Sci Eng C 32:18–23

Carpentier R, Loranger C, Chartrand J, Purcell M (1991) Photoelectrochemical cell containing chloroplast membranes as a biosensor for phytotoxicity measurement. Anal Chim Acta 249:55–60

Cesarino I, Moraes FC, Lanza MRV, Machado SAS (2012) Electrochemical detection of carbamate pesticides in fruit and vegetables with a biosensor based on acetylcholinesterase immobilised on a composite of polyaniline–carbon nanotubes. Food Chem 135:873–879

Chai Y, Niu X, Chen C, Zhao H, Lan M (2013) Carbamate insecticide sensing based on acetylcholinesterase/Prussian blue-multi-walled carbon nanotubes/screen-printed electrodes. Anal Lett 46:803–817

Chang IS, Jang JK, Gil GC, Kim M, Kim HJ, Cho BW, Kim BH (2004) Continuous determination of biochemical oxygen demand using microbial fuel cell type biosensor. Biosens Bioelectron 19:607–613

Charrier T, Chapeau C, Bendria L, Picart P, Daniel P, Thouand G (2011a) A multi-channel bioluminescent bacterial biosensor for the on-line detection of metals and toxicity. Part II: technical development and proof of concept of the biosensor. Anal Bioanal Chem 400:1061–1070

Charrier T, Durand M-J, Jouanneau S, Dion M, Pernetti M, Poncelet D, Thouand G (2011b) A multi-channel bioluminescent bacterial biosensor for the on-line detection of metals and toxicity. Part I: design and optimization of bioluminescent bacterial strains. Anal Bioanal Chem 400:1051–1060
Chen D, Cao Y, Liu B, Kong J (2002) A BOD biosensor based on a microorganism immobilized on an Al2O3 sol–gel matrix. Anal Bioanal Chem 372:737–739
Chen Q, Wu X, Wang D, Tang W, Li N, Liu F (2011a) Oligonucleotide-functionalized gold nanoparticles-enhanced QCM-D sensor for mercury(ii) ions with high sensitivity and tunable dynamic range. Analyst 136:2572–2577
Chen S, Huang J, Du D, Li J, Tu H, Liu D, Zhang A (2011b) Methyl parathion hydrolase based nanocomposite biosensors for highly sensitive and selective determination of methyl parathion. Biosens Bioelectron 26:4320–4325
Chen W, Cai S, Ren Q-Q, Wen W, Zhao Y-D (2012) Recent advances in electrochemical sensing for hydrogen peroxide: a review. Analyst 137:49–58
Chen G-H, Chen W-Y, Yen Y-C, Wang C-W, Chang H-T, Chen C-F (2014) Detection of mercury(II) ions using colorimetric gold nanoparticles on paper-based analytical devices. Anal Chem 86:6843–6849
Chen S, Chen X, Zhang L, Gao J, Ma Q (2017) Electrochemiluminescence detection of *Escherichia coli* O157:H7 based on a novel polydopamine surface imprinted polymer biosensor. ACS Appl Mater Interfaces 9:5430–5436
Chen Y, Li H, Gao T, Zhang T, Xu L, Wang B, Wang J, Pei R (2018) Selection of DNA aptamers for the development of light-up biosensor to detect Pb(II). Sensors Actuators B Chem 254:214–221
Choudhary MK, Singh M, Saharan W (2015) Application of nanobiosensors in agriculture. Popular Kheti 3:130–135
Costa SPF, Cunha E, Azevedo AMO, Pereira SAP, Neves AFDC, Vilaranda AG, Araujo ARTS, Passos MLC, Pinto PCAG, Saraiva MLMFS (2018) Microfluidic chemiluminescence system with yeast saccharomyces cerevisiae for rapid biochemical oxygen demand measurement. ACS Sustain Chem Eng 6:6094–6101
Crew A, Lonsdale D, Byrd N, Pittson R, Hart JP (2011) A screen-printed, amperometric biosensor array incorporated into a novel automated system for the simultaneous determination of organophosphate pesticides. Biosens Bioelectron 26:2847–2851
Dai Y, Liu CC (2017) Detection of 17 β-estradiol in environmental samples and for health care using a single-use, cost-effective biosensor based on differential pulse voltammetry (DPV). Biosensors 7:15
Damborský P, Švitel J, Katrlík J (2016) Optical biosensors. Essays Biochem 60:91–100
Darbha GK, Singh AK, Rai US, Yu E, Yu H, Chandra Ray P (2008) Selective detection of mercury (II) ion using nonlinear optical properties of gold nanoparticles. J Am Chem Soc 130:8038–8043
de Mora K, Joshi N, Balint BL, Ward FB, Elfick A, French CE (2011) A pH-based biosensor for detection of arsenic in drinking water. Anal Bioanal Chem 400:1031–1039
Dehnad A, Hamedi* J, Derakhshan-Khadivi F, Abuşov R (2015) Green synthesis of gold nanoparticles by a metal resistant arthrobacter nitroguajacolicus isolated from gold mine. IEEE Trans NanoBiosci 14:393–396
Deng H-H, Hong G-L, Lin F-L, Liu A-L, Xia X-H, Chen W (2016a) Colorimetric detection of urea, urease, and urease inhibitor based on the peroxidase-like activity of gold nanoparticles. Anal Chim Acta 915:74–80
Deng Y, Liu K, Liu Y, Dong H, Li S (2016b) An novel acetylcholinesterase biosensor based on nano-porous pseudo carbon paste electrode modified with gold nanoparticles for detection of methyl parathion. J Nanosci Nanotechnol 16:9460–9467
Deo RP, Wang J, Block I, Mulchandani A, Joshi KA, Trojanowicz M, Scholz F, Chen W, Lin Y (2005) Determination of organophosphate pesticides at a carbon nanotube/organophosphorus hydrolase electrochemical biosensor. Anal Chim Acta 530:185–189
DeRosa MC, Monreal C, Schnitzer M, Walsh R, Sultan Y (2010) Nanotechnology in fertilizers. Nat Nanotechnol 5:91

Deshpande K, Mishra RK, Bhand S (2010) A high sensitivity micro format chemiluminescence enzyme inhibition assay for determination of Hg(II). Sensors (Basel, Switzerland) 10:6377–6394

Dhall P, Kumar A, Joshi A, Saxsena TK, Manoharan A, Makhijani SD, Kumar R (2008) Quick and reliable estimation of BOD load of beverage industrial wastewater by developing BOD biosensor. Sensors Actuators B Chem 133:478–483

Di Lorenzo M (2016) 11 – use of microbial fuel cells in sensors. In: Scott K, Yu EH (eds) Microbial electrochemical and fuel cells. Woodhead Publishing, Boston, pp 341–356

Domínguez-Renedo O, Alonso-Lomillo MA, Ferreira-Gonçalves L, Arcos-Martínez MJ (2009) Development of urease based amperometric biosensors for the inhibitive determination of Hg (II). Talanta 79:1306–1310

Dong J, Hou J, Jiang J, Ai S (2015) Innovative approach for the electrochemical detection of non-electroactive organophosphorus pesticides using oxime as electroactive probe. Anal Chim Acta 885:92–97

Duan J, Guo ZY (2012) Development of a test strip based on DNA-functionalized gold nanoparticles for rapid detection of mercury (II) ions. Chin Chem Lett 23:225–228

Duhan JS, Kumar R, Kumar N, Kaur P, Nehra K, Duhan S (2017) Nanotechnology: the new perspective in precision agriculture. Biotechnol Rep 15:11–23

Durmuş NG, Lin RL, Kozberg M, Dermici D, Khademhosseini A, Demirci U (2015) Acoustic-based biosensors. In: Li D (ed) Encyclopedia of microfluidics and nanofluidics. Springer, New York, pp 28–40

Durrieu C, Tran-Minh C (2002) Optical algal biosensor using alkaline phosphatase for determination of heavy metals. Ecotoxicol Environ Saf 51:206–209

Dwevedi A, Kumar P, Kumar P, Kumar Y, Sharma YK, Kayastha AM (2017) New pesticidies and soil sensors: nanotechnology in the agri-food industry. In: Grumezescu AM (ed) Soil sensors: detailed insight into research updates, significance, and future prospects. Elsevier Inc., London

Economou A, Karapetis SK, Nikoleli G, Nikolelis DP, Bratakou S, Varzakas TH (2017) Enzyme-based sensors. In: Nollet FTALM (ed) Advances in food diagnostics. Wiley-Blackwell, Chichester, pp 231–250

Ejeian F, Etedali P, Mansouri-Tehrani H-A, Soozanipour A, Low Z-X, Asadnia M, Taheri-Kafrani A, Razmjou A (2018) Biosensors for wastewater monitoring: a review. Biosens Bioelectron 118:66–79

Elmer W, White JC (2018) The future of nanotechnology in plant pathology. Annu Rev Phytopathol 56:11–33

Etefagh R, Azhir E, Shahtahmasebi N (2013) Synthesis of CuO nanoparticles and fabrication of nanostructural layer biosensors for detecting Aspergillus niger fungi. Sci Iran 20:1055–1058

Eun AJ-C, Wong S-M (2000) Molecular beacons: a new approach to plant virus detection. Phytopathology 90:269–275

Eun AJ-C, Huang L, Chew F-T, Li SF-Y, Wong S-M (2002) Detection of two orchid viruses using quartz crystal microbalance (QCM) immunosensors. J Virol Methods 99:71–79

Fan L, Zhao G, Shi H, Liu M, Wang Y, Ke H (2014) A femtomolar level and highly selective 17β-estradiol photoelectrochemical aptasensor applied in environmental water samples analysis. Environ Sci Technol 48:5754–5761

Fang Y, Ramasamy RP (2015) Current and prospective methods for plant disease detection. Biosensors 5:537–561

Fang Y, Umasankar Y, Ramasamy RP (2014) Electrochemical detection of p-ethylguaiacol, a fungi infected fruit volatile using metal oxide nanoparticles. Analyst 139:3804–3810

Fang D, Gao G, Shen J, Yu Y, Zhi J (2016) A reagentless electrochemical biosensor based on thionine wrapped E. coli and chitosan-entrapped carbon nanodots film modified glassy carbon electrode for wastewater toxicity assessment. Electrochim Acta 222:303–311

Farré M, Barceló D (2009) Biosensors for aquatic toxicology evaluation. In: Barceló D, Hansen P-D (eds) Biosensors for environmental monitoring of aquatic systems: bioanalytical and chemical methods for endocrine disruptors. Springer, Berlin/Heidelberg, pp 115–160

Fei A, Liu Q, Huan J, Qian J, Dong X, Qiu B, Mao H, Wang K (2015) Label-free impedimetric aptasensor for detection of femtomole level acetamiprid using gold nanoparticles decorated multiwalled carbon nanotube-reduced graphene oxide nanoribbon composites. Biosens Bioelectron 70:122–129

Fogel R, Limson J, Seshia AA (2016) Acoustic biosensors. Essays Biochem 60:101–110

Foudeh AM, Trigui H, Mendis N, Faucher SP, Veres T, Tabrizian M (2015) Rapid and specific SPRi detection of *L. pneumophila* in complex environmental water samples. Anal Bioanal Chem 407:5541–5545

Frense D, Müller A, Beckmann D (1998) Detection of environmental pollutants using optical biosensor with immobilized algae cells. Sensors Actuators B Chem 51:256–260

Firraro G, Moretti M, Ruiz Rosquete M, Gobbi E, R L (2005) Nanobiotransducer for detecting flavescence Doree phytoplasma. J Plant Pathol 87:101–107

Gan N, Yang X, Xie D, Wu Y, Wen W (2010) A disposable organophosphorus pesticides enzyme biosensor based on magnetic composite nano-particles modified screen printed carbon electrode. Sensors (Basel, Switzerland) 10:625–638

Gao G, Qian J, Fang D, Yu Y, Zhi J (2016) Development of a mediated whole cell-based electrochemical biosensor for joint toxicity assessment of multi-pollutants using a mixed microbial consortium. Anal Chim Acta 924:21–28

Gao G, Fang D, Yu Y, Wu L, Wang Y, Zhi J (2017) A double-mediator based whole cell electrochemical biosensor for acute biotoxicity assessment of wastewater. Talanta 167:208–216

Gayathri S, Braganca J (2009) Biosensors for monitoring environmental pollutants: a review. Environ Sci Indian J 4:297–305

Giannoudi L, Piletska EV, Piletsky SA (2006) Development of biosensors for the detection of hydrogen peroxide. In: Biotechnological applications of photosynthetic proteins: biochips, biosensors and biodevices. Springer, Boston, pp 175–191

Giardi MT, Pace E (2006) Photosystem II-based biosensors for the detection of photosynthetic herbicides. In: Biotechnological applications of photosynthetic proteins: biochips, biosensors and biodevices. Springer, Boston, pp 147–154

Giardi MT, Kobľizek M, Masojídek J (2001) Photosystem II-based biosensors for the detection of pollutants. Biosens Bioelectron 16:1027–1033

Giardi MT, Guzzella L, Euzet P, Rouillon R, Esposito D (2005) Detection of herbicide subclasses by an optical multibiosensor based on an array of photosystem II mutants. Environ Sci Technol 39:5378–5384

Giardi MT, Scognamiglio V, Rea G, Rodio G, Antonacci A, Lambreva M, Pezzotti G, Johanningmeier U (2009) Optical biosensors for environmental monitoring based on computational and biotechnological tools for engineering the photosynthetic D1 protein of Chlamydomonas reinhardtii. Biosens Bioelectron 25:294–300

Girigoswami K, Akhtar N (2019) Nanobiosensors and fluorescence based biosensors: an overview. Int J Nano Dimens 10:1–17

Gong Z, Guo Y, Sun X, Cao Y, Wang X (2014) Acetylcholinesterase biosensor for carbaryl detection based on interdigitated array microelectrodes. Bioprocess Biosyst Eng 37:1929–1934

González-Techera A, Zon MA, Molina PG, Fernández H, González-Sapienza G, Arévalo FJ (2015) Development of a highly sensitive noncompetitive electrochemical immunosensor for the detection of atrazine by phage anti-immunocomplex assay. Biosens Bioelectron 64:650–656

Grieshaber D, MacKenzie R, Vörös J, Reimhult E (2008) Electrochemical biosensors – sensor principles and architectures. Sensors (Basel, Switzerland) 8:1400–1458

Gruzina TG, Zadorozhnyaya AM, Gutnik GA, Vember VV, Ulberg ZR, Kanyuk NI, Starodub NF (2007) A bacterial multisensor for determination of the contents of heavy metals in water. J Water Chem Technol 29:50–53

Gui Q, Lawson T, Shan S, Yan L, Liu Y (2017) The application of whole cell-based biosensors for use in environmental analysis and in medical diagnostics. Sensors (Basel, Switzerland) 17:1623

Guo Z, Duan J, Yang F, Li M, Hao T, Wang S, Wei D (2012) A test strip platform based on DNA-functionalized gold nanoparticles for on-site detection of mercury (II) ions. Talanta 93:49–54

Guo L, Li Z, Chen H, Wu Y, Chen L, Song Z, Lin T (2017) Colorimetric biosensor for the assay of paraoxon in environmental water samples based on the iodine-starch color reaction. Anal Chim Acta 967:59–63

Haigh-Flórez D, de la Hera C, Costas E, Orellana G (2014) Microalgae dual-head biosensors for selective detection of herbicides with fiber-optic luminescent O2 transduction. Biosens Bioelectron 54:484–491

Han S, Zhu M, Yuan Z, Li X (2001) A methylene blue-mediated enzyme electrode for the determination of trace mercury(II), mercury(I), methylmercury, and mercury–glutathione complex. Biosens Bioelectron 16:9–16

Hashimoto Y, Nakamura H, Asaga K, Karube I (2008) A new diagnostic method for soil-borne disease using a microbial biosensor. Microbes Environ 23:35–39

Hayman RB (2008) Fiber optic biosensors for bacterial detection. In: Zourob M, Elwary S, Turner A (eds) Principles of bacterial detection: biosensors, recognition receptors and microsystems. Springer, New York, pp 125–137

He S, Feng Y (2017) The impact of engineered nanomaterials on crops and soil microorganisms. In: Prasad R, Kumar V, Kumar M (eds) Nanotechnology: food and environmental paradigm. Springer, Singapore, pp 191–208

He Y, Zhang X, Zeng K, Zhang S, Baloda M, Gurung AS, Liu G (2011) Visual detection of Hg^{2+} in aqueous solution using gold nanoparticles and thymine-rich hairpin DNA probes. Biosens Bioelectron 26:4464–4470

He M-Q, Wang K, Wang J, Yu Y-L, He R-H (2017) A sensitive aptasensor based on molybdenum carbide nanotubes and label-free aptamer for detection of bisphenol A. Anal Bioanal Chem 409:1797–1803

Hleli S, Martelet C, Abdelghani A, Burais N, Jaffrezic-Renault N (2006) Atrazine analysis using an impedimetric immunosensor based on mixed biotinylated self-assembled monolayer. Sensors Actuators B Chem 113:711–717

Hollis RP, Killham K, Glover LA (2000) Design and application of a biosensor for monitoring toxicity of compounds to eukaryotes. Appl Environ Microbiol 66:1676–1679

Husu I, Rodio G, Touloupakis E, Lambreva MD, Buonasera K, Litescu SC, Giardi MT, Rea G (2013) Insights into photo-electrochemical sensing of herbicides driven by Chlamydomonas reinhardtii cells. Sensors Actuators B Chem 185:321–330

Ilangovan R, Daniel D, Krastanov A, Zachariah C, Elizabeth R (2006) Enzyme based biosensor for heavy metal ions determination. Biotechnol Biotechnol Equip 20:184–189

Ispas CR, Crivat G, Andreescu S (2012) Review: recent developments in enzyme-based biosensors for biomedical analysis. Anal Lett 45:168–186

Istamboulie G, Andreescu S, Marty J-L, Noguer T (2007) Highly sensitive detection of organophosphorus insecticides using magnetic microbeads and genetically engineered acetylcholinesterase. Biosens Bioelectron 23:506–512

Istamboulie G, Fournier D, Marty J-L, Noguer T (2009) Phosphotriesterase: a complementary tool for the selective detection of two organophosphate insecticides: chlorpyrifos and chlorfenvinfos. Talanta 77:1627–1631

Jaffrezic-Renault N, Dzyadevych SV (2008) Conductometric microbiosensors for environmental monitoring. Sensors (Basel, Switzerland) 8:2569–2588

Jatav GK, Nirmal DE (2013) Application of nano-technology in soil-plant system. Asian J Soil Sci 8:176–184

Jeyapragasam T, Saraswathi R (2014) Electrochemical biosensing of carbofuran based on acetylcholinesterase immobilized onto iron oxide–chitosan nanocomposite. Sensors Actuators B Chem 191:681–687

Ji X, Zheng J, Xu J, Rastogi VK, Cheng T-C, DeFrank JJ, Leblanc RM (2005) (CdSe)ZnS quantum dots and organophosphorus hydrolase bioconjugate as biosensors for detection of paraoxon. J Phys Chem B 109:3793–3799

Jiang D, Du X, Liu Q, Zhou L, Dai L, Qian J, Wang K (2015) Silver nanoparticles anchored on nitrogen-doped graphene as a novel electrochemical biosensing platform with enhanced sensitivity for aptamer-based pesticide assay. Analyst 140:6404–6411

Jiao Y, Hou W, Fu J, Guo Y, Sun X, Wang X, Zhao J (2017) A nanostructured electrochemical aptasensor for highly sensitive detection of chlorpyrifos. Sensors Actuators B Chem 243:1164–1170

Joe M-H, Lee K-H, Lim S-Y, Im S-H, Song H-P, Lee IS, Kim D-H (2012) Pigment-based whole-cell biosensor system for cadmium detection using genetically engineered Deinococcus radiodurans. Bioprocess Biosyst Eng 35:265–272

Justino CIL, Duarte AC, Rocha-Santos TAP (2017) Recent progress in biosensors for environmental monitoring: a review. Sensors (Basel, Switzerland) 17:2918

Kabariya JH, Ramani VM (2017) Nanobiosensors, as a next-generation diagnostic device for quality & safety of food and dairy product. In: Prasad R, Kumar V, Kumar M (eds) Nanotechnology: food and environmental paradigm. Springer, Singapore, pp 115–129

Kang T, Yoo SM, Yoon I, Lee S, Choo J, Lee SY, Kim B (2011) Au nanowire-on-film SERRS sensor for ultrasensitive Hg2+ detection. Chem Eur J 17:2211–2214

Kanjana D (2017) Advancement of nanotechnology applications on plant nutrients management and soil improvement. In: Prasad R, Kumar V, Kumar M (eds) Nanotechnology: food and environmental paradigm. Springer, Singapore, pp 209–234

Karnati C, Du H, Ji H-F, Xu X, Lvov Y, Mulchandani A, Mulchandani P, Chen W (2007) Organophosphorus hydrolase multilayer modified microcantilevers for organophosphorus detection. Biosens Bioelectron 22:2636–2642

Kaushal M, Wani S (2017) Nanosensors: frontiers in precision agriculture. In: Nanotechnolgy. Springer, Singapore, pp 279–291

Kaushik A, Solanki PR, Ansari AA, Malhotra BD, Ahmad S (2009) Iron oxide-chitosan hybrid nanobiocomposite based nucleic acid sensor for pyrethroid detection. Biochem Eng J 46:132–140

Kavita V (2017) DNA biosensors-a review. J Bioeng Biomed Sci 7:222

Khadivi Derakhshan F, Dehnad A, Salouti M (2012) Extracellular biosynthesis of gold nanoparticles by metal resistance bacteria: *Streptomyces griseus*. Synth React Inorg Met-Org Nano-Metal Chem 42:868–871

Khaledian S, Nikkhah M, Shams-bakhsh M, Hoseinzadeh S (2017) A sensitive biosensor based on gold nanoparticles to detect Ralstonia solanacearum in soil. J Gen Plant Pathol 83:231–239

Khan M, Fatima T (2014) Nanotechnology: scope and application in plant disease management. Plant Pathol J 13:214–231

Khiyami MA, Almoammar H, Awad YM, Alghuthaymi MA, Abd-Elsalam KA (2014) Plant pathogen nanodiagnostic techniques: forthcoming changes? Biotechnol Biotechnol Equip 28:775–785

Khor BH, Ismail AK, Ahamad R, Shahir S (2015) A redox mediated UME biosensor using immobilized Chromobacterium violaceum strain R1 for rapid biochemical oxygen demand measurement. Electrochim Acta 176:777–783

Kim M, Lim JW, Kim HJ, Lee SK, Lee SJ, Kim T (2015) Chemostat-like microfluidic platform for highly sensitive detection of heavy metal ions using microbial biosensors. Biosens Bioelectron 65:257–264

Kim HJ, Lim JW, Jeong H, Lee S-J, Lee D-W, Kim T, Lee SJ (2016) Development of a highly specific and sensitive cadmium and lead microbial biosensor using synthetic CadC-T7 genetic circuitry. Biosens Bioelectron 79:701–708

Knecht MR, Sethi M (2009) Bio-inspired colorimetric detection of Hg2+ and Pb2+ heavy metal ions using Au nanoparticles. Anal Bioanal Chem 394:33–46

Koblizek M, Masojidek J, Komenda J, Kucera T, Pilloton R, Mattoo AK, Giardi MT (1998) A sensitive photosystem II-based biosensor for detection of a class of herbicides. Biotechnol Bioeng 20:664–669

Koblížek M, Malý J, Masojídek J, Komenda J, Kučera T, Giardi MT, Mattoo AK, Pilloton R (2002) A biosensor for the detection of triazine and phenylurea herbicides designed using Photosystem II coupled to a screen-printed electrode. Biotechnol Bioeng 78:110–116

Koedrith P, Thasiphu T, Weon J-I, Boonprasert R, Tuitemwong K, Tuitemwong P (2015) Recent trends in rapid environmental monitoring of pathogens and toxicants: potential of nanoparticle-based biosensor and applications. Sci World J 2015:12

Kulagina NV, Shankar L, Michael AC (1999) Monitoring glutamate and ascorbate in the extracellular space of brain tissue with electrochemical microsensors. Anal Chem 71:5093–5100

Kuswandi B (2003) Simple optical fibre biosensor based on immobilised enzyme for monitoring of trace heavy metal ions. Anal Bioanal Chem 376:1104–1110

Kwak S-Y, Wong MH, Thomas Salim Lew T, Bisker G, Lee MA, Kaplan A, Dong J, Tianxiang Liu A, Koman VB, Sinclair R, Hamann C, Strano MS (2017) Nanosensor technology applied to living plant systems. Annu Rev Anal Chem 13:113–140

Kwok N-Y, Dong S, Lo W, Wong K-Y (2005) An optical biosensor for multi-sample determination of biochemical oxygen demand (BOD). Sensors Actuators B Chem 110:289–298

Lammers F, Scheper T (1999) Thermal biosensors in biotechnology. In: Bhatia PK, Danielsson B, Gemeiner P, Grabley S, Lammers F, Mukhopadhyay A, Ramanathan K, Saleemuddin M, Scheper T, Stefuca V, Thiericke R, Xie B (eds) Thermal biosensors, bioactivity, bioaffinitty. Springer, Berlin/Heidelberg, pp 35–67

Lang Q, Han L, Hou C, Wang F, Liu A (2016) A sensitive acetylcholinesterase biosensor based on gold nanorods modified electrode for detection of organophosphate pesticide. Talanta 156–157:34–41

Lara S, Perez-Potti A (2018) Review: applications of nanomaterials for immunosensing. Biosens Bioelectron 8:104–125

Lee MM, Russell DA (2003) Novel determination of cadmium ions using an enzyme self-assembled monolayer with surface plasmon resonance. Anal Chim Acta 500:119–125

Leth S, Maltoni S, Simkus R, Mattiasson B, Corbisier P, Klimant I, Wolfbeis O, Csöregi E (2002) Engineered bacteria based biosensors for monitoring bioavailable heavy metals. Electroanalysis 14:35–42

Li J, Chen L, Lou T, Wang Y (2011) Highly sensitive SERS detection of As3+ ions in aqueous media using glutathione functionalized silver nanoparticles. ACS Appl Mater Interfaces 3:3936–3941

Li M, Zhou X, Guo S, Wu N (2013) Detection of lead (II) with a “turn-on” fluorescent biosensor based on energy transfer from CdSe/ZnS quantum dots to graphene oxide. Biosens Bioelectron 43:69–74

Li M, Kong Q, Bian Z, Ma C, Ge S, Zhang Y, Yu J, Yan M (2015) Ultrasensitive detection of lead ion sensor based on gold nanodendrites modified electrode and electrochemiluminescent quenching of quantum dots by electrocatalytic silver/zinc oxide coupled structures. Biosens Bioelectron 65:176–182

Li Y, Shi L, Han G, Xiao Y, Zhou W (2017a) Electrochemical biosensing of carbaryl based on acetylcholinesterase immobilized onto electrochemically inducing porous graphene oxide network. Sensors Actuators B Chem 238:945–953

Li Z, Qu S, Cui L, Zhang S (2017b) Detection of carbofuran pesticide in seawater by using an enzyme biosensor. J Coast Res 80:1–5

Lim JW, Ha D, Lee J, Lee SK, Kim T (2015a) Review of micro/nanotechnologies for microbial biosensors. Front Bioeng Biotechnol 3:1–13

Lim JW, Ha D, Lee J, Lee SK, Kim T (2015b) Review of micro/nanotechnologies for microbial biosensors. Front Bioeng Biotechnol 3:61–61

Lin T-J, Chung M-F (2009) Detection of cadmium by a fiber-optic biosensor based on localized surface plasmon resonance. Biosens Bioelectron 24:1213–1218

Lin T-J, Huang K-T, Liu C-Y (2006) Determination of organophosphorous pesticides by a novel biosensor based on localized surface plasmon resonance. Biosens Bioelectron 22:513–518

Lin H-Y, Huang C-H, Lu S-H, Kuo IT, Chau L-K (2014) Direct detection of orchid viruses using nanorod-based fiber optic particle plasmon resonance immunosensor. Biosens Bioelectron 51:371–378

Lisa M, Chouhan RS, Vinayaka AC, Manonmani HK, Thakur MS (2009) Gold nanoparticles based dipstick immunoassay for the rapid detection of dichlorodiphenyltrichloroethane: an organochlorine pesticide. Biosens Bioelectron 25:224–227

Liu Z, Liu Y, Yang H, Yang Y, Shen G, Yu R (2005) A phenol biosensor based on immobilizing tyrosinase to modified core-shell magnetic nanoparticles supported at a carbon paste electrode. Anal Chim Acta 533:3–9

Liu P, Huang Q, Chen W (2012) Construction and application of a zinc-specific biosensor for assessing the immobilization and bioavailability of zinc in different soils. Environ Pollut 164:66–72

Liu X, Li W-J, Li L, Yang Y, Mao L-G, Peng Z (2014) A label-free electrochemical immunosensor based on gold nanoparticles for direct detection of atrazine. Sensors Actuators B Chem 191:408–414

Liu Q, Fei A, Huan J, Mao H, Wang K (2015) Effective amperometric biosensor for carbaryl detection based on covalent immobilization acetylcholinesterase on multiwall carbon nanotubes/graphene oxide nanoribbons nanostructure. J Electroanal Chem 740:8–13

Long F, Zhu A, Shi H (2013a) Recent advances in optical biosensors for environmental monitoring and early warning. Sensors (Basel, Switzerland) 13:13928–13948

Long F, Zhu A, Shi H, Wang H, Liu J (2013b) Rapid on-site/in-situ detection of heavy metal ions in environmental water using a structure-switching DNA optical biosensor. Sci Rep 3:2308

Lopez MM, Llop P, Olmos A, Marco-Noales E, Cambra M, Bertolini E (2009) Are molecular tools solving the challenges posed by detection of plant pathogen bacteria and viruses? Curr Issues Mol Biol 11:13–46

Lu L, Guo L, Li J, Kang T, Cheng S (2016) Electrochemiluminescent detection of Pb2+ by graphene/gold nanoparticles and CdSe quantum dots. Appl Surf Sci 388:431–436

Luan E, Zheng Z, Li X, Gu H, Liu S (2016) Inkjet-assisted layer-by-layer printing of quantum dot/enzyme microarrays for highly sensitive detection of organophosphorous pesticides. Anal Chim Acta 916:77–83

Lukyanenko KA, Denisov IA, Sorokin VV, Yakimov AS, Esimbekova EN, Belobrov PI (2019) Handheld enzymatic luminescent biosensor for rapid detection of heavy metals in water samples. Chemosensors 7:16

Ma W, Sun M, Xu L, Wang L, Kuang H, Xu C (2013) A SERS active gold nanostar dimer for mercury ion detection. Chem Commun 49:4989–4991

Madasamy T, Pandiaraj M, Balamurugan M, Bhargava K, Sethy NK, Karunakaran C (2014) Copper, zinc superoxide dismutase and nitrate reductase coimmobilized bienzymatic biosensor for the simultaneous determination of nitrite and nitrate. Biosens Bioelectron 52:209–215

Maderova L, Paton GI (2013) Deployment of microbial sensors to assess zinc bioavailability and toxicity in soils. Soil Biol Biochem 66:222–228

Madianos L, Tsekenis G, Skotadis E, Patsiouras L, Tsoukalas D (2018) A highly sensitive impedimetric aptasensor for the selective detection of acetamiprid and atrazine based on microwires formed by platinum nanoparticles. Biosens Bioelectron 101:268–274

Malik P, Katyal V, Malik V, Asatkar A, Inwati G, Mukherjee TK (2013) Nanobiosensors: concepts and variations. ISRN Nanomater 2013:9

Mallat E, Barzen C, Abuknesha R, Gauglitz G, Barceló D (2001) Fast determination of paraquat residues in water by an optical immunosensor and validation using capillary electrophoresis-ultraviolet detection. Anal Chim Acta 427:165–171

Maly J, Masojidek J, Masci A, Ilie M, Cianci E, Foglietti V, Vastarella W, Pilloton R (2005) Direct mediatorless electron transport between the monolayer of photosystem II and poly(mercapto-p-benzoquinone) modified gold electrode—new design of biosensor for herbicide detection. Biosens Bioelectron 21:923–932

Margarita S, Velizar G, Zdravka V (2016) Electrochemical biosensors for direct determination of organophosphorus pesticides: a review. Curr Anal Chem 12:37–42

Marques I, da Costa JP, Justino C, Santos P, Duarte K, Freitas A, Cardoso S (2017) Carbon nanotube field effect biosensor for the detection of toxins in seawater. Anal Chem 97:597–605

Martín M, Salazar P, Jiménez C, Lecuona M, Ramos MJ, Ode J, Alcoba J, Roche R, Villalonga R, Campuzano S, Pingarrón JM, González-Mora JL (2015) Rapid Legionella pneumophila determination based on a disposable core–shell Fe3O4@poly(dopamine) magnetic nanoparticles immunoplatform. Anal Chim Acta 887:51–58

Martinazzo J, Muenchen DK, de Cezaro A, Nava A, Rigo AM, Leite F, Steffens C, Steffens J (2018) Pesticide detection in soil using biosensors and nanobiosensors. Biointerface Res Appl Chem 6(6):1659–1675

Martinelli F, Scalenghe R, Davino S, Panno S, Scuderi G, Ruisi P, Villa P, Stroppiana D, Boschetti M, Goulart LR, Davis CE, Dandekar AM (2015) Advanced methods of plant disease detection. A review. Agron Sustain Dev 35:1–25

Masojídek J, Souček P, Máchová J, Frolík J, Klem K, Malý J (2011) Detection of photosynthetic herbicides: algal growth inhibition test vs. electrochemical photosystem II biosensor. Ecotoxicol Environ Saf 74:117–122

Mayorga-Martinez CC, Pino F, Kurbanoglu S, Rivas L, Ozkan SA, Merkoçi A (2014) Iridium oxide nanoparticle induced dual catalytic/inhibition based detection of phenol and pesticide compounds. J Mater Chem B 2:2233–2239

McGrath SP, Knight B, Killham K, Preston S, Paton GI (1999) Assessment of the toxicity of metals in soils amended with sewage sludge using a chemical speciation technique and a lux-based biosensor. Environ Toxicol Chem 18:659–663

McPartlin DA, Lochhead MJ, Connell LB, Doucette GJ, O'Kennedy RJ (2016) Use of biosensors for the detection of marine toxins. Essays Biochem 60:49–58

Mehrotra P (2016) Biosensors and their applications – a review. J Oral Biol Craniofacial Res 6:153–159

Meng X, Wei J, Ren X, Ren J, Tang F (2013) A simple and sensitive fluorescence biosensor for detection of organophosphorus pesticides using H2O2-sensitive quantum dots/bi-enzyme. Biosens Bioelectron 47:402–407

Mishra A, Kumar J, Melo JS (2017) An optical microplate biosensor for the detection of methyl parathion pesticide using a biohybrid of Sphingomonas sp. cells-silica nanoparticles. Biosens Bioelectron 87:332–338

Mogha NK, Sahu V, Sharma M, Sharma RK, Masram DT (2016) Biocompatible ZrO2- reduced graphene oxide immobilized AChE biosensor for chlorpyrifos detection. Mater Des 111:312–320

Mohammadi H, Amine A, Cosnier S, Mousty C (2005) Mercury–enzyme inhibition assays with an amperometric sucrose biosensor based on a trienzymatic-clay matrix. Anal Chim Acta 543:143–149

Mohammadi-Aloucheh R, Alaee Mollabashi Y, Asadi A, Baris O, Golamzadeh S (2018) The role of nanobiosensors in identifying pathogens and environmental hazards. Anthropogenic Pollut J 2:16–25

Mohanty SP, Kougianos E (2006) Biosensors: a tutorial review. IEEE Potentials 25:35–40

Moon H, Chang IS, Kang KH, Jang JK, Kim BH (2004) Improving the dynamic response of a mediator-less microbial fuel cell as a biochemical oxygen demand (BOD) sensor. Biotechnol Lett 26:1717–1721

Morgan CL, Newman DJ, Price CP (1996) Immunosensors: technology and opportunities in laboratory medicine. Clin Chem 42:193–209

Moro L, Pezzotti G, Turemis M, Sanchís J, Farré M, Denaro R, Giacobbe MG, Crisafi F, Giardi MT (2018) Fast pesticide pre-screening in marine environment using a green microalgae-based optical bioassay. Mar Pollut Bull 129:212–221

Mourzina IG, Yoshinobu T, Ermolenko YE, Vlasov YG, Schöning MJ, Iwasaki H (2004) Immobilization of urease and cholinesterase on the surface of semiconductor transducer for the development of light-addressable potentiometric sensors. Microchim Acta 144:41–50

Mulchandani A, Chen W, Mulchandani P, Wang J, Rogers KR (2001) Biosensors for direct determination of organophosphate pesticides. Biosens Bioelectron 16:225–230

Mura S, Greppi G, Roggero PP, Musu E, Pittalis D, Carletti A, Ghiglieri G, Irudayaraj J (2015) Functionalized gold nanoparticles for the detection of nitrates in water. Int J Environ Sci Technol 12:1021–1028

Naessens M, Leclerc J, Tran-Minh C (2000) Fiber optic biosensor using Chlorella vulgaris for determination of toxic compounds. Ecotoxicol Environ Saf 46:181–185

Nakamura H, Karube I (2003) Current research activity in biosensors. Anal Bioanal Chem 377:446–468

Nakamura H, Kobayashi S, Hirata Y, Suzuki K, Mogi Y, Karube I (2007) A spectrophotometric biochemical oxygen demand determination method using 2,6-dichlorophenolindophenol as the redox color indicator and the eukaryote Saccharomyces cerevisiae. Anal Biochem 369:168–174

Narsaiah K, Jha SN, Bhardwaj R, Sharma R, Kumar R (2012) Optical biosensors for food quality and safety assurance-a review. J Food Sci Technol 49:383–406

Neethirajan S, Jayas DS (2007) Sensors for grain storage. ASABE Annual International Meeting, Technical Papers. https://doi.org/10.13031/2013.23510

Nepomuscene N, Daniel D, Krastanov A (2007) Biosensor to detect chromium in wastewater. Biotechnol Biotechnol Equip 21:377–381

Niu X, Zhong Y, Chen R, Wang F, Liu Y, Luo D (2018) A "turn-on" fluorescence sensor for Pb2+ detection based on graphene quantum dots and gold nanoparticles. Sensors Actuators B Chem 255:1577–1581

Niyomdecha S, Limbut W, Numnuam A, Asawatreratanakul P, Kanatharana P, Thavarungkul P (2017) A novel BOD biosensor based on entrapped activated sludge in a porous chitosan-albumin cryogel incorporated with graphene and methylene blue. Sensors Actuators B Chem 241:473–481

Nsibande SA, Forbes PBC (2016) Fluorescence detection of pesticides using quantum dot materials – a review. Anal Chim Acta 945:9–22

Nunes GS, Montesinos T, Marques PBO, Fournier D, Marty JL (2001) Acetylcholine enzyme sensor for determining methamidophos insecticide: evaluation of some genetically modified acetylcholinesterases from Drosophila melanogaster. Anal Chim Acta 434:1–8

Nunes GS, Lins JAP, Silva FGS, Araujo LC, Silva FEPS, Mendonça CD, Badea M, Hayat A, Marty JL (2014) Design of a macroalgae amperometric biosensor; application to the rapid monitoring of organophosphate insecticides in an agroecosystem. Chemosphere 111:623–630

Paitan Y, Biran D, Biran I, Shechter N, Babai R, Rishpon J, Ron EZ (2003) On-line and in situ biosensors for monitoring environmental pollution. Biotechnol Adv 22:27–33

Pal P, Bhattacharyay D, Mukhopadhyay A, Sarkar P (2009) The detection of mercury, cadium, and arsenic by the deactivation of urease on rhodinized carbon. Environ Eng Sci 26:25–32

Pan Y, Zhou J, Su K, Hu N, Wang P (2017) A novel quantum dot fluorescence immunosensor based on magnetic beads and portable flow cytometry for detection of okadaic acid. Proced Technol 27:214–216

Pan J, Li Q, Zhou D, Chen J (2018) Ultrasensitive aptamer biosensor for arsenic (III) detection based on label-free triple-helix molecular switch and fluorescence sensing platform. Talanta 189:370–376

Pandit S, Dasgupta D, Dewan N, Ahmed A (2016) Nanotechnology based biosensors and its application. Pharma Innov J 5:18–25

Pang H-L, Kwok N-Y, Chan P-H, Yeung C-H, Lo W, Wong K-Y (2007) High-throughput determination of biochemical oxygen demand (BOD) by a microplate-based biosensor. Environ Sci Technol 41:4038–4044

Park EJ, Lee J-Y, Kim JH, Lee CJ, Kim HS, Min NK (2010) Investigation of plasma-functionalized multiwalled carbon nanotube film and its application of DNA sensor for Legionella pneumophila detection. Talanta 82:904–911

Park J, You X, Jang Y, Nam Y, Kim MJ, Min NK, Pak JJ (2014) ZnO nanorod matrix based electrochemical immunosensors for sensitivity enhanced detection of Legionella pneumophila. Sensors Actuators B Chem 200:173–180

Park M, Ha HD, Kim YT, Jung JH, Kim S-H, Kim DH, Seo TS (2015) Combination of a sample pretreatment microfluidic device with a photoluminescent graphene oxide quantum dot sensor for trace lead detection. Anal Chem 87:10969–10975

Patel N, Desai P, Patel NN, Jha A, Gautam HK (2014) Agronanotechnology for plant fungal disease management : a review. Int J Curr Microbiol Appl Sci 3:71–84

Pedrosa VA, Paliwal S, Balasubramanian S, Nepal D, Davis V, Wild J, Ramanculov E, Simonian A (2010) Enhanced stability of enzyme organophosphate hydrolase interfaced on the carbon nanotubes. Colloids Surf B: Biointerfaces 77:69–74

Peng L, Dong S, Wei W, Yuan X, Huang T (2017) Synthesis of reticulated hollow spheres structure NiCo2S4 and its application in organophosphate pesticides biosensor. Biosens Bioelectron 92:563–569

Piletska EV, Piletsky SA, Rouillon R (2006) Application of chloroplast D1 protein in biosensors for monitoring photosystem II-inhibiting herbicides. In: Biotechnological applications of photosynthetic proteins: biochips, biosensors and biodevices. Springer, Boston, pp 130–146

Pogačnik L, Franko M (1999) Determination of organophosphate and carbamate pesticides in spiked samples of tap water and fruit juices by a biosensor with photothermal detection. Biosens Bioelectron 14:569–578

Pospíšilová M, Kuncová G, Trögl J (2015) Fiber-optic chemical sensors and fiber-optic biosensors. Sensors (Basel, Switzerland) 15:25208–25259

Prabhakar N, Arora K, Singh SP, Pandey MK, Singh H, Malhotra BD (2007) Polypyrrole-polyvinyl sulphonate film based disposable nucleic acid biosensor. Anal Chim Acta 589:6–13

Prakash O, Talat M, Hasan SH, Pandey RK (2008) Enzymatic detection of mercuric ions in ground-water from vegetable wastes by immobilizing pumpkin (Cucumis melo) urease in calcium alginate beads. Bioresour Technol 99:4524–4528

Prasad R, Bhattacharyya A, Nguyen QD (2017) Nanotechnology in sustainable agriculture: recent developments, challenges, and perspectives. Front Microbiol 8:1014–1014

Přibyl J, Hepel M, Halámek J, Skládal P (2003) Development of piezoelectric immunosensors for competitive and direct determination of atrazine. Sensors Actuators B Chem 91:333–341

Qi Y, Xiu F-R, Yu G, Huang L, Li B (2017) Simple and rapid chemiluminescence aptasensor for Hg2+ in contaminated samples: a new signal amplification mechanism. Biosens Bioelectron 87:439–446

Rad F, Mohsenifar A, Tabatabaei M, Safarnejad MR, Shahryari F, Safarpour H, Foroutan A, Mardi M, Davoudi D, Fotokian M (2012) Detection of Candidatus Phytoplasma aurantifolia with a quantum dots fret-based biosensor. J Plant Pathol 94:525–534

Ragavan KV, Selvakumar LS, Thakur MS (2013) Functionalized aptamers as nano-bioprobes for ultrasensitive detection of bisphenol-A. Chem Commun 49:5960–5962

Rai V, Acharya S, Dey N (2012) Implications of nanobiosensors in agriculture. J Biomater Nanobiotechnol 03(02):10

Rajkumar P, Ramprasath T, Selvam GS (2017) 12 – A simple whole cell microbial biosensors to monitor soil pollution. In: Grumezescu AM (ed) New pesticides and soil sensors. Academic, New York, pp 437–481

Ramanathan K, Rank M, Svitel J, Dzgoev A, Danielsson B (1999) The development and applications of thermal biosensors for bioprocess monitoring. Trends Biotechnol 17:499–505

Rasmussen M, Minteer SD (2013) Self-powered herbicide biosensor utilizing thylakoid membranes. Anal Methods 5:1140–1144

Rasmussen M, Wingersky A, Minteer SD (2014) Comparative study of thylakoids from higher plants for solar energy conversion and herbicide detection. Electrochim Acta 140:304–308

Ravikumar S, Ganesh I, Yoo I-k, Hong SH (2012) Construction of a bacterial biosensor for zinc and copper and its application to the development of multifunctional heavy metal adsorption bacteria. Process Biochem 47:758–765

Ravikumar A, Panneerselvam P, Radhakrishnan K, Morad N, Anuradha CD, Sivanesan S (2017) DNAzyme based amplified biosensor on ultrasensitive fluorescence detection of Pb (II) ions from aqueous system. J Fluoresc 27:2101–2109

Reshetilov A, Arlyapov V, Alferov V, Reshetilova T (2013) BOD biosensors: application of novel technologies and prospects for the development. In: Rinken T (ed) State of the art in biosensors – environmental and medical applications. IntechOpen, Rijeka

Ricci F, Adornetto G, Palleschi G (2012) A review of experimental aspects of electrochemical immunosensors. Electrochim Acta 84:74–83

Rizzuto M, Polcaro C, Desiderio C, Koblizek M, Pilloton R, Giardi MT (2000) Herbicide monitoring in surface water samples with a Photosystem-II based biosensor. In: Proceedings of the second workshop on chemical sensors and biosensors. ENEA, Rome, pp 346–357

Rodriguez M, Sanders CA, Greenbaum E (2002) Biosensors for rapid monitoring of primary-source drinking water using naturally occurring photosynthesis. Biosens Bioelectron 17:843–849

Rodriguez BB, Bolbot JA, Tothill IE (2004) Urease–glutamic dehydrogenase biosensor for screening heavy metals in water and soil samples. Anal Bioanal Chem 380:284–292

Rodríguez-Mozaz S, Marco MP, Lopez de Alda M, BarcelÃ D (2004) Biosensors for environmental applications: future development trends. Pure Appl Chem 76(4):723–752

Ronkainen NJ, Halsall HB, Heineman WR (2010) Electrochemical biosensors. Chem Soc Rev 39:1747–1763

Rouillon R, Piletsky SA, Breton F, Piletska EV, Carpentier R (2006) Photosystem II biosensors for heavy metals monitoring. In: Biotechnological applications of photosynthetic proteins: biochips, biosensors and biodevices. Springer, Boston, pp 166–174

Sadeghi SJ (2013) Amperometric biosensors. In: Roberts GCK (ed) Encyclopedia of biophysics. Springer, Berlin/Heidelberg, pp 61–67

Safarpour H, Safarnejad MR, Tabatabaei M, Mohsenifar A, Rad F, Basirat M, Shahryari F, Hasanzadeh F (2012) Development of a quantum dots FRET-based biosensor for efficient detection of Polymyxa betae. Can J Plant Pathol 34:507–515

Sakaguchi T, Morioka Y, Yamasaki M, Iwanaga J, Beppu K, Maeda H, Morita Y, Tamiya E (2007) Rapid and onsite BOD sensing system using luminous bacterial cells-immobilized chip. Biosens Bioelectron 22:1345–1350

Salgado AM, Silva LM, Melo AF (2011) Biosensor for environmental applications. In: Somerset V (ed) Environmental biosensors. IntechOpen, Rijeka, pp 4–14

Samendra PS, Masaaki K, Charles PG, Ian LP (2014) Rapid detection technologies for monitoring microorganisms in water. Biosens J 3:009

Sang S, Zhang W, Zhao Y (2013) Review on the design art of biosensors. In: Rinken T (ed) State of the art in biosensors-general aspects. IntechOpen, Rijeka

Santos CS, Mossanha R, Pessôa CA (2015) Biosensor for carbaryl based on gold modified with PAMAM-G4 dendrimer. J Appl Electrochem 45:325–334

Sawant SN (2017) 13 – development of biosensors from biopolymer composites. In: Sadasivuni KK, Ponnamma D, Kim J, Cabibihan JJ, AlMaadeed MA (eds) Biopolymer composites in electronics. Elsevier, Cambridge, pp 353–383

Schöning MJ, Krause R, Block K, Musahmeh M, Mulchandani A, Wang J (2003) A dual amperometric/potentiometric FIA-based biosensor for the distinctive detection of organophosphorus pesticides. Sensors Actuators B Chem 95:291–296

Scognamiglio V, Raffi D, Lambreva M, Rea G, Tibuzzi A, Pezzotti G, Johanningmeier U, Giardi MT (2009) *Chlamydomonas reinhardtii* genetic variants as probes for fluorescence sensing system in detection of pollutants. Anal Bioanal Chem 394:1081

Scognamiglio V, Pezzotti I, Pezzotti G, Cano J, Manfredonia I, Buonasera K, Arduini F, Moscone D, Palleschi G, Giardi MT (2012) Towards an integrated biosensor array for simultaneous and rapid multi-analysis of endocrine disrupting chemicals. Anal Chim Acta 751:161–170

Scognamiglio V, Pezzotti I, Pezzotti G, Cano J, Manfredonia I, Buonasera K, Rodio G, Giardi MT (2013) A new embedded biosensor platform based on micro-electrodes array (MEA) technology. Sensors Actuators B Chem 176:275–283

Sekhon BS (2014) Nanotechnology in agri-food production: an overview. Nanotechnol Sci Appl 7:31–53

Seo S, Dobozi-King M, Young RF, Kish LB, Cheng M (2008) Patterning a nanowell sensor biochip for specific and rapid detection of bacteria. Microelectron Eng 85:1484–1489

Shahbazi R, Salouti M, Amini B, Jalilvand A, Naderlou E, Amini A, Shams A (2018) Highly selective and sensitive detection of Staphylococcus aureus with gold nanoparticle-based core-shell nano biosensor. Mol Cell Probes 41:8–13

Shao CY, Howe CJ, Porter AJR, Glover LA (2002) Novel Cyanobacterial Biosensor For Detection Of Herbicides. Appl Environ Microbiol 68:5026–5033

Shi H, Zhao G, Liu M, Fan L, Cao T (2013) Aptamer-based colorimetric sensing of acetamiprid in soil samples: sensitivity, selectivity and mechanism. J Hazard Mater 260:754–761

Shi L, Wang Y, Chu Z, Yin Y, Jiang D, Luo J, Ding S, Jin W (2017) A highly sensitive and reusable electrochemical mercury biosensor based on tunable vertical single-walled carbon nanotubes and a target recycling strategy. J Mater Chem B 5:1073–1080

Shitanda I, Takamatsu S, Watanabe K, Itagaki M (2009) Amperometric screen-printed algal biosensor with flow injection analysis system for detection of environmental toxic compounds. Electrochim Acta 54:4933–4936

Shofiul Azam MD, Tanver Rahman MDR, Lou Z, Jothi JS (2014) Review: advancements and application of immunosensors in the analysis of food contaminants. Nusantara Biosci 6:186–195

Shtenberg G, Massad-Ivanir N, Segal E (2015) Detection of trace heavy metal ions in water by nanostructured porous Si biosensors. Analyst 140:4507–4514

Shyuan LK, Heng LY, Ahmad M, Aziz SA, Ishak Z (2008) Evaluation of pesticide and heavy metal toxicity using immobilized enzyme alkaline phosphatase with an electrochemical biosensor. Asian J Biochem 3:359–365

Siddiquee S, Rovina K, Yusof NA, Rodrigues KF, Suryani S (2014) Nanoparticle-enhanced electrochemical biosensor with DNA immobilization and hybridization of Trichoderma harzianum gene. Sens Bio-Sens Res 2:16–22

Simonian AL, Grimsley JK, Flounders AW, Schoeniger JS, Cheng T-C, DeFrank JJ, Wild JR (2001) Enzyme-based biosensor for the direct detection of fluorine-containing organophosphates. Anal Chim Acta 442:15–23

Singh S, Singh M, Agrawal VV, Kumar A (2010) An attempt to develop surface plasmon resonance based immunosensor for Karnal bunt (Tilletia indica) diagnosis based on the experience of nano-gold based lateral flow immuno-dipstick test. Thin Solid Films 519:1156–1159

Singh S, Gupta AK, Gupta S, Gupta S, Kumar A (2014) Surface Plasmon Resonance (SPR) and cyclic voltammetry based immunosensor for determination of teliosporic antigen and diagnosis of Karnal Bunt of wheat using anti-teliosporic antibody. Sensors Actuators B Chem 191:866–873

Skottrup P, Hearty S, Frøkiær H, Leonard P, Hejgaard J, O'Kennedy R, Nicolaisen M, Justesen AF (2007) Detection of fungal spores using a generic surface plasmon resonance immunoassay. Biosens Bioelectron 22:2724–2729

Skottrup PD, Nicolaisen M, Justesen AF (2008) Towards on-site pathogen detection using antibody-based sensors. Biosens Bioelectron 24:339–348

Song T, Zhu X, Zhou S, Yang G, Gan W, Yuan Q (2015) DNA derived fluorescent bio-dots for sensitive detection of mercury and silver ions in aqueous solution. Appl Surf Sci 347:505–513

Song L, Mao K, Zhou X, Hu J (2016) A novel biosensor based on Au@Ag core–shell nanoparticles for SERS detection of arsenic (III). Talanta 146:285–290

Songa EA, Okonkwo JO (2016) Recent approaches to improving selectivity and sensitivity of enzyme-based biosensors for organophosphorus pesticides: a review. Talanta 155:289–304

Sotiropoulou S, Fournier D, Chaniotakis NA (2005) Genetically engineered acetylcholinesterase-based biosensor for attomolar detection of dichlorvos. Biosens Bioelectron 20:2347–2352

Srivastava AK, Dev A, Karmakar SJECL (2018) Nanosensors and nanobiosensors in food and agriculture. Environ Chem Lett 16:161–182

Steffens C, Steffens J, Marcia Graboski A, Manzoli A, Lima Leite F (2017) 16 – nanosensors for detection of pesticides in water. In: Grumezescu AM (ed) New pesticides and soil sensors. Academic, Cambridge, pp 595–635

Stiner L, Halverson LJ (2002) Development and characterization of a green fluorescent protein-based bacterial biosensor for bioavailable toluene and related compounds. Appl Environ Microbiol 68:1962–1971

Strosnider HM (2003) Whole-cell bacterial biosensors and the detection of bioavailable arsenic. In: National network of environmental management studies. U.S. Environmental Protection Agency, Washington, DC

Su Y-T, Lan G-Y, Chen W-Y, Chang H-T (2010) Detection of copper ions through recovery of the fluorescence of DNA-templated copper/silver nanoclusters in the presence of mercaptopropionic acid. Anal Chem 82:8566–8572

Su L, Jia W, Hou C, Lei Y (2011) Microbial biosensors: a review. Biosens Bioelectron 26:1788–1799

Sun C, Sun R, Chen Y, Tong Y, Zhu J, Bai H, Zhang S, Zheng H, Ye H (2018) Utilization of aptamer-functionalized magnetic beads for highly accurate fluorescent detection of mercury (II) in environment and food. Sensors Actuators B Chem 255:775–780

Sundarmurugasan R, Gumpu MB, Ramachandra BL, Nesakumar N, Sethuraman S, Krishnan UM, Rayappan JBB (2016) Simultaneous detection of monocrotophos and dichlorvos in orange samples using acetylcholinesterase–zinc oxide modified platinum electrode with linear regression calibration. Sensors Actuators B Chem 230:306–313

Suresh S, Periasamy M (2014) Recent trends in nanobiosensors and their applications -a review. Rev Adv Mater Sci 36:62–69

Suri CR, Kaur J, Gandhi S, Shekhawat GS (2008) Label-free ultra-sensitive detection of atrazine based on nanomechanics. Nanotechnology 19(23):235502

Tahirbegi IB, Ehgartner J, Sulzer P, Zieger S, Kasjanow A, Paradiso M, Strobl M, Bouwes D, Mayr T (2017) Fast pesticide detection inside microfluidic device with integrated optical pH, oxygen sensors and algal fluorescence. Biosens Bioelectron 88:188–195

Talan A, Mishra A, Eremin SA, Narang J, Kumar A, Gandhi S (2018) Ultrasensitive electrochemical immuno-sensing platform based on gold nanoparticles triggering chlorpyrifos detection in fruits and vegetables. Biosens Bioelectron 105:14–21

Tan L, Chen Z, Zhao Y, Wei X, Li Y, Zhang C, Wei X, Hu X (2016) Dual channel sensor for detection and discrimination of heavy metal ions based on colorimetric and fluorescence response of the AuNPs-DNA conjugates. Biosens Bioelectron 85:414–421

Tecon R, Van der Meer JR (2008) Bacterial Biosensors for measuring availability of environmental pollutants. Sensors (Basel, Switzerland) 8:4062–4080

Tekaya N, Saiapina O, Ben Ouada H, Lagarde F, Namour P, Ben Ouada H, Jaffrezic-Renault N (2014) Bi-enzymatic conductometric biosensor for detection of heavy metal ions and pesticides in water samples based on enzymatic inhibition in arthrospira platensis. J Environ Prot 5:441–453

Thevenot DR, Toth K, Durst RA, Wilson GS (2001) Electrochemical biosensors: recommended definitions and classification. Biosens Bioelectron 16:121–131

Thomas SG, Phillips AL, Hedden P (1999) Molecular cloning and functional expression of gibberellin 2- oxidases, multifunctional enzymes involved in gibberellin deactivation. Proc Natl Acad Sci U S A 96:4698–4703

Touhami A (2014) Biosensors and nanobiosensors: design and applications. In: Seifalian PA (ed) Nanomedicine. One Central Press (OCP), University College London, Achala de Mel University College London, Deepak M. Kalaskar University College London, London, pp 374–403

Touloupakis E, Giannoudi L, Piletsky SA, Guzzella L, Pozzoni F, Giardi MT (2005) A multi-biosensor based on immobilized Photosystem II on screen-printed electrodes for the detection of herbicides in river water. Biosens Bioelectron 20:1984–1992

Touloupakis E, Boutopoulos C, Buonasera K, Zergioti I, Giardi MT (2012) A photosynthetic biosensor with enhanced electron transfer generation realized by laser printing technology. Anal Bioanal Chem 402:3237–3244

Tripathi D, Ahmad P, Sharma S, Chauhan D, Dubey NK (2017) Nanomaterials in plants, algae, and microorganisms. Academic Press, Elsevier Inc., pp 1–534. https://doi.org/10.1016/C2016-0-00176-6

Tsai H-C, Doong R-A (2005) Simultaneous determination of pH, urea, acetylcholine and heavy metals using array-based enzymatic optical biosensor. Biosens Bioelectron 20:1796–1804

Tsopela A, Lale A, Vanhove E, Reynes O, Séguy I, Temple-Boyer P, Juneau P, Izquierdo R, Launay J (2014) Integrated electrochemical biosensor based on algal metabolism for water toxicity analysis. Biosens Bioelectron 61:290–297

Tsopela A, Laborde A, Salvagnac L, Ventalon V, Bedel-Pereira E, Séguy I, Temple-Boyer P, Juneau P, Izquierdo R, Launay J (2016) Development of a lab-on-chip electrochemical biosensor for water quality analysis based on microalgal photosynthesis. Biosens Bioelectron 79:568–573
Tucci M, Grattieri M, Schievano A, Cristiani P, Minteer SD (2019) Microbial amperometric biosensor for online herbicide detection: photocurrent inhibition of Anabaena variabilis. Electrochim Acta 302:102–108
Turdean GL (2011) Design and development of biosensors for the detection of heavy metal toxicity. Int J Electrochem 2011:343125
Turner APF (2000) Biosensors--sense and sensitivity. Science 290:1315–1317
Umasankar Y, Ramasamy RP (2013) Highly sensitive electrochemical detection of methyl salicylate using electroactive gold nanoparticles. Analyst 138:6623–6631
Uniyal S, Sharma RK (2018) Technological advancement in electrochemical biosensor based detection of Organophosphate pesticide chlorpyrifos in the environment: a review of status and prospects. Biosens Bioelectron 116:37–50
Upadhyay S, Rao GR, Sharma MK, Bhattacharya BK, Rao VK, Vijayaraghavan R (2009) Immobilization of acetylcholineesterase–choline oxidase on a gold–platinum bimetallic nanoparticles modified glassy carbon electrode for the sensitive detection of organophosphate pesticides, carbamates and nerve agents. Biosens Bioelectron 25:832–838
Valera E, Ramón-Azcón J, Sanchez FJ, Marco MP, Rodríguez Á (2008) Conductimetric immunosensor for atrazine detection based on antibodies labelled with gold nanoparticles. Sensors Actuators B Chem 134:95–103
Vamvakaki V, Chaniotakis NA (2007) Pesticide detection with a liposome-based nano-biosensor. Biosens Bioelectron 22:2848–2853
Védrine C, Leclerc J-C, Durrieu C, Tran-Minh C (2003) Optical whole-cell biosensor using Chlorella vulgaris designed for monitoring herbicides. Biosens Bioelectron 18:457–463
Ventrella A, Catucci L, Placido T, Longobardi F, Agostiano A (2011) Biomaterials based on photosynthetic membranes as potential sensors for herbicides. Biosens Bioelectron 26:4747–4752
Vinayaka AC, Basheer S, Thakur MS (2009) Bioconjugation of CdTe quantum dot for the detection of 2,4-dichlorophenoxyacetic acid by competitive fluoroimmunoassay based biosensor. Biosens Bioelectron 24:1615–1620
Viswanathan S, Radecka H, Radecki J (2009) Electrochemical biosensor for pesticides based on acetylcholinesterase immobilized on polyaniline deposited on vertically assembled carbon nanotubes wrapped with ssDNA. Biosens Bioelectron 24:2772–2777
Vopálenská I, Váchová L, Palková Z (2015) New biosensor for detection of copper ions in water based on immobilized genetically modified yeast cells. Biosens Bioelectron 72:160–167
Wade Elmer, White JC (2018) The future of nanotechnology in plant pathology. Annu Rev Phytopathol 56:111–133
Wan Jusoh W, Ling Shing W (2014) Exploring the potential of whole cell biosensor: a review in environmental applications. Int J Chem Environ Biol Sci 2:52–56
Wang J, Chen L, Mulchandani A, Mulchandani P, Chen W (1999) Remote biosensor for in-situ monitoring of organophosphate nerve agents. Electroanalysis 11:866–869
Wang J, Krause R, Block K, Musameh M, Mulchandani A, Schöning MJ (2003) Flow injection amperometric detection of OP nerve agents based on an organophosphorus–hydrolase biosensor detector. Biosens Bioelectron 18:255–260
Wang G, Lu Y, Yan C, Lu Y (2015) DNA-functionalization gold nanoparticles based fluorescence sensor for sensitive detection of Hg2+ in aqueous solution. Sensors Actuators B Chem 211:1–6
Wasito H, Fatoni A, Hermawan D, Susilowati SS (2019) Immobilized bacterial biosensor for rapid and effective monitoring of acute toxicity in water. Ecotoxicol Environ Saf 170:205–209
Webster DP, TerAvest MA, Doud DFR, Chakravorty A, Holmes EC, Radens CM, Sureka S, Gralnick JA, Angenent LT (2014) An arsenic-specific biosensor with genetically engineered Shewanella oneidensis in a bioelectrochemical system. Biosens Bioelectron 62:320–324
Wei M, Wang J (2015) A novel acetylcholinesterase biosensor based on ionic liquids-AuNPs-porous carbon composite matrix for detection of organophosphate pesticides. Sensors Actuators B Chem 211:290–296

Wei M, Zeng G, Lu Q (2014) Determination of organophosphate pesticides using an acetylcholinesterase-based biosensor based on a boron-doped diamond electrode modified with gold nanoparticles and carbon spheres. Microchim Acta 181:121–127

Wen Y, Xing F, He S, Song S, Wang L, Long Y, Li D, Fan C (2010) A graphene-based fluorescent nanoprobe for silver(i) ions detection by using graphene oxide and a silver-specific oligonucleotide. Chem Commun 46:2596–2598

Willardson BM, Wilkins JF, Rand TA, Schupp JM, Hill KK, Keim P, Jackson PJ (1998) Development and testing of a bacterial biosensor for toluene-based environmental contaminants. Appl Environ Microbiol 64:1006–1012

Wong ELS, Chow E, Justin Gooding J (2007) The electrochemical detection of cadmium using surface-immobilized DNA. Electrochem Commun 9:845–849

Woutersen M, Belkin S, Brouwer B, van Wezel AP, Heringa MB (2011) Are luminescent bacteria suitable for online detection and monitoring of toxic compounds in drinking water and its sources? Anal Bioanal Chem 400:915–929

Wu C-S, Khaing Oo MK, Fan X (2010) Highly sensitive multiplexed heavy metal detection using quantum-dot-labeled DNAzymes. ACS Nano 4:5897–5904

Wu Y, Liu L, Zhan S, Wang F, Zhou P (2012) Ultrasensitive aptamer biosensor for arsenic(III) detection in aqueous solution based on surfactant-induced aggregation of gold nanoparticles. Analyst 137:4171–4178

Wu L, Wang Z, Zhao S, Meng X, Song X, Feng J, Song S, Zhang H (2016) A metal–organic framework/DNA hybrid system as a novel fluorescent biosensor for mercury(II) ion detection. Chem Eur J 22:477–480

Xu L, Yin H, Ma W, Kuang H, Wang L, Xu C (2015) Ultrasensitive SERS detection of mercury based on the assembled gold nanochains. Biosens Bioelectron 67:472–476

Xu G, Huo D, Hou C, Zhao Y, Bao J, Yang M, Fa H (2018) A regenerative and selective electrochemical aptasensor based on copper oxide nanoflowers-single walled carbon nanotubes nanocomposite for chlorpyrifos detection. Talanta 178:1046–1052

Yamashita T, Ookawa N, Ishida M, Kanamori H, Sasaki H, Katayose Y, Yokoyama H (2016) A novel open-type biosensor for the in-situ monitoring of biochemical oxygen demand in an aerobic environment. Sci Rep 6:38552

Yang L, Wang G, Liu Y, Wang M (2013) Development of a biosensor based on immobilization of acetylcholinesterase on NiO nanoparticles–carboxylic graphene–nafion modified electrode for detection of pesticides. Talanta 113:135–141

Yang W, Wei X, Fraiwan A, Coogan CG, Lee H, Choi S (2016) Fast and sensitive water quality assessment: a μL-scale microbial fuel cell-based biosensor integrated with an air-bubble trap and electrochemical sensing functionality. Sensors Actuators B Chem 226:191–195

Yang X, He Y, Wang X, Yuan R (2017) A SERS biosensor with magnetic substrate CoFe2O4@Ag for sensitive detection of Hg2+. Appl Surf Sci 416:581–586

Yao KS, Li SJ, Tzeng KC, Cheng TC, Chang CY, Chiu CY, Liao CY, Hsu JJ, Lin ZP (2009) Fluorescence silica nanoprobe as a biomarker for rapid detection of plant pathogens. Adv Mater Res 79–82:513–516

Yildirim N (2016) Next generation biosensor systems for environmental water quality monitoring. Northeastern University, Boston

Yildirim N, Long F, He M, Shi H-C, Gu AZ (2014) A portable optic fiber aptasensor for sensitive, specific and rapid detection of bisphenol-A in water samples. Environ Sci Process Impacts 16:1379–1386

Yoo M-S, Shin M, Kim Y, Jang M, Choi Y-E, Park SJ, Choi J, Lee J, Park C (2017) Development of electrochemical biosensor for detection of pathogenic microorganism in Asian dust events. Chemosphere 175:269–274

Yüce M, Nazır H, Dönmez G (2010a) Using of Rhizopus arrhizus as a sensor modifying component for determination of Pb(II) in aqueous media by voltammetry. Bioresour Technol 101:7551–7555

Yüce M, Nazır H, Dönmez G (2010b) A voltammetric Rhodotorula mucilaginosa modified microbial biosensor for Cu(II) determination. Bioelectrochemistry 79:66–70

Yunus S, Jonas AM, Lakard B (2013) Potentiometric biosensors. In: Roberts GCK (ed) Encyclopedia of biophysics. Springer, Berlin/Heidelberg, pp 1941–1946

Zamaleeva AI, Sharipova IR, Shamagsumova RV, Ivanov AN, Evtugyn GA, Ishmuchametova DG, Fakhrullin RF (2011) A whole-cell amperometric herbicide biosensor based on magnetically functionalised microalgae and screen-printed electrodes. Anal Methods 3:509–513

Zhai J, Cui H, Yang R (1997) DNA based biosensors. Biotechnol Adv 15:43–58

Zhang X, Li Y, Su H, Zhang S (2010) Highly sensitive and selective detection of Hg2+ using mismatched DNA and a molecular light switch complex in aqueous solution. Biosens Bioelectron 25:1338–1343

Zhang Y, Arugula MA, Wales M, Wild J, Simonian AL (2015) A novel layer-by-layer assembled multi-enzyme/CNT biosensor for discriminative detection between organophosphorus and non-organophosphrus pesticides. Biosens Bioelectron 67:287–295

Zhang C, Zhou Y, Tang L, Zeng G, Zhang J, Peng B, Xie X, Lai C (2016) Determination of Cd2+ and Pb2+ based on mesoporous carbon nitride/self-doped polyaniline nanofibers and square wave anodic stripping voltammetry. Nanomaterials 6:7

Zhang Y, Chen M, Li H, Yan F, Pang P, Wang H, Wu Z, Yang W (2017) A molybdenum disulfide/gold nanorod composite-based electrochemical immunosensor for sensitive and quantitative detection of microcystin-LR in environmental samples. Sensors Actuators B Chem 244:606–615

Zhao X, Hilliard LR, Mechery SJ, Wang Y, Bagwe RP, Jin S, Tan W (2004) A rapid bioassay for single bacterial cell quantitation using bioconjugated nanoparticles. Proc Natl Acad Sci U S A 101:15027–15032

Zhao X-H, Kong R-M, Zhang X-B, Meng H-M, Liu W-N, Tan W, Shen G-L, Yu R-Q (2011) Graphene–DNAzyme based biosensor for amplified fluorescence "Turn-On" detection of Pb2+ with a high selectivity. Anal Chem 83:5062–5066

Zhao Y, Zhang W, Lin Y, Du D (2013) The vital function of Fe3O4@Au nanocomposites for hydrolase biosensor design and its application in detection of methyl parathion. Nanoscale 5:1121–1126

Zhao Y, Liu L, Kong D, Kuang H, Wang L, Xu C (2014) Dual amplified electrochemical immunosensor for highly sensitive detection of Pantoea stewartii sbusp. stewartii. ACS Appl Mater Interfaces 6:21178–21183

Zheng Y, Yang C, Pu W, Zhang J (2009) Carbon nanotube-based DNA biosensor for monitoring phenolic pollutants. Microchim Acta 166:21–26

Zhou B, Yang X-Y, Wang Y-S, Yi J-C, Zeng Z, Zhang H, Chen Y-T, Hu X-J, Suo Q-L (2019) Label-free fluorescent aptasensor of Cd2+ detection based on the conformational switching of aptamer probe and SYBR green I. Microchem J 144:377–382

Zhu Y, Zeng G-m, Zhang Y, Tang L, Chen J, Cheng M, Zhang L-h, He L, Guo Y, He X-x, Lai M-y, He Y-b (2014) Highly sensitive electrochemical sensor using a MWCNTs/GNPs-modified electrode for lead (II) detection based on Pb2+-induced G-rich DNA conformation. Analyst 139:5014–5020

Zhu Y-F, Wang Y-S, Zhou B, Yu J-H, Peng L-L, Huang Y-Q, Li X-J, Chen S-H, Tang X, Wang X-F (2017) A multifunctional fluorescent aptamer probe for highly sensitive and selective detection of cadmium(II). Anal Bioanal Chem 409:4951–4958

Zhu Q, Liu L, Xing Y, Zhou X (2018) Duplex functional G-quadruplex/NMM fluorescent probe for label-free detection of lead(II) and mercury(II) ions. J Hazard Mater 355:50–55

Biogenic Synthesis of Metal Nanoparticles by Plants

27

Mousa Solgi and Mina Taghizadeh

Abstract

Progressing in time proved development in technology that showed the ability of metals of nanoscale to perform specific utilities better than the bulk form of metals. Nanotechnology by means of specific traits of nanoparticles can be an identical valuable knowledge in various industry and science divisions. The noble metals like silver, gold, platinum, palladium, copper, zinc, selenium, titanium, and iron were used in synthesis of particles of nano-size. Chemical, physical, and biological ways have been used toward synthesis of various types of metal nanoparticles. The extensive potential applications of these nanoparticles made the green (biological or biogenic) synthesis by using bacteria, algae, actinomycetes, fungi, and plants. In the plant-based synthesis, several extracts (leaves, bark, stem, shoots, seeds, latex, secondary metabolites, roots, twigs, peel, fruit, seedlings, essential oils, tissue cultures, gum) are used. Therefore, the current review especially focuses on synthesis particularly plant-intermediated biosynthesis of metal nanoparticles and their classification.

Keywords

Green nanotechnology · Green nanoparticles · Biogenic nanoparticles · Nanoparticles production

M. Solgi (✉) · M. Taghizadeh
Department of Horticultural Science, Faculty of Agriculture and Natural Resources, Arak University, Arak, Iran
e-mail: M-solgi@araku.ac.ir

M. Ghorbanpour et al. (eds.), *Biogenic Nano-Particles and their Use in Agro-ecosystems*, https://doi.org/10.1007/978-981-15-2985-6_27

27.1 Introduction

27.1.1 Definition of Nanotechnology and Its Background

The term "nano" in the Greek language means small and dwarf. Each nanometer is numerically 10^{-9} or 1 billionth or as large as the three atomic widths that lie next to each other. In comparison, the size of DNA is over 2, proteins 50, influenza virus 100, and human hair diameter 10,000 nm. Nanotechnology is the foundation of many new technologies and innovations in the twenty-first century. New science of nanotechnology with Richard Feynman's famous speech, entitled "There are plenty of spaces around it" at the annual conference of the Physics Society of America in 1959, was established. Over the past few years, nanotechnology has come into various research areas and even human lives. Research and development is growing rapidly around the world. Nanotechnology refers to the ordering of atoms and molecules to the extent that new buildings form and lead to the production of materials and tools with new or even completely different properties.

Nanotechnology promotes the living standards of human beings and has great effects on the improvement and development of human security, welfare, and human health. Increasing the productivity using limited resources and energy sources will be the result of the application of this knowledge. Nanotechnology will discourse issues at the scale of disease causation and has a great potential for identifying and eliminating pathogens. Nanotechnology allows the use of drug release systems that can remain active over a period of time. Nanotechnology as a powerful technology enables humans to have a molecular and atomic attitude and can build nanoscale structures. Given the potential of nanotechnology, most countries use this technology as a tool for advancement in the world and taking the ground for "development leap" and counting it in line with their economic and national interests (Solgi et al. 2009, 2011; Dubey et al. 2010; Kaushik et al. 2010; Solgi 2012, 2014; Mukhopadhay 2014).

The income produced through worldwide nanotechnology increased quickly, presently around \$39.2 billion, and is estimated to reach \$90.5 billion in 2021 (McWilliams 2016; Rai et al. 2018). Nanotechnology has developed one of the most favorable technologies functional in all areas of knowledge. Metal nanoparticles created by nanotechnology have established universal attention as a result of their widespread applications in the physiochemical and biomedical grounds. Lately, producing metal nanoparticles by plants and microorganisms has been widely studied and has been documented as a green (biological) and competent way for additional using of microorganisms as suitable nano-factories (Singh et al. 2016).

27.1.2 Type of Nanostructures

Nanotechnology includes various subunits like nanotubes such as carbon nanotubes (CNTs), nanosensors, and nanomaterials. Each of them has many applications in industry, medicine, biosciences, agriculture, and natural resources. Nanomaterials

(NMs) also include subunits, one of the most important and most practical ones being metal nanoparticles. In fact, nanoparticles of metals are the most important product of nanotechnology. Nanoparticles are called the particles which have all the same dimensions, and their sizes are less than 100 nm. Metal nanoparticles have been very much considered for optical, catalytic, magnetic, and electrical properties (Dubey et al. 2010; Kaushik et al. 2010; Solgi et al. 2011; Solgi 2012).

Nanometals like gold, silver, platinum, copper, zinc, palladium, and iron were used in synthesis of particles of nano-size. The nanoparticle properties such as shape, size, structure, and crystalline nature determine their applications. The nanoparticles (NPs) are metal atom clusters with a range of 1–100 nm, extremely favorable due to their extensive range of requests in profitable products. The metal nanoparticles are synthesized by physical, chemical, and biological approaches. The biological synthesis of nanoparticles involves plants, bacteria, fungi, algae, and actinomycetes (Haleemkhan et al. 2015).

27.1.3 Nanoparticle Synthesis

27.1.3.1 Physical Procedures

Laser ablation, condensation evaporation, diffusion, electrolysis, pyrolysis, and high-energy ball milling are the components of metal nanoparticle manufacturing (Iravani et al. 2014). In laser ablation, colloidal nanoparticles are commonly produced using several solvents. The pulsed laser ablation in liquid (PLAL) is done within the chamber under vacuum along with a number of inert gases (Khodashenas and Ghorbani 2014). The lack of chemical reagents in solutions is the main benefit of PLAL than other methods for production of metal colloids (Iravani et al. 2014). Many nanoparticles, including Au and Ag, have been produced by evaporation-condensation method. However, it is associated with several disadvantages; for example, it occupies a huge space and takes time to gain thermal stability and also consumes a large amount of energy while raising the environmental temperature around the source material (Hong and Han 2006; Korbekandi et al. 2015).

Spray pyrolysis for the production of nanoparticles has been developed recently, and its flexibility in synthesis of particles with different appearances, sizes, and compositions has been approved (Hongwang and Swihart 2007). High-impact collisions are applied in high-energy ball milling for reducing macroscale or microscale materials into nano-crystalline structures with no chemical changes (Vijayaraghavan and Ashokkumar 2017).

27.1.3.2 Chemical Methods

Chemical reduction, micro-emulsion/colloidal, and electrochemical and thermal decomposition are the available chemical systems to manufacture nanoparticles. Chemical reduction using organic and inorganic reducing agents has been shown as the commonly used method for producing colloidal metal particles, since it is equipped with simple tools and also its simple function. Potassium bitartrate, sodium borohydride, methoxy polyethylene glycol, trisodium citrate dihydrate, ascorbate, and elemental hydrogen are the most commonly used reducing mediators (Tan et al. 2003; Kim et al. 2007; Mallick et al. 2004; Rivas et al. 2001; Iravani et al. 2014; Merga et al. 2007). Such chemical materials can reduce metallic ions like gold, silver, and lead for producing corresponding metallic nanoparticles.

Micro-emulsion technique is an adaptable and repeatable technique to manage particle's feature, including shape, size, surface area, and homogeneity (Malik et al. 2012). Size and morphology of the nanoparticles have been widely tried to be controlled via micro-emulsion method (Martínez-Rodríguez et al. 2014). Electricity is employed as a controlling force to produce nanoparticles via electrochemical methods. Accordingly, passing an electric current among two electrodes divided by an electrolyte and also nanoparticle production occurred at the electrode/electrolyte interface (Starowicz et al. 2006). Electrochemical method was employed by Rodríguez-Sánchez et al. (2000) according to the dissolution of a metallic anode in an aprotic solvent for preparing Ag nanoparticles (2–7 nm). In addition, they stated that different Ag particle sizes can be obtained via changing the current density. Thermal decomposition technique is widely used for synthesis of stable monodisperse suspensions with self-assembly (Simeonidis et al. 2007).

In general, size and the composition of the obtained nanoparticles are associated with temperature, reaction time, and surfactant molecule length. Although chemical synthesis technique has many advantages, using extreme surfactants, solvents, and other chemicals prevents the application aspects of produced nanoparticles (Vijayaraghavan and Ashokkumar 2017).

27.1.3.3 Biological Methods

Biological resources have been considered to produce metallic nanoparticles for developing cost-effective and eco-friendly method. Green (biological) synthesis includes the reduction of metal ions by biological mass/extract as the resultants. Moreover, eco-friendliness and cost-effectiveness as the advantages of biological method than the traditional chemical and physical methods indicate its efficacy for catalyzing reactions in aqueous media at a standard temperature and pressure and also the flexibility of the process (Schrofel et al. 2014). The reduction occurred by components available in biological materials, and it is mostly activated by several compounds seen in the cell, like carbonyl, phenolic, amine, proteins, amide groups, pigments, flavonoids, terpenoids, alkaloids, and other reducing materials (Asmathunisha and Kathiresan 2013). Due to the varied structure in these groups, the exact mechanism for biosynthesis of nanoparticles is not easy to explain and has not yet been entirely identified. Bacteria, fungi, yeast, virus, algae, and plant extract/biomass are crucial biological compounds applied to form metallic nanoparticles (Lombardi and Garcia Jr 1999).

Bacteria are the main group of unicellular living organisms (from prokaryotes), found in water and soil (Vijayaraghavan and Yun 2008; Vijayaraghavan and Balasubramanian 2015). Different bacterial genera (*Bacillus* and *Pseudomonas*) have been studied for biosynthesis of nanomaterials (Kalishwaralal et al. 2009; Phadke and Patel 2012). Nanometals Au and Ag were obtained by Nair and Pradeep (2002) through the reaction of the corresponding metal ions within cells of lactic acid bacteria found in buttermilk. In addition, Husseiny et al. (2007) evaluated extracellular production of Au nanoparticles by *Pseudomonas aeruginosa*. Kalimuthu et al. (2008) also produced Ag nanoparticles through *Bacillus licheniformis*. Inactive and/or dead bacterial biomasses have been found to reduce metal ions to nanoparticles due to the certain organic functional groups on the cell wall.

Fungi are eukaryotic and non-phototrophic microorganisms characterized by a firm cell wall. Their cell wall contains polysaccharides and glycoproteins, in which chitin and glucan are commonly found (Yadav et al. 2015). Production of metal nanoparticles is done extracellularly or intracellularly by fungi (Mukherjee et al. 2001). Extracellular synthesis is much rapid compared to the intracellular route (Narayanan and Sakthivel 2010a). Several studies have demonstrated fungi synthesizing nanoparticles extracellularly, namely, *Fusarium solani* (Ingle et al. 2009), *Penicillium fellutanum* (Kathiresan et al. 2009), *Phoma glomerata* (Birla et al. 2009), *Aspergillus oryzae* (Binupriya et al. 2010), *Aspergillus terreus* (Baskar et al. 2013), and *Rhizopus nigricans* (Ravindra and Rajasab 2013). On the contrary, limited investigations have been done on the intracellular synthesis of nanoparticles by fungal species (Mukherjee et al. 2001).

Algae are simple organisms, in which several specific structures and organs in earthly plants cannot be found. Using algae for production of metal nanoparticle has not widely been considered. Application of microalgae in producing nanoparticles has been negligibly reported (Sudha et al. 2013; Jena et al. 2014). The brown marine algae (*Sargassum wightii*) have shown useful for synthesizing gold nanoparticles extracellularly. Additional brown seaweeds, including *Turbinaria conoides*, green seaweeds, as well as red seaweeds have also been investigated for nanoparticle production (Rajeshkumar et al. 2013; Sangeetha et al. 2013; Priyadharshini et al. 2014).

27.1.3.3.1 Advantage of Biological Nanoparticles

Reducing metal cytotoxicity is crucial for metal nanoparticles for biomedical uses. Metal nanoparticles obtained by green (biogenic) routes are free from toxicity of by-products than the physicochemical-derived nanoparticles (Solgi and Taghizadeh 2012; Baker et al. 2013).

The green or biological production of nanoparticles is associated with several advantages, such as eco-friendly and rapid production methods and the cost-effective and biocompatible type of produced nanoparticles. In addition, there is no need for more stabilizing agents, since microorganism and plant components act as stabilizing agents (Makarov et al. 2014). Biological nanoparticles are more active due to the binding of biologically active components on the surface of synthesized nanoparticles from the biological sources, like microorganisms and plants. Many abundant metabolites with pharmacological effects are available and found to bind to the synthesized nanoparticles, providing supplementary profit through the increased effectiveness of the nanoparticles, especially in medicinal plants (Makarov et al. 2014; Singh et al. 2016). Moreover, by biological synthesis the needed steps can be reduced, for example, the attachment of some functional groups to the nanoparticle surface for making them biologically active, which is an additional step for physiochemical production (Baker et al. 2013). Time of biosynthesizing nanoparticles is also lower than that of the physiochemical methods. High-speed synthetic methods have developed with high yields via development of different plant sources. For example, silver nanoparticles (SNPs) have been synthesized via different plant extracts through 2, 5, and 45 min. Gold nanoparticles have also been shown to be produced within 3 and 5 min, indicating the simple and fast synthesis of nanoparticles by plant extracts (Lombardi and Garcia Jr 1999; Vijayaraghavan and Yun 2008; Priyadharshini et al. 2014; Singh et al. 2016).

27.1.3.3.2 Plant-Mediated Synthesis (Phytosynthesis) of Nanoparticles

Using plants or their extracts for synthesis of nanoparticles has been considered in nanotechnology as an environmentally friendly method. The basic green (biological) chemistry principles make cleaner synthesis of nanoparticles. Biological synthesis (phytosynthesis) applies molecular tolerance mechanisms and metabolomics to form nanoparticles (Rai et al. 2018).

Phytonanotechnology has recently offered new methods to synthesize nanoparticles, which are cost-effective, simple, high speed, eco-friendly, and stable. Biological synthesis using plants is associated with several benefits, including scalability, biocompatibility, and the medical use of synthesizing nanoparticles using the water, common solvent, as a reducing medium (Noruzi 2015). Therefore, plant-mediated nanoparticles derived from readily accessible plant materials and the plants' safety are both effective to achieve the high demand for nanoparticles to use in the environmental and biomedical settings. Accordingly, it has been tried to investigate various plant species to assess their potential to synthesize nanoparticles. Different plant parts, such as roots, stems, leaves, fruits, flowers, and their extracts, have been applied to produce metal nanoparticles. Table 27.1 shows some of these sections/extracts. The underlying mechanism and the components essential for plant-mediated synthetic nanoparticles have not yet been demonstrated. Proteins, amino acids, organic acid, and vitamins and also secondary metabolites, including flavonoids, alkaloids, polyphenols, terpenoids, heterocyclic compounds, and polysaccharides, have been shown effective in metal salt reduction. They also act as capping and stabilizing factors to form nanoparticles (Duan et al. 2015). In this regard, Solgi (2014) found that saffron petal extract includes phenolic compounds, including flavonoids (kaempferol) and anthocyanins (anthocyanidin, delphinidin, and pelargonidin). In addition, Solgi indicated that pomegranate peels have phenolic compounds, including ellagic acid and quercetin, quercitrin, rutin, luteolin, gallic acid, and myricetin, found in fresh flowers of Damask rose (Solgi and Taghizadeh 2012). Their hydroxyl groups are able to attach silver ions and affect the biosynthesis of SNPs and also act as reducing agent for the reduction of silver ions (Ag^{+}) to SNPs (Ag^{0}) (Solgi and Taghizadeh 2012; Solgi 2014). Furthermore, El-Kassas and El-Sheekh (2014) reported that the hydroxyl functional group of the polyphenols as well as the carbonyl group of proteins of *Corallina officinalis* extract are associated with producing and stabilizing gold nanoparticles. Philip et al. (2011) demonstrated formation and stabilization of silver and gold nanoparticles via biomolecule attachment in leaf extract of *Murraya koenigii*. It has stated that various mechanisms to synthesize nanoparticles are found in several plant species (Baker et al. 2013).

Table 27.1 Biological synthesis of several metal/metal-oxide nanoparticles by different plant parts

Plant name	Plant section	Nanoparticle name	Size (nm)	References	Year
Geranium graveolens	Leaves	Silver	27	Shankar et al.	2003
Aloe vera	Leaves	Silver and gold	15–20	Chandran et al.	2006
Carica papaya	Fruits	Silver	15	Jain et al.	2009
Rosa rugosa	Leaves	Silver and gold	Silver, 11; gold, 12	Dubey et al.	2010
Coleus amboinicus	Leaves	Gold	4.6–55	Narayanan and Sakthivel	2010a, b
Capsicum annuum	Fruits	Silver	2–6	Jha and Prasad	2011
Rosa damascena	Petals	Silver	13–28	Solgi	2012
Punica granatum	Peel	Silver	19–29	Solgi	2012
Anogeissus latifolia	Gum powder	Silver	5.5–5.9	Kora et al.	2012
Banana	Peel	Cadmium sulfide	1.48	Zhou et al.	2014
Crocus sativus	Petals	Silver	2–3.5	Solgi	2014
Euphorbia prostrata	Leaves	Silver and titanium dioxide (TiO_2)	Silver, 10–15; TiO_2, 81.7–84.7	Zahir et al.	2015
Ginkgo biloba	Leaves	Copper	15–20	Nasrollahzadeh and Sajadi	2015
Panax ginseng	Root	Silver and gold	Silver, 10–30; gold, 10–40	Singh et al.	2015
Azadirachta indica	Leaves	Silver	41–60	Poopathi et al.	2015
Cocos nucifera	Leaves	Lead	47	Elango and Roopan	2015
Pistacia atlantica	Seeds	Silver	27	Sadeghi et al.	2015
Citrus medica	Fruits	Copper	20	Shende et al.	2015
Lawsonia inermis	Leaves	Iron	21	Naseem and Farrukh	2015
Origanum vulgare	Leaves	Titanium dioxide (TiO_2)	2–15	Shiak et al.	2018
Thymus vulgaris	Waste extract (leaves and stems)	Zinc oxide (ZnO)	10–35	Abolghasemi et al.	2019

Eugenol as the certain terpenoid in *Cinnamomum zeylanisum* plays a crucial role in the synthesis of gold and SNPs (Makarov et al. 2014). It should be noted that dicot plants have different secondary metabolites possibly effective for nanoparticle synthesis (Singh et al. 2016).

27.2 Characterization of Nanoparticle

Several experimental methods have been employed to observe, form, and characterize metallic nanoparticles.

27.2.1 UV-VIS Spectroscopy

UV-visible spectroscopy (UV-Vis) method is applied to quantify the light absorbed or scattered by a sample. It has widely accepted that UV-Vis spectra can be applied for evaluating the size and shape of the controlled nanoparticles in aqueous suspensions. UV-Vis is employed for determining the concentration of the elements in solutions, namely, silver, gold, and copper. In this respect, it is used for detecting and evaluating the possibility of producing nanoparticles of metals, including silver, and its concentration in a watery environment. A wavelength of 200–700 nm is commonly applied for determination of the nanoparticle production from metals. For example, the exact wavelength peak for absorption of silver and gold is 450 and 550 nm, respectively. Indeed, UV-Vis is a tool for measuring the absorption spectra of the samples (Jain et al. 2009; Jha et al. 2009; Dubey et al. 2010; Bankar et al. 2010; Krishnaraj et al. 2010; Solgi and Taghizadeh 2012; Solgi 2014).

27.2.2 *FTIR* Spectroscopy

Fourier-transform infrared (FTIR) spectroscopy is developed for measuring the chemical bonds in surface atoms of plant samples and functional atoms involved in the recovery of nanomaterials and their production as well. FTIR with infrared light irradiation makes the molecular bands vibrate leading to design the graph. The obtained graph includes different absorption peaks, each of them shows specific chemical bonds. The infrared spectroscopy (IR spectroscopy) device to detect chemical groups is involved in the formation of SNPs. For instance, the chemical carbonyl group has a peak of approximately 11,700 cm (Jain et al. 2009; Jha et al. 2009; Dubey et al. 2010; Bankar et al. 2010; Krishnaraj et al. 2010; Solgi and Taghizadeh 2012; Solgi 2014).

27.2.3 Transmission Electron Microscopy (TEM)

Transmission electron microscopy (TEM) method is the commonly used procedure for measuring the morphology, size, and size distribution of the metal nanoparticles.

27.2.4 Scanning Electron Microscopy(SEM)

A scanning electron microscope is applied for determining the shape or precision for evaluating the size of nanomaterials formed in the nanometer size and also for assessing their shapes. For TEM, nearly 25 μl of the samples is removed coating on copper rod, followed by transmitting SNPs using an electron microscope (Jain et al. 2009; Jha et al. 2009; Dubey et al. 2010; Bankar et al. 2010; Krishnaraj et al. 2010; Solgi and Taghizadeh 2012; Solgi 2014).

27.2.5 X-Ray Diffraction (XRD)

XRD system is a useful tool for analyzing the crystal structure as well as mean particle size of the nanoparticles. It is used for the quantitative and qualitative measurement of the solid and liquid phases. Accordingly, dried SNPs are covered on the roll of the apparatus, and the spectrum is then set at 40 kV voltages and 30 mAh current and the element's radiation is transmitted. Scherrer equation was developed for calculation of the crystallite size from XRD diffraction pattern for nanoparticles:

$$d = \mathrm{K}\lambda / B \cos\,\theta$$

where

D = mean dimension of crystallites (nm)
λ = X-ray radiation wavelength
K = Scherrer constant (morphology, commonly 0.94)
B = the line full width at half maximum (FWHM) height in radians
θ = Bragg angle (the position of the diffraction peak maximum)

(Jain et al. 2009; Jha et al. 2009; Dubey et al. 2010; Bankar et al. 2010; Krishnaraj et al. 2010; Solgi and Taghizadeh 2012; Solgi 2014).

27.2.6 Energy Dispersive X-Ray Spectroscopy (EDS or EDX)

EDX spectroscopy is an appropriate method for identification, purity, and the elemental composition of the formed nanoparticles. The shape and chemical composition of the resulted nanoparticles are assessed through scanning electron microscopy (SEM), which is equipped with an energy-dispersive X-ray spectrometer (EDX or EDS) (Jain et al. 2009; Jha et al. 2009; Dubey et al. 2010; Bankar et al. 2010; Krishnaraj et al. 2010; Solgi and Taghizadeh 2012; Solgi 2014).

27.3 Nanoparticles and Their Applications

Nanotechnology has known as one of the most crucial technologies in all academic fields. Nanomaterials have been applied unknowingly for a long period of time. For instance, gold nanoparticles used for staining drinking glasses also have treated several disorders. In recent years, the different uses of metal nanoparticles have been considered in many fields, including biomedical, agricultural, environmental, and physiochemical fields (Solgi et al. 2009, 2011; Solgi 2014). Gold nanoparticles have shown to be used for the specific delivery of mediations, like methotrexate. They have also been found effective to diagnose genetic disorders, detect tumors, and also use for photoimaging. It has been shown that iron oxide nanoparticles are useful for drug delivery, treatment of cancer, tissue repair, cell labeling, targeting and immunoassays, detoxification of biological fluids, magnetic resonance imaging, and magnetically responsive drug delivery therapy (Singh et al. 2016). SNPs recently have been extensively considered due to their increasing application in various areas, including textiles, electronics, pharmaceutics, cosmetics, and environmental remediation. They have also been employed for many antimicrobial applications. Anticancer, anti-inflammatory, and wound healing are other uses of SNPs (Solgi et al. 2009, 2011; Ahamed et al. 2010). Considering their nontoxic, biocompatible, self-cleansing, skin-compatible, antimicrobial, and dermatological effects, zinc and titanium nanoparticles have been applied in cosmetic, biomedical, and ultraviolet (UV)-blocking agents (Zahir et al. 2015; Singh et al. 2016). In addition, metal nanoparticle has been used in the spatial analysis of different biomolecules, such as peptides, nucleic acids, lipids, fatty acids, glycosphingolipids, and molecules of drug to image these molecules with higher sensitivity and spatial resolution (Nasrollahzadeh et al 2014).

References

Abolghasemi R, Haghighi M, Solgi M, Mobinikhaledi A (2019) Rapid synthesis of ZnO nanoparticles by waste thyme (*Thymus vulgaris* L.). Int J Environ Sci Technol 16:6985–6990. https://doi.org/10.1007/s13762-018-2112-1

Ahamed M, Alsalhi MS, Siddiqui MK (2010) Silver nanoparticle applications and human health. Clin Chem Acta 411:1841–1848

Asmathunisha N, Kathiresan K (2013) A review on biosynthesis of nanoparticles by marine organisms. Colloids Surf B: Bionterfaces 103:283–287

Baker S, Rakshith D, Kavitha KS, Santosh P, Kavitha HU, Rao Y, Satish S (2013) Plants: emerging as nanofactories towards facile route in synthesis of nanoparticles. Bioimpacts 3:111–117

Bankar A, Joshi B, Kumar AR, Zinjarde S (2010) Banana peel extract mediated novel rout for the synthesis of silver nanoparticles. Colloids Surf A Physicochem Eng Asp 368:58–63

Baskar G, Chandhuru J, Sheraz Fahad K, Praveen AS (2013) Mycological synthesis, characterization and antifungal activity of zinc oxide nanoparticles. Asian J Pharm Technol 3:142–146

Binupriya AR, Sathishkumar M, Yun SI (2010) Myco-crystallization of silver ions to nanosized particles by live and dead cell filtrates of Aspergillus oryzae var. viridis and its bactericidal activity toward Staphylococcus aureus KCCM 12256. Ind Eng Chem Res 49:852–858

Birla SS, Tiwari VV, Gade AK, Ingle AP, Yadav AP, Rai MK (2009) Fabrication of silver nanoparticles by Phoma glomerata and its combined effect against Escherichia coli Pseudomonas aeruginosa and Staphylococcus aureus. Lett Appl Microbiol 48:173–179

Chandran SP, Chaudhary M, Pasricha R, Ahmad A, Sastry M (2006) Synthesis of gold nanotriangles and silver nanoparticles using Aloe vera plant extract. Biotechnol Prog 22(2):577–583

Duan H, Wand D, Li Y (2015) Green chemistry for nanoparticle synthesis. Chem Soc Rev 44:5778–5792

Dubey SP, Lahtineb M, Sillanpaa M (2010) Green synthesis and characterization of silver and gold nanoparticles using leaf extract of *Rosa rugosa*. Colloids Surf A Physiochem Eng Asp 364:34–41

Elango G, Roopan SM (2015) Green synthesis, spectroscopic investigation and photocatalytic activity of lead nanoparticles. Spectrochim Acta A Mol Biomol Spectrosc 139:367–373

El-Kassas HY, El-Sheekh MM (2014) Cytotoxic activity of biosynthesized gold nanoparticles with an extract of the red seaweed Corallina officinalis on the MCF-7 human breast cancer cell line. Asian Pac J Cancer Prev 15:4311–4317

Haleemkhan AA, Naseem, Vidya Vardhini B (2015) Synthesis of nanoparticles from plant extracts. Int J Mod Chem Appl Sci 2(3):195–203

Hong SJ, Han JI (2006) Synthesis and characterization of indium tin oxide (ITO) nanoparticle using gas evaporation process. J Electroceram 17:821–826

Hongwang Z, Swihart MT (2007) Synthesis of tellurium dioxide nanoparticles by spray pyrolysis. Chem Mater 19:1290–1301

Husseiny MI, El-Aziz MA, Badr Y, Mahmoud MA (2007) Biosynthesis of gold nanoparticles using Pseudomonas aeruginosa. Spectrochim Acta A Mol Biomol Spectrosc 67:1003–1006

Ingle A, Rai M, Gade A, Bawaskar M (2009) Fusarium solani: a novel biological agent for the extracellular synthesis of silver nanoparticles. J Nanopart Res 11:2079–2085

Iravani S, Korbekandi H, Mirmohammadi SV, Zolfaghari B (2014) Synthesis of silver nanoparticles: chemical, physical and biological methods. Res Pharm Sci 9:385–406

Jain D, Kumar Daima H, Kachhwaha S, Kothari SL (2009) Synthesis of plant-mediated silver nanoparticles using papaya fruit extract and evaluation of their antimicrobial activities. Dig Nanomater Biostruct 4(3):557–563

Jena J, Pradhan N, Nayak RR, Dash BP, Sukla LB, Panda PK, Mishra BK (2014) Microalga Scenedesmus sp.: a potential low-cost green machine for silver nanoparticle synthesis. J Microbiol Biotechnol 24:522–533

Jha AK, Prasad K, Prasad K, Kulkarni AR (2009) Plant system: nature's nanofactory. Colloids Surf B: Biointerfaces 73:219–223

Jha AK, Prasad K (2011) Green fruit of chili (capsicum annum L.) synthesizes nano silver! Dig J Nanomater Biostruct 6(4):1717–1723

Kalimuthu K, Babu RS, Venkataraman D, Bilal M, Gurunathan S (2008) Biosynthesis of silver nanoparticles by Bacillus licheniformis. Colloids Surf B: Biointerfaces 65:150–153

Kalishwaralal K, Deepak VS, Pandian RK, Gurunathan S (2009) Biological synthesis of gold nanocubes from Bacillus licheniformis. Bioresour Technol 100:5356–5358

Kathiresan K, Manivannan S, Nabeel MA, Dhivya B (2009) Studies on silver nanoparticles synthesized by a marine fungus, Penicillium fellutanum isolated from coastal mangrove sediment. Colloids Surf B: Bionterfaces 71:133–137

Kaushik N, Thakkar MS, Snehit S, Mhatre MS, Rasesh Y, Parikh MS (2010) Biological synthesis of metallic nanoparticles. Nanomedicine 6:257–262

Khodashenas B, Ghorbani HR (2014) Synthesis of copper nanoparticles: an overview of the various methods. Korean J Chem Eng 31:1105–1109
Kim JS, Kuk E, Yu KN, Kim JH, Park SJ, Lee HJ (2007) Antimicrobial effects of silver nanoparticles. Nanomedicine 3:95–101
Kora AJ, Beedu SR, Jayaraman A (2012) Size-controlled green synthesis of silver nanoparticles mediated by gum ghatti (*Anogeissus latifolia*) and its biological activity. Org Med Chem Lett 2:17
Korbekandi H, Iravani S, Abbass Hashim A (eds) (2015) Silver nanoparticles, nanotechnology and nanomaterials, the delivery of nanoparticles, ISBN 978-953-51-0615-9
Krishnaraj C, Jagan EG, Rajasekar S, Selvakumar P, Kalaichelvan PT, Mohan N (2010) Synthesis of silver nanoparticles using *Acalypha indica* extracts and its antimicrobial activity against water pathogens. Colloid Surf B: Biointerfaces 76:50–56
Lombardi AT, Garcia O Jr (1999) An evaluation into the potential of biological processing for the removal of metals from sewage sludges. Crit Rev Microbiol 25:275–288
Makarov VV, Love AJ, Sinitsyna OV, Makarova SS, Yaminsky IV, Taliansky ME, Kalinina NO (2014) "Green" nanotechnologies: synthesis of metal nanoparticles using plants. Acta Nat 6:35–44
Malik MA, Wani MY, Hashim MA (2012) Microemulsion method: a novel route to synthesize organic and inorganic nanomaterials: 1st nano update. Arab J Chem 5:397–417
Mallick K, Witcomb MJ, Scurell MS (2004) Polymer stabilized silver nanoparticles: a photochemical synthesis route. J Mater Sci 39:4459–4463
Martínez-Rodríguez RA, Vidal-Iglesias FJ, Solla-Gullón J, Cabrera CR, Feliu JM (2014) Synthesis of Pt nanoparticles in water-in-oil microemulsion: effect of HCl on their surface structure. J Am Chem Soc 136:1280–1283
McWilliams A (2016) The maturing nanotechnology market: products and applications. NAN031G, Global Markets, BBC Research Report. Prabhat
Merga G, Wilson R, Lynn G, Milosavljevic B, Meisel D (2007) Redox catalysis on naked silver nanoparticles. J Phys Chem C 111:12220–12226
Mukherjee P, Ahmad A, Mandal D, Senapati S, Sainkar SR, Khan MI (2001) Fungus mediated synthesis of silver nanoparticles and their immobilization in the mycelia matrix: a novel biological approach to nanoparticle synthesis. Nano Lett 1:515–519
Mukhopadhay SS (2014) Nanotechnology in agriculture: prospects and constrains. Nanotechnol Sci Appl 7:63–71
Nair B, Pradeep T (2002) Coalescence of nanoclusters and formation of submicron crystallites assisted by Lactobacillus strains. Cryst Growth Des 2:293–298
Narayanan KB, Sakthivel N (2010a) Biological synthesis of metal nanoparticles by microbes. Adv Colloid Interf Sci 156:1–13
Narayanan KB, Sakthivel N (2010b) Phytosynthesis of gold nanoparticles using leaf extract of Coleus amboinicus Lour. Mater Charact 61(11):1232–1238
Naseem T, Farrukh MA (2015) Antibacterial activity of green synthesis of iron nanoparticles using *Lawsonia inermis* and Gardenia jasminoides leaves extract. J Chem 2015:1–7. https://doi.org/10.1155/2015/912342
Nasrollahzadeh M, Sajadi SM (2015) Green synthesis of copper nano-particles using *Ginkgo biloba* L. leaf extract and their catalytic activity for the Huisgen [3+2] cycloaddition of azides and alkynes at room temperature. J Colloid Interface Sci 457:141–147
Nasrollahzadeh M, Sajadib SM, Khalajc M (2014) Green synthesis of copper nanoparticles using aqueous extract of the leaves of Euphorbia esula L and their catalytic activity for ligand-free Ullmann-coupling reaction and reduction of 4-nitrophenol. RSC Adv 4:47313–47318
Noruzi M (2015) Biosynthesis of gold nanoparticles using plant extracts. Bioprocess Biosyst Eng 38:1–14
Phadke M, Patel J (2012) Biological synthesis of silver nano particles using Pseudomonas aeruginosa. J Pure Appl Microbiol 66:1917–1924
Philip D, Unni C, Aromal SA, Vidhu VK (2011) Murraya Koenigii leaf-assisted rapid green synthesis of silver and gold nanoparticles. Spectrochim Acta A Mol Biomol Spectrosc 78:899–904

Poopathi S, De Britto LJ, Praba VL, Mani C, Praveen M (2015) Synthesis of silver nanoparticles from *Azadirachta indica*–a most effective method for mosquito control. Environ Sci Pollut Res Int 22:2956–2963

Priyadharshini RI, Prasannaraj G, Geetha N, Venkatachalam P (2014) Microwavemediated extracellular synthesis of metallic silver and zinc oxide nanoparticles using macro-algae (Gracilaria edulis) extracts and its anticancer activity against. Appl Biochem Biotechnol 174(8):2777–2790

Rai K, Kumar V, Lee SS, Raza N, Kim KH, Ok YS, Tsang DCW (2018) Nanoparticle-plant interaction: implications in energy, environment, and agriculture. Environ Int 119:1–19

Rajeshkumar S, Malarkodi C, Gnanajobitha G, Paulkumar K, Vanaja M, Kannan C, Annadurai G (2013) Seaweed-mediated synthesis of gold nanoparticles using Turbinaria conoides and its characterization. J Nanostruc Chem 3:44

Ravindra BK, Rajasab AHA (2013) A comparative study on biosynthesis of silver nanoparticles using four different fungal species. Int J Pharm Pharm Sci 6:372–376

Rivas L, Sanchez-Cortes S, Garcia-Ramos JV, Morcillo G (2001) Growth of silver colloidal particles obtained by citrate reduction to increase the Raman enhancement factor. Langmuir 17:574–577

Rodríguez-Sánchez L, Blanco MC, Lpez-Quintela MA (2000) Electrochemical synthesis of silver nanoparticles. J Phys Chem B 104(41):9683–9688

Sadeghi B, Rostami A, Momeni SS (2015) Facile green synthesis of silver nano-particles using seed aqueous extract of *Pistacia atlantica* and its antibacterial activity. Spectrochim Acta A Mol Biomol Spectrosc 134:326–332

Sangeetha N, Priyenka Devi KS, Saravanan K (2013) Synthesis and characterization of silver nanoparticles using the marine algae Ulva lactuca. Pollut Res 32:135–139

Schrofel A, Kratosova G, Safarik I, Safarikova M, Raska I, Shor LM (2014) Applications of biosynthesized metallic nanoparticles – a review. Acta Biomater 10:4023–4042

Shaik MR, Khan M, Kuniyil M, Al-Warthan A, Alkhathlan HZ, Siddiqui MRH, Shaik JP, Ahamed A, Mahmood A, Khan M, Adil SF (2018) Plant-extract-assisted green synthesis of silver nanoparticles using Origanum vulgare L. extract and their microbicidal activities. Sustainability 10:913. https://doi.org/10.3390/su10040913

Shankar SS, Ahmad A, Sastry M (2003) Geranium leaf assisted biosynthesis of silver nanoparticles. Biotechnol Prog 19(6):1627–1631

Shende S, Ingle AP, Gade A, Rai M (2015) Green synthesis of copper nanoparticles by *Citrus medica* Linn. (Idilimbu) juice and its antimicrobial activity. World J Microbiol Biotechnol 31:865–873

Simeonidis K, Mourdikoudis S, Moulla M, Tsiaoussis I, Martinez-Boubeta C, Angelakeris M, Dendrinou-Samara C, Kalogirou O (2007) Controlled synthesis and phase characterization of Fe-based nanoparticles obtained by thermal decomposition. J Magn Magn Mater 316:1–4

Singh P, Kim YJ, Yang DC (2015) A strategic approach for rapid synthesis of gold and silver nanoparticles by *Panax ginseng* leaves. Artif Cells Nanomed Biotechnol 44:1949–1957. https://doi.org/10.3109/21691401.2015.1115410

Singh P, Kim YJ, Zhang D, Yang DC (2016) Biological synthesis of nanoparticles from plants and microorganisms. Trends Biotechnol 34(7):589–599

Solgi M (2012) Application of nanotechnology and "Smart Packaging" in marketing and postharvest of cut flowers. First nanotechnology and its application in agriculture and natural resources conference. May 15–6, Karaj, Iran

Solgi M (2014) Evaluation of plant-mediated silver nanoparticles synthesis and its application in postharvest physiology of cut flowers. Physiol Mol Biol Plants 20(30):279–285

Solgi M, Taghizadeh M (2012) Silver nanoparticles ecofriendly synthesis by two medicinal plants. Int J Nanomater Biostruct 2(4):60–64

Solgi M, Kafi M, Taghavi TS, Naderi R (2009) Essential oils and silver nanoparticles (SNP) as novel agents to extend vase life of gerbera (*Gerbera jameso*nii cv. 'Dune'). Postharvest Biol Technol 53:155–158

Solgi M, Kafi M, Taghavi TS, Naderi R, Eyre J, Joyce DC (2011) Effects of silver nanoparticles (SNP) on *Gerbera jamesonii* cut flowers. Int J Postharvest Innov 2(3):274–285

Starowicz M, Stypuła B, Banas J (2006) Electrochemical synthesis of silver nanoparticles. Electrochem Commun 8:227–230

Sudha SS, Rajamanickam K, Rengaramanujam J (2013) Microalgae mediated synthesis of silver nanoparticles and their antibacterial activity against pathogenic bacteria. Indian J Exp Biol 51:393–399

Tan Y, Dai Y, Li Y, Zhua D (2003) Preparation of gold platinum, palladium and silver nanoparticles by the reduction of their salts with a weak reductant-potassium bitartrate. J Mater Chem A 13:1069–1075

Vijayaraghavan K, Ashokkumar T (2017) Plant-mediated biosynthesis of metallic nanoparticles: a review of literature, factors affecting synthesis, characterization techniques and applications. J Environ Chem Eng 5:4866–4883

Vijayaraghavan K, Balasubramanian R (2015) Is biosorption suitable for decontamination of metal-bearing wastewaters? A critical review on the state-of-the-art of biosorption processes and future directions. J Environ Manag 160:283–296

Vijayaraghavan K, Yun YS (2008) Bacterial biosorbents and biosorption. Biotechnol Adv 26:266–291

Yadav A, Kon K, Kratosova G, Duran N, Ingle AP, Rai M (2015) Fungi as an efficient mycosystem for the synthesis of metal nanoparticles: progress and key aspects of research. Biotechnol Lett 37:2099–2120

Zahir AA, Chauhan IS, Bagavan A, Kamaraj C, Elango G, Shankar J, Arjaria N, Roopan SM, Rahuman AA, Singh N (2015) Green synthesis of silver and titanium dioxide nanoparticles using *Euphorbia prostrata* extract shows shift from apoptosis to G0/G1 arrest followed by necrotic cell death in *Leishmania donovani*. Antimicrob Agents Chemother 59:4782–4799

Zhou GJ, Li SH, Zhang YC, Fu YZ (2014) Biosynthesis of CdS nanoparticles in banana peel extract. J Nanosci Nanotechnol 14:4437–4442